Why WileyPLUS for Anatomy and Physiology?

WileyPLUS: *See + Hear + Do = Success in Anatomy & Physiology!*

WileyPLUS simplifies both the teaching and learning experience, by providing a more effective and efficient environment for both student and professor success.

WileyPLUS helps students organize their resources and maximize their time and learning. Instructors use a variety of presentation, grading, and assessment tools to streamline their own course, share their course with other instructors to standardize, and provide students with a course of directed study.

A wealth of resources supporting multiple or blended learning styles are included for each chapter. These are organized around the simple principles of "See, Hear, and Do". A typical chapter within *WileyPLUS* includes the following:

SEE

- **Interactive Anatomy Review**
- **Animations of Key Physiological Processes**
- **Cadaver Lab Videos**

www.wileyplus.com

HEAR

- **Ipod/MP3 downloads keyed to significant illustrations within the text**

- **Mastering Vocabulary with audio glossary and flashcards**

DO

- **Practice Quizzes**

- **Interactive and Animated Exercises**

- **Interactive Concept Maps**

- **Review Sheets linked to Animations**

- **Illustrated Anatomy Practice and Review**

- **Cadaver Practical Practice and Review**

- **Histology Practice and Review**

All of these resources work to create a course that motivates students every step of the way. And *WileyPLUS* also gives the instructor the tools to monitor student progress and engagement throughout the course. Interested? View details or take an interactive tour: www.wileyplus.com/experience

WILEY PLUS

www.wileyplus.com

PRINCIPLES OF ANATOMY AND PHYSIOLOGY

Twelfth Edition

Volume 2—Maintenance and Continuity of the Human Body

Gerard J. Tortora

Bergen Community College

Bryan H. Derrickson

Valencia Community College

WILEY

John Wiley & Sons, Inc.

Principles of Anatomy and Physiology: Organization, Support and Movement, and Control Systems of the Human Body, Twelfth Edition, Volume 1 ISBN: 978-0470-39495-3

Principles of Anatomy and Physiology: Maintenance and Continuity of the Human Body, Twelfth Edition, Volume 2 ISBN: 978-0470-39234-8

Principles of Anatomy and Physiology, Twelfth Edition with Atlas and Registration Card, 2 Volume Set, International Student Version ISBN: 978-0470-23347-4

Printed in Asia
10 9 8 7 6 5 4 3 2

ABOUT THE AUTHORS

Gerard J. Tortora is Professor of Biology and former Biology Coordinator at Bergen Community College in Paramus, New Jersey, where he teaches human anatomy and physiology as well as microbiology. He received his bachelor's degree in biology from Fairleigh Dickinson University and his master's degree in science education from Montclair State College. He is a member of many professional organizations, including the Human Anatomy and Physiology Society (HAPS), the American Society of Microbiology (ASM), American Association for the Advancement of Science (AAAS), National Education Association (NEA), and the Metropolitan Association of College and University Biologists (MACUB).

Above all, Jerry is devoted to his students and their aspirations. In recognition of this commitment, Jerry was the recipient of MACUB's 1992 President's Memorial Award. In 1996, he received a National Institute for Staff and Organizational Development (NISOD) excellent award from the University of Texas and was selected to represent Bergen Community College in a campaign to increase awareness of the contributions of community colleges to higher education.

Jerry is the author of several best-selling science textbooks and laboratory manuals, a calling that often requires an additional 40 hours per week beyond his teaching responsibilities. Nevertheless, he still makes time for four or five weekly aerobic workouts that include biking and running. He also enjoys attending college basketball and professional hockey games and performances at the Metropolitan Opera House.

To my mother,
Angelina M. Tortora. Her love, guidance, faith,
support, and example continue to be the cornerstone
of my personal and professional life.

G.J.T.

Bryan Derrickson is Professor of Biology at Valencia Community College in Orlando, Florida, where he teaches human anatomy and physiology as well as general biology and human sexuality. He received his bachelor's degree in biology from Morehouse College and his Ph.D. in Cell Biology from Duke University. Bryan's study at Duke was in the Physiology Division within the Department of Cell Biology, so while his degree is in Cell Biology, his training focused on physiology. At Valencia, he frequently serves on faculty hiring committees. He has served as a member of the Faculty Senate, which is the governing body of the college, and as a member of the Faculty Academy Committee (now called the Teaching and Learning Academy), which sets the standards for the acquisition of tenure by faculty members. Nationally, he is a member of the Human Anatomy and Physiology Society (HAPS) and the National Association of Biology Teachers (NABT). Bryan has always wanted to teach. Inspired by several biology professors while in college, he decided to pursue physiology with an eye to teaching at the college level. He is completely dedicated to the success of his students. He particularly enjoys the challenges of his diverse student population, in terms of their age, ethnicity, and academic ability, and finds being able to reach all of them, despite their differences, a rewarding experience. His students continually recognize Bryan's efforts and care by nominating him for a campus award known as the "Valencia Professor Who Makes Valencia a Better Place to Start." Bryan has received this award three times.

To my family: Rosalind, Hurley, Cherie, and Robb.
Your support and motivation have been invaluable.

B.H.D.

PREFACE

An anatomy and physiology course can be the gateway to a gratifying career in a host of health-related professions. As active teachers of the course, we recognize both the rewards and challenges in providing a strong foundation for understanding the complexities of the human body to an increasingly diverse population of students. The twelfth edition of *Principles of Anatomy and Physiology* continues to offer a balanced presentation of content under the umbrella of our primary and unifying theme of homeostasis, supported by relevant discussions of disruptions to homeostasis. In addition, years of student feedback have convinced us that readers learn anatomy and physiology more readily when they remain mindful of the relationship between structure and function. As a writing team—an anatomist and a physiologist—our very different specializations offer practical advantages in fine-tuning the balance between anatomy and physiology.

Most importantly, our students continue to remind us of their needs for—and of the power of—simplicity, directness, and clarity. To meet these needs each chapter has been written and revised to include:

- clear, compelling, and up-to-date discussions of anatomy and physiology
- expertly executed and generously sized art
- classroom-tested pedagogy
- outstanding student study support.

As we revised the content for this edition, we kept our focus on these important criteria for success in the anatomy and physiology classroom and have refined or added new elements to enhance the teaching and learning process.

NEW TO THIS EDITION

◉ TEXT UPDATES

Every chapter in this edition of *Principles of Anatomy and Physiology* incorporates a host of improvements to both the text and the art developed by ourselves and suggested by reviewers, educators, or students. Some noteworthy text changes include the revision of the section on transport across the plasma membrane, which now begins with a discussion of passive processes (simple diffusion, facilitated diffusion, and osmosis) followed by a discussion of active processes (primary active transport, secondary active transport, and transport in vesicles, which includes endocytosis, exocytosis, and transcytosis) in Chapter 3. Chapter 12 is completely rewritten in order to provide a clearer understanding of nervous tissue structure and function. This updated narrative is supported by nine new illustrations, several revised illustrations, and a new table. Chapter 16 is rewritten in order to clarify how the brain and spinal cord process sensory and motor information and includes five new figures. Chapter 22 includes significantly revised sections on adaptive immunity, cell-mediated immunity, and antibody-mediated immunity along with updated illustrations. Chapter 26 offers revised sections on tubular reabsorption and tubular secretion, and the production of dilute and concentrated urine, which clarifies the concepts of countercurrent multiplication and countercurrent exchange, accompanied by simplified illustrations.

All clinical applications have been reviewed for currency and have been redesigned into **Clinical Connection** boxes, to be more easily recognizable within the chapter content. Many of the entries in the **Disorders: Homeostatic Imbalances** sections at chapters' ends now have new illustrations. All **Medical Terminology** sections, also at the ends of chapters, have been updated.

◉ ART AND DESIGN

The simple redesign of the twelfth edition allows the illustrations to be the focal point on each page. Each page is carefully laid out to place related text, figures, and tables near one another, minimizing the need for page turning while reading a topic. You'll notice the redesign for the updated Clinical Connection boxes within each chapter.

An outstanding illustration program has always been a signature feature of this text. Beautiful artwork, carefully chosen photographs and photomicrographs, and unique pedagogical enhancements all combine to make the visual appeal and usefulness of the illustration program in *Principles of Anatomy and Physiology* distinctive.

Continuing in this tradition, you will find exciting new three-dimensional illustrations gracing the pages of nearly every chapter in the text. Significantly, all of the illustrations in Chapters 7, 8, and 9 on the skeleton and joints are new, as well as all of the illustrations in Chapter 11 on muscles. These new illustrations are among the best that we have ever seen in any anatomy and physiology textbook and truly support the visual learner in meeting the challenge of learning so many anatomical structures. Equally important are the numerous new illustrations depicting and clarifying physiological processes. See, for example, the nine new figures in Chapter 12 on membrane potentials, or new figures in Chapter 16 on sensory and motor pathways.

Thoughtful revisions have been made to many of the figures depicting both anatomy and physiology throughout the text.

These revisions include enhanced use of color for visual impact and to better engage students, and clarifying details for better understanding of processes. All figures showing transverse sections of the spinal cord have been recolored to better reflect gray and white matter (see Figures 13.3–13.18 for example). Other examples are Figures 1.6–1.9 on body planes and cavities; Figure 4.6 on connective tissue; Figure 10.2 on skeletal muscle tissue; Figures 14.17–14.26 on cranial nerves; Figures 21.11, 21.15, 21.16. and 21.18 on immune processes; and Figures 26.18–26.19 on countercurrent multiplication and countercurrent exchange.

◉ CADAVER PHOTOGRAPHS

The number of cadaver photos in this edition has been increased, and most previously existing photos have been replaced. These distinctive images were photographed by Mark Nielsen in his laboratory at the University of Utah. Many of the meticulous dissections are the work of his colleague (and former student) Shawn Miller. Others are dissected by other students under Mark's guidance. The matching of these photographs with the line art brings your students that much closer to experiencing an actual cadaver lab.

(g) Posterior view

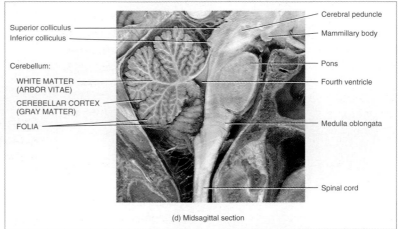

(d) Midsagittal section

◉ PHOTOMICROGRAPHS

Mark Nielsen is also responsible for most of the new photomicrographs included in this edition. Some show exploded segments at higher magnification, allowing students to clearly see specific anatomical details.

◉ MP3 DOWNLOADS

An exciting new feature has been added to the illustration program for this edition. MP3 downloads, linked to identified illustrations in each chapter give the students the opportunity to hear while they study—as they would in lecture—about the importance and relevance of the structures or concepts that are depicted. These illustrations are identified in each chapter by a distinctive icon.

(c) Photomicrographs

◉ COMPLETE TEACHING AND LEARNING PACKAGE

The twelfth edition of Principles of Anatomy and Physiology is accompanied by a host of dynamic resources designed to help you and your students maximize your time and energies. Please contact your Wiley representative for details about these and other resources or visit our website at www.wiley.com/college/sc/totora and click on the text cover to explore these assets more fully.

REAL ANATOMY

◉ **NEW! REAL ANATOMY.** Mark Nielsen and Shawn Miller of the University of Utah, led a team of media and anatomical experts in the creation of this powerful new DVD, *Real Anatomy*. Their extensive experience in undergraduate anatomy classrooms and cadaver laboratories as well as their passion for the subject matter shine through in this new, user-friendly program with its intuitive interface. The 3-D imaging software allows students to dissect through numerous layers of a real three-dimensional human body to study and learn the anatomical structures of all body systems from multiple perspectives. Histology is viewed via a virtual microscope at varied levels of magnification. Professors can use the program to capture and customize images from a large database of stunning cadaver photographs and clear histology photomicrographs for presentations, quizzing, or testing.

◉ **INTERACTIONS: EXPLORING THE FUNCTIONS OF THE HUMAN BODY 3.0** From Lancraft et. al. Covering all body systems, this dynamic and highly acclaimed program includes anatomical overviews linking form and function, rich animations of complex physiological processes, a variety of creative interactive exercises, concept maps to help students make the connections, and animated clinical case studies. The 3.0 release boasts enhancements based on user feedback, including coverage of ATP, the building blocks of proteins, and dermatomes; a new overview on Special Senses; cardiac muscle; and a revised animation on muscle contraction. Interactions is available in one DVD or in a web-based version, and is fully integrated into *WileyPLUS*.

◉ **POWERPHYS** by Allen, Harper, Ivlev, and Lancraft. Ten self-contained lab modules for exploring physiological principles. Each module contains objectives with illustrated and animated review material, prelab quizzes, prelab reporting, data collection and analysis, and a full lab report with discussion and application questions. Experiments contain randomly generated data, allowing users to experiment multiple times but still arrive at the same conclusions. Available as a stand-alone product, PowerPhys is also bundled with every new copy of the Allen and Harper Laboratory Manual and integrated into *WileyPlus*.

◉ **POWERANATOMY** by Allen, Harper and Baxley. Developed in conjunction with Primal Pictures U.K., this is an online human anatomy laboratory manual, combining beautiful 3-D images of the human body along with text, exercises, and review questions focused on the undergraduate students in anatomy or anatomy and physiology. Users can rotate the images, click on linked terms to see structures, and then answer self-assessing questions to test their knowledge.

 ◉ **WILEY*PLUS*** is a powerful online tool that provides students and instructors with an integrated suite of teaching and learning resources in one easy-to-use website. With *WileyPLUS*, students will come to class better prepared for lectures, get immediate feedback and context-sensitive help on assignments and quizzes, and have access to a full range of interactive learning resources, including a complete online version of their text. A description of some of the resources available to students within *WileyPLUS* appears on the front endpapers of this text. Instructors benefit as well with *WileyPLUS*, with all the tools and resources included to prepare and present dynamic lectures as well as assess student progress and learning. New within *WileyPLUS*, ***Quickstart*** is an organizing tool that makes it possible for you to spend less time preparing lectures and grading quizzes and more time teaching and interacting with students. Ask your sales representative to set you up with a test drive, or view a demo online.

◉ **VISUAL LIBRARY FOR ANATOMY AND PHYSIOLOGY 4.0** A cross-platform DVD includes all of the illustrations from the textbook in labeled, unlabeled, and unlabeled with leader lines formats. In addition, many illustrations and photographs not included in the text, but which could easily be added to enhance lecture or lab, are included. Search for images by chapter or by using keywords.

◉ **COMPANION WEBSITES.** A dynamic website for students, rich with many activities for review and exploration includes self-quizzes for each chapter, Visual Anatomy review exercises, and weblinks. An access code is bundled with each new text. A dedicated companion website for instructors provides many resources for preparing and presenting lectures. Additionally, this website provides a web version of the Visual Library for Anatomy and Physiology, additional critical thinking questions with answers, an editable test bank, a computerized test bank, transparencies on demand, and clicker questions. These websites can be accessed through www.wiley.com/college/tortora.

◉ **A BRIEF ATLAS OF THE SKELETON, SURFACE ANATOMY, AND SELECTED MEDICAL IMAGES** Packaged with every new copy of the text, this atlas of stunning photographs provides a visual reference for both lecture and lab.

◉ **LEARNING GUIDE** by Kathleen Schmidt Prezbindowski, College of Mount St. Joseph. Designed specifically to fit the needs of students with different learning styles, this well-received guide helps students to examine more closely important concepts through a variety of activities and exercises. The 29 chapters in the *Learning Guide* parallel those of the textbook and include many activities, quizzes, and tests for review and study.

◉ **ILLUSTRATED NOTEBOOK.** A true companion to the text, this unique notebook is a tool for organized notetaking in class and for review during study. Following the sequence in the textbook, each left-hand page displays an unlabeled black and white copy of every text figure. Students can fill in the labels during lecture or lab at the instructor's direction and take additional notes on the lined right-hand pages.

◘ **HUMAN ANATOMY AND PHYSIOLOGY LABORATORY MANUAL 3E** by Allen and Harper. This newly revised laboratory manual includes multiple activities to enhance student laboratory experience. Illustrations and terminology closely match the text, making this manual the perfect companion. Each copy of the lab manual includes a CD with the *PowerPhys* simulation software for the laboratory. Wiley*Plus*, with a wealth of integrated resources including cat, fetal pig, and rat dissection videos, is also available for adoption with this laboratory. The Cat Dissection Laboratory Guide and a Fetal Pig Laboratory Guide, depending upon your dissection needs, are available to package at no additional cost with the main laboratory manual or as standalone dissection guides.

◘ **PHOTOGRAPHIC ATLAS OF THE HUMAN BODY SECOND EDITION** by Tortora. Loaded with excellent cadaver photographs and micrographs, the high quality imagery can be used in the classroom, laboratory or for study and review.

ACKNOWLEDGMENTS

We wish to especially thank several academic colleagues for their helpful contributions to this edition.

Thanks to Marg Olfert and Linda Hardy of Saskatchewan Institute of Applied Science and Technology, who revised the end-of-chapter Self-Quiz and Critical Thinking Questions. We are grateful to Tom Lancraft of St. Petersburg College for all of his contributions to the *QuickStart Wiley*PLUS course for this textbook. Special thanks go to Kathleen Schmidt Prezbindowski, who has authored the Learning Guide for so many editions.

A talented group of educators have contributed to the high quality of the diverse supplementary materials that accompany this text. We wish to acknowledge each and thank them for their work. Special thanks to Connie Allen of Edison College, Gary Allen, Dalhousie University; Lois Borek, Georgia State University; Laura Branagan, Foothill College; Valerie Harper; Donald Ferruzzi, Suffolk Community College; Candace Francis, Palomar College; Chaya Gopalan, St. Louis Community College; Jacqueline Jordan, Clayton State University Community College; Brenda Leady, University of Toledo; Daniel Mark Pavuk, Bowling Green State University; Lynn Preston, Tarrant County College; Saeed Rahmanian, Roane State Community College; Claudia Stanescu, University of Arizona and Eric Sun of Macon State University; Lynn Wandrey, Mott Community College; and Terry Wiseth, Northland College.

We wish to thank to James Witte and Prasanthi Pallapu of Auburn University and the Institute for Learning Styles Research for their collaboration with us in developing questions and tools for students to assess, understand, and apply their learning style preferences.

This beautiful textbook would not be possible without the talent and skill of several outstanding medical illustrators. Kevin Sommerville has contributed many illustrations for us over numerous editions. For this edition, many new drawings are the work of his talented hands. We so value the long relationship we have with Kevin. We also welcome two new illustrators to our "team". John Gibb is responsible for all of the new skeletal art and most of the outstanding new muscle illustrations. Richard Coombs contributed several new illustrations for Chapters 1, 22, and 24. And we thank the artists of Imagineering Media Services for all they do to enhance the visuals within this text. Mark Nielsen and Shawn Miller of the University of Utah have our gratitude for ex-

cellent dissections in the cadaver photographs as well as the many new histological photomicrographs they provided.

We are also extremely grateful to our colleagues who have reviewed the manuscript or participated in focus groups and offered numerous suggestions for improvement: Doris Benfer, Delaware County Community College; Franklyn F. Bolander, Jr., University of South Carolina Columbia; Carolyn Bunde, Idaho State University; Brian Carver, Freed-Harman University; Bruce A. Fisher, Roane State Community College; Purti Gadkari, Wharton County Junior College; Ron Hackney, Volunteer State Community College; Clare Hays, Metropolitan State College of Denver; Catherine Hurlbut, Florida Community College Jacksonville; Leonard Jago, Northampton Community College; Wilfredo Lopez-Ojeda, University of Central Florida; Jackie Reynolds, Richland College; Benita Sabie, Jefferson Community & Technical College; Leo B. Stouder, Broward Community College; Andrew M. Scala, Dutchess Community College; R. Bruce Sundrud, Harrisburg Area Community College; Cynthia Surmacz, Bloomsburg University; Harry Womack, Salisbury University and Mark Womble, Youngstown State University.

Finally, our hats are off to everyone at Wiley. We enjoy collaborating with this enthusiastic, dedicated, and talented team of publishing professionals. Our thanks to the entire team: Bonnie Roesch, Executive Editor; Karen Trost, Developmental Editor; Lorraina Raccuia, Project Editor; Lauren Morris, Program Assistant; Lisa Wojcik, Senior Production Editor; Hilary Newman, Photo Manager; Anna Melhorn, Senior Illustration Editor; Madelyn Lesure, Designer; Karin Kincheloe, Page Make-up; Linda Muriello, Senior. Media Editor; and Clay Stone, Executive Marketing Manager.

Gerard J. Tortora
Department of Science and Health, S229
Bergen Community College
400 Paramus Road
Paramus, NJ 07652

Bryan Derrickson
Department of Science, PO Box 3028
Valencia Community College
Orlando, FL 32802
bderrickson@valenciacc.edu

NOTE TO STUDENTS

Your book has a variety of special features that will make your time studying anatomy a more rewarding experience. These features have been developed based on feedback from students like you who have used previous editions of the text.

As you start to read each section of a chapter, be sure to take note of the **Objectives** at the beginning of the section to help you focus on what is important as you read. At the end of the section, take time to try and answer the **Checkpoint** questions placed there. If you can answer, then you are ready to move on. If you have trouble answering the questions, you may want to re-read the section before continuing.

Studying the figures (illustrations that include artwork and photographs) in this book is as important as reading the text. To get the most out of the visual parts of this book, use the tools we have added to the figures to help you understand the concepts being presented. Start by reading the **Legend,** which explains what the figure is about. Next, study the **Key Concept Statement,** indicated by a "key" icon, which reveals a basic idea portrayed in the figure. Added to many figures you will also find an **Orientation Diagram** to help you understand the perspective from which you are viewing a particular piece of anatomical art. Finally, at the bottom of each figure you will find a **Figure Question,** accompanied by a "question mark" icon. If you try to answer these questions as you go along, they will serve as self-checks to help you understand the material. Often it will be possible to answer a question by examining the figure itself. Other questions will encourage you to integrate the knowledge you've gained by carefully reading the text associated with the figure. Still other questions may prompt you to think critically about the topic at hand or predict a consequence in advance of its description in the text. You will find the answer to each figure question at the end of the chapter in which the figure appears. Selected figures include **Functions** boxes, brief summaries of the functions of the anatomical structure of the system shown.

In each chapter you will find that several illustrations are marked with an icon that looks like an MP3 player. This is an indication that a download which narrates and discusses the important elements of that particular illustration is available for your study. You can access these downloads on the student companion website.

⊙ **OBJECTIVES**

• Outline the steps involved in the sliding filament mechanism of muscle contraction.
• Describe how muscle action potentials arise at the neuromuscular junction.

⊙ **CHECKPOINT**

7. What roles do contractile, regulatory, and structural proteins play in muscle contraction and relaxation?
8. How do calcium ions and ATP contribute to muscle contraction and relaxation?
9. How does sarcomere length influence the maximum tension that is possible during muscle contraction?
10. How is the motor end plate different from other parts of the sarcolemma?

Figure 24.11 External and internal anatomy of the stomach. (See Tortora, *A Photographic Atlas of the Human Body, Second Edition,* Figure 12.9.)

The four regions of the stomach are the cardia, fundus, body, and pylorus.

Esophagus
FUNDUS
Lower esophageal sphincter
Serosa
Muscularis:
CARDIA
Longitudinal layer
BODY
Lesser curvature
Circular layer
PYLORUS
Oblique layer
Greater curvature
Duodenum
Pyloric sphincter
Rugae of mucosa
PYLORIC CANAL
PYLORIC ANTRUM

(a) Anterior view of regions of stomach

Functions of the Stomach

1. Mixes saliva, food, and gastric juice to form chyme.
2. Serves as a reservoir for food before release into small intestine.
3. Secretes gastric juice, which contains HCl (kills bacteria and denatures protein), pepsin (begins the digestion of proteins), intrinsic factor (aids absorption of vitamin B_{12}), and gastric lipase (aids digestion of triglycerides).
4. Secretes gastrin into blood.

Esophagus
FUNDUS
CARDIA
Rugae of mucosa
Lesser curvature
Duodenum
PYLORUS
PYLORIC CANAL
BODY
Pyloric sphincter
PYLORIC ANTRUM
Greater curvature

(b) Anterior view of internal anatomy

? After a very large meal, does your stomach still have rugae?

Studying physiology requires an understanding of the sequence of processes. Correlation of sequential processes in text and art is achieved through the use of special numbered lists in the narrative that correspond to numbered segments in the accompanying figure. This approach is used extensively throughout the book to lend clarity to the flow of complex processes.

Learning the complex anatomy and all of the terminology involved for each body system can be a daunting task. For many topics, including the bones, joints, skeletal muscles, surface anatomy, blood vessels, and nerves, we have created special **Exhibits** which organize the material into manageable segments. Each Exhibit consists of an objective, an overview, a tabular summary of the relevant anatomy, an associated group of illustrations or photographs, and a checkpoint question. Some Exhibits also contain a relevant Clinical Connection.

Figure 12.23 Signal transmission at a chemical synapse. Through exocytosis of synaptic vesicles, a presynaptic neuron releases neurotransmitter molecules. After diffusing across the synaptic cleft, the neurotransmitter binds to receptors in the plasma membrane of the postsynaptic neuron and produces a postsynaptic potential.

At a chemical synapse, a presynaptic neuron converts an electrical signal (nerve impulse) into a chemical signal (neurotransmitter release). The postsynaptic neuron then converts the chemical signal back into an electrical signal (postsynaptic potential).

? Why may electrical synapses work in two directions, but chemical synapses can transmit a signal in only one direction?

mitter receptor is called an *ionotropic receptor*. Not all neurotransmitters bind to ionotropic receptors; some bind to *metabotropic receptors* (described shortly).

5 Binding of neurotransmitter molecules to their receptors on ligand-gated channels opens the channels and allows particular ions to flow across the membrane.

6 As ions flow through the opened channels, the voltage across the membrane changes. This change in membrane voltage is a **postsynaptic potential**. Depending on which ions the channels admit, the postsynaptic potential may be a depolarization or a hyperpolarization. For example, opening of Na^+ channels allows inflow of Na^+, which causes de-

lows K^+ to move out—in either event, the inside of the cell becomes more negative.

7 When a depolarizing postsynaptic potential reaches threshold, it triggers an action potential in the axon of the postsynaptic neuron.

At most chemical synapses, only *one-way information transfer* can occur—from a presynaptic neuron to a postsynaptic neuron or an effector, such as a muscle fiber or a gland cell. For example, synaptic transmission at a neuromuscular junction (NMJ) proceeds from a somatic motor neuron to a skeletal muscle fiber (but not in the opposite direction). Only synaptic end bulbs of presynaptic neurons can release neurotransmitter, and only the

---

EXHIBIT 11.2 — Muscles That Move the Eyeballs (Extrinsic Eye Muscles) and Upper Eyelids

OBJECTIVE
• Describe the origin, insertion, action, and innervation of the extrinsic eye muscles.

Muscles that move the eyeballs are called **extrinsic eye muscles** because they originate outside the eyeballs (in the orbit) and insert on the outer surface of the sclera ("white of the eye") (Figure 11.5). The extrinsic eye muscles are some of the fastest contracting and most precisely controlled skeletal muscles in the body.

Three pairs of extrinsic eye muscles control movements of the eyeballs: (1) superior and inferior recti, (2) lateral and medial recti, and (3) superior and inferior oblique. The four recti muscles (superior, inferior, lateral, and medial) arise from a tendinous ring in the orbit and insert into the sclera of the eye. As their names imply, the **superior** and **inferior recti** move the eyeballs superiorly and inferiorly; the **lateral** and **medial recti** move the eyeballs laterally and medially.

The actions of the oblique muscles cannot be deduced from their names. The **superior oblique** muscle originates posteriorly near the tendinous ring, then passes anteriorly, and ends in a round tendon. The tendon extends through a pulleylike loop of fibrocartilaginous tissue called the *trochlea* (= pulley) in the anterior and medial part of the roof of the orbit. Finally, the tendon turns and inserts on the posterolateral aspect of the eyeballs. Accordingly, the superior oblique muscle moves the eyeballs inferiorly and laterally. The **inferior oblique** muscle originates on the maxilla at the anteromedial aspect of the floor of the orbit. It then passes posteriorly and laterally and inserts on the pos-

terolateral aspect of the eyeballs. Because of this arrangement, the inferior oblique muscle moves the eyeballs superiorly and laterally.

The **levator palpebrae superioris**, unlike the recti and oblique muscles, does not move the eyeballs. Rather, it raises the upper eyelids, that is, opens the eyes. It is therefore an antagonist to the orbicularis oculi, which closes the eyes.

• CLINICAL CONNECTION Strabismus

Strabismus (stra-BIZ-mus; *strabismos* = squinting) is a condition in which the two eyeballs are not properly aligned. This can be hereditary or it can be due to birth injuries, poor attachments of the muscles, problems with the brain's control center, or localized disease. Strabismus can be constant or intermittent. In strabismus, each eye sends an image to a different area of the brain and because the brain usually ignores the messages sent by one of the eyes, the ignored eye becomes weaker, hence "lazy eye" or *amblyopia*, develops. *External strabismus* results when a lesion in the oculomotor (III) nerve causes the eyeball to move laterally when at rest, and results in an inability to move the eyeball medially and inferiorly. A lesion in the abducens (VI) nerve results in *internal strabismus*, a condition in which the eyeball moves medially when at rest and cannot move laterally.

Treatment options for strabismus depend on the specific type of problem and include surgery, visual therapy (retraining the brain's control center), and orthoptics (eye muscle training to straighten the eyes). •

Relating Muscles to Movements

Arrange the muscles in this exhibit according to their actions on the eyeballs: (1) elevation, (2) depression, (3) abduction, (4) adduction, (5) medial rotation, and (6) lateral rotation. The same muscle may be mentioned more than once.

CHECKPOINT
Which muscles contract and relax in each eye as you gaze to your left without moving your head?

Figure 11.5 Muscles of the head that move the eyeballs (extrinsic eye muscles) and upper eyelid.

The extrinsic muscles of the eyeball are among the fastest contracting and most precisely controlled skeletal muscles in the body.

(a) Lateral view of right eyeball

(b) Movements of right eyeball in response to contraction of extrinsic muscles

? How does the inferior oblique muscle move the eyeball superiorly and laterally?

MUSCLE	ORIGIN	INSERTION	ACTION	INNERVATION
Superior rectus (*rectus* = fascicles parallel to midline)	Common tendinous ring (attached to orbit around optic foramen).	Superior and central part of eyeballs.	Moves eyeballs superiorly (elevation) and medially (adduction), and rotates them medially.	Oculomotor (III) nerve.
Inferior rectus	Same as above.	Inferior and central part of eyeballs.	Moves eyeballs inferiorly (depression) and medially (adduction), and rotates them medially.	Oculomotor (III) nerve.
Lateral rectus	Same as above.	Lateral side of eyeballs.	Moves eyeballs laterally (abduction).	Abducens (VI) nerve.
Medial rectus	Same as above.	Medial side of eyeballs.	Moves eyeballs medially (adduction).	Oculomotor (III) nerve.
Superior oblique (*oblique* = fascicles diagonal to midline)	Sphenoid bone, superior and medial to the tendinous ring in the orbit.	Eyeball between superior and lateral recti. The muscle inserts into the superior and lateral surfaces of the eyeball via a tendon that passes through the trochlea.	Moves eyeballs inferiorly (depression) and laterally (abduction), and rotates them medially.	Trochlear (IV) nerve.
Inferior oblique	Maxilla in floor of orbit.	Eyeballs between inferior and lateral recti.	Moves eyeballs superiorly (elevation) and laterally (abduction) and rotates them laterally.	Oculomotor (III) nerve.
Levator palpebrae superioris (le-VĀ-tor PAL-pe-brē soo-per'-ē-OR-is; *palpebrae* = eyelids)	Roof of orbit (lesser wing of sphenoid bone).	Skin and tarsal plate of upper eyelids.	Elevates upper eyelids (opens eyes).	Oculomotor (III) nerve.

At the end of each chapter are other resources that you will find useful. The **Disorders: Homeostatic Imbalances** sections at the end of most chapters include concise discussions of major diseases and disorders that illustrate departures from normal homeostasis. They provide answers to many of your questions about medical problems. The **Medical Terminology** section includes selected terms dealing with both normal and pathological conditions. The **Study Outline** is a concise statement of important topics discussed in the chapter. Page numbers are listed next to key concepts so that you can refer easily to specific passages in the text for clarification or amplification. The **Self-Quiz Questions** are designed to help you evaluate your understanding of the chapter contents. **Critical Thinking Questions** are word prob-

lems that allow you to apply the concepts you have studied in the chapter to specific situations. Answers to the Self-Quiz Questions and suggested answers to the Critical Thinking Quesitons (some of which have no one right answer) appear in an appendix at the end of the book so you can check your progress.

At times, you may require extra help to learn specific anatomical features of the various body systems. One way to do this is through the use of **Mnemonics**, aids to help memory. Mnemonics are included throughout the text, some displayed in figures, tables, or exhibits, and some included within the text discussion. We encourage you to use not only the mnemonics provided, but also to create your own to help you learn the multitude of terms involved in your study of human anatomy.

STUDY OUTLINE

Brain Organization, Protection, and Blood Supply (p. 496)

1. The major parts of the brain are the brain stem, cerebellum, diencephalon, and cerebrum.
2. The brain is protected by cranial bones and the cranial meninges.
3. The cranial meninges are continuous with the spinal meninges. From superficial to deep they are the dura mater, arachnoid mater, and pia mater.
4. Blood flow to the brain is mainly via the internal carotid and vertebral arteries.
5. Any interruption of the oxygen or glucose supply to the brain can result in weakening
6. The blood–brain move between the and prevents the the brain.

Cerebrospinal Fluid

1. Cerebrospinal fluid circulates through ventricle, subarach is absorbed into th sagittal sinus.

4. A large part of the brain stem consists of small areas of gray matter and white matter called the reticular formation, which helps maintain consciousness, causes awakening from sleep, and contributes to regulating muscle tone.

The Cerebellum (p. 507)

1. The cerebellum occupies the inferior and posterior aspects of the cranial cavity. It consists of two lateral hemispheres and a medial, constricted vermis.
2. It connects to the brain stem by three pairs of cerebellar peduncles.
3. The cerebellum smoothes and coordinates the contractions of

SELF-QUIZ QUESTIONS

Fill in the blanks in the following statements.

1. The cerebral hemispheres are connected internally by a broad band of white matter known as the _____.
2. List the five lobes of the cerebrum: _____, _____, _____, _____, _____.
3. The _____ separates the cerebrum into right and left halves.

Indicate whether the following statements are true or false.

4. The brain stem consists of the medulla oblongata, pons, and diencephalon.
5. You are the greatest student of anatomy and physiology, and you are well-prepared for your exam on the brain. As you confidently answer the questions, your brain is exhibiting beta waves.

ANSWERS TO FIGURE QUESTIONS **545**

CRITICAL THINKING QUESTIONS

1. An elderly relative suffered a CVA (stroke) and now has difficulty moving her right arm, and she also has speech problems. What areas of the brain were damaged by the stroke?
2. Nicky has recently had a viral infection and now she cannot move the muscles on the right side of her face. In addition, she is experiencing a loss of taste and a dry mouth, and she cannot close her right eye. What cra

3. You have been hired by a pharmaceutical company to develop a drug to regulate a specific brain disorder. What is a major physiological roadblock to developing such a drug and how can you design a drug to bypass that roadblock so that the drug is delivered to the brain where it is needed?

ANSWERS TO FIGURE QUESTIONS

14.1 The largest part of the brain is the cerebrum.
14.2 From superficial to deep, the three cranial meninges are the dura mater, arachnoid, and pia mater.
14.3 The brain stem is anterior to the fourth ventricle, and the cerebellum is posterior to it.
14.4 Cerebrospinal fluid is reabsorbed by the arachnoid villi that project into the dural venous sinuses.
14.5 The medulla oblongata contains the pyramids; the midbrain contains the cerebral peduncles; "pons" means "bridge."
14.6 Decussation means crossing to the opposite side. The functional consequence of decussation of the pyramids is that each side of the cerebrum controls muscles on the opposite side of the body.
14.7 The cerebral peduncles are the main sites through which tracts extend and nerve impulses are conducted between the superior parts of the brain and the inferior parts of the brain and the

14.15 The somatosensory association area allows you to recognize an object by simply touching it; Broca's speech area translates thoughts into speech; the premotor area serves as a memory bank for learned motor activities that are complex and sequential; the auditory association area allows you to recognize a particular sound as speech, music, or noise.
14.16 In an EEG, theta waves indicate emotional stress.
14.17 Axons in the olfactory tracts terminate in the primary olfactory area in the temporal lobe of the cerebral cortex.
14.18 Most axons in the optic tracts terminate in the lateral geniculate nucleus of the thalamus.
14.19 The superior branch of the oculomotor nerve is distributed to the superior rectus muscle; the trochlear nerve is the smallest cranial nerve.
14.20 The trigeminal nerve is the largest cranial nerve.

Throughout the text we have included **Pronunciations** and, sometimes, **Word Roots**, for many terms that may be new to you. These appear in parentheses immediately following the new words, and the pronunciations are repeated in the glossary at the back of the book. Look at the words carefully and say them out loud several times. Learning to pronounce a new word will help you remember it and make it a useful part of your medical vocabulary. Take a few minutes now to read the following pronunciation key, so it will be familiar as you encounter new words. The key is repeated at the beginning of the Glossary, page G-1.

◉ PRONUNCIATION KEY

1. The most strongly accented syllable appears in capital letters, for example, bilateral (bī-LAT-er-al) and diagnosis (dī-ag-NŌ-sis).

2. If there is a secondary accent, it is noted by a prime ('), for example, constitution (kon′-sti-TOO-shun) and physiology (fiz′-ē-OL-ō-jē). Any additional secondary accents are also noted by a prime, for example, decarboxylation (dē′-kar-bok′-si-LĀ-shun).

3. Vowels marked by a line above the letter are pronounced with the long sound, as in the following common words:
   ā as in *māke*   ō as in *pōle*
   ē as in *bē*   ū as in *cūte*
   ī as in *īvy*

4. Vowels not marked by a line above the letter are pronounced with the short sound, as in the following words:
   a as in *above* or *at*   o as in *not*
   e as in *bet*   u as in *bud*
   i as in *sip*

5. Other vowel sounds are indicated as follows:
   oy as in *oil*
   oo as in *root*

6. Consonant sounds are pronounced as in the following words:
   b as in *bat*   m as in *mother*
   ch as in *chair*   n as in *no*
   d as in *dog*   p as in *pick*
   f as in *father*   r as in *rib*
   g as in *get*   s as in *so*
   h as in *hat*   t as in *tea*
   j as in *jump*   v as in *very*
   k as in *can*   w as in *welcome*
   ks as in *tax*   z as in *zero*
   kw as in *quit*   zh as in *lesion*
   l as in *let*

BRIEF TABLE OF CONTENTS

VOLUME 1

BRIEF TABLE OF CONTENTS

VOLUME 2

VOLUME 2 CONTENTS

Volume 1 Chapters 1-18
Volume 2 Chapters 19-29

THE CARDIOVASCULAR SYSTEM: THE BLOOD

BLOOD AND HOMEOSTASIS Blood contributes to homeostasis by transporting oxygen, carbon dioxide, nutrients, and hormones to and from your body's cells. It helps regulate body pH and temperature, and provides protection against disease through phagocytosis and the production of antibodies. •

The **cardiovascular system** (*cardio-* = heart; *vascular* = blood vessels) consists of three interrelated components: blood, the heart, and blood vessels. The focus of this chapter is blood; the next two chapters will examine the heart and blood vessels, respectively. Blood transports various substances, helps regulate several life processes, and affords protection against disease. For all of its similarities in origin, composition, and functions, blood is as unique from one person to another as are skin, bone, and hair. Health-care professionals routinely examine and analyze its differences through various blood tests when trying to determine the cause of different diseases. The branch of science concerned with the study of blood, blood-forming tissues, and the disorders associated with them is **hematology** (hēm-a-TOL-ō-jē; *hema-* or *hemato-* = blood; *-logy* = study of).

FUNCTIONS AND PROPERTIES OF BLOOD

◉ **OBJECTIVES**

- Describe the functions of blood.
- Describe the physical characteristics and principal components of blood.

Most cells of a multicellular organism cannot move around to obtain oxygen and nutrients or eliminate carbon dioxide and other wastes. Instead, these needs are met by two fluids: blood and interstitial fluid. **Blood** is a connective tissue composed of a liquid extracellular matrix called blood plasma that dissolves and suspends various cells and cell fragments. **Interstitial fluid** is the fluid that bathes body cells (see Figure 27.1 on page 1063) and is constantly renewed by the blood. Blood transports oxygen from the lungs and nutrients from the gastrointestinal tract, which diffuse from the blood into the interstitial fluid and then into body cells. Carbon dioxide and other wastes move in the reverse direction, from body cells to interstitial fluid to blood. Blood then transports the wastes to various organs—the lungs, kidneys, and skin—for elimination from the body.

Functions of Blood

Blood, which is a liquid connective tissue, has three general functions:

1. *Transportation.* As you just learned, blood transports oxygen from the lungs to the cells of the body and carbon dioxide from the body cells to the lungs for exhalation. It carries nutrients from the gastrointestinal tract to body cells and hormones from endocrine glands to other body cells. Blood also transports heat and waste products to various organs for elimination from the body.

2. *Regulation.* Circulating blood helps maintain homeostasis of all body fluids. Blood helps regulate pH through the use of buffers. It also helps adjust body temperature through the heat-absorbing and coolant properties of the water (see page 40) in blood plasma and its variable rate of flow through the skin, where excess heat can be lost from the blood to the environment. In addition, blood osmotic pressure influences the water content of cells, mainly through interactions of dissolved ions and proteins.

3. *Protection.* Blood can clot, which protects against its excessive loss from the cardiovascular system after an injury. In addition, its white blood cells protect against disease by carrying on phagocytosis. Several types of blood proteins, including antibodies, interferons, and complement, help protect against disease in a variety of ways.

Physical Characteristics of Blood

Blood is denser and more viscous (thicker) than water and feels slightly sticky. The temperature of blood is 38°C (100.4°F), about 1°C higher than oral or rectal body temperature, and it has a slightly alkaline pH ranging from 7.35 to 7.45. The color of blood varies with its oxygen content. When it has a high oxygen content, it is bright red. When it has a low oxygen content, it is dark red. Blood constitutes about 20% of extracellular fluid, amounting to 8% of the total body mass. The blood volume is 5 to 6 liters (1.5 gal) in an average-sized adult male and 4 to 5 liters (1.2 gal) in an average-sized adult female. The difference in volume is due to differences in body size. Several hormones, regulated by negative feedback, ensure that blood volume and osmotic pressure remain relatively constant. Especially important are the hormones aldosterone, antidiuretic hormone, and atrial natriuretic peptide, which regulate how much water is excreted in the urine (see pages 1065–1066).

● CLINICAL **CONNECTION** | **Withdrawing Blood**

Blood samples for laboratory testing may be obtained in several ways. The most common procedure is **venipuncture,** withdrawal of blood from a vein using a needle and collecting tube, which contains various additives. A tourniquet is wrapped around the arm above the venipuncture site, which causes blood to accumulate in the vein. This increased blood volume makes the vein stand out. Opening and closing the fist further causes it to stand out, making the venipuncture more successful. A common site for venipuncture is the median cubital vein anterior to the elbow (see Figure 21.25b on page 808). Another method of withdrawing blood is through a **finger** or **heel stick.** Diabetic patients who monitor their daily blood sugar typically perform a finger stick, and it is often used for drawing blood from infants and children. In an **arterial stick,** blood is withdrawn from an artery; this test is used to determine the level of oxygen in oxygenated blood. •

Components of Blood

Blood has two components: (1) blood plasma, a watery liquid extracellular matrix that contains dissolved substances, and (2) formed elements, which are cells and cell fragments. If a sample of blood is centrifuged (spun) in a small glass tube, the cells sink to the bottom of the tube while the lighter-weight plasma forms a layer on top (Figure 19.1a). Blood is about 45% formed elements and 55% blood plasma. Normally, more than 99% of the formed elements are cells named for their red color—red blood cells (RBCs). Pale, colorless white blood cells (WBCs) and platelets occupy less than 1% of the formed elements. Because they are less dense than red blood cells but more dense than blood plasma, they form a very thin **buffy coat** layer between the packed RBCs and plasma in centrifuged blood. Figure 19.1b shows the composition of blood plasma and the numbers of the various types of formed elements in blood.

Blood Plasma

When the formed elements are removed from blood, a straw-colored liquid called **blood plasma** (or simply **plasma**) is left. Blood plasma is about 91.5% water and 8.5% solutes, most of which (7% by weight) are proteins. Some of the proteins in blood plasma are also found elsewhere in the body, but those

Figure 19.1 Components of blood in a normal adult.

Blood is a connective tissue that consists of blood plasma (liquid) plus formed elements (red blood cells, white blood cells, and platelets).

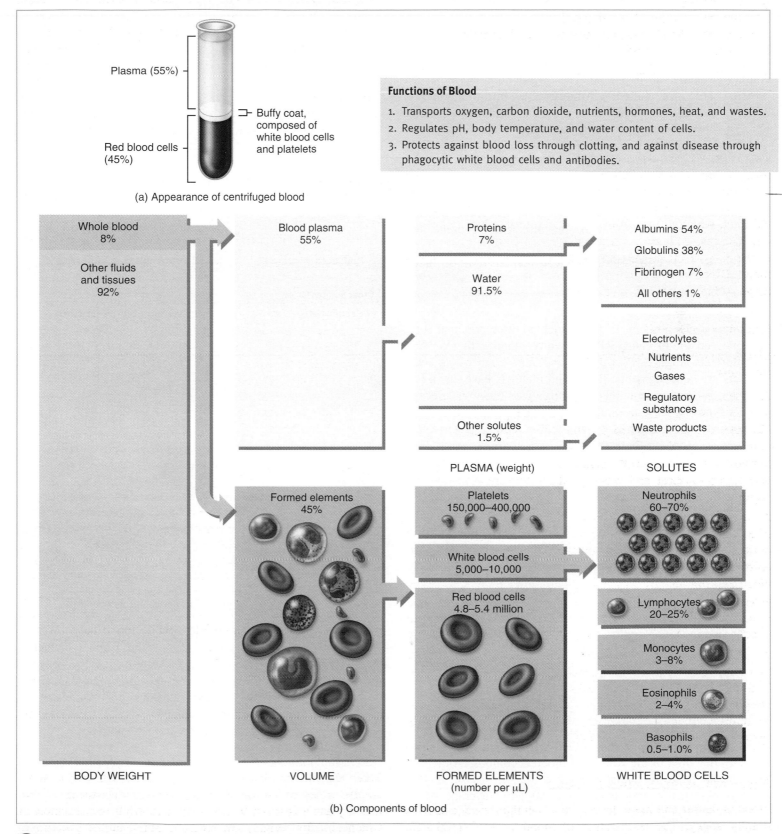

Functions of Blood

1. Transports oxygen, carbon dioxide, nutrients, hormones, heat, and wastes.
2. Regulates pH, body temperature, and water content of cells.
3. Protects against blood loss through clotting, and against disease through phagocytic white blood cells and antibodies.

Plasma (55%)

Buffy coat, composed of white blood cells and platelets

Red blood cells (45%)

(a) Appearance of centrifuged blood

Whole blood 8%

Other fluids and tissues 92%

Blood plasma 55%

Proteins 7%

Albumins 54%

Globulins 38%

Water 91.5%

Fibrinogen 7%

All others 1%

Electrolytes

Nutrients

Gases

Regulatory substances

Other solutes 1.5%

Waste products

PLASMA (weight)

SOLUTES

Formed elements 45%

Platelets 150,000–400,000

Neutrophils 60–70%

White blood cells 5,000–10,000

Red blood cells 4.8–5.4 million

Lymphocytes 20–25%

Monocytes 3–8%

Eosinophils 2–4%

Basophils 0.5–1.0%

BODY WEIGHT

VOLUME

FORMED ELEMENTS (number per μL)

WHITE BLOOD CELLS

(b) Components of blood

? **What is the approximate volume of blood in your body?**

confined to blood are called **plasma proteins.** Among other functions, these proteins play a role in maintaining proper blood osmotic pressure, which is an important factor in the exchange of fluids across capillary walls (discussed in Chapter 21).

Hepatocytes (liver cells) synthesize most of the plasma proteins, which include the **albumins** (54% of plasma proteins), **globulins** (38%), and **fibrinogen** (7%). Their functions are given in Table 19.1. Certain blood cells develop into cells that produce gamma globulins, an important type of globulin. These plasma proteins are also called **antibodies** or **immunoglobulins** because they are produced during certain immune responses. Foreign substances (antigens) such as bacteria and viruses stimulate production of millions of different antibodies. An antibody binds specifically to the antigen that stimulated its production and thus disables the invading antigen.

Besides proteins, other solutes in plasma include electrolytes, nutrients, regulatory substances such as enzymes and hormones, gases, and waste products such as urea, uric acid, creatinine, ammonia, and bilirubin.

Table 19.1 describes the chemical composition of blood plasma.

Formed Elements

The **formed elements** of the blood include three principal components: **red blood cells (RBCs), white blood cells (WBCs),** and **platelets** (Figure 19.2). RBCs and WBCs are whole cells; platelets are cell fragments. RBCs and platelets have just a few roles, but WBCs have a number of specialized functions. Several distinct types of WBCs—neutrophils, lymphocytes, monocytes, eosinophils, and basophils—each with a unique microscopic appearance, carry out these functions, which are discussed later in this chapter.

I. Red blood cells
II. White blood cells
    A. Granular leukocytes (contain conspicuous granules that are visible under a light microscope after staining)
    1. Neutrophils
    2. Eosinophils
    3. Basophils
    B. Agranular leukocytes (no granules are visible under a light microscope after staining)
    1. T and B lymphocytes and natural killer (NK) cells
    2. Monocytes
III. Platelets

The percentage of total blood volume occupied by RBCs is called the **hematocrit** (he-MAT-ō-krit); a hematocrit of 40 indicates that 40% of the volume of blood is composed of RBCs. The normal range of hematocrit for adult females is 38–46% (average = 42); for adult males, it is 40–54% (average = 47). The hormone testosterone, present in much higher concentration in males than in females, stimulates synthesis of erythropoietin

TABLE 19.1

Substances in Blood Plasma

CONSTITUENT	DESCRIPTION
Water (91.5%)	Liquid portion of blood. Acts as solvent and suspending medium for components of blood; absorbs, transports, and releases heat.
Plasma Proteins (7.0%)	Exert colloid osmotic pressure, which helps maintain water balance between blood and tissues and regulates blood volume.
Albumins	Smallest and most numerous blood plasma proteins; produced by liver. Function as transport proteins for several steroid hormones and for fatty acids.
Globulins	Produced by liver and by plasma cells, which develop from B lymphocytes. Antibodies (immunoglobulins) help attack viruses and bacteria. Alpha and beta globulins transport iron, lipids, and fat-soluble vitamins.
Fibrinogen	Produced by liver. Plays essential role in blood clotting.
Other Solutes (1.5%)	
Electrolytes	Inorganic salts. Positively charged ions (cations) include Na^+, K^+, Ca^{2+}, Mg^{2+}; negatively charged ions (anions) include Cl^-, HPO_4^{2-}, SO_4^{2-}, and HCO_3^-. Help maintain osmotic pressure and play essential roles in the function of cells.
Nutrients	Products of digestion pass into blood for distribution to all body cells. Include amino acids (from proteins), glucose (from carbohydrates), fatty acids and glycerol (from triglycerides), vitamins, and minerals.
Gases	Oxygen (O_2), carbon dioxide (CO_2), and nitrogen (N_2). More O_2 is associated with hemoglobin inside red blood cells; more CO_2 is dissolved in plasma. N_2 is present but has no known function in the body.
Regulatory substances	Enzymes, produced by body cells, catalyze chemical reactions. Hormones, produced by endocrine glands, regulate metabolism, growth, and development. Vitamins are cofactors for enzymatic reactions.
Waste products	Most are breakdown products of protein metabolism and are carried by blood to organs of excretion. Include urea, uric acid, creatine, creatinine, bilirubin, and ammonia.

(EPO), the hormone that in turn stimulates production of RBCs. Thus, testosterone contributes to higher hematocrits in males. Lower values in women during their reproductive years also may be due to excessive loss of blood during menstruation. A significant drop in hematocrit indicates *anemia,* a lower-than-

Figure 19.2 Scanning electron micrograph and photomicrograph of the formed elements of blood.

The formed elements of blood are red blood cells (RBCs), white blood cells (WBCs), and platelets.

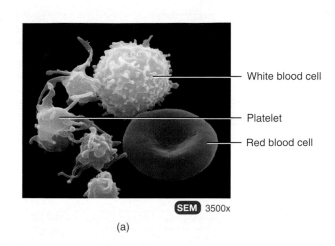

White blood cell

Platelet

Red blood cell

SEM 3500x

(a)

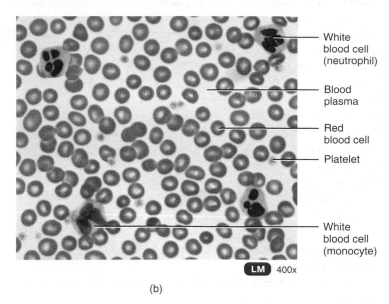

White blood cell (neutrophil)

Blood plasma

Red blood cell

Platelet

White blood cell (monocyte)

LM 400x

(b)

Which formed elements of the blood are cell fragments?

normal number of RBCs. In *polycythemia* the percentage of RBCs is abnormally high, and the hematocrit may be 65% or higher. This raises the viscosity of blood, which increases the resistance to flow and makes the blood more difficult for the heart to pump. Increased viscosity also contributes to high blood pressure and increased risk of stroke. Causes of polycythemia include abnormal increases in RBC production, tissue hypoxia, dehydration, and blood doping or the use of EPO by athletes.

◉ CHECKPOINT
1. In what ways is blood plasma similar to interstitial fluid? How does it differ?
2. What substances does blood transport?
3. How many kilograms or pounds of blood are there in your body?
4. How does the volume of blood plasma in your body compare to the volume of fluid in a two-liter bottle of Coke?
5. List the formed elements in blood plasma and describe their functions.
6. What is the significance of lower-than-normal or higher-than-normal hematocrit?

FORMATION OF BLOOD CELLS

◉ OBJECTIVE
• Explain the origin of blood cells.

Although some lymphocytes have a lifetime measured in years, most formed elements of the blood last only hours, days, or weeks, and must be replaced continually. Negative feedback

systems regulate the total number of RBCs and platelets in circulation, and their numbers normally remain steady. The abundance of the different types of WBCs, however, varies in response to challenges by invading pathogens and other foreign antigens.

The process by which the formed elements of blood develop is called **hemopoiesis** (hē-mō-poy-Ē-sis; *-poiesis* = making) or *hematopoiesis*. Before birth, hemopoiesis first occurs in the yolk sac of an embryo and later in the liver, spleen, thymus, and lymph nodes of a fetus. Red bone marrow becomes the primary site of hemopoiesis in the last three months before birth, and continues as the source of blood cells after birth and throughout life.

Red bone marrow is a highly vascularized connective tissue located in the microscopic spaces between trabeculae of spongy bone tissue. It is present chiefly in bones of the axial skeleton, pectoral and pelvic girdles, and the proximal epiphyses of the humerus and femur. About 0.05–0.1% of red bone marrow cells are derived from mesenchyme and are called **pluripotent stem cells** (ploo-RIP-ō-tent; *pluri-* = several) or *hemocytoblasts*. These cells have the capacity to develop into many different types of cells (Figure 19.3). In newborns all bone marrow is red and thus active in blood cell production. As an individual grows and in adulthood, the rate of blood cell formation decreases; the red bone marrow in the medullary (marrow) cavity of long bones becomes inactive and is replaced by yellow bone marrow, which is largely fat cells. Under certain conditions, such as severe bleeding, yellow bone marrow can revert to red bone marrow by extension of red bone marrow into yellow bone marrow and repopulation of yellow bone marrow by pluripotent stem cells.

Figure 19.3 Origin, development, and structure of blood cells. Some of the generations of some cell lines have been omitted.

Blood cell production, called hemopoiesis, occurs mainly in red bone marrow after birth.

Key:
- Progenitor cells
- Precursor cells or "blasts"
- Formed elements of circulating blood
- Tissue cells

Key:
- CFU–E — Colony-forming unit–erythrocyte
- CFU–Meg — Colony-forming unit–megakaryocyte
- CFU–GM — Colony-forming unit–granulocyte macrophage

? **From which connective tissue cells do pluripotent stem cells develop?**

• CLINICAL CONNECTION | Bone Marrow Examination

Sometimes a sample of red bone marrow must be obtained in order to diagnose certain blood disorders, such as leukemia and severe anemias. **Bone marrow examination** may involve *bone marrow aspiration* (withdrawal of a small amount of red bone marrow with a fine needle and syringe) or a *bone marrow biopsy* (removal of a core of red bone marrow with a larger needle).

Both types of samples are usually taken from the iliac crest of the hip bone, although samples are sometimes aspirated from the sternum. In young children, bone marrow samples are taken from a vertebra or tibia (shin bone). The tissue or cell sample is then sent to a pathology lab for analysis. Specifically, laboratory technicians look for signs of neoplastic (cancer) cells or other diseased cells to assist in diagnosis. •

Stem cells in red bone marrow reproduce themselves, proliferate, and differentiate into cells that give rise to blood cells, macrophages, reticular cells, mast cells, and adipocytes. Some of the stem cells can also form osteoblasts, chondroblasts, and muscle cells, and someday may be used as a source of bone, cartilage, and muscular tissue for tissue and organ replacement. The reticular cells produce reticular fibers, which form the stroma (framework) that supports red bone marrow cells. Once blood cells are produced in red bone marrow, they enter the bloodstream through *sinusoids* (also called *sinuses*), enlarged and leaky capillaries that surround red bone marrow cells and fibers. With the exception of lymphocytes, formed elements do not divide once they leave red bone marrow.

In order to form blood cells, pluripotent stem cells in red bone marrow produce two further types of stem cells, which have the capacity to develop into several types of cells. These stem cells are called *myeloid stem cells* and *lymphoid stem cells.* Myeloid stem cells begin their development in red bone marrow and give rise to red blood cells, platelets, monocytes, neutrophils, eosinophils, and basophils. Lymphoid stem cells begin their development in red bone marrow but complete it in lymphatic tissues; they give rise to lymphocytes. Although the various stem cells have distinctive cell identity markers in their plasma membranes, they cannot be distinguished histologically and resemble lymphocytes.

During hemopoiesis, some of the myeloid stem cells differentiate into **progenitor cells** (prō-JEN-i-tor). Other myeloid stem cells and the lymphoid stem cells develop directly into precursor cells (described shortly). Progenitor cells are no longer capable of reproducing themselves and are committed to giving rise to more specific elements of blood. Some progenitor cells are known as *colony-forming units (CFUs).* Following the CFU designation is an abbreviation that indicates the mature elements in blood that they will produce: CFU–E ultimately produces erythrocytes (red blood cells), CFU–Meg produces megakaryocytes, the source of platelets, and CFU–GM ultimately produces granulocytes (specifically, neutrophils) and monocytes (see Figure 19.3). Progenitor cells, like stem cells, resemble lymphocytes and cannot be distinguished by their microscopic appearance alone.

In the next generation, the cells are called **precursor cells,** also known as **blasts.** Over several cell divisions they develop into the actual formed elements of blood. For example, monoblasts develop into monocytes, eosinophilic myeloblasts develop into eosinophils, and so on. Precursor cells have recognizable microscopic appearances.

Several hormones called **hemopoietic growth factors** regulate the differentiation and proliferation of particular progenitor cells. **Erythropoietin** (e-rith′-rō-POY-e-tin) or **EPO** increases the number of red blood cell precursors. EPO is produced primarily by cells in the kidneys that lie between the kidney tubules (peritubular interstitial cells). With renal failure, EPO release slows and RBC production is inadequate. **Thrombopoietin** (throm′-bō-POY-ē-tin) or **TPO** is a hormone produced by the liver that stimulates the formation of platelets (thrombocytes) from megakaryocytes. Several different cytokines regulate development of different blood cell types. **Cytokines** are small glycoproteins that are typically produced by cells such as red bone marrow cells, leukocytes, macrophages, fibroblasts, and endothelial cells. They generally act as local hormones

(autocrines or paracrines; see Chapter 18). Cytokines stimulate proliferation of progenitor cells in red bone marrow and regulate the activities of cells involved in nonspecific defenses (such as phagocytes) and immune responses (such as B cells and T cells). Two important families of cytokines that stimulate white blood cell formation are **colony-stimulating factors (CSFs)** and **interleukins.**

• CLINICAL CONNECTION | Medical Uses of Hemopoietic Growth Factors

Hemopoietic growth factors made available through recombinant DNA technology hold tremendous potential for medical uses when a person's natural ability to form new blood cells is diminished or defective. The artificial form of erythropoietin (Epoetin alfa) is very effective in treating the diminished red blood cell production that accompanies end-stage kidney disease. Granulocyte-macrophage colony-stimulating factor and granulocyte CSF are given to stimulate white blood cell formation in cancer patients who are undergoing chemotherapy, which kills red bone marrow cells as well as cancer cells because both cell types are undergoing mitosis. (Recall that white blood cells help protect against disease.) Thrombopoietin shows great promise for preventing the depletion of platelets, which are needed to help blood clot, during chemotherapy. CSFs and thrombopoietin also improve the outcome of patients who receive bone marrow transplants. Hemopoietic growth factors are also used to treat thrombocytopenia in neonates, other clotting disorders, and various types of anemia. Research on these medications is ongoing and shows a great deal of promise. •

◉ CHECKPOINT

7. Which hemopoietic growth factors regulate differentiation and proliferation of CFU–E and formation of platelets from megakaryocytes?
8. Describe the formation of platelets from pluripotent stem cells, including the influence of hormones.

RED BLOOD CELLS

◉ OBJECTIVE

• Describe the structure, functions, life cycle, and production of red blood cells.

Red blood cells (RBCs) or **erythrocytes** (e-RITH-rō-sīts; *erythro-*= red; *-cyte* = cell) contain the oxygen-carrying protein **hemoglobin,** which is a pigment that gives whole blood its red color. A healthy adult male has about 5.4 million red blood cells per microliter (μL) of blood,* and a healthy adult female has about 4.8 million. (One drop of blood is about 50 μL.) To maintain normal numbers of RBCs, new mature cells must enter the circulation at the astonishing rate of at least 2 million per second, a pace that balances the equally high rate of RBC destruction.

*1 μL = 1 mm^3 = 10^{-6} liter.

RBC Anatomy

RBCs are biconcave discs with a diameter of 7–8 μm (Figure 19.4a). Mature red blood cells have a simple structure. Their plasma membrane is both strong and flexible, which allows them to deform without rupturing as they squeeze through narrow capillaries. As you will see later, certain glycolipids in the plasma membrane of RBCs are antigens that account for the various blood groups such as the ABO and Rh groups. RBCs lack a nucleus and other organelles and can neither reproduce nor carry on extensive metabolic activities. The cytosol of RBCs contains hemoglobin molecules; these important molecules are synthesized before loss of the nucleus during RBC production and constitute about 33% of the cell's weight.

RBC Physiology

Red blood cells are highly specialized for their oxygen transport function. Because mature RBCs have no nucleus, all their internal space is available for oxygen transport. Because RBCs lack mitochondria and generate ATP anaerobically (without oxygen), they do not use up any of the oxygen they transport. Even the shape of an RBC facilitates its function. A biconcave disc has a much greater surface area for the diffusion of gas molecules into and out of the RBC than would, say, a sphere or a cube.

Each RBC contains about 280 million hemoglobin molecules. A hemoglobin molecule consists of a protein called **globin,** composed of four polypeptide chains (two alpha and two beta chains); a ringlike nonprotein pigment called a **heme** (Figure 19.4b) is bound to each of the four chains. At the center of each heme ring is an iron ion (Fe^{2+}) that can combine reversibly with

one oxygen molecule (Figure 19.4c), allowing each hemoglobin molecule to bind four oxygen molecules. Each oxygen molecule picked up from the lungs is bound to an iron ion. As blood flows through tissue capillaries, the iron–oxygen reaction reverses. Hemoglobin releases oxygen, which diffuses first into the interstitial fluid and then into cells.

Hemoglobin also transports about 23% of the total carbon dioxide, a waste product of metabolism. Blood flowing through tissue capillaries picks up carbon dioxide, some of which combines with amino acids in the globin part of hemoglobin. As blood flows through the lungs, the carbon dioxide is released from hemoglobin and then exhaled.

In addition to its key role in transporting oxygen and carbon dioxide, hemoglobin also plays a role in the regulation of blood flow and blood pressure. The gaseous hormone **nitric oxide (NO),** produced by the endothelial cells that line blood vessels, binds to hemoglobin. Under some circumstances, hemoglobin releases NO. The released NO causes *vasodilation,* an increase in blood vessel diameter that occurs when the smooth muscle in the vessel wall relaxes. Vasodilation improves blood flow and enhances oxygen delivery to cells near the site of NO release.

RBC Life Cycle

Red blood cells live only about 120 days because of the wear and tear their plasma membranes undergo as they squeeze through blood capillaries. Without a nucleus and other organelles, RBCs cannot synthesize new components to replace damaged ones. The plasma membrane becomes more fragile with age, and the cells are more likely to burst, especially as they squeeze through narrow channels in the spleen. Ruptured

Figure 19.4 The shapes of a red blood cell (RBC) and a hemoglobin molecule. In (b), note that each of the four polypetide chains of a hemoglobin molecule (blue) has one heme group (gold), which contains an iron ion (Fe^{2+}), shown in red.

The iron portion of a heme group binds oxygen for transport by hemoglobin.

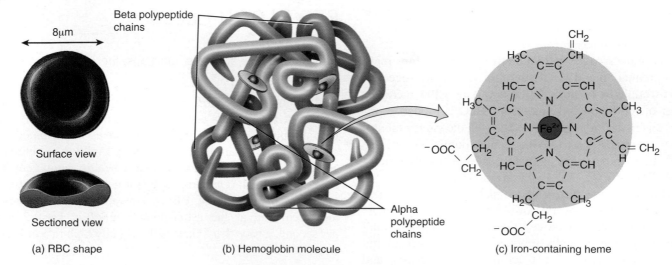

| (a) RBC shape | (b) Hemoglobin molecule | (c) Iron-containing heme |

? How many molecules of O_2 can one hemoglobin molecule transport?

red blood cells are removed from circulation and destroyed by fixed phagocytic macrophages in the spleen and liver, and the breakdown products are recycled, as follows (Figure 19.5):

1 Macrophages in the spleen, liver, or red bone marrow phagocytize ruptured and worn-out red blood cells.

2 The globin and heme portions of hemoglobin are split apart.

3 Globin is broken down into amino acids, which can be reused to synthesize other proteins.

4 Iron is removed from the heme portion in the form of Fe^{3+}, which associates with the plasma protein **transferrin** (trans-FER-in; *trans-* = across; *ferr-* = iron), a transporter for Fe^{3+} in the bloodstream.

5 In muscle fibers, liver cells, and macrophages of the spleen and liver, Fe^{3+} detaches from transferrin and attaches to an iron-storage protein called **ferritin.**

6 Upon release from a storage site or absorption from the gastrointestinal tract, Fe^{3+} reattaches to transferrin.

7 The Fe^{3+}–transferrin complex is then carried to red bone marrow, where RBC precursor cells take it up through receptor-mediated endocytosis (see Figure 3.12 on page 74) for use in hemoglobin synthesis. Iron is needed for the heme portion of the hemoglobin molecule, and amino acids are needed for the globin portion. Vitamin B_{12} is also needed for the synthesis of hemoglobin.

8 Erythropoiesis in red bone marrow results in the production of red blood cells, which enter the circulation.

9 When iron is removed from heme, the non-iron portion of heme is converted to **biliverdin** (bil'-i-VER-din), a green pigment, and then into **bilirubin** (bil' i ROO bin), a yellow orange pigment.

10 Bilirubin enters the blood and is transported to the liver.

11 Within the liver, bilirubin is released by liver cells into bile, which passes into the small intestine and then into the large intestine.

12 In the large intestine, bacteria convert bilirubin into **urobilinogen** (ūr-ō-bī-LIN-ō-jen).

13 Some urobilinogen is absorbed back into the blood, converted to a yellow pigment called **urobilin** (ūr-ō-BĪ-lin), and excreted in urine.

14 Most urobilinogen is eliminated in feces in the form of a brown pigment called **stercobilin** (ster'-kō-BĪ-lin), which gives feces its characteristic color.

Figure 19.5 Formation and destruction of red blood cells, and the recycling of hemoglobin components. RBCs circulate for about 120 days after leaving red bone marrow before they are phagocytized by macrophages.

 The rate of RBC formation by red bone marrow equals the rate of RBC destruction by macrophages.

? What is the function of transferrin?

Because free iron ions (Fe^{2+} and Fe^{3+}) bind to and damage molecules in cells or in the blood, transferrin and ferritin act as protective "protein escorts" during transport and storage of iron ions. As a result, plasma contains virtually no free iron. Furthermore, only small amounts are available inside body cells for use in synthesis of iron-containing molecules such as the cytochrome pigments needed for ATP production in mitochondria (see Figure 25.9 on page 987). In cases of **iron overload**, the amount of iron present in the body builds up. Because we have no method for eliminating excess iron, any condition that increases dietary iron absorption can cause iron overload. At some point, the proteins transferrin and ferritin become saturated with iron ions, and free iron level rises. Common consequences of iron overload are diseases of the liver, heart, pancreatic islets, and gonads. Iron overload also allows certain iron-dependent microbes to flourish. Such microbes normally are not pathogenic, but they multiply rapidly and can cause lethal effects in a short time when free iron is present. ●

Erythropoiesis: Production of RBCs

Erythropoiesis (e-rith′-rō-poy-Ē-sis), the production of RBCs, starts in the red bone marrow with a precursor cell called a **proerythroblast** (see Figure 19.3). The proerythroblast divides several times, producing cells that begin to synthesize hemoglobin. Ultimately, a cell near the end of the development sequence ejects its nucleus and becomes a **reticulocyte** (re-TIK-ū-lō-sīt). Loss of the nucleus causes the center of the cell to indent, producing the red blood cell's distinctive biconcave shape. Reticulocytes retain some mitochondria, ribosomes, and endoplasmic reticulum. They pass from red bone marrow into the bloodstream by squeezing between the endothelial cells of blood capillaries. Reticulocytes develop into mature red blood cells within 1 to 2 days after their release from red bone marrow.

Normally, erythropoiesis and red blood cell destruction proceed at roughly the same pace. If the oxygen-carrying capacity of the blood falls because erythropoiesis is not keeping up with RBC destruction, a negative feedback system steps up RBC production (Figure 19.6). The controlled condition is the amount of oxygen delivered to body tissues. Cellular oxygen deficiency, called **hypoxia** (hī-POKS-ē-a), may occur if too little oxygen enters the blood. For example, the lower oxygen content of air at high altitudes reduces the amount of oxygen in the blood. Oxygen delivery may also fall due to anemia, which has many causes: Lack of iron, lack of certain amino acids, and lack of vitamin B_{12} are but a few (see page 711). Circulatory problems that reduce blood flow to tissues may also reduce oxygen delivery. Whatever the cause, hypoxia stimulates the kidneys to step up the release of erythropoietin, which speeds the development of proerythroblasts into reticulocytes in the red bone marrow. As the number of circulating RBCs increases, more oxygen can be delivered to body tissues.

Premature newborns often exhibit anemia, due in part to inadequate production of erythropoietin. During the first weeks after birth, the liver, not the kidneys, produces most EPO.

Figure 19.6 Negative feedback regulation of erythropoiesis (red blood cell formation). Lower oxygen content of air at high altitudes, anemia, and circulatory problems may reduce oxygen delivery to body tissues.

The main stimulus for erythropoiesis is hypoxia, a decrease in the oxygen-carrying capacity of the blood.

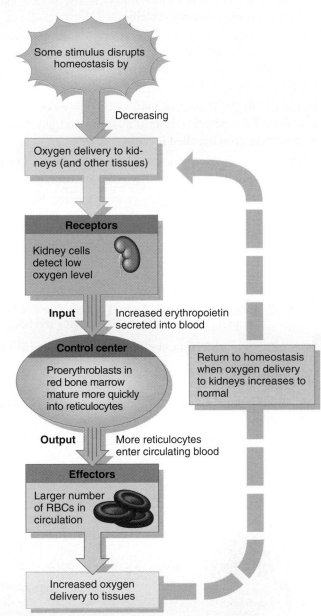

Some stimulus disrupts homeostasis by

Decreasing

Oxygen delivery to kidneys (and other tissues)

Receptors

Kidney cells detect low oxygen level

Input — Increased erythropoietin secreted into blood

Control center

Proerythroblasts in red bone marrow mature more quickly into reticulocytes

Return to homeostasis when oxygen delivery to kidneys increases to normal

Output — More reticulocytes enter circulating blood

Effectors

Larger number of RBCs in circulation

Increased oxygen delivery to tissues

❓ **How might your hematocrit change if you moved from a town at sea level to a high mountain village?**

Because the liver is less sensitive than the kidneys to hypoxia, newborns have a smaller EPO response to anemia than do adults. Because fetal hemoglobin (hemoglobin present at birth) carries up to 30% more oxygen, the loss of fetal hemoglobin, due to insufficient erythropoietin production, makes the anemia worse.

• CLINICAL **CONNECTION** | Reticulocyte Count

The rate of erythropoiesis is measured by a **reticulocyte count.** Normally, a little less than 1% of the oldest RBCs are replaced by newcomer reticulocytes on any given day. It then takes 1 to 2 days for the reticulocytes to lose the last vestiges of endoplasmic reticulum and become mature RBCs. Thus, reticulocytes account for about 0.5–1.5% of all RBCs in a normal blood sample. A low "retic" count in a person who is anemic might indicate a shortage of erythropoietin or an inability of the red bone marrow to respond to EPO, perhaps because of a nutritional deficiency or leukemia. A high "retic" count might indicate a good red bone marrow response to previous blood loss or to iron therapy in someone who had been iron deficient. It could also point to illegal use of Epoetin alfa by an athlete. •

◉ CHECKPOINT

9. Describe the size, microscopic appearance, and functions of RBCs.
10. How is hemoglobin recycled?
11. What is erythropoiesis? How does erythropoiesis affect hematocrit? What factors speed up and slow down erythropoiesis?

WHITE BLOOD CELLS

◉ OBJECTIVE

• Describe the structure, functions, and production of white blood cells (WBCs).

Types of WBCs

Unlike red blood cells, white blood cells or **leukocytes** (LOO-kō-sīts; *leuko-* = white) have nuclei and do not contain hemoglobin (Figure 19.7). WBCs are classified as either granular or agranular, depending on whether they contain conspicuous chemical-filled cytoplasmic granules (vesicles) that are made visible by staining when viewed through a light microscope. *Granular leukocytes* include neutrophils, eosinophils, and basophils; *agranular leukocytes* include lymphocytes and mono-

cytes. As shown in Figure 19.3, monocytes and granular leukocytes develop from a myeloid stem cell, and lymphocytes develop from a lymphoid stem cell.

Granular Leukocytes

After staining, each of the three types of granular leukocytes displays conspicuous granules with distinctive coloration that can be recognized under a light microscope. The large, uniform-sized granules within an **eosinophil** (ē-ō-SIN-ō-fil) are *eosinophilic* (= eosin-loving)—they stain red-orange with acidic dyes (Figure 19.7a). The granules usually do not cover or obscure the nucleus, which most often has two lobes connected by a thick strand of chromatin. The round, variable-sized granules of a **basophil** (BĀ-sō-fil) are *basophilic* (= basic loving)—they stain blue-purple with basic dyes (Figure 19.7b). The granules commonly obscure the nucleus, which has two lobes. The granules of a **neutrophil** (NOO-trō-fil) are smaller, evenly distributed, and pale lilac in color (Figure 19.7c); the nucleus has two to five lobes, connected by very thin strands of chromatin. As the cells age, the number of nuclear lobes increases. Because older neutrophils have several differently shaped nuclear lobes, they are often called *polymorphonuclear leukocytes (PMNs),* polymorphs, or "polys." Younger neutrophils are often called *bands* because their nucleus is more rod-shaped.

Agranular Leukocytes

Even though so-called agranular leukocytes possess cytoplasmic granules, the granules are not visible under a light microscope because of their small size and poor staining qualities.

The nucleus of a **lymphocyte** (LIM-fō-sīt) is round or slightly indented and stains darkly. The cytoplasm stains sky blue and forms a rim around the nucleus (Figure 19.7d). The larger the cell, the more cytoplasm is visible. Lymphocytes may be as small as 6–9 μm in diameter or as large as 10–14 μm in diameter. Although the functional significance of the size difference between small and large lymphocytes is unclear, the distinction is still clinically useful because an increase in the number of large

Figure 19.7 Types of white blood cells.

  The shapes of their nuclei and the staining properties of their cytoplasmic granules distinguish white blood cells from one another.

LM all 1600x

(a) Eosinophil   (b) Basophil   (c) Neutrophil   (d) Lymphocyte   (e) Monocyte

? Which WBCs are called granular leukocytes? Why?

lymphocytes has diagnostic significance in acute viral infections and in some immunodeficiency diseases. As noted earlier, there are three types of lymphocytes: T lymphocytes (T cells), B lymphocytes (B cells), and natural killer (NK) cells.

Monocytes (MON-ō-sīts) are 12–20 μm in diameter (Figure 19.7e). The nucleus of a monocyte is usually kidney shaped or horseshoe shaped, and the cytoplasm is blue-gray and has a foamy appearance. The color and appearance are due to very fine *azurophilic granules* (az′-ū-rō-FIL-ik; *azur-* = blue; *-philic* = loving), which are lysosomes. The blood transports monocytes from the blood into the tissues, where they enlarge and differentiate into **macrophages** (= large eaters). Some become **fixed macrophages,** which means they reside in a particular tissue; examples are alveolar macrophages in the lungs, macrophages in the spleen, and stellate reticuloendothelial (Kupffer) cells in the liver. Others become **wandering macrophages,** which roam the tissues and gather at sites of infection or inflammation.

White blood cells and all other nucleated cells in the body have proteins, called **major histocompatibility (MHC) antigens,** protruding from their plasma membrane into the extracellular fluid. These "cell identity markers" are unique for each person (except identical twins). Although RBCs possess blood group antigens, they lack the MHC antigens.

Functions of WBCs

In a healthy body, some WBCs, especially lymphocytes, can live for several months or years, but most live only a few days. During a period of infection, phagocytic WBCs may live only a few hours. WBCs are far less numerous than red blood cells; at about 5000–10,000 cells per μL of blood, they are outnumbered by RBCs by about 700 : 1. **Leukocytosis** (loo′-kō-sī-TŌ-sis), an increase in the number of WBCs above 10,000/μL, is a normal, protective response to stresses such as invading microbes, strenuous exercise, anesthesia, and surgery. An abnormally low level of white blood cells (below 5000/μL) is termed **leukopenia** (loo′-kō-PĒ-nē-a). It is never beneficial and may be caused by radiation, shock, and certain chemotherapeutic agents.

The skin and mucous membranes of the body are continuously exposed to microbes and their toxins. Some of these microbes can invade deeper tissues to cause disease. Once pathogens enter the body, the general function of white blood cells is to combat them by phagocytosis or immune responses. To accomplish these tasks, many WBCs leave the bloodstream and collect at sites of pathogen invasion or inflammation. Once granular leukocytes and monocytes leave the bloodstream to fight injury or infection, they never return to it. Lymphocytes, on the other hand, continually recirculate—from blood to interstitial spaces of tissues to lymphatic fluid and back to blood. Only 2% of the total lymphocyte population is circulating in the blood at any given time; the rest are in lymphatic fluid and organs such as the skin, lungs, lymph nodes, and spleen.

WBCs leave the bloodstream by a process termed **emigration** (em′-i-GRĀ-shun; *e-* = out; *migra-* = wander), formerly called *diapedesis* (dī-a-pe-DĒ-sis), in which they roll along the endothelium, stick to it, and then squeeze between endothelial cells (Figure 19.8). The precise signals that stimulate emigration through a particular blood vessel vary for the different types of WBCs. Molecules known as **adhesion molecules** help WBCs stick to the endothelium. For example, endothelial cells display adhesion molecules called *selectins* in response to nearby injury and inflammation. Selectins stick to carbohydrates on the surface of neutrophils, causing them to slow down and roll along the endothelial surface. On the neutrophil surface are other adhesion molecules called *integrins,* which tether neutrophils to the endothelium and assist their movement through the blood vessel wall and into the interstitial fluid of the injured tissue.

Neutrophils and macrophages are active in **phagocytosis;** they can ingest bacteria and dispose of dead matter (see Figure 3.13 on page 75). Several different chemicals released by microbes and inflamed tissues attract phagocytes, a phenomenon called **chemotaxis.** The substances that provide stimuli for chemotaxis include toxins produced by microbes; kinins, which are specialized products of damaged tissues; and some of the colony-stimulating factors (CSFs). The CSFs also enhance the phagocytic activity of neutrophils and macrophages.

Among WBCs, neutrophils respond most quickly to tissue destruction by bacteria. After engulfing a pathogen during phago-

Figure 19.8 Emigration of white blood cells.

Adhesion molecules (selectins and integrins) assist the emigration of WBCs from the bloodstream into interstitial fluid.

Interstitial fluid

Blood flow

Neutrophil

Endothelial cell

Rolling

Sticking

Squeezing between endothelial cells

Key:

§ Selectins on endothelial cells

◼ Integrins on neutrophil

❓ **In what way is the "traffic pattern" of lymphocytes in the body different from that of other WBCs?**

cytosis, a neutrophil unleashes several chemicals to destroy the pathogen. These chemicals include the enzyme **lysozyme,** which destroys certain bacteria, and **strong oxidants,** such as the superoxide anion (O_2^-), hydrogen peroxide (H_2O_2), and the hypochlorite anion (OCl^-), which is similar to household bleach. Neutrophils also contain **defensins,** proteins that exhibit a broad range of antibiotic activity against bacteria and fungi. Within a neutrophil, vesicles containing defensins merge with phagosomes containing microbes. Defensins form peptide "spears" that poke holes in microbe membranes; the resulting loss of cellular contents kills the invader.

Monocytes take longer to reach a site of infection than neutrophils, but they arrive in larger numbers and destroy more microbes. Upon their arrival monocytes enlarge and differentiate into wandering macrophages, which clean up cellular debris and microbes by phagocytosis after an infection.

At sites of inflammation, basophils leave capillaries, enter tissues, and release granules that contain heparin, histamine, and serotonin. These substances intensify the inflammatory reaction and are involved in hypersensitivity (allergic) reactions. Basophils are similar in function to mast cells, connective tissue cells that originate from pluripotent stem cells in red bone marrow. Like basophils, mast cells release substances involved in inflammation, including heparin, histamine, and proteases. Mast cells are widely dispersed in the body, particularly in connective tissues of the skin and mucous membranes of the respiratory and gastrointestinal tracts.

Eosinophils leave the capillaries and enter tissue fluid. They are believed to release enzymes, such as histaminase, that combat the effects of histamine and other substances involved in inflammation during allergic reactions. Eosinophils also phagocytize antigen–antibody complexes and are effective against certain parasitic worms. A high eosinophil count often indicates an allergic condition or a parasitic infection.

Lymphocytes are the major soldiers in immune system battles (described in detail in Chapter 22). Most lymphocytes continually move among lymphoid tissues, lymph, and blood, spending only a few hours at a time in blood. Thus, only a small proportion of the total lymphocytes are present in the blood at any given time. Three main types of lymphocytes are B cells, T cells, and natural killer (NK) cells. B cells are particularly effective in destroying bacteria and inactivating their toxins. T cells attack viruses, fungi, transplanted cells, cancer cells, and some bacteria, and are responsible for transfusion reactions, allergies, and the rejection of transplanted organs. Immune responses carried out by both B cells and T cells help combat infection and provide protection against some diseases. Natural killer cells attack a wide variety of infectious microbes and certain spontaneously arising tumor cells.

As you have already learned, an increase in the number of circulating WBCs usually indicates inflammation or infection. A physician may order a **differential white blood cell count,** a count of each of the five types of white blood cells, to detect infection or inflammation, determine the effects of possible poisoning by chemicals or drugs, monitor blood disorders (for example, leukemia) and the effects of chemotherapy, or detect

allergic reactions and parasitic infections. Because each type of white blood cell plays a different role, determining the *percentage* of each type in the blood assists in diagnosing the condition. Table 19.2 lists the significance of both high and low WBC counts.

TABLE 19.2

Significance of High and Low White Blood Cell Counts

WBC TYPE	HIGH COUNT MAY INDICATE	LOW COUNT MAY INDICATE
Neutrophils	Bacterial infection, burns, stress, inflammation.	Radiation exposure, drug toxicity, vitamin B_{12} deficiency, or systemic lupus erythematosus (SLE).
Lymphocytes	Viral infections, some leukemias.	Prolonged illness, immunosuppression, or treatment with cortisol.
Monocytes	Viral or fungal infections, tuberculosis, some leukemias, other chronic diseases.	Bone marrow suppression, treatment with cortisol.
Eosinophils	Allergic reactions, parasitic infections, autoimmune diseases.	Drug toxicity, stress.
Basophils	Allergic reactions, leukemias, cancers, hypothyroidism.	Pregnancy, ovulation, stress, or hyperthyroidism.

◉ CHECKPOINT

12. What is the importance of emigration, chemotaxis, and phagocytosis in fighting bacterial invaders?
13. How are leukocytosis and leukopenia different?
14. What is a differential white blood cell count?
15. What functions do granular leukocytes, macrophages, B cells, T cells, and natural killer cells perform?

PLATELETS

◉ OBJECTIVE
• Describe the structure, function, and origin of platelets.

Besides the immature cell types that develop into erythrocytes and leukocytes, hemopoietic stem cells also differentiate into cells that produce platelets. Under the influence of the hormone **thrombopoietin,** myeloid stem cells develop into megakaryocyte-colony-forming cells that in turn develop into precursor cells called *megakaryoblasts* (see Figure 19.3). Megakaryoblasts transform into megakaryocytes, huge cells that splinter into 2000 to 3000 fragments. Each fragment, enclosed by a piece of the plasma membrane, is a **platelet (thrombocyte).** Platelets break off from the megakaryocytes in red bone marrow and then enter the blood circulation. Between 150,000 and 400,000 platelets are present in each μL of blood. Each is disc-shaped, 2–4 μm in diameter, and has many vesicles but no nucleus.

Platelets help stop blood loss from damaged blood vessels by forming a platelet plug. Their granules also contain chemicals that, once released, promote blood clotting. Platelets have a short life span, normally just 5 to 9 days. Aged and dead platelets are removed by fixed macrophages in the spleen and liver.

Table 19.3 summarizes the formed elements in blood.

●CLINICAL CONNECTION | Complete Blood Count

A complete blood count (CBC) is a very valuable test that screens for anemia and various infections. Usually included are counts of RBCs, WBCs, and platelets per μL of whole blood; hematocrit; and differential white blood cell count. The amount of hemoglobin in grams per milliliter of blood also is determined. Normal hemoglobin ranges are: infants, 14–20 g/100 mL of blood; adult females, 12–16 g/100 mL of blood; and adult males, 13.5–18 g/100 mL of blood. ●

TABLE 19.3

Summary of Formed Elements in Blood

NAME AND APPEARANCE	NUMBER	CHARACTERISTICS*	FUNCTIONS
Red Blood Cells (RBCs) or Erythrocytes	4.8 million/μL in females; 5.4 million/μL in males	7–8 μm diameter, biconcave discs, without nuclei; live for about 120 days.	Hemoglobin within RBCs transports most of the oxygen and part of the carbon dioxide in the blood.
White Blood Cells (WBCs) or Leukocytes	5000–10,000/μL	Most live for a few hours to a few days[†].	Combat pathogens and other foreign substances that enter the body.
Granular leukocytes			
Neutrophils	60–70% of all WBCs	10–12 μm diameter; nucleus has 2–5 lobes connected by thin strands of chromatin; cytoplasm has very fine, pale lilac granules.	Phagocytosis. Destruction of bacteria with lysozyme, defensins, and strong oxidants, such as superoxide anion, hydrogen peroxide, and hypochlorite anion.
Eosinophils	2–4% of all WBCs	10–12 μm diameter; nucleus usually has 2 lobes connected by a thick strand of chromatin; large, red-orange granules fill the cytoplasm.	Combat the effects of histamine in allergic reactions, phagocytize antigen–antibody complexes, and destroy certain parasitic worms.
Basophils	0.5–1% of all WBCs	8–10 μm diameter; nucleus has 2 lobes; large cytoplasmic granules appear deep blue-purple.	Liberate heparin, histamine, and serotonin in allergic reactions that intensify the overall inflammatory response.
Agranular leukocytes			
Lymphocytes (T cells, B cells, and natural killer cells)	20–25% of all WBCs	Small lymphocytes are 6–9 μm in diameter; large lymphocytes are 10–14 μm in diameter; nucleus is round or slightly indented; cytoplasm forms a rim around the nucleus that looks sky blue; the larger the cell, the more cytoplasm is visible.	Mediate immune responses, including antigen–antibody reactions. B cells develop into plasma cells, which secrete antibodies. T cells attack invading viruses, cancer cells, and transplanted tissue cells. Natural killer cells attack a wide variety of infectious microbes and certain spontaneously arising tumor cells.
Monocytes	3–8% of all WBCs	12–20 μm diameter; nucleus is kidney shaped or horseshoe shaped; cytoplasm is blue-gray and has foamy appearance.	Phagocytosis (after transforming into fixed or wandering macrophages).
Platelets (Thrombocytes)	150,000–400,000/μL	2–4 μm diameter cell fragments that live for 5–9 days; contain many vesicles but no nucleus.	Form platelet plug in hemostasis; release chemicals that promote vascular spasm and blood clotting.

*Colors are those seen when using Wright's stain.

[†]Some lymphocytes, called T and B memory cells, can live for many years once they are established.

⊙ **CHECKPOINT**

16. How do RBCs, WBCs, and platelets compare with respect to size, number per μL of blood, and life span?

STEM CELL TRANSPLANTS FROM BONE MARROW AND CORD-BLOOD

⊙ **OBJECTIVE**

- Explain the importance of bone marrow transplants and stem cell transplants.

A **bone marrow transplant** is the replacement of cancerous or abnormal red bone marrow with healthy red bone marrow in order to establish normal blood cell counts. In patients with cancer or certain genetic diseases, the defective red bone marrow is destroyed by high doses of chemotherapy and whole body radiation just before the transplant takes place. These treatments kill the cancer cells and destroy the patient's immune system in order to decrease the chance of transplant rejection.

Healthy red bone marrow for transplanting may be supplied by a donor or by the patient when the underlying disease is inactive, as when leukemia is in remission. The red bone marrow from a donor is usually removed from the iliac crest of the hip bone under general anesthesia with a syringe and is then injected into the recipient's vein, much like a blood transfusion. The injected marrow migrates to the recipient's red bone marrow cavities, where the donor's stem cells multiply. If all goes well, the recipient's red bone marrow is replaced entirely by healthy, noncancerous cells.

Bone marrow transplants have been used to treat aplastic anemia, certain types of leukemia, severe combined immuno-deficiency disease (SCID), Hodgkin's disease, non-Hodgkin's lymphoma, multiple myeloma, thalassemia, sickle-cell disease, breast cancer, ovarian cancer, testicular cancer, and hemolytic anemia. However, there are some drawbacks. Since the recipient's white blood cells have been completely destroyed by chemotherapy and radiation, the patient is extremely vulnerable to infection. (It takes about 2–3 weeks for transplanted bone marrow to produce enough white blood cells to protect against infection.) In addition, transplanted red bone marrow may produce T cells that attack the recipient's tissues, a reaction called *graft-versus-host disease.* Similarly, any of the recipient's T cells that survived the chemotherapy and radiation can attack donor transplant cells. Another drawback is that patients must take immunosuppressive drugs for life. Because these drugs reduce the level of immune system activity, they increase the risk of infection. Immunosuppressive drugs also have side effects such as fever, muscle aches, headache, nausea, fatigue, depression, high blood pressure, and kidney and liver damage.

A more recent advance for obtaining stem cells involves a **cord-blood transplant.** The connection between the mother and embryo (and later the fetus) is the umbilical cord. Stem cells may be obtained from the umbilical cord shortly after birth. The stem cells are removed from the cord with a syringe and then frozen. Stem cells from the cord have several advantages over those obtained from red bone marrow:

1. They are easily collected following permission of the newborn's parents.

2. They are more abundant than stem cells in red bone marrow.

3. They are less likely to cause graft-versus-host disease, so the match between donor and recipient does not have to be as close as in a bone marrow transplant. This provides a larger number of potential donors.

4. They are less likely to transmit infections.

5. They can be stored indefinitely in cord-blood banks.

⊙ **CHECKPOINT**

17. How are cord-blood transplants and bone marrow transplants similar? How do they differ?

HEMOSTASIS

⊙ **OBJECTIVES**

- Describe the three mechanisms that contribute to hemostasis.
- Identify the stages of blood clotting and explain the various factors that promote and inhibit blood clotting.

Hemostasis (hē-mō-STĀ-sis), not to be confused with the very similar term homeostasis, is a sequence of responses that stops bleeding. When blood vessels are damaged or ruptured, the hemostatic response must be quick, localized to the region of damage, and carefully controlled in order to be effective. Three mechanisms reduce blood loss: (1) vascular spasm, (2) platelet plug formation, and (3) blood clotting (coagulation). When successful, hemostasis prevents **hemorrhage** (HEM-o-rij; *-rhage* = burst forth), the loss of a large amount of blood from the vessels. Hemostatic mechanisms can prevent hemorrhage from smaller blood vessels, but extensive hemorrhage from larger vessels usually requires medical intervention.

Vascular Spasm

When arteries or arterioles are damaged, the circularly arranged smooth muscle in their walls contracts immediately, a reaction called **vascular spasm.** This reduces blood loss for several minutes to several hours, during which time the other hemostatic mechanisms go into operation. The spasm is probably caused by damage to the smooth muscle, by substances released from activated platelets, and by reflexes initiated by pain receptors.

Platelet Plug Formation

Considering their small size, platelets store an impressive array of chemicals. Within many vesicles are clotting factors, ADP, ATP, Ca^{2+}, and serotonin. Also present are enzymes that produce thromboxane A2, a prostaglandin; *fibrin-stabilizing factor,* which helps to strengthen a blood clot; lysosomes; some mitochondria; membrane systems that take up and store calcium and provide channels for release of the contents of granules; and glycogen. Also within platelets is **platelet-derived growth factor (PDGF),** a hormone that can cause proliferation of vascular endothelial cells, vascular smooth muscle fibers, and fibroblasts to help repair damaged blood vessel walls.

Platelet plug formation occurs as follows (Figure 19.9):

Figure 19.9 Platelet plug formation.

A platelet plug can stop blood loss completely if the hole in a blood vessel is small enough.

1 Platelet adhesion

2 Platelet release reaction

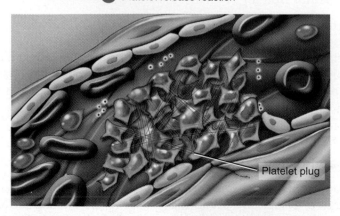

3 Platelet aggregation

? Along with platelet plug formation, which two mechanisms contribute to hemostasis?

1 Initially, platelets contact and stick to parts of a damaged blood vessel, such as collagen fibers of the connective tissue underlying the damaged endothelial cells. This process is called **platelet adhesion.**

2 Due to adhesion, the platelets become activated, and their characteristics change dramatically. They extend many projections that enable them to contact and interact with one another, and they begin to liberate the contents of their vesicles. This phase is called the **platelet release reaction.** Liberated ADP and thromboxane A2 play a major role by activating nearby platelets. Serotonin and thromboxane A2 function as vasoconstrictors, causing and sustaining contraction of vascular smooth muscle, which decreases blood flow through the injured vessel.

3 The release of ADP makes other platelets in the area sticky, and the stickiness of the newly recruited and activated platelets causes them to adhere to the originally activated platelets. This gathering of platelets is called **platelet aggregation.** Eventually, the accumulation and attachment of large numbers of platelets form a mass called a **platelet plug.**

A platelet plug is very effective in preventing blood loss in a small vessel. Although initially the platelet plug is loose, it becomes quite tight when reinforced by fibrin threads formed during clotting (see Figure 19.10). A platelet plug can stop blood loss completely if the hole in a blood vessel is not too large.

Blood Clotting

Normally, blood remains in its liquid form as long as it stays within its vessels. If it is drawn from the body, however, it thickens and forms a gel. Eventually, the gel separates from the liquid. The straw-colored liquid, called **serum,** is simply blood plasma minus the clotting proteins. The gel is called a **clot.** It consists of a network of insoluble protein fibers called fibrin in which the formed elements of blood are trapped (Figure 19.10).

The process of gel formation, called **clotting** or **coagulation,** is a series of chemical reactions that culminates in formation of fibrin threads. If blood clots too easily, the result can be **thrombosis**—clotting in an undamaged blood vessel. If the blood takes too long to clot, hemorrhage can occur.

Clotting involves several substances known as **clotting (coagulation) factors.** These factors include calcium ions (Ca^{2+}), several inactive enzymes that are synthesized by hepatocytes (liver cells) and released into the bloodstream, and various molecules associated with platelets or released by damaged tissues. Most clotting factors are identified by Roman numerals that indicate the order of their discovery (not necessarily the order of their participation in the clotting process).

Clotting is a complex cascade of enzymatic reactions in which each clotting factor activates many molecules of the next one in a fixed sequence. Finally, a large quantity of product (the insoluble protein fibrin) is formed. Clotting can be divided into three stages (Figure 19.11):

① Two pathways, called the extrinsic pathway (Figure 19.11a) and the intrinsic pathway (Figure 19.11b), which will be described shortly, lead to the formation of prothrombinase. Once prothrombinase is formed, the steps involved in the next two stages of clotting are the same for both the extrinsic and intrinsic pathways, and together these two stages are referred to as the common pathway.

② Prothrombinase converts prothrombin (a plasma protein formed by the liver) into the enzyme thrombin.

③ Thrombin converts soluble fibrinogen (another plasma protein formed by the liver) into insoluble fibrin. Fibrin forms the threads of the clot.

Figure 19.11 The blood-clotting cascade.

  In blood clotting, coagulation factors are activated in sequence, resulting in a cascade of reactions that includes positive feedback cycles.

Figure 19.10 Blood clot formation.   Notice the platelet and red blood cells entrapped in fibrin threads.

  A blood clot is a gel that contains formed elements of the blood entangled in fibrin threads.

SEM 900x

(a) Early stage

— Platelet

— Fibrin threads

SEM 900x

(b) Intermediate stage

SEM 900x

(c) Late stage

SEM 1600x

— Red blood cell

— Fibrin threads

(d) Red blood cells trapped in fibrin threads

? What is serum?

? What is the outcome of the first stage of blood clotting?

The Extrinsic Pathway

The **extrinsic pathway** of blood clotting has fewer steps than the intrinsic pathway and occurs rapidly—within a matter of seconds if trauma is severe. It is so named because a tissue protein called **tissue factor (TF),** also known as **thromboplastin,** leaks into the blood from cells *outside (extrinsic to)* blood vessels and initiates the formation of prothrombinase. TF is a complex mixture of lipoproteins and phospholipids released from the surfaces of damaged cells. In the presence of Ca^{2+}, TF begins a sequence of reactions that ultimately activates clotting factor X (Figure 19.11a). Once factor X is activated, it combines with factor V in the presence of Ca^{2+} to form the active enzyme prothrombinase, completing the extrinsic pathway.

The Intrinsic Pathway

The **intrinsic pathway** of blood clotting is more complex than the extrinsic pathway, and it occurs more slowly, usually requiring several minutes. The intrinsic pathway is so named because its activators are either in direct contact with blood or contained *within (intrinsic to)* the blood; outside tissue damage is not needed. If endothelial cells become roughened or damaged, blood can come in contact with collagen fibers in the connective tissue around the endothelium of the blood vessel. In addition, trauma to endothelial cells causes damage to platelets, resulting in the release of phospholipids by the platelets. Contact with collagen fibers (or with the glass sides of a blood collection tube) activates clotting factor XII (Figure 19.11b), which begins a sequence of reactions that eventually activates clotting factor X. Platelet phospholipids and Ca^{2+} can also participate in the activation of factor X. Once factor X is activated, it combines with factor V to form the active enzyme prothrombinase (just as occurs in the extrinsic pathway), completing the intrinsic pathway.

The Common Pathway

The formation of prothrombinase marks the beginning of the common pathway. In the second stage of blood clotting (Figure 19.11c), prothrombinase and Ca^{2+} catalyze the conversion of prothrombin to thrombin. In the third stage, thrombin, in the presence of Ca^{2+}, converts fibrinogen, which is soluble, to loose fibrin threads, which are insoluble. Thrombin also activates factor XIII (fibrin stabilizing factor), which strengthens and stabilizes the fibrin threads into a sturdy clot. Plasma contains some factor XIII, which is also released by platelets trapped in the clot.

Thrombin has two positive feedback effects. In the first positive feedback loop, which involves factor V, it accelerates the formation of prothrombinase. Prothrombinase in turn accelerates the production of more thrombin, and so on. In the second positive feedback loop, thrombin activates platelets, which reinforces their aggregation and the release of platelet phospholipids.

Clot Retraction

Once a clot is formed, it plugs the ruptured area of the blood vessel and thus stops blood loss. **Clot retraction** is the consolidation or tightening of the fibrin clot. The fibrin threads attached to the damaged surfaces of the blood vessel gradually contract as platelets pull on them. As the clot retracts, it pulls the edges of the damaged vessel closer together, decreasing the risk of further damage. During retraction, some serum can escape between the fibrin threads, but the formed elements in blood cannot. Normal retraction depends on an adequate number of platelets in the clot, which release factor XIII and other factors, thereby strengthening and stabilizing the clot. Permanent repair of the blood vessel can then take place. In time, fibroblasts form connective tissue in the ruptured area, and new endothelial cells repair the vessel lining.

Role of Vitamin K in Clotting

Normal clotting depends on adequate levels of vitamin K in the body. Although vitamin K is not involved in actual clot formation, it is required for the synthesis of four clotting factors. Normally produced by bacteria that inhabit the large intestine, vitamin K is a fat-soluble vitamin that can be absorbed through the lining of the intestine and into the blood if absorption of lipids is normal. People suffering from disorders that slow absorption of lipids (for example, inadequate release of bile into the small intestine) often experience uncontrolled bleeding as a consequence of vitamin K deficiency.

The various clotting factors, their sources, and the pathways of activation are summarized in Table 19.4.

Hemostatic Control Mechanisms

Many times a day little clots start to form, often at a site of minor roughness or at a developing atherosclerotic plaque inside a blood vessel. Because blood clotting involves amplification and positive feedback cycles, a clot has a tendency to enlarge, creating the potential for impairment of blood flow through undamaged vessels. The **fibrinolytic system** (fī-bri-nō-LIT-ik) dissolves small, inappropriate clots; it also dissolves clots at a site of damage once the damage is repaired. Dissolution of a clot is called **fibrinolysis** (fī-bri-NOL-i-sis). When a clot is formed, an inactive plasma enzyme called **plasminogen** is incorporated into the clot. Both body tissues and blood contain substances that can activate plasminogen to **plasmin (fibrinolysin),** an active plasma enzyme. Among these substances are thrombin, activated factor XII, and tissue plasminogen activator (t-PA), which is synthesized in endothelial cells of most tissues and liberated into the blood. Once plasmin is formed, it can dissolve the clot by digesting fibrin threads and inactivating substances such as fibrinogen, prothrombin, and factors V and XII.

Even though thrombin has a positive feedback effect on blood clotting, clot formation normally remains localized at the site of damage. A clot does not extend beyond a wound site into the general circulation, in part because fibrin absorbs thrombin into the clot. Another reason for localized clot formation is that because of the dispersal of some of the clotting factors by the blood, their concentrations are not high enough to bring about widespread clotting.

Several other mechanisms also control blood clotting. For example, endothelial cells and white blood cells produce a

TABLE 19.4

Clotting (Coagulation) Factors

NUMBER*	NAME(S)	SOURCE	PATHWAY(S) OF ACTIVATION
I	Fibrinogen.	Liver.	Common.
II	Prothrombin.	Liver.	Common.
III	Tissue factor (thromboplastin).	Damaged tissues and activated platelets.	Extrinsic.
IV	Calcium ions (Ca^{2+}).	Diet, bones, and platelets.	All.
V	Proaccelerin, labile factor, or accelerator globulin (AcG).	Liver and platelets.	Extrinsic and intrinsic.
VII	Serum prothrombin conversion accelerator (SPCA), stable factor, or proconvertin.	Liver.	Extrinsic.
VIII	Antihemophilic factor (AHF), antihemophilic factor A, or antihemophilic globulin (AHG).	Liver.	Intrinsic.
IX	Christmas factor, plasma thromboplastin component (PTC), or antihemophilic factor B.	Liver.	Intrinsic.
X	Stuart factor, Prower factor, or thrombokinase.	Liver.	Extrinsic and intrinsic.
XI	Plasma thromboplastin antecedent (PTA) or antihemophilic factor C.	Liver.	Intrinsic.
XII	Hageman factor, glass factor, contact factor, or antihemophilic factor D.	Liver.	Intrinsic.
XIII	Fibrin-stabilizing factor (FSF).	Liver and platelets.	Common.

*There is no factor VI. Prothrombinase (prothrombin activator) is a combination of activated factors V and X.

prostaglandin called **prostacyclin** (pros-ta-SĪ-klin) that opposes the actions of thromboxane A2. Prostacyclin is a powerful inhibitor of platelet adhesion and release.

In addition, substances that delay, suppress, or prevent blood clotting, called **anticoagulants,** are present in blood. These include **antithrombin,** which blocks the action of several factors, including XII, X, and II (prothrombin). **Heparin,** an anticoagulant that is produced by mast cells and basophils, combines with antithrombin and increases its effectiveness in blocking thrombin. Another anticoagulant, **activated protein C (APC),** inactivates the two major clotting factors not blocked by antithrombin and enhances activity of plasminogen activators. Babies that lack the ability to produce APC due to a genetic mutation usually die of blood clots in infancy.

clots may also form when blood flows too slowly (stasis), allowing clotting factors to accumulate locally in high enough concentrations to initiate coagulation. Clotting in an unbroken blood vessel (usually a vein) is called **thrombosis** (throm-BŌ-sis; *thromb-* = clot; *-osis* = a condition of). The clot itself, called a **thrombus,** may dissolve spontaneously. If it remains intact, however, the thrombus may become dislodged and be swept away in the blood. A blood clot, bubble of air, fat from broken bones, or a piece of debris transported by the bloodstream is called an **embolus** (*em-* = in; *bolus* = a mass). An embolus that breaks away from an arterial wall may lodge in a smaller-diameter artery downstream and block blood flow to a vital organ. When an embolus lodges in the lungs, the condition is called **pulmonary embolism.**

● CLINICAL CONNECTION | Anticoagulants

Patients who are at increased risk of forming blood clots may receive anticoagulants. Examples are heparin or warfarin. Heparin is often administered during hemodialysis and open-heart surgery. **Warfarin (Coumadin®)** acts as an antagonist to vitamin K and thus blocks synthesis of four clotting factors. Warfarin is slower acting than heparin. To prevent clotting in donated blood, blood banks and laboratories often add substances that remove Ca^{2+}; examples are EDTA (ethylene diamine tetraacetic acid) and CPD (citrate phosphate dextrose). ●

Intravascular Clotting

Despite the anticoagulating and fibrinolytic mechanisms, blood clots sometimes form within the cardiovascular system. Such clots may be initiated by roughened endothelial surfaces of a blood vessel resulting from atherosclerosis, trauma, or infection. These conditions induce adhesion of platelets. Intravascular

● CLINICAL CONNECTION | Aspirin and Thrombolytic Agents

In patients with heart and blood vessel disease, the events of hemostasis may occur even without external injury to a blood vessel. At low doses, **aspirin** inhibits vasoconstriction and platelet aggregation by blocking synthesis of thromboxane A2. It also reduces the chance of thrombus formation. Due to these effects, aspirin reduces the risk of transient ischemic attacks (TIA), strokes, myocardial infarction, and blockage of peripheral arteries.

Thrombolytic agents are chemical substances that are injected into the body to dissolve blood clots that have already formed to restore circulation. They either directly or indirectly activate plasminogen. The first thrombolytic agent, approved in 1982 for dissolving clots in the coronary arteries of the heart, was **streptokinase,** which is produced by streptococcal bacteria. A genetically engineered version of human **tissue plasminogen activator (t-PA)** is now used to treat victims of both heart attacks and brain attacks (strokes) that are caused by blood clots. ●

◉ CHECKPOINT

18. What is hemostasis?
19. How do vascular spasm and platelet plug formation occur?
20. What is fibrinolysis? Why does blood rarely remain clotted inside blood vessels?
21. How do the extrinsic and intrinsic pathways of blood clotting differ?
22. Define each of the following terms: anticoagulant, thrombus, embolus, and thrombolytic agent.

BLOOD GROUPS AND BLOOD TYPES

◉ OBJECTIVES

• Distinguish between the ABO and Rh blood groups.
• Explain why it is so important to match donor and recipient blood types before administering a transfusion.

The surfaces of erythrocytes contain a genetically determined assortment of **antigens** composed of glycoproteins and glycolipids. These antigens, called **agglutinogens** (a-gloo-TIN-ō-jens), occur in characteristic combinations. Based on the presence or absence of various antigens, blood is categorized into different **blood groups.** Within a given blood group, there may be two or more different **blood types.** There are at least 24 blood groups and more than 100 antigens that can be detected on the surface of red blood cells. Here we discuss two major blood groups—ABO and Rh. Other blood groups include the Lewis, Kell, Kidd, and Duffy systems. The incidence of ABO and Rh blood types varies among different population groups, as indicated in Table 19.5.

TABLE 19.5

Blood Types in the United States

POPULATION GROUP	BLOOD TYPE (PERCENTAGE)				
	O	A	B	AB	RH⁺
European-American	45	40	11	4	85
African-American	49	27	20	4	95
Korean-American	32	28	30	10	100
Japanese-American	31	38	21	10	100
Chinese-American	42	27	25	6	100
Native American	79	16	4	1	100

ABO Blood Group

The **ABO blood group** is based on two glycolipid antigens called A and B (Figure 19.12). People whose RBCs display *only antigen A* have **type A** blood. Those who have *only antigen B* are **type B.** Individuals who have *both A and B antigens* are **type AB;** those who have *neither antigen A nor B* are **type O.**

Blood plasma usually contains **antibodies** called **agglutinins** (a-GLOO-ti-nins) that react with the A or B antigens if the two are mixed. These are the **anti-A antibody,** which reacts with antigen A, and the **anti-B antibody,** which reacts with antigen B. The antibodies present in each of the four blood types are shown in Figure 19.12. You do not have antibodies that react with the antigens of your own RBCs, but you do have antibodies for any antigens that your RBCs lack. For example, if your blood type is B, you have B antigens on your red blood cells, and you have anti-A antibodies in your blood plasma. Although agglutinins start to appear in the blood within a few months after birth, the reason for their presence is not clear. Perhaps they are formed in response to bacteria that normally inhabit the gastrointestinal tract. Because the antibodies are large IgM-type antibodies (see Table 22.3 on page 860) that do not cross the placenta, ABO incompatibility between a mother and her fetus rarely causes problems.

Transfusions

Despite the differences in RBC antigens reflected in the blood group systems, blood is the most easily shared of human tissues, saving many thousands of lives every year through transfusions. A **transfusion** (trans-FŪ-shun) is the transfer of whole blood or blood components (red blood cells only or blood plasma only) into the bloodstream or directly into the red bone marrow. A transfusion is most often given to alleviate anemia, to increase blood volume (for example, after a severe hemorrhage), or to improve immunity. However, the normal components of one person's RBC plasma membrane can trigger damaging antigen–antibody responses in a transfusion recipient. In an incompatible blood transfusion, antibodies in the recipient's plasma bind to the antigens on the donated RBCs, which causes **agglutination** (a-gloo-ti-NĀ-shun), or clumping, of the RBCs. Agglutination is an antigen–antibody response in which RBCs become cross-linked to one another. (Note that agglutination is not the same as blood clotting.) When these antigen–antibody complexes form, they activate plasma proteins of the complement family (described on page 860). In essence, complement molecules make the plasma membrane of the donated RBCs leaky, causing **hemolysis** (rupture) of the RBCs and the release of hemoglobin into the blood plasma. The liberated hemoglobin may cause kidney damage by clogging the filtration membranes. Although quite rare, it is possible for the viruses that cause AIDS and hepatitis B and C to be transmitted through transfusion of contaminated blood products.

Consider what happens if a person with type A blood receives a transfusion of type B blood. The recipient's blood (type A) contains A antigens on the red blood cells and anti-B antibodies in the plasma. The donor's blood (type B) contains B antigens and anti-A antibodies. In this situation, two things can happen. First, the anti-B antibodies in the recipient's plasma can bind to the B antigens on the donor's erythrocytes, causing agglutination and hemolysis of the red blood cells. Second, the anti-A antibodies in the donor's plasma can bind to the A antigens on

Figure 19.12 Antigens and antibodies of the ABO blood types.

The antibodies in your plasma do not react with the antigens on your red blood cells.

? Which antibodies are usually present in type O blood?

the recipient's red blood cells, a less serious reaction because the donor's anti-A antibodies become so diluted in the recipient's plasma that they do not cause significant agglutination and hemolysis of the recipient's RBCs.

Table 19.6 summarizes the interactions of the four blood types of the ABO system.

People with type AB blood do not have anti-A or anti-B antibodies in their blood plasma. They are sometimes called *universal recipients* because theoretically they can receive blood from donors of all four blood types. They have no antibodies to attack antigens on donated RBCs (Table 19.6). People with type O blood have neither A nor B antigens on their RBCs and are sometimes called *universal donors* because theoretically they can donate blood to all four ABO blood types. Type O persons requiring blood may receive only type O blood (Table 19.6). In practice, use of the terms universal recipient and universal donor is misleading and dangerous. Blood contains antigens and antibodies other than those associated with the ABO system that can cause transfusion problems. Thus, blood should be carefully cross-matched or screened before transfusion. In about 80% of the population, soluble antigens of the ABO type appear in saliva and other body fluids, in which case blood type can be identified from a sample of saliva.

Rh Blood Group

The **Rh blood group** is so named because the antigen was discovered in the blood of the *Rhesus* monkey. The alleles of three genes may code for the Rh antigen. People whose RBCs have Rh antigens are designated Rh$^+$ (Rh positive); those who lack Rh antigens are designated Rh$^-$ (Rh negative). Table 19.5 shows the incidence

of Rh$^+$ and Rh$^-$ in various populations. Normally, blood plasma does not contain anti-Rh antibodies. If an Rh$^-$ person receives an Rh$^+$ blood transfusion, however, the immune system starts to make anti-Rh antibodies that will remain in the blood. If a second transfusion of Rh$^+$ blood is given later, the previously formed anti-Rh antibodies will cause agglutination and hemolysis of the RBCs in the donated blood, and a severe reaction may occur.

TABLE 19.6

Summary of ABO Blood Group Interactions

CHARACTERISTIC	BLOOD TYPE			
	A	B	AB	O
Agglutinogen (antigen) on RBCs	A	B	Both A and B	Neither A nor B
Agglutinin (antibody) in plasma	anti-B	anti-A	Neither anti-A nor anti-B	Both anti-A and anti-B
Compatible donor blood types (no hemolysis)	A, O	B, O	A, B, AB, O	O
Incompatible donor blood types (hemolysis)	B, AB	A, AB	—	A, B, AB

• CLINICAL CONNECTION | Hemolytic Disease of the Newborn

The most common problem with Rh incompatibility, **hemolytic disease of the newborn (HDN),** may arise during pregnancy (Figure 19.13). Normally, no direct contact occurs between maternal and fetal blood while a woman is pregnant. However, if a small amount of Rh⁺ blood leaks from the fetus through the placenta into the bloodstream of an Rh⁻ mother, the mother will start to make anti-Rh antibodies. Because the greatest possibility of fetal blood leakage into the maternal circulation occurs at delivery, the firstborn baby usually is not affected. If the mother becomes pregnant again, however, her anti-Rh antibodies can cross the placenta and enter the bloodstream of the fetus. If the fetus is Rh⁻, there is no problem, because Rh⁻ blood does not have the Rh antigen. If the fetus is Rh⁺, however, agglutination and hemolysis brought on by fetal–maternal incompatibility may occur in the fetal blood.

An injection of anti-Rh antibodies called anti-Rh gamma globulin (RhoGAM®) can be given to prevent HDN. All Rh⁻ women should receive RhoGAM® soon after every delivery, miscarriage, or abortion. These antibodies bind to and inactivate the fetal Rh antigens before the mother's immune system can respond to the foreign antigens by producing her own anti-Rh antibodies. •

Figure 19.13 Development of hemolytic disease of the newborn (HDN). (a) At birth, a small quantity of fetal blood usually leaks across the placenta into the maternal bloodstream. A problem can arise when the mother is Rh⁻ and the baby is Rh⁺, having inherited an allele for one of the Rh antigens from the father. (b) Upon exposure to Rh antigen, the mother's immune system responds by making anti-Rh antibodies. (c) During a subsequent pregnancy, the maternal antibodies cross the placenta into the fetal blood. If the second fetus is Rh⁺, the ensuing antigen–antibody reaction causes agglutination and hemolysis of fetal RBCs. The result is HDN.

HDN occurs when maternal anti-Rh antibodies cross the placenta and cause hemolysis of fetal RBCs.

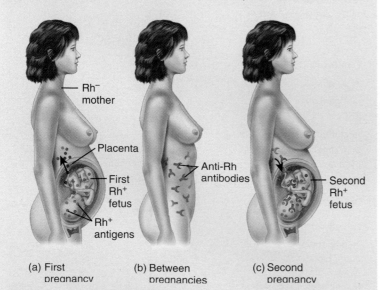

(a) First pregnancy   (b) Between pregnancies   (c) Second pregnancy

? Why is the firstborn baby unlikely to have HDN?

Typing and Cross-Matching Blood for Transfusion

To avoid blood-type mismatches, laboratory technicians type the patient's blood and then either cross-match it to potential donor blood or screen it for the presence of antibodies. In the procedure for ABO blood typing, single drops of blood are mixed with different *antisera,* solutions that contain antibodies (Figure 19.14). One drop of blood is mixed with anti-A serum, which contains anti-A antibodies that will agglutinate red blood cells that possess A antigens. Another drop is mixed with anti-B serum, which contains anti-B antibodies that will agglutinate red blood cells that possess B antigens. If the red blood cells aggluti-

Figure 19.14 ABO blood typing.

In the procedure for ABO blood typing, blood is mixed with anti-A serum and anti-B serum.

? What is agglutination?

nate only when mixed with anti-A serum, the blood is type A. If the red blood cells agglutinate only when mixed with anti-B serum, the blood is type B. The blood is type AB if both drops agglutinate; if neither drop agglutinates, the blood is type O.

In the procedure for determining Rh factor, a drop of blood is mixed with antiserum containing antibodies that will agglutinate RBCs displaying Rh antigens. If the blood agglutinates, it is Rh$^+$; no agglutination indicates Rh$^-$.

Once the patient's blood type is known, donor blood of the same ABO and Rh type is selected. In a **cross-match,** the possible donor RBCs are mixed with the recipient's serum. If agglutination does not occur, the recipient does not have antibodies that will attack the donor RBCs. Alternatively, the recipient's serum can be **screened** against a test panel of RBCs having antigens known to cause blood transfusion reactions to detect any antibodies that may be present.

◉ CHECKPOINT

23. What precautions must be taken before giving a blood transfusion?
24. What is hemolysis, and how can it occur after a mismatched blood transfusion?
25. Explain the conditions that may cause hemolytic disease of the newborn.

DISORDERS: HOMEOSTATIC IMBALANCES

Anemia

Anemia is a condition in which the oxygen-carrying capacity of blood is reduced. All of the many types of anemia are characterized by reduced numbers of RBCs or a decreased amount of hemoglobin in the blood. The person feels fatigued and is intolerant of cold, both of which are related to lack of oxygen needed for ATP and heat production. Also, the skin appears pale, due to the low content of red-colored hemoglobin circulating in skin blood vessels. Among the most important causes and types of anemia are the following:

- *Inadequate absorption of iron, excessive loss of iron, increased iron requirement, or insufficient intake of iron* causes **iron-deficiency anemia,** the most common type of anemia. Women are at greater risk for iron-deficiency anemia due to menstrual blood losses and increased iron demands of the growing fetus during pregnancy. Gastrointestinal losses, such as those that occur with malignancy or ulceration, also contribute to this type of anemia.

- *Inadequate intake of vitamin B$_{12}$ or folic acid* causes **megaloblastic anemia** in which red bone marrow produces large, abnormal red blood cells (megaloblasts). It may also be caused by drugs that alter gastric secretion or are used to treat cancer.

- *Insufficient hemopoiesis* resulting from an inability of the stomach to produce intrinsic factor, which is needed for absorption of vitamin B$_{12}$ in the small intestine, causes **pernicious anemia.**

- *Excessive loss of RBCs* through bleeding resulting from large wounds, stomach ulcers, or especially heavy menstruation leads to **hemorrhagic anemia.**

- *RBC plasma membranes rupture prematurely* in **hemolytic anemia.** The released hemoglobin pours into the plasma and may damage the filtering units (glomeruli) in the kidneys. The condition may result from inherited defects such as abnormal red blood cell enzymes, or from outside agents such as parasites, toxins, or antibodies from incompatible transfused blood.

- *Deficient synthesis of hemoglobin* occurs in **thalassemia** (thal'-a-SĒ-mē-a), a group of hereditary hemolytic anemias. The RBCs are small (microcytic), pale (hypochromic), and short-lived. Thalassemia occurs primarily in populations from countries bordering the Mediterranean Sea.

- *Destruction of red bone marrow* results in **aplastic anemia.** It is caused by toxins, gamma radiation, and certain medications that inhibit enzymes needed for hemopoiesis.

Sickle-Cell Disease

The RBCs of a person with **sickle-cell disease (SCD)** contain Hb-S, an abnormal kind of hemoglobin. When Hb-S gives up oxygen to the interstitial fluid, it forms long, stiff, rodlike structures that bend the erythrocyte into a sickle shape (Figure 19.15). The sickled cells rupture easily. Even though erythropoiesis is stimulated by the loss of the cells, it cannot keep pace with hemolysis. People with sickle-cell disease always have some degree of anemia and mild jaundice and may experience joint or bone pain, breathlessness, rapid heart rate, abdominal pain, fever, and fatigue as a result of tissue damage caused by prolonged recovery oxygen uptake (oxygen debt). Any activity that reduces the amount of oxygen in the blood, such as vigorous exercise, may produce a sickle-cell crisis (worsening of the anemia, pain in the abdomen and long bones of the limbs, fever, and shortness of breath).

Sickle-cell disease is inherited. People with two sickle-cell genes have severe anemia; those with only one defective gene have minor problems. Sickle-cell genes are found primarily among populations, or

Figure 19.15 Red blood cells from a person with sickle-cell disease.

 The red blood cells of a person with sickle-cell disease contain an abnormal type of hemoglobin called Hb-S.

Normal red blood cell

Sickled red blood cell

SEM 7000x

Red blood cells

❓ What are some symptoms of sickle-cell disease?

descendants of populations, that live in the malaria belt around the world, including parts of Mediterranean Europe, sub-Saharan Africa, and tropical Asia. The gene responsible for the tendency of the RBCs to sickle also alters the permeability of the plasma membranes of sickled cells, causing potassium ions to leak out. Low levels of potassium kill the malaria parasites that may infect sickled cells. Because of this effect, a person with one normal gene and one sickle-cell gene has higher-than-average resistance to malaria. The possession of a single sickle-cell gene thus confers a survival advantage.

Treatment of SCD consists of administration of analgesics to relieve pain, fluid therapy to maintain hydration, oxygen to reduce the possibility of oxygen debt, antibiotics to counter infections, and blood transfusions. People who suffer from SCD have normal fetal hemoglobin (Hb-F), a slightly different form of hemoglobin that predominates at birth and is present in small amounts thereafter. In some patients with sickle-cell disease, a drug called hydroxyurea promotes transcription of the normal Hb-F gene, elevates the level of Hb-F, and reduces the chance that the RBCs will sickle. Unfortunately, this drug also has toxic effects on the bone marrow; thus, its safety for long-term use is questionable.

Hemophilia

Hemophilia (hē-mō-FIL-ē-a; -*philia* = loving) is an inherited deficiency of clotting in which bleeding may occur spontaneously or after only minor trauma. It is the oldest known hereditary bleeding disorder; descriptions of the disease are found as early as the second century A.D. Hemophilia usually affects males and is sometimes referred to as "the royal disease" because many descendants of Queen Victoria, beginning with one of her sons, were affected by the disease. Different types of hemophilia are due to deficiencies of different blood clotting factors and exhibit varying degrees of severity, ranging from mild to severe bleeding tendencies. Hemophilia is characterized by sponta-

neous or traumatic subcutaneous and intramuscular hemorrhaging, nosebleeds, blood in the urine, and hemorrhages in joints that produce pain and tissue damage. Treatment involves transfusions of fresh blood plasma or concentrates of the deficient clotting factor to relieve the tendency to bleed. Another treatment is the drug desmopressin (DDAVP), which can boost the levels of the clotting factors.

Leukemia

The term **leukemia** (loo-KĒ-mē-a; *leuko*- = white) refers to a group of red bone marrow cancers in which abnormal white blood cells multiply uncontrollably. The accumulation of the cancerous white blood cells in red bone marrow interferes with the production of red blood cells, white blood cells, and platelets. As a result the oxygen-carrying capacity of the blood is reduced, an individual is more susceptible to infection, and blood clotting is abnormal. In most leukemias, the cancerous white blood cells spread to the lymph nodes, liver, and spleen, causing them to enlarge. All leukemias produce the usual symptoms of anemia (fatigue, intolerance to cold, and pale skin). In addition, weight loss, fever, night sweats, excessive bleeding, and recurrent infections may occur.

In general, leukemias are classified as acute (symptoms develop rapidly) and chronic (symptoms may take years to develop). Adults may have either type, but children usually have the acute type.

The cause of most types of leukemia is unknown. However, certain risk factors have been implicated. These include exposure to radiation or chemotherapy for other cancers, genetics (some genetic disorders such as Down syndrome), environmental factors (smoking and benzene), and microbes such as the human T cell leukemia-lymphoma virus-1 (HTLV-1) and the Epstein-Barr virus.

Treatment options include chemotherapy, radiation, stem cell transplantation, interferon, antibodies, and blood transfusion.

MEDICAL TERMINOLOGY

Acute normovolemic hemodilution (nor-mō-vō-LĒ-mik hē-mō-di-LOO-shun) Removal of blood immediately before surgery and its replacement with a cell-free solution to maintain sufficient blood volume for adequate circulation. At the end of surgery, once bleeding has been controlled, the collected blood is returned to the body.

Autologous preoperative transfusion (aw-TOL-o-gus trans-FŪ-zhun; *auto*- = self) Donating one's own blood; can be done up to 6 weeks before elective surgery. Also called **predonation**. This procedure eliminates the risk of incompatibility and blood-borne disease.

Blood bank A facility that collects and stores a supply of blood for future use by the donor or others. Because blood banks have additional and diverse functions (immunohematology reference work, continuing medical education, bone and tissue storage, and clinical consultation), they are more appropriately referred to as **centers of transfusion medicine.**

Cyanosis (sī-a-NŌ-sis; *cyano*- = blue) Slightly bluish/dark-purple skin discoloration, most easily seen in the nail beds and mucous membranes, due to an increased quantity of reduced hemoglobin (hemoglobin not combined with oxygen) in systemic blood.

Gamma globulin (GLOB-ū-lin) Solution of immunoglobulins from blood consisting of antibodies that react with specific pathogens, such as viruses. It is prepared by injecting the specific virus into animals, removing blood from the animals after antibodies have accumulated, isolating the antibodies, and injecting them into a human to provide short-term immunity.

Hemochromatosis (hē-mō-krō-ma-TŌ-sis; *chroma* = color) Disorder of iron metabolism characterized by excessive absorption of ingested iron and excess deposits of iron in tissues (especially the liver, heart, pituitary gland, gonads, and pancreas) that result in bronze discoloration of the skin, cirrhosis, diabetes mellitus, and bone and joint abnormalities.

Hemorrhage (HEM-or-ij; *rhegnynai* = bursting forth) Loss of a large amount of blood; can be either internal (from blood vessels into tissues) or external (from blood vessels directly to the surface of the body).

Jaundice (*jaund*- = yellow) An abnormal yellowish discoloration of the sclerae of the eyes, skin, and mucous membranes due to excess bilirubin (yellow-orange pigment) in the blood. The three main categories of jaundice are *prehepatic jaundice,* due to excess production of bilirubin; *hepatic jaundice,* abnormal bilirubin processing by the liver caused by congenital liver disease, cirrhosis (scar tissue formation) of the liver, or hepatitis (liver inflammation); and *extrahepatic jaundice,* due to blockage of bile drainage by gallstones or cancer of the bowel or pancreas.

Phlebotomist (fle-BOT-ō-mist; *phlebo*- = vein; -*tom* = cut) A technician who specializes in withdrawing blood.

Septicemia (sep′-ti-SĒ-mē-a; *septic*- = decay; -*emia* = condition of blood) Toxins or disease-causing bacteria in the blood. Also called "blood poisoning."

Thrombocytopenia (throm′-bō-sī-tō-PĒ-nē-a; -*penia* = poverty) Very low platelet count that results in a tendency to bleed from capillaries.

Venesection (vē′-ne-SEK-shun; *ven-* = vein) Opening of a vein for withdrawal of blood. Although **phlebotomy** (fle-BŌT-ō-mē) is a synonym for venesection, in clinical practice phlebotomy refers to therapeutic bloodletting, such as the removal of some blood to lower its viscosity in a patient with polycythemia.

Whole blood Blood containing all formed elements, plasma, and plasma solutes in natural concentrations.

STUDY OUTLINE

Introduction (p. 689)

1. The cardiovascular system consists of the blood, heart, and blood vessels.
2. Blood is a connective tissue composed of blood plasma (liquid portion) and formed elements (cells and cell fragments).

Functions and Properties of Blood (p. 690)

1. Blood transports oxygen, carbon dioxide, nutrients, wastes, and hormones.
2. It helps regulate pH, body temperature, and water content of cells.
3. It provides protection through clotting and by combating toxins and microbes through certain phagocytic white blood cells or specialized blood plasma proteins.
4. Physical characteristics of blood include a viscosity greater than that of water; a temperature of 38°C (100.4°F); and a pH of 7.35–7.45.
5. Blood constitutes about 8% of body weight, and its volume is 4–6 liters in adults.
6. Blood is about 55% blood plasma and 45% formed elements.
7. The hematocrit is the percentage of total blood volume occupied by red blood cells.
8. Blood plasma consists of 91.5% water and 8.5% solutes. Principal solutes include proteins (albumins, globulins, fibrinogen), nutrients, vitamins, hormones, respiratory gases, electrolytes, and waste products.
9. The formed elements in blood include red blood cells (erythrocytes), white blood cells (leukocytes), and platelets.

Formation of Blood Cells (p. 693)

1. Hemopoiesis is the formation of blood cells from hemopoietic stem cells in red bone marrow.
2. Myeloid stem cells form RBCs, platelets, granulocytes, and monocytes. Lymphoid stem cells give rise to lymphocytes.
3. Several hemopoietic growth factors stimulate differentiation and proliferation of the various blood cells.

Red Blood Cells (p. 695)

1. Mature RBCs are biconcave discs that lack nuclei and contain hemoglobin.
2. The function of the hemoglobin in red blood cells is to transport oxygen and some carbon dioxide.
3. RBCs live about 120 days. A healthy male has about 5.4 million RBCs/μL of blood; a healthy female has about 4.8 million/μL.
4. After phagocytosis of aged RBCs by macrophages, hemoglobin is recycled.

5. RBC formation, called erythropoiesis, occurs in adult red bone marrow of certain bones. It is stimulated by hypoxia, which stimulates the release of erythropoietin by the kidneys.
6. A reticulocyte count is a diagnostic test that indicates the rate of erythropoiesis.

White Blood Cells (p. 699)

1. WBCs are nucleated cells. The two principal types are granulocytes (neutrophils, eosinophils, and basophils) and agranulocytes (lymphocytes and monocytes).
2. The general function of WBCs is to combat inflammation and infection. Neutrophils and macrophages (which develop from monocytes) do so through phagocytosis.
3. Eosinophils combat the effects of histamine in allergic reactions, phagocytize antigen–antibody complexes, and combat parasitic worms. Basophils liberate heparin, histamine, and serotonin in allergic reactions that intensify the inflammatory response.
4. B lymphocytes, in response to the presence of foreign substances called antigens, differentiate into plasma cells that produce antibodies. Antibodies attach to the antigens and render them harmless. This antigen–antibody response combats infection and provides immunity. T lymphocytes destroy foreign invaders directly. Natural killer cells attack infectious microbes and tumor cells.
5. Except for lymphocytes, which may live for years, WBCs usually live for only a few hours or a few days. Normal blood contains 5000–10,000 WBCs/μL.

Platelets (p. 701)

1. Platelets (thrombocytes) are disc-shaped cell fragments that splinter from megakaryocytes. Normal blood contains 150,000–400,000 platelets/μL.
2. Platelets help stop blood loss from damaged blood vessels by forming a platelet plug.

Stem Cell Transplants from Bone Marrow and Cord-Blood (p. 703)

1. Bone marrow transplants involve removal of red bone marrow as a source of stem cells from the iliac crest.
2. In a cord-blood transplant, stem cells from the placenta are removed from the umbilical cord.
3. Cord-blood transplants have several advantages over bone marrow transplants.

Hemostasis (p. 703)

1. Hemostasis refers to the stoppage of bleeding.
2. It involves vascular spasm, platelet plug formation, and blood clotting (coagulation).

3. In vascular spasm, the smooth muscle of a blood vessel wall contracts, which slows blood loss.
4. Platelet plug formation involves the aggregation of platelets to stop bleeding.
5. A clot is a network of insoluble protein fibers (fibrin) in which formed elements of blood are trapped.
6. The chemicals involved in clotting are known as clotting (coagulation) factors.
7. Blood clotting involves a cascade of reactions that may be divided into three stages: formation of prothrombinase, conversion of prothrombin into thrombin, and conversion of soluble fibrinogen into insoluble fibrin.
8. Clotting is initiated by the interplay of the extrinsic and intrinsic pathways of blood clotting.
9. Normal coagulation requires vitamin K and is followed by clot retraction (tightening of the clot) and ultimately fibrinolysis (dissolution of the clot).

10. Clotting in an unbroken blood vessel is called thrombosis. A thrombus that moves from its site of origin is called an embolus.

Blood Groups and Blood Types (p. 708)

1. ABO and Rh blood groups are genetically determined and based on antigen–antibody responses.
2. In the ABO blood group, the presence or absence of A and B antigens on the surface of RBCs determines blood type.
3. In the Rh system, individuals whose RBCs have Rh antigens are classified as Rh^+; those who lack the antigen are Rh^-.
4. Hemolytic disease of the newborn (HDN) can occur when an Rh^- mother is pregnant with an Rh^+ fetus.
5. Before blood is transfused, a recipient's blood is typed and then either cross-matched to potential donor blood or screened for the presence of antibodies.

SELF-QUIZ QUESTIONS

Fill in the blanks in the following statements.

1. Plasma minus its clotting proteins is termed _____.
2. _____ is the consolidation or tightening of the fibrin clot that helps to bring the edges of a damaged vessel closer together.

Indicate whether the following statements are true or false.

3. Hemoglobin functions in transporting both oxygen and carbon dioxide and in regulating blood pressure.
4. The most numerous white blood cells in a differential white blood cell count of a healthy individual are the neutrophils.

Choose the one best answer to the following questions.

5. Which of the following are *not* required for clot formation? (1) vitamin K, (2) calcium, (3) prostacyclin, (4) plasmin, (5) fibrinogen.
   (a) 1, 2, and 5 (b) 3, 4, and 5 (c) 4 and 5
   (d) 1, 2, and 3 (e) 3 and 4

6. Place the steps involved in hemostasis in the correct order. (1) conversion of fibrinogen into fibrin, (2) conversion of prothrombin into thrombin, (3) adhesion and aggregation of platelets on damaged vessel, (4) prothrombinase formed by extrinsic or intrinsic pathway, (5) reduction of blood loss by initiation of a vascular spasm.
   (a) 5, 3, 4, 2, 1 (b) 5, 4, 3, 1, 2 (c) 3, 5, 4, 2, 1
   (d) 5, 3, 2, 1, 4 (e) 5, 3, 2, 4, 1

7. Which of the following statements explain why red blood cells (RBCs) are highly specialized for oxygen transport? (1) RBCs contain hemoglobin. (2) RBCs lack a nucleus. (3) RBCs have many mitochondria and thus generate ATP aerobically. (4) The biconcave shape of RBCs provides a large surface area for the inward and outward diffusion of gas molecules. (5) RBCs can carry up to four oxygen molecules for each hemoglobin molecule.
   (a) 1, 2, 3, and 5 (b) 1, 2, 4, and 5 (c) 2, 3, 4, and 5
   (d) 1, 3, and 5 (e) 2, 4, and 5

8. Which of the following are *true?* (1) White blood cells leave the bloodstream by emigration. (2) Adhesion molecules help white blood cells stick to the endothelium, which aids emigration. (3)

Neutrophils and macrophages are active in phagocytosis. (4) The attraction of phagocytes to microbes and inflamed tissue is termed chemotaxis. (5) Leukopenia is an increase in white blood cell count that occurs during infection.
   (a) 1, 2, 4, and 5 (b) 2, 3, 4, and 5 (c) 1, 2, 3, and 4
   (d) 1, 3, and 5 (e) 1, 2, and 4

9. A person with type A Rh^- blood can receive a blood transfusion from which of the following types? (1) A Rh^+, (2) B Rh^-, (3) AB Rh^-, (4) O Rh^-, (5) A Rh^-.
   (a) 1 only (b) 3 only (c) 4 only
   (d) 4 and 5 (e) 1 and 5

10. A person with type B positive blood receives a transfusion of type AB positive blood. What will happen?
    (a) The recipient's antibodies will react with the donor's red blood cells.
    (b) The donor's antigens will destroy the recipient's antibodies.
    (c) The donor's antibodies will react with and destroy all of the recipient's red blood cells.
    (d) The recipient's blood type will change from Rh^+ to Rh^-.
    (e) These blood types are compatible, and the transfusion will be accepted.

11. What happens to the iron (Fe^{3+}) that is released during the breakdown of damaged red blood cells?
    (a) It is used to synthesize proteins.
    (b) It is transported to the liver where it becomes part of bile.
    (c) It is converted into urobilin and excreted in urine.
    (d) It attaches to transferrin and is transported to bone marrow for use in hemoglobin synthesis.
    (e) It is utilized by intestinal bacteria to convert bilirubin into urobilinogen.

12. Which of the following would *not* cause an increase in erythropoietin?
    (a) anemia (b) high altitude
    (c) hemorrhage (d) donating blood to a blood bank
    (e) polycythemia

13. Match the following:

_____(a) contain hemoglobin and function in gas transport

_____(b) cell fragments enclosed by a piece of the cell membrane of megakaryocytes; contain clotting factors

_____(c) individual forms of progenitor cells; named on the basis of the mature elements in blood they will ultimately produce

_____(d) white blood cell showing a kidney-shaped nucleus; capable of phagocytosis

_____(e) monocytes that roam the tissues and gather at sites of infection or inflammation

_____(f) occur as B cells, T cells, and natural killer cells

_____(g) give rise to red blood cells, monocytes, neutrophils, eosinophils, basophils, and platelets

_____(h) combat the effects of histamine and other mediators of inflammation in allergic reactions; also phagocytize antigen–antibody complexes

_____(i) respond to tissue destruction by bacteria; release lysozyme, strong oxidants, and defensins

_____(j) older neutrophils with several differently shaped nuclear lobes

_____(k) released from the red bone marrow, they develop into mature red blood cells

_____(l) give rise to lymphocytes

_____(m) cells no longer capable of replenishing themselves; can only give rise to more specific formed elements of blood

_____(n) hormone that stimulates formation of platelets

_____(o) monocytes that leave the blood and reside in a particular tissue such as alveolar macrophages in the lungs

_____(p) involved in inflammatory and allergic reactions; are involved in hypersensitivity reactions

_____(q) stimulate white blood cell formation

_____(r) cells that give rise to all the formed elements of blood; derived from mesenchyme

_____(s) hormone that increases the numbers of red blood cell precursors

(1) neutrophils
(2) lymphocytes
(3) monocytes
(4) eosinophils
(5) basophils
(6) pluripotent stem cells
(7) colony-forming units
(8) red blood cells
(9) reticulocytes
(10) polymorphs
(11) myeloid stem cells
(12) lymphoid stem cells
(13) progenitor cells
(14) platelets
(15) fixed macrophages
(16) wandering macrophages
(17) erythropoietin
(18) thrombopoietin
(19) cytokines

14. Match the following:

_____(a) tissue protein that leaks into the blood from cells outside blood vessels and initiates the formation of prothrombinase

_____(b) an anticoagulant

_____(c) platelet hormone that stimulates repair of damaged vessel walls

_____(d) its formation is initiated by either the extrinsic or intrinsic pathway or both; catalyzes the conversion of prothrombin to thrombin

_____(e) glycoproteins and glycolipids on the surfaces of red blood cells that can act as antigens

_____(f) forms the threads of a clot; produced from fibrinogen

_____(g) can dissolve a clot by digesting fibrin threads

_____(h) serves as the catalyst to form fibrin; formed from prothrombin

(1) prothrombinase
(2) thrombin
(3) fibrin
(4) thromboplastin
(5) plasmin
(6) heparin
(7) agglutinogens
(8) platelet-derived growth factor

15. Match the following:

_____(a) the percentage of total blood volume occupied by red blood cells

_____(b) the percentage of each type of white blood cell

_____(c) measures numbers of RBCs, WBCs, platelets per μ of blood; hematocrit; and differential WBC count

_____(d) measures the rate of erythropoiesis

_____(e) withdrawal of blood from a vein using a needle and collecting tube

_____(f) withdrawal of a small amount of red bone marrow with a fine needle and syringe

_____(g) removal of a core of red bone marrow with a large needle

(1) reticulocyte count
(2) bone marrow biopsy
(3) venipuncture
(4) hematocrit
(5) bone marrow aspiration
(6) complete blood count
(7) differential white blood cell count

CRITICAL THINKING QUESTIONS

1. Shilpa has recently been on broad spectrum antibiotics for a recurrent urinary bladder infection. While slicing vegetables, she cut herself and had difficulty stopping the bleeding. How could the antibiotics have played a role in her bleeding?

2. Mrs. Brown is in renal failure. Her recent blood tests indicated a hematocrit of 22. Why is her hematocrit low? What can she be given to raise her hematocrit?

3. Thomas has hepatitis, which is disrupting his liver functions. What kinds of symptoms would he be experiencing based on the role(s) of the liver related to blood?

? ANSWERS TO FIGURE QUESTIONS

19.1 Blood volume is about 8% of your body mass, roughly 5–6 liters in males and 4–5 liters in females. For instance, a 70-kg (150-lb.) person has a blood volume of 5.6 liters (70 kg × 8% × 1 liter/kg).

19.2 Platelets are cell fragments.

19.3 Pluripotent stem cells develop from mesenchyme.

19.4 One hemoglobin molecule can transport a maximum of four O_2 molecules, one O_2 bound to each heme group.

19.5 Transferrin is a plasma protein that transports iron in the blood.

19.6 Once you moved to high altitude, your hematocrit would increase due to increased secretion of erythropoietin.

19.7 Neutrophils, eosinophils, and basophils are called granular leukocytes because all have cytoplasmic granules that are visible through a light microscope when stained.

19.8 Lymphocytes recirculate from blood to tissues and back to blood. After leaving the blood, other WBCs remain in the tissues until they die.

19.9 Along with platelet plug formation, vascular spasm and blood clotting contribute to hemostasis.

19.10 Serum is blood plasma minus the clotting proteins.

19.11 The outcome of the first stage of clotting is the formation of prothrombinase.

19.12 Type O blood usually contains both anti-A and anti-B antibodies.

19.13 Because the mother is most likely to start making anti-Rh antibodies after the first baby is already born, that baby suffers no damage.

19.14 Agglutination refers to clumping of red blood cells.

19.15 Some symptoms of sickle-cell disease are anemia, mild jaundice, joint pain, shortness of breath, rapid heart rate, abdominal pain, fever, and fatigue.

THE CARDIOVASCULAR SYSTEM: THE HEART

THE HEART AND HOMEOSTASIS The heart pumps blood through blood vessels to all body tissues. •

As you learned in the previous chapter, the **cardiovascular system** consists of the blood, the heart, and blood vessels. We also examined the composition and functions of blood, and in this chapter you will learn about the pump that circulates it throughout the body—the heart. Blood must be constantly pumped through the body's blood vessels so that it can reach body cells and exchange materials with them. To accomplish this, the heart beats about 100,000 times every day, which adds up to 35 million beats in a year and about 2.5 billion times in an average lifetime. Even while you are sleeping, your heart pumps 30 times its own weight (5 L or 5.3 qt) each minute, which amounts to more than 14,000 liters (3600 gal) of blood in a day and 10 million liters (2.6 million gal) in a year. Because you don't spend all your time sleeping, and your heart pumps more vigorously when you are active, the actual blood volume the heart pumps in a single day is even larger.

The scientific study of the normal heart and the diseases associated with it is known as **cardiology** (kar′-dē-OL-ō-jē; *cardio-* = heart; *-logy* = study of). This chapter explores the structure of the heart and the unique properties that permit it to pump for a lifetime without rest. •

ANATOMY OF THE HEART

◉ **OBJECTIVES**

- Describe the location of the heart.
- Describe the structure of the pericardium and the heart wall.
- Discuss the external and internal anatomy of the chambers of the heart.

Location of the Heart

For all its might, the heart is relatively small, roughly the same size (but not the same shape) as your closed fist. It is about 12 cm (5 in.) long, 9 cm (3.5 in.) wide at its broadest point, and

6 cm (2.5 in.) thick, with an average mass of 250 g (8 oz) in adult females and 300 g (10 oz) in adult males. The heart rests on the diaphragm, near the midline of the thoracic cavity. It lies in the **mediastinum** (mē-dē-a-STĪ-num), an anatomical region that extends from the sternum to the vertebral column, the first rib to the diaphragm, and between the lungs (Figure 20.1a). About two-thirds of the mass of the heart lies to the left of the body's midline (Figure 20.1b). You can visualize the heart as a cone lying on its side. The pointed **apex** is formed by the tip of the left ventricle (a lower chamber of the heart) and rests on the diaphragm. It is directed anteriorly, inferiorly, and to the left. The **base** of the heart is its posterior surface. It is formed by the atria (upper chambers) of the heart, mostly the left atrium.

Figure 20.1 Position of the heart and associated structures in the mediastinum (dashed outline). (See Tortora, *A Photographic Atlas of the Human Body, Second Edition,* Figures 6.5 and 6.6.)

🔑 The heart is located in the mediastinum, with two-thirds of its mass to the left of the midline.

(a) Inferior view of transverse section of thoracic cavity showing the heart in the mediastinum

(b) Anterior view of the heart in the thoracic cavity

❓ **What is the mediastinum?**

In addition to the apex and base, the heart has several distinct surfaces and borders (margins). The **anterior surface** is deep to the sternum and ribs. The **inferior surface** is the part of the heart between the apex and right border and rests mostly on the diaphragm (Figure 20.1b). The **right border** faces the right lung and extends from the inferior surface to the base. The **left border,** also called the *pulmonary border,* faces the left lung and extends from the base to the apex.

● CLINICAL **CONNECTION** | Cardiopulmonary Resuscitation

Because the heart lies between two rigid structures—the vertebral column and the sternum (Figure 20.1a)—external pressure on the chest (compression) can be used to force blood out of the heart and into the circulation. In cases in which the heart suddenly stops beating, **cardiopulmonary resuscitation (CPR)**—properly applied cardiac compressions, performed with artificial ventilation of the lungs via mouth-to-mouth respiration—saves lives. CPR keeps oxygenated blood circulating until the heart can be restarted.

In a 2007 Japanese study, researchers found that chest compressions alone are equally as effective as, if not better than, traditional CPR with lung ventilation. This is good news because it is easier for an emergency dispatcher to give instructions limited to chest compressions to frightened, nonmedical bystanders. As public fear of contracting contagious diseases such as hepatitis, HIV, and tuberculosis continues to rise, bystanders are much more likely to perform chest compressions alone than treatment involving mouth-to-mouth rescue breathing. ●

Pericardium

The membrane that surrounds and protects the heart is the **pericardium** (*peri-* = around). It confines the heart to its position in the mediastinum, while allowing sufficient freedom of movement for vigorous and rapid contraction. The pericardium consists of two main parts: (1) the fibrous pericardium and (2) the serous pericardium (Figure 20.2a). The superficial **fibrous pericardium** is composed of tough, inelastic, dense irregular connective tissue. It resembles a bag that rests on and attaches to the diaphragm; its open end is fused to the connective tissues of the blood vessels entering and leaving the heart. The fibrous pericardium prevents overstretching of the heart, provides protection, and anchors the heart in the mediastinum.

The deeper **serous pericardium** is a thinner, more delicate membrane that forms a double layer around the heart (Figure 20.2a). The outer **parietal layer** of the serous pericardium is fused to the fibrous pericardium. The inner **visceral layer** of the serous pericardium, also called the **epicardium** (*epi-* = on top of), is one of the layers of the heart wall and adheres tightly to the surface of the heart. Between the parietal and visceral layers of the serous pericardium is a thin film of lubricating serous fluid. This slippery secretion of the pericardial cells, known as **pericardial fluid,** reduces friction between the layers of the serous pericardium as the heart moves. The space that contains the few milliliters of pericardial fluid is called the **pericardial cavity.**

Figure 20.2 **Pericardium and heart wall.**

🔒 ⟳ The pericardium is a triple-layered sac that surrounds and protects the heart.

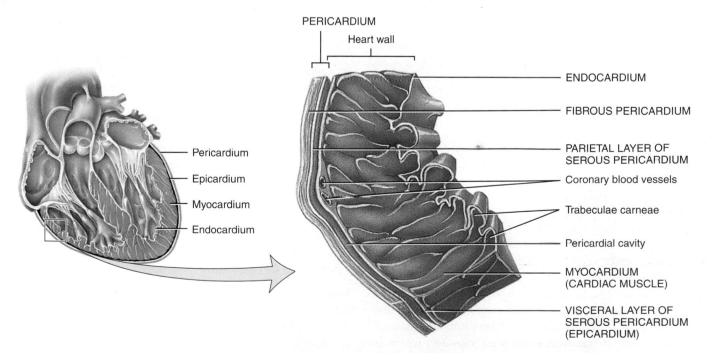

(a) Portion of pericardium and right ventricular heart wall showing the divisions of the pericardium and layers of the heart wall

 FIGURE 20.2 CONTINUES ▶

FIGURE 20.2 CONTINUED ▶

(b) Simplified relationship of the serous pericardium to the heart

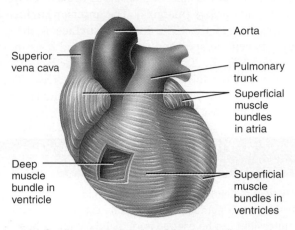

(c) Cardiac muscle bundles of the myocardium

? **Which layer is both a part of the pericardium and a part of the heart wall?**

•CLINICAL CONNECTION | **Pericarditis**

Inflammation of the pericardium is called **pericarditis** (per′-i-kar-DĪ-tis). The most common type, *acute pericarditis,* begins suddenly and has no known cause in most cases but is sometimes linked to a viral infection. As a result of irritation to the pericardium, there is chest pain that may extend to the left shoulder and down the left arm (often mistaken for a heart attack) and pericardial friction rub (a scratchy or creaking sound heard through a stethoscope as the visceral layer of the serous pericardium rubs against the parietal layer of the serous pericardium). Acute pericarditis usually lasts for about one week and is treated with drugs that reduce inflammation and pain, such as ibuprofen or aspirin.

   Chronic pericarditis begins gradually and is long-lasting. In one form of this condition, there is a buildup of pericardial fluid. If a great deal of fluid accumulates, this is a life-threatening condition because the fluid compresses the heart, a condition called *cardiac tamponade* (tam′-pon-ĀD). As a result of the compression, ventricular filling is decreased, cardiac output is reduced, venous return to the heart is diminished, blood pressure falls, and breathing is difficult. Most causes of chronic pericarditis involving cardiac tamponade are unknown, but it is sometimes caused by conditions such as cancer and tuberculosis. Treatment consists of draining the excess fluid through a needle passed into the pericardial cavity. •

Layers of the Heart Wall

The wall of the heart consists of three layers (Figure 20.2a): the epicardium (external layer), the myocardium (middle layer), and the endocardium (inner layer). As noted earlier, the outermost **epicardium,** the thin, transparent outer layer of the heart wall, is also called the *visceral layer of the serous pericardium.* It is composed of mesothelium and delicate connective tissue that imparts a smooth, slippery texture to the outermost surface of the heart. The middle **myocardium** (*myo-* = muscle), which is cardiac muscle tissue, makes up about 95% of the heart and is responsible for its pumping action. Although it is striated like skeletal muscle, cardiac muscle is involuntary like smooth

muscle. The cardiac muscle fibers swirl diagonally around the heart in bundles (Figure 20.2c). The innermost **endocardium** (*endo-* = within) is a thin layer of endothelium overlying a thin layer of connective tissue. It provides a smooth lining for the chambers of the heart and covers the valves of the heart. The endocardium is continuous with the endothelial lining of the large blood vessels attached to the heart, and it minimizes surface friction as blood passes through the heart and blood vessels.

•CLINICAL CONNECTION | **Myocarditis and Endocarditis**

Myocarditis (mī-ō-kar-DĪ-tis) is an inflammation of the myocardium that usually occurs as a complication of a viral infection, rheumatic fever, or exposure to radiation or certain chemicals or medications. Myocarditis often has no symptoms. However, if they do occur, they may include fever, fatigue, vague chest pain, irregular or rapid heartbeat, joint pain, and breathlessness. Myocarditis is usually mild and recovery occurs within two weeks. Severe cases can lead to cardiac failure and death. Treatment consists of avoiding vigorous exercise, a low-salt diet, electrocardiographic monitoring, and treatment of the cardiac failure. **Endocarditis** (en′-dō-kar-DĪ-tis) refers to an inflammation of the endocardium and typically involves the heart valves. Most cases are caused by bacteria (bacterial endocarditis). Signs and symptoms of endocarditis include fever, heart murmur, irregular or rapid heartbeat, fatigue, loss of appetite, night sweats, and chills. Treatment is with intravenous antibiotics. •

Chambers of the Heart

The heart has four chambers. The two superior receiving chambers are the **atria** (= entry halls or chambers), and the two inferior pumping chambers are the **ventricles** (= little bellies). On the anterior surface of each atrium is a wrinkled pouchlike structure called an **auricle** (OR-i-kul; *auri-* = ear), so named because of its resemblance to a dog's ear (Figure 20.3). Each auricle slightly increases the capacity of an atrium so that it can hold a

Figure 20.3 Structure of the heart: surface features. Throughout this book, blood vessels that carry oxygenated blood (which looks bright red) are colored red, whereas those that carry deoxygenated blood (which looks dark red) are colored blue.

 Sulci are grooves that contain blood vessels and fat and mark the external boundaries between the various chambers.

Brachiocephalic trunk

Superior vena cava

Ascending aorta

Right pulmonary artery

Fibrous pericardium (cut)

Right pulmonary veins

RIGHT AURICLE OF RIGHT ATRIUM

Right coronary artery

RIGHT ATRIUM

CORONARY SULCUS

RIGHT VENTRICLE

Inferior vena cava

Left common carotid artery

Left subclavian artery

Arch of aorta

Ligamentum arteriosum

Left pulmonary artery

Pulmonary trunk

Left pulmonary veins

LEFT AURICLE OF LEFT ATRIUM

Branch of left coronary artery

LEFT VENTRICLE

ANTERIOR INTERVENTRICULAR SULCUS

Descending aorta

(a) Anterior external view showing surface features

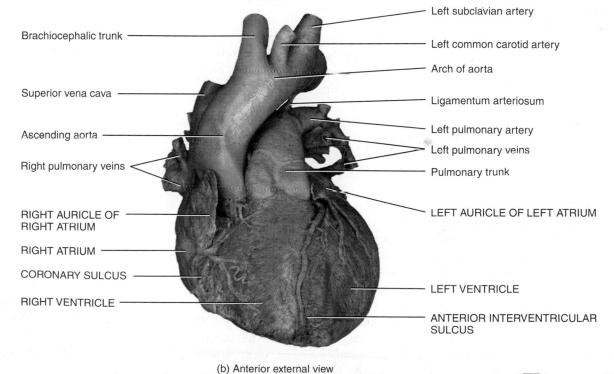

Brachiocephalic trunk

Superior vena cava

Ascending aorta

Right pulmonary veins

RIGHT AURICLE OF RIGHT ATRIUM

RIGHT ATRIUM

CORONARY SULCUS

RIGHT VENTRICLE

Left subclavian artery

Left common carotid artery

Arch of aorta

Ligamentum arteriosum

Left pulmonary artery

Left pulmonary veins

Pulmonary trunk

LEFT AURICLE OF LEFT ATRIUM

LEFT VENTRICLE

ANTERIOR INTERVENTRICULAR SULCUS

(b) Anterior external view

FIGURE 20.3 CONTINUES ▶

FIGURE 20.3 CONTINUED ▶

Left common carotid artery

Left subclavian artery

Arch of aorta

Descending aorta

Left pulmonary artery

AURICLE OF LEFT ATRIUM

Left pulmonary veins

LEFT ATRIUM

Coronary sinus
(in the coronary sulcus)

LEFT VENTRICLE

POSTERIOR
INTERVENTRICULAR SULCUS

Brachiocephalic trunk

Superior vena cava

Ascending aorta

Right pulmonary artery

Right pulmonary veins

RIGHT ATRIUM

Right coronary artery

Inferior vena cava

Middle cardiac vein

RIGHT VENTRICLE

(c) Posterior external view showing surface features

? **The coronary sulcus forms an external boundary between which chambers of the heart?**

greater volume of blood. Also on the surface of the heart are a series of grooves, called **sulci** (SUL-sē), that contain coronary blood vessels and a variable amount of fat. Each *sulcus* (SUL-kus) marks the external boundary between two chambers of the heart. The deep **coronary sulcus** (*coron-* = resembling a crown) encircles most of the heart and marks the external boundary between the superior atria and inferior ventricles. The **anterior interventricular sulcus** is a shallow groove on the anterior surface of the heart that marks the external boundary between the right and left ventricles. This sulcus continues around to the posterior surface of the heart as the **posterior interventricular sulcus,** which marks the external boundary between the ventricles on the posterior aspect of the heart (Figure 20.3c).

Right Atrium

The **right atrium** forms the right border of the heart and receives blood from three veins: the *superior vena cava, inferior vena cava,* and *coronary sinus* (Figure 20.4a) (Veins always return blood to the heart). The right atrium is about 2–3 mm (0.08–0.12 in.) in average thickness. The anterior and posterior walls of the right atrium are very different. The posterior wall is smooth; the anterior wall is rough due to the presence of muscular ridges called **pectinate muscles** (PEK-tin-āt; *pectin* = comb), which also extend into the auricle (Figure 20.4b). Between the right atrium and left atrium is a thin partition called

the **interatrial septum** (*inter-* = between; *septum* = a dividing wall or partition). A prominent feature of this septum is an oval depression called the **fossa ovalis,** the remnant of the *foramen ovale,* an opening in the interatrial septum of the fetal heart that normally closes soon after birth (see Figure 21.30 on page 822). Blood passes from the right atrium into the right ventricle through a valve that is called the **tricuspid valve** (trī-KUS-pid; *tri-* = three; *cuspid* = point) because it consists of three leaflets or cusps (Figure 20.4a). It is also called the **right atrioventricular valve.** The valves of the heart are composed of dense connective tissue covered by endocardium.

Right Ventricle

The **right ventricle** is about 4–5 mm (0.16–0.2 in.) in average thickness and forms most of the anterior surface of the heart. The inside of the right ventricle contains a series of ridges formed by raised bundles of cardiac muscle fibers called **trabeculae carneae** (tra-BEK-ū-lē KAR-nē-ē; *trabeculae* = little beams; *carneae* = fleshy; see Figure 20.2a). Some of the trabeculae carneae convey part of the conduction system of the heart, which you will learn about later in this chapter (see page 732). The cusps of the tricuspid valve are connected to tendonlike cords, the **chordae tendineae** (KOR-dē ten-DIN-ē-ē; *chord-* = cord; *tend-* = tendon), which in turn are connected to cone-shaped trabeculae carneae called **papillary muscles** (*papill-* =

Figure 20.4 Structure of the heart: internal anatomy.

Blood flows into the right atrium through the superior vena cava, inferior vena cava, and coronary sinus and into the left atrium through four pulmonary veins.

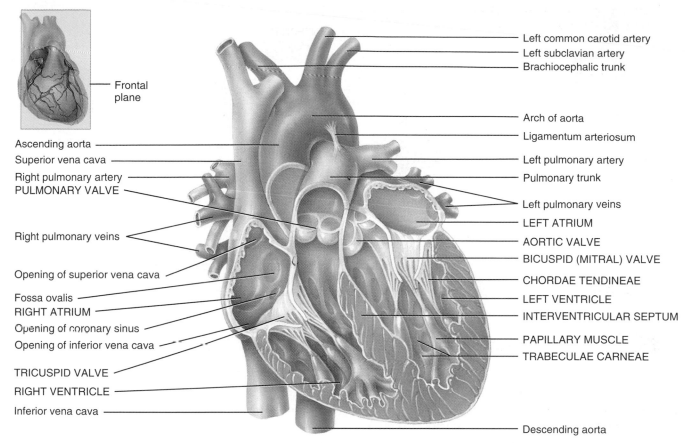

Frontal plane

Ascending aorta
Superior vena cava
Right pulmonary artery
PULMONARY VALVE

Right pulmonary veins

Opening of superior vena cava
Fossa ovalis
RIGHT ATRIUM
Opening of coronary sinus
Opening of inferior vena cava

TRICUSPID VALVE
RIGHT VENTRICLE
Inferior vena cava

Left common carotid artery
Left subclavian artery
Brachiocephalic trunk

Arch of aorta
Ligamentum arteriosum
Left pulmonary artery
Pulmonary trunk
Left pulmonary veins
LEFT ATRIUM
AORTIC VALVE
BICUSPID (MITRAL) VALVE
CHORDAE TENDINEAE
LEFT VENTRICLE
INTERVENTRICULAR SEPTUM
PAPILLARY MUSCLE
TRABECULAE CARNEAE

Descending aorta

(a) Anterior view of frontal section showing internal anatomy

Brachiocephalic trunk

Superior vena cava

Right pulmonary vein

Ascending aorta

RIGHT AURICLE
(cut open)
Pectinate muscles

RIGHT ATRIUM

Cusp of tricuspid valve
Chordae tendineae
Papillary muscle
RIGHT VENTRICLE

Left subclavian artery
Left common carotid artery
Arch of aorta
Ligamentum anteriosum
Pulmonary trunk
Left pulmonary vein
LEFT AURICLE
LEFT VENTRICLE
INTERVENTRICULAR SEPTUM
TRABECULAE CARNEAE

(b) Anterior view of partially sectioned heart

FIGURE 20.4 CONTINUES ▶

FIGURE 20.4 CONTINUED ▶

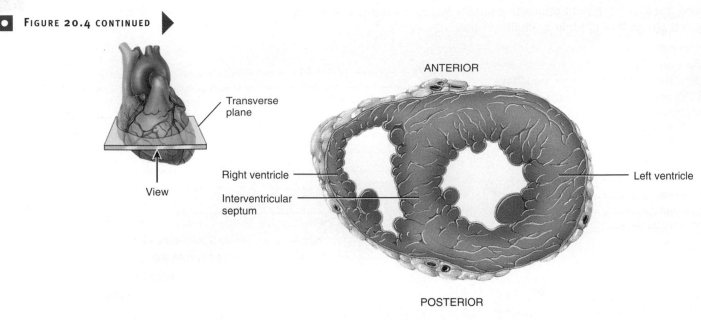

(c) Inferior view of transverse section showing differences in thickness of ventricular walls

? How does thickness of the myocardium relate to the workload of a cardiac chamber?

nipple). Internally, the right ventricle is separated from the left ventricle by a partition called the **interventricular septum.** Blood passes from the right ventricle through the **pulmonary valve** (*pulmonary semilunar valve*) into a large artery called the *pulmonary trunk,* which divides into right and left *pulmonary arteries.* Arteries always take blood away from the heart.

Left Atrium

The **left atrium** is about the same thickness as the right atrium and forms most of the base of the heart (see Figure 20.1b). It receives blood from the lungs through four *pulmonary veins.* Like the right atrium, the inside of the left atrium has a smooth posterior wall. Because pectinate muscles are confined to the auricle of the left atrium, the anterior wall of the left atrium also is smooth. Blood passes from the left atrium into the left ventricle through the **bicuspid (mitral) valve** (*bi-* = two), which, as its name implies, has two cusps. The term *mitral* refers to the resemblance of the bicuspid valve to a bishop's miter (hat), which is two-sided. It is also called the **left atrioventricular valve.**

Left Ventricle

The **left ventricle** is the thickest chamber of the heart, averaging 10–15 mm (0.4–0.6 in.) and forms the apex of the heart (see Figure 20.1b). Like the right ventricle, the left ventricle contains trabeculae carneae and has chordae tendinae that anchor the cusps of the bicuspid valve to papillary muscles. Blood passes from the left ventricle through the **aortic valve** (*aortic semilunar valve*) into the *ascending aorta* (*aorte* = to suspend, because the aorta once was believed to lift up the heart). Some of the blood

in the aorta flows into the *coronary arteries,* which branch from the ascending aorta and carry blood to the heart wall. The remainder of the blood passes into the *arch of the aorta* and *descending aorta* (*thoracic aorta* and *abdominal aorta*). Branches of the arch of the aorta and descending aorta carry blood throughout the body.

During fetal life, a temporary blood vessel, called the *ductus arteriosus,* shunts blood from the pulmonary trunk into the aorta. Hence, only a small amount of blood enters the nonfunctioning fetal lungs (see Figure 21.30 on page 822). The ductus arteriosus normally closes shortly after birth, leaving a remnant known as the **ligamentum arteriosum,** which connects the arch of the aorta and pulmonary trunk (Figure 20.4a).

Myocardial Thickness and Function

The thickness of the myocardium of the four chambers varies according to each chamber's function. The thin-walled atria deliver blood under less pressure into the adjacent ventricles. Because the ventricles pump blood under higher pressure over greater distances, their walls are thicker (Figure 20.4a). Although the right and left ventricles act as two separate pumps that simultaneously eject equal volumes of blood, the right side has a much smaller workload. It pumps blood a short distance to the lungs at lower pressure, and the resistance to blood flow is small. The left ventricle pumps blood great distances to all other parts of the body at higher pressure, and the resistance to blood flow is larger. Therefore, the left ventricle works much harder than the right ventricle to maintain the same rate of blood flow. The anatomy of the two ventricles confirms this functional difference—the muscular wall of the left ventricle is considerably thicker than the wall of the right ventricle (Figure 20.4c). Note also that the perimeter of the lumen (space) of the left ventricle is roughly circular in contrast to that of the right ventricle, which is somewhat crescent shaped.

Fibrous Skeleton of the Heart

In addition to cardiac muscle tissue, the heart wall also contains dense connective tissue that forms the **fibrous skeleton of the heart** (Figure 20.5). Essentially, the fibrous skeleton consists of four dense connective tissue rings that surround the valves of the heart, fuse with one another, and merge with the interventricular septum. In addition to forming a structural foundation for the heart valves, the fibrous skeleton prevents overstretching of the valves as blood passes through them. It also serves as a point of insertion for bundles of cardiac muscle fibers and acts as an electrical insulator between the atria and ventricles.

◉ CHECKPOINT

1. Define each of the following external features of the heart: auricle, coronary sulcus, anterior interventricular sulcus, and posterior interventricular sulcus.
2. Describe the structure of the pericardium and the layers of the wall of the heart.
3. What are the characteristic internal features of each chamber of the heart?
4. Which blood vessels deliver blood to the right and left atria?
5. What is the relationship between wall thickness and function among the various chambers of the heart?
6. What type of tissue composes the fibrous skeleton of the heart? What functions does this tissue perform?

HEART VALVES AND CIRCULATION OF BLOOD

◉ OBJECTIVES
- Describe the structure and function of the valves of the heart.
- Outline the flow of blood through the chambers of the heart and through the systemic and pulmonary circulations.
- Discuss the coronary circulation.

As each chamber of the heart contracts, it pushes a volume of blood into a ventricle or out of the heart into an artery. Valves open and close in response to *pressure changes* as the heart contracts and relaxes. Each of the four valves helps ensure the one-way flow of blood by opening to let blood through and then closing to prevent its backflow.

Operation of the Atrioventricular Valves

Because they are located between an atrium and a ventricle, the tricuspid and bicuspid valves are termed **atrioventricular (AV) valves.** When an AV valve is open, the rounded ends of the cusps project into the ventricle. When the ventricles are relaxed, the papillary muscles are relaxed, the chordae tendineae are slack, and blood moves from a higher pressure in the atria to a lower pressure in the ventricles through open AV valves (Figure 20.6a, d). When the ventricles contract, the pressure of the blood drives the cusps upward until their edges meet and close the opening (Figure 20.6b, e). At the same time, the papillary muscles contract, which pulls on and tightens the chordae tendineae. This prevents the valve cusps from everting (opening into the atria) in response to the high ventricular pressure. If the AV valves or chordae tendineae are damaged, blood may regurgitate (flow back) into the atria when the ventricles contract.

Figure 20.5 **Fibrous skeleton of the heart.** Elements of the fibrous skeleton are shown in capital letters.

 Fibrous rings support the four valves of the heart and are fused to one another.

Superior view (the atria have been removed)

❓ **In what two ways does the fibrous skeleton contribute to the functioning of heart valves?**

Figure 20.6 **Responses of the valves to the pumping of the heart.**

Heart valves prevent the backflow of blood.

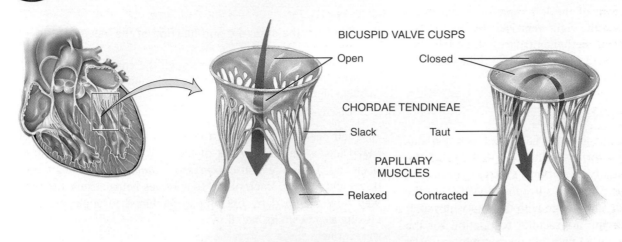

BICUSPID VALVE CUSPS

Open   Closed

CHORDAE TENDINEAE

Slack   Taut

PAPILLARY MUSCLES

Relaxed   Contracted

(a) Bicuspid valve open   (b) Bicuspid valve closed

Cusp of tricuspid valve

Chordae tendineae

Papillary muscle

(c) Tricuspid valve open

ANTERIOR

Pulmonary valve (closed)

Left coronary artery

Bicuspid valve (open)

Tricuspid valve (open)

Aortic valve (closed)

Right coronary artery

POSTERIOR

(d) Superior view with atria removed: pulmonary and aortic valves closed, bicuspid and tricuspid valves open.

ANTERIOR

Pulmonary valve (open)

Aortic valve (open)

Bicuspid valve (closed)

Tricuspid valve (closed)

POSTERIOR

(e) Superior view with atria removed: pulmonary and aortic valves open, bicuspid and tricuspid valves closed.

Operation of the Semilunar Valves

The aortic and pulmonary valves are known as the **semilunar (SL) valves** (*semi-* = half; *lunar* = moon-shaped) because they are made up of three crescent moon-shaped cusps (Figure 20.6d). Each cusp attaches to the arterial wall by its convex outer margin. The SL valves allow ejection of blood from the heart into arteries but prevent backflow of blood into the ventricles. The free borders of the cusps project into the lumen of the artery. When the ventricles contract, pressure builds up within the chambers. The semilunar valves open when pressure in the ventricles exceeds the pressure in the arteries, permitting ejection of blood from the ventricles into the pulmonary trunk and aorta (Figure 20.6e). As the ventricles relax, blood starts to flow back toward the heart. This backflowing blood fills the valve cusps, which causes the semilunar valves to close tightly (Figure 20.6d).

Surprisingly perhaps, there are no valves guarding the junctions between the venae cavae and the right atrium or the pulmonary veins and the left atrium. As the atria contract, a small amount of blood does flow backward from the atria into these vessels. However, backflow is minimized by a different mechanism; as the atrial muscle contracts, it compresses and nearly collapses the venous entry points.

• CLINICAL CONNECTION | Heart Valve Disorders

When heart valves operate normally, they open fully and close completely at the proper times. A narrowing of a heart valve opening that restricts blood flow is known as **stenosis** (ste-NŌ-sis = a narrowing); failure of a valve to close completely is termed **insufficiency** or **incompetence.** In **mitral stenosis,** scar formation or a congenital defect causes narrowing of the mitral valve. One cause of **mitral insufficiency,** in which there is backflow of blood from the left ventricle into the left atrium, is **mitral valve prolapse (MVP).** In MVP one or both cusps of the mitral valve protrude into the left atrium during ventricular contraction. Mitral valve prolapse is one of the most common valvular disorders, affecting as much as 30% of the population. It is more prevalent in women than in men, and does not always pose a serious threat. In **aortic stenosis** the aortic valve is narrowed, and in **aortic insufficiency** there is backflow of blood from the aorta into the left ventricle.

Certain infectious diseases can damage or destroy the heart valves. One example is **rheumatic fever,** an acute systemic inflammatory disease that usually occurs after a streptococcal infection of the throat. The bacteria trigger an immune response in which antibodies produced to destroy the bacteria instead attack and inflame the connective tissues in joints, heart valves, and other organs. Even though rheumatic fever may weaken the entire heart wall, most often it damages the mitral and aortic valves. •

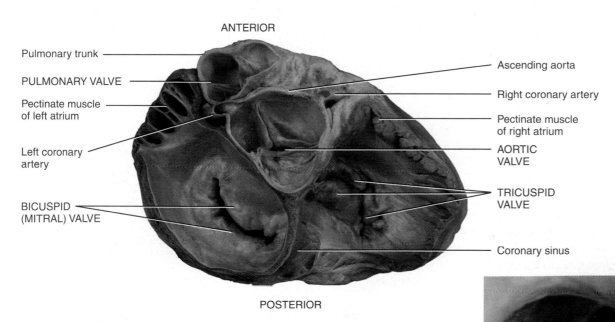

(f) Superior view of atrioventricular and semilunar valves

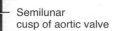

(g) Superior view of aortic valve

 How do papillary muscles prevent atrioventricular valve cusps from everting (swinging upward) into the atria?

Systemic and Pulmonary Circulations

In postnatal (after birth) circulation, the heart pumps blood into two closed circuits with each beat—**systemic circulation** and **pulmonary circulation** (*pulmon-* = lung). The two circuits are arranged in series: The output of one becomes the input of the other, as would happen if you attached two garden hoses (see Figure 21.17 on page 000). The left side of the heart is the pump for systemic circulation; it receives bright red, oxygen-rich blood from the lungs. The left ventricle ejects blood into the *aorta* (Figure 20.7). From the aorta, the blood divides into separate streams, entering progressively smaller *systemic arteries* that carry it to all organs throughout the body—except for the air sacs (alveoli) of the lungs, which are supplied by pulmonary circulation. In systemic tissues, arteries give rise to smaller-diameter *arterioles,* which finally lead into extensive beds of *systemic capillaries.* Exchange of nutrients and gases occurs across the thin capillary walls. Blood unloads O_2 (oxygen) and picks up CO_2 (carbon dioxide). In most cases, blood flows through only one capillary and then enters a *systemic venule.* Venules carry deoxygenated (oxygen-poor) blood away from tissues and merge to form larger *systemic veins.* Ultimately the blood flows back to the right atrium.

The right side of the heart is the pump for pulmonary circulation; it receives all the dark red, deoxygenated blood returning from systemic circulation. Blood ejected from the right ventricle flows into the *pulmonary trunk,* which branches into *pulmonary arteries* that carry blood to the right and left lungs. In pulmonary capillaries, blood unloads CO_2, which is exhaled, and picks up inhaled O_2. The freshly oxygenated blood then flows into pulmonary veins and returns to the left atrium.

Coronary Circulation

Nutrients are not able to diffuse quickly enough from blood in the chambers of the heart to supply all the layers of cells that make up the heart wall. For this reason, the myocardium has its own network of blood vessels, the **coronary** or **cardiac circulation** (*coron-* = crown). The **coronary arteries** branch from the ascending aorta and encircle the heart like a crown encircles the head (Figure 20.8a). While the heart is contracting, little blood flows in the coronary arteries because they are squeezed shut. When the heart relaxes, however, the high pressure of blood in the aorta propels blood through the coronary arteries, into capillaries, and then into **coronary veins** (Figure 20.8b).

Coronary Arteries

Two coronary arteries, the right and left coronary arteries, branch from the ascending aorta and supply oxygenated blood to the myocardium (Figure 20.8a). The **left coronary artery** passes inferior to the left auricle and divides into the anterior interventricular and circumflex branches. The **anterior interventricular branch** or *left anterior descending (LAD) artery* is in the anterior interventricular sulcus and supplies oxygenated blood to the walls of both ventricles. The **circumflex branch**

Figure 20.7 Systemic and pulmonary circulations.

The left side of the heart pumps oxygenated blood into the systemic circulation to all tissues of the body except the air sacs (alveoli) of the lungs. The right side of the heart pumps deoxygenated blood into the pulmonary circulation to the air sacs (alveoli) of the lungs.

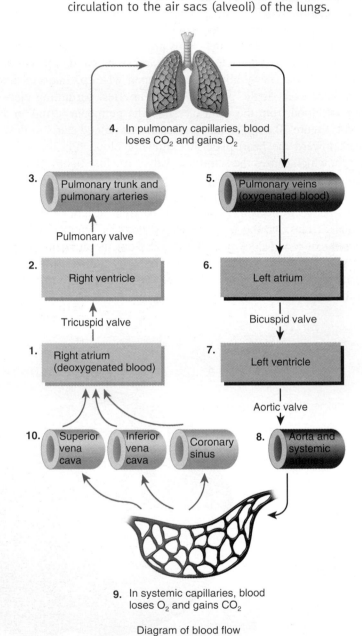

Diagram of blood flow

? Which numbers constitute the pulmonary circulation? Which constitute the systemic circulation?

lies in the coronary sulcus and distributes oxygenated blood to the walls of the left ventricle and left atrium.

The **right coronary artery** supplies small branches (*atrial branches*) to the right atrium. It continues inferior to the right auricle and ultimately divides into the posterior interventricular and marginal branches. The **posterior interventricular branch** follows the posterior interventricular sulcus and supplies the

Figure 20.8 **The coronary circulation.** The views of the heart from the anterior aspect in (a) and (b) are drawn as if the heart were transparent to reveal blood vessels on the posterior aspect. (See Tortora, *A Photographic Atlas of the Human Body, Second Edition,* Figures 6.8 and 6.9.)

The right and left coronary arteries deliver blood to the heart; the coronary veins drain blood from the heart into the coronary sinus.

(a) Anterior view of coronary arteries

(b) Anterior view of coronary veins

(c) Anterior view

Which coronary blood vessel delivers oxygenated blood to the walls of the left atrium and left ventricle?

walls of the two ventricles with oxygenated blood. The **marginal branch** in the coronary sulcus transports oxygenated blood to the myocardium of the right ventricle.

Most parts of the body receive blood from branches of more than one artery, and where two or more arteries supply the same region, they usually connect. These connections, called **anastomoses** (a-nas′-tō-MŌ-sēs), provide alternate routes called **collateral circuits,** for blood to reach a particular organ or tissue. The myocardium contains many anastomoses that connect branches of a given coronary artery or extend between branches of different coronary arteries. They provide detours for arterial blood if a main route becomes obstructed. Thus, heart muscle may receive sufficient oxygen even if one of its coronary arteries is partially blocked.

Coronary Veins

After blood passes through the arteries of the coronary circulation, it flows into capillaries, where it delivers oxygen and nutrients to the heart muscle and collects carbon dioxide and waste, and then moves into coronary veins. Most of the deoxygenated blood from the myocardium drains into a large *vascular sinus* in the coronary sulcus on the posterior surface of the heart, called the **coronary sinus** (Figure 20.8b). (A *vascular sinus* is a thin-walled vein that has no smooth muscle to alter its diameter.) The deoxygenated blood in the coronary sinus empties into the right atrium. The principal tributaries carrying blood into the coronary sinus are the following:

- **Great cardiac vein** in the anterior interventricular sulcus, which drains the areas of the heart supplied by the left coronary artery (left and right ventricles and left atrium)

- **Middle cardiac vein** in the posterior interventricular sulcus, which drains the areas supplied by the posterior interventricular branch of the right coronary artery (left and right ventricles)

- **Small cardiac vein** in the coronary sulcus, which drains the right atrium and right ventricle

- **Anterior cardiac veins,** which drain the right ventricle and open directly into the right atrium

When blockage of a coronary artery deprives the heart muscle of oxygen, **reperfusion,** the reestablishment of blood flow, may damage the tissue further. This surprising effect is due to the formation of oxygen **free radicals** from the reintroduced oxygen. As you learned in Chapter 2, free radicals are electrically charged molecules that have an unpaired electron (see Figure 2.3b on page 000). These unstable, highly reactive molecules cause chain reactions that lead to cellular damage and death. To counter the effects of oxygen free radicals, body cells produce enzymes that convert free radicals to less reactive substances. Two such enzymes are *superoxide dismutase* and *catalase*. In addition, nutrients such as vitamin E, vitamin C, beta-carotene, zinc, and selenium serve as antioxidants, which remove oxygen free radicals from circulation. Drugs that lessen reperfusion damage after a heart attack or stroke are currently under development.

• CLINICAL CONNECTION | **Myocardial Ischemia and Infarction**

Partial obstruction of blood flow in the coronary arteries may cause **myocardial ischemia** (is-KĒ-mē-a; *ische-* = to obstruct; *-emia* = in the blood), a condition of reduced blood flow to the myocardium. Usually, ischemia causes **hypoxia** (reduced oxygen supply), which may weaken cells without killing them. **Angina pectoris** (an-JĪ-na, or AN-ji-na, PEK-to-ris), which literally means "strangled chest," is a severe pain that usually accompanies myocardial ischemia. Typically, sufferers describe it as a tightness or squeezing sensation, as though the chest were in a vise. The pain associated with angina pectoris is often referred to the neck, chin, or down the left arm to the elbow. **Silent myocardial ischemia,** ischemic episodes without pain, is particularly dangerous because the person has no forewarning of an impending heart attack.

A complete obstruction to blood flow in a coronary artery may result in a **myocardial infarction** (in-FARK-shun), or **MI,** commonly called a *heart attack. Infarction* means the death of an area of tissue because of interrupted blood supply. Because the heart tissue distal to the obstruction dies and is replaced by noncontractile scar tissue, the heart muscle loses some of its strength. Depending on the size and location of the infarcted (dead) area, an infarction may disrupt the conduction system of the heart and cause sudden death by triggering ventricular fibrillation. Treatment for a myocardial infarction may involve injection of a thrombolytic (clot-dissolving) agent such as streptokinase or t-PA, plus heparin (an anticoagulant), or performing coronary angioplasty or coronary artery bypass grafting. Fortunately, heart muscle can remain alive in a resting person if it receives as little as 10–15% of its normal blood supply. •

◉ CHECKPOINT

7. What causes the heart valves to open and to close? What supporting structures ensure that the valves operate properly?

8. In correct sequence, which heart chambers, heart valves, and blood vessels would a drop of blood encounter as it flows from the right atrium to the aorta?

9. Which arteries deliver oxygenated blood to the myocardium of the left and right ventricles?

CARDIAC MUSCLE TISSUE AND THE CARDIAC CONDUCTION SYSTEM

◉ OBJECTIVES

- Describe the structural and functional characteristics of cardiac muscle tissue and the conduction system of the heart.
- Explain how an action potential occurs in cardiac contractile fibers.
- Describe the electrical events of a normal electrocardiogram (ECG).

Histology of Cardiac Muscle Tissue

Compared with skeletal muscle fibers, cardiac muscle fibers are shorter in length and less circular in transverse section (Figure 20.9). They also exhibit branching, which gives individual cardiac muscle fibers a "stair-step" appearance (see Table 4.5B on page 138). A typical cardiac muscle fiber is 50–100 μm long and

Figure 20.9 Histology of cardiac muscle tissue. (See Table 4.5B on page 138 for a light micrograph of cardiac muscle.)

 Cardiac muscle fibers connect to neighboring fibers by intercalated discs, which contain desmosomes and gap junctions.

(a) Cardiac muscle fibers

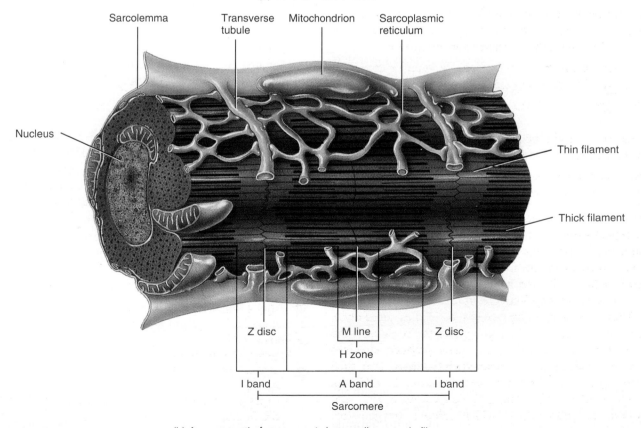

(b) Arrangement of components in a cardiac muscle fiber

? **What are the functions of intercalated discs in cardiac muscle fibers?**

has a diameter of about 14 μm. Usually one centrally located nucleus is present, although an occasional cell may have two nuclei. The ends of cardiac muscle fibers connect to neighboring fibers by irregular transverse thickenings of the sarcolemma called **intercalated discs** (in-TER-kā-lāt-ed; *intercalat-* = to insert between). The discs contain **desmosomes,** which hold the fibers together, and **gap junctions,** which allow muscle action potentials to conduct from one muscle fiber to its neighbors. Gap junctions allow the entire myocardium of the atria or the ventricles to contract as a single, coordinated unit.

Mitochondria are larger and more numerous in cardiac muscle fibers than in skeletal muscle fibers. In a cardiac muscle fiber, they take up 25% of the cytosolic space; in a skeletal muscle fiber only 2% of the cytosolic space is occupied by mitochondria. Cardiac muscle fibers have the same arrangement of actin and myosin, and the same bands, zones, and Z discs, as skeletal muscle fibers. The transverse tubules of cardiac muscle are wider but less abundant than those of skeletal muscle; the one transverse tubule per sarcomere is located at the Z disc. The sarcoplasmic reticulum of cardiac muscle fibers is somewhat smaller than the SR of skeletal muscle fibers. As a result, cardiac muscle has a smaller intracellular reserve of Ca^{2+}.

• CLINICAL CONNECTION | Regeneration of Heart Cells

As noted earlier in the chapter, the heart of a heart attack survivor often has regions of infarcted (dead) cardiac muscle tissue that typically are replaced with noncontractile fibrous scar tissue over time. Our inability to repair damage from a heart attack has been attributed to a lack of stem cells in cardiac muscle and to the absence of mitosis in mature cardiac muscle fibers. A recent study of heart transplant recipients by American and Italian scientists, however, provides evidence for significant replacement of heart cells. The researchers studied men who had received a heart from a female, and then looked for the presence of a Y chromosome in heart cells. (All female cells except gametes have two X chromosomes and lack the Y chromosome.) Several years after the transplant surgery, between 7% and 16% of the heart cells in the transplanted tissue, including cardiac muscle fibers and endothelial cells in coronary arterioles and capillaries, had been replaced by the recipient's own cells, as evidenced by the presence of a Y chromosome. The study also revealed cells with some of the characteristics of stem cells in both transplanted hearts and control hearts. Evidently, stem cells can migrate from the blood into the heart and differentiate into functional muscle and endothelial cells. The hope is that researchers can learn how to "turn on" such regeneration of heart cells to treat people with heart failure or cardiomyopathy (diseased heart). •

Autorhythmic Fibers: The Conduction System

An inherent and rhythmical electrical activity is the reason for the heart's lifelong beat. The source of this electrical activity is a network of specialized cardiac muscle fibers called **autorhythmic fibers** (*auto-* = self) because they are self-excitable. Autorhyth-

mic fibers repeatedly generate action potentials that trigger heart contractions. They continue to stimulate a heart to beat even after it is removed from the body—for example, to be transplanted into another person—and all of its nerves have been cut. (Note: Surgeons do not attempt to reattach heart nerves during heart transplant operations. For this reason, it has been said that heart surgeons are better "plumbers" than they are "electricians.")

During embryonic development, only about 1% of the cardiac muscle fibers become autorhythmic fibers; these relatively rare fibers have two important functions.

1. They act as a **pacemaker,** setting the rhythm of electrical excitation that causes contraction of the heart.

2. They form the **conduction system,** a network of specialized cardiac muscle fibers that provide a path for each cycle of cardiac excitation to progress through the heart. The conduction system ensures that cardiac chambers become stimulated to contract in a coordinated manner, which makes the heart an effective pump. As you will see later in the chapter, problems with autorhythmic fibers can result in arrhythmias (abnormal rhythms) in which the heart beats irregularly, too fast, or too slow.

Cardiac action potentials propagate through the conduction system in the following sequence (Figure 20.10a):

1 Cardiac excitation normally begins in the **sinoatrial (SA) node,** located in the right atrial wall just inferior and lateral to the opening of the superior vena cava. SA node cells do not have a stable resting potential. Rather, they repeatedly depolarize to threshold spontaneously. The spontaneous depolarization is a **pacemaker potential.** When the pacemaker potential reaches threshold, it triggers an action potential (Figure 20.10b). Each action potential from the SA node propagates throughout both atria via gap junctions in the intercalated discs of atrial muscle fibers. Following the action potential, the atria contract.

2 By conducting along atrial muscle fibers, the action potential reaches the **atrioventricular (AV) node,** located in the interatrial septum, just anterior to the opening of the coronary sinus (Figure 20.10a).

3 From the AV node, the action potential enters the **atrioventricular (AV) bundle** (also known as the **bundle of His**). This bundle is the only site where action potentials can conduct from the atria to the ventricles. (Elsewhere, the fibrous skeleton of the heart electrically insulates the atria from the ventricles.)

4 After propagating along the AV bundle, the action potential enters both the **right** and **left bundle branches.** The bundle branches extend through the interventricular septum toward the apex of the heart.

5 Finally, the large-diameter **Purkinje fibers** rapidly conduct the action potential beginning at the apex of the heart upward to the remainder of the ventricular myocardium. Then the ventricles contract, pushing the blood upward toward the semilunar valves.

Figure 20.10 **The conduction system of the heart.** Autorhythmic fibers in the SA node, located in the right atrial wall (a), act as the heart's pacemaker, initiating cardiac action potentials (b) that cause contraction of the heart's chambers.

🔑 The conduction system ensures that the chambers of the heart contract in a coordinated manner.

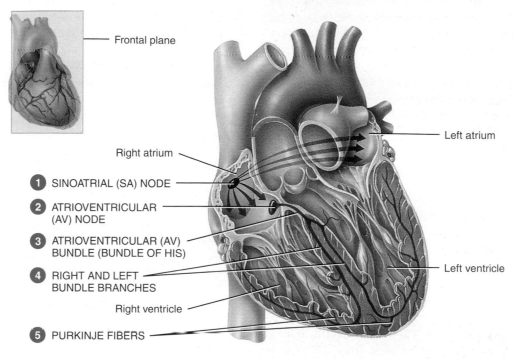

Frontal plane

Left atrium

Right atrium

1 SINOATRIAL (SA) NODE

2 ATRIOVENTRICULAR (AV) NODE

3 ATRIOVENTRICULAR (AV) BUNDLE (BUNDLE OF HIS)

4 RIGHT AND LEFT BUNDLE BRANCHES

Left ventricle

Right ventricle

5 PURKINJE FIBERS

(a) Anterior view of frontal section

+ 10 mV

Membrane potential

Action potential

Threshold

Pacemaker potential

− 60 mV

0   0.8   1.6   2.4

Time (sec) ➝

(b) Pacemaker potentials and action potentials in autorhythmic fibers of SA node

❓ **Which component of the conduction system provides the only electrical connection between the atria and the ventricles?**

On their own, autorhythmic fibers in the SA node would initiate an action potential about every 0.6 second, or 100 times per minute. This rate is faster than that of any other autorhythmic fibers. Because action potentials from the SA node spread through the conduction system and stimulate other areas before the other areas are able to generate an action potential at their own, slower rate, the SA node acts as the natural pacemaker of the heart. Nerve impulses from the autonomic nervous system (ANS) and blood-borne hormones (such as epinephrine) *modify the timing and strength* of each heartbeat, but they *do not establish the fundamental rhythm*. In a person at rest, for example, acetylcholine released by the parasympathetic division of the ANS slows SA node pacing to about every 0.8 second or 75 action potentials per minute (Figure 20.10b).

If the SA node becomes damaged or diseased, the slower AV node can pick up the pacemaking task. Its rate of spontaneous depolarization is 40 to 60 times per minute. If the activity of both nodes is suppressed, the heartbeat may still be maintained by autorhythmic fibers in the ventricles—the AV bundle, a bundle branch, or Purkinje fibers. However, the pacing rate is so slow (20–35 beats per minute) that blood flow to the brain is inadequate. When this condition occurs, normal heart rhythm can be restored and maintained by surgically implanting an **artificial pacemaker,** a device that sends out small electrical currents to stimulate the heart to contract. A pacemaker consists of a battery and impulse generator and is usually implanted beneath the skin just inferior to the clavicle. The pacemaker is connected to one or two flexible leads (wires) that are threaded through the superior vena cava and then passed into the various chambers of the heart. Many of the newer pacemakers, referred to as *activity-adjusted pacemakers,* automatically speed up the heartbeat during exercise. •

Action Potential and Contraction of Contractile Fibers

The action potential initiated by the SA node travels along the conduction system and spreads out to excite the "working" atrial and ventricular muscle fibers, called **contractile fibers.** An action potential occurs in a contractile fiber as follows (Figure 20.11):

1 *Depolarization.* Unlike autorhythmic fibers, contractile fibers have a stable resting membrane potential that is close to −90 mV. When a contractile fiber is brought to threshold by an action potential from neighboring fibers, its **voltage-gated fast Na⁺ channels** open. These sodium ion channels are referred to as "fast" because they open very rapidly in response to a threshold-level depolarization. Opening of these channels allows Na⁺ inflow because the cytosol of contractile fibers is electrically more negative than interstitial fluid and Na⁺ concentration is higher in interstitial fluid. Inflow of Na⁺ down the electrochemical gradient produces a **rapid depolarization.** Within a few milliseconds, the fast Na⁺ channels automatically inactivate and Na⁺ inflow decreases.

2 *Plateau.* The next phase of an action potential in a contractile fiber is the **plateau,** a period of maintained depolarization. It is due in part to opening of **voltage-gated slow Ca²⁺ channels** in the sarcolemma. When these channels open, calcium ions move from the interstitial fluid (which has a higher Ca²⁺ concentration) into the cytosol. This inflow of Ca²⁺ causes even more Ca²⁺ to pour out of the sarcoplasmic reticulum into the cytosol through additional Ca²⁺ channels in the sarcoplasmic reticulum membrane. The increased Ca²⁺ concentration in the cytosol ultimately triggers contraction. Several different types of **voltage-gated K⁺ channels** are also found in the sarcolemma of a contractile fiber. Just before the plateau phase begins, some of these K⁺ channels open, allowing potassium ions to leave the contractile fiber. Therefore, depolarization is sustained during the plateau phase because Ca²⁺ inflow just balances K⁺ outflow. The plateau phase lasts for about 0.25 sec, and the membrane potential of the contractile fiber is close to 0 mV. By comparison, depolarization in a neuron or skeletal muscle

Figure 20.11 Action potential in a ventricular contractile fiber. The resting membrane potential is about −90 mV.

🔑 A long refractory period prevents tetanus in cardiac muscle fibers.

❓ How does the duration of an action potential in a ventricular contractile fiber compare with that in a skeletal muscle fiber?

fiber is much briefer, about 1 msec (0.001 sec), because it lacks a plateau phase.

❸ *Repolarization.* The recovery of the resting membrane potential during the **repolarization** phase of a cardiac action potential resembles that in other excitable cells. After a delay (which is particularly prolonged in cardiac muscle), additional voltage-gated K^+ channels open. Outflow of K^+ restores the negative resting membrane potential (-90 mV). At the same time, the calcium channels in the sarcolemma and the sarcoplasmic reticulum are closing, which also contributes to repolarization.

The mechanism of contraction is similar in cardiac and skeletal muscle: The electrical activity (action potential) leads to the mechanical response (contraction) after a short delay. As Ca^{2+} concentration rises inside a contractile fiber, Ca^{2+} binds to the regulatory protein troponin, which allows the actin and myosin filaments to begin sliding past one another, and tension starts to develop. Substances that alter the movement of Ca^{2+} through slow Ca^{2+} channels influence the strength of heart contractions. Epinephrine, for example, increases contraction force by enhancing Ca^{2+} flow into the cytosol.

In muscle, the **refractory period** is the time interval during which a second contraction cannot be triggered. The refractory period of a cardiac muscle fiber lasts longer than the contraction itself (Figure 20.11). As a result, another contraction cannot begin until relaxation is well underway. For this reason, tetanus (maintained contraction) cannot occur in cardiac muscle as it can in skeletal muscle. The advantage is apparent if you consider how the ventricles work. Their pumping function depends on alternating contraction (when they eject blood) and relaxation (when they refill). If heart muscle could undergo tetanus, blood flow would cease.

ATP Production in Cardiac Muscle

In contrast to skeletal muscle, cardiac muscle produces little of the ATP it needs by anaerobic cellular respiration (see Figure 10.12 on page 319). Instead, it relies almost exclusively on aerobic cellular respiration in its numerous mitochondria. The needed oxygen diffuses from blood in the coronary circulation and is released from myoglobin inside cardiac muscle fibers. Cardiac muscle fibers use several fuels to power mitochondrial ATP production. In a person at rest, the heart's ATP comes mainly from oxidation of fatty acids (60%) and glucose (35%), with smaller contributions from lactic acid, amino acids, and ketone bodies. During exercise, the heart's use of lactic acid, produced by actively contracting skeletal muscles, rises.

Like skeletal muscle, cardiac muscle also produces some ATP from creatine phosphate. One sign that a myocardial infarction (heart attack, see page 730) has occurred is the presence in blood of creatine kinase (CK), the enzyme that catalyzes transfer of a phosphate group from creatine phosphate to ADP to make ATP. Normally, CK and other enzymes are confined within cells. Injured or dying cardiac or skeletal muscle fibers release CK into the blood.

Electrocardiogram

As action potentials propagate through the heart, they generate electrical currents that can be detected at the surface of the body. An **electrocardiogram** (e-lek′-trō-KAR-dē-ō-gram), abbreviated either **ECG** or **EKG** (from the German word *Elektrokardiogram*), is a recording of these electrical signals. The ECG is a composite record of action potentials produced by all the heart muscle fibers during each heartbeat. The instrument used to record the changes is an **electrocardiograph.**

In clinical practice, electrodes are positioned on the arms and legs (limb leads) and at six positions on the chest (chest leads) to record the ECG. The electrocardiograph amplifies the heart's electrical signals and produces 12 different tracings from different combinations of limb and chest leads. Each limb and chest electrode records slightly different electrical activity because of the difference in its position relative to the heart. By comparing these records with one another and with normal records, it is possible to determine (1) if the conducting pathway is abnormal, (2) if the heart is enlarged, (3) if certain regions of the heart are damaged, and (4) the cause of chest pain.

In a typical record, three clearly recognizable waves appear with each heartbeat (Figure 20.12). The first, called the **P wave,** is a small upward deflection on the ECG. The P wave represents **atrial depolarization,** which spreads from the SA

Figure 20.12 Normal electrocardiogram or ECG (Lead II).
P wave = atrial depolarization; QRS complex = onset of ventricular depolarization; T wave = ventricular repolarization.

 An ECG is a recording of the electrical activity that initiates each heartbeat.

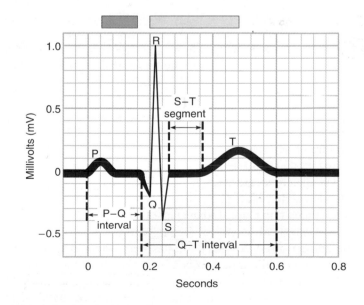

Key:

Atrial contraction

Ventricular contraction

 What is the significance of an enlarged Q wave?

node through contractile fibers in both atria. The second wave, called the **QRS complex,** begins as a downward deflection, continues as a large, upright, triangular wave, and ends as a downward wave. The QRS complex represents **rapid ventricular depolarization,** as the action potential spreads through ventricular contractile fibers. The third wave is a dome-shaped upward deflection called the **T wave.** It indicates **ventricular repolarization** and occurs just as the ventricles are starting to relax. The T wave is smaller and wider than the QRS complex because repolarization occurs more slowly than depolarization. During the plateau period of steady depolarization, the ECG tracing is flat.

In reading an ECG, the size of the waves can provide clues to abnormalities. Larger P waves indicate enlargement of an atrium; an enlarged Q wave may indicate a myocardial infarction; and an enlarged R wave generally indicates enlarged ventricles. The T wave is flatter than normal when the heart muscle is receiving insufficient oxygen—as, for example, in coronary artery disease. The T wave may be elevated in hyperkalemia (high blood K^+ level).

Analysis of an ECG also involves measuring the time spans between waves, which are called **intervals** or **segments.** For example, the **P-Q interval** is the time from the beginning of the P wave to the beginning of the QRS complex. It represents the conduction time from the beginning of atrial excitation to the beginning of ventricular excitation. Put another way, the P-Q interval is the time required for the action potential to travel through the atria, atrioventricular node, and the remaining fibers of the conduction system. As the action potential is forced to detour around scar tissue caused by disorders such as coronary artery disease and rheumatic fever, the P-Q interval lengthens.

The **S-T segment,** which begins at the end of the S wave and ends at the beginning of the T wave, represents the time when the ventricular contractile fibers are depolarized during the plateau phase of the action potential. The S-T segment is elevated (above the baseline) in acute myocardial infarction and depressed (below the baseline) when the heart muscle receives insufficient oxygen. The **Q-T interval** extends from the start of the QRS complex to the end of the T wave. It is the time from the beginning of ventricular depolarization to the end of ventricular repolarization. The Q-T interval may be lengthened by myocardial damage, myocardial ischemia (decreased blood flow), or conduction abnormalities.

Sometimes it is helpful to evaluate the heart's response to the stress of physical exercise (stress testing) (see page 751). Although narrowed coronary arteries may carry adequate oxygenated blood while a person is at rest, they will not be able to meet the heart's increased need for oxygen during strenuous exercise. This situation creates changes that can be seen on an electrocardiogram.

Abnormal heart rhythms and inadequate blood flow to the heart may occur only briefly or unpredictably. To detect these problems, **continuous ambulatory electrocardiographs** are used. With this procedure, a person wears a battery-operated monitor (Holter monitor) that records an ECG continuously for 24 hours. Electrodes attached to the chest are connected to the monitor, and information on the heart's activity is stored in the monitor and retrieved later by medical personnel.

Correlation of ECG Waves with Atrial and Ventricular Systole

As we have seen, the atria and ventricles depolarize and then contract at different times because the conduction system routes cardiac action potentials along a specific pathway. The term **systole** (SIS-tō-lē = contraction) refers to the phase of contraction; the phase of relaxation is **diastole** (dī-AS-tō-lē = dilation or expansion). The ECG waves predict the timing of atrial and ventricular systole and diastole. At a heart rate of 75 beats per minute, the timing is as follows (Figure 20.13):

1 A cardiac action potential arises in the SA node. It propagates throughout the atrial muscle and down to the AV node in about 0.03 sec. As the atrial contractile fibers depolarize, the P wave appears in the ECG.

2 After the P wave begins, the atria contract (atrial systole). Conduction of the action potential slows at the AV node because the fibers there have much smaller diameters and fewer gap junctions. (Traffic slows in a similar way where a four-lane highway narrows to one lane in a construction zone!) The resulting 0.1-sec delay gives the atria time to contract, thus adding to the volume of blood in the ventricles, before ventricular systole begins.

3 The action potential propagates rapidly again after entering the AV bundle. About 0.2 sec after onset of the P wave, it has propagated through the bundle branches, Purkinje fibers, and the entire ventricular myocardium. Depolarization progresses down the septum, upward from the apex, and outward from the endocardial surface, producing the QRS complex. At the same time, atrial repolarization is occurring, but it is not usually evident in an ECG because the larger QRS complex masks it.

4 Contraction of ventricular contractile fibers (ventricular systole) begins shortly after the QRS complex appears and continues during the S-T segment. As contraction proceeds from the apex toward the base of the heart, blood is squeezed upward toward the semilunar valves.

5 Repolarization of ventricular contractile fibers begins at the apex and spreads throughout the ventricular myocardium. This produces the T wave in the ECG about 0.4 sec after the onset of the P wave.

6 Shortly after the T wave begins, the ventricles start to relax (ventricular diastole). By 0.6 sec, ventricular repolarization is complete and ventricular contractile fibers are relaxed.

During the next 0.2 sec, contractile fibers in both the atria and ventricles are relaxed. At 0.8 sec, the P wave appears again in the ECG, the atria begin to contract, and the cycle repeats.

As you have just learned, events in the heart occur in cycles that repeat for as long as you live. Next, we will see how the

Figure 20.13 **Timing and route of action potential depolarization and repolarization through the conduction system and myocardium.** Green indicates depolarization and red indicates repolarization.

Depolarization causes contraction and repolarization causes relaxation of cardiac muscle fibers.

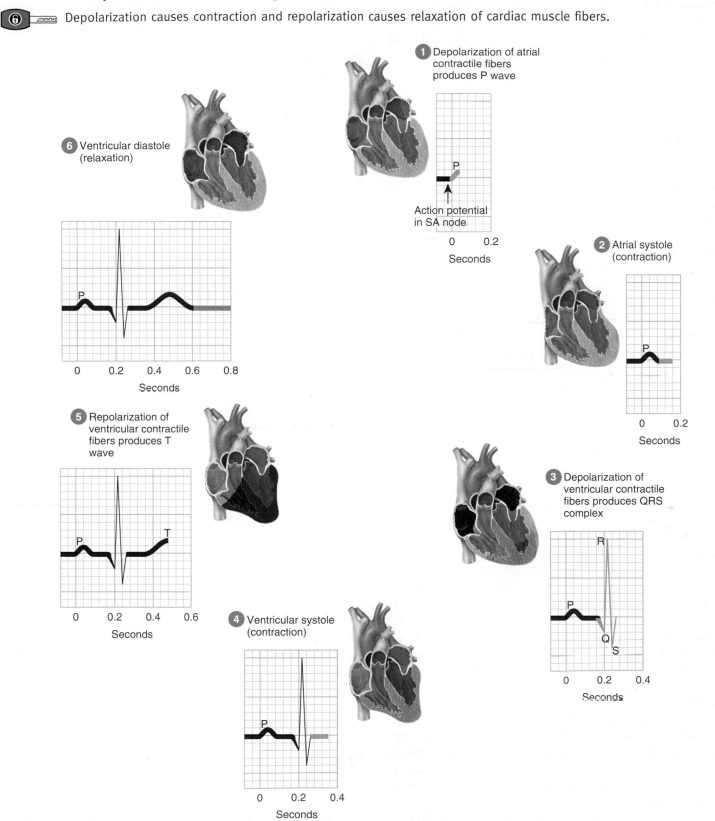

? Where in the conduction system do action potentials propagate most slowly?

pressure changes associated with relaxation and contraction of the heart chambers allow the heart to alternately fill with blood and then eject blood into the aorta and pulmonary trunk.

◉ CHECKPOINT

10. How do cardiac muscle fibers differ structurally and functionally from skeletal muscle fibers?
11. In what ways are autorhythmic fibers similar to and different from contractile fibers?
12. What happens during each of the three phases of an action potential in ventricular contractile fibers?
13. In what ways are ECGs helpful in diagnosing cardiac problems?
14. How does each ECG wave, interval, and segment relate to contraction (systole) and relaxation (diastole) of the atria and ventricles?

THE CARDIAC CYCLE

◉ OBJECTIVES

• Describe the pressure and volume changes that occur during a cardiac cycle.
• Relate the timing of heart sounds to the ECG waves and pressure changes during systole and diastole.

A single **cardiac cycle** includes all the events associated with one heartbeat. Thus, a cardiac cycle consists of systole and diastole of the atria plus systole and diastole of the ventricles.

Pressure and Volume Changes During the Cardiac Cycle

In each cardiac cycle, the atria and ventricles alternately contract and relax, forcing blood from areas of higher pressure to areas of lower pressure. As a chamber of the heart contracts, blood pressure within it increases. Figure 20.14 shows the relation between the heart's electrical signals (ECG) and changes in atrial pressure, ventricular pressure, aortic pressure, and ventricular volume during the cardiac cycle. The pressures given in Figure 20.14 apply to the left side of the heart; pressures on the right side are considerably lower. Each ventricle, however, expels the same volume of blood per beat, and the same pattern exists for both pumping chambers. When heart rate is 75 beats/min, a cardiac cycle lasts 0.8 sec. To examine and correlate the events taking place during a cardiac cycle, we will begin with atrial systole.

Atrial Systole

During **atrial systole,** which lasts about 0.1 sec, the atria are contracting. At the same time, the ventricles are relaxed.

❶ Depolarization of the SA node causes atrial depolarization, marked by the P wave in the ECG.
❷ Atrial depolarization causes atrial systole. As the atria contract, they exert pressure on the blood within, which forces blood through the open AV valves into the ventricles.
❸ Atrial systole contributes a final 25 mL of blood to the vol-

ume already in each ventricle (about 105 mL). The end of atrial systole is also the end of ventricular diastole (relaxation). Thus, each ventricle contains about 130 mL at the end of its relaxation period (diastole). This blood volume is called the **end-diastolic volume (EDV).**
❹ The QRS complex in the ECG marks the onset of ventricular depolarization.

Ventricular Systole

During **ventricular systole,** which lasts about 0.3 sec, the ventricles are contracting. At the same time, the atria are relaxed in **atrial diastole.**

❺ Ventricular depolarization causes ventricular systole. As ventricular systole begins, pressure rises inside the ventricles and pushes blood up against the atrioventricular (AV) valves, forcing them shut. For about 0.05 seconds, both the SL (semilunar) and AV valves are closed. This is the period of **isovolumetric contraction** (*iso-* = same). During this interval, cardiac muscle fibers are contracting and exerting force but are not yet shortening. Thus, the muscle contraction is isometric (same length). Moreover, because all four valves are closed, ventricular volume remains the same (isovolumic).
❻ Continued contraction of the ventricles causes pressure inside the chambers to rise sharply. When left ventricular pressure surpasses aortic pressure at about 80 millimeters of mercury (mmHg) and right ventricular pressure rises above the pressure in the pulmonary trunk (about 20 mmHg), both SL valves open. At this point, ejection of blood from the heart begins. The period when the SL valves are open is **ventricular ejection** and lasts for about 0.25 sec. The pressure in the left ventricle continues to rise to about 120 mmHg, whereas the pressure in the right ventricle climbs to about 25–30 mmHg.
❼ The left ventricle ejects about 70 mL of blood into the aorta and the right ventricle ejects the same volume of blood into the pulmonary trunk. The volume remaining in each ventricle at the end of systole, about 60 mL, is the **end-systolic volume (ESV). Stroke volume,** the volume ejected per beat from each ventricle, equals end-diastolic volume minus end-systolic volume: SV = EDV − ESV. At rest, the stroke volume is about 130 mL − 60 mL = 70 mL (a little more than 2 oz).
❽ The T wave in the ECG marks the onset of ventricular repolarization.

Relaxation Period

During the **relaxation period,** which lasts about 0.4 sec, the atria and the ventricles are both relaxed. As the heart beats faster and faster, the relaxation period becomes shorter and shorter, whereas the durations of atrial systole and ventricular systole shorten only slightly.

❾ Ventricular repolarization causes **ventricular diastole.** As the ventricles relax, pressure within the chambers falls, and

Figure 20.14 **Cardiac cycle.** (a) ECG. (b) Changes in left atrial pressure (green line), left ventricular pressure (blue line), and aortic pressure (red line) as they relate to the opening and closing of heart valves. (c) Heart sounds. (d) Changes in left ventricular volume. (e) Phases of the cardiac cycle.

🔑 A cardiac cycle is composed of all the events associated with one heartbeat.

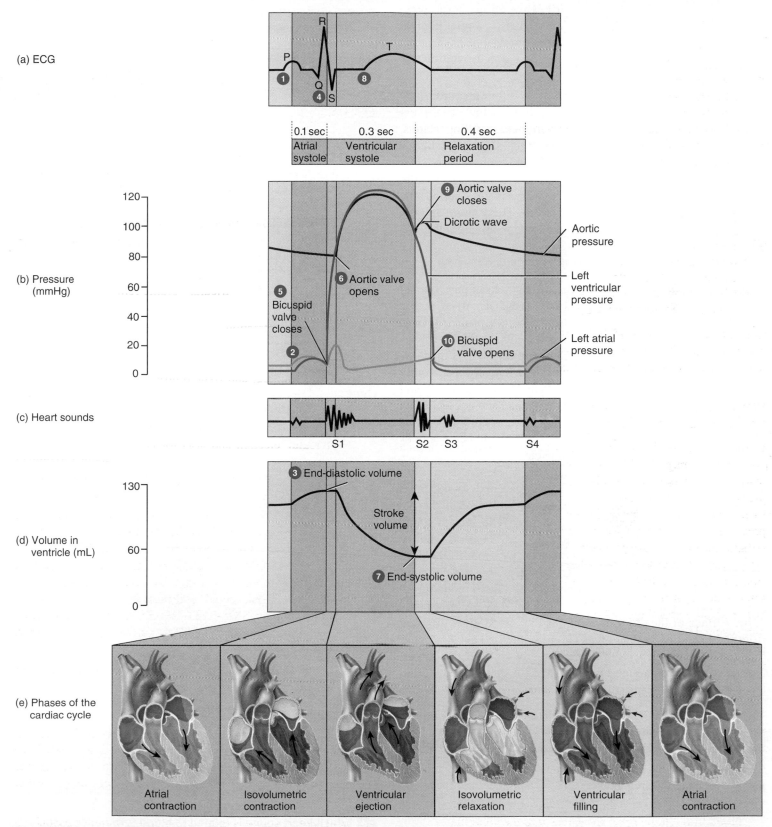

(a) ECG

(b) Pressure (mmHg)

(c) Heart sounds

(d) Volume in ventricle (mL)

(e) Phases of the cardiac cycle

Atrial contraction | Isovolumetric contraction | Ventricular ejection | Isovolumetric relaxation | Ventricular filling | Atrial contraction

❓ How much blood remains in each ventricle at the end of ventricular diastole in a resting person? What is this volume called?

blood in the aorta and pulmonary trunk begins to flow backward toward the regions of lower pressure in the ventricles. Backflowing blood catches in the valve cusps and closes the SL valves. The aortic valve closes at a pressure of about 100 mmHg. Rebound of blood off the closed cusps of the aortic valve produces the **dicrotic wave** on the aortic pressure curve. After the SL valves close, there is a brief interval when ventricular blood volume does not change because all four valves are closed. This is the period of **isovolumetric relaxation.**

10 As the ventricles continue to relax, the pressure falls quickly. When ventricular pressure drops below atrial pressure, the AV valves open, and **ventricular filling** begins. The major part of ventricular filling occurs just after the AV valves open. Blood that has been flowing into and building up in the atria during ventricular systole then rushes rapidly into the ventricles. At the end of the relaxation period, the ventricles are about three-quarters full. The P wave appears in the ECG, signaling the start of another cardiac cycle.

Heart Sounds

Auscultation (aws-kul-TĀ-shun; *ausculta-* = listening), the act of listening to sounds within the body, is usually done with a stethoscope. The sound of the heartbeat comes primarily from blood turbulence caused by the closing of the heart valves. Smoothly flowing blood is silent. Compare the sounds made by white-water rapids or a waterfall with the silence of a smoothly flowing river. During each cardiac cycle, there are four **heart sounds,** but in a normal heart only the first and second heart sounds (S1 and S2) are loud enough to be heard through a stethoscope. Figure 20.14c shows the timing of heart sounds relative to other events in the cardiac cycle.

The first sound (S1), which can be described as a **lubb** sound, is louder and a bit longer than the second sound. S1 is caused by blood turbulence associated with closure of the AV valves soon after ventricular systole begins. The second sound (S2), which is shorter and not as loud as the first, can be described as a **dupp** sound. S2 is caused by blood turbulence associated with closure of the SL valves at the beginning of ventricular diastole. Although S1 and S2 are due to blood turbulence associated with the closure of valves, they are best heard at the surface of the chest in locations that are slightly different from the locations of the valves (Figure 20.15). This is because the sound is carried by the blood flow away from the valves. Normally not loud enough to be heard, S3 is due to blood turbulence during rapid ventricular filling, and S4 is due to blood turbulence during atrial systole.

Heart sounds provide valuable information about the mechanical operation of the heart. A **heart murmur** is an abnormal sound consisting of a clicking, rushing, or gurgling noise that either is heard before, between, or after the normal heart sounds, or may mask the normal heart sounds. Heart murmurs in children are extremely common and usually do not represent a health condition. Murmurs are most frequently discovered in children between the ages of two and four. These types of heart murmurs are referred to as *innocent* or *functional heart murmurs;* they often subside or disappear with growth. Although some heart murmurs in adults are innocent, most often an adult murmur indicates a valve disorder. When a heart valve exhibits stenosis, the heart murmur is heard while the valve should be fully open but is not. For example, mitral stenosis (see page 727) produces a murmur during the relaxation period, between S2 and the next S1. An incompetent heart valve, by contrast, causes a murmur to appear when the valve should be fully closed but is not. So, a murmur due to mitral incompetence (see page 727) occurs during ventricular systole, between S1 and S2. •

◉ CHECKPOINT

15. Why must left ventricular pressure be greater than aortic pressure during ventricular ejection?
16. Does more blood flow through the coronary arteries during ventricular diastole or ventricular systole? Explain your answer.
17. During which two periods of the cardiac cycle do the heart muscle fibers exhibit isometric contractions?
18. What events produce the four normal heart sounds? Which ones can usually be heard through a stethoscope?

Figure 20.15 Heart sounds. Location of valves (purple) and auscultation sites (red) for heart sounds.

Listening to sounds within the body is called auscultation; it is usually done with a stethoscope.

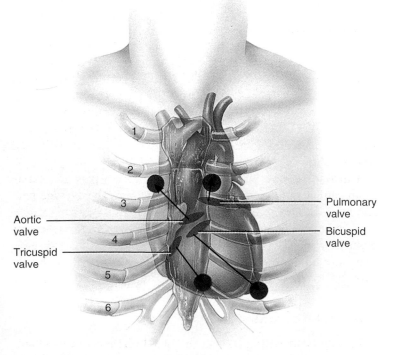

Aortic valve

Tricuspid valve

Pulmonary valve

Bicuspid valve

Anterior view of heart valve locations and auscultation sites

? Which heart sound is related to blood turbulence associated with closure of the atrioventricular valves?

CARDIAC OUTPUT

◉ **OBJECTIVES**
- Define cardiac output.
- Describe the factors that affect regulation of stroke volume.
- Outline the factors that affect the regulation of heart rate.

Although the heart has autorhythmic fibers that enable it to beat independently, its operation is governed by events occurring throughout the body. All body cells must receive a certain amount of oxygenated blood each minute to maintain health and life. When cells are metabolically active, as during exercise, they take up even more oxygen from the blood. During rest periods, cellular metabolic need is reduced, and the workload of the heart decreases.

Cardiac output (CO) is the volume of blood ejected from the left ventricle (or the right ventricle) into the aorta (or pulmonary trunk) each minute. Cardiac output equals the **stroke volume (SV),** the volume of blood ejected by the ventricle during each contraction, multiplied by the **heart rate (HR),** the number of heartbeats per minute:

$$\underset{\text{(mL/min)}}{CO} = \underset{\text{(mL/beat)}}{SV} \times \underset{\text{(beats/min)}}{HR}$$

In a typical resting adult male, stroke volume averages 70 mL/beat, and heart rate is about 75 beats/min. Thus, average cardiac output is

$$
\begin{aligned}
CO &= 70 \text{ mL/beat} \times 75 \text{ beats/min} \\
&= 5250 \text{ mL/min} \\
&= 5.25 \text{ L/min}
\end{aligned}
$$

This volume is close to the total blood volume, which is about 5 liters in a typical adult male. Thus, your entire blood volume flows through your pulmonary and systemic circulations each minute. Factors that increase stroke volume or heart rate normally increase CO. During mild exercise, for example, stroke volume may increase to 100 mL/beat, and heart rate to 100 beats/min. Cardiac output then would be 10 L/min. During intense (but still not maximal) exercise, the heart rate may accelerate to 150 beats/min, and stroke volume may rise to 130 mL/beat, resulting in a cardiac output of 19.5 L/min.

Cardiac reserve is the difference between a person's maximum cardiac output and cardiac output at rest. The average person has a cardiac reserve of four or five times the resting value. Top endurance athletes may have a cardiac reserve seven or eight times their resting CO. People with severe heart disease may have little or no cardiac reserve, which limits their ability to carry out even the simple tasks of daily living.

Regulation of Stroke Volume

A healthy heart will pump out the blood that entered its chambers during the previous diastole. In other words, if more blood returns to the heart during diastole, then more blood is ejected during the next systole. At rest, the stroke volume is 50–60% of the end-diastolic volume because 40–50% of the blood remains in the ventricles after each contraction (end-systolic volume). Three factors regulate stroke volume and ensure that the left and right ventricles pump equal volumes of blood: (1) **preload,** the degree of stretch on the heart before it contracts; (2) **contractility,** the forcefulness of contraction of individual ventricular muscle fibers; and (3) **afterload,** the pressure that must be exceeded before ejection of blood from the ventricles can occur.

Preload: Effect of Stretching

A greater preload (stretch) on cardiac muscle fibers prior to contraction increases their force of contraction. Preload can be compared to the stretching of a rubber band. The more the rubber band is stretched, the more forcefully it will snap back. Within limits, the more the heart fills with blood during diastole, the greater the force of contraction during systole. This relationship is known as the **Frank–Starling law of the heart.** The preload is proportional to the end-diastolic volume (EDV) (the volume of blood that fills the ventricles at the end of diastole). Normally, the greater the EDV, the more forceful the next contraction.

Two key factors determine EDV: (1) the duration of ventricular diastole and (2) **venous return,** the volume of blood returning to the right ventricle. When heart rate increases, the duration of diastole is shorter. Less filling time means a smaller EDV, and the ventricles may contract before they are adequately filled. By contrast, when venous return increases, a greater volume of blood flows into the ventricles, and the EDV is increased.

When heart rate exceeds about 160 beats/min, stroke volume usually declines due to the short filling time. At such rapid heart rates, EDV is less, and the preload is lower. People who have slow resting heart rates usually have large resting stroke volumes because filling time is prolonged and preload is larger.

The Frank–Starling law of the heart equalizes the output of the right and left ventricles and keeps the same volume of blood flowing to both the systemic and pulmonary circulations. If the left side of the heart pumps a little more blood than the right side, the volume of blood returning to the right ventricle (venous return) increases. The increased EDV causes the right ventricle to contract more forcefully on the next beat, bringing the two sides back into balance.

Contractility

The second factor that influences stroke volume is myocardial **contractility,** the strength of contraction at any given preload. Substances that increase contractility are **positive inotropic agents;** those that decrease contractility are **negative inotropic agents.** Thus, for a constant preload, the stroke volume increases when a positive inotropic substance is present. Positive inotropic agents often promote Ca^{2+} inflow during cardiac action potentials, which strengthens the force of the next contraction. Stimulation of the sympathetic division of the autonomic nervous system (ANS), hormones such as epinephrine and norepinephrine, increased Ca^{2+} level in the interstitial fluid, and the drug digitalis all have positive inotropic effects. In contrast, inhibition of the sympathetic division of the ANS, anoxia, acidosis, some anesthetics, and increased K^+ level in the interstitial fluid have negative inotropic effects. *Calcium channel blockers* are drugs that can have a negative inotropic effect by reducing Ca^{2+} inflow, thereby decreasing the strength of the heartbeat.

Afterload

Ejection of blood from the heart begins when pressure in the right ventricle exceeds the pressure in the pulmonary trunk (about 20 mmHg), and when the pressure in the left ventricle exceeds the pressure in the aorta (about 80 mmHg). At that point, the higher pressure in the ventricles causes blood to push the semilunar valves open. The pressure that must be overcome before a semilunar valve can open is termed the **afterload.** An increase in afterload causes stroke volume to decrease, so that more blood remains in the ventricles at the end of systole. Conditions that can increase afterload include hypertension (elevated blood pressure) and narrowing of arteries by atherosclerosis (see page 750).

● CLINICAL CONNECTION | Congestive Heart Failure

In **congestive heart failure (CHF),** there is a loss of pumping efficiency by the heart. Causes of CHF include coronary artery disease (see page 000), congenital defects, long-term high blood pressure (which increases the afterload), myocardial infarctions (regions of dead heart tissue due to a previous heart attack), and valve disorders. As the pump becomes less effective, more blood remains in the ventricles at the end of each cycle, and gradually the end-diastolic volume (preload) increases. Initially, increased preload may promote increased force of contraction (the Frank–Starling law of the heart), but as the preload increases further, the heart is overstretched and contracts less forcefully. The result is a potentially lethal positive feedback loop: Less-effective pumping leads to even lower pumping capability.

   Often, one side of the heart starts to fail before the other. If the left ventricle fails first, it can't pump out all the blood it receives. As a result, blood backs up in the lungs and causes *pulmonary edema,* fluid accumulation in the lungs that can cause suffocation if left untreated. If the right ventricle fails first, blood backs up in the systemic veins and, over time, the kidneys cause an increase in blood volume. In this case, the resulting *peripheral edema* usually is most noticeable in the feet and ankles. ●

Regulation of Heart Rate

As you have just learned, cardiac output depends on both heart rate and stroke volume. Adjustments in heart rate are important in the short-term control of cardiac output and blood pressure. The sinoatrial (SA) node initiates contraction and, if left to itself, would set a constant heart rate of about 100 beats/min. However, tissues require different volumes of blood flow under different conditions. During exercise, for example, cardiac output rises to supply working tissues with increased amounts of oxygen and nutrients. Stroke volume may fall if the ventricular myocardium is damaged or if blood volume is reduced by bleeding. In these cases, homeostatic mechanisms maintain adequate cardiac output by increasing the heart rate and contractility. Among the several factors that contribute to regulation of heart rate, the most important are the autonomic nervous system and hormones released by the adrenal medullae (epinephrine and norepinephrine).

Autonomic Regulation of Heart Rate

Nervous system regulation of the heart originates in the **cardiovascular center** in the medulla oblongata. This region of the brain stem receives input from a variety of sensory receptors and from higher brain centers, such as the limbic system and cerebral cortex. The cardiovascular center then directs appropriate output by increasing or decreasing the frequency of nerve impulses in both the sympathetic and parasympathetic branches of the ANS (Figure 20.16).

Even before physical activity begins, especially in competitive situations, heart rate may climb. This anticipatory increase occurs because the limbic system sends nerve impulses to the cardiovascular center in the medulla. As physical activity begins, **proprioceptors** that are monitoring the position of limbs and muscles send nerve impulses at an increased frequency to the cardiovascular center. Proprioceptor input is a major stimulus for the quick rise in heart rate that occurs at the onset of physical activity. Other sensory receptors that provide input to the cardiovascular center include **chemoreceptors,** which monitor chemical changes in the blood, and **baroreceptors,** which monitor the stretching of major arteries and veins caused by the pressure of the blood flowing through them. Important baroreceptors located in the arch of the aorta and in the carotid arteries (see Figure 21.13 on page 777) detect changes in blood pressure and provide input to the cardiovascular center when it changes. The role of baroreceptors in the regulation of blood pressure is discussed in detail in Chapter 21. Here we focus on the innervation of the heart by the sympathetic and parasympathetic branches of the ANS.

Sympathetic neurons extend from the medulla oblongata into the spinal cord. From the thoracic region of the spinal cord, sympathetic **cardiac accelerator nerves** extend out to the SA node, AV node, and most portions of the myocardium. Impulses in the cardiac accelerator nerves trigger the release of norepinephrine, which binds to beta-1 receptors on cardiac muscle fibers. This interaction has two separate effects: (1) In SA (and AV) node fibers, norepinephrine speeds the rate of spontaneous depolarization so that these pacemakers fire impulses more rapidly and heart rate increases; (2) in contractile fibers throughout the atria and ventricles, norepinephrine enhances Ca^{2+} entry through the voltage-gated slow Ca^{2+} channels, thereby increasing contractility. As a result, a greater volume of blood is ejected during systole. With a moderate increase in heart rate, stroke volume does not decline because the increased contractility offsets the decreased preload. With maximal sympathetic stimulation, however, heart rate may reach 200 beats/min in a 20-year-old person. At such a high heart rate, stroke volume is lower than at rest due to the very short filling time. The maximal heart rate declines with age; as a rule, subtracting your age from 220 provides a good estimate of your maximal heart rate in beats per minute.

Parasympathetic nerve impulses reach the heart via the right and left **vagus (X) nerves.** Vagal axons terminate in the SA node, AV node, and atrial myocardium. They release acetylcholine, which decreases heart rate by slowing the rate of spontaneous depolarization in autorhythmic fibers. As only a few vagal fibers innervate ventricular muscle, changes in parasympathetic activity have little effect on contractility of the ventricles.

A continually shifting balance exists between sympathetic and parasympathetic stimulation of the heart. At rest, parasympathetic stimulation predominates. The resting heart rate—about 75 beats/min—is usually lower than the autorhythmic rate of the SA node (about 100 beats/min). With maximal stimulation by the parasympathetic division, the heart can slow to 20 or 30 beats/min, or can even stop momentarily.

Chemical Regulation of Heart Rate

Certain chemicals influence both the basic physiology of cardiac muscle and the heart rate. For example, hypoxia (lowered oxygen level), acidosis (low pH), and alkalosis (high pH) all depress cardiac activity. Several hormones and cations have major effects on the heart:

1. Hormones. Epinephrine and norepinephrine (from the adrenal medullae) enhance the heart's pumping effectiveness. These hormones affect cardiac muscle fibers in much the same way as does norepinephrine released by cardiac accelerator nerves—they increase both heart rate and contractility. Exercise, stress, and excitement cause the adrenal medullae to release more hormones. Thyroid hormones also enhance cardiac contractility and increase heart rate. One sign of hyperthyroidism (excessive thyroid hormone) is **tachycardia,** an elevated resting heart rate.

2. Cations. Given that differences between intracellular and extracellular concentrations of several cations (for example, Na^+ and K^+) are crucial for the production of action potentials in all nerve and muscle fibers, it is not surprising that ionic imbalances can quickly compromise the pumping effectiveness of the heart. In particular, the relative concentrations of three cations—K^+, Ca^{2+}, and Na^+—have a large effect on cardiac function. Elevated blood levels of K^+ or Na^+ decrease heart rate and contractility. Excess Na^+ blocks Ca^{2+} inflow during cardiac action potentials, thereby decreasing the force of contraction, whereas excess K^+ blocks generation of action potentials. A moderate increase in interstitial (and thus intracellular) Ca^{2+} level speeds heart rate and strengthens the heartbeat.

Other Factors in Heart Rate Regulation

Age, gender, physical fitness, and body temperature also influence resting heart rate. A newborn baby is likely to have a resting heart rate over 120 beats/min; the rate then gradually declines throughout life. Adult females often have slightly higher resting heart rates than adult males, although regular exercise tends to bring resting heart rate down in both sexes. A physically fit person may even exhibit **bradycardia** (brad-ē-KAR-dē-a; *bradys-* = slow), a resting heart rate under 50 beats/min. This is a beneficial effect of endurance-type training because a slowly beating heart is more energy efficient than one that beats more rapidly.

Increased body temperature, as occurs during a fever or strenuous exercise, causes the SA node to discharge impulses more quickly, thereby increasing heart rate. Decreased body temperature decreases heart rate and strength of contraction.

Figure 20.16 Nervous system control of the heart.

The cardiovascular center in the medulla oblongata controls both sympathetic and parasympathetic nerves that innervate the heart.

INPUT TO CARDIOVASCULAR CENTER

From higher brain centers: cerebral cortex, limbic system, and hypothalamus

From sensory receptors:
Proprioceptors–monitor movements
Chemoreceptors–monitor blood chemistry
Baroreceptors–monitor blood pressure

Cardiovascular (CV) center

Cardiac accelerator **nerves** (sympathetic)

Vagus nerves (cranial nerve X, parasympathetic)

OUTPUT TO HEART

Increased rate of spontaneous depolarization in SA node (and AV node) increases heart rate

Increased contractility of atria and ventricles increases stroke volume

Decreased rate of spontaneous depolarization in SA node (and AV node) decreases heart rate

? What region of the heart is innervated by the sympathetic division but not by the parasympathetic division?

During surgical repair of certain heart abnormalities, it is helpful to slow a patient's heart rate by **hypothermia** (hī-pō-THER-mē-a), in which the person's body is deliberately cooled to a low core temperature. Hypothermia slows metabolism, which reduces the oxygen needs of the tissues, allowing the heart and brain to withstand short periods of interrupted or reduced blood flow during the procedure.

Figure 20.17 summarizes the factors that can increase stroke volume and heart rate to achieve an increase in cardiac output.

◉ CHECKPOINT

19. How is cardiac output calculated?
20. Define stroke volume (SV), and explain the factors that regulate it.
21. What is the Frank-Starling law of the heart? What is its significance?
22. Define cardiac reserve. How does it change with training or with heart failure?
23. How do the sympathetic and parasympathetic divisions of the autonomic nervous system adjust heart rate?

Figure 20.17 Factors that increase cardiac output.

🔑 Cardiac output equals stroke volume multiplied by heart rate.

? When you are exercising, contraction of skeletal muscles helps return blood to the heart more rapidly. Would this tend to increase or decrease stroke volume?

EXERCISE AND THE HEART

◉ OBJECTIVE
• Explain the relationship between exercise and the heart.

Regardless of current fitness level, a person's cardiovascular fitness can be improved at any age with regular exercise. Some types of exercise are more effective than others for improving the health of the cardiovascular system. **Aerobics,** any activity that works large body muscles for at least 20 minutes, elevates cardiac output and accelerates metabolic rate. Three to five such sessions a week are usually recommended for improving the health of the cardiovascular system. Brisk walking, running, bicycling, cross-country skiing, and swimming are examples of aerobic activities.

Sustained exercise increases the oxygen demand of the muscles. Whether the demand is met depends mainly on the adequacy of cardiac output and proper functioning of the respiratory system. After several weeks of training, a healthy person increases maximal cardiac output, thereby increasing the maximal rate of oxygen delivery to the tissues. Oxygen delivery also rises because skeletal muscles develop more capillary networks in response to long-term training.

During strenuous activity, a well-trained athlete can achieve a cardiac output double that of a sedentary person, in part because training causes hypertrophy (enlargement) of the heart. Even though the heart of a well-trained athlete is larger, *resting* cardiac output is about the same as in a healthy untrained person, because stroke volume is increased while heart rate is decreased. The resting heart rate of a trained athlete often is only 40–60 beats per minute *(resting bradycardia).* Regular exercise also helps to reduce blood pressure, anxiety, and depression; control weight; and increase the body's ability to dissolve blood clots by increasing fibrinolytic activity.

◉ CHECKPOINT
24. What are some of the cardiovascular benefits of regular exercise?

HELP FOR FAILING HEARTS

◉ OBJECTIVE
• Describe several techniques used for failing hearts.

As the heart fails, a person has decreasing ability to exercise or even to move around. A variety of surgical techniques and medical devices exist to aid a failing heart. For some patients, even a 10% increase in the volume of blood ejected from the ventricles can mean the difference between being bedridden and having limited mobility.

A **cardiac (heart) transplant** is the replacement of a severely damaged heart with a normal heart from a brain-dead or recently deceased donor. Cardiac transplants are performed on patients with end-stage heart failure or severe coronary artery disease. Once a suitable heart is located, the chest cavity is exposed through a midsternal cut. After the patient is placed on a heart-lung bypass machine, which oxygenates and circulates blood, the pericardium is cut to expose the heart. Next, the diseased heart is removed (usually except for the posterior wall of the left atrium) (Figure 20.18) and the donor heart is trimmed and sutured into position so that the remaining left atrium and great vessels are connected to the donor heart. The new heart is

Figure 20.18 Cardiac transplantation.

🔑 Cardiac transplantation is the replacement of a severely damaged heart with a normal heart from a brain-dead or recently deceased donor.

(a) The donor's left atrium is sutured to the recipient's left atrium

◉ FIGURE 20.18 CONTINUES ▶

FIGURE 20.18 CONTINUED ▶

Recipient's superior vena cava

Donor's right atrium

Recipient's inferior vena cava

(b) The donor's right atrium is sutured to the recipient's superior and inferior venae cavae

(c) Transplanted heart with sutures

? **Which patients are candidates for cardiac transplantation?**

started as blood flows through it (an electrical shock may be used to correct an abnormal rhythm), the patient is weaned from the heart-lung bypass machine, and the chest is closed. The patient must remain on immunosuppressant drugs for a lifetime to prevent rejection. Since the vagus (X) nerve is severed during the surgery, the new heart will beat at about 100 times per minute until nerve regrowth occurs.

Usually, a donor heart is perfused with a cold solution and then preserved in sterile ice. This can keep the heart viable for about 4–5 hours. In May 2007, surgeons in the United States performed the first beating-heart transplant. The donor heart was maintained at normal body temperature and hooked up to an

organ care system that allowed it to keep beating with warm, oxygenated blood flowing through it. This greatly prolongs the time between removal of the heart from the donor and transplantation into a recipient and also decreases injury to the heart while being deprived of blood, which can lead to rejection. The safety and benefits of the oxygen care system are currently being evaluated.

Cardiac transplants are common today and produce good results, but the availability of donor hearts is very limited. Another approach is the use of cardiac assist devices and other surgical procedures that assist heart function without removing the heart. Table 20.1 describes several of these.

TABLE 20.1

Cardiac Assist Devices and Procedures

DEVICE	DESCRIPTION
Intra-aortic balloon pump (IABP)	A 40-mL polyurethane balloon mounted on a catheter is inserted into an artery in the groin and threaded through the femoral artery into the thoracic aorta. See (a). An external pump inflates the balloon with helium gas at the beginning of ventricular diastole. As the balloon inflates, it pushes blood both backward toward the heart, which improves coronary blood flow, and forward toward peripheral tissues. The balloon then is rapidly deflated just before the next ventricular systole, sucking blood out of the left ventricle, making it easier for the left ventricle to eject blood. Because the balloon is inflated between heartbeats, this technique is called *intra-aortic balloon counterpulsation.*
Ventricular assist device (VAD)	A mechanical pump that helps a weakened ventricle to pump blood throughout the body so that the heart does not have to work as hard. A VAD is sometimes called a *bridge to transplant* if it is used to help a patient survive until a heart transplant can be performed. A VAD can also provide an alternative to heart transplantation. Such permanent support using VAD is called *destination therapy.*

Thoracic aorta

Catheter

Anterior view

Balloon

Thoracic aorta (cut)

Thoracic aorta (intact)

Catheter

Posterior view

(a) Intra-aortic balloon pump

Aorta

Left ventricle

Implanted left ventricular assist device (LVAD)

Outflow tube

Inflow tube

Outflow one-way valve

Inflow one-way valve

Pump unit

Parts of left ventricular assist device (LVAD)

Driveline

(b) Left ventricular assist device (LVAD)

VADs are classified according to the ventricle that requires support. A *left ventricular assist device (LVAD)* helps the left ventricle pump blood into the aorta. See (b). This is the most commonly used VAD. A *right ventricular assist device (RVAD)* helps the right ventricle to pump blood into the pulmonary trunk. A *bi-ventricular assist device (BVAD)* helps both the left and right ventricles perform their functions.

There are several VADs available, and which one is used depends on the specific needs of the patient. In order to help you understand how a VAD works, we will illustrate an LVAD to the right. An inflow tube attached to the apex of the left ventricle takes blood from the ventricle through a one-way valve into the pump unit. Once the pump fills with blood, an external control system triggers pumping and blood flows through a one-way valve into an outflow tube that delivers blood into the aorta. The external control system is on a belt around the waist or on a shoulder strap. Some VADs pump at a constant rate; others are coordinated with the person's heartbeat.

Cardiomyoplasty

A large piece of the patient's own skeletal muscle (left latissimus dorsi) is partially freed from its connective tissue attachments and wrapped around the heart, leaving the blood and nerve supply intact. An implanted pacemaker stimulates the skeletal muscle's motor neurons to cause contraction 10–20 times per minute, in synchrony with some of the heartbeats.

Skeletal muscle assist device

A piece of the patient's own skeletal muscle is used to fashion a pouch that is inserted between the heart and the aorta, functioning as a booster heart. A pacemaker stimulates the muscle's motor neurons to elicit contraction.

◉ CHECKPOINT
25. Describe how a heart transplant is performed.
26. Explain four different cardiac arrest devices and procedures.

DEVELOPMENT OF THE HEART

◉ OBJECTIVE

• Describe the development of the heart.

Listening to a fetal heartbeat for the first time is an exciting moment for prospective parents, but it is also an important diagnostic tool. The cardiovascular system is one of the first systems to form in an embryo, and the heart is the first functional organ.

This order of development is essential because of the need of the rapidly growing embryo to obtain oxygen and nutrients and get rid of wastes. As you will learn shortly, the development of the heart is a complex process, and any disruptions along the way can result in congenital (present at birth) disorders of the heart. Such disorders, described on page 752, are responsible for almost half of all deaths from birth defects.

The *heart* begins its development from **mesoderm** on day 18 or 19 following fertilization. In the head end of the embryo, the heart develops from a group of mesodermal cells called the **cardiogenic area** (kar-dē-ō-JEN-ik; *cardio-* = heart; *-genic* = producing) (Figure 20.19a). In response to signals from the underlying endoderm, the mesoderm in the cardiogenic area forms a pair of elongated strands called **cardiogenic cords.** Shortly after, these cords develop a hollow center and then become known as **endocardial tubes** (Figure 20.19b). With lateral folding of the

Figure 20.19 Development of the heart. Arrows within the structures indicate the direction of blood flow.

🔑 The heart begins its development from a group of mesodermal cells called the cardiogenic area during the third week after fertilization.

(a) Location of cardiogenic area — 19 days

(b) Formation of endocardial tubes — 20 days

(c) Formation of primitive heart tube — 21 days

(d) Development of regions in the primitive heart tube — 22 days

(e) Bending of the primitive heart — 23 days, 24 days

(f) Orientation of atria and ventricles to their final adult position — 28 days

❓ When during embryonic development does the primitive heart begin to contract?

embryo, the paired endocardial tubes approach each other and fuse into a single tube called the **primitive heart tube** on day 21 following fertilization (Figure 20.19c).

On the twenty-second day, the primitive heart tube develops into five distinct regions and begins to pump blood. From tail end to head end (and the direction of blood flow) they are the (1) **sinus venosus,** (2) **atrium,** (3) **ventricle,** (4) **bulbus cordis,** and (5) **truncus arteriosus.** The sinus venosus initially receives blood from all the veins in the embryo; contractions of the heart begin in this region and follow sequentially in the other regions. Thus, at this stage, the heart consists of a series of unpaired regions. The fates of the regions are as follows:

1. The sinus venosus develops into part of the *right atrium, coronary sinus,* and *sinoatrial (SA) node.*

2. The atrium develops into part of the *right atrium, right auricle, left atrium,* and *left auricle.*

3. The ventricle gives rise to the *left ventricle.*

4. The bulbus cordis develops into the *right ventricle.*

5. The truncus arteriosus gives rise to the *ascending aorta* and *pulmonary trunk.*

On day 23, the primitive heart tube elongates. Because the bulbus cordis and ventricle grow more rapidly than other parts of the tube and because the atrial and venous ends of the tube are confined by the pericardium, the tube begins to loop and fold. At first, the primitive heart tube assumes a U-shape; later it becomes S-shaped (Figure 20.19e). As a result of these movements, which are completed by day 28, the atria and ventricles of the future heart are reoriented to assume their final adult positions. The remainder of heart development consists of reconstruction of the chambers and the formation of septa and valves to form a four-chambered heart.

On about day 28, thickenings of mesoderm of the inner lining of the heart wall, called **endocardial cushions,** appear (Figure 20.20). They grow toward each other, fuse, and divide the single **atrioventricular canal** (region between atria and ventricles) into smaller, separate left and right atrioventricular canals. Also, the *interatrial septum* begins its growth toward fused endocardial cushions. Ultimately, the interatrial septum and endocardial cushions unite and an opening in the septum, the **foramen ovale,** develops. The interatrial septum divides the atrial region into a *right atrium* and a *left atrium.* Before birth, the foramen ovale allows most blood entering the right atrium to pass into the left atrium. After birth, it normally closes so that the interatrial septum is a complete partition. The remnant of the foramen ovale is the fossa ovalis (Figure 20.20). Formation of the *interventricular septum* partitions the ventricular region into a *right ventricle* and a *left ventricle.* Partitioning of the atrioventricular canal, atrial region, and ventricular region is basically complete by the end of the fifth week. The *atrioventricular valves* form between the fifth and eighth weeks. The *semilunar valves* form between the fifth and ninth weeks.

Figure 20.20 **Partitioning of the heart into four chambers.**

Partitioning the heart begins on about the 28th day after fertilization.

About 28 days

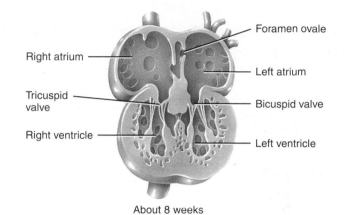

About 8 weeks

❓ **When is the partitioning of the heart complete?**

◉ CHECKPOINT

27. Why is the cardiovascular system one of the first systems to develop?

28. From which tissue does the heart develop?

DISORDERS: HOMEOSTATIC IMBALANCES

Coronary Artery Disease

Coronary artery disease (CAD) is a serious medical problem that affects about 7 million people annually. Responsible for nearly three-quarters of a million deaths in the United States each year, it is the leading cause of death for both men and women. CAD results from the effects of the accumulation of atherosclerotic plaques (described shortly) in coronary arteries, which leads to a reduction in blood flow to the myocardium. Some individuals have no signs or symptoms; others experience angina pectoris (chest pain), and still others suffer heart attacks.

Risk Factors for CAD

People who possess combinations of certain risk factors are more likely to develop CAD. *Risk factors* (characteristics, symptoms, or signs present in a disease-free person that are statistically associated with a greater chance of developing a disease) include smoking, high blood pressure, diabetes, high cholesterol levels, obesity, "type A" personality, sedentary lifestyle, and a family history of CAD. Most of these can be modified by changing diet and other habits or can be controlled by taking medications. However, other risk factors are un-modifiable (beyond our control), including genetic predisposition (family history of CAD at an early age), age, and gender. For example, adult males are more likely than adult females to develop CAD; after age 70 the risks are roughly equal. Smoking is undoubtedly the number-one risk factor in all CAD-associated diseases, roughly doubling the risk of morbidity and mortality.

Development of Atherosclerotic Plaques

Although the following discussion applies to coronary arteries, the process can also occur in arteries outside the heart. Thickening of the walls of arteries and loss of elasticity are the main characteristics of a group of diseases called **arteriosclerosis** (ar-tē-rē-ō-skle-RŌ-sis; *sclero-* = hardening). One form of arteriosclerosis is **atherosclerosis** (ath-er-ō-skle-RŌ-sis), a progressive disease characterized by the formation in the walls of large and medium-sized arteries of lesions called **atherosclerotic plaques** (Figure 20.21a,b).

To understand how atherosclerotic plaques develop, you will need to learn about the role of molecules produced by the liver and small intestine called **lipoproteins.** These spherical particles consist of an inner core of triglycerides and other lipids and an outer shell of proteins, phospholipids, and cholesterol. Like most lipids, cholesterol does not dissolve in water and must be made water-soluble in order to be transported in the blood. This is accomplished by combining it with lipoproteins. Two major lipoproteins are **low-density lipoproteins (LDLs)** and **high-density lipoproteins (HDLs).** LDLs transport cholesterol from the liver to body cells for use in cell membrane repair and the production of steroid hormones and bile salts. However, excessive amounts of LDLs promote atherosclerosis, so the cholesterol in these particles is commonly known as "bad cholesterol." HDLs, on the other hand, remove excess cholesterol from body cells and transport it to the liver for elimination. Because HDLs decrease blood cholesterol level, the cholesterol in HDLs is commonly referred to as "good cholesterol." Basically, you want your LDL concentration to be low and your HDL concentration to be high.

It has been learned recently that inflammation, a defensive response of the body to tissue damage, plays a key role in the development of atherosclerotic plaques. As a result of tissue damage, blood vessels dilate and increase their permeability, and phagocytes, including macrophages, appear in large numbers. The formation of atherosclerotic plaques begins when excess LDLs from the blood accumulate in the inner layer of an artery wall (layer closest to the bloodstream) and the lipids and proteins in the LDLs undergo oxidation and the proteins also bind to sugars. In response, endothelial and smooth muscle cells of the artery secrete substances that attract monocytes from the blood and convert them into macrophages. The macrophages then ingest and become so filled with oxidized LDL particles that they have a foamy appearance when viewed microscopically **(foam cells).** T cells (lymphocytes) follow monocytes into the inner lining of an artery where they release chemicals that intensify the inflammatory response. Together, the foam cells, macrophages, and T cells form a fatty streak, the beginning of an atherosclerotic plaque.

Macrophages secrete chemicals that cause smooth muscle cells of the middle layer of an artery to migrate to the top of the atherosclerotic plaque, forming a cap over it and thus walling it off from the blood.

Figure 20.21 Photomicrographs of a transverse section of (a) a normal artery and (b) one partially obstructed by an atherosclerotic plaque.

 Inflammation plays a key role in the development of atherosclerotic plaques.

Partially obstructed lumen (space through which blood flows)

Atherosclerotic plaque

LM 20x

LM 20x

(a) Normal artery

(b) Obstructed artery

 What is the role of HDL?

Because most atherosclerotic plaques expand away from the bloodstream rather than into it, blood can still flow through the affected artery with relative ease, often for decades. Relatively few heart attacks occur when plaque in a coronary artery expands into the bloodstream and restricts blood flow. Most heart attacks occur when the cap over the plaque breaks open in response to chemicals produced by foam cells. In addition, T cells induce foam cells to produce tissue factor (TF), a chemical that begins the cascade of reactions that result in blood clot formation. If the clot in a coronary artery is large enough, it can significantly decrease or stop the flow of blood and result in a heart attack.

In recent years, a number of new risk factors (all modifiable) have been identified as significant predictors of CAD. **C-reactive proteins (CRPs)** are proteins produced by the liver or present in blood in an inactive form that are converted to an active form during inflammation. CRPs may play a direct role in the development of atherosclerosis by promoting the uptake of LDLs by macrophages. **Lipoprotein (a)** is an LDL-like particle that binds to endothelial cells, macrophages, and blood platelets, may promote the proliferation of smooth muscle fibers, and inhibits the breakdown of blood clots. **Fibrinogen** is a glycoprotein involved in blood clotting that may help regulate cellular proliferation, vasoconstriction, and platelet aggregation. **Homocysteine** is an amino acid that may induce blood vessel damage by promoting platelet aggregation and smooth muscle fiber proliferation.

Diagnosis of CAD

Many procedures may be employed to diagnose CAD; the specific procedure used will depend on the signs and symptoms of the individual.

A resting electrocardiogram (see page 735) is the standard test employed to diagnose CAD. **Stress testing** can also be performed. In an *exercise stress test,* the functioning of the heart is monitored when placed under physical stress by exercising using a treadmill, an exercise bicycle, or arm exercises. During the procedure, ECG recordings are monitored continuously and blood pressure is monitored at intervals. A *nonexercise (pharmacologic) stress test* is used for individuals who cannot exercise due to conditions such as arthritis. A medication is injected that stresses the heart to mimic the effects of exercise. During both exercise and nonexercise stress testing, **radionuclide imaging** may be performed to evaluate blood flow through heart muscle (see page 23).

Diagnosis of CAD may also involve **echocardiography** (ek'-ō-kar-dē-OG-ra-fē), a technique that uses ultrasound waves to image the interior of the heart. Echocardiography allows the heart to be seen in motion and can be used to determine the size, shape, and functions of the chambers of the heart; the volume and velocity of blood pumped from the heart; the status of heart valves; the presence of birth defects; and abnormalities of the pericardium. A fairly recent technique for evaluating CAD is **electron beam computerized tomography (EBCT),** which detects calcium deposits in coronary arteries. These calcium deposits are indicators of atherosclerosis.

Cardiac catheterization (kath'-e-ter-i-ZĀ-shun) is an invasive procedure used to visualize the heart's chambers, valves, and great vessels in order to diagnose and treat disease not related to abnormalities of the coronary arteries. It may also be used to measure pressure in the heart and great vessels; to assess cardiac output; to measure the flow of blood through the heart and great vessels; to identify the location of septal and valvular defects; and to take tissue and blood samples. The basic procedure involves inserting a long, flexible, radiopaque **catheter** (plastic tube) into a peripheral vein (for *right heart catheterization*) or a peripheral artery (for *left heart catheterization*) and guiding it under fluoroscopy (x-ray observation).

Coronary angiography (an'-jē-OG-ra-fē; *angio-* = blood vessel; *-grapho* = to write) is an invasive procedure used to obtain information about the coronary arteries. In the procedure, a catheter is inserted into an artery in the groin or wrist and threaded under fluoroscopy toward the heart and then into the coronary arteries. After the tip of the catheter is in place, a radiopaque contrast medium is injected into the coronary arteries. The radiographs of the arteries, called *angiograms,* appear in motion on a monitor and the information is recorded on a videotape or computer disc. Coronary angiography may be used to visualize coronary arteries (see page 22) and to inject clot-dissolving drugs, such as streptokinase or tissue plasminogen activator (t-PA), into a coronary artery to dissolve an obstructing thrombus.

Treatment of CAD

Treatment options for CAD include **drugs** (antihypertensives, nitroglycerine, beta blockers, cholesterol-lowering drugs, and clot-dissolving agents) and various surgical and nonsurgical procedures designed to increase the blood supply to the heart.

Coronary artery bypass grafting (CABG) is a surgical procedure in which a blood vessel from another part of the body is attached ("grafted") to a coronary artery to bypass an area of blockage. A piece of the grafted blood vessel is sutured between the aorta and the unblocked portion of the coronary artery (Figure 20.22a).

A nonsurgical procedure used to treat CAD is **percutaneous transluminal coronary angioplasty (PTCA)** (percutaneous = through the skin; *trans-* = across; *lumen* = an opening or channel in a tube; *angio-* = blood vessel; *-plasty* = to mold or to shape). In one variation of this procedure, a balloon catheter is inserted into an artery of an arm or

Figure 20.22 **Procedures for reestablishing blood flow in occluded coronary arteries.**

 Treatment options for CAD include drugs and various nonsurgical and surgical procedures.

Ascending aorta

Grafted vessel

Obstruction

(a) Coronary artery bypass grafting (CABG)

● **FIGURE 20.22 CONTINUES** ▶

FIGURE 20.22 CONTINUED

Balloon catheter with uninflated balloon is threaded to obstructed area in artery

When balloon is inflated, it stretches arterial wall and squashes atherosclerotic plaque

After lumen is widened, balloon is deflated and catheter is withdrawn

(b) Percutaneous transluminal coronary angioplasty (PTCA)

(c) Stent in an artery

(d) Angiogram showing a stent in the circumflex artery

? **Which diagnostic procedure for CAD is used to visualize coronary blood vessels?**

leg and gently guided into a coronary artery (Figure 20.22b). While dye is released, angiograms (x-rays of blood vessels) are taken to locate the plaques. Next, the catheter is advanced to the point of obstruction, and a balloonlike device is inflated with air to squash the plaque against the blood vessel wall. Because 30–50% of PTCA-opened arteries fail due to restenosis (renarrowing) within six months after the procedure is done, a stent may be inserted via a catheter. A **stent** is a metallic, fine wire tube that is permanently placed in an artery to keep the artery *patent* (open), permitting blood to circulate (Figure 20.22c, d). Restenosis may be due to damage from the procedure itself, for PTCA may damage the arterial wall, leading to platelet activation, proliferation of smooth muscle fibers, and plaque formation. Recently, *drug-coated (drug-eluting) coronary stents* have been used to prevent restenosis. The stents are coated with one of several antiproliferative drugs (drugs that inhibit the proliferation of smooth muscle fibers of the middle layer of an artery) and anti-inflammatory drugs. It has been shown that drug-coated stents reduce the rate of restenosis when compared to bare-metal (noncoated) stents. In addition to balloon and stent angioplasty, laser-emitting catheters are used to vaporize plaques (excimer laser coronary angioplasy or ELCA) and small blades inside catheters are used to remove part of the plaque (directional coronary atherectomy).

One area of current research involves cooling the body's core temperature during procedures such as coronary artery bypass grafting (CABG). There have been some promising results from the application of cold therapy during a cerebral vascular accident (CVA or stroke). This research stemmed from observations of people who had suffered a hypothermic incident (such as cold-water drowning) and recovered with relatively minimal neurologic deficits.

Congenital Heart Defects

A defect that is present at birth, and usually before, is called a **congenital defect.** Many such defects are not serious and may go unnoticed for a lifetime. Others are life threatening and must be surgically repaired. Among the several congenital defects that affect the heart are the following (Figure 20.23):

- **Coarctation** (kō'-ark-TĀ-shun) **of the aorta.** In this condition, a segment of the aorta is too narrow, and thus the flow of oxygenated blood to the body is reduced, the left ventricle is forced to pump harder, and high blood pressure develops. Coarctation is usually repaired surgically by removing the area of obstruction. Surgical interventions that are done in childhood may require revisions in adulthood. Another surgical procedure is balloon dilation, insertion and inflation of a device in the aorta to stretch the vessel. A stent can be inserted and left in place to hold the vessel open.

- **Patent ductus arteriosus (PDA).** In some babies, the ductus arteriosus, a temporary blood vessel between the aorta and the pulmonary trunk, remains open rather than closing shortly after birth. As a result, aortic blood flows into the lower-pressure pulmonary trunk, thus increasing the pulmonary trunk blood pressure and overworking both ventricles. In uncomplicated PDA, medication can be used to facilitate the closure of the defect. In more severe cases, surgical intervention may be required.

- **Septal defect.** A septal defect is an opening in the septum that separates the interior of the heart into left and right sides. In an **atrial septal defect** the fetal foramen ovale between the two atria fails to close after birth. A **ventricular septal defect** is caused by incomplete development of the interventricular septum. In such cases, oxygenated blood flows directly from the left ventricle into

Figure 20.23 Congenital heart defects.

A congenital defect is one that is present at birth, and usually before.

(a) Coarctation of the aorta

(b) Patent ductus arteriosus

(c) Atrial septal defect

(d) Ventricular septal defect

(e) Tetralogy of Fallot

? Which four developmental defects occur in tetralogy of Fallot?

the right ventricle, where it mixes with deoxygenated blood. The condition is treated surgically.

- **Tetralogy of Fallot** (tet-RAL-Ō-jē of fal-Ō). This condition is a combination of four developmental defects: an interventricular septal defect, an aorta that emerges from both ventricles instead of from the left ventricle only, a stenosed pulmonary valve, and an enlarged right ventricle. There is a decreased flow of blood to the lungs and mixing of blood from both sides of the heart. This causes cyanosis, the bluish discoloration most easily seen in nail beds and mucous membranes when the level of deoxygenated hemoglobin is high; in infants, this condition is referred to as "blue baby." Despite the apparent complexity of this condition, surgical repair is usually successful.

Arrhythmias

The usual rhythm of heartbeats, established by the SA node, is called **normal sinus rhythm.** The term **arrhythmia** (a-RITH-mē-a) or **dysrhythmia** refers to an abnormal rhythm as a result of a defect in the conduction system of the heart. The heart may beat irregularly, too quickly, or too slowly. Symptoms include chest pain, shortness of breath, lightheadedness, dizziness, and fainting. Arrhythmias may be caused by factors that stimulate the heart such as stress, caffeine, alcohol, nicotine, cocaine, and certain drugs that contain caffeine or other stimulants. Arrhythmias may also be caused by a congenital defect, coronary artery disease, myocardial infarction, hypertension, defective heart valves, rheumatic heart disease, hyperthyroidism, and potassium deficiency.

Arrhythmias are categorized by their speed, rhythm, and origination of the problem. **Bradycardia** (brād-e-KAR-dē-a; *brady-* = slow) refers to a slow heart rate (below 50 beats per minute); **tachycardia** (tak-i-KAR-dē-a; *tachy-* = quick) refers to a rapid heart rate (over 100 beats per minute); and **fibrillation** (fi-bri-LA-shun) refers to rapid, uncoordinated heartbeats. Arrhythmias that begin in the atria are called **supraventricular** or **atrial arrhythmias;** those that originate in the ventricles are called **ventricular arrhythmias.**

- **Supraventricular tachycardia (SVT)** is a rapid but regular heart rate (160–200 beats) per minute that originates in the atria. The episodes begin and end suddenly and may last from a few minutes to many hours. SVTs can sometimes be stopped by maneuvers that stimulate the vagus (X) nerve and decrease heart rate. These include straining as if having a difficult bowel movement, rubbing the area over the carotid artery in the neck to stimulate the carotid sinus (not recommended for people over 50 since it may cause a stroke), and plunging the face into a bowl of ice-cold water. Treatment may also involve antiarrhythmic drugs and destruction of the abnormal pathway by radiofrequency ablation.

- **Heart block** is an arrhythmia that occurs when the electrical pathways between the atria and ventricles are blocked, slowing the transmission of nerve impulses. The most common site of blockage is the atrioventricular node, a condition called *atrioventricular (AV) block*. In *first-degree AV block,* the P-Q interval is prolonged, usually because conduction through the AV node is slower than normal (Figure 20.24b). In *second-degree AV block,* some of the action potentials from the SA node are not conducted through the AV node. The result is "dropped" beats because excitation doesn't always reach the ventricles. Consequently, there are fewer QRS complexes than P waves on the ECG. In *third-degree (complete) AV block,* no SA node action potentials get through the AV node. Autorhythmic fibers in the atria and ventricles pace the upper and lower chambers separately. With complete AV block, the ventricular contraction rate is less than 40 beats/min.

- **Atrial premature contraction (APC)** is a heartbeat that occurs earlier than expected and briefly interupts the normal heart rhythm. It often causes a sensation of a skipped heartbeat followed by a more forceful heartbeat. APCs originate in the atrial myocardium and are common in healthy individuals.

- **Atrial flutter** consists of rapid, regular atrial contractions (240–360 beats/min) accompanied by an atrioventricular (AV) block in which some of the nerve impulses from the SA node are not conducted through the AV node.

- **Atrial fibrillation (AF)** is a common arrhythmia, affecting mostly older adults, in which contraction of the atrial fibers is asynchronous (not in unison) so that atrial pumping ceases altogether. The atria may beat 300–600 beats/min. The ventricles may also speed up, resulting in a rapid heartbeat (up to 160 beats/min). The ECG of an individual with atrial fibrillation typically has no clearly defined P waves and irregularly spaced QRS complexes (and P-R intervals) (Figure 20.24c). Since the atria and ventricles do not beat in rhythm, the heartbeat is irregular in timing and strength. In an otherwise strong heart, atrial fibrillation reduces the pumping effectiveness of the heart by 20–30%. The most dangerous complication of atrial fibrillation is stroke since blood may stagnate in the atria and form blood clots. A stroke occurs when part of a blood clot occludes an artery supplying the brain.

- **Ventricular premature contraction.** Another form of arrhythmia arises when an *ectopic focus* (ek-TOP-ik), a region of the heart other than the conduction system, becomes more excitable than

Figure 20.24 Representative arrhythmias.

An arrhythmia is an abnormal rhythm as a result of a defect in the cardiac conduction system.

(a) Normal electrocardiogram (ECG)

(b) First-degree AV block

(c) Atrial fibrillation

(d) Ventricular tachycardia

(e) Ventricular fibrillation

? Why is ventricular fibrillation such a serious arrhythmia?

normal and causes an occasional abnormal action potential to occur. As a wave of depolarization spreads outward from the ectopic focus, it causes a **ventricular premature contraction (beat).** The contraction occurs early in diastole before the SA node is normally scheduled to discharge its action potential. Ventricular premature contractions may be relatively benign and may be caused by emotional stress, excessive intake of stimulants such as caffeine, alcohol, or nicotine, and lack of sleep. In other cases, the premature beats may reflect an underlying pathology.

- **Ventricular tachycardia (VT or V-tach)** is an arrhythmia that originates in the ventricles and is characterized by 4 or more ventricular premature contractions. It causes the ventricles to beat too fast (at least 120 beats/min) (Figure 20.24d). VT is almost always associated with heart disease or a recent myocardial infarction and may develop into a very serious arrhythmia called ventricular fibrillation (described shortly). Sustained VT is dangerous because the ventricles do not fill properly and thus do not pump sufficient blood. The result may be low blood pressure and heart failure.

- **Ventricular fibrillation (VF or V-fib)** is the most deadly arrhythmia, in which contractions of the ventricular fibers are completely asynchronous so that the ventricles quiver rather than contract in a coordinated way. As a result, ventricular pumping stops, blood ejection ceases, and circulatory failure and death occur unless there is immediate medical intervention. During ventricular fibrillation, the ECG has no clearly defined P waves, QRS complexes, or T waves (Figure 20.24e). The most common cause of ventricular fibrillation is inadequate blood flow to the heart due to coronary artery disease, as occurs during a myocardial infarction. Other causes are cardiovascular shock, electrical shock, drowning, and very low potassium levels. Ventricular fibrillation causes unconsciousness in seconds and, if untreated, seizures occur and irreversible brain damage may occur after five minutes. Death soon follows. Treatment involves cardiopulmonary resuscitation (CPR) and defibrillation. In **defibrillation** (dē-fib-re-LA-shun), also called **cardioversion** (kar′-dē-ō-VER-shun), a strong, brief electrical current is passed to the heart and often can stop the ventricular fibrillation. The electrical shock is generated by a device called a **defibrillator** (de-FIB-ri-lā-tor) and applied via two large paddle-shaped electrodes pressed against the skin of the chest. Patients who face a high risk of dying from heart rhythm disorders now can receive an **automatic implantable cardioverter defibrillator (AICD),** an implanted device that monitors their heart rhythm and delivers a small shock directly to the heart when a life-threatening rhythm disturbance occurs. Thousands of patients around the world have AICDs, including Dick Cheney, Vice President of the United States, who received a combination pacemaker-defibrillator in 2001. Also available are **automated external defibrillators (AEDs)** that function like AICDs, except that they are external devices. About the size of a laptop computer, AEDs are used by emergency response teams and are found increasingly in public places such as stadiums, casinos, airports, hotels, and shopping malls. Defibrillation may also be used as an emergency treatment for cardiac arrest.

MEDICAL TERMINOLOGY

Asystole (ā-SIS-tō-lē; *a-* = without) Failure of the myocardium to contract.

Cardiac arrest (KAR-dē-ak a-REST) A clinical term meaning cessation of an effective heartbeat. The heart may be completely stopped or in ventricular fibrillation.

Cardiomegaly (kar′-dē-ō-MEG-a-lē; *mega* = large) Heart enlargement.

Cardiac rehabilitation (rē-ha-bil-i-TĀ-shun) A supervised program of progressive exercise, psychological support, education, and training to enable a patient to resume normal activities following a myocardial infarction.

Cardiomyopathy (kar′-dē-ō-mī-OP-a-thē; *myo-* = muscle; *-pathos* = disease) A progressive disorder in which ventricular structure or function is impaired. In dilated cardiomyopathy, the ventricles enlarge (stretch) and become weaker and reduce the heart's pumping action. In hypertrophic cardiomyopathy, the ventricular walls thicken and the pumping efficiency of the ventricles is reduced.

Commotio cordis (kō-MŌ-shē-ō KOR-dis; *commotio-* = disturbance; *cordis* = heart) Damage to the heart, frequently fatal, as a result of a sharp, nonpenetrating blow to the chest while the ventricles are repolarizing.

Cor pulmonale (CP) (kor pul-mōn-ALE; *cor-* = heart; *pulmon-* = lung) A term referring to right ventricular hypertrophy from disorders that bring about hypertension (high blood pressure) in the pulmonary circulation.

Ejection fraction The fraction of the end-diastolic volume (EDV) that is ejected during an average heartbeat. Equal to stroke volume (SV) divided by EDV.

Electrophysiological testing (e-lek′-trō-fiz′-ē-OL-ō-ji-kal) A procedure in which a catheter with an electrode is passed through blood vessels and introduced into the heart. It is used to detect the exact locations of abnormal electrical conduction pathways. Once an abnormal pathway is located, it can be destroyed by sending a current through the electrode, a procedure called *radiofrequency ablation.*

Palpitation (pal′-pi-TĀ-shun) A fluttering of the heart or an abnormal rate or rhythm of the heart about which an individual is aware.

Paroxysmal tachycardia (par′-ok-SIZ-mal tak′-e-KAR-dē-a; *tachy-* = quick) A period of rapid heartbeats that begins and ends suddenly.

Sick sinus syndrome An abnormally functioning SA node that initiates heartbeats too slowly or rapidly, pauses too long between heartbeats, or stops producing heartbeats. Symptoms include lightheadedness, shortness of breath, loss of consciousness, and palpitations. It is caused by degeneration of cells in the SA node and is common in elderly persons. It is also related to coronary artery disease. Treatment consists of drugs to speed up or slow down the heart or implantation of an artificial pacemaker.

Sudden cardiac death The unexpected cessation of circulation and breathing due to an underlying heart disease such as ischemia, myocardial infarction, or a disturbance in cardiac rhythm.

STUDY OUTLINE

Anatomy of the Heart (p. 718)

1. The heart is located in the mediastinum; about two-thirds of its mass is to the left of the midline.
2. The heart is shaped like a cone lying on its side. Its apex is the pointed, inferior part. Its base is the broad, superior part.
3. The pericardium is the membrane that surrounds and protects the heart; it consists of an outer fibrous layer and an inner serous pericardium, which is composed of a parietal and a visceral layer.
4. Between the parietal and visceral layers of the serous pericardium is the pericardial cavity, a potential space filled with a few milliliters of lubricating pericardial fluid that reduces friction between the two membranes.
5. Three layers make up the wall of the heart: epicardium (visceral layer of the serous pericardium), myocardium, and endocardium.
6. The epicardium consists of mesothelium and connective tissue, the myocardium is composed of cardiac muscle tissue, and the endocardium consists of endothelium and connective tissue.
7. The heart chambers include two superior chambers, the right and left atria, and two inferior chambers, the right and left ventricles.
8. External features of the heart include the auricles (flaps of each atrium that slightly increase their volume), the coronary sulcus between the atria and ventricles, and the anterior and posterior sulci between the ventricles on the anterior and posterior surfaces of the heart, respectively.
9. The right atrium receives blood from the superior vena cava, inferior vena cava, and coronary sinus. It is separated internally from the left atrium by the interatrial septum, which contains the fossa ovalis. Blood exits the right atrium through the tricuspid valve.
10. The right ventricle receives blood from the right atrium. It is separated internally from the left ventricle by the interventricular septum and pumps blood through the pulmonary valve into the pulmonary trunk.
11. Oxygenated blood enters the left atrium from the pulmonary veins and exits through the bicuspid (mitral) valve.
12. The left ventricle pumps oxygenated blood through the aortic valve into the aorta.
13. The thickness of the myocardium of the four chambers varies according to the chamber's function. The left ventricle, with the highest workload, has the thickest wall.
14. The fibrous skeleton of the heart is dense connective tissue that surrounds and supports the valves of the heart.

Heart Valves and Circulation of Blood (p. 725)

1. Heart valves prevent backflow of blood within the heart. The atrioventricular (AV) valves, which lie between atria and ventricles, are the tricuspid valve on the right side of the heart and the bicuspid (mitral) valve on the left. The semilunar (SL) valves are the aortic valve, at the entrance to the aorta, and the pulmonary valve, at the entrance to the pulmonary trunk.
2. The left side of the heart is the pump for systemic circulation, the circulation of blood throughout the body except for the air sacs of the lungs. The left ventricle ejects blood into the aorta, and blood then flows into systemic arteries, arterioles, capillaries, venules, and veins, which carry it back to the right atrium.
3. The right side of the heart is the pump for pulmonary circulation, the circulation of blood through the lungs. The right ventricle ejects blood into the pulmonary trunk, and blood then flows into pulmonary arteries, pulmonary capillaries, and pulmonary veins, which carry it back to the left atrium.
4. The coronary circulation provides blood flow to the myocardium. The main arteries of the coronary circulation are left and right coronary arteries; the main veins are the cardiac veins and the coronary sinus.

Cardiac Muscle Tissue and the Cardiac Conduction System (p. 730)

1. Cardiac muscle fibers usually contain a single centrally located nucleus. Compared to skeletal muscle fibers, cardiac muscle fibers have more and larger mitochondria, slightly smaller sarcoplasmic reticulum, and wider transverse tubules, which are located at Z discs.
2. Cardiac muscle fibers are connected end-to-end via intercalated discs. Desmosomes in the discs provide strength, and gap junctions allow muscle action potentials to conduct from one muscle fiber to its neighbors.
3. Autorhythmic fibers form the conduction system, cardiac muscle fibers that spontaneously depolarize and generate action potentials.
4. Components of the conduction system are the sinoatrial (SA) node (pacemaker), atrioventricular (AV) node, atrioventricular (AV) bundle (bundle of His), bundle branches, and Purkinje fibers.
5. Phases of an action potential in a ventricular contractile fiber include rapid depolarization, a long plateau, and repolarization.
6. Cardiac muscle tissue has a long refractory period, which prevents tetanus.
7. The record of electrical changes during each cardiac cycle is called an electrocardiogram (ECG). A normal ECG consists of a P wave (atrial depolarization), a QRS complex (onset of ventricular depolarization), and a T wave (ventricular repolarization).
8. The P-Q interval represents the conduction time from the beginning of atrial excitation to the beginning of ventricular excitation. The S-T segment represents the time when ventricular contractile fibers are fully depolarized.

The Cardiac Cycle (p. 738)

1. A cardiac cycle consists of the systole (contraction) and diastole (relaxation) of both atria, plus the systole and diastole of both ventricles. With an average heartbeat of 75 beats/min, a complete cardiac cycle requires 0.8 seconds.
2. The phases of the cardiac cycle are (a) atrial systole, (b) ventricular systole, and (c) relaxation period.
3. S1, the first heart sound (lubb), is caused by blood turbulence associated with the closing of the atrioventricular valves. S2, the second sound (dupp), is caused by blood turbulence associated with the closing of semilunar valves.

Cardiac Output (p. 741)

1. Cardiac output (CO) is the amount of blood ejected per minute by the left ventricle into the aorta (or by the right ventricle into the pulmonary trunk). It is calculated as follows: CO (mL/min) = stroke volume (SV) in mL/beat $\times$ heart rate (HR) in beats per minute.
2. Stroke volume (SV) is the amount of blood ejected by a ventricle during each systole.
3. Cardiac reserve is the difference between a person's maximum cardiac output and his or her cardiac output at rest.
4. Stroke volume is related to preload (stretch on the heart before it

contracts), contractility (forcefulness of contraction), and afterload (pressure that must be exceeded before ventricular ejection can begin).

5. According to the Frank-Starling law of the heart, a greater preload (end-diastolic volume) stretching cardiac muscle fibers just before they contract increases their force of contraction until the stretching becomes excessive.
6. Nervous control of the cardiovascular system originates in the cardiovascular center in the medulla oblongata.
7. Sympathetic impulses increase heart rate and force of contraction; parasympathetic impulses decrease heart rate.
8. Heart rate is affected by hormones (epinephrine, norepinephrine, thyroid hormones), ions (Na^+, K^+, Ca^{2+}), age, gender, physical fitness, and body temperature.

Exercise and the Heart (p. 745)

1. Sustained exercise increases oxygen demand on muscles.

2. Among the benefits of aerobic exercise are increased cardiac output, decreased blood pressure, weight control, and increased fibrinolytic activity.

Help for Failing Hearts (p. 745)

1. A cardiac (heart) transplant is the replacement of a severely damaged heart with a normal one.
2. Cardiac arrest devices and procedures include the intra-aortic balloon pump, ventricular arrest device, cardiomyoplasty, and a skeletal muscle arrest device.

Development of the Heart (p. 748)

1. The heart develops from mesoderm.
2. The endocardial tubes develop into the four-chambered heart and great vessels of the heart.

SELF-QUIZ QUESTIONS

Fill in the blanks in the following statements.

1. The chamber of the heart with the thickest myocardium is the _____.
2. The phase of heart contraction is called _____; the phase of relaxation is called _____.

Indicate whether the following statements are true or false.

3. In auscultation, the lubb represents closing of the semilunar valves and the dupp represents closing of the atrioventricular valves.
4. The Frank-Starling law of the heart equalizes the output of the right and left ventricles and keeps the same volume of blood flowing to both the systemic and pulmonary circulations.

Choose the one best answer to the following questions.

5. Which of the following is the correct route of blood through the heart from the systemic circulation to the pulmonary circulation and back to the systemic circulation?
   (a) right atrium, tricuspid valve, right ventricle, pulmonary semilunar valve, left atrium, mitral valve, left ventricle, aortic semilunar valve
   (b) left atrium, tricuspid valve, left ventricle, pulmonary semilunar valve, right atrium, mitral valve, right ventricle, aortic semilunar valve
   (c) left atrium, pulmonary semilunar valve, right atrium, tricuspid valve, left ventricle, aortic semilunar valve, right ventricle, mitral valve
   (d) left ventricle, mitral valve, left atrium, pulmonary semilunar valve, right ventricle, tricuspid valve, right atrium, aortic semilunar valve
   (e) right atrium, mitral valve, right ventricle, pulmonary semilunar valve, left atrium, tricuspid valve, left ventricle, aortic semilunar valve
6. Which of the following represents the correct pathway for conduction of an action potential through the heart?
   (a) AV node, AV bundle, SA node, Purkinje fibers, bundle branches
   (b) AV node, bundle branches, AV bundle, SA node, Purkinje fibers
   (c) SA node, AV node, AV bundle, bundle branches, Purkinje fibers
   (d) SA node, AV bundle, bundle branches, AV node, Purkinje fibers
   (e) SA node, AV node, Purkinje fibers, bundle branches, AV bundle

7. The external boundary between the atria and ventricles is the
   (a) anterior interventricular sulcus.
   (b) interventricular septum.
   (c) interatrial septum.
   (d) coronary sulcus.
   (e) posterior interventricular sulcus.
8. A softball player is found to have a resting cardiac output of 5.0 liters per minute and a heart rate of 50 beats per minute. What is her stroke volume?
   (a) 10 mL   (b) 100 mL   (c) 1000 mL   (d) 250 mL
   (e) The information given is insufficient to calculate stroke volume.
9. Which of the following are true? (1) ANS regulation of heart rate originates in the cardiovascular center of the medulla oblongata. (2) Proprioceptor input is a major stimulus that accounts for the rapid rise in the heart rate at the onset of physical activity. (3) The vagus nerves release norepinephrine, causing the heart rate to increase. (4) Hormones from the adrenal medulla and the thyroid gland can increase the heart rate. (5) Hypothermia increases the heart rate.
   (a) 1, 2, 3, and 4   (b) 1, 2, and 4   (c) 2, 3, 4, and 5
   (d) 3, 5, and 6   (e) 1, 2, 4, and 5
10. Which of the following are true concerning action potentials and contraction in the myocardium? (1) The refractory period in a cardiac muscle fiber is very brief. (2) The binding of Ca^{2+} to troponin allows the interaction of actin and myosin filaments, resulting in contraction. (3) Repolarization occurs when the voltage-gated K^+ channels open and calcium channels are closing. (4) Opening of voltage-gated fast Na^+ channels results in depolarization. (5) Opening of voltage-gated slow Ca^{2+} channels results in a period of maintained depolarization, known as the plateau.
    (a) 1, 3, and 5   (b) 2, 3, and 4   (c) 2 and 5
    (d) 3, 4, and 5   (e) 2, 3, 4, and 5
11. Which of the following would not increase stroke volume?
    (a) increased Ca^{2+} in the interstitial fluid
    (b) epinephrine
    (c) increased K^+ in the interstitial fluid
    (d) increase in venous return
    (e) slow resting heart rate

12. Match the following:

____ (a) indicates ventricular repolarization

____ (b) represents the time from the beginning of ventricular depolarization to the end of ventricular repolarization

____ (c) represents atrial depolarization

____ (d) represents the time when the ventricular contractile fibers are fully depolarized; occurs during the plateau phase of the action potential

____ (e) represents the onset of ventricular depolarization

____ (f) represents the conduction time from the beginning of atrial excitation to the beginning of ventricular excitation

(1) P wave
(2) QRS complex
(3) T wave
(4) P-Q interval
(5) S-T segment
(6) Q-T interval

13. Match the following:

____ (a) major branch from the ascending aorta; passes inferior to the left auricle

____ (b) lies in the posterior interventricular sulcus; supplies the walls of the ventricles with oxygenated blood

____ (c) located in the coronary sulcus on the posterior surface of the heart; receives most of the deoxygenated blood from the myocardium

____ (d) lies in the coronary sulcus; supplies oxygenated blood to the walls of the right ventricle

____ (e) lies in the coronary sulcus; drains the right atrium and right ventricle

____ (f) major branch from the ascending aorta; lies inferior to the right auricle

____ (g) lies in the posterior interventricular sulcus; drains the right and left ventricles

____ (h) lies in the anterior interventricular sulcus; supplies oxygenated blood to the walls of both ventricles

____ (i) lies in the anterior interventricular sulcus; drains the walls of both ventricles and the left atrium

____ (j) lies in the coronary sulcus; supplies oxygenated blood to the walls of the left ventricle and left atrium

____ (k) drain the right ventricle and open directly into the right atrium

(1) small cardiac vein
(2) anterior interventricular branch (left anterior descending artery)
(3) anterior cardiac veins
(4) posterior interventricular branch
(5) marginal branch
(6) circumflex branch
(7) middle cardiac vein
(8) left coronary artery
(9) right coronary artery
(10) great cardiac vein
(11) coronary sinus

14. Match the following:

____ (a) collects oxygenated blood from the pulmonary circulation

____ (b) pumps deoxygenated blood to the lungs for oxygenation

____ (c) their contraction pulls on and tightens the chordae tendineae, preventing the valve cusps from everting

____ (d) cardiac muscle tissue

____ (e) increase blood-holding capacity of the atria

____ (f) tendonlike cords connected to the atrioventricular valve cusps which, along with the papillary muscles, prevent valve eversion

____ (g) the superficial dense irregular connective tissue covering of the heart

____ (h) outer layer of the serous pericardium; is fused to the fibrous pericardium

____ (i) endothelial cells lining the interior of the heart; are continuous with the endothelium of the blood vessels

____ (j) pumps oxygenated blood to all body cells, except the air sacs of the lungs

____ (k) prevents backflow of blood from the right ventricle into the right atrium

____ (l) collects deoxygenated blood from the systemic circulation

____ (m) left atrioventricular valve

____ (n) the remnant of the foramen ovale, an opening in the interatrial septum of the fetal heart

____ (o) blood vessels that pierce the heart muscle and supply blood to the cardiac muscle fibers

____ (p) grooves on the surface of the heart which delineate the external boundaries between the chambers

____ (q) prevent backflow of blood from the arteries into the ventricles

____ (r) the gap junction and desmosome connections between individual cardiac muscle fibers

____ (s) internal wall dividing the chambers of the heart

____ (t) separate the upper and lower heart chambers, preventing backflow of blood from the ventricles back into the atria

____ (u) inner visceral layer of the pericardium; adheres tightly to the surface of the heart

____ (v) ridges formed by raised bundles of cardiac muscle fibers

(1) right atrium
(2) right ventricle
(3) left atrium
(4) left ventricle
(5) tricuspid valve
(6) bicuspid (mitral) valve
(7) chordae tendineae
(8) auricles
(9) papillary muscles
(10) trabeculae carneae
(11) fibrous pericardium
(12) parietal pericardium
(13) epicardium
(14) myocardium
(15) endocardium
(16) atrioventricular valves
(17) semilunar valves
(18) intercalated discs
(19) sulci
(20) septum
(21) fossa ovalis
(22) coronary circulation

15. Match the following:

____ (a) amount of blood contained in the ventricles at the end of ventricular relaxation

____ (b) period of time when cardiac muscle fibers are contracting and exerting force but not shortening

____ (c) amount of blood ejected per beat by each ventricle

____ (d) amount of blood remaining in the ventricles following ventricular contraction

____ (e) difference between a person's maximum cardiac output and cardiac output at rest

____ (f) period of time when semilunar valves are open and blood flows out of the ventricles

____ (g) period when all four valves are closed and ventricular blood volume does not change

(1) cardiac reserve
(2) stroke volume
(3) end-diastolic volume (EDV)
(4) isovolumetric relaxation
(5) end-systolic volume (ESV)
(6) ventricular ejection
(7) isovolumetric contraction

CRITICAL THINKING QUESTIONS

1. Gerald recently visited the dentist to have his teeth cleaned and checked. During the cleaning process, Gerald had some bleeding from his gums. A couple of days later, Gerald developed a fever, rapid heartbeat, sweating, and chills. He visited his family physician who detected a slight heart murmur. Gerald was given antibiotics and continued to have his heart monitored. How was Gerald's dental visit related to his illness?

2. Unathletic Sylvia makes a resolution to begin an exercise program. She tells you that she wants to make her heart "beat as fast as it can" during exercise. Explain to her why that may not be a good idea.

3. Mr. Perkins is a large, 62-year-old man with a weakness for sweets and fried foods. His idea of exercise is walking to the kitchen for more potato chips to eat while he is watching sports on television. Lately, he's been troubled by chest pains when he walks up stairs. His doctor told him to quit smoking and scheduled a cardiac angiography for next week. What is involved in performing this procedure? Why did the doctor order this test?

? ANSWERS TO FIGURE QUESTIONS

20.1 The mediastinum is the anatomical region that extends from the sternum to the vertebral column, from the first rib to the diaphragm, and between the lungs.

20.2 The visceral layer of the serous pericardium (epicardium) is both a part of the pericardium and a part of the heart wall.

20.3 The coronary sulcus forms a boundary between the atria and ventricles.

20.4 The greater the workload of a heart chamber, the thicker its myocardium.

20.5 The fibrous skeleton attaches to the heart valves and prevents overstretching of the valves as blood passes through them.

20.6 The papillary muscles contract, which pulls on the chordae tendineae and prevents cusps of the atrioventricular valves from everting and letting blood flow back into the atria.

20.7 Numbers 6 (right ventricle) through 10 depict the pulmonary circulation; numbers 1 (left ventricle) through 5 (right atrium) depict the systemic circulation.

20.8 The circumflex artery delivers oxygenated blood to the left atrium and left ventricle.

20.9 The intercalated discs hold the cardiac muscle fibers together and enable action potentials to propagate from one muscle fiber to another.

20.10 The only electrical connection between the atria and the ventricles is the atrioventricular bundle.

20.11 The duration of an action potential is much longer in a ventricular contractile fiber (0.3 sec = 300 msec) than in a skeletal muscle fiber (1–2 msec).

20.12 An enlarged Q wave may indicate a myocardial infarction (heart attack).

20.13 Action potentials propagate most slowly through the AV node.

20.14 The amount of blood in each ventricle at the end of ventricular diastole—called the end-diastolic volume—is about 130 mL in a resting person.

20.15 The first heart sound (S1), or lubb, is associated with closure of the atrioventricular valves.

20.16 The ventricular myocardium receives innervation from the sympathetic division only.

20.17 The skeletal muscle "pump" increases stroke volume by increasing preload (end-diastolic volume).

20.18 Individuals with end-stage heart failure or severe coronary artery disease are candidates for cardiac transplantation.

20.19 The heart begins to contract by the twenty-second day of gestation.

20.20 Partitioning of the heart is complete by the end of the fifth week.

20.21 HDL removes excess cholesterol from body cells and transports it to the liver for elimination.

20.22 Coronary angiography is used to visualize many blood vessels.

20.23 Tetralogy of Fallot involves an interventricular septal defect, an aorta that emerges from both ventricles, a stenosed pulmonary valve, and an enlarged right ventricle.

20.24 Ventricular fibrillation is such a serious arrhythmia because ventricular pumping stops, blood ejection ceases, and circulatory failure and death can occur without immediate medical intervention.

21 | THE CARDIOVASCULAR SYSTEM: BLOOD VESSELS AND HEMODYNAMICS

BLOOD VESSELS, HEMODYNAMICS, AND HOMEOSTASIS Blood vessels contribute to homeostasis by providing the structures for the flow of blood to and from the heart and the exchange of nutrients and wastes in tissues. They also play an important role in adjusting the velocity and volume of blood flow. •

The cardiovascular system contributes to homeostasis of other body systems by transporting and distributing blood throughout the body to deliver materials (such as oxygen, nutrients, and hormones) and carry away wastes. The structures involved in these important tasks are the blood vessels, which form a closed system of tubes that carries blood away from the heart, transports it to the tissues of the body, and then returns it to the heart. The left side of the heart pumps blood through an estimated 100,000 km (60,000 mi) of blood vessels. The right side of the heart pumps blood through the lungs, enabling blood to pick up oxygen and unload carbon dioxide. Chapters 19 and 20 described the composition and functions of blood and the structure and function of the heart. In this chapter, we focus on the structure and functions of the various types of blood vessels; on **hemodynamics** (hē-mō-dī-NAM-iks; *hemo-* = blood; *dynamics* = power), the forces involved in circulating blood throughout the body; and on the blood vessels that constitute the major circulatory routes.

STRUCTURE AND FUNCTION OF BLOOD VESSELS

◉ OBJECTIVES

- Contrast the structure and function of arteries, arterioles, capillaries, venules, and veins.
- Outline the vessels through which the blood moves in its passage from the heart to the capillaries and back.
- Distinguish between pressure reservoirs and blood reservoirs.

The five main types of blood vessels are arteries, arterioles, capillaries, venules, and veins. **Arteries** (AR-ter-ēz) carry blood *away from the heart* to other organs. Large, elastic arteries leave the heart and divide into medium-sized, muscular arteries that branch out into the various regions of the body. Medium-sized arteries then divide into small arteries, which in turn divide into still smaller arteries called **arterioles** (ar-TĒR-ē-ōls). As the arterioles enter a tissue, they branch into numerous tiny vessels called **capillaries** (KAP-i-lar′-ēz = hairlike). The thin walls of capillaries allow the exchange of substances between the blood and body tissues. Groups of capillaries within a tissue reunite to form small veins called **venules** (VEN-ūls). These in turn merge to form progressively larger blood vessels called veins. **Veins** (VĀNZ) are the blood vessels that convey blood from the tissues *back to the heart.*

● CLINICAL **CONNECTION** | Angiogenesis and Disease

Angiogenesis (an′-jē-ō-JEN-e-sis; *angio-* = blood vessel; *-genesis* = production) refers to the growth of new blood vessels. It is an important process in embryonic and fetal development, and in postnatal life serves important functions such as wound healing, formation of a new uterine lining after menstruation, formation of the corpus luteum after ovulation, and development of blood vessels around obstructed arteries in the coronary circulation. Several proteins (peptides) are known to promote and inhibit angiogenesis.

Clinically angiogenesis is important because cells of a malignant tumor secrete proteins called *tumor angiogenesis factors (TAFs)* that stimulate blood vessel growth to provide nourishment for the tumor cells. Scientists are seeking chemicals that would inhibit angiogenesis and thus stop the growth of tumors. In diabetic retinopathy, angiogenesis may be important in the development of blood vessels that actually cause blindness, so finding inhibitors of angiogenesis may also prevent the blindness associated with diabetes. ●

Basic Structure of a Blood Vessel

The wall of a blood vessel consists of three layers, or tunics, of different tissues: an epithelial inner lining, a middle layer consisting of a smooth muscle and elastic connective tissue, and a connective tissue outer covering. The three structural layers of a generalized blood vessel from innermost to outermost are the *tunica interna (intima), tunica media,* and *tunica externa* (Figure 21.1). Modifications of this basic design account for the five types of blood vessels and the structural and functional differ-

ences among the various vessel types. Always remember that structural variations correlate to the differences in function that occur throughout the cardiovascular system.

Tunica Interna (Intima)

The **tunica interna (intima)** (*tunic* = garment; *interna* or *intima* = innermost) forms the inner lining of a blood vessel and is in direct contact with the blood as it flows through the **lumen,** or interior opening, of the vessel (Figure 21.1a, b). Although this layer has multiple parts, these tissue components contribute minimally to the thickness of the vessel wall. Its innermost layer is a simple squamous epithelium, called *endothelium,* which is continuous with the endocardial lining of the heart. The endothelium is a thin layer of flattened cells that lines the inner surface of the entire cardiovascular system (heart and blood vessels). Until recently, endothelial cells were regarded as little more than a passive barrier between the blood and the remainder of the vessel wall. It is now known that endothelial cells are active participants in a variety of vessel-related activities, including physically influencing blood flow, secreting locally acting chemical mediators that influence the contractile state of the vessel's overlying smooth muscle, and assisting with capillary permeability.

The second component of the tunica interna is a *basement membrane* deep to the endothelium. It provides a physical support base for the epithelial layer. Its framework of collagen fibers affords the basal lamina significant tensile strength, yet its properties also provide resilience for stretching and recoil. The basal lamina anchors the endothelium to the underlying connective tissue while also regulating molecular movement. It appears to play an important role in guiding cell movements during tissue repair of blood vessel walls. The outermost part of the tunica interna, which forms the boundary between the tunica interna and tunica media, is the *internal elastic lamina* (*lamina* = thin plate). The internal elastic lamina is a thin sheet of elastic fibers with a variable number of window-like openings that give it the look of Swiss cheese. These openings facilitate diffusion of materials through the tunica interna to the thicker tunica media.

Tunica Media

The **tunica media** (*media* = middle) is a muscular and connective tissue layer that displays the greatest variation among the different vessel types (Figure 21.1a, b). In most vessels, it is a relatively thick layer comprised mainly of smooth muscle cells and substantial amounts of elastic fibers. The primary role of the smooth muscle cells, which extend circularly around the lumen like a ring encircles your finger, is to regulate the diameter of the lumen. As you will learn in more detail shortly, the rate of blood flow through different parts of the body is regulated by the extent of smooth muscle contraction in the walls of particular vessels. Furthermore, the extent of smooth muscle contraction in particular vessel types is crucial in the regulation of blood pressure.

In addition to regulating blood flow and blood pressure, smooth muscle contracts when vessels are damaged to help limit loss of blood through the injured vessel, and smooth muscle cells help

Figure 21.1 Comparative structure of blood vessels. The capillary in (c) is enlarged relative to the structures shown in parts (a) and (b).

🔑 Arteries carry blood from the heart to tissues; veins carry blood from tissues to the heart.

TUNICA INTERNA:
Endothelium

Basement membrane

Internal elastic lamina

TUNICA MEDIA:
Smooth muscle

External elastic lamina

TUNICA EXTERNA

Lumen
(a) Artery

Valve

Lumen
(b) Vein

Lumen

Basement membrane

Endothelium

(c) Capillary

Internal elastic lamina

External elastic lamina

Tunica externa

Lumen with blood cells

Tunica interna

Tunica media

Connective tissue

LM 200x

(d) Transverse section through an artery

Connective tissue

Red blood cell

LM 600x

Capillary endothelial cells

(e) Red blood cells passing through a capillary

❓ Which vessel—the femoral artery or the femoral vein—has a thicker wall? Which has a wider lumen?

produce the elastic fibers within the tunica media that allow the vessels to stretch and recoil under the applied pressure of the blood.

The tunica media is the most variable of the tunics. As you study the different types of blood vessels in the remainder of this chapter, you will see that the structural differences in this layer account for the many variations in function among the different vessel types.

Tunica Externa

The outer covering of a blood vessel, the **tunica externa** (*externa* = outermost), consists of elastic and collagen fibers (Figure 21.1a, b). Separating the tunica externa from the tunica media is a network of elastic fibers, the *external elastic lamina* which is part of the tunica media. The tunica externa contains numerous nerves and, especially in larger vessels, tiny blood vessels that supply the tissue of the vessel wall. These small vessels that supply blood to the tissues of the vessel are called **vasa vasorum** (*vas* = vessel), or vessels to the vessels. They are easily seen on large vessels such as the aorta. In addition to the important role of supplying the vessel wall with nerves and self-vessels, the tunica externa helps anchor the vessels to surrounding tissues.

Arteries

Because **arteries** (*ar-* = air; *ter-* = to carry) were found empty at death, in ancient times they were thought to contain only air. The wall of an artery has the three layers of a typical blood vessel, but has a thick muscular-to-elastic tunica media (Figure 21.1a). Due to their plentiful elastic fibers, arteries normally have high *compliance,* which means that their walls stretch easily or expand without tearing in response to a small increase in pressure.

Sympathetic fibers of the autonomic nervous system innervate the smooth muscle of blood vessels. An increase in sympathetic stimulation typically stimulates the smooth muscle to contract, squeezing the vessel wall and narrowing the lumen. Such a decrease in the diameter of the lumen of a blood vessel is called **vasoconstriction.** In contrast, when sympathetic stimulation decreases, or in the presence of certain chemicals (such as nitric oxide, H^+, and lactic acid), smooth muscle fibers relax. The resulting increase in lumen diameter is called **vasodilation.** Additionally, when an artery or arteriole is damaged, its smooth muscle contracts, producing vascular spasm of the vessel. Such a vasospasm limits blood flow through the damaged vessel and helps reduce blood loss if the vessel is small.

Elastic Arteries

Elastic arteries are the largest arteries in the body, ranging from the garden hose–sized aorta and pulmonary trunk to the finger-sized branches of the aorta. They have the largest diameter among arteries, but their vessel walls (approximately one-tenth of the vessel's total diameter) are relatively thin compared to the overall size of the vessel. These vessels are characterized by well-defined internal and external elastic laminae, along with a thick tunica media

that is dominated by elastic fibers, called the **elastic lamellae** (*lamellae* = little plate). Elastic arteries include the two major trunks that exit the heart (the aorta and the pulmonary trunk), along with the aorta's major initial branches, such as the brachiocephalic, subclavian, common carotid, and common iliac arteries (see Figure 21.19a). Elastic arteries perform an important function: They help propel blood onward while the ventricles are relaxing. As blood is ejected from the heart into elastic arteries, their walls stretch, easily accommodating the surge of blood. As they stretch, the elastic fibers momentarily store mechanical energy, functioning as a **pressure reservoir** (Figure 21.2a). Then, the elastic fibers recoil and convert stored (potential) energy in the vessel into kinetic energy of the blood. Thus, blood continues to move through the arteries even while the ventricles are relaxed (Figure 21.2b). Because they conduct blood from the heart to medium-sized, more muscular arteries, elastic arteries also are called *conducting arteries.*

Figure 21.2 **Pressure reservoir function of elastic arteries.**

Recoil of elastic arteries keeps blood flowing during ventricular relaxation (diastole).

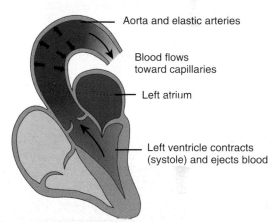

(a) Elastic aorta and arteries stretch during ventricular contraction

- Aorta and elastic arteries
- Blood flows toward capillaries
- Left atrium
- Left ventricle contracts (systole) and ejects blood

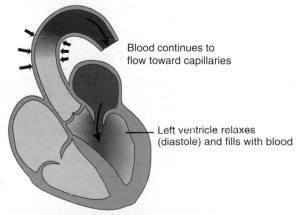

(b) Elastic aorta and arteries recoil during ventricular relaxation

- Blood continues to flow toward capillaries
- Left ventricle relaxes (diastole) and fills with blood

? In atherosclerosis, the walls of elastic arteries become less compliant (stiffer). What effect does reduced compliance have on the pressure reservoir function of arteries?

Muscular Arteries

Medium-sized arteries are called **muscular arteries** because their tunica media contains more smooth muscle and fewer elastic fibers than elastic arteries. The large amount of smooth muscle, approximately three-quarters of the total mass, makes the walls of muscular arteries relatively thick. Thus, muscular arteries are capable of greater vasoconstriction and vasodilation to adjust the rate of blood flow. Muscular arteries have a well-defined internal elastic lamina but a thin external elastic lamina. These two elastic laminae form the inner and outer boundaries of the muscular tunica media. In large arteries, the thick tunica media can have as many as 40 layers of circumferentially arranged smooth muscle cells; in smaller arteries there are as few as three layers.

Muscular arteries span a range of sizes from the pencil-sized femoral and axillary arteries to string-sized arteries that enter organs, measuring as little as 0.5 mm in diameter. Compared to elastic arteries, the vessel wall of muscular arteries comprises a larger percentage (25%) of the total vessel diameter. Because the muscular arteries continue to branch and ultimately distribute blood to each of the various organs, they are called **distributing arteries.** Examples include the brachial artery in the arm and radial artery in the forearm (see Figure 21.19a).

The tunica externa is often thicker than the tunica media in muscular arteries. This outer layer contains fibroblasts, collagen fibers, and elastic fibers all oriented longitudinally. The loose structure of this layer permits changes in the diameter of the vessel to take place but also prevents shortening or retraction of the vessel when it is cut.

Because of the reduced amount of elastic tissue in the walls of muscular arteries, these vessels do not have the ability to recoil and help propel the blood like the elastic arteries. Instead, the thick, muscular tunica media is primarily responsible for the functions of the muscular arteries. The ability of the muscle to contract and maintain a state of partial contraction is referred to as *vascular tone*. Vascular tone stiffens the vessel wall and is important in maintaining vessel pressure and efficient blood flow.

Anastomoses

Most tissues of the body receive blood from more than one artery. The union of the branches of two or more arteries supplying the same body region is called an **anastomosis** (a-nas-tō-MŌ-sis = connecting; plural is **anastomoses**) (see Figure 21.21b). Anastomoses between arteries provide alternative routes for blood to reach a tissue or organ. If blood flow stops for a short time when normal movements compress a vessel, or if a vessel is blocked by disease, injury, or surgery, then circulation to a part of the body is not necessarily stopped. The alternative route of blood flow to a body part through an anastomosis is known as **collateral circulation.** Anastomoses may also occur between veins and between arterioles and venules. Arteries that do not anastomose are known as **end arteries.** Obstruction of an end artery interrupts the blood supply to a whole segment of an organ, producing necrosis (death) of that segment. Alternative blood routes may also be provided by nonanastomosing vessels that supply the same region of the body.

Arterioles

Literally meaning small arteries, **arterioles** are abundant microscopic vessels that regulate the flow of blood into the capillary networks of the body's tissues (see Figure 21.3). The approximately 400 million arterioles have diameters that range in size from 15 μm to 300 μm. The wall thickness of arterioles is one-half of the total vessel diameter.

Arterioles have a thin tunica interna with a thin, fenestrated (with small pores) internal elastic lamina that disappears at the terminal end. The tunica media consists of one to two layers of smooth muscle cells having a circular orientation in the vessel wall. The terminal end of the arteriole, the region called the **metarteriole** (*meta* = after), tapers toward the capillary junction. At the metarteriole–capillary junction, the distal-most muscle cell forms the **precapillary sphincter** (= to bind tight), which monitors the blood flow into the capillary; the other muscle cells in the arteriole regulate the resistance (opposition) to blood flow (see Figure 21.3).

The tunica externa of the arteriole consists of loose collagenous connective tissue containing abundant unmyelinated sympathetic nerves. This sympathetic nerve supply, along with the actions of local chemical mediators, can alter the diameter of arterioles and thus vary the rate of blood flow and resistance through these vessels.

Arterioles play a key role in regulating blood flow from arteries into capillaries by regulating **resistance,** the opposition to blood flow. Because of this they are known as *resistance vessels*. In a blood vessel, resistance is due mainly to friction between blood and the inner walls of blood vessels. When the blood vessel diameter is smaller, the friction is greater, so there is more resistance. Contraction of the smooth muscle of an arteriole causes vasoconstriction, which increases resistance even more and decreases blood flow into capillaries supplied by that arteriole. By contrast, relaxtion of the smooth muscle of an arteriole causes vasodilation, which decreases resistance and increases blood flow into capillaries. A change in arteriole diameter can also affect blood pressure: Vasoconstriction of arterioles increases blood pressure, and vasodilation of arterioles decreases blood pressure.

Capillaries

Capillaries (*capillus* = little hair), the smallest of blood vessels, have diameters of 5–10 μm, and form the "U-turns" that connect the arterial outflow to the venous return (Figure 21.3). Since red blood cells have a diameter of 8 μm, they must often fold upon themselves in order to pass single file through the lumens of these vessels. Capillaries form an extensive network, approximately 20 billion in number, of short (hundreds of μm in length), branched, interconnecting vessels that course among the individual cells of the body. This network forms an enormous surface area to make contact with the body's cells. The flow of blood from a metarteriole through capillaries and into a **postcapillary venule** (venule that re-large, they acquire one or two layers of circularly arranged smooth muscle cells. These **muscular venules** (50 μm to 200 μm) have thicker walls across which exchanges with the interstitial fluid can no longer occur.

Veins

While **veins** do show structural changes as they increase in size from small to medium to large, the structural changes are not as distinct as they are in arteries. Veins, in general, have very thin walls relative to their total diameter (average thickness is less than one-tenth of the vessel diameter). They range in size from 0.5 mm in diameter for small veins to 3 cm in the large caval veins entering the heart.

Although veins are composed of essentially the same three layers as arteries, the relative thicknesses of the layers are different. The tunica interna of veins is thinner than that of arteries; the tunica media of veins is much thinner than in arteries, with relatively little smooth muscle and elastic fibers. The tunica externa of veins is the thickest layer and consists of collagen and elastic fibers. Veins lack the internal or external elastic laminae found in arteries (see Figure 22.1b). They are distensible enough to adapt to variations in the volume and pressure of blood passing through them, but are not designed to withstand high pressure. The lumen of a vein is larger than that of a comparable artery, and veins often appear collapsed (flattened) when sectioned.

The pumping action of the heart is a major factor in moving venous blood back to the heart. The contraction of skeletal muscles in the lower limbs also helps boost venous return to the heart (Figure 21.5). The average blood pressure in veins is considerably lower than in arteries. The difference in pressure can be noticed when blood flows from a cut vessel. Blood leaves a cut vein in an even, slow flow but spurts rapidly from a cut artery. Most of the structural differences between arteries and veins reflect this pressure difference. For example, the walls of veins are not as strong as those of arteries.

Many veins, especially those in the limbs, also contain **valves,** thin folds of tunica interna that form flaplike cusps. The valve cusps project into the lumen, pointing toward the heart (see Figure 21.9). The low blood pressure in veins allows blood returning to the heart to slow and even back up; the valves aid in venous return by preventing the backflow of blood.

A **vascular (venous) sinus** is a vein with a thin endothelial wall that has no smooth muscle to alter its diameter. In a vascular sinus, the surrounding dense connective tissue replaces the tunica media and tunica externa in providing support. For example, dural venous sinuses, which are supported by the dura mater, convey deoxygenated blood from the brain to the heart. Another example of a vascular sinus is the coronary sinus of the heart (see Figure 20.3c on page 722).

While veins follow paths similar to those of their arterial counterparts, they differ from arteries in a number of ways, aside from the structures of their walls. First, veins are more numerous than arteries for several reasons. Some veins are paired and accompany medium- to small-sized muscular arteries. These double sets of veins escort the arteries and connect with one another via

venous channels called **anastomotic veins.** The anastomotic veins cross the accompanying artery to form ladder-like rungs between the paired veins (see Figure 21.25c). The greatest number of paired veins occurs within the limbs. The subcutaneous layer deep to the skin is another source of veins. These veins, called **superficial veins,** course through the subcutaneous layer unaccompanied by parallel arteries. Along their course, the superficial veins form small connections (anastomoses) with the **deep veins** that travel between the skeletal muscles. These connections allow communication between the deep and superficial flow of blood. The amount of blood flow through superficial veins varies from location to location within the body. In the upper limb, the superficial veins are much larger than the deep veins and serve as the major pathways from the capillaries of the upper limb back to the heart. In the lower limb, the opposite is true; the deep veins serve as the principal return pathways. In fact, one-way valves in small anastomosing vessels allow blood to pass from the superficial to the deep veins, but prevent the blood from passing in the reverse direction. This design has important implications in the development of varicose veins.

Figure 21.5 Venous valves.

Valves in veins allow blood to flow in one direction only—toward the heart.

Transverse plane

Vein

Frontal plane

Cusps of valve

Transverse section

Cusps of valve

Longitudinally cut

Photographs of a valve in a vein

? Why are valves more important in arm veins and leg veins than in neck veins?

● CLINICAL **CONNECTION** | Varicose Veins

Leaky venous valves can cause veins to become dilated and twisted in appearance, a condition called **varicose veins** or **varices** (VAR-i-sēz; *varic-* = a swollen vein). The singular is *varix* (VAR-iks). The condition may occur in the veins of almost any body part, but it is most common in the esophagus and in superficial veins of the lower limbs. Those in the lower limbs can range from cosmetic problems to serious medical conditions. The valvular defect may be congenital or may result from mechanical stress (prolonged standing or pregnancy) or aging. The leaking venous valves allow the backflow of blood from the deep veins to the less efficient superficial veins, where the blood pools. This creates pressure that distends the vein and allows fluid to leak into surrounding tissue. As a result, the affected vein and the tissue around it may become inflamed and painfully tender. Veins close to the surface of the legs, especially the saphenous vein, are highly susceptible to varicosities; deeper veins are not as vulnerable because surrounding skeletal muscles prevent their walls from stretching excessively. Varicose veins in the anal canal are referred to as *hemorrhoids*. Esophageal varices result from dilated veins in the walls of the lower part of the

---

TABLE 21.1

Distinguishing Features of Blood Vessels

BLOOD VESSEL	SIZE	TUNICA INTERNA	TUNICA MEDIA	TUNICA EXTERNA	FUNCTION
Elastic arteries	Largest arteries in the body.	Well-defined internal elastic lamina.	Thick and dominated by elastic fibers; well-defined external elastic lamina.	Thinner than tunica media.	Conduct blood from the heart to muscular arteries.
Muscular arteries	Medium-sized arteries.	Well-defined internal elastic lamina.	Thick and dominated by smooth muscle; thin external elastic lamina.	Thicker than tunica media.	Distribute blood to arterioles.
Arterioles	Microscopic (15–300 μm in diameter).	Thin with a fenestrated internal elastic lamina that disappears distally.	One or two layers of circularly oriented smooth muscle; distalmost smooth muscle cell forms a precapillary sphincter.	Loose collagenous connective tissue and sympathetic nerves.	Deliver blood to capillaries and help regulate blood flow from arteries to capillaries.
Capillaries	Microscopic; smallest blood vessels (5–10 μm in diameter).	Endothelium and basement membrane.	None.	None.	Permit exchange of nutrients and wastes between blood and interstitial fluid; distribute blood to postcapillary venules.
Postcapillary venules	Microscopic (10–50 μm in diameter).	Endothelium and basement membrane.	None.	Sparse.	Pass blood into muscular venules; permit exchange of nutrients and wastes between blood and interstitial fluid and function in white blood cell emigration.
Muscular venules	Microscopic (50–200 μm in diameter).	Endothelium and basement membrane.	One or two layers of circularly oriented smooth muscle.	Sparse.	Pass blood into vein; reservoirs for accumulating large volumes of blood (along with postcapillary venules.
Veins	Range from 0.5 mm–3 cm in diameter.	Endothelium and basement membrane; no internal elastic lamina; contain valves; lumen is much larger than in accompanying artery.	Much thinner than in arteries; no external elastic lamina.	Thickest of the three layers.	Return blood to the heart, facilitated by valves in veins in limbs.

esophagus and sometimes the upper part of the stomach. Bleeding esophageal varices are life-threatening and are usually a result of chronic liver disease.

Several treatment options are available for varicose veins in the lower limbs. *Elastic stockings* (support hose) may be used for individuals with mild symptoms or for whom other options are not recommended. *Sclerotherapy* involves injection of a solution into varicose veins that damages the tunica interna by producing a harmless superficial thrombophlebitis (inflammation involving a blood clot). Healing of the damaged part leads to scar formation that occludes the vein. *Radiofrequency endovenous occlusion* involves the application of radiofrequency energy to heat up and close off varicose veins. *Laser occlusion* uses laser therapy to shut down veins. In a surgical procedure called *stripping*, veins may be removed. In this procedure, a flexible wire is threaded through the vein and then pulled out to strip (remove) it from the body. •

A summary of the distinguishing features of blood vessels is presented in Table 21.1 on page 768.

Blood Distribution

The largest portion of your blood volume at rest—about 64%—is in systemic veins and venules (Figure 21.6). Systemic arteries and arterioles hold about 13% of the blood volume, systemic capillaries hold about 7%, pulmonary blood vessels hold about 9%, and the heart holds about 7%. Because systemic veins and venules contain a large percentage of the blood volume, they function as **blood reservoirs** from which blood can be diverted

Figure 21.6 Blood distribution in the cardiovascular system at rest.

Because systemic veins and venules contain more than half the total blood volume, they are called blood reservoirs.

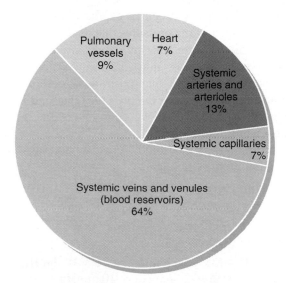

? If your total blood volume is 5 liters, what volume is in your venules and veins right now? In your capillaries?

quickly if the need arises. For example, during increased muscular activity, the cardiovascular center in the brain stem sends a larger number of sympathetic impulses to veins. The result is *venoconstriction*, constriction of veins, which reduces the volume of blood in reservoirs and allows a greater blood volume to flow to skeletal muscles, where it is needed most. A similar mechanism operates in cases of hemorrhage, when blood volume and pressure decrease; in this case, venoconstriction helps counteract the drop in blood pressure. Among the principal blood reservoirs are the veins of the abdominal organs (especially the liver and spleen) and the veins of the skin.

◉ CHECKPOINT

1. What is the function of elastic fibers and smooth muscle in the tunica media of arteries?
2. How are elastic arteries and muscular arteries different?
3. What structural features of capillaries allow the exchange of materials between blood and body cells?
4. What is the difference between pressure reservoirs and blood reservoirs? Why is each important?
5. What is the relationship between anastomoses and collateral circulation?

CAPILLARY EXCHANGE

◉ OBJECTIVE

• Discuss the pressures that cause movement of fluids between capillaries and interstitial spaces.

The mission of the entire cardiovascular system is to keep blood flowing through capillaries to allow **capillary exchange,** the movement of substances between blood and interstitial fluid. The 7% of the blood in systemic capillaries at any given time is continually exchanging materials with interstitial fluid. Substances enter and leave capillaries by three basic mechanisms: diffusion, transcytosis, and bulk flow.

Diffusion

The most important method of capillary exchange is simple diffusion. Many substances, such as oxygen (O_2), carbon dioxide (CO_2), glucose, amino acids, and hormones, enter and leave capillaries by simple diffusion. Because O_2 and nutrients normally are present in higher concentrations in blood, they diffuse down their concentration gradients into interstitial fluid and then into body cells. CO_2 and other wastes released by body cells are present in higher concentrations in interstitial fluid, so they diffuse into blood.

Substances in blood or interstitial fluid can cross the walls of a capillary by diffusing through the intercellular clefts or fenestrations or by diffusing through the endothelial cells (see Figure 21.4). Water-soluble substances such as glucose and amino acids pass across capillary walls through intercellular clefts or fenestrations. Lipid-soluble materials, such as O_2, CO_2, and steroid hormones, may pass across capillary walls directly through the lipid bilayer of endothelial cell plasma membranes.

Most plasma proteins and red blood cells cannot pass through capillary walls of continuous and fenestrated capillaries because they are too large to fit through the intercellular clefts and fenestrations.

In sinusoids, however, the intercellular clefts are so large that they allow even proteins and blood cells to pass through their walls. For example, hepatocytes (liver cells) synthesize and release many plasma proteins, such as fibrinogen (the main clotting protein) and albumin, which then diffuse into the bloodstream through sinusoids. In red bone marrow, blood cells are formed (hemopoiesis) and then enter the bloodstream through sinusoids.

In contrast to sinusoids, the capillaries of the brain allow only a few substances to move across their walls. Most areas of the brain contain continuous capillaries; however, these capillaries are very "tight." The endothelial cells of most brain capillaries are sealed together by tight junctions. The resulting blockade to movement of materials into and out of brain capillaries is known as the *blood–brain barrier* (see page 498). In brain areas that lack the blood–brain barrier, for example, the hypothalamus, pineal gland, and pituitary gland, materials undergo capillary exchange more freely.

Transcytosis

A small quantity of material crosses capillary walls by **transcytosis** (*trans-* = across; *cyt-* = cell; *-osis* = process). In this process, substances in blood plasma become enclosed within tiny pinocytic vesicles that first enter endothelial cells by endocytosis, then move across the cell and exit on the other side by exocytosis. This method of transport is important mainly for large, lipid-insoluble molecules that cannot cross capillary walls in any other way. For example, the hormone insulin (a small protein) enters the bloodstream by transcytosis, and certain antibodies (also proteins) pass from the maternal circulation into the fetal circulation by transcytosis.

Bulk Flow: Filtration and Reabsorption

Bulk flow is a passive process in which *large* numbers of ions, molecules, or particles in a fluid move together in the same direction. The substances move at rates far greater than can be accounted for by diffusion alone. Bulk flow occurs from an area of higher pressure to an area of lower pressure, and it continues as long as a pressure difference exists. Diffusion is more important for *solute exchange* between blood and interstitial fluid, but bulk flow is more important for regulation of the *relative volumes of blood and interstitial fluid.* Pressure-driven movement of fluid and solutes *from* blood capillaries *into* interstitial fluid is called **filtration.** Pressure-driven movement *from* interstitial fluid *into* blood capillaries is called **reabsorption.**

Two pressures promote filtration: **blood hydrostatic pressure (BHP),** the pressure generated by the pumping action of the heart, and **interstitial fluid osmotic pressure.** The main pressure promoting reabsorption of fluid is **blood colloid osmotic pressure.** The balance of these pressures, called **net filtration pressure (NFP),** determines whether the volumes of blood and interstitial fluid remain steady or change. Overall, the volume of fluid and solutes reabsorbed normally is almost as large as the volume filtered. This near equilibrium is known as **Starling's law of the capillaries.** Let's see how these hydrostatic and osmotic pressures balance.

Within vessels, the hydrostatic pressure is due to the pressure that water in blood plasma exerts against blood vessel walls. The **blood hydrostatic pressure (BHP)** is about 35 millimeters of mercury (mmHg) at the arterial end of a capillary, and about 16 mmHg at the capillary's venous end (Figure 21.7). BHP "pushes" fluid out of capillaries into interstitial fluid. The opposing pressure of the interstitial fluid, called **interstitial fluid hydrostatic pressure (IFHP),** "pushes" fluid from interstitial spaces back into capillaries. However, IFHP is close to zero. (IFHP is difficult to measure, and its reported values vary from small positive values to small negative values.) For our discussion we assume that IFHP equals 0 mmHg all along the capillaries.

The difference in osmotic pressure across a capillary wall is due almost entirely to the presence in blood of plasma proteins, which are too large to pass through either fenestrations or gaps between endothelial cells. **Blood colloid osmotic pressure (BCOP)** is a force caused by the colloidal suspension of these large proteins in plasma that averages 26 mmHg in most capillaries. The effect of BCOP is to "pull" fluid from interstitial spaces into capillaries. Opposing BCOP is **interstitial fluid osmotic pressure (IFOP),** which "pulls" fluid out of capillaries into interstitial fluid. Normally, IFOP is very small— 0.1–5 mmHg—because only tiny amounts of protein are present in interstitial fluid. The small amount of protein that leaks from blood plasma into interstitial fluid does not accumulate there because it passes into lymph in lymphatic capillaries and is eventually returned to the blood. For discussion, we can use a value of 1 mmHg for IFOP.

Whether fluids leave or enter capillaries depends on the balance of pressures. If the pressures that push fluid out of capillaries exceed the pressures that pull fluid into capillaries, fluid will move from capillaries into interstitial spaces (filtration). If, however, the pressures that push fluid out of interstitial spaces into capillaries exceed the pressures that pull fluid out of capillaries, then fluid will move from interstitial spaces into capillaries (reabsorption).

The net filtration pressure (NFP), which indicates the direction of fluid movement, is calculated as follows:

$$NFP = (BHP + IFOP) - (BCOP + IFHP)$$

| Pressures that promote filtration | Pressures that promote reabsorption |

At the arterial end of a capillary,

$$NFP = (35 + 1) \text{ mmHg} - (26 + 0) \text{ mmHg}$$
$$= 36 - 26 \text{ mmHg} = 10 \text{ mmHg}$$

Thus, at the arterial end of a capillary, there is a *net outward pressure* of 10 mmHg, and fluid moves out of the capillary into interstitial spaces (filtration).

At the venous end of a capillary,

$$NFP = (16 + 1)\ mmHg - (26 + 0)\ mmHg$$
$$= 17 - 26\ mmHg = -9\ mmHg$$

At the venous end of a capillary, the negative value (–9 mmHg) represents a *net inward pressure,* and fluid moves into the capillary from tissue spaces (reabsorption).

On average, about 85% of the fluid filtered out of capillaries is reabsorbed. The excess filtered fluid and the few plasma proteins that do escape from blood into interstitial fluid enter lymphatic capillaries (see Figure 22.2 on page 834). As lymph drains into the junction of the jugular and subclavian veins in the upper thorax (see Figure 22.3 on page 835), these materials return to the blood. Every day about 20 liters of fluid filter out of capillaries in tissues throughout the body. Of this fluid, 17 liters are reabsorbed and 3 liters enter lymphatic capillaries (excluding filtration during urine formation).

Figure 21.7 **Dynamics of capillary exchange (Starling's law of the capillaries).** Excess filtered fluid drains into lymphatic capillaries.

Blood hydrostatic pressure pushes fluid out of capillaries (filtration), and blood colloid osmotic pressure pulls fluid into capillaries (reabsorption).

A person who has liver failure cannot synthesize the normal amount of plasma proteins. How does a deficit of plasma proteins affect blood colloid osmotic pressure, and what is the effect on capillary filtration and reabsorption?

• CLINICAL **CONNECTION** Edema

If filtration greatly exceeds reabsorption, the result is **edema** (= swelling), an abnormal increase in interstitial fluid volume. Edema is not usually detectable in tissues until interstitial fluid volume has risen to 30% above normal. Edema can result from either excess filtration or inadequate reabsorption.

Two situations may cause excess filtration:

• *Increased capillary blood pressure* causes more fluid to be filtered from capillaries.
• *Increased permeability of capillaries* raises interstitial fluid osmotic pressure by allowing some plasma proteins to escape. Such leakiness may be caused by the destructive effects of chemical, bacterial, thermal, or mechanical agents on capillary walls.

One situation commonly causes inadequate reabsorption:

• *Decreased concentration of plasma proteins* lowers the blood colloid osmotic pressure. Inadequate synthesis or dietary intake or loss of plasma proteins is associated with liver disease, burns, malnutrition (for example, kwashiorkor, see page 1013), and kidney disease. •

◉ CHECKPOINT

6. How can substances enter and leave blood plasma?
7. How do hydrostatic and osmotic pressures determine fluid movement across the walls of capillaries?
8. Define edema and describe how it develops.

HEMODYNAMICS: FACTORS AFFECTING BLOOD FLOW

◉ OBJECTIVES
• Explain the factors that regulate the volume of blood flow.
• Explain how blood pressure changes throughout the cardiovascular system.
• Describe the factors that determine mean arterial pressure and systemic vascular resistance.
• Describe the relationship between cross-sectional area and velocity of blood flow.

Blood flow is the volume of blood that flows through any tissue in a given time period (in mL/min). Total blood flow is cardiac output (CO), the volume of blood that circulates through systemic (or pulmonary) blood vessels each minute. In Chapter 20 we saw that cardiac output depends on heart rate and stroke volume: Cardiac output (CO) = heart rate (HR) × stroke volume (SV). How the cardiac output becomes distributed into circulatory routes that serve various body tissues depends on two more factors: (1) the *pressure difference* that drives the blood flow through a tissue and (2) the *resistance* to blood flow in specific blood vessels. Blood flows from regions of higher pressure to regions of lower pressure; the greater the pressure difference,

the greater the blood flow. But the higher the resistance, the smaller the blood flow.

Blood Pressure

As you have just learned, blood flows from regions of higher pressure to regions of lower pressure; the greater the pressure difference, the greater the blood flow. Contraction of the ventricles generates **blood pressure (BP),** the hydrostatic pressure exerted by blood on the walls of a blood vessel. BP is determined by cardiac output (see page 741), blood volume, and vascular resistance (described shortly). BP is highest in the aorta and large systemic arteries; in a resting, young adult, BP rises to about 110 mmHg during systole (ventricular contraction) and drops to about 70 mmHg during diastole (ventricular relaxation). **Systolic blood pressure** is the highest pressure attained in arteries during systole, and **diastolic blood pressure** is the lowest arterial pressure during diastole (Figure 21.8). As blood leaves the aorta and flows through the systemic circulation, its pressure falls progressively as the distance from the left ventricle increases. Blood pressure decreases to about 35 mmHg as blood passes from systemic arteries through systemic arterioles and into capillaries, where the pressure fluctuations disappear. At the venous end of capillaries, blood pressure has dropped to about

Figure 21.8 Blood pressures in various parts of the cardiovascular system. The dashed line is the mean (average) blood pressure in the aorta, arteries, and arterioles.

⑥ ⎓ Blood pressure rises and falls with each heartbeat in blood vessels leading to capillaries.

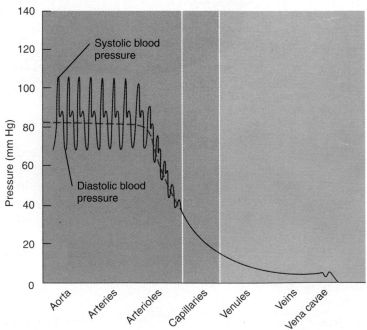

? Is the mean blood pressure in the aorta closer to systolic or to diastolic pressure?

16 mmHg. Blood pressure continues to drop as blood enters systemic venules and then veins because these vessels are farthest from the left ventricle. Finally, blood pressure reaches 0 mmHg as blood flows into the right ventricle.

Mean arterial pressure (MAP), the average blood pressure in arteries, is roughly one-third of the way between the diastolic and systolic pressures. It can be estimated as follows:

MAP = diastolic BP + 1/3 (systolic BP − diastolic BP)

Thus, in a person whose BP is 110/70 mmHg, MAP is about 83 mmHg (70 + 1/3(110 − 70)).

We have already seen that cardiac output equals heart rate multiplied by stroke volume. Another way to calculate cardiac output is to divide mean arterial pressure (MAP) by resistance (R): CO = MAP ÷ R. By rearranging the terms of this equation, you can see that MAP = CO × R. If cardiac output rises due to an increase in stroke volume or heart rate, then the mean arterial pressure rises as long as resistance remains steady. Likewise, a decrease in cardiac output causes a decrease in mean arterial pressure if resistance does not change.

Blood pressure also depends on the total volume of blood in the cardiovascular system. The normal volume of blood in an adult is about 5 liters (5.3 qt). Any decrease in this volume, as from hemorrhage, decreases the amount of blood that is circulated through the arteries each minute. A modest decrease can be compensated for by homeostatic mechanisms that help maintain blood pressure (described on page 755), but if the decrease in blood volume is greater than 10% of the total, blood pressure drops. Conversely, anything that increases blood volume, such as water retention in the body, tends to increase blood pressure.

Vascular Resistance

As noted earlier, **vascular resistance** is the opposition to blood flow due to friction between blood and the walls of blood vessels. Vascular resistance depends on (1) size of the blood vessel lumen, (2) blood viscosity, and (3) total blood vessel length.

1. *Size of the lumen.* The smaller the lumen of a blood vessel, the greater its resistance to blood flow. Resistance is inversely proportional to the fourth power of the diameter (d) of the blood vessel's lumen ($R \propto 1/d^4$). The smaller the diameter of the blood vessel, the greater the resistance it offers to blood flow. For example, if the diameter of a blood vessel decreases by one-half, its resistance to blood flow increases 16 times. Vasoconstriction narrows the lumen, and vasodilation widens it. Normally, moment-to-moment fluctuations in blood flow through a given tissue are due to vasoconstriction and vasodilation of the tissue's arterioles. As arterioles dilate, resistance decreases, and blood pressure falls. As arterioles constrict, resistance increases, and blood pressure rises.

2. *Blood viscosity.* The viscosity (thickness) of blood depends mostly on the ratio of red blood cells to plasma (fluid) volume, and to a smaller extent on the concentration of proteins in plasma. The higher the blood's viscosity, the higher the resistance. Any condition that increases the viscosity of blood, such as dehydration or polycythemia (an unusually high number of red blood cells), thus increases blood pressure. A depletion of plasma proteins or red blood cells, due to anemia or hemorrhage, decreases viscosity and thus decreases blood pressure.

3. *Total blood vessel length.* Resistance to blood flow through a vessel is directly proportional to the length of the blood vessel. The longer a blood vessel, the greater the resistance. Obese people often have hypertension (elevated blood pressure) because the additional blood vessels in their adipose tissue increase their total blood vessel length. An estimated 650 km (about 400 miles) of additional blood vessels develop for each extra kilogram (2.2 lb) of fat.

Systemic vascular resistance (SVR), also known as *total peripheral resistance (TPR),* refers to all the vascular resistances offered by systemic blood vessels. The diameters of arteries and veins are large, so their resistance is very small because most of the blood does not come into physical contact with the walls of the blood vessel. The smallest vessels—arterioles, capillaries, and venules—contribute the most resistance. A major function of arterioles is to control SVR—and therefore blood pressure and blood flow to particular tissues—by changing their diameters. Arterioles need to vasodilate or vasoconstrict only slightly to have a large effect on SVR. The main center for regulation of SVR is the vasomotor center in the brain stem (described shortly).

Venous Return

Venous return, the volume of blood flowing back to the heart through the systemic veins, occurs due to the pressure generated by contractions of the heart's left ventricle. The pressure difference from venules (averaging about 16 mmHg) to the right ventricle (0 mmHg), although small, normally is sufficient to cause venous return to the heart. If pressure increases in the right atrium or ventricle, venous return will decrease. One cause of increased pressure in the right atrium is an incompetent (leaky) tricuspid valve, which lets blood regurgitate (flow backward) as the ventricles contract. The result is decreased venous return and buildup of blood on the venous side of the systemic circulation.

When you stand up, for example, at the end of an anatomy and physiology lecture, the pressure pushing blood up the veins in your lower limbs is barely enough to overcome the force of gravity pushing it back down. Besides the heart, two other mechanisms "pump" blood from the lower body back to the heart: (1) the skeletal muscle pump, and (2) the respiratory pump. Both pumps depend on the presence of valves in veins.

The **skeletal muscle pump** operates as follows (Figure 21.9):

1 While you are standing at rest, both the venous valve closer to the heart (proximal valve) and the one farther from the heart (distal valve) in this part of the leg are open, and blood flows upward toward the heart.

Figure 21.9 Action of the skeletal muscle pump in returning blood to the heart. ❶ At rest, both proximal and distal venous valves are open and blood flows toward the heart. ❷ Contraction of leg muscles pushes blood through the proximal valve while closing the distal valve. ❸ As the leg muscles relax, the proximal valve closes and the distal valve opens. When the vein fills with blood from the foot, the proximal valve reopens.

🔑 Milking refers to skeletal muscle contractions that drive venous blood toward the heart.

Proximal valve

Distal valve

❶   ❷   ❸

❓ **Aside from cardiac contractions, what mechanisms act as pumps to boost venous return?**

❷ Contraction of leg muscles, such as when you stand on tip-toes or take a step, compresses the vein. The compression pushes blood through the proximal valve, an action called *milking*. At the same time, the distal valve in the uncompressed segment of the vein closes as some blood is pushed against it. People who are immobilized through injury or disease lack these contractions of leg muscles. As a result, their venous return is slower and they may develop circulation problems.

❸ Just after muscle relaxation, pressure falls in the previously compressed section of vein, which causes the proximal valve to close. The distal valve now opens because blood pressure in the foot is higher than in the leg, and the vein fills with blood from the foot.

The **respiratory pump** is also based on alternating compression and decompression of veins. During inhalation, the diaphragm moves downward, which causes a decrease in pressure in the thoracic cavity and an increase in pressure in the abdominal cavity. As a result, abdominal veins are compressed, and a greater volume of blood moves from the compressed abdominal veins into the decompressed thoracic veins and then into the right atrium. When the pressures reverse during exhalation, the valves in the veins prevent backflow of blood from the thoracic veins to the abdominal veins.

Figure 21.10 summarizes the factors that increase blood pressure through increasing cardiac output or systemic vascular resistance.

Velocity of Blood Flow

Earlier we saw that blood flow is the *volume* of blood that flows through any tissue in a given time period (in mL/min). The speed or *velocity* of blood flow (in cm/sec) is inversely related to the cross-sectional area. Velocity is slowest where the total cross-sectional area is greatest (Figure 21.11). Each time an artery branches, the total cross-sectional area of all its branches is greater than the cross-sectional area of the original vessel, so blood flow becomes slower and slower as blood moves further away from the heart, and is slowest in the capillaries. Conversely, when venules unite to form veins, the total cross-sectional area becomes smaller and flow becomes faster. In an adult, the cross-sectional area of the aorta is only 3–5 cm², and the average velocity of the blood there is 40 cm/sec. In capillaries, the total cross-sectional area is 4500–6000 cm², and the velocity of blood flow is less than 0.1 cm/sec. In the two venae cavae combined, the cross-sectional area is about 14 cm², and the velocity is about 15 cm/sec. Thus, the velocity of blood flow decreases as blood flows from the aorta to arteries to arterioles to capillaries, and increases as it leaves capillaries and returns to the heart. The relatively slow rate of flow through capillaries aids the exchange of materials between blood and interstitial fluid.

Circulation time is the time required for a drop of blood to pass from the right atrium, through the pulmonary circulation, back to the left atrium, through the systemic circulation down to the foot, and back again to the right atrium. In a resting person, circulation time normally is about 1 minute.

●CLINICAL CONNECTION | Syncope

Syncope (SIN-kō-pē), or fainting, is a sudden, temporary loss of consciousness that is not due to head trauma, followed by spontaneous recovery. It is most commonly due to cerebral ischemia, lack of sufficient blood flow to the brain. Syncope may occur for several reasons:

- *Vasodepressor syncope* is due to sudden emotional stress or real, threatened, or fantasized injury.
- *Situational syncope* is caused by pressure stress associated with urination, defecation, or severe coughing.
- *Drug-induced syncope* may be caused by drugs such as antihypertensives, diuretics, vasodilators, and tranquilizers.
- *Orthostatic hypotension*, an excessive decrease in blood pressure that occurs upon standing up, may cause fainting. •

Figure 21.10 Summary of factors that increase blood pressure. Changes noted within green boxes increase cardiac output; changes noted within blue boxes increase systemic vascular resistance.

Increases in cardiac output and increases in systemic vascular resistance will increase mean arterial pressure.

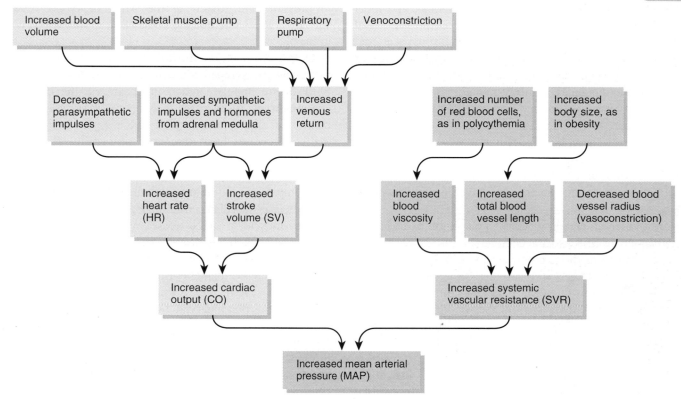

Which type of blood vessel exerts the major control of systemic vascular resistance, and how does it achieve this?

Figure 21.11 Relationship between velocity (speed) of blood flow and total cross-sectional area in different types of blood vessels.

 Velocity of blood flow is slowest in the capillaries because they have the largest total cross-sectional area.

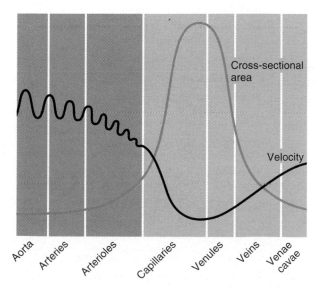

In which blood vessels is the velocity of flow fastest?

CHECKPOINT

9. Explain how blood pressure and resistance determine volume of blood flow.
10. What is systemic vascular resistance and what factors contribute to it?
11. How is the return of venous blood to the heart accomplished?
12. Why is the velocity of blood flow faster in arteries and veins than in capillaries?

CONTROL OF BLOOD PRESSURE AND BLOOD FLOW

OBJECTIVE

• Describe how blood pressure is regulated.

Several interconnected negative feedback systems control blood pressure by adjusting heart rate, stroke volume, systemic vascular resistance, and blood volume. Some systems allow rapid adjustments to cope with sudden changes, such as the drop in blood pressure in the brain that occurs when you get out of bed; others act more slowly to provide long-term regulation of blood pressure. The body may also require adjustments to the distribution of blood flow. During exercise, for example, a greater percentage of the total blood flow is diverted to skeletal muscles.

Role of the Cardiovascular Center

In Chapter 20, we noted how the **cardiovascular (CV) center** in the medulla oblongata helps regulate heart rate and stroke volume. The CV center also controls neural, hormonal, and local negative feedback systems that regulate blood pressure and blood flow to specific tissues. Groups of neurons scattered within the CV center regulate heart rate, contractility (force of contraction) of the ventricles, and blood vessel diameter. Some neurons stimulate the heart (cardiostimulatory center); others inhibit the heart (cardioinhibitory center). Still others control blood vessel diameter by causing constriction (vasoconstrictor center) or dilation (vasodilator center); these neurons are referred to collectively as the vasomotor center. Because the CV center neurons communicate with one another, function together, and are not clearly separated anatomically, we discuss them here as a group.

The cardiovascular center receives input both from higher brain regions and from sensory receptors (Figure 21.12). Nerve impulses descend from the cerebral cortex, limbic system, and hypothalamus to affect the cardiovascular center. For example, even before you start to run a race, your heart rate may increase due to nerve impulses conveyed from the limbic system to the CV center. If your body temperature rises during a race, the hypothalamus sends nerve impulses to the CV center. The resulting vasodilation of skin blood vessels allows heat to dissipate more rapidly from the surface of the skin. The three main types of sensory receptors that provide input to the cardiovascular center are proprioceptors, baroreceptors, and chemoreceptors.

Proprioceptors monitor movements of joints and muscles and provide input to the cardiovascular center during physical activity. Their activity accounts for the rapid increase in heart rate at the beginning of exercise. *Baroreceptors* monitor changes in pressure and stretch in the walls of blood vessels, and *chemoreceptors* monitor the concentration of various chemicals in the blood.

Output from the cardiovascular center flows along sympathetic and parasympathetic neurons of the ANS (Figure 21.12). Sympathetic impulses reach the heart via the **cardiac accelerator nerves.** An increase in sympathetic stimulation increases heart rate and contractility; a decrease in sympathetic stimulation decreases heart rate and contractility. Parasympathetic stimulation, conveyed along the **vagus (X) nerves,** decreases heart rate. Thus, opposing sympathetic (stimulatory) and parasympathetic (inhibitory) influences control the heart.

The cardiovascular center also continually sends impulses to smooth muscle in blood vessel walls via **vasomotor nerves.** These sympathetic neurons exit the spinal cord through all thoracic and the first one or two lumbar spinal nerves and then pass into the sympathetic trunk ganglia (see Figure 15.2 on page 550). From there, impulses propagate along sympathetic neurons that innervate blood vessels in viscera and peripheral areas. The vasomotor region of the cardiovascular center continually sends impulses over these routes to arterioles throughout the body, but especially to those in the skin and abdominal viscera. The result is a moderate state of tonic contraction or vasoconstriction, called **vasomotor tone,** that sets the resting level of systemic vascular resistance. Sympathetic stimu-

Figure 21.12 Location and function of the cardiovascular (CV) center in the medulla oblongata. The CV center receives input from higher brain centers, proprioceptors, baroreceptors, and chemoreceptors. Then, it provides output to the sympathetic and parasympathetic divisions of the autonomic nervous system (ANS).

The cardiovascular center is the main region for nervous system regulation of the heart and blood vessels.

? What types of effector tissues are regulated by the cardiovascular center?

lation of most veins causes constriction that moves blood out of venous blood reservoirs and increases blood pressure.

Neural Regulation of Blood Pressure

The nervous system regulates blood pressure via negative feedback loops that occur as two types of reflexes: baroreceptor reflexes and chemoreceptor reflexes.

Baroreceptor Reflexes

Baroreceptors, pressure-sensitive sensory receptors, are located in the aorta, internal carotid arteries (arteries in the neck that supply blood to the brain), and other large arteries in the neck and chest. They send impulses to the cardiovascular center to help regulate blood pressure. The two most important **baroreceptor reflexes** are the carotid sinus reflex and the aortic reflex.

Baroreceptors in the wall of the carotid sinuses initiate the **carotid sinus reflex,** which helps regulate blood pressure in the brain. The **carotid sinuses** are small widenings of the right and left internal carotid arteries just above the point where they branch from the common carotid arteries (Figure 21.13). Blood pressure stretches the wall of the carotid sinus, which stimulates the baroreceptors. Nerve impulses propagate from the carotid sinus baroreceptors over sensory axons in the **glossopharyngeal (IX) nerves** to the cardiovascular center in the medulla oblongata. Baroreceptors in the wall of the ascending aorta and arch of the aorta initiate the **aortic reflex,** which regulates systemic blood pressure. Nerve impulses from aortic baroreceptors reach the cardiovascular center via sensory axons of the **vagus (X) nerves.**

When blood pressure falls, the baroreceptors are stretched less, and they send nerve impulses at a slower rate to the cardiovascular center (Figure 21.14). In response, the CV center decreases parasympathetic stimulation of the heart by way of motor axons of the vagus nerves and increases sympathetic stimulation of the heart via cardiac accelerator nerves. Another consequence of increased sympathetic stimulation is increased secretion of epinephrine and norepinephrine by the adrenal medulla. As the heart beats faster and more forcefully, and as systemic vascular resistance increases, cardiac output and systemic vascular resistance rise, and blood pressure increases to the normal level.

Conversely, when an increase in pressure is detected, the baroreceptors send impulses at a faster rate. The CV center responds by increasing parasympathetic stimulation and decreasing sympathetic stimulation. The resulting decreases in heart rate and force of contraction reduce the cardiac output. The cardiovascular center also slows the rate at which it sends sympathetic impulses along vasomotor neurons that normally cause vasoconstriction. The resulting vasodilation lowers systemic vascular resistance. Decreased cardiac output and decreased systemic vascular resistance both lower systemic arterial blood pressure to the normal level.

Figure 21.13 ANS innervation of the heart and the baroreceptor reflexes that help regulate blood pressure.

Baroreceptors are pressure-sensitive neurons that monitor stretching.

Key:
Sensory (afferent) neurons ⟵
Motor (efferent) neurons ⟶

? Which cranial nerves conduct impulses to the cardiovascular center from baroreceptors in the carotid sinuses and the arch of the aorta?

Figure 21.14 Negative feedback regulation of blood pressure via baroreceptor reflexes.

When blood pressure decreases, heart rate increases.

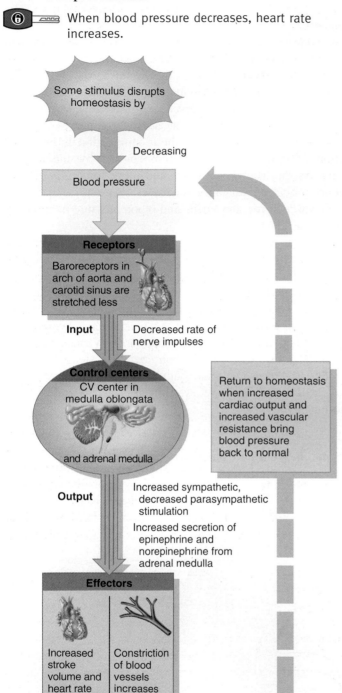

Does this negative feedback cycle represent the changes that occur when you lie down or when you stand up?

Moving from a prone (lying down) to an erect position decreases blood pressure and blood flow in the head and upper part of the body. The baroreceptor reflexes, however, quickly counteract the drop in pressure. Sometimes these reflexes operate more slowly than normal, especially in the elderly, in which case a person can faint due to reduced brain blood flow upon standing up too quickly.

• CLINICAL CONNECTION | **Carotid Sinus Massage and Carotid Sinus Syncope**

Because the carotid sinus is close to the anterior surface of the neck, it is possible to stimulate the baroreceptors there by putting pressure on the neck. Physicians sometimes use **carotid sinus massage,** which involves carefully massaging the neck over the carotid sinus, to slow heart rate in a person who has paroxysmal supraventricular tachycardia, a type of tachycardia that originates in the atria. Anything that stretches or puts pressure on the carotid sinus, such as hyperextension of the head, tight collars, or carrying heavy shoulder loads, may also slow heart rate and can cause **carotid sinus syncope,** fainting due to inappropriate stimulation of the carotid sinus baroreceptors. •

Chemoreceptor Reflexes

Chemoreceptors, sensory receptors that monitor the chemical composition of blood, are located close to the baroreceptors of the carotid sinus and arch of the aorta in small structures called **carotid bodies** and **aortic bodies,** respectively. These chemoreceptors detect changes in blood level of O_2, CO_2, and H^+. *Hypoxia* (lowered O_2 availability), *acidosis* (an increase in H^+ concentration), or *hypercapnia* (excess CO_2) stimulates the chemoreceptors to send impulses to the cardiovascular center. In response, the CV center increases sympathetic stimulation to arterioles and veins, producing vasoconstriction and an increase in blood pressure. These chemoreceptors also provide input to the respiratory center in the brain stem to adjust the rate of breathing.

Hormonal Regulation of Blood Pressure

As you learned in Chapter 18, several hormones help regulate blood pressure and blood flow by altering cardiac output, changing systemic vascular resistance, or adjusting the total blood volume:

1. Renin–angiotensin–aldosterone (RAA) system. When blood volume falls or blood flow to the kidneys decreases, juxtaglomerular cells in the kidneys secrete **renin** into the bloodstream. In sequence, renin and angiotensin converting enzyme (ACE) act on their substrates to produce the active hormone **angiotensin II,** which raises blood pressure in two ways. First, angiotensin II is a potent vasoconstrictor; it raises blood pressure by increasing systemic vascular resistance. Second, it stimulates secretion of **aldosterone,** which increases reabsorption of sodium ions (Na^+) and water by the kidneys. The water reabsorption increases total blood volume, which increases blood pressure. (See page 783.)

2. *Epinephrine and norepinephrine.* In response to sympathetic stimulation, the adrenal medulla releases epinephrine and norepinephrine. These hormones increase cardiac output by increasing the rate and force of heart contractions. They also cause vasoconstriction of arterioles and veins in the skin and abdominal organs and vasodilation of arterioles in cardiac and skeletal muscle, which helps increase blood flow to muscle during exercise. (See Figure 18.20 on page 677.)

3. *Antidiuretic hormone (ADH).* ADH is produced by the hypothalamus and released from the posterior pituitary in response to dehydration or decreased blood volume. Among other actions, ADH causes vasoconstriction, which increases blood pressure. For this reason ADH is also called **vasopressin.** (See Figure 18.9 on page 657.)

4. *Atrial natriuretic peptide (ANP).* Released by cells in the atria of the heart, ANP lowers blood pressure by causing vasodilation and by promoting the loss of salt and water in the urine, which reduces blood volume.

Table 21.2 summarizes the regulation of blood pressure by hormones.

Autoregulation of Blood Pressure

In each capillary bed, local changes can regulate vasomotion. When vasodilators produce local dilation of arterioles and relaxation of precapillary sphincters, blood flow into capillary networks is increased, which increases O_2 level. Vasoconstrictors have the opposite effect. The ability of a tissue to automatically adjust its blood flow to match its metabolic demands is called **autoregulation.** In tissues such as the heart and skeletal muscle, where the demand for O_2 and nutrients and for the removal of wastes can increase as much as tenfold during physical activity, autoregulation is an important contributor to increased blood flow through the tissue. Autoregulation also controls regional blood flow in the brain; blood distribution to various parts of the brain changes dramatically for different mental and physical activities. During a conversation, for example, blood flow increases to your motor speech areas when you are talking and increases to the auditory areas when you are listening.

Two general types of stimuli cause autoregulatory changes in blood flow:

1. *Physical changes.* Warming promotes vasodilation, and cooling causes vasoconstriction. In addition, smooth muscle in arteriole walls exhibits a **myogenic response**—it contracts more forcefully when it is stretched and relaxes when stretching lessens. If, for example, blood flow through an arteriole decreases, stretching of the arteriole walls decreases. As a result, the smooth muscle relaxes and produces vasodilation, which increases blood flow.

2. *Vasodilating and vasoconstricting chemicals.* Several types of cells—including white blood cells, platelets, smooth muscle fibers, macrophages, and endothelial cells—release a wide variety of chemicals that alter blood-vessel diameter. Vasodilating chemicals released by metabolically active tissue cells include K^+, H^+, lactic acid (lactate), and adenosine (from ATP). Another important vasodilator released by endothelial cells is nitric oxide (NO). Tissue trauma or inflammation causes release of vasodilating kinins and histamine. Vasoconstrictors include thromboxane A2, superoxide radicals, serotonin (from platelets), and endothelins (from endothelial cells).

An important difference between the pulmonary and systemic circulations is their autoregulatory response to changes in O_2 level. The walls of blood vessels in the systemic circulation *dilate* in response to low O_2. With vasodilation, O_2 delivery increases, which restores the normal O_2 level. By contrast, the walls of blood vessels in the pulmonary circulation *constrict* in response to low levels of O_2. This response ensures that blood mostly bypasses those alveoli (air sacs) in the lungs that are poorly ventilated by fresh air. Thus, most blood flows to better-ventilated areas of the lung.

◉ **CHECKPOINT**

13. What are the principal inputs to and outputs from the cardiovascular center?
14. Explain the operation of the carotid sinus reflex and the aortic reflex.
15. What is the role of chemoreceptors in the regulation of blood pressure?
16. How do hormones regulate blood pressure?
17. What is autoregulation, and how does it differ in the systemic and pulmonary circulations?

TABLE 21.2		
Blood Pressure Regulation by Hormones		
FACTOR INFLUENCING BLOOD PRESSURE	**HORMONE**	**EFFECT ON BLOOD PRESSURE**
CARDIAC OUTPUT		
Increased heart rate and contractility	Norepinephrine Epinephrine	Increase
SYSTEMIC VASCULAR RESISTANCE		
Vasoconstriction	Angiotensin II	Increase
	Antidiuretic hormone (vasopressin)	
	Norepinephrine*	
	Epinephrine†	
Vasodilation	Atrial natriuretic peptide	Decrease
	Epinephrine†	
	Nitric oxide	
BLOOD VOLUME		
Blood volume increase	Aldosterone Antidiuretic hormone	Increase
Blood volume decrease	Atrial natriuretic peptide	Decrease

*Acts at α_1 receptors in arterioles of abdomen and skin.

†Acts at β_2 receptors in arterioles of cardiac and skeletal muscle; norepinephrine has a much smaller vasodilating effect.

CHECKING CIRCULATION

• Define pulse, and define systolic, diastolic, and pulse pressures.

Pulse

The alternate expansion and recoil of elastic arteries after each systole of the left ventricle creates a traveling pressure wave that is called the **pulse.** The pulse is strongest in the arteries closest to the heart, becomes weaker in the arterioles, and disappears altogether in the capillaries. The pulse may be felt in any artery that lies near the surface of the body that can be compressed against a bone or other firm structure. Table 21.3 depicts some common pulse points.

The pulse rate normally is the same as the heart rate, about 70 to 80 beats per minute at rest. **Tachycardia** (tak′-i-KAR-dē-a; *tachy-* = fast) is a rapid resting heart or pulse rate over 100 beats/min. **Bradycardia** (brād′-i-KAR-dē-a; *brady-* = slow) is a slow resting heart or pulse rate under 50 beats/min. Endurance-trained athletes normally exhibit bradycardia.

Measuring Blood Pressure

In clinical use, the term **blood pressure** usually refers to the pressure in arteries generated by the left ventricle during systole and the pressure remaining in the arteries when the ventricle is in diastole. Blood pressure is usually measured in the brachial artery in the left arm (Table 21.3). The device used to measure blood pressure is a **sphygmomanometer** (sfig′-mō-ma-NOM-e-ter; *sphygmo-* = pulse; *manometer* = instrument used to measure pressure). It consists of a rubber cuff connected to a rubber bulb that is used to inflate the cuff and a meter that registers the pressure in the cuff. With the arm resting on a table so that it is about the same level as the heart, the cuff of the sphygmo-

TABLE 21.3

Pulse Points

STRUCTURE	LOCATION	STRUCTURE	LOCATION
Superficial temporal artery	Lateral to orbit of eye.	Femoral artery	Inferior to inguinal ligament.
Facial artery	Mandible (lower jawbone) on a line with the corners of the mouth.	Popliteal artery	Posterior to knee.
		Radial artery	Distal aspect of wrist.
Common carotid artery	Lateral to larynx (voice box).	Dorsal artery of the foot (dorsalis pedis artery)	Superior to instep of foot.
Brachial artery	Medial side of biceps brachii muscle.		

Superficial temporal artery
Facial artery
Common carotid artery
Brachial artery
Radial artery

Femoral artery
Popliteal artery
Dorsal artery of the foot (dorsalis pedis artery)

manometer is wrapped around a bared arm. The cuff is inflated by squeezing the bulb until the brachial artery is compressed and blood flow stops, about 30 mmHg higher than the person's usual systolic pressure. The technician places a stethoscope below the cuff on the brachial artery, and slowly deflates the cuff. When the cuff is deflated enough to allow the artery to open, a spurt of blood passes through, resulting in the first sound heard through the stethoscope. This sound corresponds to **systolic blood pressure (SBP),** the force of blood pressure on arterial walls just after ventricular contraction (Figure 21.15). As the cuff is deflated further, the sounds suddenly become too faint to be heard through the stethoscope. This level, called the **diastolic blood pressure (DBP),** represents the force exerted by the blood remaining in arteries during ventricular relaxation. At pressures below diastolic blood pressure, sounds disappear altogether. The various sounds that are heard while taking blood pressure are called **Korotkoff sounds** (kō-ROT-kof).

The normal blood pressure of an adult male is less than 120 mmHg systolic and less than 80 mmHg diastolic. For example, "110 over 70" (written as 110/70) is a normal blood pressure. In young adult females, the pressures are 8 to 10 mmHg less. People who exercise regularly and are in good physical condition may have even lower blood pressures. Thus, blood pressure slightly lower than 120/80 may be a sign of good health and fitness.

The difference between systolic and diastolic pressure is called **pulse pressure.** This pressure, normally about 40 mmHg, provides information about the condition of the cardiovascular system. For example, conditions such as atherosclerosis and patent (open) ductus arteriosus greatly increase pulse pressure. The normal ratio of systolic pressure to diastolic pressure to pulse pressure is about 3 : 2 : 1.

Figure 21.15 Relationship of blood pressure changes to cuff pressure.

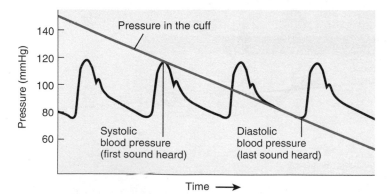

As the cuff is deflated, sounds first occur at the systolic blood pressure; the sounds suddenly become faint at the diastolic blood pressure.

If a blood pressure is reported as "142 over 95," what are the diastolic, systolic, and pulse pressures? Does this person have hypertension as defined on page 825?

Does this person have hypertension as defined on page 825?

CHECKPOINT

18. Where may the pulse be felt?
19. What do tachycardia and bradycardia mean?
20. How are systolic and diastolic blood pressures measured with a sphygmomanometer?

SHOCK AND HOMEOSTASIS

OBJECTIVES

- Define shock, and describe the four types of shock.
- Explain how the body's response to shock is regulated by negative feedback.

Shock is a failure of the cardiovascular system to deliver enough O_2 and nutrients to meet cellular metabolic needs. The causes of shock are many and varied, but all are characterized by inadequate blood flow to body tissues. With inadequate oxygen delivery, cells switch from aerobic to anaerobic production of ATP, and lactic acid accumulates in body fluids. If shock persists, cells and organs become damaged, and cells may die unless proper treatment begins quickly.

Types of Shock

Shock can be of four different types: (1) **hypovolemic shock** (hī-pō-vō-LĒ-mik; *hypo-* = low; *-volemic* = volume) due to decreased blood volume, (2) **cardiogenic shock** due to poor heart function, (3) **vascular shock** due to inappropriate vasodilation, and (4) **obstructive shock** due to obstruction of blood flow.

A common cause of hypovolemic shock is acute (sudden) hemorrhage. The blood loss may be external, as occurs in trauma, or internal, as in rupture of an aortic aneurysm. Loss of body fluids through excessive sweating, diarrhea, or vomiting also can cause hypovolemic shock. Other conditions—for instance, diabetes mellitus—may cause excessive loss of fluid in the urine. Sometimes, hypovolemic shock is due to inadequate intake of fluid. Whatever the cause, when the volume of body fluids falls, venous return to the heart declines, filling of the heart lessens, stroke volume decreases, and cardiac output decreases. Replacing fluid volume as quickly as possible is essential in managing hypovolemic shock.

In cardiogenic shock, the heart fails to pump adequately, most often because of a myocardial infarction (heart attack). Other causes of cardiogenic shock include poor perfusion of the heart (ischemia), heart valve problems, excessive preload or afterload, impaired contractility of heart muscle fibers, and arrhythmias.

Even with normal blood volume and cardiac output, shock may occur if blood pressure drops due to a decrease in systemic vascular resistance. A variety of conditions can cause inappropriate dilation of arterioles or venules. In *anaphylactic shock,* a severe allergic reaction—for example, to a bee sting—releases histamine and other mediators that cause vasodilation. In *neurogenic shock,* vasodilation may occur following trauma to the head that causes malfunction of the cardiovascular center in the medulla. Shock stemming from certain bacterial toxins that produce vasodilation is termed *septic shock.* In the United States, septic shock causes more than 100,000 deaths per year and is the most common cause of death in hospital critical care units.

Obstructive shock occurs when blood flow through a portion of the circulation is blocked. The most common cause is *pulmonary embolism,* a blood clot lodged in a blood vessel of the lungs.

Homeostatic Responses to Shock

The major mechanisms of compensation in shock are *negative feedback systems* that work to return cardiac output and arterial blood pressure to normal. When shock is mild, compensation by homeostatic mechanisms prevents serious damage. In an otherwise healthy person, compensatory mechanisms can maintain

Figure 21.16 **Negative feedback systems that can restore normal blood pressure during hypovolemic shock.**

Homeostatic mechanisms can compensate for an acute blood loss of as much as 10% of total blood volume.

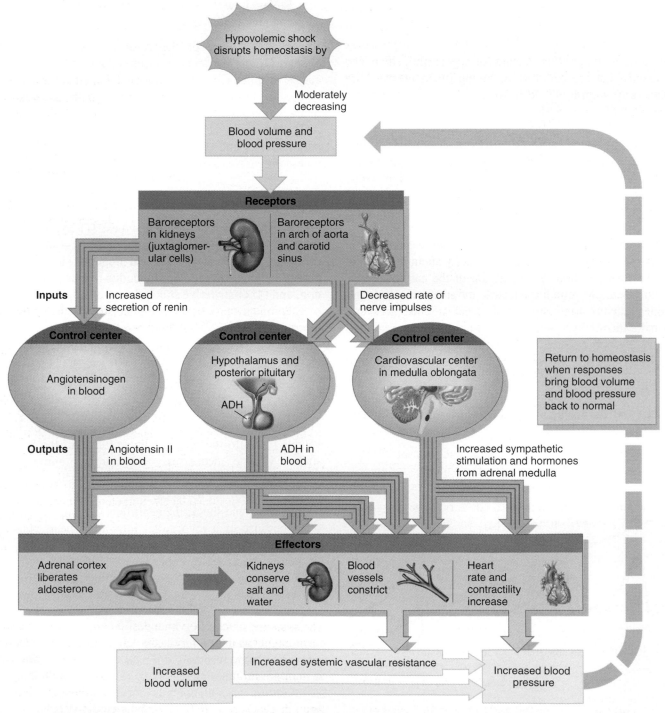

? **Does almost-normal blood pressure in a person who has lost blood indicate that the patient's tissues are receiving adequate perfusion (blood flow)?**

adequate blood flow and blood pressure despite an acute blood loss of as much as 10% of total volume. Figure 21.16 on page 782 shows several of the negative feedback systems that respond to hypovolemic shock.

1. *Activation of the renin–angiotensin–aldosterone system.* Decreased blood flow to the kidneys causes the kidneys to secrete renin and initiates the renin–angiotensin–aldosterone system (see Figure 18.16 on page 667). Recall that angiotensin II causes vasoconstriction and stimulates the adrenal cortex to secrete aldosterone, a hormone that increases reabsorption of Na^+ and water by the kidneys. The increases in systemic vascular resistance and blood volume help raise blood pressure.

2. *Secretion of antidiuretic hormone.* In response to decreased blood pressure, the posterior pituitary releases more antidiuretic hormone (ADH). ADH enhances water reabsorption by the kidneys, which conserves remaining blood volume. It also causes vasoconstriction, which increases systemic vascular resistance. (See Figure 18.9 on page 657.)

3. *Activation of the sympathetic division of the ANS.* As blood pressure decreases, the aortic and carotid baroreceptors initiate powerful sympathetic responses throughout the body. One result is marked vasoconstriction of arterioles and veins of the skin, kidneys, and other abdominal viscera. (Vasoconstriction does not occur in the brain or heart.) The constriction of arterioles increases systemic vascular resistance, and the constriction of veins increases venous return. Both effects help maintain an adequate blood pressure. Sympathetic stimulation also increases heart rate and contractility and increases secretion of epinephrine and norepinephrine by the adrenal medulla. These hormones intensify vasoconstriction and increase heart rate and contractility, all of which help raise blood pressure.

4. *Release of local vasodilators.* In response to *hypoxia,* cells liberate vasodilators—including K^+, H^+, lactic acid, adenosine, and nitric oxide—that dilate arterioles and relax precapillary sphincters. Such vasodilation increases local blood flow and may restore O_2 level to normal in part of the body. However, vasodilation also has the potentially harmful effect of decreasing systemic vascular resistance and thus lowering the blood pressure.

If blood volume drops more than 10–20%, or if the heart cannot bring blood pressure up sufficiently, compensatory mechanisms may fail to maintain adequate blood flow to tissues. At this point, shock becomes life-threatening as damaged cells start to die.

Signs and Symptoms of Shock

Even though the signs and symptoms of shock vary with the severity of the condition, most can be predicted in light of the responses generated by the negative feedback systems that attempt to correct the problem. Among the signs and symptoms of shock are the following:

- Systolic blood pressure is lower than 90 mmHg.
- Resting heart rate is rapid due to sympathetic stimulation

and increased blood levels of epinephrine and norepinephrine.
- Pulse is weak and rapid due to reduced cardiac output and fast heart rate.
- Skin is cool, pale, and clammy due to sympathetic constriction of skin blood vessels and sympathetic stimulation of sweating.
- Mental state is altered due to reduced oxygen supply to the brain.
- Urine formation is reduced due to increased levels of aldosterone and antidiuretic hormone (ADH).
- The person is thirsty due to loss of extracellular fluid.
- The pH of blood is low (acidosis) due to buildup of lactic acid.
- The person may have nausea because of impaired blood flow to the digestive organs from sympathetic vasoconstriction.

◉ CHECKPOINT
21. Which symptoms of hypovolemic shock relate to actual body fluid loss, and which relate to the negative feedback systems that attempt to maintain blood pressure and blood flow?
22. Describe the types of shock and their causes and how a person in hypovolemic shock should be treated.

CIRCULATORY ROUTES

◉ OBJECTIVES
- Describe and compare the major routes that blood takes through various regions of the body.

Blood vessels are organized into **circulatory routes** that carry blood to specific organs in the body (Figure 21.17). The routes are parallel—in most cases a portion of the cardiac output flows separately to each tissue of the body, so that each organ receives its own supply of freshly oxygenated blood. The two main circulatory routes, the systemic circulation and pulmonary circulation, differ in two important ways. First, blood in the pulmonary circulation need not be pumped as far as blood in the systemic circulation. Second, compared to systemic arteries, pulmonary arteries have larger diameters, thinner walls, and less elastic tissue. As a result, the resistance to pulmonary blood flow is very low, which means that less pressure is needed to move blood through the lungs. The peak systolic pressure in the right ventricle is only 20% of that in the left ventricle.

The Systemic Circulation

The **systemic circulation** includes the arteries and arterioles that carry oxygenated blood from the left ventricle to systemic capillaries, plus the veins and venules that return deoxygenated blood to the right atrium. Blood leaving the aorta and flowing through the systemic arteries is a bright red color. As blood flows through capillaries, it loses some of its oxygen and picks up carbon dioxide, becoming a dark red color. All systemic arteries branch from the **aorta.** Completing the circuit, all the veins of

Figure 21.17 Circulatory routes. Long black arrows indicate the systemic circulation (detailed in Exhibits 21.3–21.12), short black arrows the pulmonary circulation (detailed in Figure 21.29), and red arrows the hepatic portal circulation (detailed in Figure 21.28). Refer to Figure 20.8 on page 729 for details of the coronary circulation, and to Figure 21.30 for details of the fetal circulation.

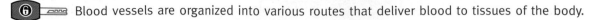 Blood vessels are organized into various routes that deliver blood to tissues of the body.

■ = Oxygenated blood
■ = Deoxygenated blood

? What are the two main circulatory routes?

the systemic circulation drain into the **superior vena cava,** the **inferior vena cava,** or the **coronary sinus,** which in turn empty into the right atrium. The bronchial arteries, which carry nutrients to the lungs, also are part of the systemic circulation.

Exhibits 21.1–21.12 and Figures 21.18–21.27 describe the main arteries and veins of the systemic circulation. The blood vessels are organized in the exhibits according to body regions. Figure 21.18a provides an overview of the major arteries, and Figure 21.23 provides an overview of the major veins. As you study the various blood vessels in the exhibits, refer to these two figures to see the relationships of the blood vessels under consideration to other regions of the body.

EXHIBIT 21.1   The Aorta and its Branches

OBJECTIVES

- Identify the four principal divisions of the aorta.
- Locate the major arterial branches arising from each division.

The **aorta** (= to lift up) is the largest artery of the body, with a diameter of 2–3 cm (about 1 in.). Its four principal divisions are the ascending aorta, arch of the aorta, thoracic aorta, and abdominal aorta (Figure 21.18). The portion of the aorta that emerges from the left ventricle posterior to the pulmonary trunk is the **ascending aorta.** The beginning of the aorta contains the aortic valve (see Figure 20.4a on page 723). The ascending aorta gives off two coronary artery branches that supply the myocardium of the heart. Then the ascending aorta arches to the left, forming the **arch of the aorta,** which descends and ends at the level of the intervertebral disc between the fourth and fifth thoracic vertebrae. As the aorta continues to descend, it lies close to the vertebral bodies, passes through the aortic hiatus of the diaphragm, and divides at the level of the fourth lumbar vertebra into two **common iliac arteries,** which carry blood to the lower limbs. The section of the aorta between the arch of the aorta and the diaphragm is called the **thoracic aorta;** the section between the diaphragm and the common iliac arteries is the **abdominal aorta.** Each division of the aorta gives off arteries that branch into distributing arteries that lead to various organs. Within the organs, the arteries divide into arterioles and then into capillaries that service the systemic tissues (all tissues except the alveoli of the lungs).

CHECKPOINT

What general regions do each of the four principal divisions of the aorta supply?

DIVISION AND BRANCHES	REGION SUPPLIED
ASCENDING AORTA	
Right and left coronary arteries	Heart.
ARCH OF THE AORTA	
Brachiocephalic trunk (brā′-kē-ō-se-FAL-ik)	
Right common carotid artery (ka-ROT-id)	Right side of head and neck.
Right subclavian artery (sub-KLĀ-vē-an)	Right upper limb.
Left common carotid artery	Left side of head and neck.
Left subclavian artery	Left upper limb.
THORACIC AORTA (*Thorac-* = Chest)	
Pericardial arteries (per-i-KAR-dē-al)	Pericardium.
Bronchial arteries (BRONG-kē-al)	Bronchi of lungs.
Esophageal arteries (e-sof′-a-JĒ-al)	Esophagus.
Mediastinal arteries (mē′-dē-as-TĪ-nal)	Structures in mediastinum.
Posterior intercostal arteries (in′-ter-KOS-tal)	Intercostal and chest muscles.
Subcostal arteries (sub-KOS-tal)	Same as posterior intercostals.
Superior phrenic arteries (FREN-ik)	Superior and posterior surfaces of diaphragm.
ABDOMINAL AORTA	
Inferior phrenic arteries (FREN-ik)	Inferior surface of diaphragm.
Celiac trunk (SĒ-lē-ak)	
Common hepatic artery (he-PAT-ik)	Liver, stomach, duodenum, and pancreas.
Left gastric artery (GAS-trik)	Stomach and esophagus.
Splenic artery (SPLĒN-ik)	Spleen, pancreas, and stomach.
Superior mesenteric artery (MES-en-ter′-ik)	Small intestine, cecum, ascending and transverse colons, and pancreas.
Suprarenal arteries (soo-pra-RĒ-nal)	Adrenal (suprarenal) glands.
Renal arteries (RĒ-nal)	Kidneys.
Gonadal arteries (gō-NAD-al)	
Testicular arteries (tes-TIK-ū-lar)	Testes (male).
Ovarian arteries (ō-VAR-ē-an)	Ovaries (female).
Inferior mesenteric artery	Transverse, descending, and sigmoid colons; rectum.
Common iliac arteries (IL-ē-ak)	
External iliac arteries	Lower limbs.
Internal iliac arteries	Uterus (female), prostate (male), muscles of buttocks, and urinary bladder.

EXHIBIT 21.1 CONTINUES ▶

EXHIBIT 21.1   **785**

EXHIBIT 21.1 The Aorta and Its Branches CONTINUED

Figure 21.18 Aorta and its principal branches.

All systemic arteries branch from the aorta.

Right internal carotid

Right vertebral

Right common carotid

Right subclavian

Brachiocephalic trunk

ASCENDING AORTA

Right brachial

ABDOMINAL AORTA

Celiac trunk

Common hepatic

Right radial

Right renal

Right ulnar

Right deep palmar arch

Right superficial palmar arch

Right deep femoral

Right external carotid

Left common carotid

Left subclavian

ARCH OF AORTA

Left axillary

THORACIC AORTA

Diaphragm

Left gastric

Splenic

Left renal

Superior mesenteric

Left gonadal (testicular or ovarian)

Inferior mesenteric

Left common iliac

Left external iliac

Left internal iliac

Left common palmar digital

Left proper palmar digital

Left femoral

Left deep femoral

Left popliteal

Left anterior tibial

Left posterior tibial

Left fibular (peroneal)

Left dorsal artery of foot (dorsalis pedis)

Left arcuate

Left dorsal metatarsal

Left dorsal digital

(a) Overall anterior view of the principal branches of the aorta

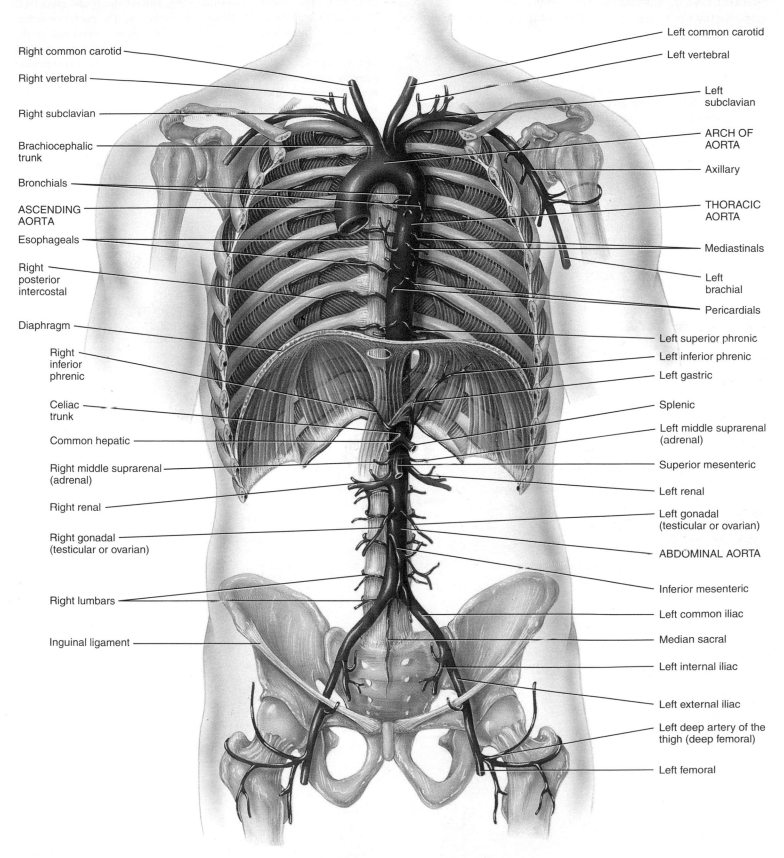

Right common carotid

Right vertebral

Right subclavian

Brachiocephalic trunk

Bronchials

ASCENDING AORTA

Esophageals

Right posterior intercostal

Diaphragm

Right inferior phrenic

Celiac trunk

Common hepatic

Right middle suprarenal (adrenal)

Right renal

Right gonadal (testicular or ovarian)

Right lumbars

Inguinal ligament

Left common carotid

Left vertebral

Left subclavian

ARCH OF AORTA

Axillary

THORACIC AORTA

Mediastinals

Left brachial

Pericardials

Left superior phronic

Left inferior phrenic

Left gastric

Splenic

Left middle suprarenal (adrenal)

Superior mesenteric

Left renal

Left gonadal (testicular or ovarian)

ABDOMINAL AORTA

Inferior mesenteric

Left common iliac

Median sacral

Left internal iliac

Left external iliac

Left deep artery of the thigh (deep femoral)

Left femoral

(b) Detailed anterior view of the principal branches of the aorta

What are the four subdivisions of the aorta?

EXHIBIT 21.1 **787**

EXHIBIT 21.2   Ascending Aorta

◉ OBJECTIVE

- Identify the two primary arterial branches of the ascending aorta.

The **ascending aorta** is about 5 cm (2 in.) in length and begins at the aortic valve (see Figure 20.8 on page 729). It is directed superiorly, slightly anteriorly, and to the right. It ends at the level of the sternal angle, where it becomes the arch of the aorta. The beginning of the ascending aorta is posterior to the pulmonary trunk and right auricle; the right pulmonary artery is posterior to it. At its origin, the ascending aorta contains three dilations called *aortic sinuses*. Two of these, the right and left sinuses, give rise to the right and left coronary arteries, respectively.

The right and left **coronary arteries** (*coron-* = crown) arise from the ascending aorta just superior to the aortic valve (see Figure 20.8 on page 729). They form a crownlike ring around the heart, giving off branches to the atrial and ventricular myocardium. The **posterior interventricular branch** (in-ter-ven-TRIK-ū-lar; *inter-* = between) of the right coronary artery supplies both ventricles, and the **marginal branch** supplies the right ventricle. The **anterior interventricular branch,** also known as the **left anterior descending (LAD) branch,** of the left coronary artery supplies both ventricles, and the **circumflex branch** (SER-kum-flex; *circum-* = around; *-flex* = to bend) supplies the left atrium and left ventricle.

◉ CHECKPOINT

Which branches of the coronary arteries supply the left ventricle? Why does the left ventricle have such an extensive arterial blood supply?

SCHEME OF DISTRIBUTION

EXHIBIT 21.3 — The Arch of the Aorta

◉ **OBJECTIVE**

- Identify the three principal arteries that branch from the arch of the aorta.

The **arch of the aorta** is 4–5 cm (almost 2 in.) in length and is the continuation of the ascending aorta. It emerges from the pericardium posterior to the sternum at the level of the sternal angle (Figure 21.19). The arch of the aorta is directed superiorly and posteriorly to the left and then inferiorly; it ends at the intervertebral disc between the fourth and fifth thoracic vertebrae, where it becomes the thoracic aorta. Three major arteries branch from the superior aspect of the arch of the aorta: the brachiocephalic trunk, the left common carotid, and the left subclavian.

BRANCH	DESCRIPTION AND REGION SUPPLIED
BRACHIOCEPHALIC TRUNK	The **brachiocephalic trunk** divides to form the right subclavian artery and right common carotid artery (Figure 21.19a).
Right subclavian artery (sub-KLĀ-vē-an)	The **right subclavian artery** extends from the brachiocephalic trunk to the first rib and then passes into the armpit (axilla). The general distribution of the artery is to the brain and spinal cord, neck, shoulder, thoracic viscera and wall, and scapular muscles.
Internal thoracic or mammary artery (thor-AS-ik; *thorac-* = chest)	The **internal thoracic artery** arises from the first part of the subclavian artery and descends posterior to the costal cartilages of the superior six ribs. It terminates at the sixth intercostal space. It supplies the anterior thoracic wall and structures in the mediastinum. In coronary artery bypass grafting, if only a single vessel is obstructed, the internal thoracic (usually the left) is used to create the bypass. The upper end of the artery is left attached to the subclavian artery and the cut end is connected to the coronary artery at a point distal to the blockage. The lower end of the internal thoracic artery is tied off. Artery grafts are preferred over vein grafts because arteries can withstand the greater pressure of blood flowing through coronary arteries and are less likely to become obstructed over time.
Vertebral artery (VER-te-bral)	Before passing into the axilla, the right subclavian artery gives off a major branch to the brain called the **right vertebral artery** (Figure 21.19b). The right vertebral artery passes through the foramina of the transverse processes of the sixth through first cervical vertebrae and enters the skull through the foramen magnum to reach the inferior surface of the brain. Here it unites with the left vertebral artery to form the **basilar** (BAS-i-lar) **artery.** The vertebral artery supplies the posterior portion of the brain with blood. The basilar artery passes along the midline of the anterior aspect of the brain stem. It gives off several branches (**posterior cerebral** and **cerebellar arteries**) that supply the cerebellum and pons of the brain and the inner ear.
Axillary artery (AK-sil-ār-ē = armpit)	The continuation of the right subclavian artery into the axilla is called the **axillary artery.** (Note that the right subclavian artery, which passes deep to the clavicle, is a good example of the practice of giving the same vessel different names as it passes through different regions.) Its general distribution is the shoulder, thoracic and scapular muscles, and humerus.
Brachial artery (BRĀ-kē-al = arm)	The **brachial artery** is the continuation of the axillary artery into the arm. The brachial artery provides the main blood supply to the arm and is superficial and palpable along its course. It begins at the tendon of the teres major muscle and ends just distal to the bend of the elbow. At first, the brachial artery is medial to the humerus, but as it descends it gradually curves laterally and passes through the cubital fossa, a triangular depression anterior to the elbow where you can easily detect the pulse of the brachial artery and listen to the various sounds when taking a person's blood pressure. Just distal to the bend in the elbow, the brachial artery divides into the radial artery and ulnar artery. Blood pressure is usually measured in the brachial artery. In order to control hemorrhage, the best place to compress the brachial artery is near the middle of the arm.
Radial artery (RĀ-dē-al = radius)	The **radial artery** is the smaller branch and is a direct continuation of the brachial artery. It passes along the lateral (radial) aspect of the forearm and then through the wrist and hand, supplying these structures with blood. At the wrist, the radial artery makes contact with the distal end of the radius, where it is covered only by fascia and skin. Because of its superficial location at this point, it is a common site for measuring the radial pulse.
Ulnar artery (UL-nar = ulna)	The **ulnar artery,** the larger branch of the brachial artery, passes along the medial (ulnar) aspect of the forearm and then into the wrist and hand, supplying these structures with blood. In the palm, branches of the radial and ulnar arteries anastomose to form the superficial palmar arch and the deep palmar arch.
Superficial palmar arch (*palma* = palm)	The **superficial palmar arch** is formed mainly by the ulnar artery, with a contribution from a branch of the radial artery. The arch is superficial to the long flexor tendons of the fingers and extends across the palm at the bases of the metacarpals. It gives rise to **common palmar digital arteries,** which supply the palm. Each divides into a pair of **proper palmar digital arteries,** which supply the fingers.
Deep palmar arch	Mainly the radial artery forms the **deep palmar arch,** with a contribution from a branch of the ulnar artery. The arch is deep to the long flexor tendons of the fingers and extends across the palm, just distal to the bases of the metacarpals. Arising from the deep palmar arch are **palmar metacarpal arteries,** which supply the palm and anastomose with the common palmar digital arteries of the superficial palmar arch.
Right common carotid artery	The **right common carotid artery** begins at the bifurcation (division into two branches) of the brachiocephalic trunk, posterior to the right sternoclavicular joint, and passes superiorly in the neck to supply structures in the head (Figure 21.19b). At the superior border of the larynx (voice box), it divides into the right external and right internal carotid arteries. Pulse may be detected in the common carotid artery, just lateral to the larynx. It is convenient to detect a carotid pulse when exercising or when administering cardiopulmonary resuscitation.
External carotid artery	The **external carotid artery** begins at the superior border of the larynx and terminates near the temporomandibular joint of the parotid gland, where it divides into two branches: the superficial temporal and maxillary arteries. The carotid pulse can be detected in the external carotid artery just anterior to the sternocleidomastoid muscle at the superior border of the larynx. The general distribution of the external carotid artery is to structures external to the skull.

● **EXHIBIT 21.3 CONTINUES** ▶

EXHIBIT 21.3 **789**

EXHIBIT 21.3   The Arch of the Aorta CONTINUED

BRANCH	DESCRIPTION AND REGION SUPPLIED
Internal carotid artery	The **internal carotid artery** has no branches in the neck and supplies structures internal to the skull. It enters the cranial cavity through the carotid foramen in the temporal bone. The internal carotid artery supplies blood to the eyeball and other orbital structures, ear, most of the cerebrum of the brain, pituitary gland, and external nose. The terminal branches of the internal carotid artery are the **anterior cerebral artery,** which supplies most of the medial surface of the cerebrum and deep masses of gray matter within the cerebrum, and the **middle cerebral artery,** which supplies most of the lateral surface of the cerebrum (Figure 21.19c). Inside the cranium, anastomoses of the left and right internal carotid arteries along with the basilar artery form an arrangement of blood vessels at the base of the brain near the hypophyseal fossa called the **cerebral arterial circle (circle of Willis).** From this circle (Figure 21.19c) arise arteries supplying most of the brain. Essentially, the cerebral arterial circle is formed by the union of the **anterior cerebral arteries** (branches of internal carotids) and **posterior cerebral arteries** (branches of basilar artery). The posterior cerebral arteries supply the inferolateral surface of the temporal lobe and lateral and medial

SCHEME OF DISTRIBUTION

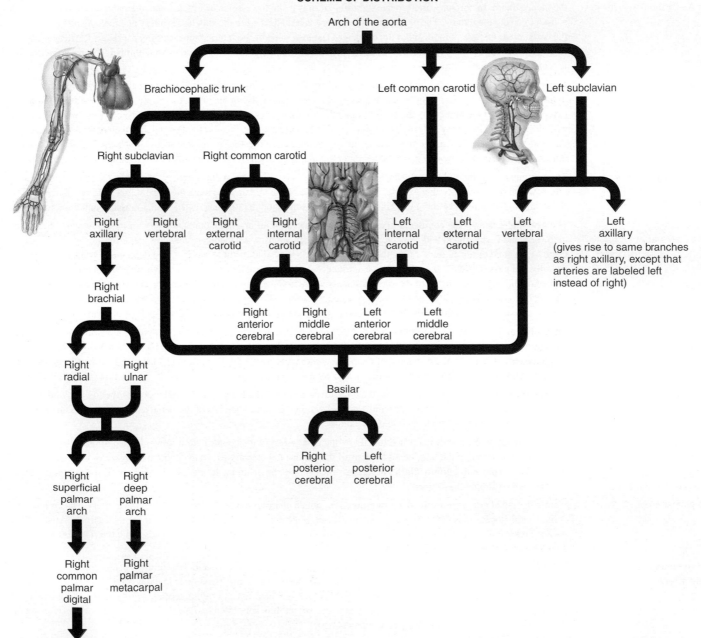

surfaces of the occipital lobe of the cerebrum, deep masses of gray matter within the cerebrum, and midbrain. The posterior cerebral arteries are connected with the internal carotid arteries by the **posterior communicating arteries.** The **anterior communicating arteries** connect the anterior cerebral arteries. The **internal carotid arteries** are also considered part of the cerebral arterial circle. The functions of the cerebral arterial circle are to equalize blood pressure to the brain and provide alternate routes for blood flow to the brain, should the arteries become damaged.

LEFT COMMON CAROTID ARTERY	See description in the introduction to this exhibit.
LEFT SUBCLAVIAN ARTERY	See description in the introduction to this exhibit.

Figure 21.19 Arch of the aorta and its branches. Note in (c) the arteries that constitute the cerebral arterial circle (circle of Willis).

The arch of the aorta ends at the level of the intervertebral disc between the fourth and fifth thoracic vertebrae.

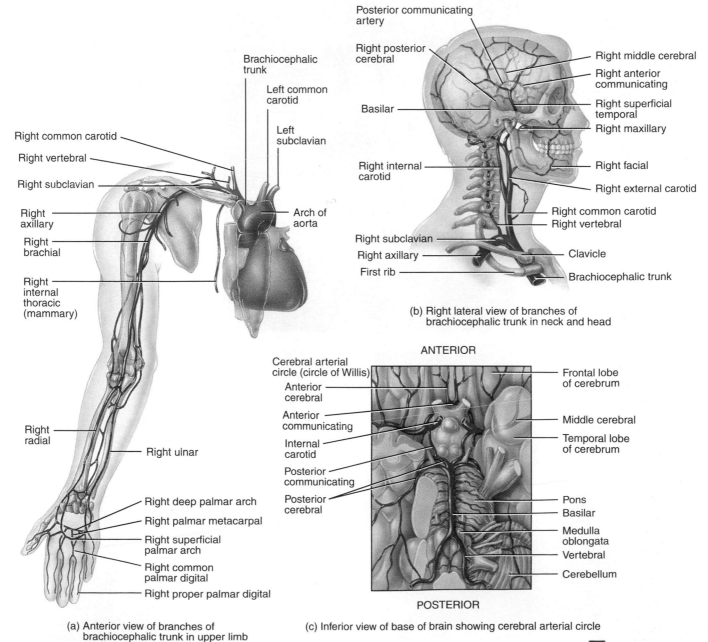

(a) Anterior view of branches of brachiocephalic trunk in upper limb

(b) Right lateral view of branches of brachiocephalic trunk in neck and head

(c) Inferior view of base of brain showing cerebral arterial circle

EXHIBIT 21.3 CONTINUES

EXHIBIT 21.3 **791**

EXHIBIT 21.3 The Arch of the Aorta CONTINUED

The first and largest branch from the arch of the aorta is the **brachio-cephalic trunk** (brā′-kē-ō-se-FAL-ik; *brachio-* = arm; *-cephalic* = head). It extends superiorly, bending slightly to the right, and divides at the right sternoclavicular joint to form the right subclavian artery and right common carotid artery. The second branch from the arch of the aorta is the **left common carotid artery** (ka-ROT-id), which divides into the same branches with the same names as the right common carotid artery. The third branch from the arch of the aorta is the **left subclavian artery** (sub-KLĀ-vē-an), which distributes blood to the left vertebral artery and vessels of the left upper limb. Arteries branching from the left subclavian artery are similar in distribution and name to those branching from the right subclavian artery. The following description focuses on the principal arteries originating from the brachiocephalic trunk.

◉ **CHECKPOINT**

What general regions do the arteries that arise from the arch of the aorta supply?

● **FIGURE 21.19 CONTINUED** ▶

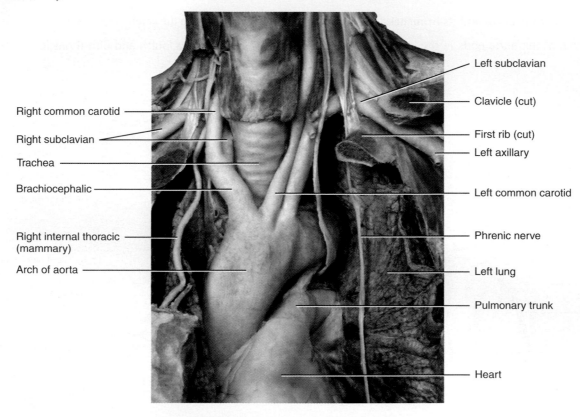

(d) Anterior view, branches of arch of aorta

? **What are the three major branches of the arch of the aorta, in order of their origination?**

EXHIBIT 21.4 Thoracic Aorta

OBJECTIVE

- Identify the visceral and parietal branches of the thoracic aorta.

The **thoracic aorta** is about 20 cm (8 in.) long and is a continuation of the arch of the aorta (Figure 21.20). It begins at the level of the intervertebral disc between the fourth and fifth thoracic vertebrae, where it lies to the left of the vertebral column. As it descends, it moves closer to the midline and extends through an opening in the diaphragm (aortic hiatus), which is located anterior to the vertebral column at the level of the intervertebral disc between the twelfth thoracic and first lumbar vertebrae.

Along its course, the thoracic aorta sends off numerous small arteries, **visceral branches** to viscera, and **parietal branches** to body wall structures.

CHECKPOINT

What general regions do the visceral and parietal branches of the thoracic aorta supply?

BRANCH	DESCRIPTION AND REGION SUPPLIED
VISCERAL	
Pericardial arteries (per'-i-KAR-dē-al; *peri-* = around; *cardia-* = heart)	Two or three tiny **pericardial arteries** supply blood to the pericardium.
Bronchial arteries (BRONG-kē-al = windpipe)	One right and two left **bronchial arteries** supply the bronchial tubes, pleurae, bronchial lymph nodes, and esophagus. (The right bronchial artery arises from the third posterior intercostal artery; the two left bronchial arteries arise from the thoracic aorta.)
Esophageal arteries (e-sof'-a-JĒ-al; *eso-* = to carry; *phage-* = food)	Four or five **esophageal arteries** supply the esophagus.
Mediastinal arteries (mē'-dē-as-TĪ-nal)	Numerous small **mediastinal arteries** supply blood to structures in the mediastinum.
PARIETAL	
Posterior intercostal arteries (in'-ter-KOS-tal; *inter-* = between; *costa* = rib)	Nine pairs of **posterior intercostal arteries** supply the intercostal, pectoralis major and minor, and serratus anterior muscles; overlying subcutaneous tissue and skin; mammary glands; and vertebrae, meninges, and spinal cord.
Subcostal arteries (sub-KOS-tal; *sub-* = under)	The left and right **subcostal arteries** have a distribution similar to that of the posterior intercostals.
Superior phrenic arteries (FREN-ik = pertaining to the diaphragm)	Small **superior phrenic arteries** supply the superior and posterior surfaces of the diaphragm.

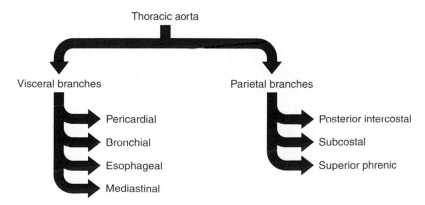

SCHEME OF DISTRIBUTION

Thoracic aorta

Visceral branches
- Pericardial
- Bronchial
- Esophageal
- Mediastinal

Parietal branches
- Posterior intercostal
- Subcostal
- Superior phrenic

EXHIBIT 21.4 CONTINUES

EXHIBIT 21.4 **793**

EXHIBIT 21.4 Thoracic Aorta CONTINUED

Figure 21.20 Thoracic aorta and abdominal aorta and their principal branches.

The thoracic aorta is the continuation of the ascending aorta.

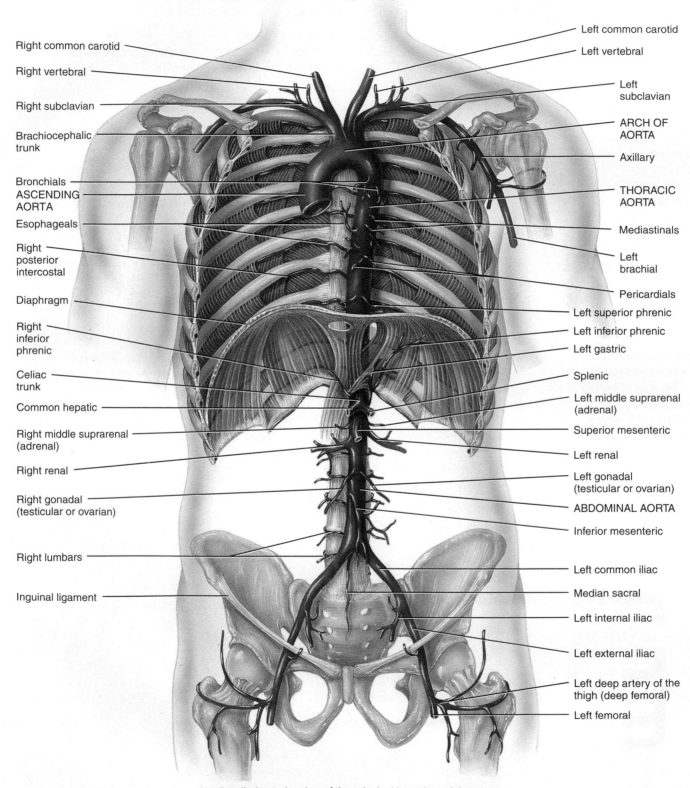

Right common carotid

Right vertebral

Right subclavian

Brachiocephalic trunk

Bronchials
ASCENDING
AORTA

Esophageals

Right posterior intercostal

Diaphragm

Right inferior phrenic

Celiac trunk

Common hepatic

Right middle suprarenal (adrenal)

Right renal

Right gonadal (testicular or ovarian)

Right lumbars

Inguinal ligament

Left common carotid

Left vertebral

Left subclavian

ARCH OF AORTA

Axillary

THORACIC AORTA

Mediastinals

Left brachial

Pericardials

Left superior phrenic

Left inferior phrenic

Left gastric

Splenic

Left middle suprarenal (adrenal)

Superior mesenteric

Left renal

Left gonadal (testicular or ovarian)

ABDOMINAL AORTA

Inferior mesenteric

Left common iliac

Median sacral

Left internal iliac

Left external iliac

Left deep artery of the thigh (deep femoral)

Left femoral

(b) Detailed anterior view of the principal branches of the aorta

Where does the thoracic aorta begin?

EXHIBIT 21.5

Abdominal Aorta

EXHIBIT 21.5

OBJECTIVE

• Identify the visceral and parietal branches of the abdominal aorta.

The **abdominal aorta** is the continuation of the thoracic aorta (Figure 21.21 on pages 797–798). It begins at the aortic hiatus in the diaphragm and ends at about the level of the fourth lumbar vertebra, where it divides into the right and left common iliac arteries. The abdominal aorta lies anterior to the vertebral column.

As with the thoracic aorta, the abdominal aorta gives off **visceral** and **parietal branches.** The unpaired visceral branches arise from the anterior surface of the aorta and include the **celiac trunk** and the **superior mesenteric** and **inferior mesenteric arteries** (see Figure 21.20).

BRANCH	DESCRIPTION AND REGION SUPPLIED
UNPAIRED VISCERAL BRANCHES	
Celiac trunk (SĒ-lē-ak)	The **celiac trunk (artery)** is the first visceral branch from the aorta inferior to the diaphragm, at about the level of the twelfth thoracic vertebra (Figure 21.21a). Almost immediately the celiac trunk divides into three branches: the left gastric, splenic, and common hepatic arteries (Figure 21.21a). 1. The **left gastric artery** (GAS-trik = stomach) is the smallest of the three branches. It passes superiorly to the left toward the esophagus and then turns to follow the lesser curvature of the stomach. It supplies the stomach and esophagus. 2. The **splenic artery** (SPLĒN-ik = spleen) is the largest branch of the celiac trunk. It arises from the left side of the celiac trunk distal to the left gastric artery and passes horizontally to the left along the pancreas. Before reaching the spleen, it gives rise to three arteries: • **Pancreatic artery** (pan-krē-AT-ik), which supplies the pancreas. • **Left gastroepiploic artery** (gas′-trō-ep′-i-PLŌ-ik; *epiplo-* = omentum), which supplies the stomach and greater omentum. • **Short gastric artery,** which supplies the stomach. 3. The **common hepatic artery** (he-PAT-ik = liver) is intermediate in size between the left gastric and splenic arteries. Unlike the other two branches of the celiac trunk, the common hepatic artery arises from the right side. It gives rise to three arteries: • **Proper hepatic artery,** which supplies the liver, gallbladder, and stomach. • **Right gastric artery,** which supplies the stomach. • **Gastroduodenal artery** (gas′-trō-doo′-ō-DĒ-nal), which supplies the stomach, duodenum of the small intestine, pancreas, and greater omentum.
Superior mesenteric artery (MES-en-ter′-ik; *meso-* = middle; *enteric* = pertaining to the intestines)	The **superior mesenteric artery** (Figure 21.21b) arises from the anterior surface of the abdominal aorta about 1 cm inferior to the celiac trunk at the level of the first lumbar vertebra. It extends inferiorly and anteriorly and between the layers of mesentery, which is a portion of the peritoneum that attaches the small intestine to the posterior abdominal wall. It anastomoses extensively and has five branches: 1. The **inferior pancreaticoduodenal artery** (pan′-krē-at′-i-kō-doo′-ō-DĒ-nal) supplies the pancreas and duodenum. 2. The **jejunal** (je-JOO-nal) and **ileal arteries** (IL-ē-al) supply the jejunum and ileum of the small intestine, respectively. 3. The **ileocolic artery** (il′ē-ō-KŌL-ik) supplies the ileum and ascending colon of the large intestine. 4. The **right colic artery** (KŌL-ik) supplies the ascending colon. 5. The **middle colic artery** supplies the transverse colon of the large intestine.
Inferior mesenteric artery	The **inferior mesenteric artery** (Figure 21.21c) arises from the anterior aspect of the abdominal aorta at the level of the third lumbar vertebra and then passes inferiorly to the left of the aorta. It anastomoses extensively and has three branches: 1. The **left colic artery** supplies the transverse colon and descending colon of the large intestine. 2. The **sigmoid arteries** (SIG-moyd) supply the descending colon and sigmoid colon of the large intestine. 3. The **superior rectal artery** (REK-tal) supplies the rectum of the large intestine.
PAIRED VISCERAL BRANCHES	
Suprarenal arteries (soo′-pra-RĒ-nal; *supra-* = above; *ren-* = kidney)	Although there are three pairs of **suprarenal (adrenal) arteries** that supply the adrenal (suprarenal) glands (superior, middle, and inferior), only the middle pair originates directly from the abdominal aorta (see Figure 21.20). The middle suprarenal arteries arise at the level of the first lumbar vertebra at or superior to the renal arteries. The superior suprarenal arteries arise from the inferior phrenic artery, and the inferior suprarenal arteries originate from the renal arteries.
Renal arteries (RĒ-nal; *ren-* = kidney)	The right and left **renal arteries** usually arise from the lateral aspects of the abdominal aorta at the superior border of the second lumbar vertebra, about 1 cm inferior to the superior mesenteric artery (see Figure 21.20). The right renal artery, which is longer than the left, arises slightly lower than the left and passes posterior to the right renal vein and inferior vena cava. The left renal artery is posterior to the left renal vein and is crossed by the inferior mesenteric vein. The renal arteries carry blood to the kidneys, adrenal (suprarenal) glands, and ureters. Their distribution within the kidneys is discussed in Chapter 26.
Gonadal (gō-NAD-al; *gon-* = seed) [testicular (test-TIK-ū-lar) or ovarian arteries (ō-VAR-ē-an)]	The **gonadal arteries** arise from the abdominal aorta at the level of the second lumbar vertebra just inferior to the renal arteries (see Figure 21.20). In males, the gonadal arteries are specifically referred to as the **testicular arteries.** They pass through the inguinal canal and supply the testes, epididymis, and ureters. In females, the gonadal arteries are called the **ovarian arteries.** They are much shorter than the testicular arteries and supply the ovaries, uterine (fallopian) tubes, and ureters.

EXHIBIT 21.5 CONTINUES ▶

EXHIBIT 21.5   **795**

EXHIBIT 21.5 Abdominal Aorta CONTINUED

The paired visceral branches arise from the lateral surfaces of the aorta and include the **suprarenal, renal,** and **gonadal arteries.** The unpaired parietal branch is the **median sacral artery.** The paired parietal branches arise from the posterolateral surfaces of the aorta and include the **inferior phrenic** and **lumbar arteries.**

⦿ **CHECKPOINT**
Name the paired visceral and parietal branches and the unpaired visceral and parietal branches of the abdominal aorta, and indicate the general regions they supply.

UNPAIRED PARIETAL BRANCH	
Median sacral artery (SĀ-kral = pertaining to the sacrum)	The **median sacral artery** arises from the posterior surface of the abdominal aorta about 1 cm superior to the bifurcation (division into two branches) of the aorta into the right and left common iliac arteries (see Figure 21.20). The median sacral artery supplies the sacrum and coccyx.

PAIRED PARIETAL BRANCHES	
Inferior phrenic arteries (FREN-ik = pertaining to the diaphragm)	The **inferior phrenic arteries** are the first paired branches of the abdominal aorta, immediately superior to the origin of the celiac trunk (see Figure 21.20). (They may also arise from the renal arteries.) The inferior phrenic arteries are distributed to the inferior surface of the diaphragm and adrenal (suprarenal) glands.
Lumbar arteries (LUM-bar = pertaining to the loin)	The four pairs of **lumbar arteries** arise from the posterolateral surface of the abdominal aorta (see Figure 21.20). They supply the lumbar vertebrae, spinal cord and its meninges, and the muscles and skin of the lumbar region of the back.

SCHEME OF DISTRIBUTION

Figure 21.21 Abdominal aorta and its principal branches.

🔓 The abdominal aorta is the continuation of the thoracic aorta.

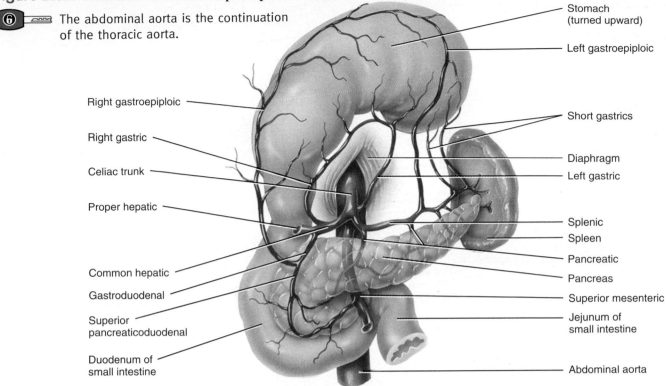

Stomach (turned upward)
Left gastroepiploic
Right gastroepiploic
Short gastrics
Right gastric
Diaphragm
Celiac trunk
Left gastric
Proper hepatic
Splenic
Spleen
Pancreatic
Common hepatic
Pancreas
Gastroduodenal
Superior mesenteric
Superior pancreaticoduodenal
Jejunum of small intestine
Duodenum of small intestine
Abdominal aorta

(a) Anterior view of celiac trunk and its branches

Transverse colon of large intestine
Superior mesenteric
Middle colic
Inferior pancreaticoduodenal
Jejunum
Right colic
Jejunals
Ileocolic
Inferior mesenteric
Ileals
Ascending colon of large intestine
Ileum of small intestine
Appendix

(b) Anterior view of superior mesenteric artery and its branches

⬤ **EXHIBIT 21.5 CONTINUES** ▶

EXHIBIT 21.5   **797**

EXHIBIT 21.5   Abdominal Aorta CONTINUED

Transverse colon
of large intestine

Superior mesenteric

Inferior
mesenteric

Left colic

Abdominal aorta

Common iliac

Descending colon
of large intestine

Sigmoids

Superior rectal

Sigmoid colon

Rectum of
large intestine

(c) Anterior view of inferior mesenteric artery and its branches

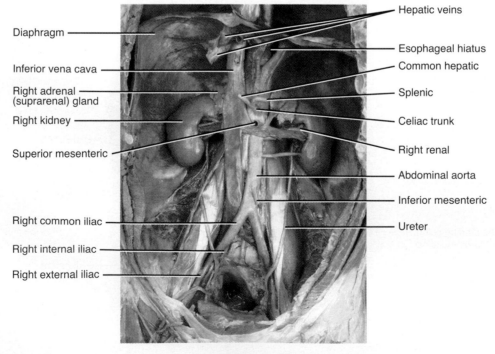

Diaphragm

Hepatic veins

Inferior vena cava

Esophageal hiatus

Right adrenal
(suprarenal) gland

Common hepatic

Splenic

Right kidney

Celiac trunk

Superior mesenteric

Right renal

Abdominal aorta

Inferior mesenteric

Right common iliac

Ureter

Right internal iliac

Right external iliac

(d) Anterior view of arteries of abdomen and pelvis

? Where does the abdominal aorta begin?

EXHIBIT 21.6   Arteries of the Pelvis and Lower Limbs

◉ OBJECTIVE

• Identify the two major branches of the common iliac arteries.

The abdominal aorta ends by dividing into the right and left **common iliac arteries** (Figure 21.22 on page 801). These, in turn, divide into the **internal** and **external iliac arteries.** In sequence, the external iliacs be- come the **femoral arteries** in the thighs, the **popliteal arteries** poste- rior to the knee, and the **anterior** and **posterior tibial arteries** in the legs.

◉ CHECKPOINT

What general regions do the internal and external iliac arteries supply?

BRANCH	DESCRIPTION AND REGION SUPPLIED
COMMON ILIAC ARTERIES (IL-ē-AK = pertaining to the ilium)	At about the level of the fourth lumbar vertebra, the abdominal aorta divides into the right and left **common iliac arteries,** the terminal branches of the abdominal aorta. Each passes inferiorly about 5 cm (2 in.) and gives rise to two branches: internal iliac and external iliac arteries. The general distribution of the common iliac arteries is to the pelvis, external genitals, and lower limbs.
Internal iliac arteries	The **internal iliac (hypogastric) arteries** are the primary arteries of the pelvis. They begin at the bifurcation (division into two branches) of the common iliac arteries anterior to the sacroiliac joint at the level of the lumbosacral intervertebral disc. They pass posteromedially as they descend in the pelvis and divide into anterior and posterior divisions. The general distribution of the internal iliac arteries is to the pelvis, buttocks, external genitals, and thigh.
External iliac arteries	The **external iliac arteries** are larger than the internal iliac arteries. Like the internal iliac arteries, they begin at the bifurcation of the common iliac arteries. They descend along the medial border of the psoas major muscles following the pelvic brim, pass posterior to the midportion of the inguinal ligaments, and become the femoral arteries. The general distribution of the external iliac arteries is to the lower limbs. Specifically, branches of the external iliac arteries supply the muscles of the anterior abdominal wall, the cremaster muscle in males and the round ligament of the uterus in females, and the lower limbs.
Femoral arteries (FEM-o-ral = pertaining to the thigh)	The **femoral arteries** descend along the anteromedial aspects of the thighs to the junction of the middle and lower third of the thighs. Here they pass through an opening in the tendon of the adductor magnus muscle, then emerge posterior to the femurs as the popliteal arteries. A pulse may be felt in the femoral artery just inferior to the inguinal ligament. Recall from Chapter 11 that the femoral artery, along with the femoral vein and nerve and deep inguinal lymph nodes, are located in the *femoral triangle* (see Figure 11.20a on page 395). The general distribution of the femoral arteries is to the lower abdominal wall, groin, external genitals, and muscles of the thigh. A major branch of the femoral artery, the **deep artery of the thigh (deep femoral),** supplies most of the muscles of the thigh: quadriceps femoris, adductors, and hamstrings. Recall that in cardiac catheterization a catheter is inserted through a blood vessel and advanced into the major vessels and heart chamber. A catheter often contains a measuring instrument or other device at its tip. To reach the left side of the heart, the catheter is inserted into the femoral artery and passed into the aorta to the coronary arteries or heart chamber.
Popliteal arteries (pop'-li-TĒ-al = posterior surface of the knee)	The **popliteal arteries** are the continuation of the femoral arteries through the popliteal fossa (space behind the knee). They descend to the inferior border of the popliteus muscles, where they divide into the anterior and posterior tibial arteries. A pulse may be detected in the popliteal arteries. In addition to supplying the adductor magnus and hamstring muscles and the skin on the posterior aspect of the legs, branches of the popliteal arteries also supply the gastrocnemius, soleus, and plantaris muscles of the calf, knee joint, femur, patella, and fibula.
Anterior tibial arteries (TIB-ē-al = pertaining to the shin bone)	The **anterior tibial arteries** descend from the bifurcation of the popliteal arteries. They are smaller than the posterior tibial arteries. The anterior tibial arteries descend through the anterior muscular compartment of the leg. They pass through the interosseous membrane that connects the tibia and fibula, lateral to the tibia. The anterior tibial arteries supply the knee joints, anterior compartment muscles of the legs, skin over the anterior aspects of the legs, and ankle joints. At the ankles, the anterior tibial arteries become the **dorsal arteries of the foot (dorsalis pedis arteries).** A pulse in this artery may be taken to evaluate the peripheral vascular system. The dorsal arteries of the foot supply the muscles, skin, and joints on the dorsal aspects of the feet. On the dorsum of the feet, the dorsal arteries of the foot give off a transverse branch at the first (medial) cuneiform bone called the **arcuate arteries** (*arcuat-* = bowed) that run laterally over the bases of the metatarsals. From the arcuate arteries branch the **dorsal metatarsal arteries,** which supply the feet. The dorsal metatarsal arteries terminate by dividing into the **dorsal digital arteries,** which supply the toes.
Posterior tibial arteries	The **posterior tibial arteries,** the direct continuations of the popliteal arteries, descend from the bifurcation of the popliteal arteries. They pass down the posterior muscular compartment of the legs posterior to the medial malleolus of the tibia. They terminate by dividing into the medial and lateral plantar arteries. Their general distribution is to the muscles, bones, and joints of the leg and foot. Major branches of the posterior tibial arteries are the **fibular (peroneal) arteries,** which supply the fibularis, soleus, tibialis posterior, and flexor hallucis muscles. They also supply the fibula, tarsus, and lateral aspect of the heel. The bifurcation of the posterior tibial arteries into the medial and lateral plantar arteries occurs deep to the flexor retinaculum on the medial side of the feet. The **medial plantar arteries** (PLAN-tar = sole) supply the abductor hallucis and flexor digitorum brevis muscles and the toes. The **lateral plantar arteries** unite with a branch of the dorsal arteries of the foot to form the **plantar arch.** The arch begins at the base of the fifth metatarsal and extends medially across the metacarpals. As the arch crosses the foot, it gives off **plantar metatarsal arteries,** which supply the feet. These terminate by dividing into **plantar digital arteries,** which supply the toes.

◉ **EXHIBIT 21.6 CONTINUES** ▶

EXHIBIT 21.6   **799**

EXHIBIT 21.6 Arteries of the Pelvis and Lower Limbs CONTINUED

SCHEME OF DISTRIBUTION

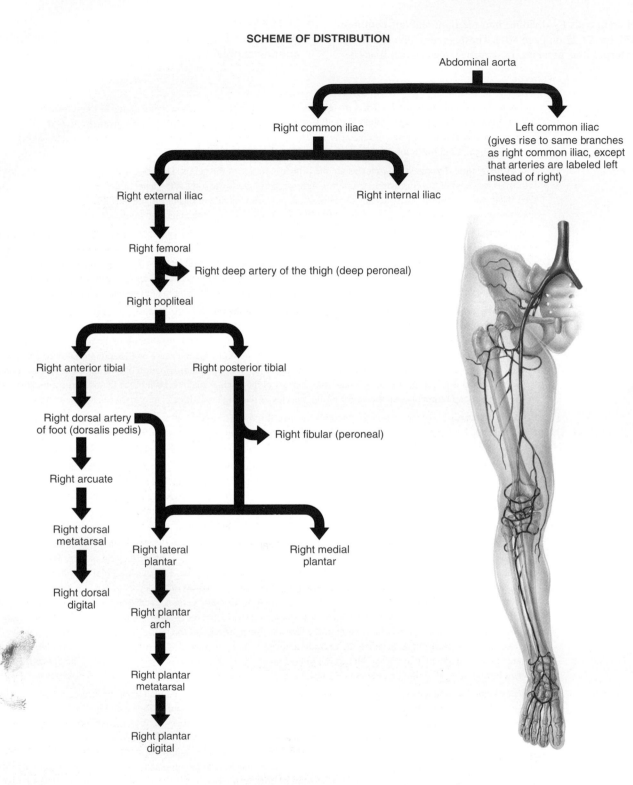

Abdominal aorta

Right common iliac

Left common iliac
(gives rise to same branches
as right common iliac, except
that arteries are labeled left
instead of right)

Right external iliac

Right internal iliac

Right femoral

Right deep artery of the thigh (deep peroneal)

Right popliteal

Right anterior tibial

Right posterior tibial

Right dorsal artery
of foot (dorsalis pedis)

Right fibular (peroneal)

Right arcuate

Right dorsal
metatarsal

Right lateral
plantar

Right medial
plantar

Right dorsal
digital

Right plantar
arch

Right plantar
metatarsal

Right plantar
digital

Figure 21.22 **Arteries of the pelvis and right lower limb.**

The internal iliac arteries carry most of the blood supply to the pelvic viscera and wall.

Right common iliac

Right internal iliac

Right external iliac

Abdominal aorta

Left common iliac

Right deep artery of the thigh (deep femoral)

Right femoral

Right popliteal

Right anterior tibial

Right posterior tibial

Right fibular (peroneal)

Right arcuate

Right dorsal artery of foot (dorsalis pedis)

Right dorsal metatarsal

Right dorsal digital

Right lateral plantar

Right medial plantar

Right plantar arch

Right plantar metatarsal

Right plantar digital

(a) Anterior view

(b) Posterior view

? **At what point does the abdominal aorta divide into the common iliac arteries?**

EXHIBIT 21.6 **801**

EXHIBIT 21.7   Veins of the Systemic Circulation

◉ OBJECTIVE

- Identify the three systemic veins that return deoxygenated blood to the heart.

As you have already learned, arteries distribute blood to various parts of the body, and veins drain blood away from them. For the most part, arteries are deep; veins may be superficial or deep. Superficial veins are located just beneath the skin and can be seen easily. Because there are no large superficial arteries, the names of superficial veins do not correspond to those of arteries. Superficial veins are clinically important as sites for withdrawing blood or giving injections. Deep veins generally travel alongside arteries and usually bear the same name. Arteries usually follow definite pathways; veins are more difficult to follow because they connect in irregular networks in which many tributaries merge to form a large vein. Although only one systemic artery, the aorta, takes oxygenated blood away from the heart (left ventricle), three systemic veins, the **coronary sinus, superior vena cava,** and **inferior vena cava,** return deoxygenated blood to the heart (right atrium) (Figure 21.23). The coronary sinus receives blood from the cardiac veins; the superior vena cava receives blood from other veins superior to the diaphragm, except the air sacs (alveoli) of the lungs; the inferior vena cava receives blood from veins inferior to the diaphragm.

◉ CHECKPOINT

What are the three tributaries of the coronary sinus?

VEIN	DESCRIPTION AND REGION DRAINED
Coronary sinus (KOR-ō-nar-ē; *corona* = crown)	The **coronary sinus** is the main vein of the heart; it receives almost all venous blood from the myocardium. It is located in the coronary sulcus (see Figure 20.3c on page 722) and opens into the right atrium between the orifice of the inferior vena cava and the tricuspid valve. It is a wide venous channel into which three veins drain. It receives the **great cardiac vein** (in the anterior interventricular sulcus) into its left end, and the **middle cardiac vein** (in the posterior interventricular sulcus) and the **small cardiac vein** into its right end. Several **anterior cardiac veins** drain directly into the right atrium.
Superior vena cava (SVC) (VĒ-na CĀ-va; *vena* = vein; *cava* = cavelike)	The **superior vena cava** is about 7.5 cm (3 in.) long and 2 cm (1 in.) in diameter and empties its blood into the superior part of the right atrium. It begins posterior to the right first costal cartilage by the union of the right and left brachiocephalic veins and ends at the level of the right third costal cartilage, where it enters the right atrium. The SVC drains the head, neck, chest, and free upper limbs.
Inferior vena cava (IVC)	The **inferior vena cava** is the largest vein in the body, about 3.5 cm (1.4 in.) in diameter. It begins anterior to the fifth lumbar vertebra by the union of the common iliac veins, ascends behind the peritoneum to the right of the midline, pierces the caval opening of the diaphragm at the level of the eighth thoracic vertebra, and enters the inferior part of the right atrium. The IVC drains the abdomen, pelvis, and free lower limbs. The inferior vena cava is commonly compressed during the later stages of pregnancy by the enlarging uterus, producing edema of the ankles and feet and temporary varicose veins.

Figure 21.23 Principal veins.

Deoxygenated blood returns to the heart via the superior vena cava, inferior vena cava, and the coronary sinus.

Superior sagittal sinus
Inferior sagittal sinus
Straight sinus
Right transverse sinus
Sigmoid sinus

Right internal jugular
Right external jugular
Right subclavian
Right brachiocephalic
Superior vena cava
Right axillary
Right cephalic
Right hepatic
Right brachial
Right median cubital
Right basilic
Right radial
Right median antebrachial
Right ulnar
Right palmar venous plexus
Right palmar digital
Right proper palmar digital

Pulmonary trunk
Coronary sinus
Great cardiac
Hepatic portal
Splenic
Superior mesenteric
Left renal
Inferior mesenteric
Inferior vena cava
Left common iliac
Left internal iliac
Left external iliac

Left femoral
Left great saphenous
Left popliteal
Left small saphenous
Left anterior tibial
Left posterior tibial
Left dorsal venous arch
Left dorsal metatarsal
Left dorsal digital

Overall anterior view of the principal veins

? **Which general regions of the body are drained by the superior vena cava and the inferior vena cava?**

EXHIBIT 21.7 **803**

EXHIBIT 21.8 Veins of the Head and Neck

- Identify the three major veins that drain blood from the head.

Most blood draining from the head passes into three pairs of veins: the **internal jugular, external jugular,** and **vertebral veins** (Figure 21.24)**.** Within the brain, all veins drain into dural venous sinuses and then into the internal jugular veins. Dural venous sinuses are endothelial-lined venous channels between layers of the cranial dura mater.

◉ **CHECKPOINT**

Which general areas are drained by the internal jugular, external jugular, and vertebral veins?

VEIN	DESCRIPTION AND REGION DRAINED
Internal jugular veins (JUG-ū-lar; *jugular* = throat)	The flow of blood from the dural venous sinuses into the internal jugular veins is as follows (Figure 21.24): The **superior sagittal sinus** (SAJ-i-tal = straight) begins at the frontal bone, where it receives a vein from the nasal cavity, and passes posteriorly to the occipital bone. Along its course, it receives blood from the superior, medial, and lateral aspects of the cerebral hemispheres, meninges, and cranial bones. The superior sagittal sinus usually turns to the right and drains into the right transverse sinus. The **inferior sagittal sinus** is much smaller than the superior sagittal sinus; it begins posterior to the attachment of the falx cerebri and receives the great cerebral vein to become the straight sinus. The great cerebral vein drains the deeper parts of the brain. Along its course the inferior sagittal sinus also receives tributaries from the superior and medial aspects of the cerebral hemispheres.
	The **straight sinus** runs in the tentorium cerebelli and is formed by the union of the inferior sagittal sinus and the great cerebral vein. The straight sinus also receives blood from the cerebellum and usually drains into the left transverse sinus.
	The **transverse sinuses** begin near the occipital bone, pass laterally and anteriorly, and become the sigmoid sinuses near the temporal bone. The transverse sinuses receive blood from the cerebrum, cerebellum, and cranial bones.
	The **sigmoid sinuses** (SIG-moyd = S-shaped) are located along the temporal bone. They pass through the jugular foramina, where they terminate in the internal jugular veins. The sigmoid sinuses drain the transverse sinuses.
	The **cavernous sinuses** (KAV-er-nus = cavelike) are located on either side of the sphenoid bone. They receive blood from the ophthalmic veins from the orbits, and from the cerebral veins from the cerebral hemispheres. They ultimately empty into the transverse sinuses and internal jugular veins. The cavernous sinuses are unique because they have nerves and a major blood vessel passing through them on their way to the orbit and face. The oculomotor (III) nerve, trochlear (IV) nerve, and ophthalmic and maxillary branches of the trigeminal (V) nerve as well as the internal carotid arteries pass through the cavernous sinuses.
	The right and left **internal jugular veins** pass inferiorly on either side of the neck lateral to the internal carotid and common carotid arteries. They then unite with the subclavian veins posterior to the clavicles at the sternoclavicular joints to form the right and left brachiocephalic veins (brā′-kē-ō-se-FAL-ik; *brachio-* = arm; *cephal-* = head). From here blood flows into the superior vena cava. The general structures drained by the internal jugular veins are the brain (through the dural venous sinuses), face, and neck.
External jugular veins	The right and left **external jugular veins** begin in the parotid glands near the angle of the mandible. They are superficial veins that descend through the neck across the sternocleidomastoid muscles. They terminate at a point opposite the middle of the clavicles, where they empty into the subclavian veins. The general structures drained by the external jugular veins are external to the cranium, such as the scalp and superficial and deep regions of the face. When venous pressure rises, for example, during heavy coughing or straining or in cases of heart failure, the external jugular veins become very prominent along the side of the neck.
Vertebral veins (VER-te-bral; *vertebra* = vertebrae)	The right and left **vertebral veins** originate inferior to the occipital condyles. They descend through successive transverse foramina of the first six cervical vertebrae and emerge from the foramina of the sixth cervical vertebra to enter the brachiocephalic veins in the root of the neck. The vertebral veins drain deep structures in the neck such as the cervical vertebrae, cervical spinal cord, and some neck muscles.

SCHEME OF DRAINAGE

Figure 21.24 Principal veins of the head and neck.

Blood draining from the head passes into the internal jugular, external jugular, and vertebral veins.

Right lateral view

❓ Into which veins in the neck does all venous blood in the brain drain?

EXHIBIT 21.8 **805**

EXHIBIT 21.9   Veins of the Upper Limbs

■ OBJECTIVE

• Identify the principal veins that drain the upper limbs.

Both superficial and deep veins return blood from the upper limbs to the heart (Figure 21.25 on page 808–809). **Superficial veins** are located just deep to the skin and are often visible. They anastomose ex-

tensively with one another and with deep veins, and they do not accompany arteries. Superficial veins are larger than deep veins and return most of the blood from the upper limbs. **Deep veins** are located deep in the body. They usually accompany arteries and have the same names as the corresponding arteries. Both superficial and deep veins have valves, but valves are more numerous in the deep veins.

VEIN	DESCRIPTION AND REGION DRAINED
SUPERFICIAL	
Cephalic veins (se-FAL-ik = pertaining to the head)	The principal superficial veins that drain the upper limbs are the cephalic and basilic veins. They originate in the hand and convey blood from the smaller superficial veins into the axillary veins. The **cephalic veins** begin on the lateral aspect of the **dorsal venous networks of the hands (dorsal venous arches),** networks of veins on the dorsum of the hands formed by the **dorsal metacarpal veins** (Figure 21.25a). These veins, in turn, drain the **dorsal digital veins,** which pass along the sides of the fingers. Following their formation from the dorsal venous networks of the hands, the cephalic veins arch around the radial side of the forearms to the anterior surface and ascend through the entire limbs along the anterolateral surface. The cephalic veins end where they join the axillary veins, just inferior to the clavicles. **Accessory cephalic veins** originate either from a venous plexus on the dorsum of the forearms or from the medial aspects of the dorsal venous networks of the hands, and unite with the cephalic veins just inferior to the elbow. The cephalic veins drain blood from the lateral aspect of the upper limbs.
Basilic veins (ba-SIL-ik = royal, of prime importance)	The **basilic veins** begin on the medial aspects of the dorsal venous networks of the hands and ascend along the posteromedial surface of the forearm and anteromedial surface of the arm (Figure 21.25b). They drain blood from the medial aspects of the upper limbs. Anterior to the elbow, the basilic veins are connected to the cephalic veins by the **median cubital veins** (*cubital* = pertaining to the elbow), which drain the forearm. If veins must be punctured for an injection, transfusion, or removal of a blood sample, the median cubital veins are preferred. After receiving the median cubital veins, the basilic veins continue ascending until they reach the middle of the arm. There they penetrate the tissues deeply and run alongside the brachial arteries until they join the brachial veins. As the basilic and brachial veins merge in the axillary area, they form the axillary veins.
Median antebrachial veins (an′-tē-BRĀ-kē-al; *ante-* = before, in front of; *brachi-* = arm)	The **median antebrachial veins (median veins of the forearm)** begin in the **palmar venous plexuses,** networks of veins on the palms. The plexuses drain the **palmar digital veins** in the fingers. The median antebrachial veins ascend anteriorly in the forearms to join the basilic or median cubital veins, sometimes both. They drain the palms and forearms.
DEEP	
Radial veins (RĀ-dē-al = pertaining to the radius)	The paired **radial veins** begin at the **deep palmar venous arches** (Figure 21.25c). These arches drain the **palmar metacarpal veins** in the palms. The radial veins drain the lateral aspects of the forearms and pass alongside the radial arteries. Just inferior to the elbow joint, the radial veins unite with the ulnar veins to form the brachial veins.
Ulnar veins (UL-nar = pertaining to the ulna)	The paired **ulnar veins,** which are larger than the radial veins, begin at the **superficial palmar venous arches.** These arches drain the **common palmar digital veins** and the **proper palmar digital veins** in the fingers. The ulnar veins drain the medial aspect of the forearms, pass alongside the ulnar arteries, and join with the radial veins to form the brachial veins.
Brachial veins (BRĀ-kē-al; *brachi-* = arm)	The paired **brachial veins** accompany the brachial arteries. They drain the forearms, elbow joints, arms, and humerus. They pass superiorly and join with the basilic veins to form the axillary veins.
Axillary veins (AK-sil-ār-ē; *axilla* = armpit)	The **axillary veins** ascend to the outer borders of the first ribs, where they become the subclavian veins. The axillary veins receive tributaries that correspond to the branches of the axillary arteries. The axillary veins drain the arms, axillas, and superolateral chest wall.
Subclavian veins (sub-KLĀ-vē-an; *sub-* = under; *clavian* = pertaining to the clavicle)	The **subclavian veins** are continuations of the axillary veins that terminate at the sternal end of the clavicles, where they unite with the internal jugular veins to form the brachiocephalic veins. The subclavian veins drain the arms, neck, and thoracic wall. The thoracic duct of the lymphatic system delivers lymph into the junction between the left subclavian and the left internal jugular veins. The right lymphatic duct delivers lymph into the junction between the right subclavian and right internal jugular veins (see Figure 22.3 on page 835). In a procedure called *central line placement,* the right subclavian vein is frequently used to administer nutrients and medication and measure venous pressure.

Where do the cephalic, basilic, median antebrachial, radial,
and ulnar veins originate?

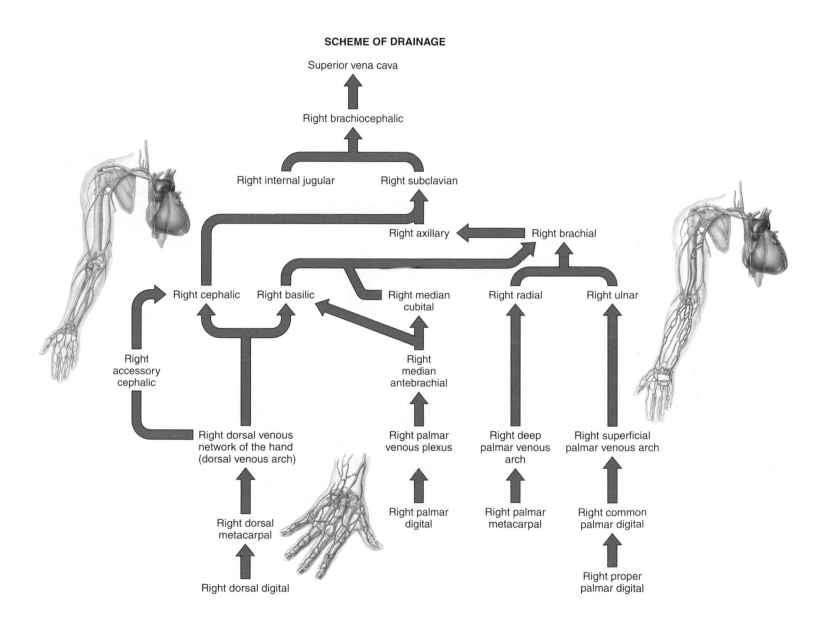

SCHEME OF DRAINAGE

Superior vena cava

Right brachiocephalic

Right internal jugular   Right subclavian

Right axillary   Right brachial

Right cephalic   Right basilic   Right median cubital   Right radial   Right ulnar

Right accessory cephalic

Right median antebrachial

Right dorsal venous network of the hand (dorsal venous arch)   Right palmar venous plexus   Right deep palmar venous arch   Right superficial palmar venous arch

Right dorsal metacarpal   Right palmar digital   Right palmar metacarpal   Right common palmar digital

Right dorsal digital   Right proper palmar digital

EXHIBIT 21.9   Veins of the Upper Limbs CONTINUED

Figure 21.25 **Principal veins of the right upper limb.**

⑥ 🔑 Deep veins usually accompany arteries that have similar names.

Right cephalic

Right dorsal venous network of the hand (dorsal venous arch)

Right dorsal metacarpal

Right dorsal digital

(a) Posterior view of superficial veins of the hand

Right external jugular

Right subclavian

Right internal jugular

Right brachiocephalic

Right axillary

Right basilic

Right cephalic

Superior vena cava

Sternum

Right accessory cephalic

Right median cubital

Right cephalic

Right basilic

Right median antebrachial

Right palmar venous plexus

Right palmar digital

(b) Anterior view of superficial veins

Right external jugular

Right subclavian

Right brachiocephalic

Right axillary

Right brachial

Right internal jugular

Superior vena cava

Right radials

Right ulnars

Right deep palmar venous arch

Right superficial palmar venous arch

Right common palmar digital

Right proper palmar digital

Right palmar metacarpal

(c) Anterior view of deep veins

Biceps brachii muscle

Brachial

Median cubital

Basilic

Cephalic

Median antebrachial

(d) Anteromedial view of superficial veins of arm and forearm

❓ **From which vein in the upper limb is a blood sample often taken?**

EXHIBIT 21.9   **809**

EXHIBIT 21.10 | Veins of the Thorax

OBJECTIVE
• Identify the components of the azygos system of veins.

Although the brachiocephalic veins drain some portions of the thorax, most thoracic structures are drained by a network of veins, called the **azygos system,** that runs on either side of the vertebral column (Figure 21.26). The system consists of three veins—the **azygos, hemiazygos,** and **accessory hemiazygos veins**—that show considerable variation in origin, course, tributaries, anastomoses, and termination. Ultimately they empty into the superior vena cava.

CHECKPOINT
What is the importance of the azygos system relative to the inferior vena cava?

VEIN	DESCRIPTION AND REGION DRAINED
Brachiocephalic vein (brā′-kē-ō-se-FAL-ik; *brachio-* = arm; *cephalic* = pertaining to the head)	The right and left **brachiocephalic veins,** formed by the union of the subclavian and internal jugular veins, drain blood from the head, neck, upper limbs, mammary glands, and superior thorax. The brachiocephalic veins unite to form the superior vena cava. Because the superior vena cava is to the right of the body's midline, the left brachiocephalic vein is longer than the right. The right brachiocephalic vein is anterior and to the right of the brachiocephalic trunk. The left brachiocephalic vein is anterior to the brachiocephalic trunk, the left common carotid and left subclavian arteries, the trachea, the left vagus (X) nerve, and the phrenic nerve.
Azygos system (az-Ī-gos = unpaired)	The **azygos system,** besides collecting blood from the thorax and abdominal wall, may serve as a bypass for the inferior vena cava that drains blood from the lower body. Several small veins directly link the azygos system with the inferior vena cava. Large veins that drain the lower limbs and abdomen conduct blood into the azygos system. If the inferior vena cava or hepatic portal vein becomes obstructed, the azygos system can return blood from the lower body to the superior vena cava.
Azygos vein	The **azygos vein** is anterior to the vertebral column, slightly to the right of the midline. It usually begins at the junction of the right ascending lumbar and right subcostal veins near the diaphragm. At the level of the fourth thoracic vertebra, it arches over the root of the right lung to end in the superior vena cava. Generally, the azygos vein drains the right side of the thoracic wall, thoracic viscera, and abdominal wall. Specifically, the azygos vein receives blood from most of the **right posterior intercostal, hemiazygos, accessory hemiazygos, esophageal, mediastinal, pericardial, and bronchial veins.**
Hemiazygos vein (HEM-ē-az-ī-gos; *hemi-* = half)	The **hemiazygos vein** is anterior to the vertebral column and slightly to the left of the midline. It often begins at the junction of the left ascending lumbar and left subcostal veins. It terminates by joining the azygos vein at about the level of the ninth thoracic vertebra. Generally, the hemiazygos vein drains the left side of the thoracic wall, thoracic viscera, and abdominal wall. Specifically, the hemiazygos vein receives blood from the ninth through eleventh **left posterior intercostal, esophageal, mediastinal,** and sometimes the **accessory hemiazygos veins.**
Accessory hemiazygos vein	The **accessory hemiazygos vein** is also anterior to the vertebral column and to the left of the midline. It begins at the fourth or fifth intercostal space and descends from the fifth to the eighth thoracic vertebra or ends in the hemiazygos vein. It terminates by joining the azygos vein at about the level of the eighth thoracic vertebra. The accessory hemiazygos vein drains the left side of the thoracic wall. It receives blood from the fourth through eighth **left posterior intercostal veins** (the first through third left posterior intercostal veins open into the left brachiocephalic vein), **left bronchial,** and **mediastinal veins.**

SCHEME OF DRAINAGE

Figure 21.26 Principal veins of the thorax, abdomen, and pelvis.

Most thoracic structures are drained by the azygos system of veins.

Right internal jugular
Right external jugular
Right brachiocephalic
Right superior intercostal
Superior vena cava
Right posterior intercostal
Azygos
Mediastinals
Bronchial
Pericardial
Diaphragm
Hepatics
Right suprarenal
Right subcostal
Right renal
Right ascending lumbar
Right gonadal (testicular or ovarian)
Right lumbar
Right common iliac
Right internal iliac
Right external iliac

Left internal jugular
Left external jugular
Left subclavian
Left brachiocephalic
Left superior intercostal
Left axillary
Left cephalic
Left posterior intercostal
Left brachial
Accessory hemiazygos
Left basilic
Esophageals
Left inferior phrenics
Hemiazygos
Left suprarenal
Left renal
Left ascending lumbar
Left gonadal (testicular or ovarian)
Inferior vena cava
Left common iliac
Inguinal ligament
Middle sacral
Left internal iliac
Left external iliac
Left femoral

(a) Anterior view

● **Exhibit 21.10 continues** ▶

EXHIBIT 21.10 **811**

EXHIBIT 21.10   Veins of the Thorax CONTINUED

FIGURE **21.26** CONTINUED

Right superior intercostal vein

Junction of azygos with
superior vena cava

Azygos vein

Right posterior intercostals

Vertebral column

Accessory hemiazygos

Left posterior intercostal

Hemiazygos

Diaphragm (cut)

(b) Anterior view of posterior thoracic wall

? Which vein returns blood from the abdominopelvic viscera to the heart?

EXHIBIT 21.11   Veins of the Abdomen and Pelvis

● OBJECTIVE

• Identify the principal veins that drain the abdomen and pelvis.

Blood from the abdominal and pelvic viscera and abdominal wall returns to the heart via the inferior vena cava. Many small veins enter the inferior vena cava. Most carry return flow from parietal branches of the abdominal aorta, and their names correspond to the names of the arteries (see also Figure 21.26 on page 811).

The inferior vena cava does not receive veins directly from the gastrointestinal tract, spleen, pancreas, and gallbladder. These organs pass their blood into a common vein, the **hepatic portal vein,** which delivers the blood to the liver. The superior mesenteric and splenic veins unite to form the hepatic portal vein (see Figure 21.28). This special flow of venous blood, called the **hepatic portal circulation,** is described shortly. After passing through the liver for processing, blood drains into the hepatic veins, which empty into the inferior vena cava.

VEIN	DESCRIPTION AND REGION DRAINED
Inferior vena cava (VĒ-na CĀ-va; *vena* = vein; *cava* = cavelike)	The two common iliac veins that drain the lower limbs, pelvis, and abdomen unite to form the inferior vena cava. The inferior vena cava extends superiorly through the abdomen and thorax to the right atrium.
Common iliac veins (IL-ē-ak = pertaining to the ilium)	The **common iliac veins** are formed by the union of the internal and external iliac veins anterior to the sacroiliac joint and represent the distal continuation of the inferior vena cava at their bifurcation. The right common iliac vein is much shorter than the left and is also more vertical. Generally, the common iliac veins drain the pelvis, external genitals, and lower limbs.
Internal iliac veins	The **internal iliac veins** begin near the superior portion of the greater sciatic notch and run medial to their corresponding arteries. Generally, the veins drain the thigh, buttocks, external genitals, and pelvis.
External iliac veins	The **external iliac veins** are companions of the internal iliac arteries and begin at the inguinal ligaments as continuations of the femoral veins. They end anterior to the sacroiliac joint where they join with the internal iliac veins to form the common iliac veins. The external iliac veins drain the lower limbs, cremaster muscle in males, and the abdominal wall.
Lumbar veins (LUM-bar = pertaining to the loin)	A series of parallel **lumbar veins,** usually four on each side, drain blood from both sides of the posterior abdominal wall, vertebral canal, spinal cord, and meninges. The lumbar veins run horizontally with the lumbar arteries. The lumbar veins connect at right angles with the right and left **ascending lumbar veins,** which form the origin of the corresponding azygos or hemiazygos vein. The lumbar veins drain blood into the ascending lumbars and then run to the inferior vena cava, where they release the remainder of the flow.
Gonadal veins (gō-NAD-al; *gono* = seed) [**testicular** (tes-TIK-ū-lar) or **ovarian** (ō-VAR-ē-an)]	The **gonadal veins** ascend with the gonadal arteries along the posterior abdominal wall. In the male, the gonadal veins are called the testicular veins. The **testicular veins** drain the testes (the left testicular vein empties into the left renal vein, and the right testicular vein drains into the inferior vena cava). In the female, the gonadal veins are called the ovarian veins. The **ovarian veins** drain the ovaries. The left ovarian vein empties into the left renal vein, and the right ovarian vein drains into the inferior vena cava.
Renal veins (RĒ-nal = *ren-* = kidney)	The large **renal veins** pass anterior to the renal arteries. The left renal vein is longer than the right renal vein and passes anterior to the abdominal aorta. It receives the left testicular (or ovarian), left inferior phrenic, and usually left suprarenal veins. The right renal vein empties into the inferior vena cava posterior to the duodenum. The renal veins drain the kidneys.
Suprarenal veins (soo'-pra-RĒ-nal; *supra-* = above)	The **suprarenal veins** drain the adrenal (suprarenal) glands (the left suprarenal vein empties into the left renal vein, and the right suprarenal vein empties into the inferior vena cava).
Inferior phrenic veins (FREN-ik = pertaining to the diaphragm)	The **inferior phrenic veins** drain the diaphragm (the left inferior phrenic vein usually sends one tributary to the left suprarenal vein, which empties into the left renal vein, and another tributary that empties into the inferior vena cava; the right inferior phrenic vein empties into the inferior vena cava).
Hepatic veins (he-PAT-ik = pertaining to the liver)	The **hepatic veins** drain the liver.

■ **EXHIBIT 21.11 CONTINUES** ▶

EXHIBIT 21.11   **813**

EXHIBIT 21.11   Veins of the Abdomen and Pelvis   CONTINUED

◉ **CHECKPOINT**

What structures do the lumbar, gonadal, renal, suprarenal, inferior phrenic, and hepatic veins drain?

SCHEME OF DRAINAGE

Heart

Inferior phrenic

Hepatic ← Hepatic portal

Suprarenal

Renal

Gonadal

Lumbar

Inferior vena cava

Right common iliac   Left common iliac

Right external iliac   Right internal iliac   Left internal iliac   Left external iliac

EXHIBIT 21.12 Veins of the Lower Limbs

- Identify the principal superficial and deep veins that drain the lower limbs.

As with the upper limbs, blood from the lower limbs is drained by both **superficial** and **deep veins.** The superficial veins often anastomose with one another and with deep veins along their length. Deep veins, for the most part, have the same names as corresponding arteries (Figure 21.27 on page 817). All veins of the lower limbs have valves, which are more numerous than in veins of the upper limbs.

◉ CHECKPOINT
What is the clinical importance of the great saphenous veins?

VEIN	DESCRIPTION AND REGION DRAINED
SUPERFICIAL VEINS	
Great saphenous veins (sa-FĒ-nus; *saphen-* = clearly visible)	The **great (long) saphenous veins,** the longest veins in the body, ascend from the foot to the groin in the subcutaneous layer. They begin at the medial end of the dorsal venous arches of the foot. The **dorsal venous arches** (VĒ-nus) are networks of veins on the dorsum of the foot formed by the **dorsal digital veins,** which collect blood from the toes, and then unite in pairs to form the **dorsal metatarsal veins,** which parallel the metatarsals. As the dorsal metatarsal veins approach the foot, they combine to form the dorsal venous arches. The great saphenous veins pass anterior to the medial malleolus of the tibia and then superiorly along the medial aspect of the leg and thigh just deep to the skin. They receive tributaries from superficial tissues and connect with the deep veins as well. They empty into the femoral veins at the groin. Generally, the great saphenous veins drain mainly the medial side of the leg and thigh, the groin, external genitals, and abdominal wall._x000D_ _x000D_ Along their length, the great saphenous veins have from 10 to 20 valves, with more located in the leg than the thigh. These veins are more likely to be subject to varicosities than other veins in the lower limbs because they must support a long column of blood and are not well supported by skeletal muscles._x000D_ _x000D_ The great saphenous veins are often used for prolonged administration of intravenous fluids. This is particularly important in very young children and in patients of any age who are in shock and whose veins are collapsed. In coronary artery bypass grafting, if multiple blood vessels need to be grafted, sections of the great saphenous vein are used along with at least one artery as a graft. After the great saphenous vein is removed and divided into sections, the sections are used to bypass the blockages. The vein grafts are reversed so that the valves do not obstruct the flow of blood.
Small saphenous veins	The **small (short) saphenous veins** begin at the lateral aspect of the dorsal venous arches of the foot. They pass posterior to the lateral malleolus of the fibula and ascend deep to the skin along the posterior aspect of the leg. They empty into the popliteal veins in the popliteal fossa, posterior to the knee. Along their length, the small saphenous veins have from 9 to 12 valves. The small saphenous veins drain the foot and posterior aspect of the leg. They may communicate with the great saphenous veins in the proximal thigh.
DEEP VEINS	
Posterior tibial veins (TIB-ē-al)	The **plantar digital veins** on the plantar surfaces of the toes unite to form the **plantar metatarsal veins,** which parallel the metatarsals. They in turn unite to form the **deep plantar venous arches.** From each arch emerges the **medial** and **lateral plantar veins.**_x000D_ _x000D_ The medial and lateral plantar veins, posterior to the medial malleolus of the tibia, form the paired **posterior tibial veins,** which sometimes merge into a single vessel. They accompany the posterior tibial artery through the leg. They ascend deep to the muscles in the posterior aspect of the leg and drain the foot and posterior compartment muscles. About two-thirds of the way up the leg, the posterior tibial veins drain blood from the **fibular (peroneal) veins,** which drain the lateral and posterior leg muscles. The posterior tibial veins unite with the anterior tibial veins just inferior to the popliteal fossa to form the popliteal veins.
Anterior tibial veins	The paired **anterior tibial veins** arise in the dorsal venous arch and accompany the anterior tibial artery. They ascend in the interosseous membrane between the tibia and fibula and unite with the posterior tibial veins to form the popliteal vein. The anterior tibial veins drain the ankle joint, knee joint, tibiofibular joint, and anterior portion of the leg.
Popliteal veins (pop'-li-TĒ-al = pertaining to the hollow behind knee)	The **popliteal veins,** formed by the union of the anterior and posterior tibial veins, also receive blood from the small saphenous veins and tributaries that correspond to branches of the popliteal artery. The popliteal veins drain the knee joint and the skin, muscles, and bones of portions of the calf and thigh around the knee joint.
Femoral veins (FEM-o-ral)	The **femoral veins** accompany the femoral arteries and are the continuations of the popliteal veins just superior to the knee. The femoral veins extend up the posterior surface of the thighs and drain the muscles of the thighs, femurs, external genitals, and superficial lymph nodes. The largest tributaries of the femoral veins are the **deep veins of the thigh (deep femoral veins).** Just before penetrating the abdominal wall, the femoral veins receive the deep femoral veins and the great saphenous veins. The veins formed from this union penetrate the body wall and enter the pelvic cavity. Here they are known as the **external iliac veins.** In order to take blood samples or pressure recordings from the right side of the heart, a catheter is inserted into the femoral vein as it passes through the femoral triangle. The catheter passes through the external and common iliac veins and inferior vena cava into the right atrium.

◉ **EXHIBIT 21.12 CONTINUES** ▶

EXHIBIT 21.12   **815**

EXHIBIT 21.12   Veins of the Lower Limbs CONTINUED

SCHEME OF DRAINAGE

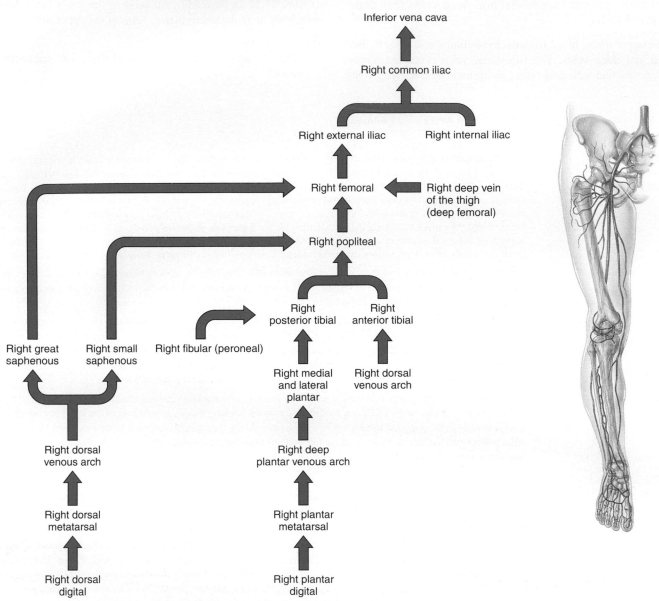

Figure 21.27 Principal veins of the pelvis and lower limbs.

Deep veins usually bear the names of their companion arteries.

Inferior vena cava

Right common iliac

Right internal iliac

Right external iliac

Left common iliac

Right deep vein of the thigh (deep femoral)

Right femoral

Right accessory saphenous

Right great saphenous

Right popliteal

Right small saphenous

Right anterior tibial

Right great saphenous

Right fibular (peroneal)

Right small saphenous

Right posterior tibial

Right dorsal venous arch

Right medial plantar

Right lateral plantar

Right dorsal metatarsal

Right deep plantar venous arch

Right plantar metatarsal

Right dorsal digital

Right plantar digital

(a) Anterior view

(b) Posterior view

Which veins of the lower limb are superficial?

EXHIBIT 21.12   **817**

The Hepatic Portal Circulation

The **hepatic portal circulation** carries venous blood from the gastrointestinal organs and spleen to the liver. A vein that carries blood from one capillary network to another is called a **portal vein.** The **hepatic portal vein** (*hepat-* = liver) receives blood from capillaries of gastrointestinal organs and the spleen and delivers it to the sinusoids of the liver (Figure 21.28). After a meal, hepatic portal blood is rich in nutrients absorbed from the gastrointestinal tract. The liver stores some of them and modifies others before they pass into the general circulation. For example, the liver converts glucose into glycogen for storage, reducing blood glucose level shortly after a meal. The liver also detoxifies harmful substances, such as alcohol, that have been absorbed from the gastrointestinal tract and destroys bacteria by phagocytosis.

The superior mesenteric and splenic veins unite to form the hepatic portal vein. The **superior mesenteric vein** drains blood from the small intestine and portions of the large intestine, stomach, and pancreas through the *jejunal, ileal, ileocolic, right colic, middle colic, pancreaticoduodenal,* and *right gastroepiploic veins.* The **splenic vein** drains blood from the stomach,

Figure 21.28 Hepatic portal circulation. A schematic diagram of blood flow through the liver, including arterial circulation, is shown in (b). As usual, deoxygenated blood is indicated in blue, oxygenated blood in red.

The hepatic portal circulation delivers venous blood from the organs of the gastrointestinal tract and spleen to the liver.

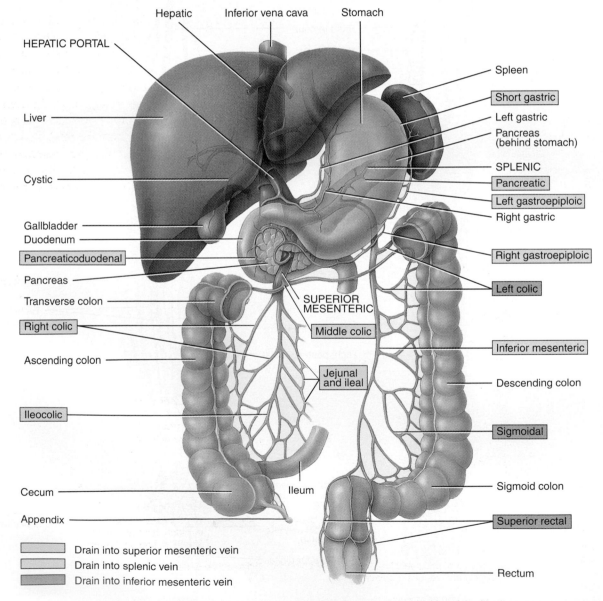

Drain into superior mesenteric vein
Drain into splenic vein
Drain into inferior mesenteric vein

(a) Anterior view of veins draining into the hepatic portal vein pancreas, and portions of the large intestine through the *short gastric, left gastroepiploic, pancreatic,* and *inferior mesenteric veins.* The inferior mesenteric vein, which passes into the splenic vein, drains portions of the large intestine through the superior *rectal, sigmoidal,* and *left colic veins.* The *right* and *left gastric veins,* which open directly into the hepatic portal vein, drain the stomach. The *cystic vein,* which also opens into the hepatic portal vein, drains the gallbladder.

At the same time the liver is receiving nutrient-rich but deoxygenated blood via the hepatic portal vein, it is also receiving oxygenated blood via the hepatic artery, a branch of the celiac trunk. The oxygenated blood mixes with the deoxygenated blood in sinusoids. Eventually, blood leaves the sinusoids of the liver through the **hepatic veins,** which drain into the inferior vena cava.

The Pulmonary Circulation

The **pulmonary circulation** (*pulmo-* = lung) carries deoxygenated blood from the right ventricle to the air sacs (alveoli) within the lungs and returns oxygenated blood from the air sacs to the left atrium (Figure 21.29). The **pulmonary trunk** emerges from the right ventricle and passes superiorly, posteriorly, and to the left. It then divides into two branches: the **right pulmonary artery** to the right lung and the **left pulmonary artery** to the left lung. After birth, the pulmonary arteries are the only arteries that carry deoxygenated blood. On entering the lungs, the branches divide and subdivide until finally they form capillaries around the air sacs (alveoli) within the lungs. CO_2 passes from the blood into the air sacs and is exhaled. Inhaled O_2

passes from the air within the lungs into the blood. The pulmonary capillaries unite to form venules and eventually **pulmonary veins,** which exit the lungs and carry the oxygenated blood to the left atrium. Two left and two right pulmonary veins enter the left atrium. After birth, the pulmonary veins are the only veins that carry oxygenated blood. Contractions of the left ventricle then eject the oxygenated blood into the systemic circulation.

The Fetal Circulation

The circulatory system of a fetus, called the **fetal circulation,** exists only in the fetus and contains special structures that allow the developing fetus to exchange materials with its mother (Figure 21.30 on page 821). It differs from the postnatal (after birth) circulation because the lungs, kidneys, and gastrointestinal organs do not begin to function until birth. The fetus obtains O_2 and nutrients from and eliminates CO_2 and other wastes into the maternal blood.

The exchange of materials between fetal and maternal circulations occurs through the **placenta** (pla-SEN-ta), which forms inside the mother's uterus and attaches to the umbilicus (navel) of the fetus by the **umbilical cord** (um-BIL-i-kal). The placenta communicates with the mother's cardiovascular system through many small blood vessels that emerge from the uterine wall. The umbilical cord contains blood vessels that branch into capillaries in the placenta. Wastes from the fetal blood diffuse out of the capillaries, into spaces containing maternal blood (intervillous spaces) in the placenta, and finally into the mother's uterine veins. Nutrients travel the opposite route—from the

(b) Scheme of principal blood vessels of hepatic portal circulation and arterial supply and venous drainage of liver

? **Which veins carry blood away from the liver?**

maternal blood vessels to the intervillous spaces to the fetal capillaries. Normally, there is no direct mixing of maternal and fetal blood because all exchanges occur by diffusion through capillary walls.

Blood passes from the fetus to the placenta via two **umbilical arteries** (Figure 21.30a, c). These branches of the internal iliac (hypogastric) arteries are within the umbilical cord. At the placenta, fetal blood picks up O_2 and nutrients and eliminates CO_2 and wastes. The oxygenated blood returns from the placenta via a single **umbilical vein.** This vein ascends to the liver of the fetus, where it divides into two branches. Some blood flows through the branch that joins the hepatic portal vein and enters the liver, but most of the blood flows into the second branch, the

ductus venosus (DUK-tus ve-NŌ-sus), which drains into the inferior vena cava.

Deoxygenated blood returning from lower body regions of the fetus mingles with oxygenated blood from the ductus venosus in the inferior vena cava. This mixed blood then enters the right atrium. Deoxygenated blood returning from upper body regions of the fetus enters the superior vena cava and also passes into the right atrium.

Most of the fetal blood does not pass from the right ventricle to the lungs, as it does in postnatal circulation, because an opening called the **foramen ovale** (fō-RĀ-men ō-VAL-ē) exists in the septum between the right and left atria. Most of the blood that enters the right atrium passes through the foramen ovale

Figure 21.29 Pulmonary circulation.

The pulmonary circulation brings deoxygenated blood from the right ventricle to the lungs and returns oxygenated blood from the lungs to the left atrium.

Superior vena cava
Right pulmonary artery
Pulmonary trunk
Right pulmonary veins
Right lung
Inferior vena cava

Arch of aorta
Ascending aorta
Left pulmonary artery
Left pulmonary veins
Left lung
Diaphragm
Abdominal aorta

(a) Anterior view

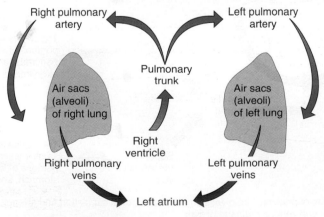

Right pulmonary artery
Left pulmonary artery
Pulmonary trunk
Air sacs (alveoli) of right lung
Air sacs (alveoli) of left lung
Right ventricle
Right pulmonary veins
Left pulmonary veins
Left atrium

(b) Scheme of pulmonary circulation

? **After birth, which are the only arteries that carry deoxygenated blood?**

into the left atrium and joins the systemic circulation. The blood that does pass into the right ventricle is pumped into the pulmonary trunk, but little of this blood reaches the nonfunctioning fetal lungs. Instead, most is sent through the **ductus arteriosus** (ar-tē-rē-Ō-sus), a vessel that connects the pulmonary trunk with the aorta. The blood in the aorta is carried to all fetal tissues through the systemic circulation. When the common iliac arteries branch into the external and internal iliacs, part of the blood flows into the internal iliacs, into the umbilical arteries, and back to the placenta for another exchange of materials.

Figure 21.30 **Fetal circulation and changes at birth.** The gold boxes between parts (a) and (b) describe the fate of certain fetal structures once postnatal circulation is established.

The lungs and gastrointestinal organs do not begin to function until birth.

DUCTUS ARTERIOSUS becomes Ligamentum arteriosum

FORAMEN OVALE becomes Fossa ovalis

DUCTUS VENOSUS becomes Ligamentum venosum

UMBILICAL VEIN becomes Ligamentum teres

UMBILICAL ARTERIES become Medial umbilical ligaments

Right atrium
Arch of aorta
Superior vena cava
Left atrium
Lung
Pulmonary artery
Pulmonary veins
Heart
Right ventricle
Left ventricle
Liver
Hepatic portal vein
Umbilicus
Inferior vena cava
Abdominal aorta
Common iliac artery
Urinary bladder
Urethra
UMBILICAL CORD
Placenta

(a) Fetal circulation

(b) Circulation at birth

Oxygenated blood

Mixed oxygenated and deoxygenated blood

Deoxygenated blood

FIGURE 21.30 CONTINUES

⬛ FIGURE 21.30 CONTINUED ▶

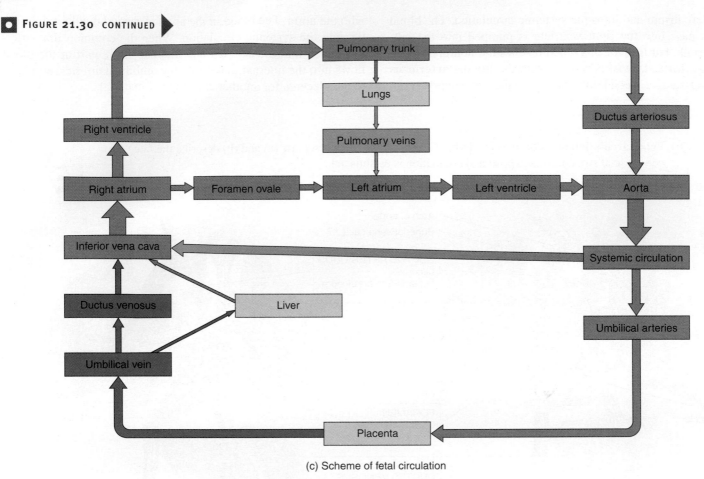

(c) Scheme of fetal circulation

❓ **Which structure provides for exchange of materials between mother and fetus?**

After birth, when pulmonary (lung), renal (kidney), and digestive functions begin, the following vascular changes occur (Figure 21.30b):

1. When the umbilical cord is tied off, blood no longer flows through the umbilical arteries, they fill with connective tissue, and the distal portions of the umbilical arteries become fibrous cords called the **medial umbilical ligaments.** Although the arteries are closed functionally only a few minutes after birth, complete obliteration of the lumens may take 2 to 3 months.

2. The umbilical vein collapses but remains as the **ligamentum teres (round ligament),** a structure that attaches the umbilicus to the liver.

3. The ductus venosus collapses but remains as the **ligamentum venosum,** a fibrous cord on the inferior surface of the liver.

4. The placenta is expelled as the **"afterbirth."**

5. The foramen ovale normally closes shortly after birth to become the **fossa ovalis,** a depression in the interatrial septum. When an infant takes its first breath, the lungs expand and blood flow to the lungs increases. Blood returning from the lungs to the heart increases pressure in the left atrium. This closes the foramen ovale by pushing the valve that guards it against the interatrial septum. Permanent closure occurs in about a year.

6. The ductus arteriosus closes by vasoconstriction almost immediately after birth and becomes the **ligamentum arteriosum.** Complete anatomical obliteration of the lumen takes 1 to 3 months.

◉ **CHECKPOINT**

23. Diagram the hepatic portal circulation. Why is this route important?

24. Diagram the route of the pulmonary circulation.

25. Discuss the anatomy and physiology of the fetal circulation. Indicate the function of the umbilical arteries, umbilical vein, ductus venosus, foramen ovale, and ductus arteriosus.

DEVELOPMENT OF BLOOD VESSELS AND BLOOD

◉ **OBJECTIVE**

• Describe the development of blood vessels and blood.

The development of blood cells and the formation of blood vessels begins outside the embryo as early as 15 to 16 days in the **mesoderm** of the wall of the yolk sac, chorion, and connecting stalk. About two days later, blood vessels form within the embryo. The early formation of the cardiovascular system is linked to the small amount of yolk in the ovum and yolk sac. As the embryo develops rapidly during the third week, there is a greater need to develop a cardiovascular system to supply sufficient nutrients to the embryo and remove wastes from it.

Blood vessels and blood cells develop from the same precursor cell, called a **hemangioblast** (hē-MAN-jē-ō-blast; *hema-* = blood; *-blast* = immature stage). Once mesenchyme develops into hemangioblasts, they can give rise to cells that produce blood vessels (angioblasts) or cells that produce blood cells (pluripotent stem cells).

Blood vessels develop from **angioblasts,** which are derived from hemangioblasts (Figure 21.31). Angioblasts aggregate to form isolated masses and cords throughout the embryonic discs called **blood islands** (Figure 21.31). Spaces soon appear in the islands and become the lumens of the blood vessels. Some of the angioblasts immediately around the spaces give rise to the *endothelial lining of the blood vessels.* Angioblasts around the endothelium form the *tunics* (interna, media, and externa) of the larger blood vessels. Growth and fusion of blood islands form an extensive network of blood vessels throughout the embryo. By continuous branching, blood vessels outside the embryo connect with those inside the embryo, linking the embryo with the placenta.

Blood cells develop from **pluripotent stem cells** derived from hemangioblasts. This development occurs in the walls of blood vessels in the yolk sac, chorion, and allantois at about three weeks after fertilization. Blood formation in the embryo itself begins at about the fifth week in the liver and the twelfth week in the spleen, red bone marrow, and thymus.

◉ CHECKPOINT

26. What are the sites of blood cell production outside the embryo and within the embryo?

AGING AND THE CARDIOVASCULAR SYSTEM

◉ OBJECTIVE

• Explain the effects of aging on the cardiovascular system.

General changes in the cardiovascular system associated with aging include decreased compliance (distensibility) of the aorta, reduction in cardiac muscle fiber size, progressive loss of cardiac muscular strength, reduced cardiac output, a decline in maximum heart rate, and an increase in systolic blood pressure. Total blood cholesterol tends to increase with age, as does low-density lipoprotein (LDL); high-density lipoprotein (HDL) tends to decrease. There is an increase in the incidence of coronary artery disease (CAD), the major cause of heart disease and death in older Americans. Congestive heart failure, a set of symptoms associated with impaired pumping of the heart, is also prevalent in older individuals. Changes in blood vessels that serve brain

Figure 21.31 Development of blood vessels and blood cells from blood islands.

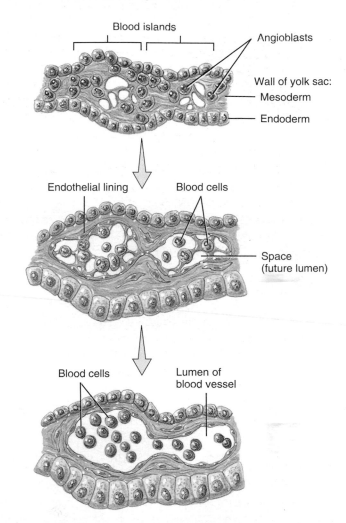

⊙ ⚿ Blood vessel development begins in the embryo on about the 15th or 16th day.

❓ From which germ cell layer are blood vessels and blood derived?

tissue—for example, atherosclerosis—reduce nourishment to the brain and result in malfunction or death of brain cells. By age 80, cerebral blood flow is 20% less and renal blood flow is 50% less than in the same person at age 30.

◉ CHECKPOINT

27. How does aging affect the heart?

• • •

To appreciate the many ways the blood, heart, and blood vessels contribute to homeostasis of other body systems, examine *Focus on Homeostasis: The Cardiovascular System.*

For all body systems

The heart pumps blood through blood vessels to body tissues, delivering oxygen and nutrients and removing wastes by means of capillary exchange. Circulating blood keeps body tissues at a proper temperature.

Integumentary system

Blood delivers clotting factors and white blood cells that aid in hemostasis when skin is damaged and contribute to repair of injured skin. Changes in skin blood flow contribute to body temperature regulation by adjusting the amount of heat loss via the skin. Blood flowing in skin may give skin a pink hue.

Skeletal system

Blood delivers calcium and phosphate ions that are needed for building bone extracellular matrix, hormones that govern building and breakdown of bone extracellular matrix, and erythropoietin that stimulates production of red blood cells by red bone marrow.

Muscular system

Blood circulating through exercising muscle removes heat and lactic acid.

Nervous system

Endothelial cells lining choroid plexuses in brain ventricles help produce cerebrospinal fluid (CSF) and contribute to the blood–brain barrier.

Endocrine system

Circulating blood delivers most hormones to their target tissues. Atrial cells secrete atrial natriuretic peptide.

Lymphatic system and immunity

Circulating blood distributes lymphocytes, antibodies, and macrophages that carry out immune functions. Lymph forms from excess interstitial fluid, which filters from blood plasma due to blood pressure generated by the heart.

Respiratory system

Circulating blood transports oxygen from the lungs to body tissues and carbon dioxide to the lungs for exhalation.

Digestive system

Blood carries newly absorbed nutrients and water to liver. Blood distributes hormones that aid digestion.

Urinary system

Heart and blood vessels deliver 20% of the resting cardiac output to the kidneys, where blood is filtered, needed substances are reabsorbed, and unneeded substances remain as part of urine, which is excreted.

Reproductive systems

Vasodilation of arterioles in penis and clitoris cause erection during sexual intercourse. Blood distributes hormones that regulate reproductive functions.

The Cardiovascular System

Hypertension

About 50 million Americans have **hypertension,** or persistently high blood pressure. It is the most common disorder affecting the heart and blood vessels and is the major cause of heart failure, kidney disease, and stroke. In May 2003, the Joint National Committee on Prevention, Detection, Evaluation, and Treatment of High Blood Pressure published new guidelines for hypertension because clinical studies have linked what were once considered fairly low blood pressure readings to an increased risk of cardiovascular disease. The new guidelines are as follows:

Category	Systolic (mmHg)	Diastolic (mmHg)
Normal	Less than 120 *and*	Less than 80
Prehypertension	120–139 *or*	80–89
Stage 1 hypertension	140–159 *or*	90–99
Stage 2 hypertension	Greater than 160 *or*	Greater than 100

Using the new guidelines, the normal classification was previously considered optimal; prehypertension now includes many more individuals previously classified as normal or high-normal; stage 1 hypertension is the same as in previous guidelines; and stage 2 hypertension now combines the previous stage 2 and stage 3 categories since treatment options are the same for the former stages 2 and 3.

Types and Causes of Hypertension

Between 90 and 95% of all cases of hypertension are **primary hypertension,** a persistently elevated blood pressure that cannot be attributed to any identifiable cause. The remaining 5–10% of cases are **secondary hypertension,** which has an identifiable underlying cause. Several disorders cause secondary hypertension:

- *Obstruction of renal blood flow* or disorders that damage renal tissue may cause the kidneys to release excessive amounts of renin into the blood. The resulting high level of angiotensin II causes vasoconstriction, thus increasing systemic vascular resistance.

- *Hypersecretion of aldosterone*—resulting, for instance, from a tumor of the adrenal cortex—stimulates excess reabsorption of salt and water by the kidneys, which increases the volume of body fluids.

- *Hypersecretion of epinephrine and norepinephrine* by a **pheochromocytoma** (fē-ō-krō′-mō-sī-TŌ-ma), a tumor of the adrenal medulla. Epinephrine and norepinephrine increase heart rate and contractility and increase systemic vascular resistance.

Damaging Effects of Untreated Hypertension

High blood pressure is known as the "silent killer" because it can cause considerable damage to the blood vessels, heart, brain, and kidneys before it causes pain or other noticeable symptoms. It is a major risk factor for the number-one (heart disease) and number-three (stroke) causes of death in the United States. In blood vessels, hypertension causes thickening of the tunica media, accelerates development of atherosclerosis and coronary artery disease, and increases systemic vascular resistance. In the heart, hypertension increases the afterload, which forces the ventricles to work harder to eject blood.

The normal response to an increased workload due to vigorous and regular exercise is hypertrophy of the myocardium, especially in the wall of the left ventricle. An increased afterload, however, leads to

myocardial hypertrophy that is accompanied by muscle damage and fibrosis (a buildup of collagen fibers between the muscle fibers). As a result, the left ventricle enlarges, weakens, and dilates. Because arteries in the brain are usually less protected by surrounding tissues than are the major arteries in other parts of the body, prolonged hypertension can eventually cause them to rupture, resulting in a stroke. Hypertension also damages kidney arterioles, causing them to thicken, which narrows the lumen; because the blood supply to the kidneys is thereby reduced, the kidneys secrete more renin, which elevates the blood pressure even more.

Lifestyle Changes to Reduce Hypertension

Although several categories of drugs (described next) can reduce elevated blood pressure, the following lifestyle changes are also effective in managing hypertension:

- *Lose weight.* This is the best treatment for high blood pressure short of using drugs. Loss of even a few pounds helps reduce blood pressure in overweight hypertensive individuals.

- *Limit alcohol intake.* Drinking in moderation may lower the risk of coronary heart disease, mainly among males over 45 and females over 55. Moderation is defined as no more than one 12-oz beer per day for females and no more than two 12-oz beers per day for males.

- *Exercise.* Becoming more physically fit by engaging in moderate activity (such as brisk walking) several times a week for 30 to 45 minutes can lower systolic blood pressure by about 10 mmHg.

- *Reduce intake of sodium (salt).* Roughly half the people with hypertension are "salt sensitive." For them, a high-salt diet appears to promote hypertension, and a low-salt diet can lower their blood pressure.

- *Maintain recommended dietary intake of potassium, calcium, and magnesium.* Higher levels of potassium, calcium, and magnesium in the diet are associated with a lower risk of hypertension.

- *Don't smoke.* Smoking has devastating effects on the heart and can augment the damaging effects of high blood pressure by promoting vasoconstriction.

- *Manage stress.* Various meditation and biofeedback techniques help some people reduce high blood pressure. These methods may work by decreasing the daily release of epinephrine and norepinephrine by the adrenal medulla.

Drug Treatment of Hypertension

Drugs having several different mechanisms of action are effective in lowering blood pressure. Many people are successfully treated with *diuretics,* agents that decrease blood pressure by decreasing blood volume because they increase elimination of water and salt in the urine. *ACE (angiotensin converting enzyme) inhibitors* block formation of angiotensin II and thereby promote vasodilation and decrease the secretion of aldosterone. *Beta blockers* reduce blood pressure by inhibiting the secretion of renin and by decreasing heart rate and contractility. *Vasodilators* relax the smooth muscle in arterial walls, causing vasodilation and lowering blood pressure by lowering systemic vascular resistance. An important category of vasodilators are the *calcium channel blockers,* which slow the inflow of Ca^{2+} into vascular smooth muscle cells. They reduce the heart's workload by slowing Ca^{2+} entry into pacemaker cells and regular myocardial fibers, thereby decreasing heart rate and the force of myocardial contraction.

MEDICAL TERMINOLOGY

Aneurysm (AN-ū-rizm) A thin, weakened section of the wall of an artery or a vein that bulges outward, forming a balloonlike sac. Common causes are atherosclerosis, syphilis, congenital blood vessel defects, and trauma. If untreated, the aneurysm enlarges and the blood vessel wall becomes so thin that it bursts. The result is massive hemorrhage with shock, severe pain, stroke, or death. Treatment may involve surgery in which the weakened area of the blood vessel is removed and replaced with a graft of synthetic material.

Aortography (ā′-or-TOG-ra-fē) X-ray examination of the aorta and its main branches after injection of a radiopaque dye.

Carotid endarterectomy (ka-ROT-id end′-ar-ter-EK-tō-mē) The removal of atherosclerotic plaque from the carotid artery to restore greater blood flow to the brain.

Claudication (klaw′-di-KĀ-shun) Pain and lameness or limping caused by defective circulation of the blood in the vessels of the limbs.

Deep vein thrombosis The presence of a thrombus (blood clot) in a deep vein of the lower limbs. It may lead to (1) pulmonary embolism, if the thrombus dislodges and then lodges within the pulmonary arterial blood flow, and (2) postphlebitic syndrome, which consists of edema, pain, and skin changes due to destruction of venous valves.

Doppler ultrasound scanning Imaging technique commonly used to measure blood flow. A transducer is placed on the skin and an image is displayed on a monitor that provides the exact position and severity of a blockage.

Femoral angiography An imaging technique in which a contrast medium is injected into the femoral artery and spreads to other arteries in the lower limb, and then a series of radiographs are taken of one or more sites. It is used to diagnose narrowing or blockage of arteries in the lower limbs.

Hypotension (hī-pō-TEN-shun) Low blood pressure; most commonly used to describe an acute drop in blood pressure, as occurs during excessive blood loss.

Normotensive (nor′-mō-TEN-siv) Characterized by normal blood pressure.

Occlusion (ō-KLOO-shun) The closure or obstruction of the lumen of a structure such as a blood vessel. An example is an atherosclerotic plaque in an artery.

Orthostatic hypotension (or′-thō-STAT-ik; *ortho-* = straight; *-static* = causing to stand) An excessive lowering of systemic blood pressure when a person assumes an erect or semi-erect posture; it is usually a sign of a disease. May be caused by excessive fluid loss, certain drugs, and cardiovascular or neurogenic factors. Also called **postural hypotension.**

Phlebitis (fle-BĪ-tis; *phleb-* = vein) Inflammation of a vein, often in a leg.

Thrombectomy (throm-BEK-tō-mē; *thrombo-* = clot) An operation to remove a blood clot from a blood vessel.

Thrombophlebitis (throm′-bō-fle-BĪ-tis) Inflammation of a vein involving clot formation. Superficial thrombophlebitis occurs in veins under the skin, especially in the calf.

Venipuncture (VEN-i-punk-chur; *vena-* = vein) The puncture of a vein, usually to withdraw blood for analysis or introduce a solution, for example, an antibiotic. The median cubital vein is frequently used.

White coat (office) hypertension A stress-induced syndrome found in patients who have elevated blood pressure when being examined by health-care personnel, but otherwise have normal blood pressure.

STUDY OUTLINE

Structure and Function of Blood Vessels (p. 761)

1. Arteries carry blood away from the heart. The wall of an artery consists of a tunica interna, a tunica media (which maintains elasticity and contractility), and a tunica externa.
2. Large arteries are termed elastic (conducting) arteries, and medium-sized arteries are called muscular (distributing) arteries.
3. Many arteries anastomose: The distal ends of two or more vessels unite. An alternative blood route from an anastomosis is called collateral circulation. Arteries that do not anastomose are called end arteries.
4. Arterioles are small arteries that deliver blood to capillaries.
5. Through constriction and dilation, arterioles assume a key role in regulating blood flow from arteries into capillaries and in altering arterial blood pressure.
6. Capillaries are microscopic blood vessels through which materials are exchanged between blood and tissue cells; some capillaries are continuous, and others are fenestrated.
7. Capillaries branch to form an extensive network throughout a tissue. This network increases surface area, allowing a rapid exchange of large quantities of materials.
8. Precapillary sphincters regulate blood flow through capillaries.
9. Microscopic blood vessels in the liver are called sinusoids.
10. Venules are small vessels that continue from capillaries and merge to form veins.
11. Veins consist of the same three tunics as arteries but have a thinner tunica interna and a thinner tunica media. The lumen of a vein is also larger than that of a comparable artery.
12. Veins contain valves to prevent backflow of blood.
13. Weak valves can lead to varicose veins.
14. Vascular (venous) sinuses are veins with very thin walls.
15. Systemic veins are collectively called blood reservoirs because they hold a large volume of blood. If the need arises, this blood can be shifted into other blood vessels through vasoconstriction of veins.
16. The principal blood reservoirs are the veins of the abdominal organs (liver and spleen) and skin.

Capillary Exchange (p. 769)

1. Substances enter and leave capillaries by diffusion, transcytosis, or bulk flow.
2. The movement of water and solutes (except proteins) through capillary walls depends on hydrostatic and osmotic pressures.
3. The near equilibrium between filtration and reabsorption in capillaries is called Starling's law of the capillaries.
4. Edema is an abnormal increase in interstitial fluid.

Hemodynamics: Factors Affecting Blood Flow (p. 772)

1. The velocity of blood flow is inversely related to the cross-sectional area of blood vessels; blood flows slowest where cross-sectional area is greatest.
2. The velocity of blood flow decreases from the aorta to arteries to capillaries and increases in venules and veins.
3. Blood pressure and resistance determine blood flow.
4. Blood flows from regions of higher to lower pressure. The higher the resistance, however, the lower the blood flow.
5. Cardiac output equals the mean arterial pressure divided by total resistance ($CO = MAP \div R$).
6. Blood pressure is the pressure exerted on the walls of a blood vessel.
7. Factors that affect blood pressure are cardiac output, blood volume, viscosity, resistance, and the elasticity of arteries.
8. As blood leaves the aorta and flows through the systemic circulation, its pressure progressively falls to 0 mmHg by the time it reaches the right ventricle.
9. Resistance depends on blood vessel diameter, blood viscosity, and total blood vessel length.
10. Venous return depends on pressure differences between the venules and the right ventricle.
11. Blood return to the heart is maintained by several factors, including skeletal muscular contractions, valves in veins (especially in the limbs), and pressure changes associated with breathing.

Control of Blood Pressure and Blood Flow (p. 775)

1. The cardiovascular (CV) center is a group of neurons in the medulla oblongata that regulates heart rate, contractility, and blood vessel diameter.
2. The cardiovascular center receives input from higher brain regions and sensory receptors (baroreceptors and chemoreceptors).
3. Output from the cardiovascular center flows along sympathetic and parasympathetic axons. Sympathetic impulses propagated along cardioaccelerator nerves increase heart rate and contractility; parasympathetic impulses propagated along vagus nerves decrease heart rate.
4. Baroreceptors monitor blood pressure, and chemoreceptors monitor blood levels of O_2, CO_2, and hydrogen ions. The carotid sinus reflex helps regulate blood pressure in the brain. The aortic reflex regulates general systemic blood pressure.
5. Hormones that help regulate blood pressure are epinephrine, norepinephrine, ADH (vasopressin), angiotensin II, and ANP.
6. Autoregulation refers to local, automatic adjustments of blood flow in a given region to meet a particular tissue's need.
7. O_2 level is the principal stimulus for autoregulation.

Checking Circulation (p. 780)

1. Pulse is the alternate expansion and elastic recoil of an artery wall with each heartbeat. It may be felt in any artery that lies near the surface or over a hard tissue.
2. A normal resting pulse (heart) rate is 70–80 beats/min.
3. Blood pressure is the pressure exerted by blood on the wall of an artery when the left ventricle undergoes systole and then diastole. It is measured by the use of a sphygmomanometer.
4. Systolic blood pressure (SBP) is the arterial blood pressure during ventricular contraction. Diastolic blood pressure (DBP) is the arterial blood pressure during ventricular relaxation. Normal blood pressure is less than 120/80.
5. Pulse pressure is the difference between systolic and diastolic blood pressure. It normally is about 40 mmHg.

Shock and Homeostasis (p. 781)

1. Shock is a failure of the cardiovascular system to deliver enough O_2 and nutrients to meet the metabolic needs of cells.
2. Types of shock include hypovolemic, cardiogenic, vascular, and obstructive.
3. Signs and symptoms of shock include systolic blood pressure less than 90 mmHg; rapid resting heart rate; weak, rapid pulse; clammy, cool, pale skin; sweating; hypotension; altered mental state; decreased urinary output; thirst; and acidosis.

Circulatory Routes (p. 783)

1. The two main circulatory routes are the systemic and pulmonary circulations.
2. Among the subdivisions of the systemic circulation are the coronary (cardiac) circulation and the hepatic portal circulation.
3. The systemic circulation carries oxygenated blood from the left ventricle through the aorta to all parts of the body, including some lung tissue, but *not* the air sacs (alveoli) of the lungs, and returns the deoxygenated blood to the right atrium.
4. The aorta is divided into the ascending aorta, the arch of the aorta, and the descending aorta. Each section gives off arteries that branch to supply the whole body.
5. Blood returns to the heart through the systemic veins. All veins of the systemic circulation drain into the superior or inferior venae cavae or the coronary sinus, which in turn empty into the right atrium.
6. The major blood vessels of the systemic circulation may be reviewed in Exhibits 21.1–21.12 on pages 785–817.
7. The hepatic portal circulation directs venous blood from the gastrointestinal organs and spleen into the hepatic portal vein of the liver before it returns to the heart. It enables the liver to utilize nutrients and detoxify harmful substances in the blood.
8. The pulmonary circulation takes deoxygenated blood from the right ventricle to the alveoli within the lungs and returns oxygenated blood from the alveoli to the left atrium.
9. Fetal circulation exists only in the fetus. It involves the exchange of materials between fetus and mother via the placenta.
10. The fetus derives O_2 and nutrients from and eliminates CO_2 and wastes into maternal blood. At birth, when pulmonary (lung), digestive, and liver functions begin, the special structures of fetal circulation are no longer needed.

Development of Blood Vessels and Blood (p. 822)

1. Blood vessels develop from mesenchyme (hemangioblasts → angioblasts → blood islands) in mesoderm called blood islands.
2. Blood cells also develop from mesenchyme (hemangioblasts → pluripotent stem cells).
3. The development of blood cells from pluripotent stem cells derived from angioblasts occurs in the walls of blood vessels in the yolk sac, chorion, and allantois at about three weeks after fertilization. Within the embryo, blood is produced by the liver at about the fifth week and in the spleen, red bone marrow, and thymus at about the twelfth week.

Aging and the Cardiovascular System (p. 823)

1. General changes associated with aging include reduced compliance (distensibility) of blood vessels, reduction in cardiac muscle size, reduced cardiac output, and increased systolic blood pressure.

2. The incidence of coronary artery disease (CAD), congestive heart failure (CHF), and atherosclerosis increases with age.

 SELF-QUIZ QUESTIONS

Fill in the blanks in the following statements.

1. The _____ reflex helps maintain normal blood pressure in the brain; the _____ reflex governs general systemic blood pressure.
2. In addition to the pressure created by contraction of the left ventricle, venous return is aided by the _____ and the _____, both of which depend on the presence of valves in the veins.

Indicate whether the following statements are true or false.

3. Baroreceptors and chemoreceptors are located in the aorta and carotid arteries.
4. The most important method of capillary exchange is simple diffusion.

Choose the one best answer to the following questions.

5. Which of the following are *not* true? (1) Muscular arteries are also known as conducting arteries. (2) Capillaries play a key role in regulating resistance. (3) The flow of blood through true capillaries is controlled by precapillary sphincters. (4) The lumen of an artery is larger than in a comparable vein. (5) Elastic arteries help propel blood. (6) The tunica media of arteries is thicker than the tunica media of veins.
   (a) 2, 3, and 6   (b) 1, 2, and 4   (c) 1, 2, 4, and 6
   (d) 3, 4, and 5   (e) 1, 2, 3, and 4

6. Which of the following are *true* concerning capillary exchange? (1) Large, lipid-insoluble molecules cross capillary walls by transcytosis. (2) The blood hydrostatic pressure promotes reabsorption of fluid into the capillaries. (3) If the pressures that promote filtration are greater than the pressures that promote reabsorption, fluid will move out of a capillary and into interstitial spaces. (4) A negative net filtration pressure results in reabsorption of fluid from interstitial spaces into a capillary. (5) The difference in osmotic pressure across a capillary wall is due primarily to red blood cells.
   (a) 1, 3, and 4   (b) 1, 2, 3, 4, and 5   (c) 1, 2, 3, and 4
   (d) 3 and 4   (e) 2, 4, and 5

7. Which of the following would *not* increase vascular resistance? (1) vasodilation, (2) polycythemia, (3) obesity, (4) dehydration, (5) anemia.
   (a) 1 and 2   (b) 1, 3, and 4   (c) 1 and 5
   (d) 1, 4, and 5   (e) 1 only

8. Capillary exchange is enhanced by (1) the slow rate of flow through the capillaries, (2) a small cross-sectional area, (3) the thinness of capillary walls, (4) the respiratory pump, (5) extensive branching, which increases the surface area.
   (a) 1, 2, 3, 4, and 5   (b) 1, 2, 3, and 5   (c) 1 and 3
   (d) 3 and 5   (e) 1, 3, and 5

9. Systemic vascular resistance depends on which of the following factors? (1) blood viscosity, (2) total blood vessel length, (3) size of the lumen, (4) type of blood vessel, (5) oxygen concentration of the blood.
   (a) 1, 2, and 3   (b) 2, 3, and 4   (c) 3, 4, and 5
   (d) 1, 3, and 5   (e) 2, 4, and 5

10. Which of the following help regulate blood pressure and help control regional blood flow? (1) baroreceptor and chemoreceptor reflexes, (2) hormones, (3) autoregulation, (4) H^+ concentration of blood, (5) oxygen concentration of the blood.
    (a) 1, 2, and 4   (b) 2, 4, and 5   (c) 1, 4, and 5
    (d) 1, 2, 3, 4, and 5   (e) 3, 4, and 5

11. For each of the following, indicate if it causes vasoconstriction or vasodilation. Use D for vasodilation and C for vasoconstriction.
    (a) atrial natriuretic peptide   (b) ADH
    (c) decrease in body temperature   (d) lactic acid
    (e) histamine   (f) hypoxia
    (g) hypercapnia   (h) angiotensin II
    (i) nitric oxide   (j) decreased sympathetic
    (k) acidosis   impulses

12. Match the following:
    ____(a) pressure generated by the pumping of the heart; pushes fluids out of capillaries
    ____(b) pressure created by proteins present in the interstitial fluid; pulls fluid out of capillaries
    ____(c) balance of pressure; determines whether blood volume and interstitial fluid remain steady or change
    ____(d) force due to presence of plasma proteins; pulls fluid into capillaries from interstitial spaces
    ____(e) pressure due to fluid in interstitial spaces; pushes fluid back into capillaries

    (1) net filtration pressure
    (2) blood hydrostatic pressure
    (3) interstitial fluid hydrostatic pressure
    (4) blood colloid osmotic pressure
    (5) interstitial fluid osmotic pressure

13. Match the following:

____(a) supplies blood to the kidney

____(b) drains blood from the small intestine, portions of the large intestine, stomach, and pancreas

____(c) main blood supply to arm; commonly used to measure blood pressure

____(d) supply blood to the free lower limbs

____(e) drain oxygenated blood from the lungs and send it to the left atrium

____(f) supplies blood to the stomach, liver, and pancreas

____(g) supply blood to the brain

____(h) supplies blood to the large intestine

____(i) drain blood from the head

____(j) detours venous blood from the gastrointestinal organs and spleen through the liver before it returns to the heart

____(k) drains most of the thorax and abdominal wall; can serve as a bypass for the inferior vena cava

____(l) a part of the venous circulation of the leg; a vessel used in heart bypass surgery

____(m) carry deoxygenated blood from the right ventricle to the lungs

(1) superior mesenteric vein
(2) inferior mesenteric artery
(3) pulmonary veins
(4) brachial artery
(5) hepatic portal circulation
(6) carotid arteries
(7) jugular veins
(8) celiac trunk
(9) common iliac arteries
(10) azygos veins
(11) renal artery
(12) great saphenous vein
(13) pulmonary arteries

14. Match the following:

____(a) a traveling pressure wave created by the alternate expansion and recoil of elastic arteries after each systole of the left ventricle

____(b) the lowest blood pressure in arteries during ventricular relaxation

____(c) a slow resting heart rate or pulse rate

____(d) an inadequate cardiac output that results in a failure of the cardio-vascular system to deliver enough oxygen and nutrients to meet the metabolic needs of body cells

____(e) a rapid resting heart rate or pulse rate

____(f) the highest force with which blood pushes against arterial walls as a result of ventricular contraction

(1) shock
(2) pulse
(3) tachycardia
(4) bradycardia
(5) systolic blood pressure
(6) diastolic blood pressure

15. Match the following (some answers will be used more than once):

____(a) returns oxygenated blood from the placenta to the fetal liver

____(b) an opening in the septum between the right and left atria

____(c) becomes the ligamentum venosum after birth

____(d) passes blood from the fetus to the placenta

____(e) bypasses the nonfunctioning lungs; becomes the ligamentum arteriosum at birth

____(f) become the medial umbilical ligaments at birth

____(g) transports oxygenated blood into the inferior vena cava

____(h) becomes the ligamentum teres at birth

____(i) becomes the fossa ovalis after birth

(1) ductus venosus
(2) ductus arteriosus
(3) foramen ovale
(4) umbilical arteries
(5) umbilical vein

CRITICAL THINKING QUESTIONS

1. Kim Sung was told that her baby was born with a hole in the upper chambers of his heart. Is this something Kim Sung should worry about?

2. Michael was brought into the emergency room suffering from a gunshot wound. He is bleeding profusely and exhibits the following: systolic blood pressure is 40 mmHg; weak pulse of 200 beats per minute; cool, pale, and clammy skin. Michael is not producing urine but is asking for water. He is confused and disoriented. What is his diagnosis and what, specifically, is causing these symptoms?

3. Maureen's job entails standing on a concrete floor for 10-hour days on an assembly line. Lately she has noticed swelling in her ankles at the end of the day and some tenderness in her calves. What do you suspect is Maureen's problem and how could she help counteract the problem?

? ANSWERS TO FIGURE QUESTIONS

21.1 The femoral artery has the thicker wall; the femoral vein has the wider lumen.

21.2 Due to atherosclerosis, less energy is stored in the less-compliant elastic arteries during systole; thus, the heart must pump harder to maintain the same rate of blood flow.

21.3 Metabolically active tissues use O_2 and produce wastes more rapidly than inactive tissues, so they require more extensive capillary networks.

21.4 Materials cross capillary walls through intercellular clefts and fenestrations, via transcytosis in pinocytic vesicles, and through the plasma membranes of endothelial cells.

21.5 Valves are more important in arm veins and leg veins than in neck veins because, when you are standing, gravity causes pooling of blood in the veins of the free limbs but aids the flow of blood in neck veins back toward the heart.

21.6 Blood volume in veins is about 64% of 5 liters, or 3.2 liters; blood volume in capillaries is about 7% of 5 liters, or 350 mL.

21.7 Blood colloid osmotic pressure is lower than normal in a person with a low level of plasma proteins, and therefore capillary reabsorption is low. The result is edema.

21.8 Mean blood pressure in the aorta is closer to diastolic than to systolic pressure.

21.9 The skeletal muscle pump and respiratory pump aid venous return.

21.10 Vasodilation and vasoconstriction of arterioles are the main regulators of systemic vascular resistance.

21.11 Velocity of blood flow is fastest in the aorta and arteries.

21.12 The effector tissues regulated by the cardiovascular center are cardiac muscle in the heart and smooth muscle in blood vessel walls.

21.13 Impulses to the cardiovascular center pass from baroreceptors in the carotid sinuses via the glossopharyngeal (IX) nerves and from baroreceptors in the arch of the aorta via the vagus (X) nerves.

21.14 It represents a change that occurs when you stand up because gravity causes pooling of blood in leg veins once you are upright, decreasing the blood pressure in your upper body.

21.15 Diastolic blood pressure = 95 mmHg; systolic blood pressure = 142 mmHg; pulse pressure = 47 mmHg. This person has stage I hypertension because the systolic blood pressure is greater than 140 mmHg and the diastolic blood pressure is greater than 90 mmHg.

21.16 Almost-normal blood pressure in a person who has lost blood does not necessarily indicate that the patient's tissues are receiving adequate blood flow; if systemic vascular resistance has increased greatly, tissue perfusion may be inadequate.

21.17 The two main circulatory routes are the systemic circulation and the pulmonary circulation.

21.18 The subdivisions of the aorta are the ascending aorta, arch of the aorta, thoracic aorta, and abdominal aorta.

21.19 Branches of the arch of aorta (in order of origination) are the brachiocephalic trunk, left common carotid artery, and left subclavian artery.

21.20 The thoracic aorta begins at the level of the intervertebral disc between T4 and T5.

21.21 The abdominal aorta begins at the aortic hiatus in the diaphragm.

21.22 The abdominal aorta divides into the common iliac arteries at about the level of L4.

21.23 The superior vena cava drains regions above the diaphragm, and the inferior vena cava drains regions below the diaphragm.

21.24 All venous blood in the brain drains into the internal jugular veins.

21.25 The median cubital vein of the free upper limb is often used for withdrawing blood.

21.26 The inferior vena cava returns blood from abdominopelvic viscera to the heart.

21.27 Superficial veins of the free lower limbs are the dorsal venous arch and the great saphenous and small saphenous veins.

21.28 The hepatic veins carry blood away from the liver.

21.29 After birth, the pulmonary arteries are the only arteries that carry deoxygenated blood.

21.30 Exchange of materials between mother and fetus occurs across the placenta.

21.31 Blood vessels and blood are derived from mesoderm.

THE LYMPHATIC SYSTEM AND IMMUNITY | 22

The lymphatic system contributes to homeostasis by draining interstitial fluid as well as providing the mechanisms for defense against disease. •

Maintaining homeostasis in the body requires continual combat against harmful agents in our internal and external environment. Despite constant exposure to a variety of **pathogens** (PATH-ō-jens), disease-producing microbes such as bacteria and viruses, most people remain healthy. The body surface also endures cuts and bumps, exposure to ultraviolet rays in sunlight, chemical toxins, and minor burns with an array of defensive ploys.

Immunity or **resistance** is the ability to ward off damage or disease through our defenses. Vulnerability or lack of resistance is termed **susceptibility.** The two general types of immunity are (1) innate and (2) adaptive. **Innate (nonspecific) immunity** refers to defenses that are present at birth. They are always present and available to provide rapid responses to protect us against disease. Innate immunity does not involve specific recognition of a microbe and acts against all microbes in the same way. In addition, innate immunity does not have a memory component; that is, it cannot recall a previous contact with a foreign molecule. Among the components of innate immunity are the first line of defense (the physical and chemical barriers of the skin and mucous membranes) and the second line of defense (antimicrobial substances, natural killer cells, phagocytes, inflammation, and fever). Innate immune responses represent immunity's early warning system and are designed to prevent microbes from gaining access into the body and to help eliminate those that do gain access.

Adaptive (specific) immunity refers to defenses that involve specific recognition of a microbe once it has breached the innate immunity defenses. Adaptive immunity is based on a specific response to a specific microbe; that is, it adapts or adjusts to handle a specific microbe. Unlike innate immunity, adaptive immunity is slower to respond, but it does have a memory component. Adaptive immunity involves lymphocytes (a type of white blood cell) called T lymphocytes (T cells) and B lymphocytes (B cells).

The body system responsible for adaptive immunity (and some aspects of innate immunity) is the lymphatic system. This system is closely allied with the cardiovascular system, and it also functions with the digestive system in the absorption of fatty foods. In this chapter, we will explore the mechanisms that provide defenses against intruders and promote the repair of damaged body tissues.

LYMPHATIC SYSTEM STRUCTURE AND FUNCTION

⊙ **OBJECTIVES**

- List the components and major functions of the lymphatic system.
- Describe the organization of lymphatic vessels.
- Explain the formation and flow of lymph.
- Compare the structure and functions of the primary and secondary lymphatic organs and tissues.

The **lymphatic system** (lim-FAT-ik) consists of a fluid called lymph, vessels called lymphatic vessels that transport the lymph, a number of structures and organs containing lymphatic tissue, and red bone marrow, where stem cells develop into the various types of blood cells, including lymphocytes (Figure 22.1). It assists in circulating body fluids and helps defend the body against disease-causing agents. As you will see shortly, most components of blood plasma filter through blood capillary walls to form interstitial fluid. After interstitial fluid passes into lymphatic vessels, it is called **lymph** (LIMF = clear fluid). The major difference between interstitial fluid and lymph is location: Interstitial fluid is found between cells, and lymph is located within lymphatic vessels and lymphatic tissue.

Lymphatic tissue is a specialized form of reticular connective tissue (see Table 4.4C on page 129) that contains large numbers of lymphocytes. Recall from Chapter 19 that lymphocytes are agranular white blood cells (see page 699). Two types of lymphocytes participate in adaptive immune responses: B cells and T cells.

Functions of the Lymphatic System

The lymphatic system has three primary functions:

1. Drains excess interstitial fluid. Lymphatic vessels drain excess interstitial fluid from tissue spaces and return it to the blood.

2. Transports dietary lipids. Lymphatic vessels transport lipids and lipid-soluble vitamins (A, D, E, and K) absorbed by the gastrointestinal tract.

3. Carries out immune responses. Lymphatic tissue initiates highly specific responses directed against particular microbes or abnormal cells.

Lymphatic Vessels and Lymph Circulation

Lymphatic vessels begin as **lymphatic capillaries.** These tiny vessels, which are located in the spaces between cells, are closed at one end (Figure 22.2 on page 834). Just as blood capillaries converge to form venules and then veins, lymphatic capillaries unite to form larger lymphatic vessels (see Figure 22.1), which resemble veins in structure but have thinner walls and more valves. At intervals along the lymphatic vessels, lymph flows through lymph nodes, encapsulated bean-shaped organs consisting of masses of B cells and T cells. In the skin, lymphatic ves-

sels lie in the subcutaneous tissue and generally follow the same route as veins; lymphatic vessels of the viscera generally follow arteries, forming plexuses (networks) around them. Tissues that lack lymphatic capillaries include avascular tissues (such as cartilage, the epidermis, and the cornea of the eye), the central nervous system, portions of the spleen, and red bone marrow.

Lymphatic Capillaries

Lymphatic capillaries are slightly larger in diameter than blood capillaries and have a unique one-way structure that permits interstitial fluid to flow into them but not out. The ends of endothelial cells that make up the wall of a lymphatic capillary overlap (Figure 22.2b). When pressure is greater in the interstitial fluid than in lymph, the cells separate slightly, like the opening of a one-way swinging door, and interstitial fluid enters the lymphatic capillary. When pressure is greater inside the lymphatic capillary, the cells adhere more closely, and lymph cannot escape back into interstitial fluid. The pressure is relieved as lymph moves further down the lymphatic capillary. Attached to the lymphatic capillaries are *anchoring filaments,* which contain elastic fibers. They extend out from the lymphatic capillary, attaching lymphatic endothelial cells to surrounding tissues. When excess interstitial fluid accumulates and causes tissue swelling, the anchoring filaments are pulled, making the openings between cells even larger so that more fluid can flow into the lymphatic capillary.

In the small intestine, specialized lymphatic capillaries called **lacteals** (LAK-tē-als; *lact-* = milky) carry dietary lipids into lymphatic vessels and ultimately into the blood. The presence of these lipids causes the lymph draining from the small intestine to appear creamy white; such lymph is referred to as **chyle** (KĪL = juice). Elsewhere, lymph is a clear, pale-yellow fluid.

Lymph Trunks and Ducts

As you have already learned, lymph passes from lymphatic capillaries into lymphatic vessels and then through lymph nodes. As lymphatic vessels exit lymph nodes in a particular region of the body, they unite to form **lymph trunks.** The principal trunks are the lumbar, intestinal, bronchomediastinal, subclavian, and jugular trunks (Figure 22.3 on page 835). The **lumbar trunks** drain lymph from the lower limbs, the wall and viscera of the pelvis, the kidneys, the adrenal glands, and the abdominal wall. The **intestinal trunk** drains lymph from the stomach, intestines, pancreas, spleen, and part of the liver. The **bronchomediastinal trunks** drain lymph from the thoracic wall, lung, and heart. The **subclavian trunks** drain the upper limbs. The **jugular trunks** drain the head and neck.

Lymph passes from lymph trunks into two main channels, the thoracic duct and the right lymphatic duct, and then drains into venous blood. The **thoracic (left lymphatic) duct** is about 38–45 cm (15–18 in.) long and begins as a dilation called the **cisterna chyli** (sis-TER-na KĪ-lē; *cisterna* = cavity or reservoir) anterior to the second lumbar vertebra. The thoracic duct is the main duct for the return of lymph to blood. The cisterna

Figure 22.1 Components of the lymphatic system.

The lymphatic system consists of lymph, lymphatic vessels, lymphatic tissues, and red bone marrow.

Palatine tonsil
Submandibular node
Cervical node
Right internal jugular vein
Right lymphatic duct
Right subclavian vein
Thymus
Lymphatic vessel
Thoracic duct
Cisterna chyli
Intestinal node
Large intestine
Appendix
Red bone marrow

Left internal jugular vein
Left subclavian vein
Thoracic duct
Axillary node
Spleen
Aggregated lymphatic follicle (Peyer's patch)
Small intestine
Iliac node
Inguinal node

Lymphatic vessel

Anterior view of principal components of lymphatic system

(b) Areas drained by right lymphatic and thoracic ducts

Area drained by right lymphatic duct

Area drained by thoracic duct

Functions

1. Drains excess interstitial fluid.
2. Transports dietary lipids from the gastrointestinal tract to the blood.
3. Protects against invasion through immune responses.

? What tissue contains stem cells that develop into lymphocytes?

Figure 22.2 Lymphatic capillaries.

Lymphatic capillaries are found throughout the body except in avascular tissues, the central nervous system, portions of the spleen, and bone marrow.

(a) Relationship of lymphatic capillaries to tissue cells and blood capillaries

(b) Details of a lymphatic capillary

? **Is lymph more similar to blood plasma or to interstitial fluid? Why?**

chyli receives lymph from the right and left lumbar trunks and from the intestinal trunk. In the neck, the thoracic duct also receives lymph from the left jugular, left subclavian, and left bronchomediastinal trunks. Therefore, the thoracic duct receives lymph from the left side of the head, neck, and chest, the left upper limb, and the entire body inferior to the ribs (see Figure 22.1b). The thoracic duct in turn drains lymph into venous blood at the junction of the left internal jugular and left subclavian veins.

The **right lymphatic duct** (Figure 22.3) is about 1.2 cm (0.5 in.) long and receives lymph from the right jugular, right subclavian, and right bronchomediastinal trunks. Thus, the right lymphatic duct receives lymph from the upper right side of the body (see Figure 22.1b). From the right lymphatic duct, lymph drains into venous blood at the junction of the right internal jugular and right subclavian veins.

Formation and Flow of Lymph

Most components of blood plasma filter freely through the capillary walls to form interstitial fluid, but more fluid filters out of blood capillaries than returns to them by reabsorption (see Figure 21.7 on page 771). The excess filtered fluid—about 3 liters per day—drains into lymphatic vessels and becomes lymph. Because most plasma proteins are too large to leave blood vessels, interstitial fluid contains only a small amount of protein. Proteins that do leave blood plasma cannot return to the blood by diffusion because the concentration gradient (high level of proteins inside blood capillaries, low level outside) opposes

such movement. The proteins can, however, move readily through the more permeable lymphatic capillaries into lymph. Thus, an important function of lymphatic vessels is to return the lost plasma proteins to the bloodstream.

Like veins, lymphatic vessels contain valves, which ensure the one-way movement of lymph. As noted previously, lymph drains into venous blood through the right lymphatic duct and the thoracic duct at the junction of the internal jugular and subclavian veins (Figure 22.3). Thus, the sequence of fluid flow is blood capillaries (blood) → interstitial spaces (interstitial fluid) → lymphatic capillaries (lymph) → lymphatic vessels (lymph) → lymphatic ducts (lymph) → junction of the internal jugular and subclavian veins (blood). Figure 22.4 on page 837 illustrates this sequence, along with the relationship of the lymphatic and cardiovascular systems.

The same two "pumps" that aid the return of venous blood to the heart maintain the flow of lymph.

1. Skeletal muscle pump. The "milking action" of skeletal muscle contractions (see Figure 21.9 on page 774) compresses lymphatic vessels (as well as veins) and forces lymph toward the junction of the internal jugular and subclavian veins.

2. Respiratory pump. Lymph flow is also maintained by pressure changes that occur during inhalation (breathing in). Lymph flows from the abdominal region, where the pressure is higher, toward the thoracic region, where it is lower. When the pressures reverse during exhalation (breathing out), the valves prevent backflow of lymph. In addition, when a lymphatic vessel

Figure 22.3 **Routes for drainage of lymph from lymph trunks into the thoracic and right lymphatic ducts.**

All lymph returns to the bloodstream through the thoracic (left) lymphatic duct and right lymphatic duct.

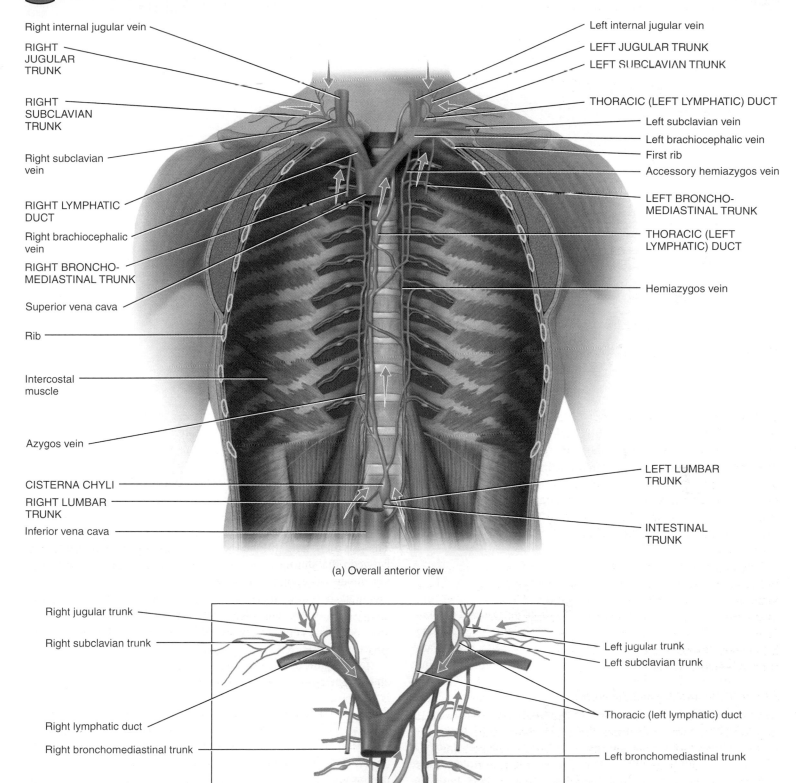

(a) Overall anterior view

(b) Detailed anterior view

Which lymphatic vessels empty into the cisterna chyli, and which duct receives lymph from the cisterna chyli?

Figure 22.4 Schematic diagram showing the relationship of the lymphatic system to the cardiovascular system. The arrows indicate the direction of flow of lymph and blood.

The sequence of fluid flow is blood capillaries (blood) → interstitial spaces (interstitial fluid) → lymphatic capillaries (lymph) → lymphatic vessels (lymph) → lymphatic ducts (lymph) → junction of the internal jugular and subclavian veins (blood).

SYSTEMIC CIRCULATION   PULMONARY CIRCULATION

? **Does inhalation promote or hinder the flow of lymph?**

distends, the smooth muscle in its wall contracts, which helps move lymph from one segment of the vessel to the next.

Lymphatic Organs and Tissues

The widely distributed lymphatic organs and tissues are classified into two groups based on their functions. **Primary lymphatic organs** are the sites where stem cells divide and become **immunocompetent,** that is, capable of mounting an immune response. The primary lymphatic organs are the red bone marrow (in flat bones and the epiphyses of long bones of adults) and the thymus. Pluripotent stem cells in red bone marrow give rise to mature, immunocompetent B cells and to pre-T cells. The pre-T cells in turn migrate to the thymus, where they become immunocompetent T cells. The **secondary lymphatic organs** and **tissues** are the sites where most immune responses occur. They include lymph nodes, the spleen, and lymphatic nodules (follicles). The thymus, lymph nodes, and spleen are considered organs because each is surrounded by a connective tissue capsule; lymphatic nodules, in contrast, are not considered organs because they lack a capsule.

Thymus

The **thymus** is a bilobed organ located in the mediastinum between the sternum and the aorta (Figure 22.5a). An enveloping layer of connective tissue holds the two lobes closely together, but a connective tissue **capsule** separates the two. Extensions of the capsule, called **trabeculae** (tra-BEK-ū-lē = little beams), penetrate inward and divide each lobe into **lobules** (Figure 22.5b).

Each thymic lobule consists of a deeply staining outer cortex and a lighter-staining central medulla (Figure 22.5b). The **cortex** is composed of large numbers of T cells and scattered dendritic cells, epithelial cells, and macrophages. Immature T cells (pre-T cells) migrate from red bone marrow to the cortex of the thymus, where they proliferate and begin to mature. **Dendritic cells** (*dendr-* = a tree), which are derived from monocytes, and so named because they have long, branched projections that resemble the dendrites of a neuron, assist the maturation process. As you will see shortly, dendritic cells in other parts of the body, such as lymph nodes, play another key role in immune responses. Each of the specialized **epithelial cells** in the cortex has several long processes that surround and serve as a framework

Figure 22.5 Thymus. (See Tortora, *A Photographic Atlas of the Human Body, Second Edition,* Figure 7.2a.)

The bilobed thymus is largest at puberty and then atrophies with age.

(a) Thymus of adolescent

(b) Thymic lobules

(c) Details of the thymic medulla

LM 30x

LM 385x

? Which type of lymphocytes mature in the thymus?

for as many as 50 T cells. These epithelial cells help "educate" the pre-T cells in a process known as positive selection (see Figure 22.22). Additionally, they produce thymic hormones that are thought to aid in the maturation of T cells. Only about 2% of developing T cells survive in the cortex. The remaining cells die via apoptosis (programmed cell death). Thymic **macrophages** help clear out the debris of dead and dying cells. The surviving T cells enter the medulla.

The **medulla** consists of widely scattered, more mature T cells, epithelial cells, dendritic cells, and macrophages (Figure 22.5c). Some of the epithelial cells become arranged into concentric layers of flat cells that degenerate and become filled with keratohyalin granules and keratin. These clusters are called **thymic (Hassall's) corpuscles.** Although their role is uncertain, they may serve as sites of T cell death in the medulla. T cells that leave the thymus via the blood migrate to lymph nodes, the spleen, and other lymphatic tissues where they colonize parts of these organs and tissues.

In infants, the thymus is large, with a mass of about 70 g (2.3 oz). After puberty, adipose and areolar connective tissue begin to replace the thymic tissue. By the time a person reaches maturity, the gland has atrophied considerably, and in old age it may weigh only 3 g (0.1 oz). Before the thymus atrophies, it populates the secondary lymphatic organs and tissues with T cells. However, some T cells continue to proliferate in the thymus throughout an individual's lifetime.

Lymph Nodes

Located along lymphatic vessels are about 600 bean-shaped **lymph nodes.** They are scattered throughout the body, both superficially and deep, and usually occur in groups (see Figure 22.1). Large groups of lymph nodes are present near the mammary glands and in the axillae and groin.

Lymph nodes are 1–25 mm (0.04–1 in.) long and, like the thymus, are covered by a **capsule** of dense connective tissue that extends into the node (Figure 22.6). The capsular extensions,

Phagocytosis occurs in five phases: chemotaxis, adherence, ingestion, digestion, and killing (Figure 22.9):

① *Chemotaxis.* Phagocytosis begins with **chemotaxis,** a chemically stimulated movement of phagocytes to a site of damage. Chemicals that attract phagocytes might come from invading microbes, white blood cells, damaged tissue cells, or activated complement proteins.

② *Adherence.* Attachment of the phagocyte to the microbe or other foreign material is termed **adherence.** The binding of complement proteins to the invading pathogen enhances adherence.

③ *Ingestion.* The plasma membrane of the phagocyte extends projections, called **pseudopods,** that engulf the microbe in a process called **ingestion.** When the pseudopods meet, they fuse, surrounding the microorganism with a sac called a **phagosome.**

④ *Digestion.* The phagosome enters the cytoplasm and merges with lysosomes to form a single, larger structure called a **phagolysosome.** The lysosome contributes lysozyme, which breaks down microbial cell walls, and other digestive enzymes that degrade carbohydrates, proteins, lipids, and nucleic acids. The phagocyte also forms lethal oxidants, such as superoxide anion (O_2^-), hypochlorite anion (OCl^-), and hydrogen peroxide (H_2O_2), in a process called an **oxidative burst.**

⑤ *Killing.* The chemical onslaught provided by lysozyme, digestive enzymes, and oxidants within a phagolysosome quickly kills many types of microbes. Any materials that cannot be degraded further remain in structures called **residual bodies.**

Inflammation

Inflammation is a nonspecific, defensive response of the body to tissue damage. Among the conditions that may produce inflammation are pathogens, abrasions, chemical irritations, distortion or disturbances of cells, and extreme temperatures. The four characteristic signs and symptoms of inflammation are **redness, pain, heat,** and **swelling.** Inflammation can also cause the **loss of function** in the injured area (for example, the inability to detect sensations), depending on the site and extent of the injury. Inflammation is an attempt to dispose of microbes, toxins, or foreign material at the site of injury, to prevent their spread to other tissues, and to prepare the site for tissue repair in an attempt to restore tissue homeostasis.

Because inflammation is one of the body's nonspecific defense mechanisms, the response of a tissue to a cut is similar to the response to damage caused by burns, radiation, or bacterial or viral invasion. In each case, the inflammatory response has three basic stages: (1) vasodilation and increased permeability of blood vessels, (2) emigration (movement) of phagocytes from the blood into interstitial fluid, and, ultimately, (3) tissue repair.

VASODILATION AND INCREASED PERMEABILITY OF BLOOD VESSELS   Two immediate changes occur in the blood vessels in a region of tissue injury: **vasodilation** (increase in the diameter) of arterioles and **increased permeability** of capillaries (Figure 22.10). Increased permeability means that substances normally retained in blood are permitted to pass from the blood vessels. Vasodilation allows more blood to flow through the damaged area, and increased permeability permits defensive pro-

Figure 22.9 Phagocytosis of a microbe.

🔑 The major types of phagocytes are neutrophils and macrophages.

(a) Phases of phagocytosis

SEM 1800x

(b) Phagocyte (white blood cell) engulfing a microbe.

❓ **What chemicals are responsible for killing ingested microbes?**

teins such as antibodies and clotting factors to enter the injured area from the blood. The increased blood flow also helps remove microbial toxins and dead cells.

Among the substances that contribute to vasodilation, increased permeability, and other aspects of the inflammatory response are the following:

- *Histamine.* In response to injury, mast cells in connective tissue and basophils and platelets in blood release histamine. Neutrophils and macrophages attracted to the site of injury also stimulate the release of histamine, which causes vasodilation and increased permeability of blood vessels.

- *Kinins.* These polypeptides, formed in blood from inactive precursors called kininogens, induce vasodilation and increased permeability and serve as chemotactic agents for phagocytes. An example of a kinin is bradykinin.

- *Prostaglandins (PGs).* These lipids, especially those of the E series, are released by damaged cells and intensify the effects of histamine and kinins. PGs also may stimulate the emigration of phagocytes through capillary walls.

- *Leukotrienes (LTs).* Produced by basophils and mast cells, LTs cause increased permeability; they also function in adherence of phagocytes to pathogens and as chemotactic agents that attract phagocytes.

- *Complement.* Different components of the complement system stimulate histamine release, attract neutrophils by chemotaxis, and promote phagocytosis; some components can also destroy bacteria.

Dilation of arterioles and increased permeability of capillaries produce three of the signs and symptoms of inflammation: heat, redness (erythema), and swelling (edema). Heat and redness result from the large amount of blood that accumulates in the damaged area. As the local temperature rises slightly, metabolic reactions proceed more rapidly and release additional heat. Edema results from increased permeability of blood vessels, which permits more fluid to move from blood plasma into tissue spaces.

Pain is a prime symptom of inflammation. It results from injury to neurons and from toxic chemicals released by microbes. Kinins affect some nerve endings, causing much of the pain associated with inflammation. Prostaglandins intensify and prolong the pain associated with inflammation. Pain may also be due to increased pressure from edema.

The increased permeability of capillaries allows leakage of blood-clotting factors into tissues. The clotting sequence is set into motion, and fibrinogen is ultimately converted to an insoluble, thick mesh of fibrin threads that localizes and traps invading microbes and blocks their spread.

EMIGRATION OF PHAGOCYTES   Within an hour after the inflammatory process starts, phagocytes appear on the scene. As large amounts of blood accumulate, neutrophils begin to stick to the inner surface of the endothelium (lining) of blood vessels (Figure 22.10). Then the neutrophils begin to squeeze through the wall of the blood vessel to reach the damaged area. This process, called **emigration,** depends on chemotaxis. Neutrophils

Figure 22.10   Inflammation.

 The three stages of inflammation are as follows: (1) vasodilation and increased permeability of blood vessels, (2) phagocyte emigration, and (3) tissue repair.

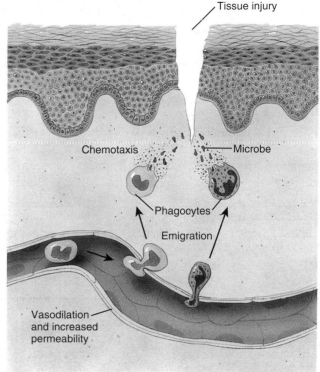

Phagocytes migrate from blood to site of tissue injury

? What causes each of the following signs and symptoms of inflammation: redness, pain, heat, and swelling?

attempt to destroy the invading microbes by phagocytosis. A steady stream of neutrophils is ensured by the production and release of additional cells from red bone marrow. Such an increase in white blood cells in the blood is termed **leukocytosis.**

Although neutrophils predominate in the early stages of infection, they die off rapidly. As the inflammatory response continues, monocytes follow the neutrophils into the infected area. Once in the tissue, monocytes transform into wandering macrophages that add to the phagocytic activity of the fixed macrophages already present. True to their name, macrophages are much more potent phagocytes than neutrophils. They are large enough to engulf damaged tissue, worn-out neutrophils, and invading microbes.

Eventually, macrophages also die. Within a few days, a pocket of dead phagocytes and damaged tissue forms; this collection of dead cells and fluid is called **pus.** Pus formation occurs in most inflammatory responses and usually continues until the infection subsides. At times, pus reaches the surface of the body or drains into an internal cavity and is dispersed; on other occasions the pus remains even after the infection is terminated. In this case, the pus is gradually destroyed over a period of days and is absorbed.

• CLINICAL CONNECTION Abscesses and Ulcers

If pus cannot drain out of an inflamed region, the result is an **abscess**—an excessive accumulation of pus in a confined space. Common examples are pimples and boils. When superficial inflamed tissue sloughs off the surface of an organ or tissue, the resulting open sore is called an **ulcer**. People with poor circulation—for instance, diabetics with advanced atherosclerosis—are susceptible to ulcers in the tissues of their legs. These ulcers, which are called stasis ulcers, develop because of poor oxygen and nutrient supply to tissues that then become very susceptible to even a very mild injury or an infection. •

Fever

Fever is an abnormally high body temperature that occurs because the hypothalamic thermostat is reset. It commonly occurs during infection and inflammation. Many bacterial toxins elevate body temperature, sometimes by triggering release of fever-causing cytokines such as interleukin-1 from macrophages. Elevated body temperature intensifies the effects of interferons, inhibits the growth of some microbes, and speeds up body reactions that aid repair.

Table 22.1 summarizes the components of innate immunity.

☉ CHECKPOINT

7. What physical and chemical factors provide protection from disease in the skin and mucous membranes?
8. What internal defenses provide protection against microbes that penetrate the skin and mucous membranes?
9. How are the activities of natural killer cells and phagocytes similar and different?
10. What are the main signs, symptoms, and stages of inflammation?

ADAPTIVE IMMUNITY

☉ OBJECTIVES

• Define adaptive immunity, and describe how T cells and B cells arise.
• Explain the relationship between an antigen and an antibody.
• Compare the functions of cell-mediated immunity and antibody-mediated immunity.

The ability of the body to defend itself against specific invading agents such as bacteria, toxins, viruses, and foreign tissues is called **adaptive (specific) immunity**. Substances that are recognized as foreign and provoke immune responses are called **antigens (Ags).** Two properties distinguish adaptive immunity from innate immunity: (1) *specificity* for particular foreign molecules (antigens), which also involves distinguishing self from nonself molecules, and (2) *memory* for most previously encountered antigens so that a second encounter prompts an even more rapid and vigorous response. The branch of science that deals with the responses of the body when challenged by antigens is called **immunology** (im′-ū-NOL-ō-jē; *immuno-* = free from service or exempt; *-logy* = study of). The **immune system** includes the cells and tissues that carry out immune responses.

TABLE 22.1

Summary of Innate Defenses

COMPONENT	FUNCTIONS
FIRST LINE OF DEFENSE: SKIN AND MUCOUS MEMBRANES	
Physical factors	
Epidermis of skin	Forms a physical barrier to the entrance of microbes.
Mucous membranes	Inhibit the entrance of many microbes, but not as effective as intact skin.
Mucus	Traps microbes in respiratory and gastrointestinal tracts.
Hairs	Filter out microbes and dust in nose.
Cilia	Together with mucus, trap and remove microbes and dust from upper respiratory tract.
Lacrimal apparatus	Tears dilute and wash away irritating substances and microbes.
Saliva	Washes microbes from surfaces of teeth and mucous membranes of mouth.
Urine	Washes microbes from urethra.
Defecation and vomiting	Expel microbes from body.
Chemical factors	
Sebum	Forms a protective acidic film over the skin surface that inhibits growth of many microbes.
Lysozyme	Antimicrobial substance in perspiration, tears, saliva, nasal secretions, and tissue fluids.
Gastric juice	Destroys bacteria and most toxins in stomach.
Vaginal secretions	Slight acidity discourages bacterial growth; flush microbes out of vagina.
SECOND LINE OF DEFENSE: INTERNAL DEFENSES	
Antimicrobial substances	
Interferons (IFNs)	Protect uninfected host cells from viral infection.
Complement system	Causes cytolysis of microbes, promotes phagocytosis, and contributes to inflammation.
Iron-binding proteins	Inhibit growth of certain bacteria by reducing the amount of available iron.
Antimicrobial proteins (AMPs)	Have broad spectrum antimicrobial activities and attract dendritic cells and mast cells.
Natural killer (NK) cells	Kill infected target cells by releasing granules that contain perforin and granzymes. Phagocytes then kill the released microbes.
Phagocytes	Ingest foreign particulate matter.
Inflammation	Confines and destroys microbes and initiates tissue repair.
Fever	Intensifies the effects of interferons, inhibits growth of some microbes, and speeds up body reactions that aid repair.

Maturation of T Cells and B Cells

Adaptive immunity involves lymphocytes called B cells and T cells. Both develop in primary lymphatic organs (red bone marrow and the thymus) from pluripotent stem cells that originate in red bone marrow (see Figure 19.3 on page 694). B cells complete their development in red bone marrow, a process that con-

tinues throughout life. T cells develop from pre-T cells that migrate from red bone marrow into the thymus, where they mature (Figure 22.11). Most T cells arise before puberty, but they continue to mature and leave the thymus throughout life. B cells and T cells are named based on where they mature. In birds, B cells mature in an organ called the *bursa of Fabricius*. Although

Figure 22.11 B cells and pre-T cells arise from pluripotent stem cells in red bone marrow. B cells and T cells develop in primary lymphatic tissues (red bone marrow and the thymus) and are activated in secondary lymphatic organs and tissues (lymph nodes, spleen, and lymphatic nodules). Once activated, each type of lymphocyte forms a clone of cells that can recognize a specific antigen. For simplicity, antigen receptors, CD4 proteins, and CD8 proteins are not shown in the plasma membranes of the cells of the lymphocyte clones.

The two types of adaptive immunity are cell-mediated immunity and antibody-mediated immunity.

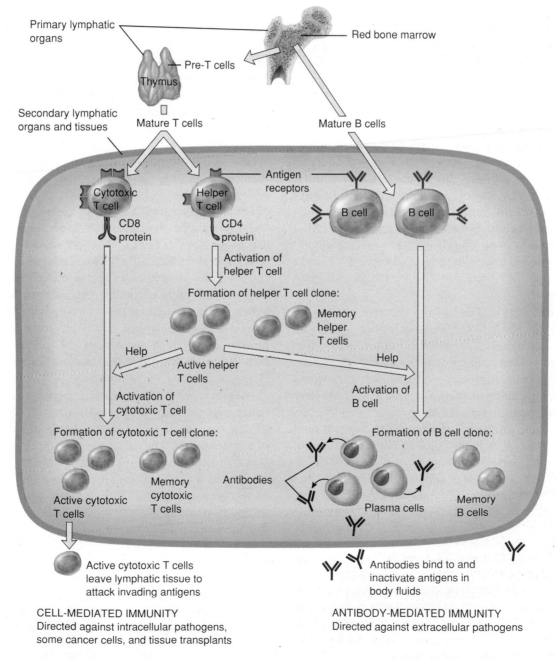

Which type of T cell participates in both cell-mediated and antibody-mediated immune responses?

this organ is not present in humans, the term *B cell* is still used, but the letter *B* stands for *bursa equivalent*, which is the red bone marrow since that is the location in humans where B cells mature. T cells are so named because they mature in the *thymus* gland.

Before T cells leave the thymus or B cells leave red bone marrow, they develop **immunocompetence,** the ability to carry out adaptive immune responses. This means that B cells and T cells begin to make several distinctive proteins that are inserted into their plasma membranes. Some of these proteins function as **antigen receptors**—molecules capable of recognizing specific antigens (Figure 22.11).

There are two major types of mature T cells that exit the thymus: **helper T cells** and **cytotoxic T cells** (Figure 22.11). Helper T cells are also known as **CD4 T cells,** which means that, in addition to antigen receptors, their plasma membranes include a protein called CD4. Cytotoxic T cells are also referred to as **CD8 T cells** because their plasma membranes not only contain antigen receptors but also a protein known as CD8. As we will see later in this chapter, these two types of T cells have very different functions.

Types of Adaptive Immunity

There are two types of adaptive immunity: cell-mediated immunity and antibody-mediated immunity. Both types of adaptive immunity are triggered by antigens. In **cell-mediated immunity,** cytotoxic T cells directly attack invading antigens. In **antibody-mediated immunity,** B cells transform into plasma cells, which synthesize and secrete specific proteins called **antibodies (Abs)** or **immunoglobulins** (im′-ū-nō-GLOB-ū-lins). A given antibody can bind to and inactivate a specific antigen. Helper T cells aid the immune responses of both cell-mediated and antibody-mediated immunity.

Cell-mediated immunity is particularly effective against (1) intracellular pathogens, which include any viruses, bacteria, or fungi that are inside cells; (2) some cancer cells; and (3) foreign tissue transplants. Thus, cell-mediated immunity always involves cells attacking cells. Antibody-mediated immunity works mainly against extracellular pathogens, which include any viruses, bacteria, or fungi that are in body fluids outside cells. Since antibody-mediated immunity involves antibodies that bind to antigens in body *humors* or fluids (such as blood and lymph), it is also referred to as *humoral immunity*.

In most cases, when a particular antigen initially enters the body, there is only a small group of lymphocytes with the correct antigen receptors to respond to that antigen; this small group of cells includes a few helper T cells, cytotoxic T cells, and B cells. Depending on its location, a given antigen can provoke both types of adaptive immune responses. This is due to the fact that when a specific antigen invades the body, there are usually many copies of that antigen spread throughout the body's tissues and fluids. Some copies of the antigen may be present inside body cells (which provokes a cell-mediated immune response by

cytotoxic T cells), while other copies of the antigen may be present in extracellular fluid (which provokes an antibody-mediated immune response by B cells). Thus, cell-mediated and antibody-mediated immune responses often work together to get rid of the large number of copies of a particular antigen from the body.

Clonal Selection: The Principle

As you just learned, when a specific antigen is present in the body, there are usually many copies of that antigen located throughout the body's tissues and fluids. The numerous copies of the antigen initially outnumber the small group of helper T cells, cytotoxic T cells, and B cells with the correct antigen receptors to respond to that antigen. Therefore, once each of these lymphocytes encounters a copy of the antigen and receives stimulatory cues, it subsequently undergoes clonal selection. **Clonal selection** is the process by which a lymphocyte **proliferates** (divides) and **differentiates** (forms more highly specialized cells) in response to a specific antigen. The result of clonal selection is the formation of a population of identical cells, called a **clone,** that can recognize the same specific antigen as the original lymphocyte (Figure 22.11). Before the first exposure to a given antigen, only a few lymphocytes are able to recognize it, but once clonal selection occurs, there are thousands of lymphocytes that can respond to that antigen. Clonal selection of lymphocytes occurs in the secondary lymphatic organs and tissues. The swollen tonsils or lymph nodes in your neck you experienced the last time you were sick were probably caused by clonal selection of lymphocytes participating in an immune response.

A lymphocyte that undergoes clonal selection gives rise to two major types of cells in the clone: effector cells and memory cells. The thousands of **effector cells** of a lymphocyte clone carry out immune responses that ultimately result in the destruction or inactivation of the antigen. Effector cells include **active helper T cells,** which are part of a helper T cell clone; **active cytotoxic T cells,** which are part of a cytotoxic T cell clone; and **plasma cells,** which are part of a B cell clone. Most effector cells eventually die after the immune response has been completed.

Memory cells do not actively participate in the initial immune response to the antigen. However, if the same antigen enters the body again in the future, the thousands of memory cells of a lymphocyte clone are available to initiate a far swifter reaction than occurred during the first invasion. The memory cells respond to the antigen by proliferating and differentiating into more effector cells and more memory cells. Consequently, the second response to the antigen is usually so fast and so vigorous that the antigen is destroyed before any signs or symptoms of disease can occur. Memory cells include **memory helper T cells,** which are part of a helper T cell clone; **memory cytotoxic T cells,** which are part of a cytotoxic T cell clone; and **memory B cells,** which are part of a B cell clone. Most memory cells do not die at the end of an immune response. Instead, they have long life spans (often lasting for decades). The functions of effector cells and memory cells are described in more detail later in this chapter.

Antigens and Antigen Receptors

Antigens have two important characteristics: immunogenicity and reactivity. **Immunogenicity** (im-ū-nō-je-NIS-i-tē; -*genic* = producing) is the ability to provoke an immune response by stimulating the production of specific antibodies, the proliferation of specific T cells, or both. The term **antigen** derives from its function as an *antibody gen*erator. **Reactivity** is the ability of the antigen to react specifically with the antibodies or cells it provoked. Strictly speaking, immunologists define antigens as substances that have reactivity; substances with both immunogenicity and reactivity are considered **complete antigens.** Commonly, however, the term *antigen* implies both immunogenicity and reactivity, and we use the word in this way.

Entire microbes or parts of microbes may act as antigens. Chemical components of bacterial structures such as flagella, capsules, and cell walls are antigenic, as are bacterial toxins. Nonmicrobial examples of antigens include chemical components of pollen, egg white, incompatible blood cells, and transplanted tissues and organs. The huge variety of antigens in the environment provides myriad opportunities for provoking immune responses. Typically, just certain small parts of a large antigen molecule act as the triggers for immune responses. These small parts are called **epitopes** (EP-i-tops), or *antigenic determinants* (Figure 22.12). Most antigens have many epitopes, each of which induces production of a specific antibody or activates a specific T cell.

Antigens that get past the innate defenses generally follow one of three routes into lymphatic tissue: (1) Most antigens that enter the bloodstream (for example, through an injured blood vessel) are trapped as they flow through the spleen. (2) Antigens that penetrate the skin enter lymphatic vessels and lodge in lymph nodes. (3) Antigens that penetrate mucous membranes are entrapped by mucosa-associated lymphatic tissue (MALT).

Chemical Nature of Antigens

Antigens are large, complex molecules. Most often, they are proteins. However, nucleic acids, lipoproteins, glycoproteins, and certain large polysaccharides may also act as antigens. Complete antigens usually have large molecular weights of 10,000 daltons or more, but large molecules that have simple, repeating subunits—for example, cellulose and most plastics—are not usually antigenic. This is why plastic materials can be used in artificial heart valves or joints.

A smaller substance that has reactivity but lacks immunogenicity is called a **hapten** (HAP-ten = to grasp). A hapten can stimulate an immune response only if it is attached to a larger carrier molecule. An example is the small lipid toxin in poison ivy, which triggers an immune response after combining with a body protein. Likewise, some drugs, such as penicillin, may combine with proteins in the body to form immunogenic complexes. Such hapten-stimulated immune responses are responsible for some allergic reactions to drugs and other substances in the environment (see page 868).

As a rule, antigens are foreign substances; they are not usually part of body tissues. However, sometimes the immune system fails to distinguish "friend" (self) from "foe" (nonself). The result is an autoimmune disorder (see page 868) in which self-molecules or cells are attacked as though they were foreign.

Diversity of Antigen Receptors

An amazing feature of the human immune system is its ability to recognize and bind to at least a billion (10^9) different epitopes. Before a particular antigen ever enters the body, T cells and B cells that can recognize and respond to that intruder are ready and waiting. Cells of the immune system can even recognize artificially made molecules that do not exist in nature. The basis for the ability to recognize so many epitopes is an equally large diversity of antigen receptors. Given that human cells contain only about 35,000 genes, how could a billion or more different antigen receptors possibly be generated?

The answer to this puzzle turned out to be simple in concept. The diversity of antigen receptors in both B cells and T cells is the result of shuffling and rearranging a few hundred versions of several small gene segments. This process is called **genetic recombination.** The gene segments are put together in different combinations as the lymphocytes are developing from stem cells in red bone marrow and the thymus. The situation is similar to shuffling a deck of 52 cards and then dealing out three cards. If you did this over and over, you could generate many more than 52 different sets of three cards. Because of genetic recombination, each B cell or T cell has a unique set of gene segments that codes for its unique antigen receptor. After transcription and translation, the receptor molecules are inserted into the plasma membrane.

Figure 22.12 Epitopes (antigenic determinants).

Most antigens have several epitopes that induce the production of different antibodies or activate different T cells.

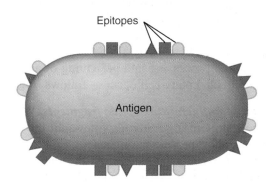

Epitopes

Antigen

? What is the difference between an epitope and a hapten?

④ *Packaging of antigen–MHC-I molecules.* From the ER, antigen–MHC-I molecules are packaged into vesicles.

⑤ *Insertion of antigen–MHC-I complexes into the plasma membrane.* The vesicles that contain antigen–MHC-I complexes undergo exocytosis. As a result, the antigen–MHC-I complexes are inserted into the plasma membrane.

Most cells of the body can process and present endogenous antigens. The display of an endogenous antigen bound to an MHC-I molecule signals that a cell has been infected and needs help.

Cytokines

Cytokines are small protein hormones that stimulate or inhibit many normal cell functions, such as cell growth and differentiation. Lymphocytes and antigen-presenting cells secrete cytokines, as do fibroblasts, endothelial cells, monocytes, hepatocytes, and kidney cells. Some cytokines stimulate proliferation of progenitor blood cells in red bone marrow. Others regulate activities of cells involved in innate defenses or adaptive immune responses, as described in Table 22.2.

•CLINICAL CONNECTION | Cytokine Therapy

Cytokine therapy is the use of cytokines to treat medical conditions. Interferons were the first cytokines shown to have limited effects against some human cancers. Alpha-interferon (Intron A®) is approved in the United States for treating Kaposi (KAP-ō-sē) sarcoma, a cancer that often occurs in patients infected with HIV, the virus that causes AIDS. Other approved uses for alpha-interferon include treating genital herpes caused by the herpes virus; treating hepatitis B and C, caused by the hepatitis B and C viruses; and treating hairy cell leukemia. A form of beta-interferon (Betaseron®) slows the progression of multiple sclerosis (MS) and lessens the frequency and severity of MS attacks. Of the interleukins, the one most widely used to fight cancer is interleukin-2. Although this treatment is effective in causing tumor regression in some patients, it also can be very toxic. Among the adverse effects are high fever, severe weakness, difficulty breathing due to pulmonary edema, and hypotension leading to shock. •

◉ CHECKPOINT

11. What is immunocompetence, and which body cells display it?
12. How do the major histocompatibility complex class I and class II self-antigens function?
13. How do antigens arrive at lymphatic tissues?
14. How do antigen-presenting cells process exogenous antigens?
15. What are cytokines, where do they arise, and how do they function?

TABLE 22.2

Summary of Cytokines Participating in Immune Responses

CYTOKINE	ORIGINS AND FUNCTIONS
Interleukin-1 (IL-1)	Produced by monocytes and macrophages; promotes proliferation of helper T cells; acts on hypothalamus to cause fever.
Interleukin-2 (IL-2) (T cell growth factor)	Secreted by helper T cells; costimulates the proliferation of helper T cells, cytotoxic T cells, and B cells; activates NK cells.
Interleukin-4 (IL-4) (B cell stimulating factor)	Produced by activated helper T cells; costimulator for B cells; causes plasma cells to secrete IgE antibodies (see Table 22.3); promotes growth of T cells.
Interleukin-5 (IL-5)	Produced by certain activated CD4 T cells and activated mast cells; costimulator for B cells; causes plasma cells to secrete IgA antibodies.
Tumor necrosis factor (TNF)	Produced mainly by macrophages; stimulates accumulation of neutrophils and macrophages at sites of inflammation and stimulates their killing of microbes; stimulates macrophages to produce IL-1; induces synthesis of colony-stimulating factors by endothelial cells and fibroblasts; exerts an interferon-like protective effect against viruses; and functions as an endogenous pyrogen to induce fever.
Transforming growth factor beta (TGF-β)	Secreted by T cells and macrophages; has some positive effects but is thought to be important for turning off immune responses; inhibits proliferation of T cells and activation of macrophages.
Gamma-interferon (γ-IFN)	Secreted by helper and cytotoxic T cells and NK cells; strongly stimulates phagocytosis by neutrophils and macrophages; activates NK cells; enhances both cell-mediated and antibody-mediated immune responses.
Alpha- and beta-interferons (α-IFN and β-IFN)	Produced by virus-infected cells to inhibit viral replication in uninfected cells; produced by antigen-stimulated macrophages to stimulate T cell growth; activate NK cells, inhibit cell growth, and suppress formation of some tumors.
Lymphotoxin (LT)	Secreted by cytotoxic T cells; kills infected target cells by activating enzymes that cause fragmentation of DNA.
Perforin	Secreted by cytotoxic T cells and by NK cells; perforates cell membranes of infected target cells, which causes cytolysis.
Macrophage migration inhibiting factor	Produced by cytotoxic T cells; prevents macrophages from leaving site of infection.
Granzymes	Secreted by cytotoxic T cells and NK cells; cause infected target cells to undergo apoptosis.
Granulysin	Secreted by cytotoxic T cells; creates holes in microbial plasma membranes that kill the microbe.

CELL-MEDIATED IMMUNITY

◉ **OBJECTIVES**

- Outline the steps in a cell-mediated immune response.
- Distinguish between the action of natural killer cells and cytotoxic T cells.
- Define immunological surveillance.

A cell-mediated immune response begins with *activation* of a small number of T cells by a specific antigen. Once a T cell has been activated, it undergoes clonal selection. Recall that **clonal selection** is the process by which a lymphocyte proliferates (divides several times) and differentiates (forms more highly specialized cells) in response to a specific antigen. The result of clonal selection is the formation of a **clone** of cells that can recognize the same antigen as the original lymphocyte (see Figure 22.11). Some of the cells of a T cell clone become effector cells, while other cells of the clone become memory cells. The effector cells of a T cell clone carry out immune responses that ultimately result in *elimination* of the intruder.

Activation of T Cells

At any given time, most T cells are inactive. As you learned in the last section, antigen receptors on the surface of T cells, called **T-cell receptors (TCRs),** recognize and bind to specific foreign antigen fragments that are presented in antigen–MHC complexes. There are millions of different T cells; each has its own unique TCRs that can recognize a specific antigen–MHC complex. When an antigen enters the body, only a few T cells have TCRs that can recognize and bind to the antigen. Antigen recognition also involves other surface proteins on T cells, the CD4 or CD8 proteins. These proteins interact with the MHC antigens and help maintain the TCR-MHC coupling. For this reason, they are referred to as *coreceptors.* Antigen recognition by a TCR with CD4 or CD8 proteins is the *first signal* in activation of a T cell.

A T cell becomes activated only if it binds to the foreign antigen and at the same time receives a *second signal,* a process known as **costimulation.** Of the more than 20 known costimulators, some are cytokines, such as **interleukin-2.** Other costimulators include pairs of plasma membrane molecules, one on the surface of the T cell and a second on the surface of an antigen-presenting cell, that enable the two cells to adhere to one another for a period of time.

The need for two signals to activate a T cell is a little like starting and driving a car: When you insert the correct key (antigen) in the ignition (TCR) and turn it, the car starts (recognition of specific antigen), but it cannot move forward until you move the gear shift into drive (costimulation). The need for costimulation may prevent immune responses from occurring accidentally. Different costimulators affect the activated T cell in different ways, just as shifting a car into reverse has a different effect than shifting it into drive. Moreover, recognition (antigen binding to a receptor) without costimulation leads to a prolonged *state of inactivity* called **anergy** in both T cells and B cells. Anergy is

rather like leaving a car in neutral gear with its engine running until it's out of gas!

Once a T cell has received these two signals (antigen recognition and costimulation), it is **activated.** An activated T cell subsequently undergoes clonal selection.

Activation and Clonal Selection of Helper T Cells

Most T cells that display CD4 develop into **helper T cells,** also known as **CD4 T cells.** Inactive (resting) helper T cells recognize exogenous antigen fragments associated with major histocompatibility complex class II (MHC-II) molecules at the surface of an APC (Figure 22.15). With the aid of the CD4 protein,

Figure 22.15 **Activation and clonal selection of a helper T cell.**

🔐 ▭ Once a helper T cell is activated, it forms a clone of active helper T cells and memory helper T cells.

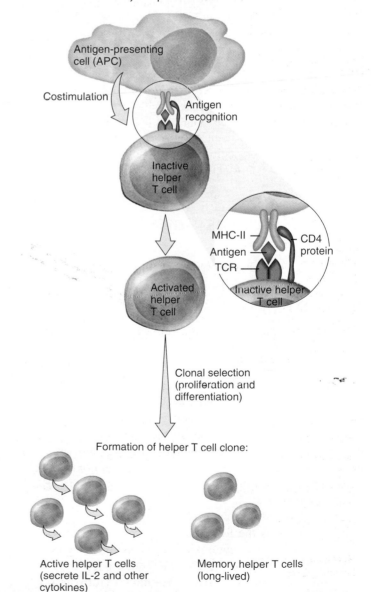

? **What are the first and second signals in activation of a T cell?**

the helper T cell and APC interact with each other (antigenic recognition), costimulation occurs, and the helper T cell becomes activated.

Once activated, the helper T cell undergoes clonal selection (Figure 22.15). The result is the formation of a clone of helper T cells that consists of active helper T cells and memory helper T cells. Within hours after costimulation, **active helper T cells** start secreting a variety of cytokines (see Table 22.2). One very important cytokine produced by helper T cells is interleukin-2 (IL-2), which is needed for virtually all immune responses and is the prime trigger of T cell proliferation. IL-2 can act as a costimulator for resting helper T cells or cytotoxic T cells, and it enhances activation and proliferation of T cells, B cells, and natural killer cells. Some actions of interleukin-2 provide a good example of a beneficial positive feedback system. As noted earlier, activation of a helper T cell stimulates it to start secreting IL-2, which then acts in an autocrine manner by binding to IL-2 receptors on the plasma membrane of the cell that secreted it. One effect is stimulation of cell division. As the helper T cells proliferate, a positive feedback effect occurs because they secrete more IL-2, which causes further cell division. IL-2 may also act in a paracrine manner by binding to IL-2 receptors on neighboring helper T cells, cytotoxic T cells, or B cells. If any of these neighboring cells have already become bound to a copy of the same antigen, IL-2 serves as a costimulator.

The **memory helper T cells** of a helper T cell clone are not active cells. However, if the same antigen enters the body again in the future, memory helper T cells can quickly proliferate and differentiate into more active helper T cells and more memory helper T cells.

Activation and Clonal Selection of Cytotoxic T Cells

Most T cells that display CD8 develop into **cytotoxic T cells,** also termed **CD8 T cells.** Cytotoxic T cells recognize foreign antigens combined with major histocompatibility complex class I (MHC-I) molecules on the surface of (1) body cells infected by microbes, (2) some tumor cells, and (3) cells of a tissue transplant (Figure 22.16). Recognition requires the TCR and CD8 protein to maintain the coupling with MHC-I. Following antigenic recognition, costimulation occurs. In order to become activated, cytotoxic T cells require costimulation by interleukin-2 or other cytokines produced by active helper T cells that have already become bound to copies of the same antigen. (Recall that helper T cells are activated by antigen associated with MHC-II molecules.) Thus, *maximal activation* of cytotoxic T cells requires presentation of antigen associated with both MHC-I and MHC-II molecules.

Once activated, the cytotoxic T cell undergoes clonal selection. The result is the formation of a clone of cytotoxic T cells that consists of active cytotoxic T cells and memory cytotoxic T cells. **Active cytotoxic T cells** attack other body cells that have been infected with the antigen. **Memory cytotoxic T cells** do not attack infected body cells. Instead, they can quickly proliferate and differentiate into more active cytotoxic T cells and more

memory cytotoxic T cells if the same antigen enters the body at a future time.

Elimination of Invaders

Cytotoxic T cells are the soldiers that march forth to do battle with foreign invaders in cell-mediated immune responses. They leave secondary lymphatic organs and tissues and migrate to seek out and destroy infected target cells, cancer cells, and transplanted cells (Figure 22.17). Cytotoxic T cells recognize and attach to target cells. Then, the cytotoxic T cells deliver a "lethal hit" that kills the target cells.

Figure 22.16 Activation and clonal selection of a cytotoxic T cell.

Once a cytotoxic T cell is activated, it forms a clone of active cytotoxic T cells and memory cytotoxic T cells.

? What is the function of the CD8 protein of a cytotoxic T cell?

Cytotoxic T cells kill infected target body cells much like natural killer cells do. The major difference is that cytotoxic T cells have receptors specific for a particular microbe and thus kill only target body cells infected with *one* particular type of microbe; natural killer cells can destroy a wide variety of microbe-infected body cells. Cytotoxic T cells have two principal mechanisms for killing infected target cells.

1. Cytotoxic T cells, using receptors on their surfaces, recognize and bind to infected target cells that have microbial antigens displayed on their surface. The cytotoxic T cell then releases **granzymes,** protein-digesting enzymes that trigger apoptosis (Figure 22.17a). Once the infected cell is destroyed, the released microbes are killed by phagocytes.

2. Alternatively, cytotoxic T cells bind to infected body cells and release two proteins from their granules: perforin and granulysin. **Perforin** inserts into the plasma membrane of the target cell and creates channels in the membrane (Figure 22.17b). As a result, extracellular fluid flows into the target cell and cytolysis (cell bursting) occurs. Other granules in cytotoxic T cells release **granulysin,** which enters through the channels and destroys the microbes by creating holes in their plasma membranes. Cytotoxic T cells may also destroy target

Figure 22.17 Activity of cytotoxic T cells. After delivering a "lethal hit," a cytotoxic T cell can detach and attack another infected target cell displaying the same antigen.

🔑 Cytotoxic T cells release granzymes that trigger apoptosis and perforin that triggers cytolysis of infected target cells.

(a) Cytotoxic T cell destruction of infected cell by release of granzymes that cause apoptosis; released microbes are destroyed by phagocyte

(b) Cytotoxic T cell destruction of infected cell by release of perforins that cause cytolysis; microbes are destroyed by granulysin

Key:
TCR   CD8 protein
Antigen–MHC-I complex

❓ **In addition to cells infected by microbes, what other types of target cells are attacked by cytotoxic T cells?**

cells by releasing a toxic molecule called **lymphotoxin,** which activates enzymes in the target cell. These enzymes cause the target cell's DNA to fragment and the cell dies. In addition, cytotoxic T cells secrete gamma-interferon, which attracts and activates phagocytic cells, and macrophage migration inhibition factor, which prevents migration of phagocytes from the infection site. After detaching from a target cell, a cytotoxic T cell can seek out and destroy another target cell.

Immunological Surveillance

When a normal cell transforms into a cancerous cell, it often displays novel cell surface components called **tumor antigens.** These molecules are rarely, if ever, displayed on the surface of normal cells. If the immune system recognizes a tumor antigen as nonself, it can destroy any cancer cells carrying that antigen. Such immune responses, called **immunological surveillance,** are carried out by cytotoxic T cells, macrophages, and natural killer cells. Immunological surveillance is most effective in eliminating tumor cells due to cancer-causing viruses. For this reason, transplant recipients who are taking immunosuppressive drugs to prevent transplant rejection have an increased incidence of virus-associated cancers. Their risk for other types of cancer is not increased.

• CLINICAL **CONNECTION** | Graft Rejection and Tissue Typing

Organ transplantation involves the replacement of an injured or diseased organ, such as the heart, liver, kidney, lungs, or pancreas, with an organ donated by another individual. Usually, the immune system recognizes the proteins in the transplanted organ as foreign and mounts both cell-mediated and antibody-mediated immune responses against them. This phenomenon is known as **graft rejection.**

The success of an organ or tissue transplant depends on **histocompatibility** (his'-tō-kom-pat-i-BIL-i-tē)—that is, the tissue compatibility between the donor and the recipient. The more similar the MHC antigens, the greater the histocompatibility, and thus the greater the probability that the transplant will not be rejected. **Tissue typing (histocompatibility testing)** is done before any organ transplant. In the United States, a nationwide computerized registry helps physicians select the most histocompatible and needy organ transplant recipients whenever donor organs become available. The closer the match between the major histocompatibility complex proteins of the donor and recipient, the weaker is the graft rejection response.

To reduce the risk of graft rejection, organ transplant recipients receive immunosuppressive drugs. One such drug is *cyclosporine,* derived from a fungus, which inhibits secretion of interleukin-2 by helper T cells but has only a minimal effect on B cells. Thus, the risk of rejection is diminished while resistance to some diseases is maintained. •

◉ CHECKPOINT

16. What are the functions of helper, cytotoxic, and memory T cells?
17. How do cytotoxic T cells kill infected target cells?
18. How is immunological surveillance useful?

ANTIBODY-MEDIATED IMMUNITY

◉ OBJECTIVES
- Describe the steps in an antibody-mediated immune response.
- List the chemical characteristics and actions of antibodies.
- Explain how the complement system operates.
- Distinguish between a primary response and a secondary response to infection.

The body contains not only millions of different T cells but also millions of different B cells, each capable of responding to a specific antigen. Cytotoxic T cells leave lymphatic tissues to seek out and destroy a foreign antigen, but B cells stay put. In the presence of a foreign antigen, a specific B cell in a lymph node, the spleen, or mucosa-associated lymphatic tissue becomes activated. Then it undergoes clonal selection, forming a clone of plasma cells and memory cells. Plasma cells are the effector cells of a B cell clone; they secrete specific antibodies, which in turn circulate in the lymph and blood to reach the site of invasion.

Activation and Clonal Selection of B Cells

During activation of a B cell, an antigen binds to **B-cell receptors (BCRs)** (Figure 22.18). These integral transmembrane proteins are chemically similar to the antibodies that eventually are secreted by plasma cells. Although B cells can respond to an unprocessed antigen present in lymph or interstitial fluid, their response is much more intense when they process the antigen. Antigen processing in a B cell occurs in the following way: The antigen is taken into the B cell, broken down into peptide fragments and combined with MHC-II self-antigens, and moved to the B cell plasma membrane. Helper T cells recognize the antigen–MHC-II complex and deliver the costimulation needed for B cell proliferation and differentiation. The helper T cell produces interleukin-2 and other cytokines that function as costimulators to activate B cells.

Once activated, a B cell undergoes clonal selection (Figure 22.18). The result is the formation of a clone of B cells that consists of plasma cells and memory B cells. **Plasma cells** secrete antibodies. A few days after exposure to an antigen, a plasma cell secretes hundreds of millions of antibodies each day for about 4 or 5 days, until the plasma cell dies. Most antibodies travel in lymph and blood to the invasion site. Interleukin-4 and interleukin-6, also produced by helper T cells, enhance B cell proliferation, B cell differentiation into plasma cells, and secretion of antibodies by plasma cells. **Memory B cells** do not secrete antibodies. Instead, they can quickly proliferate and differentiate into more plasma cells and more memory B cells should the same antigen reappear at a future time.

Different antigens stimulate different B cells to develop into plasma cells and their accompanying memory B cells. All of the B cells of a particular clone are capable of secreting only one type of antibody, which is identical to the antigen receptor displayed by the B cell that first responded. Each specific antigen

Figure 22.18 Activation and clonal selection of B cells. Plasma cells are actually much larger than B cells.

⑥ ⬜═══ Plasma cells secrete antibodies.

❓ **How many different kinds of antibodies will be secreted by the plasma cells in the clone shown here?**

activates only those B cells that are predestined (by the combination of gene segments they carry) to secrete antibody specific to that antigen. Antibodies produced by a clone of plasma cells enter the circulation and form antigen–antibody complexes with the antigen that initiated their production.

Antibodies

An **antibody (Ab)** can combine specifically with the epitope on the antigen that triggered its production. The antibody's structure matches its antigen much as a lock accepts a specific key. In theory, plasma cells could secrete as many different antibodies as there are different B-cell receptors because the same recombined gene segments code for both the BCR and the antibodies eventually secreted by plasma cells.

Antibody Structure

Antibodies belong to a group of glycoproteins called globulins, and for this reason they are also known as **immunoglobulins (Igs).** Most antibodies contain four polypeptide chains (Figure 22.19). Two of the chains are identical to each other and are called **heavy (H) chains;** each consists of about 450 amino acids. Short carbohydrate chains are attached to each heavy polypeptide chain. The two other polypeptide chains, also identical to each other, are called **light (L) chains,** and each consists of about 220 amino acids. A disulfide bond (S—S) holds each light chain to a heavy chain. Two disulfide bonds also link the midregion of the two heavy chains; this part of the antibody displays considerable flexibility and is called the

Figure 22.19 Chemical structure of the immunoglobulin G (IgG) class of antibody. Each molecule is composed of four polypeptide chains (two heavy and two light) plus a short carbohydrate chain attached to each heavy chain. In (a), each circle represents one amino acid. In (b), V_L = variable regions of light chain, C_L = constant region of light chain, V_H = variable region of heavy chain, and C_H = constant region of heavy chain.

An antibody combines only with the epitope on the antigen that triggered its production.

(a) Model of IgG molecule

(b) Diagram of IgG heavy and light chains

? What is the function of the variable regions in an antibody molecule?

hinge region. Because the antibody "arms" can move somewhat as the hinge region bends, an antibody can assume either a T shape (Figure 22.19a) or a Y shape (Figure 22.19b). Beyond the hinge region, parts of the two heavy chains form the **stem region.**

Within each H and L chain are two distinct regions. The tips of the H and L chains, called the **variable (V) regions,** constitute the **antigen-binding site.** The variable region, which is different for each kind of antibody, is the part of the antibody that recognizes and attaches specifically to a particular antigen. Because most antibodies have two antigen-binding sites, they are said to be *bivalent.* Flexibility at the hinge allows the antibody to simultaneously bind to two epitopes that are some distance apart—for example, on the surface of a microbe.

The remainder of each H and L chain, called the **constant (C) region,** is nearly the same in all antibodies of the same class and is responsible for the type of antigen–antibody reaction that occurs. However, the constant region of the H chain differs from one class of antibody to another, and its structure serves as a basis for distinguishing five different classes, designated IgG, IgA, IgM, IgD, and IgE. Each class has a distinct chemical structure and a specific biological role. Because they appear first and are relatively short-lived, IgM antibodies indicate a recent invasion. In a sick patient, the responsible pathogen may be suggested by the presence of high levels of IgM specific to a particular organism. Resistance of the fetus and newborn baby to infection stems mainly from maternal IgG antibodies that cross the placenta before birth and IgA antibodies in breast milk after birth. Table 22.3 summarizes the structures and functions of the five classes of antibodies.

Antibody Actions

The actions of the five classes of immunoglobulins differ somewhat, but all of them act to disable antigens in some way. Actions of antibodies include:

- *Neutralizing antigen.* The reaction of antibody with antigen blocks or neutralizes some bacterial toxins and prevents attachment of some viruses to body cells.

- *Immobilizing bacteria.* If antibodies form against antigens on the cilia or flagella of motile bacteria, the antigen–antibody reaction may cause the bacteria to lose their motility, which limits their spread into nearby tissues.

- *Agglutinating and precipitating antigen.* Because antibodies have two or more sites for binding to antigen, the antigen–antibody reaction may cross-link pathogens to one another, causing agglutination (clumping together). Phagocytic cells ingest agglutinated microbes more readily. Likewise, soluble antigens may come out of solution and form a more-easily phagocytized precipitate when cross-linked by antibodies.

- *Activating complement.* Antigen–antibody complexes initiate the classical pathway of the complement system (discussed shortly).

- *Enhancing phagocytosis.* The stem region of an antibody acts as a flag that attracts phagocytes once antigens have bound to the antibody's variable region. Antibodies enhance the activity of phagocytes by causing agglutination and precipitation, by activating complement, and by coating microbes so that they are more susceptible to phagocytosis.

TABLE 22.3

Classes of Immunoglobulins (Igs)

NAME AND STRUCTURE	CHARACTERISTICS AND FUNCTIONS
IgG	Most abundant, about 80% of all antibodies in the blood; found in blood, lymph, and the intestines; monomer (one-unit) structure. Protects against bacteria and viruses by enhancing phagocytosis, neutralizing toxins, and triggering the complement system. It is the only class of antibody to cross the placenta from mother to fetus, conferring considerable immune protection in newborns.
IgA	Found mainly in sweat, tears, saliva, mucus, breast milk, and gastrointestinal secretions. Smaller quantities are present in blood and lymph. Makes up 10–15% of all antibodies in the blood; occurs as monomers and dimers (two units). Levels decrease during stress, lowering resistance to infection. Provides localized protection of mucous membranes against bacteria and viruses.
IgM	About 5–10% of all antibodies in the blood; also found in lymph. Occurs as pentamers (five units); first antibody class to be secreted by plasma cells after an initial exposure to any antigen. Activates complement and causes agglutination and lysis of microbes. Also present as monomers on the surfaces of B cells, where they serve as antigen receptors. In blood plasma, the anti-A and anti-B antibodies of the ABO blood group, which bind to A and B antigens during incompatible blood transfusions, are also IgM antibodies (see Figure 19.12 on page 709).
IgD	Mainly found on the surfaces of B cells as antigen receptors, where it occurs as monomers; involved in activation of B cells. About 0.2% of all antibodies in the blood.
IgE	Less than 0.1% of all antibodies in the blood; occurs as monomers; located on mast cells and basophils. Involved in allergic and hypersensitivity reactions; provides protection against parasitic worms.

● CLINICAL **CONNECTION** | Monoclonal Antibodies

The antibodies produced against a given antigen by plasma cells can be harvested from an individual's blood. However, because an antigen typically has many epitopes, several different clones of plasma cells produce different antibodies against the antigen. If a single plasma cell could be isolated and induced to proliferate into a clone of identical plasma cells, then a large quantity of identical antibodies could be produced. Unfortunately, lymphocytes and plasma cells are difficult to grow in culture, so scientists sidestepped this difficulty by fusing B cells with tumor cells that grow easily and proliferate endlessly. The resulting hybrid cell is called a **hybridoma** (hī-bri-DŌ-ma). Hybridomas are long-term sources of large quantities of pure, identical antibodies, called **monoclonal antibodies (MAbs)** because they come from a single clone of identical cells. One clinical use of monoclonal antibodies is for measuring levels of a drug in a patient's blood. Other uses include the diagnosis of strep throat, pregnancy, allergies, and diseases such as hepatitis, rabies, and some sexually transmitted diseases. MAbs have also been used to detect cancer at an early stage and to ascertain the extent of metastasis. They may also be useful in preparing vaccines to counteract the rejection associated with transplants, to treat autoimmune diseases, and perhaps to treat AIDS. ●

Role of the Complement System in Immunity

The **complement system** is a defensive system made up of over 30 proteins produced by the liver and found circulating in blood plasma and within tissues throughout the body. Collectively, the complement proteins destroy microbes by causing phagocytosis, cytolysis, and inflammation; they also prevent excessive damage to body tissues.

Most complement proteins are designated by an uppercase letter *C*, numbered C1 through C9, named for the order in which they were discovered. The C1–C9 complement proteins are inactive and become activated only when split by enzymes into active fragments, which are indicated by lowercase letters *a* and *b*. For example, inactive complement protein C3 is split into the activated fragments, C3a and C3b. The active fragments carry out the destructive actions of the C1–C9 complement proteins. Other complement proteins are referred to as factors B, D, and P (properdin).

Complement proteins act in a *cascade*—one reaction triggers another reaction, which in turn triggers another reaction, and so on. With each succeeding reaction, more and more product is formed so that the net effect is amplified many times.

Complement activation may begin by three different pathways (described shortly), all of which activate C3. Once activated, C3 begins a cascade of reactions that brings about phagocytosis, cytolysis, and inflammation as follows (Figure 22.20):

1 Inactivated C3 splits into activated C3a and C3b.

2 C3b binds to the surface of a microbe and receptors on phagocytes attach to the C3b. Thus C3b enhances **phagocytosis** by coating a microbe, a process called **opsonization** (op-sō-ni-ZĀ-shun). Opsonization promotes attachment of a phagocyte to a microbe.

3 C3b also initiates a series of reactions that bring about cytolysis. First, C3b splits C5. The C5b fragment then binds to C6 and C7, which attach to the plasma membrane of an invading microbe. Then C8 and several C9 molecules join the other complement proteins and together form a cylinder-shaped **membrane attack complex,** which inserts into the plasma membrane.

4 The membrane attack complex creates channels in the plasma membrane that result in **cytolysis,** the bursting of the microbial cells due to the inflow of extracellular fluid through the channels.

5 C3a and C5a bind to mast cells and cause them to release histamine that increases blood vessel permeability during **inflammation.** C5a also attracts phagocytes to the site of inflammation (chemotaxis).

Figure 22.20 Complement activation and results of activation. (Adapted from Tortora, Funke, and Case, *Microbiology: An Introduction, Eighth Edition,* Figure 16.10, Pearson Benjamin-Cummings, 2004.)

When activated, complement proteins enhance phagocytosis, cytolysis, and inflammation.

PHAGOCYTOSIS:
Enhancement of phagocytosis by coating with C3b

Histamine

Mast cell

Microbe

INFLAMMATION:
Increase of blood vessel permeability and chemotactic attraction of phagocytes

Channel

Microbial plasma membrane

Membrane attack complex forms channel

CYTOLYSIS:
Bursting of microbe due to inflow of extracellular fluid through channel formed by membrane attack complex C5-C9

? Which pathway for activation of complement involves antibodies? Explain why.

C3 can be activated in three ways: (1) The **classical pathway** starts when antibodies bind to antigens (microbes). The antigen–antibody complex binds and activates C1. Eventually, C3 is activated and the C3 fragments initiate phagocytosis, cytolysis, and inflammation. (2) The **alternative pathway** does not involve antibodies. It is initiated by an interaction between lipid-carbohydrate complexes on the surface of microbes and complement protein factors B, D, and P. This interaction activates C3. (3) In the **lectin pathway,** macrophages that digest microbes release chemicals that cause the liver to produce proteins called **lectins.** Lectins bind to the carbohydrates on the surface of microbes, ultimately causing the activation of C3.

Once complement is activated, proteins in blood and on body cells such as blood cells break down activated C3. In this way, its destructive capabilities cease very quickly so that damage to body cells is minimized.

Immunological Memory

A hallmark of immune responses is memory for specific antigens that have triggered immune responses in the past. Immunological memory is due to the presence of long-lasting antibodies and very long-lived lymphocytes that arise during clonal selection of antigen-stimulated B cells and T cells.

Immune responses, whether cell-mediated or antibody-mediated, are much quicker and more intense after a second or subsequent exposure to an antigen than after the first exposure. Initially, only a few cells have the correct specificity to respond, and the immune response may take several days to build to maximum intensity. Because thousands of memory cells exist after an initial encounter with an antigen, the next time the same antigen appears they can proliferate and differentiate into helper T cells, cytotoxic T cells, or plasma cells within hours.

One measure of immunological memory is *antibody titer* (TĪ-ter), the amount of antibody in serum. After an initial contact with an antigen, no antibodies are present for a period of several days. Then, a slow rise in the antibody titer occurs, first IgM and then IgG, followed by a gradual decline in antibody titer (Figure 22.21). This is the **primary response.**

Memory cells may remain for decades. Every new encounter with the same antigen results in a rapid proliferation of memory cells. After subsequent encounters, the antibody titer is far greater than during a primary response and consists mainly of IgG antibodies. This accelerated, more intense response is called the **secondary response.** Antibodies produced during a secondary response have an even higher affinity for the antigen than those produced during a primary response, and thus they are more successful in disposing of it.

Primary and secondary responses occur during microbial infection. When you recover from an infection without taking antimicrobial drugs, it is usually because of the primary response. If the same microbe infects you later, the secondary response could be so swift that the microbes are destroyed before you exhibit any signs or symptoms of infection.

Figure 22.21 Production of antibodies in the primary (after first exposure) and secondary (after second exposure) responses to a given antigen.

 Immunological memory is the basis for successful immunization by vaccination.

? According to this graph, how much more IgG is circulating in the blood in the secondary response than in the primary response? (*Hint:* Notice that each mark on the antibody titer axis represents a 10-fold increase.)

Immunological memory provides the basis for immunization by vaccination against certain diseases (for example, polio). When you receive the vaccine, which may contain **attenuated** (weakened) or killed whole microbes or portions of microbes, your B cells and T cells are activated. Should you subsequently encounter the living pathogen as an infecting microbe, your body initiates a secondary response.

Table 22.4 summarizes the various ways to acquire adaptive immunity.

TABLE 22.4

Ways to Acquire Adaptive Immunity

METHOD	DESCRIPTION
Naturally acquired active immunity	Following exposure to a microbe, antigen recognition by B cells and T cells and costimulation lead to the formation of antibody-secreting plasma cells, cytotoxic T cells, and B and T memory cells.
Naturally acquired passive immunity	Transfer of IgG antibodies from mother to fetus across placenta, or of IgA antibodies from mother to baby in milk during breast-feeding.
Artificially acquired active immunity	Antigens introduced during a vaccination stimulate cell-mediated and antibody-mediated immune responses, leading to the production of memory cells. The antigens are pretreated to be immunogenic but not pathogenic; that is, they will trigger an immune response but not cause significant illness.
Artificially acquired passive immunity	Intravenous injection of immunoglobulins (antibodies).

23

THE RESPIRATORY SYSTEM

THE RESPIRATORY SYSTEM AND HOMEOSTASIS

The respiratory system contributes to homeostasis by providing for the exchange of gases—oxygen and carbon dioxide—between the atmospheric air, blood, and tissue cells. It also helps adjust the pH of body fluids.

Your body's cells continually use oxygen (O_2) for the metabolic reactions that release energy from nutrient molecules and produce ATP. At the same time, these reactions release carbon dioxide (CO_2). Because an excessive amount of CO_2 produces acidity that can be toxic to cells, excess CO_2 must be eliminated quickly and efficiently. The cardiovascular and respiratory systems cooperate to supply O_2 and eliminate CO_2. The respiratory system provides for gas exchange—intake of O_2 and elimination of CO_2—and the cardiovascular system transports blood containing the gases between the lungs and body cells. Failure of either system disrupts homeostasis by causing rapid death of cells from oxygen starvation and buildup of waste products. In addition to functioning in gas exchange, the respiratory system also participates in regulating blood pH, contains receptors for the sense of smell, filters inspired air, produces sounds, and rids the body of some water and heat in exhaled air. Like the digestive and urinary systems that will be covered in subsequent chapters, in the respiratory system there is an extensive area of contact between the external environment and capillary blood vessels. This area of contact allows the body to constantly renew and replenish the internal fluid environment that surrounds and nourishes every body cell.

RESPIRATORY SYSTEM ANATOMY

◉ OBJECTIVES

- Describe the anatomy and histology of the nose, pharynx, larynx, trachea, bronchi, and lungs.
- Identify the functions of each respiratory system structure.

The **respiratory system** consists of the nose, pharynx (throat), larynx (voice box), trachea (windpipe), bronchi, and lungs (Figure 23.1). Its parts can be classified according to either structure or function. *Structurally,* the respiratory system consists of two parts: (1) The **upper respiratory system** includes the nose, pharynx, and associated structures. (2) The **lower respiratory system** includes the larynx, trachea, bronchi, and lungs. *Functionally,* the respiratory system also consists of two parts: (1) The **conducting zone** consists of a series of interconnecting cavities and tubes both outside and within the lungs. These include the nose, pharynx, larynx, trachea, bronchi, bronchioles, and terminal bronchioles; their function is to filter, warm, and moisten air and conduct it into the lungs. (2) The **respiratory zone** consists of tissues within the lungs where gas exchange occurs. These include the respiratory bronchioles, alveolar ducts, alveolar sacs, and alveoli; they are the main sites of gas exchange between air and blood.

The branch of medicine that deals with the diagnosis and treatment of diseases of the ears, nose, and throat (ENT) is called **otorhinolaryngology** (ō′-tō-rī′-nō-lar′-in-GOL-ō-jē; *oto- = ear; rhino- = nose; laryngo- = voice box; -logy = study of*). A **pulmonologist** is a specialist in the diagnosis and treatment of diseases of the lungs.

Nose

The **nose** can be divided into external and internal portions. The **external nose** is the portion of the nose visible on the face and consists of a supporting framework of bone and hyaline cartilage covered with muscle and skin and lined by a mucous membrane. The frontal bone, nasal bones, and maxillae form the *bony framework* of the external nose (Figure 23.2a on page 877). The *cartilaginous framework* of the external nose consists of the **septal nasal cartilage,** which forms the anterior portion of the nasal septum; the **lateral nasal cartilages** inferior to the nasal bones; and the **alar cartilages,** which form a portion of the walls of the nostrils. Because it consists of pliable hyaline cartilage, the cartilaginous framework of the external nose is somewhat flexible. On the undersurface of the external nose are two openings called the **external nares** (NĀ-rez; singular is **naris**) or **nostrils.** Figure 23.3 on page 878 shows the surface anatomy of the nose.

The interior structures of the external nose have three functions: (1) warming, moistening, and filtering incoming air; (2) detecting olfactory stimuli; and (3) modifying speech vibrations as they pass through the large, hollow resonating chambers. *Resonance* refers to prolonging, amplifying, or modifying a sound by vibration.

The **internal nose** is a large cavity beyond the nasal vestibule in the anterior aspect of the skull that lies inferior to the nasal bone and superior to the mouth; it is lined with muscle and mucous membrane. Anteriorly, the internal nose merges with the external nose, and posteriorly it communicates with the pharynx through two openings called the **internal nares** or **choanae** (kō-Ā-nē) (see Figure 23.2b). Ducts from the *paranasal sinuses* (which drain mucus) and the *nasolacrimal ducts* (which drain tears) also open into the internal nose. Recall from Chapter 7 that the paranasal sinuses are cavities in certain cranial and facial bones lined with mucous membranes that are continuous with the lining of the nasal cavity. Skull bones containing the paranasal sinuses are the frontal, sphenoid, ethmoid, and maxillae. Besides producing mucus, the paranasal sinuses serve as resonating chambers for sound as we speak or sing. The lateral walls of the internal nose are formed by the ethmoid, maxillae, lacrimal, palatine, and inferior nasal conchae bones (see Figure 7.9 on page 210); the ethmoid bone also forms the roof. The palatine bones and palatine processes of the maxillae, which together constitute the hard palate, form the floor of the internal nose.

The space within the internal nose is called the **nasal cavity.** The anterior portion of the nasal cavity just inside the nostrils, called the **nasal vestibule,** is surrounded by cartilage; the superior part of the nasal cavity is surrounded by bone. A vertical partition, the **nasal septum,** divides the nasal cavity into right and left sides. The anterior portion of the nasal septum consists primarily of hyaline cartilage; the remainder is formed by the vomer, perpendicular plate of the ethmoid, maxillae, and palatine bones (see Figure 7.11 on page 213).

When air enters the nostrils, it passes first through the vestibule, which is lined by skin containing coarse hairs that filter out large dust particles. Three shelves formed by projections of the superior, middle, and inferior nasal conchae extend out of each lateral wall of the nasal cavity. The conchae, almost reaching the nasal septum, subdivide each side of the nasal cavity into a series of groovelike passageways—the **superior, middle,** and **inferior meatuses** (mē-Ā-tus-ēz = openings or passages; singular is **meatus**). Mucous membrane lines the cavity and its shelves. The arrangement of conchae and meatuses increases surface area in the internal nose and prevents dehydration by trapping water droplets during exhalation.

Figure 23.1 Structures of the respiratory system.

The upper respiratory system includes the nose, pharynx, and associated structures; the lower respiratory system includes the larynx, trachea, bronchi, and lungs.

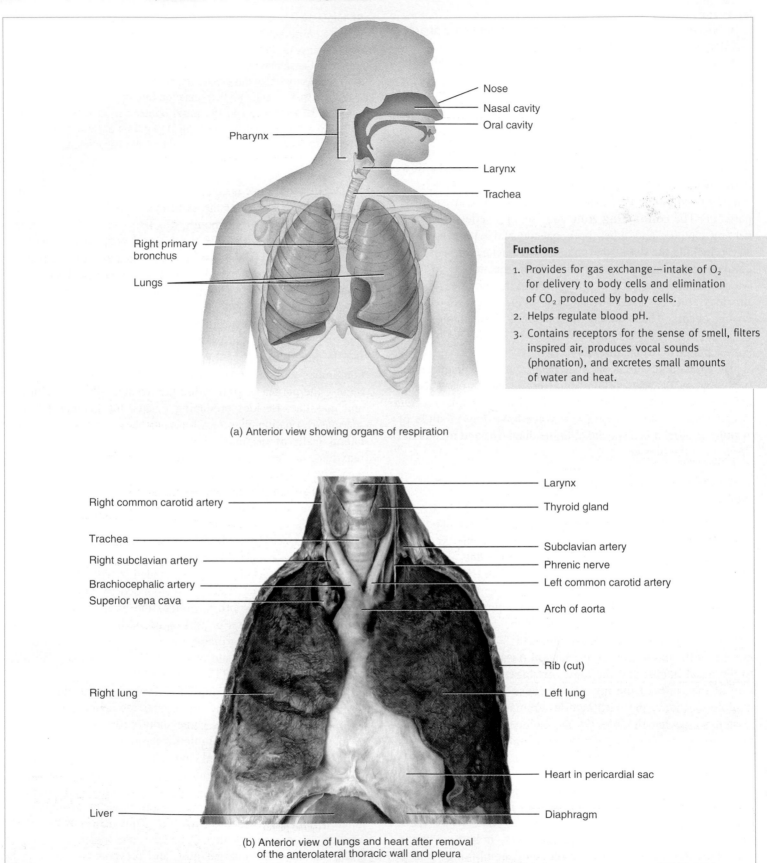

Nose
Nasal cavity
Oral cavity
Pharynx
Larynx
Trachea
Right primary bronchus
Lungs

Functions

1. Provides for gas exchange—intake of O_2 for delivery to body cells and elimination of CO_2 produced by body cells.
2. Helps regulate blood pH.
3. Contains receptors for the sense of smell, filters inspired air, produces vocal sounds (phonation), and excretes small amounts of water and heat.

(a) Anterior view showing organs of respiration

Right common carotid artery
Trachea
Right subclavian artery
Brachiocephalic artery
Superior vena cava
Right lung
Liver

Larynx
Thyroid gland
Subclavian artery
Phrenic nerve
Left common carotid artery
Arch of aorta
Rib (cut)
Left lung
Heart in pericardial sac
Diaphragm

(b) Anterior view of lungs and heart after removal of the anterolateral thoracic wall and pleura

Which structures are part of the conducting zone of the respiratory system?

Figure 23.2 Respiratory structures in the head and neck. (See Tortora, *A Photographic Atlas of the Human Body, Second Edition,* Figures 11.2 and 11.3.)

As air passes through the nose, it is warmed, filtered, moistened, and olfaction occurs.

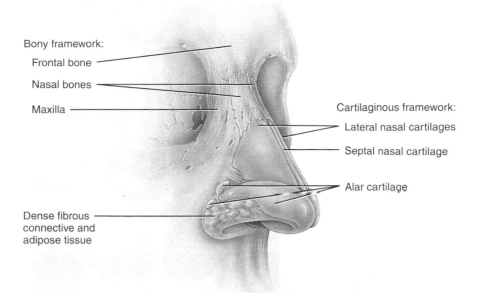

Bony framework:

Frontal bone

Nasal bones

Maxilla

Cartilaginous framework:

Lateral nasal cartilages

Septal nasal cartilage

Alar cartilage

Dense fibrous connective and adipose tissue

(a) Anterolateral view of external portion of nose showing cartilaginous and bony framework

Sagittal plane

Nasal meatuses
- Superior
- Middle
- Inferior

Frontal sinus

Frontal bone

Olfactory epithelium

Sphenoid bone

Sphenoidal sinus

Internal naris

Pharyngeal tonsil

NASOPHARYNX

Opening of auditory tube

Uvula

Palatine tonsil

Fauces

OROPHARYNX

Epiglottis

LARYNGOPHARYNX (hypopharynx)

Esophagus

Trachea

Nasal conchae
- Superior
- Middle
- Inferior

Nasal vestibule

External naris

Maxilla

Oral cavity

Tongue

Palatine bone

Soft palate

Lingual tonsil

Mandible

Hyoid bone

Ventricular fold (false vocal cord)

Vocal fold (true vocal cord)

Larynx

Thyroid cartilage

Cricoid cartilage

Thyroid gland

Nasopharynx

Oropharynx

Laryngopharynx

Regions of the pharynx

(b) Sagittal section of the left side of the head and neck showing the location of respiratory structures

FIGURE 23.2 CONTINUES

Figure 23.4 Larynx. (See Tortora, *A Photographic Atlas of the Human Body, Second Edition,* Figures 11.5 and 11.6.)

The larynx is composed of nine pieces of cartilage.

Larynx   Thyroid gland

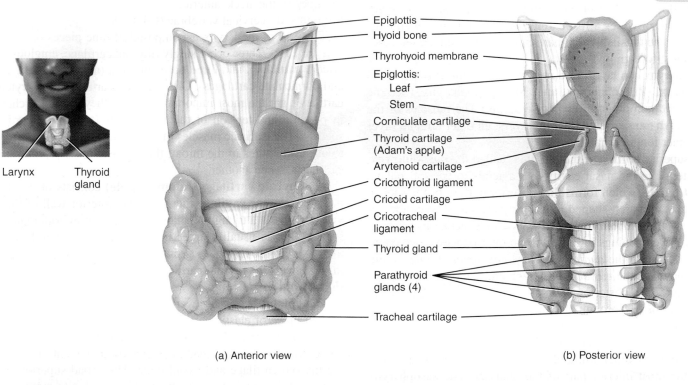

Epiglottis
Hyoid bone
Thyrohyoid membrane
Epiglottis:
   Leaf
   Stem
Corniculate cartilage
Thyroid cartilage (Adam's apple)
Arytenoid cartilage
Cricothyroid ligament
Cricoid cartilage
Cricotracheal ligament
Thyroid gland
Parathyroid glands (4)
Tracheal cartilage

(a) Anterior view

(b) Posterior view

Sagittal plane

Epiglottis
Thyrohyoid membrane
Cuneiform cartilage
Corniculate cartilage
Arytenoid cartilage
Cricoid cartilage
Tracheal cartilage

Hyoid bone
Thyrohyoid membrane
Fat body
Ventricular fold (false vocal cord)
Thyroid cartilage
Vocal fold (true vocal cord)
Cricothyroid ligament
Cricotracheal ligament

(c) Sagittal section

? How does the epiglottis prevent aspiration of foods and liquids?

cuneiform cartilages (KŪ-nē-i-form = wedge-shaped), club-shaped elastic cartilages anterior to the corniculate cartilages, support the vocal folds and lateral aspects of the epiglottis.

The lining of the larynx superior to the vocal folds is nonkeratinized stratified squamous epithelium. The lining of the larynx inferior to the vocal folds is pseudostratified ciliated columnar

epithelium consisting of ciliated columnar cells, goblet cells, and basal cells. The mucus produced by the goblet cells helps trap dust not removed in the upper passages. The cilia in the upper respiratory tract move mucus and trapped particles *down* toward the pharynx; the cilia in the lower respiratory tract move them *up* toward the pharynx.

The Structures of Voice Production

The mucous membrane of the larynx forms two pairs of folds (Figure 23.4c): a superior pair called the **ventricular folds (false vocal cords)** and an inferior pair called the **vocal folds (true vocal cords).** The space between the ventricular folds is known as the **rima vestibuli.** The **laryngeal sinus (ventricle)** is a lateral expansion of the middle portion of the laryngeal cavity inferior to the ventricular folds and superior to the vocal folds (see Figure 23.2b).

When the ventricular folds are brought together, they function in holding the breath against pressure in the thoracic cavity, such as might occur when you strain to lift a heavy object. Deep to the mucous membrane of the vocal folds, which is lined by nonkeratinized stratified squamous epithelium, bands of elastic ligaments are stretched between pieces of rigid cartilage like the strings on a guitar. Intrinsic laryngeal muscles attach to both the rigid cartilage and the vocal folds. When the muscles contract, they pull the elastic ligaments tight and stretch the vocal folds out into the airways so that the rima glottidis is narrowed. If air is directed against the vocal folds, they vibrate and produce sounds (phonation) by setting up sound waves in the column of air in the pharynx, nose, and mouth. The greater the pressure of air, the louder the sound.

When the intrinsic muscles of the larynx contract, they pull on the arytenoid cartilages, which causes them to pivot and slide. Contraction of the posterior cricoarytenoid muscles, for example, moves the vocal folds apart (abduction), thereby opening the rima glottidis (Figure 23.5a). By contrast, contraction of the lateral cricoarytenoid muscles moves the vocal folds together (adduction), thereby closing the rima glottidis (Figure 23.5b). Other intrinsic muscles can elongate (and place tension on) or shorten (and relax) the vocal folds.

Pitch is controlled by the tension on the vocal folds. If they are pulled taut by the muscles, they vibrate more rapidly, and a higher pitch results. Decreasing the muscular tension on the vocal folds causes them to vibrate more slowly and produce lower-pitch sounds. Due to the influence of androgens (male sex hormones), vocal folds are usually thicker and longer in males than in females, and therefore they vibrate more slowly. This is why a man's voice generally has a lower range of pitch than that of a woman.

Figure 23.5 **Movement of the vocal folds.**

The glottis consists of a pair of folds of mucous membrane in the larynx (the vocal folds) and the space between them (the rima glottidis).

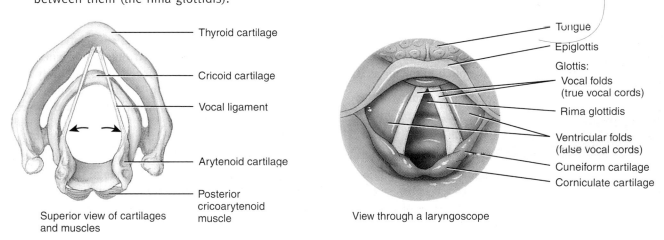

(a) Movement of vocal folds apart (abduction)

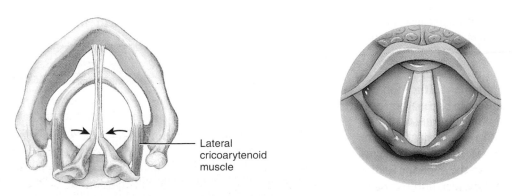

(b) Movement of vocal folds together (adduction)

FIGURE 23.5 CONTINUES

Figure 23.7 Branching of airways from the trachea: the bronchial tree. (See Tortora, *A Photographic Atlas of the Human Body, Second Edition*, Figure 11.8.)

🔑 The bronchial tree begins at the trachea and ends at the terminal bronchioles.

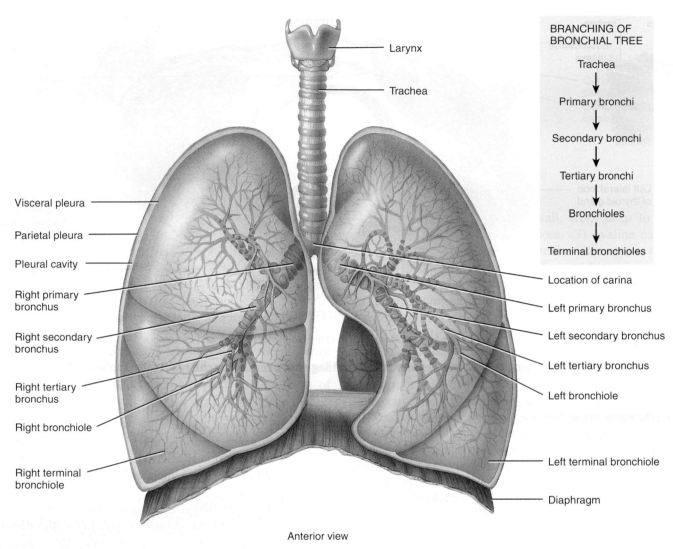

Anterior view

❓ How many lobes and secondary bronchi are present in each lung?

2. Plates of cartilage gradually replace the incomplete rings of cartilage in primary bronchi and finally disappear in the distal bronchioles.

3. As the amount of cartilage decreases, the amount of smooth muscle increases. Smooth muscle encircles the lumen in spiral bands. Because there is no supporting cartilage, however, muscle spasms can close off the airways. This is what happens during an asthma attack, which can be a life-threatening situation.

During exercise, activity in the sympathetic division of the autonomic nervous system (ANS) increases and the adrenal medulla releases the hormones epinephrine and norepinephrine; both of these events cause relaxation of smooth muscle in the bronchioles, which dilates the airways. Because air reaches the alveoli more quickly, lung ventilation improves. The parasympathetic division of the ANS and mediators of allergic reactions such as histamine have the opposite effect, causing contraction of bronchiolar smooth muscle, which results in constriction of distal bronchioles.

◉ CHECKPOINT

4. List the roles of each of the three anatomical regions of the pharynx in respiration.
5. How does the larynx function in respiration and voice production?
6. Describe the location, structure, and function of the trachea.
7. Describe the structure of the bronchial tree.

Lungs

The **lungs** (= lightweights, because they float) are paired cone-shaped organs in the thoracic cavity. They are separated from each other by the heart and other structures in the mediastinum, which divides the thoracic cavity into two anatomically distinct chambers. As a result, if trauma causes one lung to collapse, the other may remain expanded. Each lung is enclosed and protected by a double-layered serous membrane called the **pleural membrane** (PLOOR-al; *pleur-* = side). The superficial layer, called the **parietal pleura,** lines the wall of the thoracic cavity; the deep layer, the **visceral pleura,** covers the lungs themselves (Figure 23.8). Between the visceral and parietal pleurae is a small space, the **pleural cavity,** which contains a small amount of lubricating fluid secreted by the membranes. This pleural fluid reduces friction between the membranes, allowing them to slide easily over one another during breathing. Pleural fluid also causes the two membranes to adhere to one another just as a film of water causes two glass microscope slides to stick together, a phenomenon called surface tension. Separate pleural cavities surround the left and right lungs. Inflammation of the pleural membrane, called **pleurisy** or **pleuritis,** may in its early stages cause pain due to friction between the parietal and visceral layers of the pleura. If the inflammation persists, excess fluid accumulates in the pleural space, a condition known as **pleural effusion.**

• CLINICAL **CONNECTION** | Pneumothorax and Hemothorax

In certain conditions, the pleural cavities may fill with air (**pneumothorax;** *pneumo-* = air or breath), blood (**hemothorax**), or pus. Air in the pleural cavities, most commonly introduced in a surgical opening of the chest or as a result of a stab or gunshot wound, may cause the lungs to collapse. This collapse of a part of a lung, or rarely an entire lung, is called **atelectasis** (at'-e-LEK-ta-sis; *ateles-* = incomplete; *-ectasis-* = expansion). The goal of treatment is the evacuation of air (or blood) from the pleural space, which allows the lung to reinflate. A small pneumothorax may resolve on its own, but it is often necessary to insert a chest tube to assist in evacuation. •

The lungs extend from the diaphragm to just slightly superior to the clavicles and lie against the ribs anteriorly and posteriorly (Figure 23.9a). The broad inferior portion of the lung, the **base,** is concave and fits over the convex area of the diaphragm. The narrow superior portion of the lung is the **apex.** The surface of the lung lying against the ribs, the **costal surface,** matches the rounded curvature of the ribs. The **mediastinal (medial) surface** of each lung contains a region, the **hilum,** through which bronchi, pulmonary blood vessels, lymphatic vessels, and nerves enter and exit (Figure 23.9e). These structures are held together by the pleura and connective tissue and constitute the **root** of the

Figure 23.8 Relationship of the pleural membranes to the lungs.

The parietal pleura lines the thoracic cavity, and the visceral pleura covers the lungs.

Inferior view of a transverse section through the thoracic cavity showing the pleural cavity and pleural membranes

? **What type of membrane is the pleural membrane?**

Figure 23.14 Pressure changes in pulmonary ventilation. During inhalation, the diaphragm contracts, the chest expands, the lungs are pulled outward, and alveolar pressure decreases. During exhalation, the diaphragm relaxes, the lungs recoil inward, and alveolar pressure increases, forcing air out of the lungs.

Air moves into the lungs when alveolar pressure is less than atmospheric pressure, and out of the lungs when alveolar pressure is greater than atmospheric pressure.

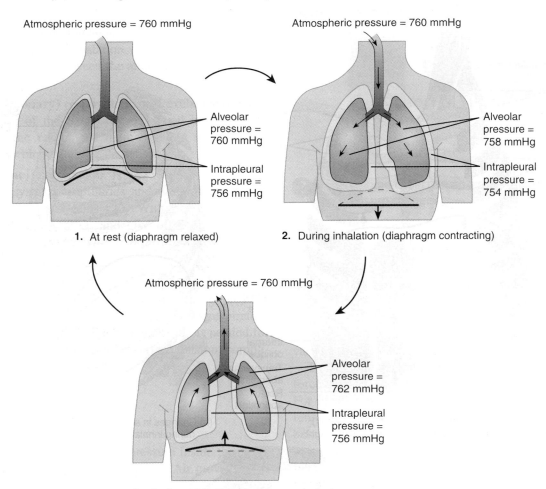

Atmospheric pressure = 760 mmHg

Alveolar pressure = 760 mmHg

Intrapleural pressure = 756 mmHg

1. At rest (diaphragm relaxed)

Atmospheric pressure = 760 mmHg

Alveolar pressure = 758 mmHg

Intrapleural pressure = 754 mmHg

2. During inhalation (diaphragm contracting)

Atmospheric pressure = 760 mmHg

Alveolar pressure = 762 mmHg

Intrapleural pressure = 756 mmHg

3. During exhalation (diaphragm relaxing)

? How does the intrapleural pressure change during a normal, quiet breath?

always flows from a region of higher pressure to a region of lower pressure, inhalation takes place. Air continues to flow into the lungs as long as a pressure difference exists. During deep, forceful inhalations, accessory muscles of inspiration also participate in increasing the size of the thoracic cavity (see Figure 23.13a). The muscles are so named because they make little, if any, contribution during normal quiet inhalation, but during exercise or forced ventilation they may contract vigorously. The accessory muscles of inhalation include the sternocleidomastoid muscles, which elevate the sternum; the scalene muscles, which elevate the first two ribs; and the pectoralis minor muscles, which elevate the third through fifth ribs. Because both normal quiet inhalation and inhalation during exercise or forced ventilation involve muscular contraction, the process of inhalation is said to be *active*.

Figure 23.15a summarizes the events of inhalation.

Exhalation

Breathing out, called **exhalation (expiration),** is also due to a pressure gradient, but in this case the gradient is in the opposite direction: The pressure in the lungs is greater than the pressure of the atmosphere. Normal exhalation during quiet breathing, unlike inhalation, is a *passive process* because no muscular contractions are involved. Instead, exhalation results from **elastic recoil** of the chest wall and lungs, both of which have a natural tendency to spring back after they have been stretched. Two inwardly directed forces contribute to elastic recoil: (1) the recoil of elastic fibers that were stretched during inhalation and (2) the inward pull of surface tension due to the film of alveolar fluid.

Exhalation starts when the inspiratory muscles relax. As the diaphragm relaxes, its dome moves superiorly owing to its elasticity. As the external intercostals relax, the ribs are

Figure 23.15 Summary of events of inhalation and exhalation.

Inhalation and exhalation are caused by changes in alveolar pressure.

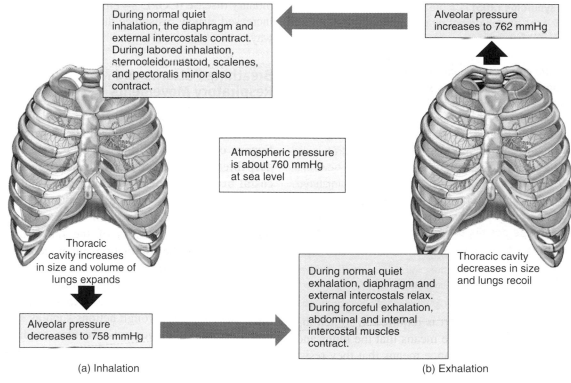

During normal quiet inhalation, the diaphragm and external intercostals contract. During labored inhalation, sternocleidomastoid, scalenes, and pectoralis minor also contract.

Alveolar pressure increases to 762 mmHg

Atmospheric pressure is about 760 mmHg at sea level

Thoracic cavity increases in size and volume of lungs expands

During normal quiet exhalation, diaphragm and external intercostals relax. During forceful exhalation, abdominal and internal intercostal muscles contract.

Thoracic cavity decreases in size and lungs recoil

Alveolar pressure decreases to 758 mmHg

(a) Inhalation

(b) Exhalation

? What is normal atmospheric pressure at sea level?

depressed. These movements decrease the vertical, lateral, and anteroposterior diameters of the thoracic cavity, which decreases lung volume. In turn, the alveolar pressure increases to about 762 mmHg. Air then flows from the area of higher pressure in the alveoli to the area of lower pressure in the atmosphere (see Figure 23.14).

Exhalation becomes active only during forceful breathing, as occurs while playing a wind instrument or during exercise. During these times, muscles of exhalation—the abdominals and internal intercostals (see Figure 23.13a)—contract, which increases pressure in the abdominal region and thorax. Contraction of the abdominal muscles moves the inferior ribs downward and compresses the abdominal viscera, thereby forcing the diaphragm superiorly. Contraction of the internal intercostals, which extend inferiorly and posteriorly between adjacent ribs, pulls the ribs inferiorly. Although intrapleural pressure is always less than alveolar pressure, it may briefly exceed atmospheric pressure during a forceful exhalation, such as during a cough.

Figure 23.15b summarizes the events of exhalation.

Other Factors Affecting Pulmonary Ventilation

As you have just learned, air pressure differences drive airflow during inhalation and exhalation. However, three other factors affect the rate of airflow and the ease of pulmonary ventilation:

surface tension of the alveolar fluid, compliance of the lungs, and airway resistance.

Surface Tension of Alveolar Fluid

As noted earlier, a thin layer of alveolar fluid coats the luminal surface of alveoli and exerts a force known as **surface tension.** Surface tension arises at all air–water interfaces because the polar water molecules are more strongly attracted to each other than they are to gas molecules in the air. When liquid surrounds a sphere of air, as in an alveolus or a soap bubble, surface tension produces an inwardly directed force. Soap bubbles "burst" because they collapse inward due to surface tension. In the lungs, surface tension causes the alveoli to assume the smallest possible diameter. During breathing, surface tension must be overcome to expand the lungs during each inhalation. Surface tension also accounts for two-thirds of lung elastic recoil, which decreases the size of alveoli during exhalation.

The surfactant (a mixture of phospholipids and lipoproteins) present in alveolar fluid reduces its surface tension below the surface tension of pure water. A deficiency of surfactant in premature infants causes *respiratory distress syndrome,* in which the surface tension of alveolar fluid is greatly increased, so that many alveoli collapse at the end of each exhalation. Great effort is then needed at the next inhalation to reopen the collapsed alveoli.

respiratory zone of the respiratory system—the respiratory bronchioles, alveolar ducts, alveolar sacs, and alveoli—and participates in external respiration. The other 30% (150 mL) remains in the conducting airways of the nose, pharynx, larynx, trachea, bronchi, bronchioles, and terminal bronchioles. Collectively, the conducting airways with air that does not undergo respiratory exchange are known as the **anatomic (respiratory) dead space.** (An easy rule of thumb for determining the volume of your anatomic dead space is that it is about the same in milliliters as your ideal weight in pounds.) Not all of the minute ventilation can be used in gas exchange because some of it remains in the anatomic dead space. The **alveolar ventilation rate** is the volume of air per minute that actually reaches the respiratory zone. In the example just given, alveolar ventilation rate would be 350 mL/breath $\times$ 12 breaths/min = 4200 mL/min.

Several other lung volumes are defined relative to forceful breathing. In general, these volumes are larger in males, taller individuals, and younger adults, and smaller in females, shorter individuals, and the elderly. Various disorders also may be diagnosed by comparison of actual and predicted normal values for a patient's gender, height, and age. The values given here are averages for young adults.

By taking a very deep breath, you can inhale a good deal more than 500 mL. This additional inhaled air, called the **inspiratory reserve volume,** is about 3100 mL in an average adult male and 1900 mL in an average adult female (Figure 23.16). Even more air can be inhaled if inhalation follows forced exhalation. If you inhale normally and then exhale as forcibly as possible, you should be able to push out considerably more air in addition to the 500 mL of tidal volume. The extra 1200 mL in males and 700 mL in females is called the **expiratory reserve volume.** The **FEV$_{1.0}$** is the **forced expiratory volume in 1 second,** the volume of air that can be exhaled from the lungs in 1 second with maximal effort following a maximal inhalation. Typically, chronic obstructive pulmonary disease (COPD) greatly reduces FEV$_{1.0}$ because COPD increases airway resistance.

Even after the expiratory reserve volume is exhaled, considerable air remains in the lungs because the subatmospheric intrapleural pressure keeps the alveoli slightly inflated, and some air also remains in the noncollapsible airways. This volume, which cannot be measured by spirometry, is called the **residual volume** and amounts to about 1200 mL in males and 1100 mL in females.

If the thoracic cavity is opened, the intrapleural pressure rises to equal the atmospheric pressure and forces out some of the residual volume. The air remaining is called the **minimal volume.** Minimal volume provides a medical and legal tool for determining whether a baby is born dead (stillborn) or died after birth. The presence of minimal volume can be demonstrated by placing a piece of lung in water and observing if it floats. Fetal lungs contain no air, so the lung of a stillborn baby will not float in water.

Lung capacities are combinations of specific lung volumes (Figure 23.16). **Inspiratory capacity** is the sum of tidal volume and inspiratory reserve volume (500 mL + 3100 mL = 3600 mL in males and 500 mL + 1900 mL = 2400 mL in females). **Functional residual capacity** is the sum of residual volume and expiratory reserve volume (1200 mL + 1200 mL = 2400 mL in males and 1100 mL + 700 mL = 1800 mL in females). **Vital capacity** is the sum of inspiratory reserve volume, tidal volume, and expiratory reserve volume (4800 mL in males and 3100 mL in females). Finally, **total lung capacity** is the sum of vital capacity and residual volume (4800 mL + 1200 mL = 6000 mL in males and 3100 mL + 1100 mL = 4200 mL in females).

CHECKPOINT

16. What is a spirometer?
17. What is the difference between a lung volume and a lung capacity?
18. How is minute ventilation calculated?
19. Define alveolar ventilation rate and FEV$_{1.0}$.

EXCHANGE OF OXYGEN AND CARBON DIOXIDE

OBJECTIVES
• Explain Dalton's law and Henry's law.
• Describe the exchange of oxygen and carbon dioxide in external and internal respiration.

The exchange of oxygen and carbon dioxide between alveolar air and pulmonary blood occurs via passive diffusion, which is governed by the behavior of gases as described by two gas laws, Dalton's law and Henry's law. Dalton's law is important for understanding how gases move down their pressure differences by diffusion, and Henry's law helps explain how the solubility of a gas relates to its diffusion.

Gas Laws: Dalton's Law and Henry's Law

According to **Dalton's law,** each gas in a mixture of gases exerts its own pressure as if no other gases were present. The pressure of a specific gas in a mixture is called its *partial pressure* (P$_x$); the subscript is the chemical formula of the gas. The total pressure of the mixture is calculated simply by adding all the partial pressures. Atmospheric air is a mixture of gases—nitrogen (N$_2$), oxygen (O$_2$), water vapor (H$_2$O), and carbon dioxide (CO$_2$), plus other gases present in small quantities. Atmospheric pressure is the sum of the pressures of all these gases:

$$\text{Atmospheric pressure (760 mmHg)}$$
$$= P_{N_2} + P_{O_2} + P_{H_2O} + P_{CO_2} + P_{\text{other gases}}$$

We can determine the partial pressure exerted by each component in the mixture by multiplying the percentage of the gas in the mixture by the total pressure of the mixture. Atmospheric air is 78.6% nitrogen, 20.9% oxygen, 0.04% carbon dioxide, and 0.06% other gases; a variable amount of water vapor is also present, about 0.4% on a cool, dry day. Thus, the partial pressures of the gases in inhaled air are as follows:

$$P_{N_2} = 0.786 \times 760 \text{ mmHg} = 597.4 \text{ mmHg}$$
$$P_{O_2} = 0.209 \times 760 \text{ mmHg} = 158.8 \text{ mmHg}$$
$$P_{H_2O} = 0.004 \times 760 \text{ mmHg} = 3.0 \text{ mmHg}$$
$$P_{CO_2} = 0.0004 \times 760 \text{ mmHg} = 0.3 \text{ mmHg}$$
$$P_{\text{other gases}} = 0.0006 \times 760 \text{ mmHg} = \underline{0.5 \text{ mmHg}}$$
$$\text{Total} = 760.0 \text{ mmHg}$$

These partial pressures determine the movement of O_2 and CO_2 between the atmosphere and lungs, between the lungs and blood, and between the blood and body cells. Each gas diffuses across a permeable membrane from the area where its partial pressure is greater to the area where its partial pressure is less. The greater the difference in partial pressure, the faster the rate of diffusion.

Compared with inhaled air, alveolar air has less O_2 (13.6% versus 20.9%) and more CO_2 (5.2% versus 0.04%) for two reasons. First, gas exchange in the alveoli increases the CO_2 content and decreases the O_2 content of alveolar air. Second, when air is inhaled it becomes humidified as it passes along the moist mucosal linings. As water vapor content of the air increases, the relative percentage that is O_2 decreases. In contrast, exhaled air contains more O_2 than alveolar air (16% versus 13.6%) and less CO_2 (4.5% versus 5.2%) because some of the exhaled air was in the anatomic dead space and did not participate in gas exchange. Exhaled air is a mixture of alveolar air and inhaled air that was in the anatomic dead space.

Henry's law states that the quantity of a gas that will dissolve in a liquid is proportional to the partial pressure of the gas and its solubility. In body fluids, the ability of a gas to stay in solution is greater when its partial pressure is higher and when it has a high solubility in water. The higher the partial pressure of a gas over a liquid and the higher the solubility, the more gas will stay in solution. In comparison to oxygen, much more CO_2 is dissolved in blood plasma because the solubility of CO_2 is 24 times greater than that of O_2. Even though the air we breathe contains mostly N_2, this gas has no known effect on bodily functions, and at sea level pressure very little of it dissolves in blood plasma because its solubility is very low.

An everyday experience gives a demonstration of Henry's law. You have probably noticed that a soft drink makes a hissing sound when the top of the container is removed, and bubbles rise to the surface for some time afterward. The gas dissolved in carbonated beverages is CO_2. Because the soft drink is bottled or canned under high pressure and capped, the CO_2 remains dissolved as long as the container is unopened. Once you remove the cap, the pressure decreases and the gas begins to bubble out of solution.

Henry's law explains two conditions resulting from changes in the solubility of nitrogen in body fluids. Even though the air we breathe contains about 79% nitrogen, this gas has no known effect on bodily functions, and very little of it dissolves in blood plasma because of its low solubility at sea level pressure. As the total air pressure increases, the partial pressures of all its gases increase. When a scuba diver breathes air under high pressure, the nitrogen in the mixture can have serious negative effects.

Because the partial pressure of nitrogen is higher in a mixture of compressed air than in air at sea level pressure, a considerable amount of nitrogen dissolves in plasma and interstitial fluid. Excessive amounts of dissolved nitrogen may produce giddiness and other symptoms similar to alcohol intoxication. The condition is called **nitrogen narcosis** or "rapture of the deep."

If a diver comes to the surface slowly, the dissolved nitrogen can be eliminated by exhaling it. However, if the ascent is too rapid, nitrogen comes out of solution too quickly and forms gas bubbles in the tissues, resulting in **decompression sickness** (the **bends**). The effects of decompression sickness typically result from bubbles in nervous tissue and can be mild or severe, depending on the number of bubbles formed. Symptoms include joint pain, especially in the arms and legs, dizziness, shortness of breath, extreme fatigue, paralysis, and unconsciousness.

●CLINICAL CONNECTION | **Hyperbaric Oxygenation**

A major clinical application of Henry's law is **hyperbaric oxygenation** (*hyper* = over; *baros* = pressure), the use of pressure to cause more O_2 to dissolve in the blood. It is an effective technique in treating patients infected by anaerobic bacteria, such as those that cause tetanus and gangrene. (Anaerobic bacteria cannot live in the presence of free O_2.) A person undergoing hyperbaric oxygenation is placed in a hyperbaric chamber, which contains O_2 at a pressure greater than one atmosphere (760 mmHg). As body tissues pick up the O_2, the bacteria are killed. Hyperbaric chambers may also be used for treating certain heart disorders, carbon monoxide poisoning, gas embolisms, crush injuries, cerebral edema, certain hard-to-treat bone infections caused by anaerobic bacteria, smoke inhalation, near-drowning, asphyxia, vascular insufficiencies, and burns. ●

External and Internal Respiration

External respiration or **pulmonary gas exchange** is the diffusion of O_2 from air in the alveoli of the lungs to blood in pulmonary capillaries and the diffusion of CO_2 in the opposite direction (Figure 23.17a). External respiration in the lungs converts **deoxygenated blood** (depleted of some O_2) coming from the right side of the heart into **oxygenated blood** (saturated with O_2) that returns to the left side of the heart (see Figure 21.29 on page 820). As blood flows through the pulmonary capillaries, it picks up O_2 from alveolar air and unloads CO_2 into alveolar air. Although this process is commonly called an "exchange" of gases, each gas diffuses independently from the area where its partial pressure is higher to the area where its partial pressure is lower.

As Figure 23.17a shows, O_2 diffuses from alveolar air, where its partial pressure is 105 mmHg, into the blood in pulmonary capillaries, where P_{O_2} is only 40 mmHg in a resting person. If you have been exercising, the P_{O_2} will be even lower because contracting muscle fibers are using more O_2. Diffusion continues until the P_{O_2} of pulmonary capillary blood increases to match the P_{O_2} of alveolar air, 105 mmHg. Because blood leaving pulmonary capillaries near alveolar air spaces mixes with a small

Figure 23.17 Changes in partial pressures of oxygen and carbon dioxide (in mmHg) during external and internal respiration.

Gases diffuse from areas of higher partial pressure to areas of lower partial pressure.

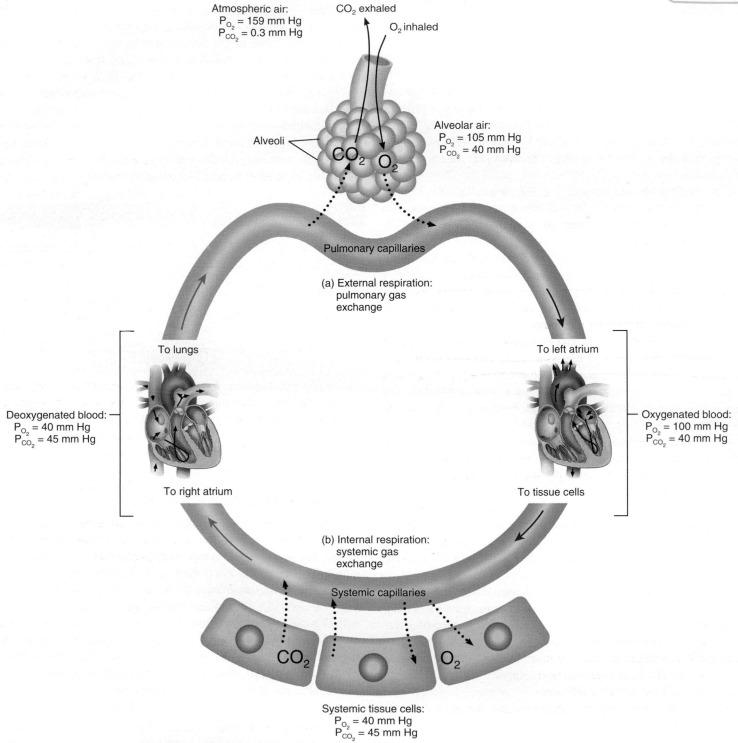

Atmospheric air:
$P_{O_2} = 159$ mm Hg
$P_{CO_2} = 0.3$ mm Hg

CO_2 exhaled

O_2 inhaled

Alveoli

CO_2 O_2

Alveolar air:
$P_{O_2} = 105$ mm Hg
$P_{CO_2} = 40$ mm Hg

Pulmonary capillaries

(a) External respiration: pulmonary gas exchange

To lungs

To left atrium

Deoxygenated blood:
$P_{O_2} = 40$ mm Hg
$P_{CO_2} = 45$ mm Hg

Oxygenated blood:
$P_{O_2} = 100$ mm Hg
$P_{CO_2} = 40$ mm Hg

To right atrium

To tissue cells

(b) Internal respiration: systemic gas exchange

Systemic capillaries

CO_2 O_2

Systemic tissue cells:
$P_{O_2} = 40$ mm Hg
$P_{CO_2} = 45$ mm Hg

? **What causes oxygen to enter pulmonary capillaries from alveoli and to enter tissue cells from systemic capillaries?**

volume of blood that has flowed through conducting portions of the respiratory system, where gas exchange does not occur, the P_{O_2} of blood in the pulmonary veins is slightly less than the P_{O_2} in pulmonary capillaries, about 100 mmHg.

While O_2 is diffusing from alveolar air into deoxygenated blood, CO_2 is diffusing in the opposite direction. The P_{CO_2} of deoxygenated blood is 45 mmHg in a resting person, and the P_{CO_2} of alveolar air is 40 mmHg. Because of this difference in

P_{CO_2}, carbon dioxide diffuses from deoxygenated blood into the alveoli until the P_{CO_2} of the blood decreases to 40 mmHg. Exhalation keeps alveolar P_{CO_2} at 40 mmHg. Oxygenated blood returning to the left side of the heart in the pulmonary veins thus has a P_{CO_2} of 40 mmHg.

The number of capillaries near alveoli in the lungs is very large, and blood flows slowly enough through these capillaries that it picks up a maximal amount of O_2. During vigorous exercise, when cardiac output is increased, blood flows more rapidly through both the systemic and pulmonary circulations. As a result, blood's transit time in the pulmonary capillaries is shorter. Still, the P_{O_2} of blood in the pulmonary veins normally reaches 100 mmHg. In diseases that decrease the rate of gas diffusion, however, the blood may not come into full equilibrium with alveolar air, especially during exercise. When this happens, the P_{O_2} declines and P_{CO_2} rises in systemic arterial blood.

The left ventricle pumps oxygenated blood into the aorta and through the systemic arteries to systemic capillaries. The exchange of O_2 and CO_2 between systemic capillaries and tissue cells is called **internal respiration** or **systemic gas exchange** (Figure 23.17b). As O_2 leaves the bloodstream, oxygenated blood is converted into deoxygenated blood. Unlike external respiration, which occurs only in the lungs, internal respiration occurs in tissues throughout the body.

The P_{O_2} of blood pumped into systemic capillaries is higher (100 mmHg) than the P_{O_2} in tissue cells (40 mmHg at rest) because the cells constantly use O_2 to produce ATP. Due to this pressure difference, oxygen diffuses out of the capillaries into tissue cells and blood P_{O_2} drops to 40 mmHg by the time the blood exits systemic capillaries.

While O_2 diffuses from the systemic capillaries into tissue cells, CO_2 diffuses in the opposite direction. Because tissue cells are constantly producing CO_2, the P_{CO_2} of cells (45 mmHg at rest) is higher than that of systemic capillary blood (40 mmHg). As a result, CO_2 diffuses from tissue cells through interstitial fluid into systemic capillaries until the P_{CO_2} in the blood increases to 45 mmHg. The deoxygenated blood then returns to the heart and is pumped to the lungs for another cycle of external respiration.

In a person at rest, tissue cells, on average, need only 25% of the available O_2 in oxygenated blood; despite its name, deoxygenated blood retains 75% of its O_2 content. During exercise, more O_2 diffuses from the blood into metabolically active cells, such as contracting skeletal muscle fibers. Active cells use more O_2 for ATP production, causing the O_2 content of deoxygenated blood to drop below 75%.

The *rate* of pulmonary and systemic gas exchange depends on several factors.

- **Partial pressure difference of the gases.** Alveolar P_{O_2} must be higher than blood P_{O_2} for oxygen to diffuse from alveolar air into the blood. The rate of diffusion is faster when the difference between P_{O_2} in alveolar air and pulmonary capillary blood is larger; diffusion is slower when the difference is smaller. The differences between P_{O_2} and P_{CO_2} in alveolar air versus pulmonary blood increase during exercise. The larger partial pressure differences accelerate the rates of gas diffusion. The partial pressures of O_2 and CO_2 in alveolar air also depend on the rate of airflow into and out of the lungs. Certain drugs (such as morphine) slow ventilation, thereby decreasing the amount of O_2 and CO_2 that can be exchanged between alveolar air and blood. With increasing altitude, the total atmospheric pressure decreases, as does the partial pressure of O_2—from 159 mmHg at sea level, to 110 mmHg at 10,000 ft, to 73 mmHg at 20,000 ft. Although O_2 still is 20.9% of the total, the P_{O_2} of inhaled air decreases with increasing altitude. Alveolar P_{O_2} decreases correspondingly, and O_2 diffuses into the blood more slowly. The common signs and symptoms of **high altitude sickness**—shortness of breath, headache, fatigue, insomnia, nausea, and dizziness—are due to a lower level of oxygen in the blood.

- **Surface area available for gas exchange.** As you learned earlier in the chapter, the surface area of the alveoli is huge (about 70 m^2 or 750 ft^2). In addition, many capillaries surround each alveolus, so many that as much as 900 mL of blood is able to participate in gas exchange at any instant. Any pulmonary disorder that decreases the functional surface area of the respiratory membranes decreases the rate of external respiration. In emphysema (page 913), for example, alveolar walls disintegrate, so surface area is smaller than normal and pulmonary gas exchange is slowed.

- **Diffusion distance.** The respiratory membrane is very thin, so diffusion occurs quickly. Also, the capillaries are so narrow that the red blood cells must pass through them in single file, which minimizes the diffusion distance from an alveolar air space to hemoglobin inside red blood cells. Buildup of interstitial fluid between alveoli, as occurs in pulmonary edema (page 914), slows the rate of gas exchange because it increases diffusion distance.

- **Molecular weight and solubility of the gases.** Because O_2 has a lower molecular weight than CO_2, it could be expected to diffuse across the respiratory membrane about 1.2 times faster. However, the solubility of CO_2 in the fluid portions of the respiratory membrane is about 24 times greater than that of O_2. Taking both of these factors into account, net outward CO_2 diffusion occurs 20 times more rapidly than net inward O_2 diffusion. Consequently, when diffusion is slower than normal, for example, in emphysema or pulmonary edema, O_2 insufficiency (hypoxia) typically occurs before there is significant retention of CO_2 (hypercapnia).

⦿ CHECKPOINT

20. Distinguish between Dalton's law and Henry's law and give a practical application of each.
21. How does the partial pressure of oxygen change as altitude changes?
22. What are the diffusion paths of oxygen and carbon dioxide during external and internal respiration?
23. What factors affect the rate of diffusion of oxygen and carbon dioxide?

TRANSPORT OF OXYGEN AND CARBON DIOXIDE

⊙ OBJECTIVE

• Describe how the blood transports oxygen and carbon dioxide.

As you have already learned, the blood transports gases between the lungs and body tissues. When O_2 and CO_2 enter the blood,

certain chemical reactions occur that aid in gas transport and gas exchange.

Oxygen Transport

Oxygen does not dissolve easily in water, so only about 1.5% of inhaled O_2 is dissolved in blood plasma, which is mostly water. About 98.5% of blood O_2 is bound to hemoglobin in red blood cells (Figure 23.18). Each 100 mL of oxygenated blood contains

Figure 23.18 Transport of oxygen (O_2) and carbon dioxide (CO_2) in the blood.

Most O_2 is transported by hemoglobin as oxyhemoglobin (Hb–O_2) within red blood cells; most CO_2 is transported in blood plasma as bicarbonate ions (HCO_3^-).

? What is the most important factor that determines how much O_2 binds to hemoglobin?

the equivalent of 20 mL of gaseous O_2. Using the percentages just given, the amount dissolved in the plasma is 0.3 mL and the amount bound to hemoglobin is 19.7 mL.

The heme portion of hemoglobin contains four atoms of iron, each capable of binding to a molecule of O_2 (see Figure 19.4b, c on page 696). Oxygen and hemoglobin bind in an easily reversible reaction to form **oxyhemoglobin:**

$$\underset{\substack{\text{Reduced hemoglobin}\\ \text{(deoxyhemoglobin)}}}{\text{Hb}} + \underset{\text{Oxygen}}{O_2} \underset{\substack{\text{Dissociation}\\ \text{of } O_2}}{\overset{\text{Binding of } O_2}{\rightleftharpoons}} \underset{\text{Oxyhemoglobin}}{\text{Hb--}O_2}$$

The 98.5% of the O_2 that is bound to hemoglobin is trapped inside RBCs, so only the dissolved O_2 (1.5%) can diffuse out of tissue capillaries into tissue cells. Thus, it is important to understand the factors that promote O_2 binding to and dissociation (separation) from hemoglobin.

The Relationship Between Hemoglobin and Oxygen Partial Pressure

The most important factor that determines how much O_2 binds to hemoglobin is the P_{O_2}; the higher the P_{O_2}, the more O_2 combines with Hb. When reduced hemoglobin (Hb) is completely converted to oxyhemoglobin (Hb--O_2), the hemoglobin is said to be **fully saturated;** when hemoglobin consists of a mixture of Hb and Hb--O_2, it is **partially saturated.** The **percent saturation of hemoglobin** expresses the average saturation of hemoglobin with oxygen. For instance, if each hemoglobin molecule has bound two O_2 molecules, then the hemoglobin is 50% saturated because each Hb can bind a maximum of four O_2.

The relationship between the percent saturation of hemoglobin and P_{O_2} is illustrated in the oxygen–hemoglobin dissociation curve in Figure 23.19. Note that when the P_{O_2} is high, hemoglobin binds with large amounts of O_2 and is almost 100% saturated. When P_{O_2} is low, hemoglobin is only partially saturated. In other words, the greater the P_{O_2}, the more O_2 will bind to hemoglobin, until all the available hemoglobin molecules are saturated. Therefore, in pulmonary capillaries, where P_{O_2} is high, a lot of O_2 binds to hemoglobin. In tissue capillaries, where the P_{O_2} is lower, hemoglobin does not hold as much O_2, and the dissolved O_2 is unloaded via diffusion into tissue cells (see Figure 23.18b). Note that hemoglobin is still 75% saturated with O_2 at a P_{O_2} of 40 mmHg, the average P_{O_2} of tissue cells in a person at rest. This is the basis for the earlier statement that only 25% of the available O_2 unloads from hemoglobin and is used by tissue cells under resting conditions.

When the P_{O_2} is between 60 and 100 mmHg, hemoglobin is 90% or more saturated with O_2 (Figure 23.19). Thus, blood picks up a nearly full load of O_2 from the lungs even when the P_{O_2} of alveolar air is as low as 60 mmHg. The Hb–P_{O_2} curve explains why people can still perform well at high altitudes or when they have certain cardiac and pulmonary diseases, even though P_{O_2} may drop as low as 60 mmHg. Note also in the curve that at a considerably lower P_{O_2} of 40 mmHg, hemoglobin is still 75% saturated with O_2. However, oxygen saturation of Hb drops

to 35% at 20 mmHg. Between 40 and 20 mmHg, large amounts of O_2 are released from hemoglobin in response to only small decreases in P_{O_2}. In active tissues such as contracting muscles, P_{O_2} may drop well below 40 mmHg. Then, a large percentage of the O_2 is released from hemoglobin, providing more O_2 to metabolically active tissues.

Other Factors Affecting the Affinity of Hemoglobin for Oxygen

Although P_{O_2} is the most important factor that determines the percent O_2 saturation of hemoglobin, several other factors influence the tightness or **affinity** with which hemoglobin binds O_2. In effect, these factors shift the entire curve either to the left (higher affinity) or to the right (lower affinity). The changing affinity of hemoglobin for O_2 is another example of how homeostatic mechanisms adjust body activities to cellular needs. Each one makes sense if you keep in mind that metabolically active tissue cells need O_2 and produce acids, CO_2, and heat as wastes. The following four factors affect the affinity of hemoglobin for O_2:

1. *Acidity (pH).* As acidity increases (pH decreases), the affinity of hemoglobin for O_2 decreases, and O_2 dissociates more readily from hemoglobin (Figure 23.20a). In other words, increasing acidity enhances the unloading of oxygen from hemoglobin. The main acids produced by metabolically active tissues are lactic acid and carbonic acid. When pH decreases, the entire

Figure 23.19 Oxygen–hemoglobin dissociation curve showing the relationship between hemoglobin saturation and P_{O_2} at normal body temperature.

As P_{O_2} increases, more O_2 combines with hemoglobin.

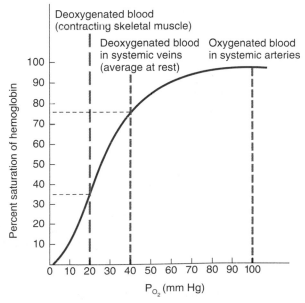

? What point on the curve represents blood in your pulmonary veins right now? In your pulmonary veins if you were jogging?

3. *Bicarbonate ions.* The greatest percentage of CO_2—about 70%—is transported in blood plasma as **bicarbonate ions** **(HCO_3^-)**. As CO_2 diffuses into systemic capillaries and enters red blood cells, it reacts with water in the presence of the enzyme carbonic anhydrase (CA) to form carbonic acid, which dissociates into H^+ and HCO_3^-:

$$\underset{\substack{\text{Carbon} \\ \text{dioxide}}}{CO_2} + \underset{\text{Water}}{H_2O} \overset{CA}{\rightleftharpoons} \underset{\substack{\text{Carbonic} \\ \text{acid}}}{H_2CO_3} \rightleftharpoons \underset{\substack{\text{Hydrogen} \\ \text{ion}}}{H^+} + \underset{\substack{\text{Bicarbonate} \\ \text{ion}}}{HCO_3^-}$$

Thus, as blood picks up CO_2, HCO_3^- accumulates inside RBCs. Some HCO_3^- moves out into the blood plasma, down its concentration gradient. In exchange, chloride ions (Cl^-) move from plasma into the RBCs. This exchange of negative ions, which maintains the electrical balance between blood plasma and RBC cytosol, is known as the **chloride shift** (Figure 23.23b). The net effect of these reactions is that CO_2 is removed from tissue cells and transported in blood plasma as HCO_3^-. As blood passes through pulmonary capillaries in the lungs, all these reactions reverse and CO_2 is exhaled.

Figure 23.23 **Summary of chemical reactions that occur during gas exchange.** (a) As carbon dioxide (CO_2) is exhaled, hemoglobin (Hb) inside red blood cells in pulmonary capillaries unloads CO_2 and picks up O_2 from alveolar air. Binding of O_2 to Hb—H releases hydrogen ions (H^+). Bicarbonate ions (HCO_3^-) pass into the RBC and bind to released H^+, forming carbonic acid (H_2CO_3). The H_2CO_3 dissociates into water (H_2O) and CO_2, and the CO_2 diffuses from blood into alveolar air. To maintain electrical balance, a chloride ion (Cl^-) exits the RBC for each HCO_3^- that enters (reverse chloride shift). (b) CO_2 diffuses out of tissue cells that produce it and enters red blood cells, where some of it binds to hemoglobin, forming carbaminohemoglobin (Hb–CO_2). This reaction causes O_2 to dissociate from oxyhemoglobin (Hb–O_2). Other molecules of CO_2 combine with water to produce bicarbonate ions (HCO_3^-) and hydrogen ions (H^+). As Hb buffers H^+, the Hb releases O_2 (Bohr effect). To maintain electrical balance, a chloride ion (Cl^-) enters the RBC for each HCO_3^- that exits (chloride shift).

Hemoglobin inside red blood cells transports O_2, CO_2, and H^+.

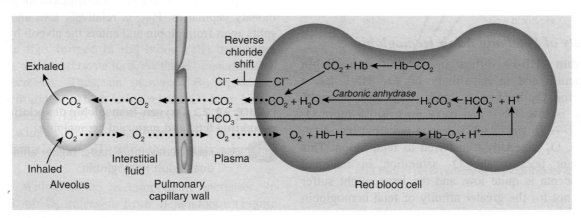

(a) Exchange of O_2 and CO_2 in pulmonary capillaries (external respiration)

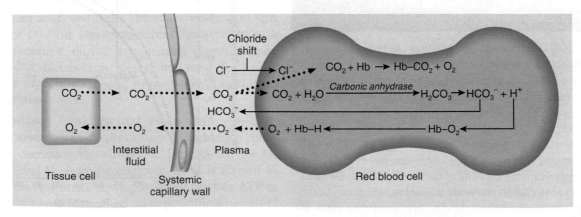

(b) Exchange of O_2 and CO_2 in systemic capillaries (internal respiration)

? Would you expect the concentration of HCO_3^- to be higher in blood plasma taken from a systemic artery or a systemic vein?

The amount of CO_2 that can be transported in the blood is influenced by the percent saturation of hemoglobin with oxygen. The lower the amount of oxyhemoglobin (Hb–O_2), the higher the CO_2 carrying capacity of the blood, a relationship known as the **Haldane effect.** Two characteristics of deoxyhemoglobin give rise to the Haldane effect: (1) Deoxyhemoglobin binds to and thus transports more CO_2 than does Hb–O_2. (2) Deoxyhemoglobin also buffers more H^+ than does Hb–O_2, thereby removing H^+ from solution and promoting conversion of CO_2 to HCO_3^- via the reaction catalyzed by carbonic anhydrase.

Summary of Gas Exchange and Transport in Lungs and Tissues

Deoxygenated blood returning to the pulmonary capillaries in the lungs (Figure 23.23a on page 904) contains CO_2 dissolved in blood plasma, CO_2 combined with globin as carbaminohemoglobin (Hb–CO_2), and CO_2 incorporated into HCO_3^- within RBCs. The RBCs have also picked up H^+, some of which binds to and therefore is buffered by hemoglobin (Hb–H). As blood passes through the pulmonary capillaries, molecules of CO_2 dissolved in blood plasma and CO_2 that dissociates from the globin portion of hemoglobin diffuse into alveolar air and are exhaled. At the same time, inhaled O_2 is diffusing from alveolar air into RBCs and is binding to hemoglobin to form oxyhemoglobin (Hb–O_2). Carbon dioxide also is released from HCO_3^- when H^+ combines with HCO_3^- inside RBCs. The H_2CO_3 formed from this reaction then splits into CO_2, which is exhaled, and H_2O. As the concentration of HCO_3^- declines inside RBCs in pulmonary capillaries, HCO_3^- diffuses in from the blood plasma, in exchange for Cl^-. In sum, oxygenated blood leaving the lungs has increased O_2 content and decreased amounts of CO_2 and H^+. In systemic capillaries, as cells use O_2 and produce CO_2, the chemical reactions reverse (Figure 23.23b).

◉ C H E C K P O I N T

24. In a resting person, how many O_2 molecules are attached to each hemoglobin molecule, on average, in blood in the pulmonary arteries? In blood in the pulmonary veins?
25. What is the relationship between hemoglobin and P_{O_2}? How do temperature, H^+, P_{CO_2}, and BPG influence the affinity of Hb for O_2?
26. Why can hemoglobin unload more oxygen as blood flows through capillaries of metabolically active tissues, such as skeletal muscle during exercise, than is unloaded at rest?

CONTROL OF RESPIRATION

◉ O B J E C T I V E S
• Explain how the nervous system controls breathing.
• List the factors that can alter the rate and depth of breathing.

At rest, about 200 mL of O_2 are used each minute by body cells. During strenuous exercise, however, O_2 use typically increases 15- to 20-fold in normal healthy adults, and as much as 30-fold in elite endurance-trained athletes. Several mechanisms help match respiratory effort to metabolic demand.

Respiratory Center

The size of the thorax is altered by the action of the respiratory muscles, which contract as a result of nerve impulses transmitted to them from centers in the brain and relax in the absence of nerve impulses. These nerve impulses are sent from clusters of neurons located bilaterally in the medulla oblongata and pons of the brain stem. This widely dispersed group of neurons, collectively called the **respiratory center,** can be divided into three areas on the basis of their functions: (1) the medullary rhythmicity area in the medulla oblongata; (2) the pneumotaxic area in the pons; and (3) the apneustic area, also in the pons (Figure 23.24).

Medullary Rhythmicity Area

The function of the **medullary rhythmicity area** (rith-MIS-i-tē) is to control the basic rhythm of respiration. There are inspiratory and expiratory areas within the medullary rhythmicity

Figure 23.24 Locations of areas of the respiratory center.

 The respiratory center is composed of neurons in the medullary rhythmicity area in the medulla oblongata plus the pneumotaxic and apneustic areas in the pons.

Sagittal plane

RESPIRATORY CENTER:

Pneumotaxic area
Apneustic area

Medullary rhythmicity area:
  Inspiratory area
  Expiratory area

Midbrain
Pons
Medulla oblongata
Spinal cord

Sagittal section of brain stem

❓ **Which area contains neurons that are active and then inactive in a repeating cycle?**

area. Figure 23.25 shows the relationships of the inspiratory and expiratory areas during normal quiet breathing and forceful breathing.

During quiet breathing, inhalation lasts for about 2 seconds and exhalation lasts for about 3 seconds. Nerve impulses generated in the **inspiratory area** establish the basic rhythm of breathing. While the inspiratory area is active, it generates nerve impulses for about 2 seconds (Figure 23.25a). The impulses propagate to the external intercostal muscles via intercostal nerves and to the diaphragm via the phrenic nerves. When the nerve impulses reach the diaphragm and external intercostal muscles, the muscles contract and inhalation occurs. Even when all incoming nerve connections to the inspiratory area are cut or blocked, neurons in this area still rhythmically discharge impulses that cause inhalation. At the end of 2 seconds, the inspiratory area becomes inactive and nerve impulses cease. With no impulses arriving, the diaphragm and external intercostal muscles relax for about 3 seconds, allowing passive elastic recoil of the lungs and thoracic wall. Then, the cycle repeats.

The neurons of the **expiratory area** remain inactive during quiet breathing. However, during forceful breathing nerve impulses from the inspiratory area activate the expiratory area (Figure 23.25b). Impulses from the expiratory area cause contraction of the internal intercostal and abdominal muscles, which decreases the size of the thoracic cavity and causes forceful exhalation.

Pneumotaxic Area

Although the medullary rhythmicity area controls the basic rhythm of respiration, other sites in the brain stem help coordinate the transition between inhalation and exhalation. One of these sites is the **pneumotaxic area** (noo-mō-TAK-sik; *pneumo-* = air or breath; *-taxic* = arrangement) in the upper pons (see Figure 23.24), which transmits inhibitory impulses to the inspiratory area. The major effect of these nerve impulses is to help turn off the inspiratory area before the lungs become too full of air. In other words, the impulses shorten the duration of inhalation. When the pneumotaxic area is more active, breathing rate is more rapid.

Apneustic Area

Another part of the brain stem that coordinates the transition between inhalation and exhalation is the **apneustic area** (ap-NOO-stik) in the lower pons (see Figure 23.24). This area sends stimulatory impulses to the inspiratory area that activate it and prolong inhalation. The result is a long, deep inhalation. When the pneumotaxic area is active, it overrides signals from the apneustic area.

Regulation of the Respiratory Center

The basic rhythm of respiration set and coordinated by the inspiratory area can be modified in response to inputs from other brain regions, receptors in the peripheral nervous system, and other factors.

Cortical Influences on Respiration

Because the cerebral cortex has connections with the respiratory center, we can voluntarily alter our pattern of breathing. We can even refuse to breathe at all for a short time. Voluntary control is protective because it enables us to prevent water or irritating

Figure 23.25 Roles of the medullary rhythmicity area in controlling (a) the basic rhythm of respiration and (b) forceful breathing.

During normal, quiet breathing, the expiratory area is inactive; during forceful breathing, the inspiratory area activates the expiratory area.

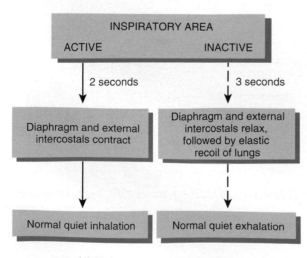

(a) During normal quiet breathing

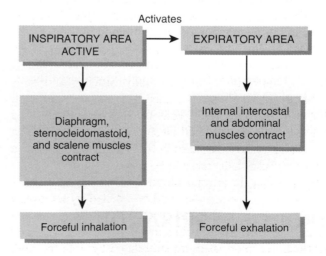

(b) During forceful breathing

? **Which nerves convey impulses from the respiratory center to the diaphragm?**

gases from entering the lungs. The ability to not breathe, however, is limited by the buildup of CO_2 and H^+ in the body. When P_{CO_2} and H^+ concentrations increase to a certain level, the inspiratory area is strongly stimulated, nerve impulses are sent along the phrenic and intercostal nerves to inspiratory muscles, and breathing resumes, whether the person wants it to or not. It is impossible for small children to kill themselves by voluntarily holding their breath, even though many have tried in order to get their way. If breath is held long enough to cause fainting, breathing resumes when consciousness is lost. Nerve impulses from the hypothalamus and limbic system also stimulate the respiratory center, allowing emotional stimuli to alter respirations as, for example, in laughing and crying.

Chemoreceptor Regulation of Respiration

Certain chemical stimuli modulate how quickly and how deeply we breathe. The respiratory system functions to maintain proper levels of CO_2 and O_2 and is very responsive to changes in the levels of these gases in body fluids. We introduced sensory neurons that are responsive to chemicals, called **chemoreceptors,** in Chapter 21. Chemoreceptors in two locations monitor levels of CO_2, H^+, and O_2 and provide input to the respiratory center (Figure 23.26). **Central chemoreceptors** are located in or near the medulla oblongata in the *central* nervous system. They respond to changes in H^+ concentration or P_{CO_2}, or both, in cerebrospinal fluid. **Peripheral chemoreceptors** are located in the **aortic bodies,** clusters of chemoreceptors located in the wall of the arch of the aorta, and in the **carotid bodies,** which are oval nodules in the wall of the left and right common carotid arteries where they divide into the internal and external carotid arteries. (The chemoreceptors of the aortic bodies are located close to the aortic baroreceptors, and the carotid bodies are located close to the carotid sinus baroreceptors. Recall from Chapter 21 that baroreceptors are sensory receptors that monitor blood pressure.) These chemoreceptors are part of the *peripheral* nervous system and are sensitive to changes in P_{O_2}, H^+, and P_{CO_2} in the blood. Axons of sensory neurons from the aortic bodies are part of the vagus (X) nerves, and those from the carotid bodies are part of the right and left glossopharyngeal (IX) nerves.

Because CO_2 is lipid-soluble, it easily diffuses into cells where in the presence of carbonic anhydrase, it combines with water, (H_2O) to form carbonic acid (H_2CO_3). Carbonic acid quickly breaks down into H^+ and HCO_3^-. Thus, an increase in CO_2 in the blood causes an increase in H^+ inside cells, and a decrease in CO_2 causes a decrease in H^+.

Normally, the P_{CO_2} in arterial blood is 40 mmHg. If even a slight increase in P_{CO_2} occurs—a condition called **hypercapnia** or **hypercarbia**—the central chemoreceptors are stimulated and respond vigorously to the resulting increase in H^+ level. The peripheral chemoreceptors also are stimulated by both the high P_{CO_2} and the rise in H^+. In addition, the peripheral chemoreceptors (but not the central chemoreceptors) respond to a deficiency of O_2. When P_{O_2} in arterial blood falls from a normal level of

Figure 23.26 Locations of peripheral chemoreceptors.

Chemoreceptors are sensory neurons that respond to changes in the levels of certain chemicals in the body.

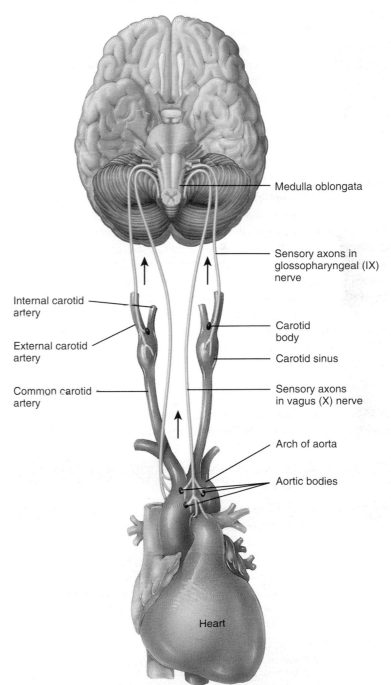

- Medulla oblongata
- Sensory axons in glossopharyngeal (IX) nerve
- Internal carotid artery
- Carotid body
- External carotid artery
- Carotid sinus
- Common carotid artery
- Sensory axons in vagus (X) nerve
- Arch of aorta
- Aortic bodies
- Heart

? Which chemicals stimulate peripheral chemoreceptors?

100 mmHg but is still above 50 mmHg, the peripheral chemoreceptors are stimulated. Severe deficiency of O_2 depresses activity of the central chemoreceptors and inspiratory area, which then do not respond well to any inputs and send fewer impulses to the muscles of inhalation. As the breathing rate decreases or breathing ceases altogether, P_{O_2} falls lower and lower, establishing a positive feedback cycle with a possibly fatal result.

The chemoreceptors participate in a negative feedback system that regulates the levels of CO_2, O_2, and H^+ in the blood (Figure 23.27). As a result of increased P_{CO_2}, decreased pH (increased H^+), or decreased P_{O_2}, input from the central and peripheral chemoreceptors causes the inspiratory area to become highly active, and the rate and depth of breathing increase. Rapid and deep breathing, called **hyperventilation,** allows the inhalation of more O_2 and exhalation of more CO_2 until P_{CO_2} and H^+ are lowered to normal.

If arterial P_{CO_2} is lower than 40 mmHg—a condition called **hypocapnia** or **hypocarbia**—the central and peripheral chemoreceptors are not stimulated, and stimulatory impulses are not sent to the inspiratory area. As a result, the area sets its own moderate pace until CO_2 accumulates and the P_{CO_2} rises to 40 mmHg. The inspiratory center is more strongly stimulated when P_{CO_2} is rising above normal than when P_{O_2} is falling below normal. As a result, people who hyperventilate voluntarily and cause hypocapnia can hold their breath for an unusually long period. Swimmers were once encouraged to hyperventilate just before diving in to compete. However, this practice is risky because the O_2 level may fall dangerously low and cause fainting before the P_{CO_2} rises high enough to stimulate inhalation. If you faint on land you may suffer bumps and bruises, but if you faint in the water you could drown.

● CLINICAL **CONNECTION** Hypoxia

Hypoxia (hī-POK-sē-a; *hypo-* = under) is a deficiency of O_2 at the tissue level. Based on the cause, we can classify hypoxia into four types, as follows:

1. **Hypoxic hypoxia** is caused by a low P_{O_2} in arterial blood as a result of high altitude, airway obstruction, or fluid in the lungs.

2. In **anemic hypoxia,** too little functioning hemoglobin is present in the blood, which reduces O_2 transport to tissue cells. Among the causes are hemorrhage, anemia, and failure of hemoglobin to carry its normal complement of O_2, as in carbon monoxide poisoning.

3. In **ischemic hypoxia,** blood flow to a tissue is so reduced that too little O_2 is delivered to it, even though P_{O_2} and oxyhemoglobin levels are normal.

4. In **histotoxic hypoxia,** the blood delivers adequate O_2 to tissues, but the tissues are unable to use it properly because of the action of some toxic agent. One cause is cyanide poisoning, in which cyanide blocks an enzyme required for the use of O_2 during ATP synthesis. ●

Figure 23.27 Regulation of breathing in response to changes in blood P_{CO_2}, P_{O_2}, and pH (H^+ concentration) via negative feedback control.

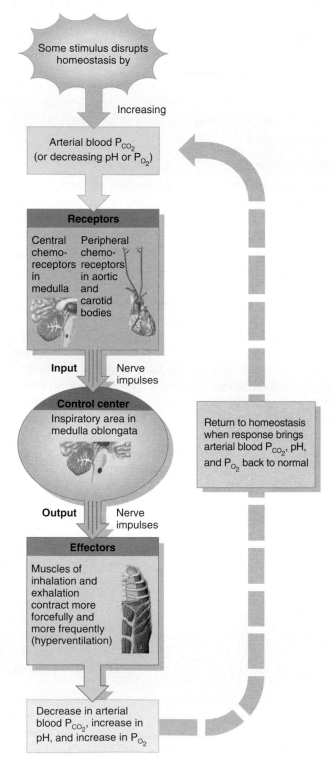

An increase in arterial blood P_{CO_2} stimulates the inspiratory center.

Some stimulus disrupts homeostasis by

Increasing

Arterial blood P_{CO_2} (or decreasing pH or P_{O_2})

Receptors

Central chemoreceptors in medulla

Peripheral chemoreceptors in aortic and carotid bodies

Input Nerve impulses

Control center

Inspiratory area in medulla oblongata

Return to homeostasis when response brings arterial blood P_{CO_2}, pH, and P_{O_2} back to normal

Output Nerve impulses

Effectors

Muscles of inhalation and exhalation contract more forcefully and more frequently (hyperventilation)

Decrease in arterial blood P_{CO_2}, increase in pH, and increase in P_{O_2}

? What is the normal arterial blood P_{CO_2}?

Proprioceptor Stimulation of Respiration

As soon as you start exercising, your rate and depth of breathing increase, even before changes in P_{O_2}, P_{CO_2}, or H^+ level occur. The main stimulus for these quick changes in respiratory effort is input from proprioceptors, which monitor movement of joints and muscles. Nerve impulses from the proprioceptors stimulate the inspiratory area of the medulla oblongata. At the same time, axon collaterals (branches) of upper motor neurons that originate in the primary motor cortex (precentral gyrus) also feed excitatory impulses into the inspiratory area.

The Inflation Reflex

Similar to those in the blood vessels, stretch-sensitive receptors called **baroreceptors** or **stretch receptors** are located in the walls of bronchi and bronchioles. When these receptors become stretched during overinflation of the lungs, nerve impulses are sent along the vagus (X) nerves to the inspiratory and apneustic areas. In response, the inspiratory area is inhibited directly, and the apneustic area is inhibited from activating the inspiratory area. As a result, exhalation begins. As air leaves the lungs during exhalation, the lungs deflate and the stretch receptors are no longer stimulated. Thus, the inspiratory and apneustic areas are no longer inhibited, and a new inhalation begins. This reflex, referred to as the **inflation (Hering–Breuer) reflex,** is mainly a protective mechanism for preventing excessive inflation of the lungs rather than a key component in the normal regulation of respiration.

Other Influences on Respiration

Other factors that contribute to regulation of respiration include the following:

- *Limbic system stimulation.* Anticipation of activity or emotional anxiety may stimulate the limbic system, which then sends excitatory input to the inspiratory area, increasing the rate and depth of ventilation.

- *Temperature.* An increase in body temperature, as occurs during a fever or vigorous muscular exercise, increases the rate of respiration. A decrease in body temperature decreases respiratory rate. A sudden cold stimulus (such as plunging into cold water) causes temporary **apnea** (AP-nē-a; *a-* = without; *pneu* = breath), an absence of breathing.

- *Pain.* A sudden, severe pain brings about brief apnea, but a prolonged somatic pain increases respiratory rate. Visceral pain may slow the rate of respiration.

- *Stretching the anal sphincter muscle.* This action increases the respiratory rate and is sometimes used to stimulate respiration in a newborn baby or a person who has stopped breathing.

- *Irritation of airways.* Physical or chemical irritation of the pharynx or larynx brings about an immediate cessation of breathing followed by coughing or sneezing.

- *Blood pressure.* The carotid and aortic baroreceptors that detect changes in blood pressure have a small effect on respiration. A sudden rise in blood pressure decreases the rate of respiration, and a drop in blood pressure increases the respiratory rate.

Table 23.2 summarizes the stimuli that affect the rate and depth of ventilation.

◉ CHECKPOINT

27. How does the medullary rhythmicity area regulate respiration?
28. How are the apneustic and pneumotaxic areas related to the control of respiration?
29. How do the cerebral cortex, levels of CO_2 and O_2, proprioceptors, inflation reflex, temperature changes, pain, and irritation of the airways modify respiration?

TABLE 23.2

Summary of Stimuli That Affect Ventilation Rate and Depth

STIMULI THAT INCREASE VENTILATION RATE AND DEPTH	STIMULI THAT DECREASE VENTILATION RATE AND DEPTH
Voluntary hyperventilation controlled by the cerebral cortex and anticipation of activity by stimulation of the limbic system.	Voluntary hypoventilation controlled by the cerebral cortex.
Increase in arterial blood P_{CO_2} above 40 mmHg (causes an increase in H^+) detected by peripheral and central chemoreceptors.	Decrease in arterial blood P_{CO_2} below 40 mmHg (causes a decrease in H^+) detected by peripheral and central chemoreceptors.
Decrease in arterial blood P_{O_2} from 105 mmHg to 50 mmHg.	Decrease in arterial blood P_{O_2} below 50 mmHg.
Increased activity of proprioceptors.	Decreased activity of proprioceptors.
Increase in body temperature.	Decrease in body temperature decreases the rate of respiration, and a sudden cold stimulus causes apnea.
Prolonged pain.	Severe pain causes apnea.
Decrease in blood pressure.	Increase in blood pressure.
Stretching the anal sphincter.	Irritation of pharynx or larynx by touch or chemicals causes brief apnea followed by coughing or sneezing.

Respiratory failure A condition in which the respiratory system either cannot supply sufficient O_2 to maintain metabolism or cannot eliminate enough CO_2 to prevent respiratory acidosis (a lower-than-normal pH in interstitial fluid).

Rhinitis (rī-NĪ-tis; *rhin-* = nose) Chronic or acute inflammation of the mucous membrane of the nose due to viruses, bacteria, or irritants. Excessive mucus production produces a runny nose, nasal congestion, and postnasal drip.

Sleep apnea (AP-nē-a; *a-* = without; *-pnea* = breath) A disorder in which a person repeatedly stops breathing for 10 or more seconds while sleeping. Most often, it occurs because loss of muscle tone in pharyngeal muscles allows the airway to collapse.

Sputum (SPŪ-tum = to spit) Mucus and other fluids from the air passages that is expectorated (expelled by coughing).

Strep throat Inflammation of the pharynx caused by the bacterium *Streptococcus pyogenes*. It may also involve the tonsils and middle ear.

Tachypnea (tak'-ip-NĒ-a; *tachy-* = rapid; *-pnea* = breath) Rapid breathing rate.

Wheeze (HWEĒZ) A whistling, squeaking, or musical high-pitched sound during breathing resulting from a partially obstructed airway.

STUDY OUTLINE

Respiratory System Anatomy (p. 875)

1. The respiratory system consists of the nose, pharynx, larynx, trachea, bronchi, and lungs. They act with the cardiovascular system to supply oxygen (O_2) and remove carbon dioxide (CO_2) from the blood.

2. The external portion of the nose is made of cartilage and skin and is lined with a mucous membrane. Openings to the exterior are the external nares.

3. The internal portion of the nose communicates with the paranasal sinuses and nasopharynx through the internal nares.

4. The nasal cavity is divided by a septum. The anterior portion of the cavity is called the vestibule. The nose warms, moistens, and filters air and functions in olfaction and speech.

5. The pharynx (throat) is a muscular tube lined by a mucous membrane. The anatomic regions are the nasopharynx, oropharynx, and laryngopharynx.

6. The nasopharynx functions in respiration. The oropharynx and laryngopharynx function both in digestion and in respiration.

7. The larynx (voice box) is a passageway that connects the pharynx with the trachea. It contains the thyroid cartilage (Adam's apple); the epiglottis, which prevents food from entering the larynx; the cricoid cartilage, which connects the larynx and trachea; and the paired arytenoid, corniculate, and cuneiform cartilages.

8. The larynx contains vocal folds, which produce sound as they vibrate. Taut folds produce high pitches, and relaxed ones produce low pitches.

9. The trachea (windpipe) extends from the larynx to the primary bronchi. It is composed of C-shaped rings of cartilage and smooth muscle and is lined with pseudostratified ciliated columnar epithelium.

10. The bronchial tree consists of the trachea, primary bronchi, secondary bronchi, tertiary bronchi, bronchioles, and terminal bronchioles. Walls of bronchi contain rings of cartilage; walls of bronchioles contain increasingly smaller plates of cartilage and increasing amounts of smooth muscle.

11. Lungs are paired organs in the thoracic cavity enclosed by the pleural membrane. The parietal pleura is the superficial layer that lines the thoracic cavity; the visceral pleura is the deep layer that covers the lungs.

12. The right lung has three lobes separated by two fissures; the left lung has two lobes separated by one fissure and a depression, the cardiac notch.

13. Secondary bronchi give rise to branches called segmental bronchi, which supply segments of lung tissue called bronchopulmonary segments.

14. Each bronchopulmonary segment consists of lobules, which contain lymphatics, arterioles, venules, terminal bronchioles, respiratory bronchioles, alveolar ducts, alveolar sacs, and alveoli.

15. Alveolar walls consist of type I alveolar cells, type II alveolar cells, and associated alveolar macrophages.

16. Gas exchange occurs across the respiratory membranes.

Pulmonary Ventilation (p. 890)

1. Pulmonary ventilation, or breathing, consists of inhalation and exhalation.

2. The movement of air into and out of the lungs depends on pressure changes governed in part by Boyle's law, which states that the volume of a gas varies inversely with pressure, assuming that temperature remains constant.

3. Inhalation occurs when alveolar pressure falls below atmospheric pressure. Contraction of the diaphragm and external intercostals increases the size of the thorax, thereby decreasing the intrapleural pressure so that the lungs expand. Expansion of the lungs decreases alveolar pressure so that air moves down a pressure gradient from the atmosphere into the lungs.

4. During forceful inhalation, accessory muscles of inhalation (sternocleidomastoids, scalenes, and pectoralis minors) are also used.

5. Exhalation occurs when alveolar pressure is higher than atmospheric pressure. Relaxation of the diaphragm and external intercostals results in elastic recoil of the chest wall and lungs, which increases intrapleural pressure; lung volume decreases and alveolar pressure increases, so air moves from the lungs to the atmosphere.

6. Forceful exhalation involves contraction of the internal intercostal and abdominal muscles.

7. The surface tension exerted by alveolar fluid is decreased by the presence of surfactant.

8. Compliance is the ease with which the lungs and thoracic wall can expand.

9. The walls of the airways offer some resistance to breathing.
10. Normal quiet breathing is termed eupnea; other patterns are costal breathing and diaphragmatic breathing. Modified respiratory movements, such as coughing, sneezing, sighing, yawning, sobbing, crying, laughing, and hiccupping, are used to express emotions and to clear the airways. (See Table 23.1 on page 895.)

Lung Volumes and Capacities (p. 894)

1. Lung volumes exchanged during breathing and the rate of respiration are measured with a spirometer.
2. Lung volumes measured by spirometry include tidal volume, minute ventilation, alveolar ventilation rate, inspiratory reserve volume, expiratory reserve volume, and $FEV_{1.0}$. Other lung volumes include anatomic dead space, residual volume, and minimal volume.
3. Lung capacities, the sum of two or more lung volumes, include inspiratory, functional, residual, vital, and total lung capacities.

Exchange of Oxygen and Carbon Dioxide (p. 896)

1. The partial pressure of a gas is the pressure exerted by that gas in a mixture of gases. It is symbolized by P_x, where the subscript is the chemical formula of the gas.
2. According to Dalton's law, each gas in a mixture of gases exerts its own pressure as if all the other gases were not present.
3. Henry's law states that the quantity of a gas that will dissolve in a liquid is proportional to the partial pressure of the gas and its solubility (given that the temperature remains constant).
4. In internal and external respiration, O_2 and CO_2 diffuse from areas of higher partial pressures to areas of lower partial pressures.
5. External respiration or pulmonary gas exchange is the exchange of gases between alveoli and pulmonary blood capillaries. It depends on partial pressure differences, a large surface area for gas exchange, a small diffusion distance across the respiratory membrane, and the rate of airflow into and out of the lungs.
6. Internal respiration or systemic gas exchange is the exchange of gases between systemic blood capillaries and tissue cells.

Transport of Oxygen and Carbon Dioxide (p. 900)

1. In each 100 mL of oxygenated blood, 1.5% of the O_2 is dissolved in blood plasma and 98.5% is bound to hemoglobin as oxyhemoglobin ($Hb–O_2$).
2. The binding of O_2 to hemoglobin is affected by P_{O_2}, acidity (pH), P_{CO_2}, temperature, and 2,3-bisphosphoglycerate (BPG).
3. Fetal hemoglobin differs from adult hemoglobin in structure and has a higher affinity for O_2.

4. In each 100 mL of deoxygenated blood, 7% of CO_2 is dissolved in blood plasma, 23% combines with hemoglobin as carbaminohemoglobin ($Hb–CO_2$), and 70% is converted to bicarbonate ions (HCO_3^-).
5. In an acidic environment, hemoglobin's affinity for O_2 is lower, and O_2 dissociates more readily from it (Bohr effect).
6. In the presence of O_2, less CO_2 binds to hemoglobin (Haldane effect).

Control of Respiration (p. 905)

1. The respiratory center consists of a medullary rhythmicity area in the medulla oblongata and a pneumotaxic area and an apneustic area in the pons.
2. The inspiratory area sets the basic rhythm of respiration.
3. The pneumotaxic and apneustic areas coordinate the transition between inhalation and exhalation.
4. Respirations may be modified by a number of factors, including cortical influences; the inflation reflex; chemical stimuli, such as O_2 and CO_2 and H+ levels; proprioceptor input; blood pressure changes; limbic system stimulation; temperature; pain; and irritation to the airways. (See Table 23.2 on page 909.)

Exercise and the Respiratory System (p. 910)

1. The rate and depth of ventilation change in response to both the intensity and duration of exercise.
2. An increase in pulmonary perfusion and O_2-diffusing capacity occurs during exercise.
3. The abrupt increase in ventilation at the start of exercise is due to neural changes that send excitatory impulses to the inspiratory area in the medulla oblongata. The more gradual increase in ventilation during moderate exercise is due to chemical and physical changes in the bloodstream.

Development of the Respiratory System (p. 910)

1. The respiratory system begins as an outgrowth of endoderm called the respiratory diverticulum.
2. Smooth muscle, cartilage, and connective tissue of the bronchial tubes and pleural sacs develop from mesoderm.

Aging and the Respiratory System (p. 911)

1. Aging results in decreased vital capacity, decreased blood level of O_2, and diminished alveolar macrophage activity.
2. Elderly people are more susceptible to pneumonia, emphysema, bronchitis, and other pulmonary disorders.

LAYERS OF THE GI TRACT

⦿ OBJECTIVE

- Describe the structure and function of the layers that form the wall of the gastrointestinal tract.

The wall of the GI tract from the lower esophagus to the anal canal has the same basic, four-layered arrangement of tissues. The four layers of the tract, from deep to superficial, are the mucosa, submucosa, muscularis, and serosa (Figure 24.2).

Mucosa

The **mucosa,** or inner lining of the GI tract, is a mucous membrane. It is composed of (1) a layer of epithelium in direct contact with the contents of the GI tract, (2) a layer of connective tissue called the lamina propria, and (3) a thin layer of smooth muscle (muscularis mucosae).

1. The **epithelium** in the mouth, pharynx, esophagus, and anal canal is mainly nonkeratinized stratified squamous epithelium that serves a protective function. Simple columnar epithelium, which functions in secretion and absorption, lines the stomach and intestines. The tight junctions that firmly seal neighboring simple columnar epithelial cells to one another restrict leakage between the cells. The rate of renewal of GI tract epithelial cells is rapid: Every 5 to 7 days they slough off and are replaced by new cells. Located among the epithelial cells are exocrine cells that secrete mucus and fluid into the lumen of the tract, and several types of endocrine cells, collectively called **enteroendocrine cells,** that secrete hormones.

2. The **lamina propria** (*lamina* = thin, flat plate; *propria* = one's own) is areolar connective tissue containing many blood and lymphatic vessels, which are the routes by which nutrients absorbed into the GI tract reach the other tissues of the body. This layer supports the epithelium and binds it to the muscularis mucosae (discussed next). The lamina propria also contains the majority of the cells of the **mucosa-associated lymphatic tissue (MALT).** These prominent lymphatic nodules contain immune system cells that protect against disease (see Chapter 22). MALT is present all along the GI tract, especially in the tonsils, small intestine, appendix, and large intestine.

3. A thin layer of smooth muscle fibers called the **muscularis mucosae** throws the mucous membrane of the stomach and small intestine into many small folds, which increase the surface area for digestion and absorption. Movements of the muscularis mucosae ensure that all absorptive cells are fully exposed to the contents of the GI tract.

Figure 24.2 Layers of the gastrointestinal tract. Variations in this basic plan may be seen in the esophagus (Figure 24.9), stomach (Figure 24.12), small intestine (Figure 24.18), and large intestine (Figure 24.23).

🔑 The four layers of the GI tract, from deep to superficial, are the mucosa, submucosa, muscularis, and serosa.

❓ What are the functions of the lamina propria?

Submucosa

The **submucosa** consists of areolar connective tissue that binds the mucosa to the muscularis. It contains many blood and lymphatic vessels that receive absorbed food molecules. Also located in the submucosa is an extensive network of neurons known as the submucosal plexus (to be described shortly). The submucosa may also contain glands and lymphatic tissue.

Muscularis

The **muscularis** of the mouth, pharynx, and superior and middle parts of the esophagus contains *skeletal muscle* that produces voluntary swallowing. Skeletal muscle also forms the external anal sphincter, which permits voluntary control of defecation. Throughout the rest of the tract, the muscularis consists of *smooth muscle* that is generally found in two sheets: an inner sheet of circular fibers and an outer sheet of longitudinal fibers. Involuntary contractions of the smooth muscle help break down food, mix it with digestive secretions, and propel it along the tract. Between the layers of the muscularis is a second plexus of neurons—the myenteric plexus (to be described shortly).

Serosa

Those portions of the GI tract that are suspended in the abdominopelvic cavity have a superficial layer called the **serosa.** As its name implies, the serosa is a serous membrane composed of areolar connective tissue and simple squamous epithelium (mesothelium). The serosa is also called the *visceral peritoneum* because it forms a portion of the peritoneum, which we examine in detail shortly. The esophagus lacks a serosa; instead only a single layer of areolar connective tissue called the *adventitia* forms the superficial layer of this organ.

◉ CHECKPOINT

4. Where along the GI tract is the muscularis composed of skeletal muscle? Is control of this skeletal muscle voluntary or involuntary?

5. Name the four layers of the gastrointestinal tract, and describe their functions.

NEURAL INNERVATION OF THE GI TRACT

◉ OBJECTIVE

• Describe the nerve supply of the GI tract.

The gastrointestinal tract is regulated by an intrinsic set of nerves known as the enteric nervous system and by an extrinsic set of nerves that are part of the autonomic nervous system.

Enteric Nervous System

We first introduced you to the **enteric nervous system (ENS),** the "brain of the gut," in Chapter 12. It consists of about 100 million neurons that extend from the esophagus to the anus.

The neurons of the ENS are arranged into two plexuses: the myenteric plexus and submucosal plexus (see Figure 24.2). The **myenteric plexus** (*myo-* = muscle), or *plexus of Auerbach,* is located between the longitudinal and circular smooth muscle layers of the muscularis. The **submucosal plexus,** or *plexus of Meissner,* is found within the submucosa. The plexuses of the ENS consist of motor neurons, interneurons, and sensory neurons (Figure 24.3). Because the motor neurons of the myenteric plexus supply the longitudinal and circular smooth muscle layers of the muscularis, this plexus mostly controls GI tract motility (movement), particularly the frequency and strength of contraction of the muscularis. The motor neurons of the submucosal plexus supply the secretory cells of the mucosal epithelium, controlling the secretions of the organs of the GI tract. The interneurons of the ENS interconnect the neurons of the myenteric and submucosal plexuses. The sensory neurons of the ENS supply the mucosal epithelium. Some of these sensory neurons function as *chemoreceptors,* receptors that are activated by the presence of certain chemicals in food located in the lumen of a GI organ. Other sensory neurons function as *stretch receptors,* receptors that are activated when food distends (stretches) the wall of a GI organ.

Autonomic Nervous System

Although the neurons of the ENS can function independently, they are subject to regulation by the neurons of the autonomic nervous system. The vagus (X) nerves supply parasympathetic fibers to most parts of the GI tract, with the exception of the

Figure 24.3 Organization of the enteric nervous system.

 The enteric nervous system consists of neurons arranged into the myenteric and submucosal plexuses.

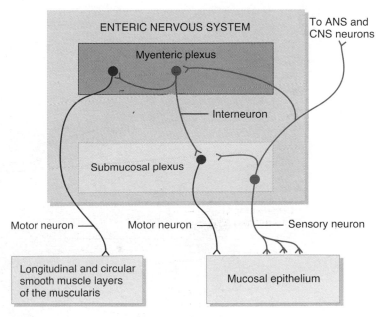

? **What are the functions of the myenteric and submucosal plexuses of the enteric nervous system?**

last half of the large intestine, which is supplied with parasympathetic fibers from the sacral spinal cord. The parasympathetic nerves that supply the GI tract form neural connections with the ENS. Parasympathetic preganglionic neurons of the vagus or pelvic splanchnic nerves synapse with parasympathetic postganglionic neurons located in the myenteric and submucosal plexuses. Some of the parasympathetic postganglionic neurons in turn synapse with neurons in the ENS; others directly innervate smooth muscle and glands within the wall of the GI tract. In general, stimulation of the parasympathetic nerves that innervate the GI tract causes an increase in GI secretion and motility by increasing the activity of ENS neurons.

Sympathetic nerves that supply the GI tract arise from the thoracic and upper lumbar regions of the spinal cord. Like the parasympathetic nerves, these sympathetic nerves form neural connections with the ENS. Sympathetic postganglionic neurons synapse with neurons located in the myenteric plexus and the submucosal plexus. In general, the sympathetic nerves that supply the GI tract cause a decrease in GI secretion and motility by inhibiting the neurons of the ENS. Emotions such as anger, fear, and anxiety may slow digestion because they stimulate the sympathetic nerves that supply the GI tract.

Gastrointestinal Reflex Pathways

Many neurons of the ENS are components of *GI (gastrointestinal) reflex pathways* that regulate GI secretion and motility in response to stimuli present in the lumen of the GI tract. The initial components of a typical GI reflex pathway are sensory receptors (such as chemoreceptors and stretch receptors) that are associated with the sensory neurons of the ENS. The axons of these sensory neurons can synapse with other neurons located in the ENS, CNS, or ANS, informing these regions about the nature of the contents and the degree of distension (stretching) of the GI tract. The neurons of the ENS, CNS, or ANS subsequently activate or inhibit GI glands and smooth muscle, altering GI secretion and motility.

◉ CHECKPOINT

6. How is the enteric nervous system regulated by the autonomic nervous system?
7. What is a gastrointestinal reflex pathway?

Figure 24.4 **Relationship of the peritoneal folds to one another and to organs of the digestive system.** The size of the peritoneal cavity in (a) is exaggerated for emphasis.

🔑 The peritoneum is the largest serous membrane in the body.

(a) Midsagittal section showing the peritoneal folds

PERITONEUM

⊙ OBJECTIVE
• Describe the peritoneum and its folds.

The **peritoneum** (per′-i-tō-NĒ-um; *peri-* = around) is the largest serous membrane of the body; it consists of a layer of simple squamous epithelium (mesothelium) with an underlying supporting layer of areolar connective tissue. The peritoneum is divided into the **parietal peritoneum,** which lines the wall of the abdominopelvic cavity, and the **visceral peritoneum,** which covers some of the organs in the cavity and is their serosa (Figure 24.4a on page 926). The slim space containing lubricating serous fluid that is between the parietal and visceral portions of the peritoneum is called the **peritoneal cavity.** In certain diseases, the peritoneal cavity may become distended by the accu-

mulation of several liters of fluid, a condition called **ascites** (a-SĪ-tēz).

As you will see shortly, some organs lie on the posterior abdominal wall and are covered by peritoneum only on their anterior surfaces; they are not in the peritoneal cavity. Such organs, including the kidneys, ascending and descending colons of the large intestine, duodenum of the small intestine, and pancreas, are said to be **retroperitoneal** (*retro-* = behind).

Unlike the pericardium and pleurae, which smoothly cover the heart and lungs, the peritoneum contains large folds that weave between the viscera. The folds bind the organs to one another and to the walls of the abdominal cavity. They also contain blood vessels, lymphatic vessels, and nerves that supply the abdominal organs. There are five major peritoneal folds: the greater omentum, falciform ligament, lesser omentum, mesentery, and mesocolon.

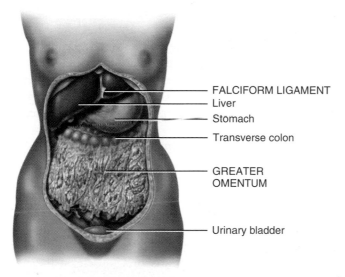

FALCIFORM LIGAMENT
Liver
Stomach
Transverse colon
GREATER OMENTUM
Urinary bladder

(b) Anterior view

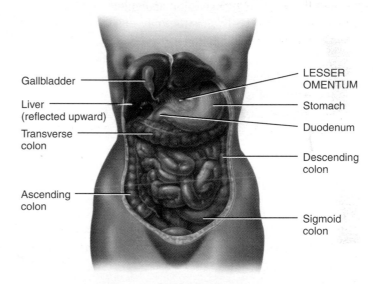

Gallbladder
Liver (reflected upward)
Transverse colon
Ascending colon

LESSER OMENTUM
Stomach
Duodenum
Descending colon
Sigmoid colon

(c) Lesser omentum, anterior view (liver and gallbladder lifted)

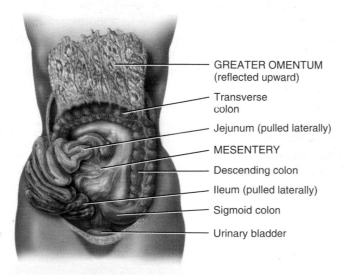

GREATER OMENTUM (reflected upward)
Transverse colon
Jejunum (pulled laterally)
MESENTERY
Descending colon
Ileum (pulled laterally)
Sigmoid colon
Urinary bladder

(d) Anterior view (greater omentum lifted and small intestine reflected to right side)

Lungs
Heart
Diaphragm
Right lobe of liver
FALCIFORM LIGAMENT
Left lobe of liver
Stomach
GREATER OMENTUM

(e) Anterior view

❓ **Which peritoneal fold binds the small intestine to the posterior abdominal wall?**

1. The **greater omentum** (ō-MEN-tum = fat skin), the largest peritoneal fold, drapes over the transverse colon and coils of the small intestine like a "fatty apron" (Figure 24.4a, d). The greater omentum is a double sheet that folds back on itself, giving it a total of four layers. From attachments along the stomach and duodenum, the greater omentum extends downward anterior to the small intestine, then turns and extends upward and attaches to the transverse colon. The greater omentum normally contains a considerable amount of adipose tissue. Its adipose tissue content can greatly expand with weight gain, giving rise to the characteristic "beer belly" seen in some overweight individuals. The many lymph nodes of the greater omentum contribute macrophages and antibody-producing plasma cells that help combat and contain infections of the GI tract.

2. The **falciform ligament** (FAL-si-form; *falc-* = sickle-shaped) attaches the liver to the anterior abdominal wall and diaphragm (Figure 24.4b). The liver is the only digestive organ that is attached to the anterior abdominal wall.

3. The **lesser omentum** arises as an anterior fold in the serosa of the stomach and duodenum, and it suspends the stomach and duodenum from the liver (Figure 24.4a, c). It is the pathway for blood vessels entering the liver and contains the hepatic portal vein, common hepatic artery, and common bile duct, along with some lymph nodes.

4. A fan-shaped fold of the peritoneum, called the **mesentery** (MEZ-en-ter′-ē; *mes-* = middle), binds the jejunum and ileum of the small intestine to the posterior abdominal wall (Figure 24.4a, d). It extends from the posterior abdominal wall to wrap around the small intestine and then returns to its origin, forming a double-layered structure. Between the two layers are blood and lymphatic vessels and lymph nodes.

5. Two separate folds of peritoneum, called the **mesocolon** (mez′-ō-KŌ-lon), bind the transverse colon (transverse mesocolon) and sigmoid colon (sigmoid mesocolon) of the large intestine to the posterior abdominal wall (Figure 24.4a). It also carries blood and lymphatic vessels to the intestines. Together, the mesentery and mesocolon hold the intestines loosely in place, allowing movement as muscular contractions mix and move the luminal contents along the GI tract.

• CLINICAL **CONNECTION** | Peritonitis

A common cause of **peritonitis,** an acute inflammation of the peritoneum, is contamination of the peritoneum by infectious microbes, which can result from accidental or surgical wounds in the abdominal wall, or from perforation or rupture of abdominal organs. If, for example, bacteria gain access to the peritoneal cavity through an intestinal perforation or rupture of the appendix, they can produce an acute, life-threatening form of peritonitis. A less serious (but still painful) form of peritonitis can result from the rubbing together of inflamed peritoneal surfaces. Peritonitis is of particularly grave concern to those who rely on peritoneal dialysis, a procedure in which the peritoneum is used to filter the blood when the kidneys do not function properly (see page 1048). •

CHECKPOINT
8. Where are the visceral peritoneum and parietal peritoneum located?
9. Describe the attachment sites and functions of the mesentery, mesocolon, falciform ligament, lesser omentum, and greater omentum.

MOUTH

OBJECTIVES
• Identify the locations of the salivary glands, and describe the functions of their secretions.
• Describe the structure and functions of the tongue.
• Identify the parts of a typical tooth, and compare deciduous and permanent dentitions.

The **mouth,** also referred to as the **oral** or **buccal cavity** (BUK-al; *bucca* = cheeks), is formed by the cheeks, hard and soft palates, and tongue (Figure 24.5). The **cheeks** form the lateral walls of the oral cavity. They are covered externally by skin and internally by a mucous membrane, which consists of nonkeratinized stratified squamous epithelium. Buccinator muscles and connective tissue lie between the skin and mucous membranes of the cheeks. The anterior portions of the cheeks end at the lips.

The **lips** or **labia** (= fleshy borders) are fleshy folds surrounding the opening of the mouth. They contain the orbicularis oris muscle and are covered externally by skin and internally by a mucous membrane. The inner surface of each lip is attached to its corresponding gum by a midline fold of mucous membrane called the **labial frenulum** (LĀ-bē-al FREN-ū-lum; *frenulum* = small bridle). During chewing, contraction of the buccinator muscles in the cheeks and orbicularis oris muscle in the lips helps keep food between the upper and lower teeth. These muscles also assist in speech.

The **oral vestibule** (= entrance to a canal) of the oral cavity is a space bounded externally by the cheeks and lips and internally by the gums and teeth. The **oral cavity proper** is a space that extends from the gums and teeth to the **fauces** (FAW-sēs = passages), the opening between the oral cavity and the oropharynx (throat).

The **palate** is a wall or septum that separates the oral cavity from the nasal cavity, forming the roof of the mouth. This important structure makes it possible to chew and breathe at the same time. The **hard palate**—the anterior portion of the roof of the mouth—is formed by the maxillae and palatine bones and is covered by a mucous membrane; it forms a bony partition between the oral and nasal cavities. The **soft palate,** which forms the posterior portion of the roof of the mouth, is an arch-shaped muscular partition between the oropharynx and nasopharynx that is lined with mucous membrane.

Hanging from the free border of the soft palate is a conical muscular process called the **uvula** (Ū-vū-la = little grape). During swallowing, the soft palate and uvula are drawn superiorly, closing off the nasopharynx and preventing swallowed foods and liquids from entering the nasal cavity. Lateral to the base of the uvula are two muscular folds that run down the lateral sides of the soft palate: Anteriorly, the **palatoglossal arch** extends to the side of the base of the tongue; posteriorly, the **palatopharyngeal arch** (PAL-a-tō-fa-rin′-jē-al) extends to the side of the pharynx. The palatine tonsils are situated between the arches, and the lingual tonsils are situated at the base of the tongue. At the posterior border of the soft palate, the mouth opens into the oropharynx through the fauces (Figure 24.5).

Salivary Glands

A **salivary gland** is a gland that releases a secretion called saliva into the oral cavity. Ordinarily, just enough saliva is secreted to keep the mucous membranes of the mouth and pharynx moist and to cleanse the mouth and teeth. When food enters the mouth, however, secretion of saliva increases, and it lubricates, dissolves, and begins the chemical breakdown of the food.

The mucous membrane of the mouth and tongue contains many small salivary glands that open directly, or indirectly via short ducts, to the oral cavity. These glands include *labial, buccal,* and *palatal glands* in the lips, cheeks, and palate, respectively, and *lingual glands* in the tongue, all of which make a small contribution to saliva.

However, most saliva is secreted by the **major salivary glands,** which lie beyond the oral mucosa, into ducts that lead to the oral cavity. There are three pairs of major salivary glands: the parotid, submandibular, and sublingual glands (Figure 24.6a). The **parotid glands** (*par-* = near; *to-* = ear) are located inferior and anterior to the ears, between the skin and the masseter muscle. Each secretes saliva into the oral cavity via a **parotid duct** that pierces the buccinator muscle to open into the vestibule opposite the second maxillary (upper) molar tooth. The **submandibular glands** are found in the floor of the mouth; they are medial and partly inferior to the body of the mandible. Their ducts, the **submandibular ducts,** run under the mucosa on either side of the midline of the floor of the mouth and enter the oral cavity proper lateral to the lingual frenulum. The **sublingual glands** are beneath the tongue and superior to the submandibular glands. Their ducts, the **lesser sublingual ducts,** open into the floor of the mouth in the oral cavity proper.

Figure 24.5 **Structures of the mouth (oral cavity).**

The mouth is formed by the cheeks, hard and soft palates, and tongue.

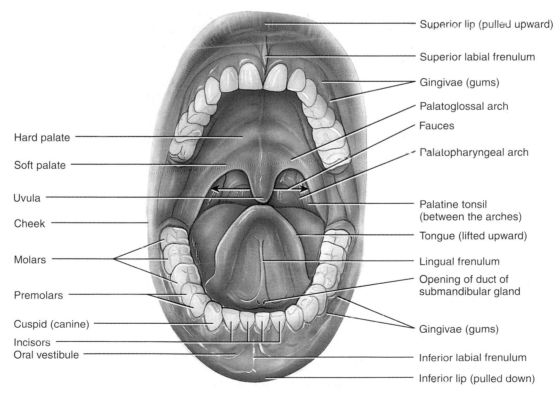

Anterior view

? What is the function of the uvula?

TABLE 24.3

Summary of Digestive Activities in the Stomach

STRUCTURE	ACTIVITY	RESULT
MUCOSA		
Chief cells	Secrete pepsinogen.	Pepsin, the activated form, breaks down proteins into peptides.
	Secrete gastric lipase.	Splits triglycerides into fatty acids and monoglycerides.
Parietal cells	Secrete hydrochloric acid.	Kills microbes in food; denatures proteins; converts pepsinogen into pepsin.
	Secrete intrinsic factor.	Needed for absorption of vitamin B_{12}, which is used in red blood cell formation (erythropoiesis).
Surface mucous cells and mucous neck cells	Secrete mucus.	Forms a protective barrier that prevents digestion of stomach wall.
	Absorption.	Small quantity of water, ions, short-chain fatty acids, and some drugs enter the bloodstream.
G cells	Secrete gastrin.	Stimulates parietal cells to secrete HCl and chief cells to secrete pepsinogen; contracts lower esophageal sphincter, increases motility of the stomach, and relaxes pyloric sphincter.
Muscularis	Mixing waves.	Macerate food and mix it with gastric juice, forming chyme.
	Peristalsis.	Forces chyme through pyloric sphincter.
Pyloric sphincter	Opens to permit passage of chyme into duodenum.	Regulates passage of chyme from stomach to duodenum; prevents backflow of chyme from duodenum to stomach.

● CLINICAL CONNECTION | Vomiting

Vomiting or *emesis* is the forcible expulsion of the contents of the upper GI tract (stomach and sometimes duodenum) through the mouth. The strongest stimuli for vomiting are irritation and distension of the stomach; other stimuli include unpleasant sights, general anesthesia, dizziness, and certain drugs such as morphine and derivatives of digitalis. Nerve impulses are transmitted to the vomiting center in the medulla oblongata, and returning impulses propagate to the upper GI tract organs, diaphragm, and abdominal muscles. Vomiting involves squeezing the stomach between the diaphragm and abdominal muscles and expelling the contents through open esophageal sphincters. Prolonged vomiting, especially in infants and elderly people, can be serious because the loss of acidic gastric juice can lead to alkalosis (higher than normal blood pH), dehydration, and damage to the esophagus and teeth. •

● CHECKPOINT

20. Compare the epithelium of the esophagus with that of the stomach. How is each adapted to the function of the organ?
21. What is the importance of rugae, surface mucous cells, mucous neck cells, chief cells, parietal cells, and G cells in the stomach?
22. What is the role of pepsin? Why is it secreted in an inactive form?
23. What are the functions of gastric lipase and lingual lipase in the stomach?

PANCREAS

● OBJECTIVE

• Describe the location, anatomy, histology, and function of the pancreas.

From the stomach, chyme passes into the small intestine. Because chemical digestion in the small intestine depends on activities of the pancreas, liver, and gallbladder, we first consider the activities of these accessory digestive organs and their contributions to digestion in the small intestine.

Anatomy of the Pancreas

The **pancreas** (*pan-* = all; *-creas* = flesh), a retroperitoneal gland that is about 12–15 cm (5–6 in.) long and 2.5 cm (1 in.) thick, lies posterior to the greater curvature of the stomach. The pancreas consists of a head, a body, and a tail and is usually connected to the duodenum by two ducts (Figure 24.14a). The **head** is the expanded portion of the organ near the curve of the duodenum; superior to and to the left of the head are the central **body** and the tapering **tail**.

Pancreatic juices are secreted by exocrine cells into small ducts that ultimately unite to form two larger ducts, the pancreatic duct and the accessory duct. These in turn convey the secretions into the small intestine. The **pancreatic duct (duct of Wirsung)** is the larger of the two ducts. In most people, the pancreatic duct joins the common bile duct from the liver and gallbladder and enters the duodenum as a dilated common duct called the **hepatopancreatic ampulla (ampulla of Vater).** The ampulla opens on an elevation of the duodenal mucosa known as the **major duodenal papilla,** which lies about 10 cm (4 in.) inferior to the pyloric sphincter of the stomach. The passage of pancreatic juice and bile through the hepatopancreatic

Figure 24.14 **Relation of the pancreas to the liver, gallbladder, and duodenum.** The inset shows details of the common bile duct and pancreatic duct forming the hepatopancreatic ampulla (ampulla of Vater) and emptying into the duodenum. (See Tortora, *A Photographic Atlas of the Human Body, Second Edition,* Figures 12.10, 12.11.)

Pancreatic enzymes digest starches (polysaccharides), proteins, triglycerides, and nucleic acids.

(a) Anterior view

(b) Details of hepatopancreatic ampulla

Key:
- Liver
- Gallbladder
- Pancreas

(c) Ducts carrying bile from liver and gallbladder and pancreatic juice from pancreas to the duodenum

(d) Anterior view

FIGURE 24.14 CONTINUES

■ **FIGURE 24.14** CONTINUED ▶

SUPERIOR

Duodenum (cut open)

Common bile duct

Major duodenal papilla

Head of pancreas

Uncinate process

MEDIAL

Tail of pancreas

Body of pancreas

Pancreatic duct (duct of Wirsung)

LATERAL

(e) Anterior view dissected to show pancreatic duct

? **What type of fluid is found in the pancreatic duct? The common bile duct? The hepatopancreatic ampulla?**

ampulla into the small intestine is regulated by a mass of smooth muscle known as the **sphincter of the hepatopancreatic ampulla (sphincter of Oddi).** The other major duct of the pancreas, the **accessory duct (duct of Santorini),** leads from the pancreas and empties into the duodenum about 2.5 cm (1 in.) superior to the hepatopancreatic ampulla.

Histology of the Pancreas

The pancreas is made up of small clusters of glandular epithelial cells. About 99% of the clusters, called **acini** (AS-i-nē), constitute the *exocrine* portion of the organ (see Figure 18.18b, c on page 670). The cells within acini secrete a mixture of fluid and digestive enzymes called **pancreatic juice.** The remaining 1% of the clusters, called **pancreatic islets (islets of Langerhans),** form the *endocrine* portion of the pancreas. These cells secrete the hormones glucagon, insulin, somatostatin, and pancreatic polypeptide. The functions of these hormones are discussed in Chapter 18.

Composition and Functions of Pancreatic Juice

Each day the pancreas produces 1200–1500 mL (about 1.2–1.5 qt) of **pancreatic juice,** a clear, colorless liquid consisting mostly of water, some salts, sodium bicarbonate, and several enzymes. The sodium bicarbonate gives pancreatic juice a slightly alkaline pH (7.1–8.2) that buffers acidic gastric juice in chyme, stops the action of pepsin from the stomach, and creates the proper pH for the action of digestive enzymes in the small intestine. The enzymes in pancreatic juice include a starch-digesting enzyme called **pancreatic amylase;** several protein-digesting enzymes called **trypsin** (TRIP-sin), **chymotrypsin** (kī′-mō-TRIP-sin), **carboxypeptidase** (kar-bok′-sē-PEP-ti-dās), and **elastase** (ē-LAS-tās); the principal triglyceride-digesting enzyme in adults, called **pancreatic lipase;** and nucleic acid–digesting enzymes called **ribonuclease** and **deoxyribonuclease.**

The protein-digesting enzymes of the pancreas are produced in an inactive form just as pepsin is produced in the stomach as pepsinogen. Because they are inactive, the enzymes do not

digest cells of the pancreas itself. Trypsin is secreted in an inactive form called **trypsinogen** (trip-SIN-ō-jen). Pancreatic acinar cells also secrete a protein called **trypsin inhibitor** that combines with any trypsin formed accidentally in the pancreas or in pancreatic juice and blocks its enzymatic activity. When trypsinogen reaches the lumen of the small intestine, it encounters an activating brush-border enzyme called **enterokinase** (en′-ter-ō-KĪ-nās), which splits off part of the trypsinogen molecule to form trypsin. In turn, trypsin acts on the inactive precursors (called **chymotrypsinogen, procarboxypeptidase,** and **proelastase**) to produce chymotrypsin, carboxypeptidase, and elastase, respectively.

● CLINICAL CONNECTION | Pancreatitis and Pancreatic Cancer

Inflammation of the pancreas, as may occur in association with alcohol abuse or chronic gallstones, is called **pancreatitis** (pan′-krē-a-TĪ-tis). In a more severe condition known as **acute pancreatitis,** which is associated with heavy alcohol intake or biliary tract obstruction, the pancreatic cells may release either trypsin instead of trypsinogen or insufficient amounts of trypsin inhibitor, and the trypsin begins to digest the pancreatic cells. Patients with acute pancreatitis usually respond to treatment, but recurrent attacks are the rule. In some people pancreatitis is idiopathic, meaning that the cause is unknown. Other causes of pancreatitis include cystic fibrosis, high levels of calcium in the blood (hypercalcemia), high levels of blood fats (hyperlipidemia or hypertriglyceridemia), some drugs, and certain autoimmune conditions. However, in roughly 70% of adults with pancreatitis, the cause is alcoholism. Often the first episode happens between ages 30 and 40.

Pancreatic cancer usually affects people over 50 years of age and occurs more frequently in males. Typically, there are few symptoms until the disorder reaches an advanced stage and often not until it has metastasized to other parts of the body such as the lymph nodes, liver, or lungs. The disease is nearly always fatal and is the fourth most common cause of death from cancer in the United States. Pancreatic cancer has been linked to fatty foods, high alcohol consumption, genetic factors, smoking, and chronic pancreatitis. ●

⦿ CHECKPOINT

24. Describe the duct system connecting the pancreas to the duodenum.
25. What are pancreatic acini? How do their functions differ from those of the pancreatic islets (islets of Langerhans)?
26. What are the digestive functions of the components of pancreatic juice?

LIVER AND GALLBLADDER

⦿ OBJECTIVE

• Describe the location, anatomy, histology, and functions of the liver and gallbladder.

The **liver** is the heaviest gland of the body, weighing about 1.4 kg (about 3 lb) in an average adult. Of all of the organs of the body, it is second only to the skin in size. The liver is inferior to the diaphragm and occupies most of the right hypochondriac and part of the epigastric regions of the abdominopelvic cavity (see Figure 1.12a on page 20).

The **gallbladder** (*gall-* = bile) is a pear-shaped sac that is located in a depression of the posterior surface of the liver. It is 7–10 cm (3–4 in.) long and typically hangs from the anterior inferior margin of the liver (Figure 24.14a).

Anatomy of the Liver and Gallbladder

The liver is almost completely covered by visceral peritoneum and is completely covered by a dense irregular connective tissue layer that lies deep to the peritoneum. The liver is divided into two principal lobes—a large **right lobe** and a smaller **left lobe**— by the **falciform ligament,** a fold of the mesentery (Figure 24.14a). Although the right lobe is considered by many anatomists to include an inferior **quadrate lobe** and a posterior **caudate lobe,** based on internal morphology (primarily the distribution of blood vessels), the quadrate and caudate lobes more appropriately belong to the left lobe. The falciform ligament extends from the undersurface of the diaphragm between the two principal lobes of the liver to the superior surface of the liver, helping to suspend the liver in the abdominal cavity. In the free border of the falciform ligament is the **ligamentum teres (round ligament),** a remnant of the umbilical vein of the fetus (see Figure 21.30a, b on page 821); this fibrous cord extends from the liver to the umbilicus. The right and left **coronary ligaments** are narrow extensions of the parietal peritoneum that suspend the liver from the diaphragm.

The parts of the gallbladder include the broad **fundus,** which projects inferiorly beyond the inferior border of the liver; the **body,** the central portion; and the **neck,** the tapered portion. The body and neck project superiorly.

Histology of the Liver and Gallbladder

Histologically, the liver is composed of several components (Figure 24.15a–c):

1. **Hepatocytes** (*hepat-* = liver; *-cytes* = cell). Hepatocytes are the major functional cells of the liver and perform a wide array of metabolic, secretory, and endocrine functions. These are specialized epithelial cells with 5 to 12 sides that make up about 80% of the volume of the liver. Hepatocytes form complex three-dimensional arrangements called **hepatic laminae.** The hepatic laminae are plates of hepatocytes one cell thick bordered on either side by the endothelial-lined vascular spaces called hepatic sinusoids. The hepatic laminae are highly branched, irregular structures. Grooves in the cell membranes between neighboring hepatocytes provide spaces for canaliculi (described next) into which the hepatocytes secrete bile. Bile, a yellow, brownish, or olive-green liquid secreted by hepatocytes, serves as both an excretory product and a digestive secretion.

2. **Bile canaliculi** (kan-a-LIK-ū-li = small canals). These are small ducts between hepatocytes that collect bile produced by the hepatocytes. From bile canaliculi, bile passes into **bile ductules** and then **bile ducts.** The bile ducts merge and eventually form the larger **right** and **left hepatic ducts,** which unite and exit the liver as the **common hepatic duct** (see Figure 24.14). The common hepatic duct joins the **cystic duct** (*cystic* = bladder) from the gallbladder to form the **common bile duct.** From here, bile enters the small intestine to participate in digestion.

3. **Hepatic sinusoids.** These are highly permeable blood capillaries between rows of hepatocytes that receive oxygenated blood from branches of the hepatic artery and nutrient-rich deoxygenated blood from branches of the hepatic portal vein. Recall that the hepatic portal vein brings venous blood from the gastrointestinal organs and spleen into the liver. Hepatic sinusoids converge and deliver blood into a **central vein.** From central veins the blood flows into the **hepatic veins,** which drain into the inferior vena cava (see Figure 21.28 on page 818). In contrast to blood which flows toward a central vein, bile flows in the opposite direction. Also present in the hepatic sinusoids are fixed phagocytes called **stellate reticuloendothelial (Kupffer) cells,** which destroy worn-out white and red blood cells, bacteria, and other foreign matter in the venous blood draining from the gastrointestinal tract.

Together, a bile duct, branch of the hepatic artery, and branch of the hepatic vein are referred to as a **portal triad** (*tri* = three).

The hepatocytes, bile duct system, and hepatic sinusoids can be organized into anatomical and functional units in three different ways:

1. **Hepatic lobule.** For years, anatomists described the hepatic lobule as the functional unit of the liver. According to this model, each hepatic lobule is shaped like a hexagon (six-sided structure). Figure 24.15e, left. At its center is the central vein, and radiating out from it are rows of hepatocytes and hepatic sinusoids. Located at three corners of the hexagon is a portal triad. This model is based on a description of the liver of adult pigs. In the human liver it is difficult to find such well-defined hepatic lobules surrounded by thick layers of connective tissue.

2. **Portal lobule.** This model emphasized the exocrine function of the liver, that is, bile secretion. Accordingly, the bile duct of a

Jaundice (JAWN-dis = yellowed) is a yellowish coloration of the sclerae (whites of the eyes), skin, and mucous membranes due to a buildup of a yellow compound called bilirubin. After bilirubin is formed from the breakdown of the heme pigment in aged red blood cells, it is transported to the liver, where it is processed and eventually excreted into bile. The three main categories of jaundice are (1) *prehepatic jaundice,* due to excess production of bilirubin; (2) *hepatic jaundice,* due to congenital liver disease, cirrhosis of the liver, or hepatitis; and (3) *extrahepatic jaundice,* due to blockage of bile drainage by gallstones or cancer of the bowel or the pancreas.

Because the liver of a newborn functions poorly for the first week or so, many babies experience a mild form of jaundice called *neonatal (physiological) jaundice* that disappears as the liver matures. Usually, it is treated by exposing the infant to blue light, which converts bilirubin into substances the kidneys can excrete.

Blood Supply of the Liver

The liver receives blood from two sources (Figure 24.16). From the hepatic artery it obtains oxygenated blood, and from the hepatic portal vein it receives deoxygenated blood containing newly absorbed nutrients, drugs, and possibly microbes and toxins from the gastrointestinal tract (see Figure 21.28 on page 818). Branches of both the hepatic artery and the hepatic portal vein carry blood into liver sinusoids, where oxygen, most of the nutrients, and certain toxic substances are taken up by the hepatocytes. Products manufactured by the hepatocytes and nutrients needed by other cells are secreted back into the blood, which then drains into the central vein and eventually passes into a hepatic vein. Because blood from the gastrointestinal tract passes through the liver as part of the hepatic portal circulation, the liver is often a site for metastasis of cancer that originates in the GI tract.

Role and Composition of Bile

Each day, hepatocytes secrete 800–1000 mL (about 1 qt) of **bile,** a yellow, brownish, or olive-green liquid. It has a pH of 7.6–8.6 and consists mostly of water, bile salts, cholesterol, a phospholipid called lecithin, bile pigments, and several ions.

The principal bile pigment is **bilirubin.** The phagocytosis of aged red blood cells liberates iron, globin, and bilirubin (derived from heme) (see Figure 19.5 on page 697). The iron and globin are recycled; the bilirubin is secreted into the bile and is eventually broken down in the intestine. One of its breakdown products—**stercobilin**—gives feces their normal brown color.

Bile is partially an excretory product and partially a digestive secretion. Bile salts, which are sodium salts and potassium salts of bile acids (mostly chenodeoxycholic acid and cholic acid), play a role in **emulsification,** the breakdown of large lipid globules into a suspension of small lipid globules. The small lipid globules present a very large surface area that allows pancreatic lipase to more rapidly accomplish digestion of triglycerides. Bile salts also aid in the absorption of lipids following their digestion.

Although hepatocytes continually release bile, they increase production and secretion when the portal blood contains more bile acids; thus, as digestion and absorption continue in the small intestine, bile release increases. Between meals, after most absorption has occurred, bile flows into the gallbladder for storage because the sphincter of the hepatopancreatic ampulla (sphincter of Oddi; see Figure 24.14) closes off the entrance to the duodenum.

Figure 24.16 Hepatic blood flow: sources, path through the liver, and return to the heart.

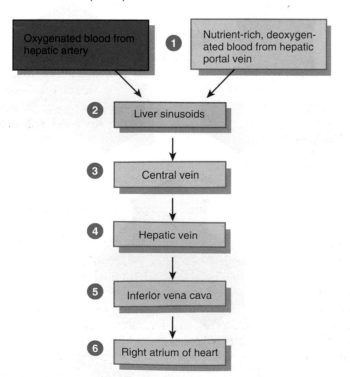

🔑 The liver receives oxygenated blood via the hepatic artery and nutrient-rich deoxygenated blood via the hepatic portal vein.

Oxygenated blood from hepatic artery

1 Nutrient-rich, deoxygenated blood from hepatic portal vein

2 Liver sinusoids

3 Central vein

4 Hepatic vein

5 Inferior vena cava

6 Right atrium of heart

❓ During the first few hours after a meal, how does the chemical composition of blood change as it flows through the liver sinusoids?

Functions of the Liver

In addition to secreting bile, which is needed for absorption of dietary fats, the liver performs many other vital functions:

- *Carbohydrate metabolism.* The liver is especially important in maintaining a normal blood glucose level. When blood glucose is low, the liver can break down glycogen to glucose and release the glucose into the bloodstream. The liver can also convert certain amino acids and lactic acid to glucose, and it can convert other sugars, such as fructose and galactose, into glucose. When blood glucose is high, as occurs just after eating a meal, the liver converts glucose to glycogen and triglycerides for storage.

- *Lipid metabolism.* Hepatocytes store some triglycerides; break down fatty acids to generate ATP; synthesize lipoproteins, which transport fatty acids, triglycerides, and cholesterol to and from body cells; synthesize cholesterol; and use cholesterol to make bile salts.

- *Protein metabolism.* Hepatocytes *deaminate* (remove the amino group, NH_2, from) amino acids so that the amino acids can be used for ATP production or converted to carbohydrates or fats. The resulting toxic ammonia (NH_3) is then converted into the much less toxic urea, which is excreted in urine. Hepatocytes also synthesize most plasma proteins, such as alpha and beta globulins, albumin, prothrombin, and fibrinogen.

- *Processing of drugs and hormones.* The liver can detoxify substances such as alcohol and excrete drugs such as penicillin, erythromycin, and sulfonamides into bile. It can also chemically alter or excrete thyroid hormones and steroid hormones such as estrogens and aldosterone.

- *Excretion of bilirubin.* As previously noted, bilirubin, derived from the heme of aged red blood cells, is absorbed by the liver from the blood and secreted into bile. Most of the bilirubin in bile is metabolized in the small intestine by bacteria and eliminated in feces.

- *Synthesis of bile salts.* Bile salts are used in the small intestine for the emulsification and absorption of lipids.

- *Storage.* In addition to glycogen, the liver is a prime storage site for certain vitamins (A, B_{12}, D, E, and K) and minerals (iron and copper), which are released from the liver when needed elsewhere in the body.

- *Phagocytosis.* The stellate reticuloendothelial (Kupffer) cells of the liver phagocytize aged red blood cells, white blood cells, and some bacteria.

- *Activation of vitamin D.* The skin, liver, and kidneys participate in synthesizing the active form of vitamin D.

The liver functions related to metabolism are discussed more fully in Chapter 25.

☉ CHECKPOINT

27. Draw and label a diagram of a liver lobule.
28. Describe the pathways of blood flow into, through, and out of the liver.
29. How are the liver and gallbladder connected to the duodenum?
30. Once bile has been formed by the liver, how is it collected and transported to the gallbladder for storage?
31. What is the function of bile?

SMALL INTESTINE

☉ OBJECTIVE

- Describe the location, anatomy, histology, and functions of the small intestine.

Most digestion and absorption of nutrients occur in a long tube called the **small intestine.** Because of this, its structure is specially adapted for these functions. Its length alone provides a large surface area for digestion and absorption, and that area is further increased by circular folds, villi, and microvilli. The small intestine begins at the pyloric sphincter of the stomach, coils through the central and inferior part of the abdominal cavity, and eventually opens into the large intestine. It averages 2.5 cm (1 in.) in diameter; its length is about 3 m (10 ft) in a living person and about 6.5 m (21 ft) in a cadaver due to the loss of smooth muscle tone after death.

Anatomy of the Small Intestine

The small intestine is divided into three regions (Figure 24.17). The **duodenum** (doo-ō-DĒ-num), the shortest region, is retroperitoneal. It starts at the pyloric sphincter of the stomach and extends about 25 cm (10 in.) until it merges with the jejunum. *Duodenum* means "12"; it is so named because it is about as long as the width of 12 fingers. The **jejunum** (je-JOO-num) is about 1 m (3 ft) long and extends to the ileum. *Jejunum* means "empty," which is how it is found at death. The final and longest region of the small intestine, the **ileum** (IL-ē-um = twisted), measures about 2 m (6 ft) and joins the large intestine at a smooth muscle sphincter called the **ileocecal sphincter** (il'-ē-ō-SĒ-kal).

Histology of the Small Intestine

The wall of the small intestine is composed of the same four layers that make up most of the GI tract: mucosa, submucosa, muscularis, and serosa (Figure 24.18a on page 951). The **mucosa** is composed of a layer of epithelium, lamina propria, and muscularis mucosae. The epithelial layer of the small intestinal mucosa consists of simple columnar epithelium that contains many types of cells (Figure 24.18b). **Absorptive cells** of the epithelium digest and absorb nutrients in small intestinal chyme. Also present in the epithelium are **goblet cells,** which secrete mucus. The small intestinal mucosa contains many deep crevices lined with glandular epithelium. Cells lining the crevices form the **intestinal glands (crypts of Lieberkühn)** and secrete intestinal juice (to be discussed shortly). Besides absorptive cells and goblet cells, the intestinal glands also contain paneth cells and

Figure 24.19 Histology of the duodenum and ileum.

Microvilli in the small intestine contain several brush-border enzymes that help digest nutrients.

Lumen of duodenum
Villi
Mucosa
Intestinal glands
Muscularis mucosae
Duodenal gland
Submucosa
Muscularis

LM 45x

(a) Wall of the duodenum

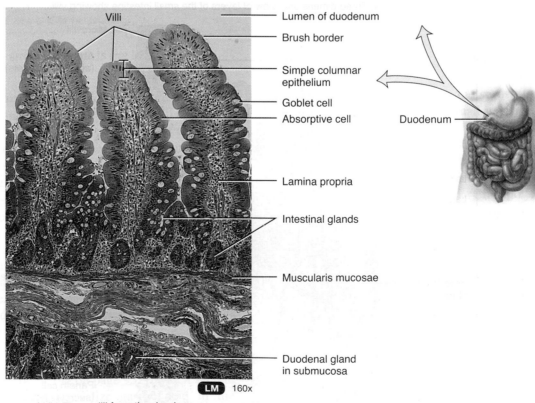

Villi
Lumen of duodenum
Brush border
Simple columnar epithelium
Goblet cell
Absorptive cell
Lamina propria
Intestinal glands
Muscularis mucosae
Duodenal gland in submucosa

Duodenum

LM 160x

(b) Three villi from the duodenum

propria are an arteriole, a venule, a blood capillary network, and a **lacteal** (LAK-tē-al = milky), which is a lymphatic capillary. Nutrients absorbed by the epithelial cells covering the villus pass through the wall of a capillary or a lacteal to enter blood or lymph, respectively.

Besides circular folds and villi, the small intestine also has **microvilli** (mī-krō-VIL-ī; *micro-* = small), which are projections of the apical (free) membrane of the absorptive cells. Each microvillus is a 1 μm-long cylindrical, membrane-covered projection that contains a bundle of 20–30 actin filaments. When viewed through a light microscope, the microvilli are too small to be seen individually; instead they form a fuzzy line, called the **brush**

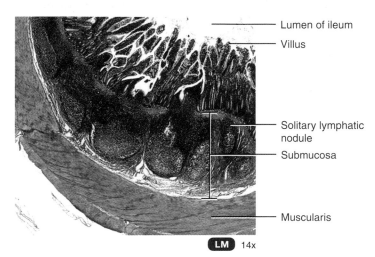

(c) Lymphatic nodules in the ileum

Lumen of ileum
Villus
Solitary lymphatic nodule
Submucosa
Muscularis
LM 14x

Microvilli
Brush border
Simple columnar epithelial cell
TEM 46,800x

(d) Several microvilli from the duodenum

❓ **What is the function of the fluid secreted by duodenal (Brunner's) glands?**

border, extending into the lumen of the small intestine (Figure 24.19d). There are an estimated 200 million microvilli per square millimeter of small intestine. Because the microvilli greatly increase the surface area of the plasma membrane, larger amounts of digested nutrients can diffuse into absorptive cells in a given period. The brush border also contains several brush-border enzymes that have digestive functions (discussed shortly).

Role of Intestinal Juice and Brush-Border Enzymes

About 1–2 liters (1–2 qt) of **intestinal juice,** a clear yellow fluid, are secreted each day. Intestinal juice contains water and mucus and is slightly alkaline (pH 7.6). Together, pancreatic and intestinal juices provide a liquid medium that aids the absorption of substances from chyme in the small intestine. The absorptive cells of the small intestine synthesize several digestive enzymes, called **brush-border enzymes,** and insert them in the plasma membrane of the microvilli. Thus, some enzymatic digestion occurs at the surface of the absorptive cells that line the villi, rather than in the lumen exclusively, as occurs in other parts of the GI tract. Among the brush-border enzymes are four carbohydrate-digesting enzymes called α-dextrinase, maltase, sucrase, and lactase; protein-digesting enzymes called peptidases (aminopeptidase and dipeptidase); and two types of nucleotide-digesting enzymes, nucleosidases and phosphatases. Also, as absorptive cells slough off into the lumen of the small intestine, they break apart and release enzymes that help digest nutrients in the chyme.

Mechanical Digestion in the Small Intestine

The two types of movements of the small intestine—segmentations and a type of peristalsis called migrating motility complexes—are governed mainly by the myenteric plexus. **Segmentations** are localized, mixing contractions that occur in portions of intestine distended by a large volume of chyme. Segmentations mix chyme with the digestive juices and bring the particles of food into contact with the mucosa for absorption; they do not push the intestinal contents along the tract. A segmentation starts with the contractions of circular muscle fibers in a portion of the small intestine, an action that constricts the intestine into segments. Next, muscle fibers that encircle the middle of each segment also contract, dividing each segment again. Finally, the fibers that first contracted relax, and each small segment unites with an adjoining small segment so that large segments are formed again. As this sequence of events repeats, the chyme sloshes back and forth. Segmentations occur most rapidly in the duodenum, about 12 times per minute, and progressively slow to about 8 times per minute in the ileum. This movement is similar to alternately squeezing the middle and then the ends of a capped tube of toothpaste.

After most of a meal has been absorbed, which lessens distension of the wall of the small intestine, segmentation stops and peristalsis begins. The type of peristalsis that occurs in the small intestine, termed a **migrating motility complex (MMC),** begins

Absorption in the Small Intestine

All the chemical and mechanical phases of digestion from the mouth through the small intestine are directed toward changing food into forms that can pass through the absorptive epithelial cells lining the mucosa and into the underlying blood and lymphatic vessels. These forms are monosaccharides (glucose, fructose, and galactose) from carbohydrates; single amino acids, dipeptides, and tripeptides from proteins; and fatty acids, glycerol, and monoglycerides from triglycerides. Passage of these digested nutrients from the gastrointestinal tract into the blood or lymph is called **absorption.**

Absorption of materials occurs via diffusion, facilitated diffusion, osmosis, and active transport. About 90% of all absorption of nutrients occurs in the small intestine; the other 10% occurs in the stomach and large intestine. Any undigested or unabsorbed material left in the small intestine passes on to the large intestine.

Absorption of Monosaccharides

All carbohydrates are absorbed as monosaccharides. The capacity of the small intestine to absorb monosaccharides is huge—an estimated 120 grams per hour. As a result, all dietary carbohydrates that are digested normally are absorbed, leaving only indigestible cellulose and fibers in the feces. Monosaccharides pass from the lumen through the apical membrane via *facilitated diffusion* or *active transport*. Fructose, a monosaccharide found in fruits, is transported via *facilitated diffusion;* glucose and galactose are transported into absorptive cells of the villi via *secondary active transport* that is coupled to the active transport of Na^+ (Figure 24.20a). The transporter has binding sites for one glucose molecule and two sodium ions; unless all three sites are filled, neither substance is transported. Galactose competes with glucose to ride the same transporter. (Because both Na^+ and glucose or galactose move in the same direction, this is a *symporter.* Monosaccharides then move out of the absorptive cells through their basolateral surfaces via *facilitated diffusion* and enter the capillaries of the villi) (see Figure 24.20b).

Absorption of Amino Acids, Dipeptides, and Tripeptides

Most proteins are absorbed as amino acids via *active transport* processes that occur mainly in the duodenum and jejunum. About half of the absorbed amino acids are present in food; the other half come from the body itself as proteins in digestive juices and dead cells that slough off the mucosal surface! Normally, 95–98% of the protein present in the small intestine is digested and absorbed. Different transporters carry different types of amino acids. Some amino acids enter absorptive cells of the villi via Na^+-dependent secondary active transport processes that are similar to the glucose transporter; other amino acids are actively transported by themselves. At least one symporter brings in dipeptides and tripeptides together with H^+; the peptides then are hydrolyzed to single amino acids inside the absorptive cells. Amino acids move out of the absorp-

tive cells via diffusion and enter capillaries of the villus (Figure 24.20a, b). Both monosaccharides and amino acids are transported in the blood to the liver by way of the hepatic portal system. If not removed by hepatocytes, they enter the general circulation.

Absorption of Lipids

All dietary lipids are absorbed via *simple diffusion.* Adults absorb about 95% of the lipids present in the small intestine; due to their lower production of bile, newborn infants absorb only about 85% of lipids. As a result of their emulsification and digestion, triglycerides are mainly broken down into monoglycerides and fatty acids, which can be either short-chain fatty acids or long-chain fatty acids. Although short-chain fatty acids are hydrophobic, they are very small in size. Because of their size, they can dissolve in the watery intestinal chyme, pass through the absorptive cells via simple diffusion, and follow the same route taken by monosaccharides and amino acids into a blood capillary of a villus (Figure 24.20a). Long-chain fatty acids and monoglycerides are large and hydrophobic and have difficulty being suspended in the watery environment of the intestinal chyme. Besides their role in emulsification, bile salts also help to make these long-chain fatty acids and monoglycerides more soluble. The bile salts in intestinal chyme surround the long-chain fatty acids and monoglycerides, forming tiny spheres called **micelles** (mī-SELZ = small morsels), each of which is 2–10 nm in diameter and includes 20–50 bile salt molecules (Figure 24.20a). Micelles are formed due to the amphipathic nature of bile salts: The hydrophobic regions of bile salts interact with the long-chain fatty acids and monoglycerides, and the hydrophilic regions of bile salts interact with the watery intestinal chyme. Once formed, the micelles move from the interior of the small intestinal lumen to the brush border of the absorptive cells. At that point, the long-chain fatty acids and monoglycerides diffuse out of the micelles into the absorptive cells, leaving the micelles behind in the chyme. The micelles continually repeat this ferrying function as they move from the brush border back through the chyme to the interior of the small intestinal lumen to pick up more long-chain fatty acids and monoglycerides. Micelles also solubilize other large hydrophobic molecules such as fat-soluble vitamins (A, D, E, and K) and cholesterol that may be present in intestinal chyme, and aid in their absorption. These fat-soluble vitamins and cholesterol molecules are packed in the micelles along with the long-chain fatty acids and monoglycerides.

Once inside the absorptive cells, long-chain fatty acids and monoglycerides are recombined to form triglycerides, which aggregate into globules along with phospholipids and cholesterol and become coated with proteins. These large spherical masses, about 80 nm in diameter, are called **chylomicrons.** Chylomicrons leave the absorptive cell via exocytosis. Because they are so large and bulky, chylomicrons cannot enter blood capillaries—the pores in the walls of blood capillaries are too small. Instead, chylomicrons enter lacteals, which have much larger pores than blood capillaries. From lacteals, chylomicrons are transported by way of lymphatic vessels to the thoracic duct and enter the blood at the left subclavian vein

Figure 24.20 Absorption of digested nutrients in the small intestine. For simplicity, all digested foods are shown in the lumen of the small intestine, even though some nutrients are digested by brush-border enzymes.

Long-chain fatty acids and monoglycerides are absorbed into lacteals; other products of digestion enter blood capillaries.

(a) Mechanisms for movement of nutrients through absorptive epithelial cells of the villi

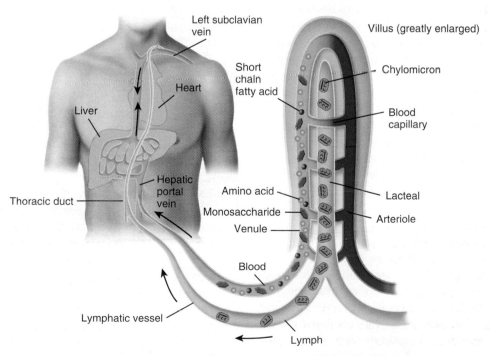

(b) Movement of absorbed nutrients into the blood and lymph

? **A monoglyceride may be larger than an amino acid. Why can monoglycerides be absorbed by simple diffusion, but amino acids cannot?**

(Figure 24.20b). The hydrophilic protein coat that surrounds each chylomicron keeps the chylomicrons suspended in blood and prevents them from sticking to each other.

Within 10 minutes after absorption, about half of the chylomicrons have already been removed from the blood as they pass through blood capillaries in the liver and adipose tissue. This removal is accomplished by an enzyme attached to the apical surface of capillary endothelial cells, called **lipoprotein lipase,** that breaks down triglycerides in chylomicrons and other lipoproteins into fatty acids and glycerol. The fatty acids diffuse into hepatocytes and adipose cells and combine with glycerol during resynthesis of triglycerides. Two or three hours after a meal, few chylomicrons remain in the blood.

After participating in the emulsification and absorption of lipids, 90–95% of the bile salts are reabsorbed by active transport in the final segment of the small intestine (ileum) and returned by the blood to the liver through the hepatic portal system for recycling. This cycle of bile salt secretion by hepatocytes into bile, reabsorption by the ileum, and resecretion into bile is called the **enterohepatic circulation.** Insufficient bile salts, due either to obstruction of the bile ducts or removal of the gallbladder, can result in the loss of up to 40% of dietary lipids in feces due to diminished lipid absorption. When lipids are not absorbed properly, the fat-soluble vitamins are not adequately absorbed.

Absorption of Electrolytes

Many of the electrolytes absorbed by the small intestine come from gastrointestinal secretions, and some are part of ingested foods and liquids. Recall that electrolytes are compounds that separate into ions in water and conduct electricity. Sodium ions are actively transported out of absorptive cells by basolateral sodium–potassium pumps (Na^+/K^+ ATPase) after they have moved into absorptive cells via diffusion and secondary active transport. Thus, most of the sodium ions (Na^+) in gastrointestinal secretions are reclaimed and not lost in the feces. Negatively charged bicarbonate, chloride, iodide, and nitrate ions can passively follow Na^+ or be actively transported. Calcium ions also are absorbed actively in a process stimulated by calcitriol. Other electrolytes such as iron, potassium, magnesium, and phosphate ions also are absorbed via active transport mechanisms.

Absorption of Vitamins

As you have just learned, the fat-soluble vitamins A, D, E, and K are included with ingested dietary lipids in micelles and are absorbed via simple diffusion. Most water-soluble vitamins, such as most B vitamins and vitamin C, also are absorbed via simple diffusion. Vitamin B_{12}, however, combines with intrinsic factor produced by the stomach, and the combination is absorbed in the ileum via an active transport mechanism.

Absorption of Water

The total volume of fluid that enters the small intestine each day—about 9.3 liters (9.8 qt)—comes from ingestion of liquids (about 2.3 liters) and from various gastrointestinal secretions

(about 7.0 liters). Figure 24.21 depicts the amounts of fluid ingested, secreted, absorbed, and excreted by the GI tract. The small intestine absorbs about 8.3 liters of the fluid; the remainder passes into the large intestine, where most of the rest of it—about 0.9 liter—is also absorbed. Only 0.1 liter (100 mL) of water is excreted in the feces each day. Most is excreted via the urinary system.

All water absorption in the GI tract occurs via *osmosis* from the lumen of the intestines through absorptive cells and into blood capillaries. Because water can move across the intestinal mucosa in both directions, the absorption of water from the small intestine depends on the absorption of electrolytes and nutrients to maintain an osmotic balance with the blood. The absorbed electrolytes, monosaccharides, and amino acids establish a concentration gradient for water that promotes water absorption via osmosis.

Table 24.5 summarizes the digestive activities of the pancreas, liver, gallbladder, and small intestine.

Figure 24.21 Daily volumes of fluid ingested, secreted, absorbed, and excreted from the GI tract.

 All water absorption in the GI tract occurs via osmosis.

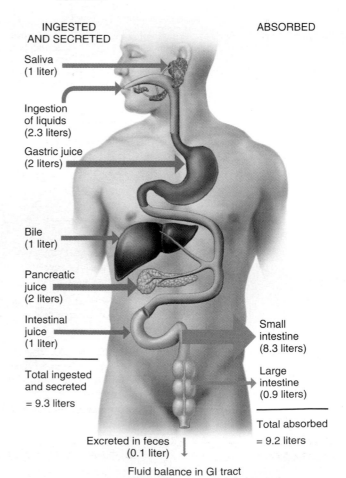

Fluid balance in GI tract

❓ **Which two organs of the digestive system secrete the most fluid?**

● CLINICAL CONNECTION | Absorption of Alcohol

The intoxicating and incapacitating effects of alcohol depend on the blood alcohol level. Because it is lipid-soluble, alcohol begins to be absorbed in the stomach. However, the surface area available for absorption is much greater in the small intestine than in the stomach, so when alcohol passes into the duodenum, it is absorbed more rapidly. Thus, the longer the alcohol remains in the stomach, the more slowly blood alcohol level rises. Because fatty acids in chyme slow gastric emptying, blood alcohol level will rise more slowly when fat-rich foods, such as pizza, hamburgers, or nachos, are consumed with alcoholic beverages. Also, the enzyme alcohol dehydrogenase, which is present in gastric mucosa cells, breaks down some of the alcohol to acetaldehyde, which is not intoxicating. When the rate of gastric emptying is slower, proportionally more alcohol will be ab-

sorbed and converted to acetaldehyde in the stomach, and thus less alcohol will reach the bloodstream. Given identical consumption of alcohol, females often develop higher blood alcohol levels (and therefore experience greater intoxication) than males of comparable size because the activity of gastric alcohol dehydrogenase is up to 60% lower in females than in males. Asian males may also have lower levels of this gastric enzyme. ●

◉ CHECKPOINT

32. List the regions of the small intestine and describe their functions.
33. In what ways are the mucosa and submucosa of the small intestine adapted for digestion and absorption?
34. Describe the types of movement that occur in the small intestine.
35. Explain the functions of pancreatic amylase, aminopeptidase, gastric lipase, and deoxyribonuclease.
36. What is the difference between digestion and absorption? How are the end products of carbohydrate, protein, and lipid digestion absorbed?
37. By what routes do absorbed nutrients reach the liver?
38. Describe the absorption of electrolytes, vitamins, and water by the small intestine.

LARGE INTESTINE

◉ OBJECTIVE

• Describe the anatomy, histology, and functions of the large intestine.

The large intestine is the terminal portion of the GI tract. The overall functions of the large intestine are the completion of absorption, the production of certain vitamins, the formation of feces, and the expulsion of feces from the body.

Anatomy of the Large Intestine

The **large intestine,** which is about 1.5 m (5 ft) long and 6.5 cm (2.5 in.) in diameter, extends from the ileum to the anus. It is attached to the posterior abdominal wall by its **mesocolon,** which is a double layer of peritoneum (see Figure 24.4a). Structurally, the four major regions of the large intestine are the cecum, colon, rectum, and anal canal (Figure 24.22a).

The opening from the ileum into the large intestine is guarded by a fold of mucous membrane called the **ileocecal sphincter (valve),** which allows materials from the small intestine to pass into the large intestine. Hanging inferior to the ileocecal valve is the **cecum,** a small pouch about 6 cm (2.4 in.) long. Attached to the cecum is a twisted, coiled tube, measuring about 8 cm (3 in.) in length, called the **appendix** or **vermiform appendix** (*vermiform* = worm-shaped; *appendix* = appendage). The mesentery of the appendix, called the **mesoappendix,** attaches the appendix to the inferior part of the mesentery of the ileum.

The open end of the cecum merges with a long tube called the **colon** (= food passage), which is divided into ascending, transverse, descending, and sigmoid portions. Both the ascending and

TABLE 24.5

Summary of Digestive Activities in the Pancreas, Liver, Gallbladder, and Small Intestine

STRUCTURE	ACTIVITY
Pancreas	Delivers pancreatic juice into the duodenum via ... pancreatic duct (see Table 24.4 for pancreatic ... functions).
Liver	
Gallbladder	
Small intestine	
Mucosa/su...	
Intestin...	
Duode... (Brunn...	
Micro...	... of ... ...at
V...	the sites ... ...se the ... ...n.
Circular ...	...ncrease the surfa... ...on.
Muscularis	
Segmentation	Consists of alternating contractions of circular smooth muscle fibers that produce segmentation and resegmentation of sections of the small intestine; mixes chyme with digestive juices and brings food into contact with the mucosa for absorption.
Migrating motility complex (MMC)	A type of peristalsis consisting of waves of contraction and relaxation of circular and longitudinal smooth muscle fibers passing the length of the small intestine; moves chyme toward ileocecal sphincter.

Figure 24.22 Anatomy of the large intestine. (See Tortora, *A Photographic Atlas of the Human Body, Second Edition,* Figure 12.13.)

6   The regions of the large intestine are the cecum, colon, rectum, and anal canal.

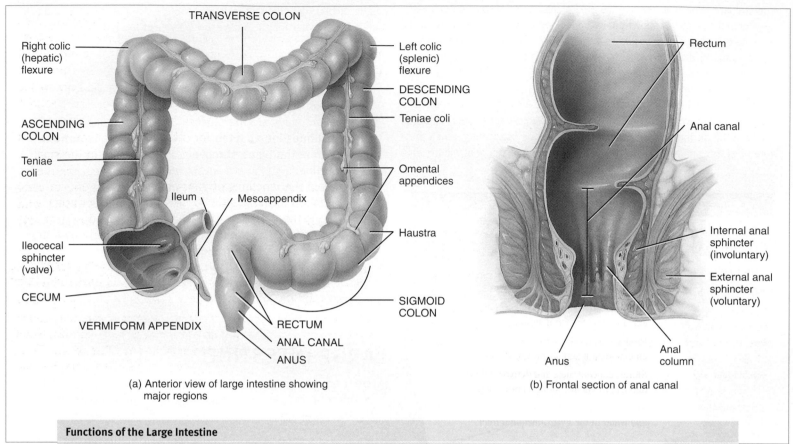

TRANSVERSE COLON

Right colic (hepatic) flexure

Left colic (splenic) flexure

ASCENDING COLON

DESCENDING COLON

Teniae coli

Teniae coli

Ileum

Mesoappendix

Omental appendices

Ileocecal sphincter (valve)

Haustra

CECUM

VERMIFORM APPENDIX

SIGMOID COLON

RECTUM

ANAL CANAL

ANUS

Rectum

Anal canal

Internal anal sphincter (involuntary)

External anal sphincter (voluntary)

Anus

Anal column

(a) Anterior view of large intestine showing major regions

(b) Frontal section of anal canal

Functions of the Large Intestine

1. Haustral churning, peristalsis, and mass peristalsis drive the contents of the colon into the rectum.
2. Bacteria in the large intestine convert proteins to amino acids, break down amino acids, and produce some B vitamins and vitamin K.
3. Absorbing some water, ions, and vitamins.
4. Forming feces.
5. Defecating (emptying the rectum).

? Which portions of the colon are retroperitoneal?

descending colon are retroperitoneal; the transverse and sigmoid colon are not. True to its name, the **ascending colon** ascends on the right side of the abdomen, reaches the inferior surface of the liver, and turns abruptly to the left to form the **right colic (hepatic) flexure.** The colon continues across the abdomen to the left side as the **transverse colon.** It curves beneath the inferior end of the spleen on the left side as the **left colic (splenic) flexure** and passes inferiorly to the level of the iliac crest as the **descending colon.** The **sigmoid colon** (*sigm-* = S-shaped) begins near the left iliac crest, projects medially to the midline, and terminates as the rectum at about the level of the third sacral vertebra.

The **rectum,** the last 20 cm (8 in.) of the GI tract, lies anterior to the sacrum and coccyx. The terminal 2–3 cm (1 in.) of the rectum is called the **anal canal** (Figure 24.22b). The mucous membrane of the anal canal is arranged in longitudinal folds

called **anal columns** that contain a network of arteries and veins. The opening of the anal canal to the exterior, called the **anus,** is guarded by an **internal anal sphincter** of smooth muscle (involuntary) and an **external anal sphincter** of skeletal muscle (voluntary). Normally these sphincters keep the anus closed except during the elimination of feces.

•CLINICAL CONNECTION | Appendicitis

Inflammation of the appendix, termed **appendicitis,** is preceded by obstruction of the lumen of the appendix by chyme, inflammation, a foreign body, a carcinoma of the cecum, stenosis, or kinking of the organ. It is characterized by high fever, elevated white blood cell count, and a neutrophil count higher than 75%. The infection that follows may result in edema and ischemia and may progress to gangrene and perforation within 24 hours. Typically, appendicitis begins with referred pain in the umbilical region of the abdomen, followed by anorexia (loss of appetite for food), nausea, and vomiting. After several hours the pain localizes in the right lower quadrant (RLQ) and is continuous, dull or severe, and intensified by coughing, sneezing, or body movements. Early appendectomy (removal of the appendix) is recommended because it is safer to operate than to risk rupture, peritonitis, and gangrene. Although it required major abdominal surgery in the past, today appendectomies are usually performed laparoscopically. •

Histology of the Large Intestine

The wall of the large intestine contains the typical four layers found in the rest of the GI tract: mucosa, submucosa, muscularis, and serosa. The **mucosa** consists of simple columnar epithelium, lamina propria (areolar connective tissue), and muscularis mucosae (smooth muscle) (Figure 24.23a). The epithelium contains mostly absorptive and goblet cells (Figure 24.23b, c). The absorptive cells function primarily in water absorption; the goblet cells secrete mucus that lubricates the passage of the colonic contents. Both absorptive and goblet cells are located in long, straight, tubular **intestinal glands (crypts of Lieberkühn)** that extend the full thickness of the mucosa. Solitary lymphatic nodules are also found in the lamina propria of the mucosa and may extend through the muscularis mucosae into the submucosa. Compared to the small intestine, the mucosa of the large intestine does not have as many structural adaptations that increase surface area. There are no circular folds or villi; however, microvilli are present on the absorptive cells.

Consequently, much more absorption occurs in the small intestine than in the large intestine.

The **submucosa** of the large intestine consists of areolar connective tissue. The **muscularis** consists of an external layer of longitudinal smooth muscle and an internal layer of circular smooth muscle. Unlike other parts of the GI tract, portions of the longitudinal muscles are thickened, forming three conspicuous bands called the **teniae coli** (TĒ-nē-ē KŌ-lī; *teniae* = flat bands) that run most of the length of the large intestine (see Figure 24.22a). The teniae coli are separated by portions of the wall with less or no longitudinal muscle. Tonic contractions of the bands gather the colon into a series of pouches called **haustra** (HAWS-tra = shaped like pouches; singular is **haustrum**), which give the colon a puckered appearance. A single layer of circular smooth muscle lies between teniae coli. The **serosa** of the large intestine is part of the visceral peritoneum. Small pouches of visceral peritoneum filled with fat are attached to teniae coli and are called **omental (fatty) appendices.**

• CLINICAL **CONNECTION** | Polyps in the Colon

Polyps in the colon are generally slow-developing benign growths that arise from the mucosa of the large intestine. Often, they do not cause symptoms. If symptoms do occur, they include diarrhea, blood in the feces, and mucus discharged from the anus. The polyps are removed by colonoscopy or surgery because some of them may become cancerous. •

Figure 24.23 Histology of the large intestine.

Intestinal glands formed by simple columnar epithelial cells and goblet cells extend the full thickness of the mucosa.

Lumen of large intestine

Openings of intestinal glands
Simple columnar epithelium
Intestinal gland
Lamina propria
Lymphatic nodule
Muscularis mucosae
Lymphatic vessel
Arteriole
Venule
Circular layer of muscle
Myenteric plexus
Longitudinal layer of muscle

MUCOSA
SUBMUCOSA
MUSCULARIS
SEROSA

(a) Three-dimensional view of layers of the large intestine

FIGURE 24.23 CONTINUES ▶

occurs, and the feces become dry and hard. Constipation may be caused by poor habits (delaying defecation), spasms of the colon, insufficient fiber in the diet, inadequate fluid intake, lack of exercise, emotional stress, and certain drugs. A common treatment is a mild laxative, such as milk of

magnesia, which induces defecation. However, many physicians maintain that laxatives are habit-forming, and that adding fiber to the diet, increasing the amount of exercise, and increasing fluid intake are safer ways of controlling this common problem.

Table 24.6 summarizes the digestive activities in the large intestine, and Table 24.7 summarizes the functions of all digestive system organs.

TABLE 24.6

Summary of Digestive Activities in the Large Intestine

STRUCTURE	ACTIVITY	FUNCTION(S)
Lumen	Bacterial activity.	Breaks down undigested carbohydrates, proteins, and amino acids into products that can be expelled in feces or absorbed and detoxified by liver; synthesizes certain B vitamins and vitamin K.
Mucosa	Secretes mucus.	Lubricates colon and protects mucosa.
	Absorption.	Water absorption solidifies feces and contributes to the body's water balance; solutes absorbed include ions and some vitamins.
Muscularis	Haustral churning.	Moves contents from haustrum to haustrum by muscular contractions.
	Peristalsis.	Moves contents along length of colon by contractions of circular and longitudinal muscles.
	Mass peristalsis.	Forces contents into sigmoid colon and rectum.
	Defecation reflex.	Eliminates feces by contractions in sigmoid colon and rectum.

● CLINICAL CONNECTION Dietary Fiber

Dietary fiber consists of indigestible plant carbohydrates—such as cellulose, lignin, and pectin—found in fruits, vegetables, grains, and beans. **Insoluble fiber,** which does not dissolve in water, includes the woody or structural parts of plants such as the skins of fruits and vegetables and the bran coating around wheat and corn kernels. Insoluble fiber passes through the GI tract largely unchanged but speeds up the passage of material through the tract. **Soluble fiber,** which does dissolve in water, forms a gel that slows the passage of material through the tract. It is found in abundance in beans, oats, barley, broccoli, prunes, apples, and citrus fruits.

People who choose a fiber-rich diet may reduce their risk of developing obesity, diabetes, atherosclerosis, gallstones, hemorrhoids, diverticulitis, appendicitis, and colorectal cancer. Soluble fiber also may help lower blood cholesterol. The liver normally converts cholesterol to bile salts, which are released into the small intestine to help fat digestion. Having accomplished their task, the bile salts are reabsorbed by the small intestine and recycled back to the liver. Since soluble fiber binds to bile salts to prevent their reabsorption, the liver makes more bile salts to replace those lost in feces. Thus, the liver uses more cholesterol to make more bile salts and blood cholesterol level is lowered. ●

TABLE 24.7

Summary of Organs of the Digestive System and Their Functions

ORGAN	FUNCTIONS
Mouth	See other listings in this table for the functions of the tongue, salivary glands, and teeth, all of which are in the mouth. Additionally, the lips and cheeks keep food between the teeth during mastication, and buccal glands lining the mouth produce saliva.
Tongue	Maneuvers food for mastication, shapes food into a bolus, maneuvers food for deglutition, detects taste and touch sensations, and initiates digestion of triglycerides.
Salivary glands	Produce saliva, which softens, moistens, and dissolves foods; cleanses mouth and teeth; and initiates the digestion of starch.
Teeth	Cut, tear, and pulverize food to reduce solids to smaller particles for swallowing.
Pharynx	Receives a bolus from the oral cavity and passes it into the esophagus.
Esophagus	Receives a bolus from the pharynx and moves it into the stomach. This requires relaxation of the upper esophageal sphincter and secretion of mucus.
Stomach	Mixing waves macerate food, mix it with secretions of gastric glands (gastric juice), and reduce food to chyme. Gastric juice activates pepsin and kills many microbes in food. Intrinsic factor aids absorption of vitamin B_{12}. The stomach serves as a reservoir for food before releasing it into the small intestine.
Pancreas	Pancreatic juice buffers acidic gastric juice in chyme (creating the proper pH for digestion in the small intestine), stops the action of pepsin from the stomach, and contains enzymes that digest carbohydrates, proteins, triglycerides, and nucleic acids.
Liver	Produces bile, which is needed for the emulsification and absorption of lipids in the small intestine.
Gallbladder	Stores and concentrates bile and releases it into the small intestine.
Small intestine	Segmentations mix chyme with digestive juices; migrating motility complexes propel chyme toward the ileocecal sphincter; digestive secretions from the small intestine, pancreas, and liver complete the digestion of carbohydrates, proteins, lipids, and nucleic acids; circular folds, villi, and microvilli increase surface area for absorption; site where about 90% of nutrients and water are absorbed.
Large intestine	Haustral churning, peristalsis, and mass peristalsis drive the contents of the colon into the rectum; bacteria produce some B vitamins and vitamin K; absorption of some water, ions, and vitamins; defecation.

CHECKPOINT

39. What are the major regions of the large intestine?
40. How does the muscularis of the large intestine differ from that of the rest of the gastrointestinal tract? What are haustra?
41. Describe the mechanical movements that occur in the large intestine.
42. What is defecation and how does it occur?
43. What activities occur in the large intestine to change its contents into feces?

PHASES OF DIGESTION

OBJECTIVES

• Describe the three phases of digestion.
• Describe the major hormones that regulate digestive activities.

Digestive activities occur in three overlapping phases: the cephalic phase, the gastric phase, and the intestinal phase.

Cephalic Phase

During the **cephalic phase** of digestion, the smell, sight, thought, or initial taste of food activates neural centers in the cerebral cortex, hypothalamus, and brain stem. The brain stem then activates the facial (VII), glossopharyngeal (IX), and vagus (X) nerves. The facial and glossopharyngeal nerves stimulate the salivary glands to secrete saliva, while the vagus nerves stimulate the gastric glands to secrete gastric juice. The purpose of the cephalic phase of digestion is to prepare the mouth and stomach for food that is about to be eaten.

Gastric Phase

Once food reaches the stomach, the **gastric phase** of digestion begins. Neural and hormonal mechanisms regulate the gastric phase of digestion to promote gastric secretion and gastric motility.

• *Neural regulation.* Food of any kind distends the stomach and stimulates stretch receptors in its walls. Chemoreceptors in the stomach monitor the pH of the stomach chyme. When the stomach walls are distended or pH increases because proteins have entered the stomach and buffered some of the stomach acid, the stretch receptors and chemoreceptors are activated, and a neural negative feedback loop is set in motion (Figure 24.24). From the stretch receptors and chemoreceptors, nerve impulses propagate to the submucosal plexus, where they activate parasympathetic and enteric neurons. The resulting nerve impulses cause waves of peristalsis and continue to stimulate the flow of gastric juice from gastric glands. The peristaltic waves mix the food with gastric juice; when the waves become strong enough, a small quantity of chyme undergoes gastric emptying into the duodenum. The pH of the stomach chyme decreases (becomes more acidic) and the distension of the stomach walls lessens because chyme has passed into the small intestine, suppressing secretion of gastric juice.

Figure 24.24 Neural negative feedback regulation of the pH of gastric juice and gastric motility during the gastric phase of digestion.

Food entering the stomach stimulates secretion of gastric juice and causes vigorous waves of peristalsis.

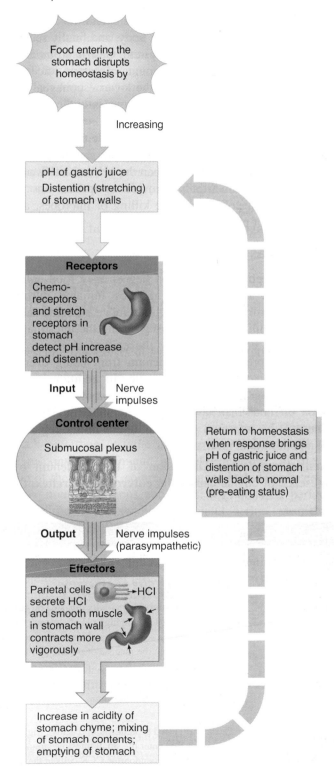

? Why does food initially cause the pH of the gastric juice to rise?

BODY SYSTEM	CONTRIBUTION OF THE DIGESTIVE SYSTEM

For all body systems
The digestive system breaks down dietary nutrients into forms that can be absorbed and used by body cells for producing ATP and building body tissues. Absorbs water, minerals, and vitamins needed for growth and function of body tissues; and eliminates wastes from body tissues in feces.

Integumentary system
Small intestine absorbs vitamin D, which skin and kidneys modify to produce the hormone calcitriol. Excess dietary calories are stored as triglycerides in adipose cells in dermis and subcutaneous layer.

Skeletal system
Small intestine absorbs dietary calcium and phosphorus salts needed to build bone extracellular matrix.

Muscular system
Liver can convert lactic acid (produced by muscles during exercise) to glucose.

Nervous system
Gluconeogenesis (synthesis of new glucose molecules) in liver plus digestion and absorption of dietary carbohydrates provide glucose, needed for ATP production by neurons.

Endocrine system
Liver inactivates some hormones, ending their activity. Pancreatic islets release insulin and glucagon. Cells in mucosa of stomach and small intestine release hormones that regulate digestive activities. Liver produces angiotensinogen.

Cardiovascular system
GI tract absorbs water that helps maintain blood volume and iron that is needed for synthesis of hemoglobin in red blood cells. Bilirubin from hemoglobin breakdown is partially excreted in feces. Liver synthesizes most plasma proteins.

Lymphatic system and immunity
Acidity of gastric juice destroys bacteria and most toxins in stomach.

Respiratory system
Pressure of abdominal organs against the diaphragm helps expel air quickly during a forced exhalation.

Urinary system
Absorption of water by GI tract provides water needed to excrete waste products in urine.

Reproductive systems
Digestion and absorption provides adequate nutrients, including fats, for normal development of reproductive structures, for production of gametes (oocytes and sperm), and for fetal growth and development during pregnancy.

THE DIGESTIVE SYSTEM

DISORDERS: HOMEOSTATIC IMBALANCES

Dental Caries

Dental caries, or tooth decay, involves a gradual demineralization (softening) of the enamel and dentin. If untreated, microorganisms may invade the pulp, causing inflammation and infection, with subsequent death of the pulp and abscess of the alveolar bone surrounding the root's apex, requiring root canal therapy (see page 932).

Dental caries begin when bacteria, acting on sugars, produce acids that demineralize the enamel. **Dextran,** a sticky polysaccharide produced from sucrose, causes the bacteria to stick to the teeth. Masses of bacterial cells, dextran, and other debris adhering to teeth constitute **dental plaque.** Saliva cannot reach the tooth surface to buffer the acid because the plaque covers the teeth. Brushing the teeth after eating removes the plaque from flat surfaces before the bacteria can produce acids. Dentists also recommend that the plaque between the teeth be removed every 24 hours with dental floss.

Periodontal Disease

Periodontal disease is a collective term for a variety of conditions characterized by inflammation and degeneration of the gingivae, alveolar bone, periodontal ligament, and cementum. In one such condition, called **pyorrhea,** initial symptoms include enlargement and inflammation of the soft tissue and bleeding of the gums. Without treatment, the soft tissue may deteriorate and the alveolar bone may be resorbed, causing loosening of the teeth and recession of the gums. Periodontal diseases are often caused by poor oral hygiene; by local irritants, such as bacteria, impacted food, and cigarette smoke; or by a poor "bite."

Peptic Ulcer Disease

In the United States, 5–10% of the population develops **peptic ulcer disease (PUD).** An **ulcer** is a craterlike lesion in a membrane; ulcers that develop in areas of the GI tract exposed to acidic gastric juice are called **peptic ulcers.** The most common complication of peptic ulcers is bleeding, which can lead to anemia if enough blood is lost. In acute cases, peptic ulcers can lead to shock and death. Three distinct causes of PUD are recognized: (1) the bacterium *Helicobacter pylori;* (2) nonsteroidal anti-inflammatory drugs (NSAIDs) such as aspirin; and (3) hypersecretion of HCl, as occurs in Zollinger–Ellison syndrome, a gastrin-producing tumor, usually of the pancreas.

Helicobacter pylori (previously named *Campylobacter pylori*) is the most frequent cause of PUD. The bacterium produces an enzyme called urease, which splits urea into ammonia and carbon dioxide. While shielding the bacterium from the acidity of the stomach, the ammonia also damages the protective mucous layer of the stomach and the underlying gastric cells. *H. pylori* also produces catalase, an enzyme that may protect the microbe from phagocytosis by neutrophils, plus several adhesion proteins that allow the bacterium to attach itself to gastric cells.

Several therapeutic approaches are helpful in the treatment of PUD. Cigarette smoke, alcohol, caffeine, and NSAIDs should be avoided because they can impair mucosal defensive mechanisms, which increases mucosal susceptibility to the damaging effects of HCl. In cases associated with *H. pylori,* treatment with an antibiotic drug often resolves the problem. Oral antacids such as Tums® or Maalox® can help temporarily by buffering gastric acid. When hypersecretion of HCl is the cause of PUD, H_2 blockers (such as Tagamet®) or proton pump inhibitors such as omeprazole (Prilosec®), which block secretion of H^+ from parietal cells, may be used.

Diverticular Disease

In **diverticular disease,** saclike outpouchings of the wall of the colon, termed **diverticula,** occur in places where the muscularis has weakened and may become inflamed. Development of diverticula is known as **diverticulosis.** Many people who develop diverticulosis have no symptoms and experience no complications. Of those people known to have diverticulosis, 10–25% eventually develop an inflammation known as **diverticulitis.** This condition may be characterized by pain, either constipation or increased frequency of defecation, nausea, vomiting, and low-grade fever. Because diets low in fiber contribute to development of diverticulitis, patients who change to high-fiber diets show marked relief of symptoms. In severe cases, affected portions of the colon may require surgical removal. If diverticula rupture, the release of bacteria into the abdominal cavity can cause peritonitis.

Colorectal Cancer

Colorectal cancer is among the deadliest of malignancies, ranking second to lung cancer in males and third after lung cancer and breast cancer in females. Genetics plays a very important role; an inherited predisposition contributes to more than half of all cases of colorectal cancer. Intake of alcohol and diets high in animal fat and protein are associated with increased risk of colorectal cancer; dietary fiber, retinoids, calcium, and selenium may be protective. Signs and symptoms of colorectal cancer include diarrhea, constipation, cramping, abdominal pain, and rectal bleeding, either visible or occult (hidden in feces). Precancerous growths on the mucosal surface, called **polyps,** also increase the risk of developing colorectal cancer. Screening for colorectal cancer includes testing for blood in the feces, digital rectal examination, sigmoidoscopy, colonoscopy, and barium enema. Tumors may be removed endoscopically or surgically.

Hepatitis

Hepatitis is an inflammation of the liver that can be caused by viruses, drugs, and chemicals, including alcohol. Clinically, several types of viral hepatitis are recognized.

Hepatitis A (infectious hepatitis) is caused by the hepatitis A virus and is spread via fecal contamination of objects such as food, clothing, toys, and eating utensils (fecal–oral route). It is generally a mild disease of children and young adults characterized by loss of appetite, malaise, nausea, diarrhea, fever, and chills. Eventually, jaundice appears. This type of hepatitis does not cause lasting liver damage. Most people recover in 4 to 6 weeks.

Hepatitis B is caused by the hepatitis B virus and is spread primarily by sexual contact and contaminated syringes and transfusion equipment. It can also be spread via saliva and tears. Hepatitis B virus can be present for years or even a lifetime, and it can produce cirrhosis and possibly cancer of the liver. Individuals who harbor the active hepatitis B virus also become carriers. Vaccines produced through recombinant DNA technology (for example, Recombivax HB®) are available to prevent hepatitis B infection.

Hepatitis C, caused by the hepatitis C virus, is clinically similar to hepatitis B. Hepatitis C can cause cirrhosis and possibly liver cancer. In developed nations, donated blood is screened for the presence of hepatitis B and C.

Hepatitis D is caused by the hepatitis D virus. It is transmitted like hepatitis B and in fact a person must have been co-infected with hepatitis B before contracting hepatitis D. Hepatitis D results in severe

Mouth (p. 928)

1. The mouth is formed by the cheeks, hard and soft palates, lips, and tongue.
2. The vestibule is the space bounded externally by the cheeks and lips and internally by the teeth and gums.
3. The oral cavity proper extends from the vestibule to the fauces.
4. The tongue, together with its associated muscles, forms the floor of the oral cavity. It is composed of skeletal muscle covered with mucous membrane.
5. The upper surface and sides of the tongue are covered with papillae, some of which contain taste buds.
6. The major portion of saliva is secreted by the major salivary glands, which lie outside the mouth and pour their contents into ducts that empty into the oral cavity.
7. There are three pairs of major salivary glands: parotid, submandibular, and sublingual glands.
8. Saliva lubricates food and starts the chemical digestion of carbohydrates.
9. Salivation is controlled by the nervous system.
10. The teeth (dentes) project into the mouth and are adapted for mechanical digestion.
11. A typical tooth consists of three principal regions: crown, root, and neck.
12. Teeth are composed primarily of dentin and are covered by enamel, the hardest substance in the body.
13. There are two dentitions: deciduous and permanent.
14. Through mastication, food is mixed with saliva and shaped into a soft, flexible mass called a bolus.
15. Salivary amylase begins the digestion of starches, and lingual lipase acts on triglycerides.

Pharynx (p. 934)

1. The pharynx is a funnel-shaped tube that extends from the internal nares to the esophagus posteriorly and to the larynx anteriorly.
2. The pharynx has both respiratory and digestive functions.

Esophagus (p. 934)

1. The esophagus is a collapsible, muscular tube that connects the pharynx to the stomach.
2. It contains an upper and a lower esophageal sphincter.

Deglutition (p. 935)

1. Deglutition, or swallowing, moves a bolus from the mouth to the stomach.
2. Swallowing consists of a voluntary stage, a pharyngeal stage (involuntary), and an esophageal stage (involuntary).

Stomach (p. 937)

1. The stomach connects the esophagus to the duodenum.
2. The principal anatomic regions of the stomach are the cardia, fundus, body, and pylorus.
3. Adaptations of the stomach for digestion include rugae; glands that produce mucus, hydrochloric acid, pepsin, gastric lipase, and intrinsic factor; and a three-layered muscularis.
4. Mechanical digestion consists of mixing waves.
5. Chemical digestion consists mostly of the conversion of proteins into peptides by pepsin.

6. The stomach wall is impermeable to most substances.
7. Among the substances the stomach can absorb are water, certain ions, drugs, and alcohol.

Pancreas (p. 942)

1. The pancreas consists of a head, a body, and a tail and is connected to the duodenum via the pancreatic duct and accessory duct.
2. Endocrine pancreatic islets (islets of Langerhans) secrete hormones, and exocrine acini secrete pancreatic juice.
3. Pancreatic juice contains enzymes that digest starch (pancreatic amylase), proteins (trypsin, chymotrypsin, carboxypeptidase, and elastase), triglycerides (pancreatic lipase), and nucleic acids (ribonuclease and deoxyribonuclease).

Liver and Gallbladder (p. 945)

1. The liver has left and right lobes; the right lobe includes a quadrate lobe and a caudate lobe. The gallbladder is a sac located in a depression on the posterior surface of the liver that stores and concentrates bile.
2. The lobes of the liver are made up of lobules that contain hepatocytes (liver cells), sinusoids, stellate reticuloendothelial (Kupffer) cells, and a central vein.
3. Hepatocytes produce bile that is carried by a duct system to the gallbladder for concentration and temporary storage.
4. Bile's contribution to digestion is the emulsification of dietary lipids.
5. The liver also functions in carbohydrate, lipid, and protein metabolism; processing of drugs and hormones; excretion of bilirubin; synthesis of bile salts; storage of vitamins and minerals; phagocytosis; and activation of vitamin D.

Small Intestine (p. 949)

1. The small intestine extends from the pyloric sphincter to the ileocecal sphincter.
2. It is divided into duodenum, jejunum, and ileum.
3. Its glands secrete fluid and mucus, and the circular folds, villi, and microvilli of its wall provide a large surface area for digestion and absorption.
4. Brush-border enzymes digest α-dextrins, maltose, sucrose, lactose, peptides, and nucleotides at the surface of mucosal epithelial cells.
5. Pancreatic and intestinal brush-border enzymes break down starches into maltose, maltotriose, and α-dextrins (pancreatic amylase), α-dextrins into glucose (α-dextrinase), maltose to glucose (maltase), sucrose to glucose and fructose (sucrase), lactose to glucose and galactose (lactase), and proteins into peptides (trypsin, chymotrypsin, and elastase). Also, enzymes break off amino acids at the carboxyl ends of peptides (carboxypeptidases) and break off amino acids at the amino ends of peptides (aminopeptidases). Finally, enzymes split dipeptides into amino acids (dipeptidases), triglycerides to fatty acids and monoglycerides (lipases), and nucleotides to pentoses and nitrogenous bases (nucleosidases and phosphatases).
6. Mechanical digestion in the small intestine involves segmentation and migrating motility complexes.
7. Absorption occurs via diffusion, facilitated diffusion, osmosis, and active transport; most absorption occurs in the small intestine.
8. Monosaccharides, amino acids, and short-chain fatty acids pass into the blood capillaries.

9. Long-chain fatty acids and monoglycerides are absorbed from micelles, resynthesized to triglycerides, and formed into chylomicrons.
10. Chylomicrons move into lymph in the lacteal of a villus.
11. The small intestine also absorbs electrolytes, vitamins, and water.

Large Intestine (p. 959)

1. The large intestine extends from the ileocecal sphincter to the anus.
2. Its regions include the cecum, colon, rectum, and anal canal.
3. The mucosa contains many goblet cells, and the muscularis consists of teniae coli and haustra.
4. Mechanical movements of the large intestine include haustral churning, peristalsis, and mass peristalsis.
5. The last stages of chemical digestion occur in the large intestine through bacterial action. Substances are further broken down, and some vitamins are synthesized.
6. The large intestine absorbs water, ions, and vitamins.
7. Feces consist of water, inorganic salts, epithelial cells, bacteria, and undigested foods.
8. The elimination of feces from the rectum is called defecation.
9. Defecation is a reflex action aided by voluntary contractions of the diaphragm and abdominal muscles and relaxation of the external anal sphincter.

Phases of Digestion (p. 965)

1. Digestive activities occur in three overlapping phases: cephalic phase, gastric phase, and intestinal phase.

2. During the cephalic phase of digestion, salivary glands secrete saliva and gastric glands secrete gastric juice in order to prepare the mouth and stomach for food that is about to be eaten.
3. The presence of food in the stomach causes the gastric phase of digestion, which promotes gastric juice secretion and gastric motility.
4. During the intestinal phase of digestion, food is digested in the small intestine. In addition, gastric motility and gastric secretion decrease in order to slow the exit of chyme from the stomach, which prevents the small intestine from being overloaded with more chyme than it can handle.
5. The activities that occur during the various phases of digestion are coordinated by neural pathways and by hormones. Table 24.8 on page 966 summarizes the major hormones that control digestion.

Development of the Digestive System (p. 967)

1. The endoderm of the primitive gut forms the epithelium and glands of most of the gastrointestinal tract.
2. The mesoderm of the primitive gut forms the smooth muscle and connective tissue of the gastrointestinal tract.

Aging and the Digestive System (p. 967)

1. General changes include decreased secretory mechanisms, decreased motility, and loss of tone.
2. Specific changes may include loss of taste, pyorrhea, hernias, peptic ulcer disease, constipation, hemorrhoids, and diverticular diseases.

SELF-QUIZ QUESTIONS

Fill in the blanks in the following statements.

1. The end products of chemical digestion of carbohydrates are _____, of proteins are _____, of lipids are _____ and _____, and of nucleic acids are _____, _____, and _____.
2. List the mechanisms of absorption of materials in the small intestine: _____, _____, _____, and _____.

Indicate whether the following statements are true or false.

3. The soft palate, uvula, and epiglottis prevent swallowed foods and liquids from entering the respiratory passages.
4. The coordinated contractions and relaxations of the muscularis, which propels materials through the GI tract, is known as peristalsis.

Choose the one best answer to the following questions.

5. Which of the following are mismatched?
   (a) chemical digestion: splitting food molecules into simple substances by hydrolysis with the assistance of digestive enzymes
   (b) motility: mechanical processes that break apart ingested food into small molecules
   (c) ingestion: taking foods and liquids into the mouth
   (d) propulsion: movement of food through GI tract due to smooth muscle contraction
   (e) absorption: passage into blood or lymph of ions, fluids and small molecules via the epithelial lining of the GI tract lumen

6. Which of the following are *true* concerning the peritoneum? (1) The kidneys and pancreas are retroperitoneal. (2) The greater omentum is the largest of the peritoneal folds. (3) The lesser omentum binds the large intestine to the posterior abdominal wall. (4) The falciform ligament attaches the liver to the anterior abdominal wall and diaphragm. (5) The mesentery is associated with the jejunum and ileum.
   (a) 1, 2, 3, and 5   (b) 1, 2, and 5   (c) 2 and 5
   (d) 1, 2, 4, and 5   (e) 3, 4, and 5

7. When a surgeon makes an incision in the small intestine, in what order would the physician encounter these structures? (1) epithelium, (2) submucosa, (3) serosa, (4) muscularis, (5) lamina propria, (6) muscularis mucosae.
   (a) 3, 4, 5, 6, 2, 1   (b) 1, 2, 3, 4, 6, 5   (c) 1, 5, 6, 2, 4, 3
   (d) 5, 1, 2, 6, 4, 3   (e) 3, 4, 2, 6, 5, 1

8. Which of the following are functions of the liver? (1) carbohydrate, lipid, and protein metabolism, (2) nucleic acid metabolism, (3) excretion of bilirubin, (4) synthesis of bile salts, (5) activation of vitamin D.
   (a) 1, 2, and 5   (b) 1, 2, 3, and 4   (c) 1, 3, 4, and 5
   (d) 2, 3, 4, and 5   (e) 1, 2, 4, and 5

? ANSWERS TO FIGURE QUESTIONS

24.1 Digestive enzymes are produced by the salivary glands, tongue, stomach, pancreas, and small intestine.

24.2 The lamina propria has the following functions: (1) It contains blood vessels and lymphatic vessels, which are the routes by which nutrients are absorbed from the GI tract; (2) it supports the mucosal epithelium and binds it to the muscularis mucosae; and (3) it contains mucosa-associated lymphatic tissue (MALT), which helps protect against disease.

24.3 The neurons of the myenteric plexus regulate GI tract motility, and the neurons of the submucosal plexus regulate GI secretion.

24.4 Mesentery binds the small intestine to the posterior abdominal wall.

24.5 The uvula helps prevent foods and liquids from entering the nasal cavity during swallowing.

24.6 Chloride ions in saliva activate salivary amylase.

24.7 The main component of teeth is connective tissue, specifically dentin.

24.8 The first, second, and third molars do not replace any deciduous teeth.

24.9 The esophageal mucosa and submucosa contain mucus-secreting glands.

24.10 Both. Initiation of swallowing is voluntary and the action is carried out by skeletal muscles. Completion of swallowing—moving a bolus along the esophagus and into the stomach—is involuntary and involves peristalsis by smooth muscle.

24.11 After a large meal, the rugae stretch and disappear as the stomach fills.

24.12 Parietal cells secrete HCl, which is a component of gastric juice. HCl kills microbes in food, denatures proteins, and converts pepsinogen into pepsin.

24.13 Hydrogen ions secreted into gastric juice are derived from carbonic acid (H_2CO_3).

24.14 The pancreatic duct contains pancreatic juice (fluid and digestive enzymes); the common bile duct contains bile; the hepatopancreatic ampulla contains pancreatic juice and bile.

24.15 The phagocytic cell in the liver is the stellate reticuloendothelial (Kupffer) cell.

24.16 While a meal is being absorbed, nutrients, O_2, and certain toxic substances are removed by hepatocytes from blood flowing through liver sinusoids.

24.17 The ileum is the longest part of the small intestine.

24.18 Nutrients being absorbed in the small intestine enter the blood via capillaries or the lymph via lacteals.

24.19 The fluid secreted by duodenal (Brunner's) glands—alkaline mucus—neutralizes gastric acid and protects the mucosal lining of the duodenum.

24.20 Because monoglycerides are hydrophobic (nonpolar) molecules, they can dissolve in and diffuse through the lipid bilayer of the plasma membrane.

24.21 The stomach and pancreas are the two digestive system organs that secrete the largest volumes of fluid.

24.22 The ascending and descending portions of the colon are retroperitoneal.

24.23 Goblet cells in the large intestine secrete mucus to lubricate colonic contents.

24.24 The pH of gastric juice rises due to the buffering action of some amino acids in food proteins.

METABOLISM AND NUTRITION | 25

METABOLISM, NUTRITION AND HOMEOSTASIS Metabolic reactions contribute to homeostasis by harvesting chemical energy from consumed nutrients to contribute to the body's growth, repair, and normal functioning. •

Plants use the green pigment chlorophyll to trap energy in sunlight. We don't have a similarly functioning pigment in our skin, so the food we eat is our only source of energy for running, walking, and even breathing. Many molecules needed to maintain cells and tissues can be made from simpler precursors by the body's metabolic reactions; others—the essential amino acids, essential fatty acids, vitamins, and minerals—must be obtained from the food we eat. As you learned in Chapter 24, carbohydrates, lipids, and proteins in food are digested by enzymes and absorbed in the gastrointestinal tract. The products of digestion that reach body cells are monosaccharides, fatty acids, glycerol, monoglycerides, and amino acids. Some minerals and many vitamins are part of enzyme systems that catalyze the breakdown and synthesis of carbohydrates, lipids, and proteins. Food molecules absorbed by the gastrointestinal (GI) tract have three main fates:

1. Most food molecules are used to *supply energy* for sustaining life processes, such as active transport, DNA replication, protein synthesis, muscle contraction, maintenance of body temperature, and mitosis.

2. Some food molecules *serve as building blocks* for the synthesis of more complex structural or functional molecules, such as muscle proteins, hormones, and enzymes.

3. Other food molecules are *stored for future use*. For example, glycogen is stored in liver cells, and triglycerides are stored in adipose cells.

In this chapter we discuss how metabolic reactions harvest the chemical energy stored in foods, how each group of food molecules contributes to the body's growth, repair, and energy needs, and how heat and energy balance is maintained in the body. Finally, we explore some aspects of nutrition to discover why you should opt for fish instead of a burger the next time you eat out.

METABOLIC REACTIONS

◉ OBJECTIVES

- Define metabolism.
- Explain the role of ATP in anabolism and catabolism.

Metabolism (me-TAB-ō-lizm; *metabol-* = change) refers to all of the chemical reactions that occur in the body. There are two types of metabolism: catabolism and anabolism. Those chemical reactions that break down complex organic molecules into simpler ones are collectively known as **catabolism** (ka-TAB-ō-lizm; *cata-* = downward). Overall, catabolic (decomposition) reactions are *exergonic;* they produce more energy than they consume, releasing the chemical energy stored in organic molecules. Important sets of catabolic reactions occur in glycolysis, the Krebs cycle, and the electron transport chain, each of which will be discussed later in the chapter.

Chemical reactions that combine simple molecules and monomers to form the body's complex structural and functional components are collectively known as **anabolism** (a-NAB-ō-lizm; *ana-* = upward). Examples of anabolic reactions are the formation of peptide bonds between amino acids during protein synthesis, the building of fatty acids into phospholipids that form the plasma membrane bilayer, and the linkage of glucose monomers to form glycogen. Anabolic reactions are *endergonic;* they consume more energy than they produce.

Metabolism is an energy-balancing act between catabolic (decomposition) reactions, and anabolic (synthesis) reactions. The molecule that participates most often in energy exchanges in living cells is **ATP (adenosine triphosphate),** which couples energy-releasing catabolic reactions to energy-requiring anabolic reactions.

The metabolic reactions that occur depend on which enzymes are active in a particular cell at a particular time, or even in a particular part of the cell. Catabolic reactions can be occurring in the mitochondria of a cell at the same time as anabolic reactions are taking place in the endoplasmic reticulum.

A molecule synthesized in an anabolic reaction has a limited lifetime. With few exceptions, it will eventually be broken down and its component atoms recycled into other molecules or excreted from the body. Recycling of biological molecules occurs continuously in living tissues, more rapidly in some than in others. Individual cells may be refurbished molecule by molecule, or a whole tissue may be rebuilt cell by cell.

Coupling of Catabolism and Anabolism by ATP

The chemical reactions of living systems depend on the efficient transfer of manageable amounts of energy from one molecule to another. The molecule that most often performs this task is ATP, the "energy currency" of a living cell. Like money, it is readily available to "buy" cellular activities; it is spent and earned over and over. A typical cell has about a billion molecules of ATP, each of which typically lasts for less than a minute before being used. Thus, ATP is not a long-term-storage form of currency,

like gold in a vault, but rather convenient cash for moment-to-moment transactions.

Recall from Chapter 2 that a molecule of ATP consists of an adenine molecule, a ribose molecule, and three phosphate groups bonded to one another (see Figure 2.25 on page 56). Figure 25.1 shows how ATP links anabolic and catabolic reactions. When the terminal phosphate group is split off ATP, adenosine diphosphate (ADP) and a phosphate group (symbolized as Ⓟ) are formed. Some of the energy released is used to drive anabolic reactions such as the formation of glycogen from glucose. In addition, energy from complex molecules is used in catabolic reactions to combine ADP and a phosphate group to resynthesize ATP:

$$ADP + Ⓟ + energy \longrightarrow ATP$$

About 40% of the energy released in catabolism is used for cellular functions; the rest is converted to heat, some of which helps maintain normal body temperature. Excess heat is lost to the environment. Compared with machines, which typically convert only 10–20% of energy into work, the 40% efficiency of the body's metabolism is impressive. Still, the body has a continuous need to take in and process external sources of energy so that cells can synthesize enough ATP to sustain life.

◉ CHECKPOINT

1. What is metabolism? Distinguish between anabolism and catabolism, and give examples of each.
2. How does ATP link anabolism and catabolism?

Figure 25.1 Role of ATP in linking anabolic and catabolic reactions. When complex molecules and polymers are split apart (catabolism, at left), some of the energy is transferred to form ATP and the rest is given off as heat. When simple molecules and monomers are combined to form complex molecules (anabolism, at right), ATP provides the energy for synthesis, and again some energy is given off as heat.

🔑 The coupling of energy-releasing and energy-requiring reactions is achieved through ATP.

❓ In a pancreatic cell that produces digestive enzymes, does anabolism or catabolism predominate?

ENERGY TRANSFER

- Describe oxidation–reduction reactions.
- Explain the role of ATP in metabolism.

Various catabolic reactions transfer energy into the "high-energy" phosphate bonds of ATP. Although the amount of energy in these bonds is not exceptionally large, it can be released quickly and easily. Before discussing metabolic pathways, it is important to understand how this transfer of energy occurs. Two important aspects of energy transfer are oxidation–reduction reactions and mechanisms of ATP generation.

Oxidation–Reduction Reactions

Oxidation is the *removal of electrons* from an atom or molecule; the result is a *decrease* in the potential energy of the atom or molecule. Because most biological oxidation reactions involve the loss of hydrogen atoms, they are called *dehydrogenation reactions.* An example of an oxidation reaction is the conversion of lactic acid into pyruvic acid:

$$
\underset{\text{Lactic acid}}{
\begin{array}{c}
COOH \\
| \\
H-C-OH \\
| \\
CH_3
\end{array}}
\xrightarrow[\text{Remove 2H (H}^+ + \text{H}^-)]{\text{Oxidation}}
\underset{\text{Pyruvic acid}}{
\begin{array}{c}
COOH \\
| \\
C=O \\
| \\
CH_3
\end{array}}
$$

In the preceding reaction, 2H ($H^+ + H^-$) means that two neutral hydrogen atoms (2H) are removed as one hydrogen ion (H^+) plus one hydride ion (H^-).

Reduction is the opposite of oxidation; it is the *addition of electrons* to a molecule. Reduction results in an *increase* in the potential energy of the molecule. An example of a reduction reaction is the conversion of pyruvic acid into lactic acid:

$$
\underset{\text{Pyruvic acid}}{
\begin{array}{c}
COOH \\
| \\
C=O \\
| \\
CH_3
\end{array}}
\xrightarrow[\text{Add 2H (H}^+ + \text{H}^-)]{\text{Reduction}}
\underset{\text{Lactic acid}}{
\begin{array}{c}
COOH \\
| \\
H-C-OH \\
| \\
CH_3
\end{array}}
$$

When a substance is oxidized, the liberated hydrogen atoms do not remain free in the cell but are transferred immediately by coenzymes to another compound. Two coenzymes are commonly used by animal cells to carry hydrogen atoms: **nicotinamide adenine dinucleotide (NAD),** a derivative of the B vitamin niacin, and **flavin adenine dinucleotide (FAD),** a derivative of vitamin B_2 (riboflavin). The oxidation and reduction states of NAD^+ and FAD can be represented as follows:

$$
\underset{\text{Oxidized}}{NAD^+}
\underset{-\text{ 2H (H}^+ + \text{H}^-)}{\overset{+\text{ 2H (H}^+ + \text{H}^-)}{\rightleftharpoons}}
\underset{\text{Reduced}}{NADH + H^+}
$$

$$
\underset{\text{Oxidized}}{FAD}
\underset{-\text{ 2H (H}^+ + \text{H}^-)}{\overset{+\text{ 2H (H}^+ + \text{H}^-)}{\rightleftharpoons}}
\underset{\text{Reduced}}{FADH_2}
$$

When NAD^+ is reduced to $NADH + H^+$, the NAD^+ gains a hydride ion (H^-), neutralizing its charge, and the H^+ is released into the surrounding solution. When NADH is oxidized to NAD^+, the loss of the hydride ion results in one less hydrogen atom and an additional positive charge. FAD is reduced to $FADH_2$ when it gains a hydrogen ion and a hydride ion, and $FADH_2$ is oxidized to FAD when it loses the same two ions.

Oxidation and reduction reactions are always coupled; each time one substance is oxidized, another is simultaneously reduced. Such paired reactions are called **oxidation–reduction** or **redox reactions.** For example, when lactic acid is *oxidized* to form pyruvic acid, the two hydrogen atoms removed in the reaction are used to *reduce* NAD^+. This coupled redox reaction may be written as follows:

$$
\begin{array}{lcr}
\underset{\text{Reduced}}{\text{Lactic acid}} & & \underset{\text{Oxidized}}{NAD^+} \\
& \diagdown\diagup & \\
& \diagup\diagdown & \\
\underset{\text{Oxidized}}{\text{Pyruvic acid}} & & \underset{\text{Reduced}}{NADH + H^+}
\end{array}
$$

An important point to remember about oxidation–reduction reactions is that oxidation is usually an exergonic (energy-releasing) reaction. Cells use multistep biochemical reactions to release energy from energy-rich, highly reduced compounds (with many hydrogen atoms) to lower-energy, highly oxidized compounds (with many oxygen atoms or multiple bonds). For example, when a cell oxidizes a molecule of glucose ($C_6H_{12}O_6$), the energy in the glucose molecule is removed in a stepwise manner. Ultimately, some of the energy is captured by transferring it to ATP, which then serves as an energy source for energy-requiring reactions within the cell. Compounds with many hydrogen atoms such as glucose contain more chemical potential energy than oxidized compounds. For this reason, glucose is a valuable nutrient.

Mechanisms of ATP Generation

Some of the energy released during oxidation reactions is captured within a cell when ATP is formed. Briefly, a phosphate group (Ⓟ) is added to ADP, with an input of energy, to form ATP. The two high-energy phosphate bonds that can be used to transfer energy are indicated by "squiggles" ($\sim$):

$$
\underset{\text{ADP}}{\text{Adenosine} - Ⓟ \sim Ⓟ} + Ⓟ + \text{energy} \longrightarrow
$$

$$
\underset{\text{ATP}}{\text{Adenosine} - Ⓟ \sim Ⓟ \sim Ⓟ}
$$

The high-energy phosphate bond that attaches the third phosphate group contains the energy stored in this reaction. The addition of a phosphate group to a molecule, called **phosphorylation** (fos'-for-i-LĀ-shun), increases its potential energy. Organisms use three mechanisms of phosphorylation to generate ATP:

1. Substrate-level phosphorylation generates ATP by transferring a high-energy phosphate group from an intermediate phosphorylated metabolic compound—a substrate—directly to ADP. In human cells, this process occurs in the cytosol.

The Krebs Cycle

Once the pyruvic acid has undergone decarboxylation and the remaining acetyl group has attached to CoA, the resulting compound (acetyl CoA) is ready to enter the Krebs cycle (Figure 25.6). The **Krebs cycle**—named for the biochemist Hans Krebs, who described these reactions in the 1930s—is also known as the **citric acid cycle,** for the first molecule formed when an acetyl group joins the cycle. The reactions occur in the matrix of mitochondria and consist of a series of oxidation–reduction reactions and decarboxylation reactions that release CO_2. In the Krebs cycle, the oxidation–reduction reactions transfer chemical energy, in the form of electrons, to two coenzymes—NAD^+ and FAD. The pyruvic acid derivatives are oxidized, and the coenzymes are reduced. In addition, one step generates ATP. Figure 25.7 shows the reactions of the Krebs cycle in more detail.

The reduced coenzymes ($NADH$ and $FADH_2$) are the most important outcome of the Krebs cycle because they contain the energy originally stored in glucose and then in pyruvic acid. Overall, for every acetyl CoA that enters the Krebs cycle, three $NADH$, three H^+, and one $FADH_2$ are produced by oxidation–reduction reactions, and one molecule of ATP is generated by substrate-level phosphorylation (see Figure 25.6). In the electron transport chain, the three $NADH$ + three H^+ will later yield nine ATP molecules, and the $FADH_2$ will later yield two ATP molecules. Thus, each "turn" of the Krebs cycle eventually generates 12 molecules of ATP. Because each glucose molecule provides two acetyl CoA molecules, glucose catabolism via the Krebs cycle and the electron transport chain yields 24 molecules of ATP per glucose molecule.

Liberation of CO_2 occurs as pyruvic acid is converted to acetyl CoA and during the two decarboxylation reactions of the Krebs cycle (see Figure 25.6). Because each molecule of glucose generates two molecules of pyruvic acid, six molecules of CO_2 are liberated from each original glucose molecule catabolized along this pathway. The molecules of CO_2 diffuse out of the mitochondria, through the cytosol and plasma membrane, and then into the blood. Blood transports the CO_2 to the lungs, where it eventually is exhaled.

Figure 25.6 After formation of acetyl coenzyme A, the next stage of cellular respiration is the Krebs cycle.

Reactions of the Krebs cycle occur in the matrix of mitochondria.

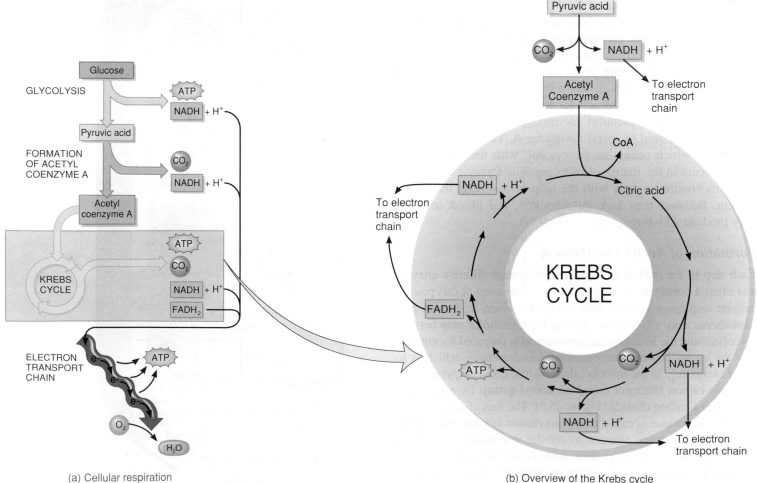

(a) Cellular respiration

(b) Overview of the Krebs cycle

? When in cellular respiration is carbon dioxide given off? What happens to this gas?

Figure 25.7 **The eight reactions of the Krebs cycle.** ❶ *Entry of the acetyl group.* The chemical bond that attaches the acetyl group to coenzyme A (CoA) breaks, and the two-carbon acetyl group attaches to a four-carbon molecule of oxaloacetic acid to form a six-carbon molecule called citric acid. CoA is free to combine with another acetyl group from pyruvic acid and repeat the process. ❷ *Isomerization.* Citric acid undergoes isomerization to isocitric acid, which has the same molecular formula as citrate. Notice, however, that the hydroxyl group (—OH) is attached to a different carbon. ❸ *Oxidative decarboxylation.* Isocitric acid is oxidized and loses a molecule of CO_2, forming alpha-ketoglutaric acid. The H^- from the oxidation is passed on to NAD^+, which is reduced to $NADH + H^+$. ❹ *Oxidative decarboxylation.* Alpha-ketoglutaric acid is oxidized, loses a molecule of CO_2, and picks up CoA to form succinyl CoA. ❺ *Substrate-level phosphorylation.* CoA is displaced by a phosphate group, which is then transferred to guanosine diphosphate (GDP) to form guanosine triphosphate (GTP). GTP can donate a phosphate group to ADP to form ATP. ❻ *Dehydrogenation.* Succinic acid is oxidized to fumaric acid as two of its hydrogen atoms are transferred to the coenzyme flavin adenine dinucleotide (FAD), which is reduced to $FADH_2$. ❼ *Hydration.* Fumaric acid is converted to malic acid by the addition of a molecule of water. ❽ *Dehydrogenation.* In the final step in the cycle, malic acid is oxidized to re-form oxaloacetic acid. Two hydrogen atoms are removed and one is transferred to NAD^+, which is reduced to $NADH + H^+$. The regenerated oxaloacetic acid can combine with another molecule of acetyl CoA, beginning a new cycle.

🔑 The three main results of the Krebs cycle are the production of reduced coenzymes ($NADH + H^+$ and $FADH_2$), which contain stored energy; the generation of GTP, a high-energy compound that is used to produce ATP; and the formation of CO_2, which is transported to the lungs and exhaled.

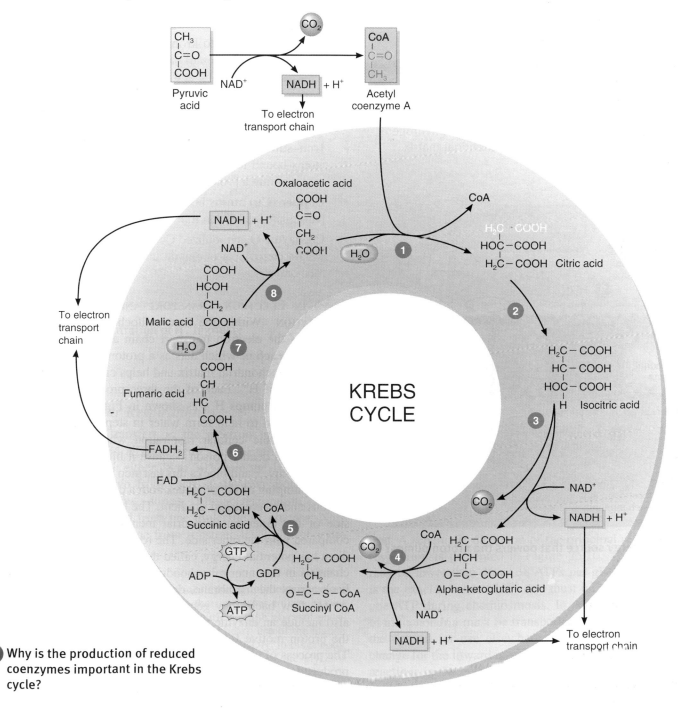

❓ Why is the production of reduced coenzymes important in the Krebs cycle?

Triglycerides stored in adipose tissue constitute 98% of all body energy reserves. They are stored more readily than glycogen, in part because triglycerides are hydrophobic and do not exert osmotic pressure on cell membranes. Adipose tissue also insulates and protects various parts of the body. Adipocytes in the subcutaneous layer contain about 50% of the stored triglycerides. Other adipose tissues account for the other half: about 12% around the kidneys, 10–15% in the omenta, 15% in genital areas, 5–8% between muscles, and 5% behind the eyes, in the sulci of the heart, and attached to the outside of the large intestine. Triglycerides in adipose tissue are continually broken down and resynthesized. Thus, the triglycerides stored in adipose tissue today are not the same molecules that were present last month because they are continually released from storage, transported in the blood, and redeposited in other adipose tissue cells.

Lipid Catabolism: Lipolysis

In order for muscle, liver, and adipose tissue to oxidize the fatty acids derived from triglycerides to produce ATP, the triglycerides must first be split into glycerol and fatty acids, a process called **lipolysis** (li-POL-i-sis). Lipolysis is catalyzed by enzymes called *lipases.* Epinephrine and norepinephrine enhance triglyceride breakdown into fatty acids and glycerol. These hormones

are released when sympathetic tone increases, as occurs, for example, during exercise. Other lipolytic hormones include cortisol, thyroid hormones, and insulinlike growth factors. By contrast, insulin inhibits lipolysis.

The glycerol and fatty acids that result from lipolysis are catabolized via different pathways (Figure 25.14). Glycerol is converted by many cells of the body to glyceraldehyde 3-phosphate, one of the compounds also formed during the catabolism of glucose. If ATP supply in a cell is high, glyceraldehyde 3-phosphate is converted into glucose, an example of gluconeogenesis. If ATP supply in a cell is low, glyceraldehyde 3-phosphate enters the catabolic pathway to pyruvic acid.

Fatty acids are catabolized differently than glycerol and yield more ATP. The first stage in fatty acid catabolism is a series of reactions, collectively called **beta oxidation,** that occurs in the matrix of mitochondria. Enzymes remove two carbon atoms at a time from the long chain of carbon atoms composing a fatty acid and attach the resulting two-carbon fragment to coenzyme A, forming acetyl CoA. Then, acetyl CoA enters the Krebs cycle (Figure 25.14). A 16-carbon fatty acid such as palmitic acid can yield as many as 129 ATPs upon its complete oxidation via beta oxidation, the Krebs cycle, and the electron transport chain.

As part of normal fatty acid catabolism, hepatocytes can take two acetyl CoA molecules at a time and condense them to form **acetoacetic acid.** This reaction liberates the bulky CoA portion, which cannot diffuse out of cells. Some acetoacetic acid is con-

Figure 25.14 Pathways of lipid metabolism. Glycerol may be converted to glyceraldehyde 3-phosphate, which can then be converted to glucose or enter the Krebs cycle for oxidation. Fatty acids undergo beta oxidation and enter the Krebs cycle via acetyl coenzyme A. The synthesis of lipids from glucose or amino acids is called lipogenesis.

Glycerol and fatty acids are catabolized in separate pathways.

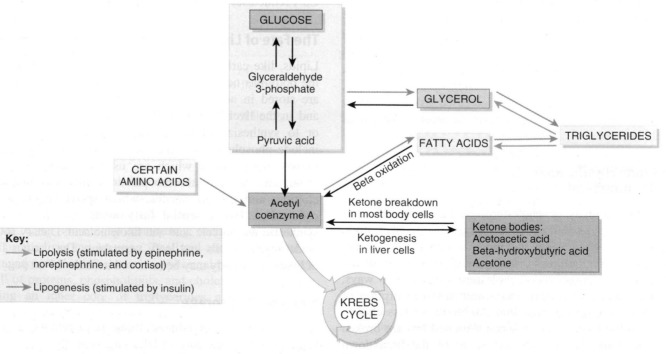

? **What types of cells can carry out lipogenesis, beta oxidation, and lipolysis? What type of cell can carry out ketogenesis?**

verted into **beta-hydroxybutyric acid** and **acetone.** The formation of these three substances, collectively known as **ketone bodies,** is called **ketogenesis** (Figure 25.14). Because ketone bodies freely diffuse through plasma membranes, they leave hepatocytes and enter the bloodstream.

Other cells take up acetoacetic acid and attach its four carbons to two coenzyme A molecules to form two acetyl CoA molecules, which can then enter the Krebs cycle for oxidation. Heart muscle and the cortex (outer part) of the kidneys use acetoacetic acid in preference to glucose for generating ATP. Hepatocytes, which make acetoacetic acid, cannot use it for ATP production because they lack the enzyme that transfers acetoacetic acid back to coenzyme A.

Lipid Anabolism: Lipogenesis

Liver cells and adipose cells can synthesize lipids from glucose or amino acids through **lipogenesis** (Figure 25.14), which is stimulated by insulin. Lipogenesis occurs when individuals consume more calories than are needed to satisfy their ATP needs. Excess dietary carbohydrates, proteins, and fats all have the same fate—they are converted into triglycerides. Certain amino acids can undergo the following reactions: amino acids → acetyl CoA → fatty acids → triglycerides. The use of glucose to form lipids takes place via two pathways: (1) glucose → glyceraldehyde 3-phosphate → glycerol and (2) glucose → glyceraldehyde 3-phosphate → acetyl CoA → fatty acids. The resulting glycerol and fatty acids can undergo anabolic reactions to become stored triglycerides, or they can go through a series of anabolic reactions to produce other lipids such as lipoproteins, phospholipids, and cholesterol.

● CLINICAL CONNECTION | Ketosis

The level of ketone bodies in the blood normally is very low because other tissues use them for ATP production as fast as they are generated from the breakdown of fatty acids in the liver. During periods of excessive beta oxidation, however, the production of ketone bodies exceeds their uptake and use by body cells. This might occur after a meal rich in triglycerides, or during fasting or starvation, because few carbohydrates are available for catabolism. Excessive beta oxidation may also occur in poorly controlled or untreated diabetes mellitus for two reasons: (1) Because adequate glucose cannot get into cells, triglycerides are used for ATP production, and (2) because insulin normally inhibits lipolysis, a lack of insulin accelerates the pace of lipolysis. When the concentration of ketone bodies in the blood rises above normal—a condition called ketosis—the ketone bodies, most of which are acids, must be buffered. If too many accumulate, they decrease the concentration of buffers, such as bicarbonate ions, and blood pH falls. Extreme or prolonged ketosis can lead to **acidosis (ketoacidosis),** an abnormally low blood pH. The decreased blood pH in turn causes depression of the central nervous system, which can result in disorientation, coma, and even death if the condition is not treated. When a diabetic becomes seriously insulin-deficient, one of the telltale signs is the sweet smell on the breath from the ketone body acetone. ●

● CHECKPOINT

13. What are the functions of the apoproteins in lipoproteins?
14. Which lipoprotein particles contain "good" and "bad" cholesterol, and why are these terms used?
15. Where are triglycerides stored in the body?
16. Explain the principal events of the catabolism of glycerol and fatty acids.
17. What are ketone bodies? What is ketosis?
18. Define lipogenesis and explain its importance.

PROTEIN METABOLISM

● OBJECTIVE

• Describe the fate, metabolism, and functions of proteins.

During digestion, proteins are broken down into amino acids. Unlike carbohydrates and triglycerides, which are stored, proteins are not warehoused for future use. Instead, amino acids are either oxidized to produce ATP or used to synthesize new proteins for body growth and repair. Excess dietary amino acids are not excreted in the urine or feces but instead are converted into glucose (gluconeogenesis) or triglycerides (lipogenesis).

The Fate of Proteins

The active transport of amino acids into body cells is stimulated by insulinlike growth factors (IGFs) and insulin. Almost immediately after digestion, amino acids are reassembled into proteins. Many proteins function as enzymes; others are involved in transportation (hemoglobin) or serve as antibodies, clotting chemicals (fibrinogen), hormones (insulin), or contractile elements in muscle fibers (actin and myosin). Several proteins serve as structural components of the body (collagen, elastin, and keratin). The various functions of proteins in the body may be reviewed in Table 2.8 on page 50.

Protein Catabolism

A certain amount of protein catabolism occurs in the body each day, stimulated mainly by cortisol from the adrenal cortex. Proteins from worn-out cells (such as red blood cells) are broken down into amino acids. Some amino acids are converted into other amino acids, peptide bonds are re-formed, and new proteins are synthesized as part of the recycling process. Hepatocytes convert some amino acids to fatty acids, ketone bodies, or glucose. Cells throughout the body oxidize a small amount of amino acids to generate ATP via the Krebs cycle and the electron transport chain. However, before amino acids can be oxidized, they must first be converted to molecules that are part of the Krebs cycle or can enter the Krebs cycle, such as acetyl CoA (Figure 25.15). Before amino acids can enter the Krebs cycle, their amino group (NH_2) must first be removed—a process called **deamination** (dē-am′-i-NĀ-shun). Deamination occurs

Figure 25.18 Principal metabolic pathways during the postabsorptive state.

The principal function of postabsorptive state reactions is to maintain a normal blood glucose level.

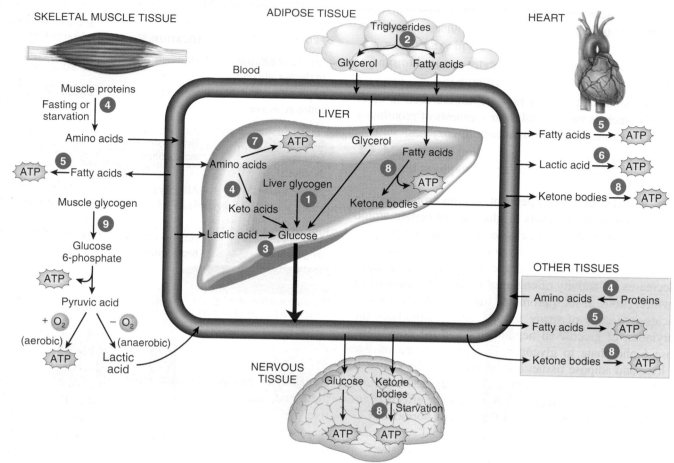

? What processes directly elevate blood glucose during the postabsorptive state, and where does each occur?

glucose. The following reactions produce ATP without using glucose:

5 *Oxidation of fatty acids.* The fatty acids released by lipolysis of triglycerides cannot be used for glucose production because acetyl CoA cannot be readily converted to pyruvic acid. But most cells can oxidize the fatty acids directly, feed them into the Krebs cycle as acetyl CoA, and produce ATP through the electron transport chain.

6 *Oxidation of lactic acid.* Cardiac muscle can produce ATP aerobically from lactic acid.

7 *Oxidation of amino acids.* In hepatocytes, amino acids may be oxidized directly to produce ATP.

8 *Oxidation of ketone bodies.* Hepatocytes also convert fatty acids to ketone bodies, which can be used by the heart, kidneys, and other tissues for ATP production.

9 *Breakdown of muscle glycogen.* Skeletal muscle cells break down glycogen to glucose 6-phosphate, which undergoes glycolysis and provides ATP for muscle contraction.

Regulation of Metabolism During the Postabsorptive State

Both hormones and the sympathetic division of the autonomic nervous system (ANS) regulate metabolism during the postabsorptive state. The hormones that regulate postabsorptive state metabolism sometimes are called anti-insulin hormones because they counter the effects of insulin during the absorptive state. As blood glucose level declines, the secretion of insulin falls and the release of anti-insulin hormones rises.

When blood glucose concentration starts to drop, the pancreatic alpha cells release glucagon at a faster rate, and the beta cells secrete insulin more slowly. The primary target tissue of glucagon is the liver; the major effect is increased release of glucose into the bloodstream due to gluconeogenesis and glycogenolysis.

Low blood glucose also activates the sympathetic branch of the ANS. Glucose-sensitive neurons in the hypothalamus detect low blood glucose and increase sympathetic output. As a result, sympathetic nerve endings release the neurotransmitter norepinephrine, and the adrenal medulla releases two catecholamine hormones—epinephrine and norepinephrine—into the bloodstream. Like glucagon, epinephrine stimulates glycogen breakdown. Epinephrine and norepinephrine are both potent stimulators of lipolysis. These actions of the catecholamines help to increase glucose and free fatty acid levels in the blood. As a result, muscle uses more fatty acids for ATP production, and more glucose is available to the nervous system. Table 25.4 summarizes the hormonal regulation of metabolism in the postabsorptive state.

Metabolism During Fasting and Starvation

The term **fasting** means going without food for many hours or a few days; **starvation** implies weeks or months of food deprivation or inadequate food intake. People can survive without food for two months or more if they drink enough water to prevent dehydration. Although glycogen stores are depleted within a few hours of beginning a fast, catabolism of stored triglycerides and structural proteins can provide energy for several weeks. The amount of adipose tissue the body contains determines the lifespan possible without food.

During fasting and starvation, nervous tissue and RBCs continue to use glucose for ATP production. There is a ready supply of amino acids for gluconeogenesis because lowered insulin and increased cortisol levels slow the pace of protein synthesis and promote protein catabolism. Most cells in the body, especially skeletal muscle cells because of their high protein content, can spare a fair amount of protein before their performance is adversely affected. During the first few days of fasting, protein catabolism outpaces protein synthesis by about 75 grams daily as some of the "old" amino acids are being deaminated and used

for gluconeogenesis and "new" (that is, dietary) amino acids are lacking.

By the second day of a fast, blood glucose level has stabilized at about 65 mg/100 mL (3.6 mmol/liter); at the same time the level of fatty acids in plasma has risen fourfold. Lipolysis of triglycerides in adipose tissue releases glycerol, which is used for gluconeogenesis, and fatty acids. The fatty acids diffuse into muscle fibers and other body cells, where they are used to produce acetyl-CoA, which enters the Krebs cycle. ATP then is synthesized as oxidation proceeds via the Krebs cycle and the electron transport chain.

The most dramatic metabolic change that occurs with fasting and starvation is the increase in the formation of ketone bodies by hepatocytes. During fasting, only small amounts of glucose undergo glycolysis to pyruvic acid, which in turn can be converted to oxaloacetic acid. Acetyl-CoA enters the Krebs cycle by combining with oxaloacetic acid (see Figure 25.16); when oxaloacetic acid is scarce due to fasting, only some of the available acetyl-CoA can enter the Krebs cycle. Surplus acetyl-CoA is used for ketogenesis, mainly in hepatocytes. Ketone body production thus increases as catabolism of fatty acids rises. Lipid-soluble ketone bodies can diffuse through plasma membranes and across the blood–brain barrier and be used as an alternative fuel for ATP production, especially by cardiac and skeletal muscle fibers and neurons. Normally, only a trace of ketone bodies (0.01 mmol/liter) are present in the blood, so they are a negligible fuel source. After two days of fasting, however, the level of ketones is 100–300 times higher and supplies roughly a third of the brain's fuel for ATP production. By 40 days of starvation, ketones provide up to two-thirds of the brain's energy needs. The presence of ketones actually reduces the use of glucose for ATP production, which in turn decreases the demand for gluconeogenesis and slows the catabolism of muscle proteins later in starvation to about 20 grams daily.

◉ CHECKPOINT

23. What are the roles of insulin, glucagon, epinephrine, insulinlike growth factors, thyroxine, cortisol, estrogen, and testosterone in regulation of metabolism?
24. Why is ketogenesis more significant during fasting or starvation than during normal absorptive and postabsorptive states?

HEAT AND ENERGY BALANCE

◉ OBJECTIVES

- Define basal metabolic rate (BMR), and explain several factors that affect it.
- Describe the factors that influence body heat production.
- Explain how normal body temperature is maintained by negative feedback loops involving the hypothalamic thermostat.

Your body produces more or less heat depending on the rates of its metabolic reactions. Because homeostasis of body temperature can be maintained only if the rate of heat loss from the body

TABLE 25.4

Hormonal Regulation of Metabolism in the Postabsorptive State

PROCESS	LOCATION(S)	MAIN STIMULATING HORMONE(S)
Glycogenolysis (glycogen breakdown)	Hepatocytes and skeletal muscle fibers.	Glucagon and epinephrine.
Lipolysis (triglyceride breakdown)	Adipocytes.	Epinephrine, norepinephrine, cortisol, insulinlike growth factors, thyroid hormones, and others.
Protein breakdown	Most body cells, but especially skeletal muscle fibers.	Cortisol.
Gluconeogenesis (synthesis of glucose from noncarbohydrates)	Hepatocytes and kidney cortex cells.	Glucagon and cortisol.

Figure 25.19 Negative feedback mechanisms that conserve heat and increase heat production.

Core temperature is the temperature in body structures deep to the skin and subcutaneous layer; shell temperature is the temperature near the body surface.

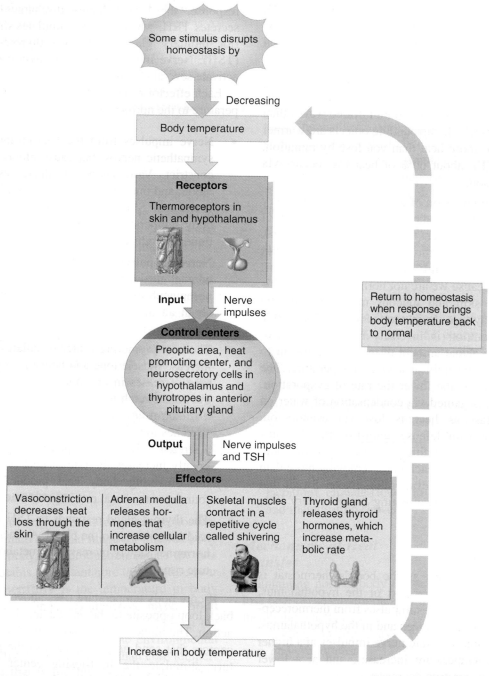

What factors can increase metabolic rate and thus increase the rate of heat production?

Hypothermia is a lowering of core body temperature to 35°C (95°F) or below. Causes of hypothermia include an overwhelming cold stress (immersion in icy water), metabolic diseases (hypoglycemia, adrenal insufficiency, or hypothyroidism), drugs (alcohol, antidepressants, sedatives, or tranquilizers), burns, and malnutrition. Hypothermia is charac-

terized by the following as core body temperature falls: sensation of cold, shivering, confusion, vasoconstriction, muscle rigidity, bradycardia, acidosis, hypoventilation, hypotension, loss of spontaneous movement, coma, and death (usually caused by cardiac arrhythmias). Because the elderly have reduced metabolic protection against a cold environment coupled with a reduced perception of cold, they are at greater risk for developing hypothermia. ●

Energy Homeostasis and Regulation of Food Intake

Most mature animals and many men and women maintain **energy homeostasis,** the precise matching of energy intake (in food) to energy expenditure over time. When the energy content of food balances the energy used by all the cells of the body, body weight remains constant (unless there is a gain or loss of water). In many people, weight stability persists despite large day-to-day variations in activity and food intake. In the more affluent nations, however, a large fraction of the population is overweight. Easy access to tasty, high-calorie foods and a "couch-potato" lifestyle promote weight gain. Being overweight increases the risk of dying from a variety of cardiovascular and metabolic disorders, including hypertension, varicose veins, diabetes mellitus, arthritis, certain cancers, and gallbladder disease.

Energy intake depends only on the amount of food consumed (and absorbed), but three components contribute to total energy expenditure:

1. The basal metabolic rate accounts for about 60% of energy expenditure.

2. Physical activity typically adds 30–35% but can be lower in sedentary people. The energy expenditure is partly from voluntary exercise, such as walking, and partly from **nonexercise activity thermogenesis (NEAT),** the energy costs for maintaining muscle tone, posture while sitting or standing, and involuntary fidgeting movements.

3. **Food-induced thermogenesis,** the heat produced while food is being digested, absorbed, and stored, represents 5–10% of total energy expenditure.

The major site of stored chemical energy in the body is adipose tissue. When energy use exceeds energy input, triglycerides in adipose tissue are catabolized to provide the extra energy, and when energy input exceeds energy expenditure, triglycerides are stored. Over time, the amount of stored triglycerides indicates the excess of energy intake over energy expenditure. Even small differences add up over time. A gain of 20 lb (9 kg) between ages 25 and 55 represents only a tiny imbalance, about 0.3% more energy intake in food than energy expenditure.

Clearly, negative feedback mechanisms are regulating both our energy intake and our energy expenditure. But no sensory receptors exist to monitor our weight or size. How, then, is food intake regulated? The answer to this question is incomplete, but important advances in understanding regulation of food intake have occurred in the past decade. It depends on many factors, including neural and endocrine signals, levels of certain nutrients in the blood, psychological elements such as stress or depression, signals from the GI tract and the special senses, and neural connections between the hypothalamus and other parts of the brain.

Within the hypothalamus are clusters of neurons that play key roles in regulating food intake. **Satiety** (sa-TĪ-i-tē) is a feeling of fullness accompanied by lack of desire to eat. Two hypothalamic areas involved in regulation of food intake are the *arcuate nucleus* and the *paraventricular nucleus* (see Figure 14.10 on page 512). In 1994, the first experiments were reported on a mouse gene, named *obese,* that causes overeating and severe obesity in its mutated form. The product of this gene is the hormone **leptin.** In both mice and humans, leptin helps decrease **adiposity,** total body-fat mass. Leptin is synthesized and secreted by adipocytes in proportion to adiposity; as more triglycerides are stored, more leptin is secreted into the bloodstream. Leptin acts on the hypothalamus to inhibit circuits that stimulate eating while also activating circuits that increase energy expenditure. The hormone insulin has a similar, but smaller, effect. Both leptin and insulin are able to pass through the blood–brain barrier.

When leptin and insulin levels are *low,* neurons that extend from the arcuate nucleus to the paraventricular nucleus release a neurotransmitter called **neuropeptide Y** that stimulates food intake. Other neurons that extend between the arcuate and paraventricular nuclei release a neurotransmitter called **melanocortin,** which is similar to melanocyte-stimulating hormone (MSH). Leptin stimulates release of melanocortin, which acts to inhibit food intake. Although leptin, neuropeptide Y, and melanocortin are key signaling molecules for maintaining energy homeostasis, several other hormones and neurotransmitters also contribute. An understanding of the brain circuits involved is still far from complete. Other areas of the hypothalamus plus nuclei in the brain stem, limbic system, and cerebral cortex take part.

Achieving energy homeostasis requires regulation of energy intake. Most increases and decreases in food intake are due to changes in meal size rather than changes in number of meals. Many experiments have demonstrated the presence of satiety signals, chemical or neural changes that help terminate eating when "fullness" is attained. For example, an increase in blood glucose level, as occurs after a meal, decreases appetite. Several hormones, such as glucagon, cholecystokinin, estrogens, and epinephrine (acting via beta receptors) act to signal satiety and to increase energy expenditure. Distension of the GI tract, particularly the stomach and duodenum, also contributes to termination of food intake. Other hormones increase appetite and decrease energy expenditure. These include growth hormone–releasing hormone (GHRH), androgens, glucocorticoids, epinephrine (acting via alpha receptors), and progesterone.

● CLINICAL **CONNECTION** | Emotional Eating

In addition to keeping us alive, eating serves countless psychological, social, and cultural purposes. We eat to celebrate, punish, comfort, defy, and deny. Eating in response to emotional drives, such as feeling stressed, bored, or tired, is called **emotional eating.** Emotional eating is so common that, within limits, it is considered well within the range of normal behavior. Who hasn't at one time or another headed for the refrigerator after a bad day? Problems arise when emotional eating becomes so excessive that it interferes with health. Physical health problems include obesity and associated disorders such as hypertension and heart disease. Psychological health problems include poor self-esteem, an inability to cope effectively with feelings of stress, and in extreme cases, eating disorders such as anorexia nervosa, bulimia, and obesity.

Eating provides comfort and solace, numbing pain and "feeding the hungry heart." Eating may provide a biochemical "fix" as well. Emotional eaters typically overeat carbohydrate foods (sweets and starches), which may raise brain serotonin levels and lead to feelings of relaxation. Food becomes a way to self-medicate when negative emotions arise. •

◉ CHECKPOINT

25. Define a kilocalorie (kcal). How is the unit used? How does it relate to a calorie?
26. Distinguish between core temperature and shell temperature.
27. In what ways can a person lose heat to or gain heat from the surroundings? How is it possible for a person to lose heat on a sunny beach when the temperature is 40°C (104°F) and the humidity is 85%?
28. What does the term energy homeostasis mean?
29. How is food intake regulated?

NUTRITION

◉ OBJECTIVES

- Describe how to select foods to maintain a healthy diet.
- Compare the sources, functions, and importance of minerals and vitamins in metabolism.

Nutrients are chemical substances in food that body cells use for growth, maintenance, and repair. The six main types of nutri-ents are water, carbohydrates, lipids, proteins, minerals, and vitamins. Water is the nutrient needed in the largest amount—about 2–3 liters per day. As the most abundant compound in the body, water provides the medium in which most metabolic reactions occur, and it also participates in some reactions (for example, hydrolysis reactions). The important roles of water in the body can be reviewed on pages 39–40. Three organic nutrients—carbohydrates, lipids, and proteins—provide the energy needed for metabolic reactions and serve as building blocks to make body structures. Some minerals and many vitamins are components of the enzyme systems that catalyze metabolic reactions. *Essential nutrients* are specific nutrient molecules that the body cannot make in sufficient quantity to meet its needs and thus must be obtained from the diet. Some amino acids, fatty acids, vitamins, and minerals are essential nutrients.

Next, we discuss some guidelines for healthy eating and the roles of minerals and vitamins in metabolism.

Guidelines for Healthy Eating

Each gram of protein or carbohydrate in food provides about 4 Calories; 1 gram of fat (lipids) provides about 9 Calories. We do not know with certainty what levels and types of carbohydrate, fat, and protein are optimal in the diet. Different populations around the world eat redically different diets that are adapted to their particular lifestyles. However, many experts recommend the following distribution of calories: 50–60% from carbohydrates, with less than 15% from simple sugars; less than 30% from fats (triglycerides are the main type of dietary fat), with no more than 10% as saturated fats; and about 12–15% from proteins.

Figure 25.20 MyPyramid.

🔑 ▭ MyPyramid is a new, personalized approach to making healthy food choices and maintaining regular physical activity.

| GRAINS | VEGETABLES | FRUITS | OILS | MILK | MEAT & BEANS |

❓ What does the wider base of each band mean?

The guidelines for healthy eating are to:

- Eat a variety of foods.
- Maintain a healthy weight.
- Choose foods low in fat, saturated fat, and cholesterol.
- Eat plenty of vegetables, fruits, and grain products.
- Use sugars in moderation only.

In 2005, the United States Department of Agriculture (USDA) introduced a new food pyramid called **MyPyramid,** which represents a *personalized* approach to making healthy food choices and maintaining regular physical activity. By consulting a chart, it is possible to determine your calorie level based on your gender, age, and activity level. Once this is determined, you can choose the type and amount of food to be consumed.

If you carefully examine the MyPyramid in Figure 25.20 on page 1006 you will note that the six color bands represent the five basic food groups plus oils. Foods from all bands are needed each day. Also note that the overall size of the bands suggests the proportion of food a person should choose on a daily basis. The wider base of each band represents foods with little or no solid fats or added sugars; these foods should be selected more often. The narrower top of each band represents foods with more added sugars and solid fats, which should be selected less frequently. The person climbing the steps is a reminder of the need for daily physical activity.

As an example of how the MyPyramid works, let's assume based upon consulting a chart that the calorie level of an 18-year-old moderately active female is 2000 Calories and that of an 18-year-old moderately active male is 2800 Calories. Accordingly, it is suggested that the following foods should be chosen in the following amounts:

Calorie Level	2000	2800
Fruits (includes all fresh, frozen, canned, and dried fruits and fruit juices)	2 cups	2.5 cups
Vegetables (includes all fresh, frozen, canned, and dried vegetables and vegetable juices)	2.5 cups	3.5 cups
Grains (includes all foods made from wheat, rice, oats, cornmeal, and barley such as bread, cereals, oatmeal, rice, pasta, crackers, tortillas, and grits)	6 oz	10 oz
Meats and beans (includes lean meat, poultry, fish, eggs, peanut butter, beans, nuts, and seeds)	5.5 oz	7 oz
Milk group (includes milk products and foods made from milk that retain their calcium content such as cheeses and yogurt)	3 cups	3 cups
Oils (Choose mostly fats that contain monounsaturated and polyunsaturated fatty acids such as fish, nuts, seeds, and vegetable oils)	6 tsp	8 tsp

In addition, you should choose and prepare foods with little salt. In fact, sodium intake should be less than 2300 mg per day. If you choose to drink alcohol, it should be consumed in moderation (no more than 1 drink per day for women and 2 drinks per day for men). A drink is defined as 12 oz of regular beer, 5 oz of wine, or 1½ oz of 80 proof distilled spirits.

Minerals

Minerals are inorganic elements that occur naturally in the Earth's crust. In the body they appear in combination with one another, in combination with organic compounds, or as ions in solution. Minerals constitute about 4% of total body mass and are concentrated most heavily in the skeleton. Minerals with known functions in the body include calcium, phosphorus, potassium, sulfur, sodium, chloride, magnesium, iron, iodide, manganese, copper, cobalt, zinc, fluoride, selenium, and chromium. Table 25.5 describes the vital functions of these minerals. Note that the body generally uses the ions of the minerals rather than the non-ionized form. Some minerals, such as chlorine, are toxic or even fatal if ingested in the non-ionized form. Other minerals—aluminum, boron, silicon, and molybdenum—are present but their functions are unclear. Typical diets supply adequate amounts of potassium, sodium, chloride, and magnesium. Some attention must be paid to eating foods that provide enough calcium, phosphorus, iron, and iodide. Excess amounts of most minerals are excreted in the urine and feces.

Calcium and phosphorus form part of the matrix of bone. Because minerals do not form long-chain compounds, they are otherwise poor building materials. A major role of minerals is to help regulate enzymatic reactions. Calcium, iron, magnesium, and manganese are constituents of some coenzymes. Magnesium also serves as a catalyst for the conversion of ADP to ATP. Minerals such as sodium and phosphorus work in buffer systems, which help control the pH of body fluids. Sodium also helps regulate the osmosis of water and, along with other ions, is involved in the generation of nerve impulses.

Vitamins

Organic nutrients required in small amounts to maintain growth and normal metabolism are called **vitamins.** Unlike carbohydrates, lipids, or proteins, vitamins do not provide energy or serve as the body's building materials. Most vitamins with known functions are coenzymes.

Most vitamins cannot be synthesized by the body and must be ingested in food. Other vitamins, such as vitamin K, are produced by bacteria in the GI tract and then absorbed. The body can assemble some vitamins if the raw materials, called **provitamins,** are provided. For example, vitamin A is produced by the body from the provitamin beta-carotene, a chemical present in yellow vegetables such as carrots and in dark green vegetables such as spinach. No single food contains all the required vitamins—one of the best reasons to eat a varied diet.

Vitamins are divided into two main groups: fat-soluble and water-soluble. The **fat-soluble vitamins,** vitamins A, D, E, and K, are absorbed along with other dietary lipids in the small intestine and packaged into chylomicrons. They cannot be absorbed in adequate quantity unless they are ingested with other lipids.

TABLE 25.5

Minerals Vital to the Body

MINERAL	COMMENTS	IMPORTANCE
Calcium	Most abundant mineral in body. Appears in combination with phosphates. About 99% is stored in bone and teeth. Blood Ca^{2+} level is controlled by parathyroid hormone (PTH). Calcitriol promotes absorption of dietary calcium. Excess is excreted in feces and urine. Sources are milk, egg yolk, shellfish, and leafy green vegetables.	Formation of bones and teeth, blood clotting, normal muscle and nerve activity, endocytosis and exocytosis, cellular motility, chromosome movement during cell division, glycogen metabolism, and release of neurotransmitters and hormones.
Phosphorus	About 80% is found in bones and teeth as phosphate salts. Blood phosphate level is controlled by parathyroid hormone (PTH). Excess is excreted in urine; small amount is eliminated in feces. Sources are dairy products, meat, fish, poultry, and nuts.	Formation of bones and teeth. Phosphates ($H_2PO_4^-$, HPO_4^-, and PO_4^{3-}) constitute a major buffer system of blood. Plays important role in muscle contraction and nerve activity. Component of many enzymes. Involved in energy transfer (ATP). Component of DNA and RNA.
Potassium	Major cation (K^+) in intracellular fluid. Excess excreted in urine. Present in most foods (meats, fish, poultry, fruits, and nuts).	Needed for generation and conduction of action potentials in neurons and muscle fibers.
Sulfur	Component of many proteins (such as insulin and chrondroitin sulfate), electron carriers in electron transport chain, and some vitamins (thiamine and biotin). Excreted in urine. Sources include beef, liver, lamb, fish, poultry, eggs, cheese, and beans.	As component of hormones and vitamins, regulates various body activities. Needed for ATP production by electron transport chain.
Sodium	Most abundant cation (Na^+) in extracellular fluids; some found in bones. Excreted in urine and perspiration. Normal intake of NaCl (table salt) supplies more than the required amounts.	Strongly affects distribution of water through osmosis. Part of bicarbonate buffer system. Functions in nerve and muscle action potential conduction.
Chloride	Major anion (Cl^-) in extracellular fluid. Excess excreted in urine. Sources include table salt (NaCl), soy sauce, and processed foods.	Plays role in acid–base balance of blood, water balance, and formation of HCl in stomach.
Magnesium	Important cation (Mg^{2+}) in intracellular fluid. Excreted in urine and feces. Widespread in various foods, such as green leafy vegetables, seafood, and whole-grain cereals.	Required for normal functioning of muscle and nervous tissue. Participates in bone formation. Constituent of many coenzymes.

Fat-soluble vitamins may be stored in cells, particularly hepatocytes. The **water-soluble vitamins,** including several B vitamins and vitamin C, are dissolved in body fluids. Excess quantities of these vitamins are not stored but instead are excreted in the urine.

Besides their other functions, three vitamins—C, E, and beta-carotene (a provitamin)—are termed **antioxidant vitamins** because they inactivate oxygen free radicals. Recall that free radicals are highly reactive ions or molecules that carry an unpaired electron in their outermost electron shell (see Figure 2.3 on page 32). Free radicals damage cell membranes, DNA, and other cellular structures and contribute to the formation of artery-narrowing atherosclerotic plaques. Some free radicals arise naturally in the body, and others come from environmental hazards such as tobacco smoke and radiation. Antioxidant vitamins are thought to play a role in protecting against some kinds of cancer, reducing the buildup of atherosclerotic plaque, delaying

some effects of aging, and decreasing the chance of cataract formation in the lens of the eyes. Table 25.6 on pages 1010–1011 lists the major vitamins, their sources, their functions, and related deficiency disorders.

● CLINICAL **CONNECTION** | **Vitamin and Mineral Supplements**

Most nutritionists recommend eating a balanced diet that includes a variety of foods rather than taking vitamin or mineral supplements, except in special circumstances. Common examples of necessary supplementations include iron for women who have excessive menstrual bleeding; iron and calcium for women who are pregnant or breast-feeding; folic acid (folate) for all women who may become pregnant, to reduce the risk of fetal neural tube defects; calcium for most adults, because they do not receive the recommended amount in their diets; and vitamin B_{12} for strict vegetarians, who eat no meat. Because most North Americans

MINERAL	COMMENTS	IMPORTANCE
Iron	About 66% found in hemoglobin of blood. Normal losses of iron occur by shedding of hair, epithelial cells, and mucosal cells, and in sweat, urine, feces, bile, and blood lost during menstruation. Sources are meat, liver, shellfish, egg yolk, beans, legumes, dried fruits, nuts, and cereals.	As component of hemoglobin, reversibly binds O_2. Component of cytochromes involved in electron transport chain.
Iodide	Essential component of thyroid hormones. Excreted in urine. Sources are seafood, iodized salt, and vegetables grown in iodine-rich soils.	Required by thyroid gland to synthesize thyroid hormones, which regulate metabolic rate.
Manganese	Some stored in liver and spleen. Most excreted in feces.	Activates several enzymes. Needed for hemoglobin synthesis, urea formation, growth, reproduction, lactation, bone formation, and possibly production and release of insulin, and inhibition of cell damage.
Copper	Some stored in liver and spleen. Most excreted in feces. Sources include eggs, whole-wheat flour, beans, beets, liver, fish, spinach, and asparagus.	Required with iron for synthesis of hemoglobin. Component of coenzymes in electron transport chain and enzyme necessary for melanin formation.
Cobalt	Constituent of vitamin B_{12}.	As part of vitamin B_{12}, required for erythropoiesis.
Zinc	Important component of certain enzymes. Widespread in many foods, especially meats.	As a component of carbonic anhydrase, important in carbon dioxide metabolism. Necessary for normal growth and wound healing, normal taste sensations and appetite, and normal sperm counts in males. As a component of peptidases, it is involved in protein digestion.
Fluoride	Components of bones, teeth, other tissues.	Appears to improve tooth structure and inhibit tooth decay.
Selenium	Important component of certain enzymes. Found in seafood, meat, chicken, tomatoes, egg yolk, milk, mushrooms, and garlic, and cereal grains grown in selenium-rich soil.	Needed for synthesis of thyroid hormones, sperm motility, and proper functioning of the immune system. Also functions as an antioxidant. Prevents chromosome breakage and may play a role in preventing certain birth defects, miscarriage, prostate cancer, and coronary artery disease.
Chromium	Found in high concentrations in brewer's yeast. Also found in wine and some brands of beer.	Needed for normal activity of insulin in carbohydrate and lipid metabolism.

do not ingest in their food the high levels of antioxidant vitamins thought to have beneficial effects, some experts recommend supplementing vitamins C and E. More is not always better, however; larger doses of vitamins or minerals can be very harmful.

Hypervitaminosis (hī-per-vī-ta-mi-NŌ-sis; *hyper-* = too much or above) refers to dietary intake of a vitamin that exceeds the ability of the body to utilize, store, or excrete the vitamin. Since water-soluble vitamins are not stored in the body, few of them cause any problems related to hypervitaminosis. However, because lipid-soluble vitamins are stored in the body, excessive consumption may cause problems. For example, excess intake of vitamin A can cause drowsiness, general weakness, irritability, headache, vomiting, dry and peeling skin, partial hair loss, joint pain, liver and spleen enlargement, coma, and even death. Excessive intake of vitamin D may result in loss of appetite, nausea, vomiting, excessive thirst, general weakness, irritability, hypertension, and kidney damage and malfunction. **Hypovitaminosis** (*hypo-* = too little or below), or vitamin deficiency, is discussed in Table 25.6 for the various vitamins. •

◉ CHECKPOINT

30. What is a nutrient?
31. Describe the food guide pyramid and give examples of foods from each food group.
32. What is a mineral? Briefly describe the functions of the following minerals: calcium, phosphorus, potassium, sulfur, sodium, chloride, magnesium, iron, iodine, copper, zinc, fluoride, manganese, cobalt, chromium, and selenium.
33. Define a vitamin. Explain how we obtain vitamins. Distinguish between a fat-soluble vitamin and a water-soluble vitamin.
34. For each of the following vitamins, indicate its principal function and the effect(s) of deficiency: A, D, E, K, B_1, B_2, niacin, B_6, B_{12}, pantothenic acid, folic acid, biotin, and C.

TABLE 25.6

The Principal Vitamins

VITAMIN	COMMENT AND SOURCE	FUNCTIONS	DEFICIENCY SYMPTOMS AND DISORDERS
Fat-soluble	All require bile salts and some dietary lipids for adequate absorption.		
A	Formed from provitamin beta-carotene (and other provitamins) in GI tract. Stored in liver. Sources of carotene and other provitamins include orange, yellow, and green vegetables; sources of vitamin A include liver and milk.	Maintains general health and vigor of epithelial cells. Beta-carotene acts as an antioxidant to inactivate free radicals. Essential for formation of light-sensitive pigments in photoreceptors of retina. Aids in growth of bones and teeth by helping to regulate activity of osteoblasts and osteoclasts.	Deficiency results in atrophy and keratinization of epithelium, leading to dry skin and hair; increased incidence of ear, sinus, respiratory, urinary, and digestive system infections; inability to gain weight; drying of cornea; and skin sores. **Night blindness** or decreased ability for dark adaptation. Slow and faulty development of bones and teeth.
D	Sunlight converts 7-dehydrocholesterol in the skin to cholecalciferol (vitamin D_3). A liver enzyme then converts cholecalciferol to 25-hydroxycholecalciferol. A second enzyme in the kidneys converts 25-hydroxycholecalciferol to calcitriol (1,25-dihydroxycalciferol), which is the active form of vitamin D. Most is excreted in bile. Dietary sources include fish-liver oils, egg yolk, and fortified milk.	Essential for absorption of calcium and phosphorus from GI tract. Works with parathyroid hormone (PTH) to maintain Ca^{2+} homeostasis.	Defective utilization of calcium by bones leads to **rickets** in children and **osteomalacia** in adults. Possible loss of muscle tone.
E (tocopherols)	Stored in liver, adipose tissue, and muscles. Sources include fresh nuts and wheat germ, seed oils, and green leafy vegetables.	Inhibits catabolism of certain fatty acids that help form cell structures, especially membranes. Involved in formation of DNA, RNA, and red blood cells. May promote wound healing, contribute to the normal structure and functioning of the nervous system, and prevent scarring. May help protect liver from toxic chemicals such as carbon tetrachloride. Acts as an antioxidant to inactivate free radicals.	May cause oxidation of monounsaturated fats, resulting in abnormal structure and function of mitochondria, lysosomes, and plasma membranes. A possible consequence is hemolytic anemia.
K	Produced by intestinal bacteria. Stored in liver and spleen. Dietary sources include spinach, cauliflower, cabbage, and liver.	Coenzyme essential for synthesis of several clotting factors by liver, including prothrombin.	Delayed clotting time results in excessive bleeding.
Water-soluble	Dissolved in body fluids. Most are not stored in body. Excess intake is eliminated in urine.		
B_1 (thiamine)	Rapidly destroyed by heat. Sources include whole-grain products, eggs, pork, nuts, liver, and yeast.	Acts as coenzyme for many different enzymes that break carbon-to-carbon bonds and are involved in carbohydrate metabolism of pyruvic acid to CO_2 and H_2O. Essential for synthesis of the neurotransmitter acetylcholine.	Improper carbohydrate metabolism leads to buildup of pyruvic and lactic acids and insufficient production of ATP for muscle and nerve cells. Deficiency leads to: (1) **beriberi,** partial paralysis of smooth muscle of GI tract, causing digestive disturbances; skeletal muscle paralysis; and atrophy of limbs; (2) **polyneuritis,** due to degeneration of myelin sheaths; impaired reflexes, impaired sense of touch, stunted growth in children, and poor appetite.

VITAMIN	COMMENT AND SOURCE	FUNCTIONS	DEFICIENCY SYMPTOMS AND DISORDERS
Water-soluble (continued)			
B_2 (riboflavin)	Small amounts supplied by bacteria of GI tract. Dietary sources include yeast, liver, beef, veal, lamb, eggs, whole-grain products, asparagus, peas, beets, and peanuts.	Component of certain coenzymes (for example, FAD and FMN) in carbohydrate and protein metabolism, especially in cells of eye, integument, mucosa of intestine, and blood.	Deficiency may lead to improper utilization of oxygen, resulting in blurred vision, cataracts, and corneal ulcerations. Also dermatitis and cracking of skin, lesions of intestinal mucosa, and one type of anemia.
Niacin (nicotinamide)	Derived from amino acid tryptophan. Sources include yeast, meats, liver, fish, whole-grain products, peas, beans, and nuts.	Essential component of NAD and NADP, coenzymes in oxidation–reduction reactions. In lipid metabolism, inhibits production of cholesterol and assists in triglyceride breakdown.	Principal deficiency is **pellagra**, characterized by dermatitis, diarrhea, and psychological disturbances.
B_6 (pyridoxine)	Synthesized by bacteria of GI tract. Stored in liver, muscle, and brain. Other sources include salmon, yeast, tomatoes, yellow corn, spinach, whole grain products, liver, and yogurt.	Essential coenzyme for normal amino acid metabolism. Assists production of circulating antibodies. May function as coenzyme in triglyceride metabolism.	Most common deficiency symptom is dermatitis of eyes, nose, and mouth. Other symptoms are retarded growth and nausea.
B_{12} (cyanocobalamin)	Only B vitamin not found in vegetables; only vitamin containing cobalt. Absorption from GI tract depends on intrinsic factor secreted by gastric mucosa. Sources include liver, kidney, milk, eggs, cheese, and meat.	Coenzyme necessary for red blood cell formation, formation of the amino acid methionine, entrance of some amino acids into Krebs cycle, and manufacture of choline (used to synthesize acetylcholine).	Pernicious anemia, neuropsychiatric abnormalities (ataxia, memory loss, weakness, personality and mood changes, and abnormal sensations), and impaired activity of osteoblasts.
Pantothenic acid	Some produced by bacteria of GI tract. Stored primarily in liver and kidneys. Other sources include kidney, liver, yeast, green vegetables, and cereal.	Constituent of coenzyme A, which is essential for transfer of acetyl group from pyruvic acid into the Krebs cycle, conversion of lipids and amino acids into glucose, and synthesis of cholesterol and steroid hormones.	Fatigue, muscle spasms, insufficient production of adrenal steroid hormones, vomiting, and insomnia.
Folic acid (folate, folacin)	Synthesized by bacteria of GI tract. Dietary sources include green leafy vegetables, broccoli, asparagus, breads, dried beans, and citrus fruits.	Component of enzyme systems synthesizing nitrogenous bases of DNA and RNA. Essential for normal production of red and white blood cells.	Production of abnormally large red blood cells (macrocytic anemia). Higher risk of neural tube defects in babies born to folate-deficient mothers.
Biotin	Synthesized by bacteria of GI tract. Dietary sources include yeast, liver, egg yolk, and kidneys.	Essential coenzyme for conversion of pyruvic acid to oxaloacetic acid and synthesis of fatty acids and purines.	Mental depression, muscular pain, dermatitis, fatigue, and nausea.
C (ascorbic acid)	Rapidly destroyed by heat. Some stored in glandular tissue and plasma. Sources include citrus fruits, tomatoes, and green vegetables.	Promotes protein synthesis including laying down of collagen in formation of connective tissue. As coenzyme, may combine with poisons, rendering them harmless until excreted. Works with antibodies, promotes wound healing, and functions as an antioxidant.	Scurvy; anemia; many symptoms related to poor collagen formation, including tender swollen gums, loosening of teeth (alveolar processes also deteriorate), poor wound healing, bleeding (vessel walls are fragile because of connective tissue degeneration), and retardation of growth.

DISORDERS: HOMEOSTATIC IMBALANCES

Fever

A **fever** is an elevation of core temperature caused by a resetting of the hypothalamic thermostat. The most common causes of fever are viral or bacterial infections and bacterial toxins; other causes are ovulation, excessive secretion of thyroid hormones, tumors, and reactions to vaccines. When phagocytes ingest certain bacteria, they are stimulated to secrete a **pyrogen** (PĪ-rō-gen; *pyro-* = fire; *-gen* = produce), a fever-producing substance. One pyrogen is interleukin-1. It circulates to the hypothalamus and induces neurons of the preoptic area to secrete prostaglandins. Some prostaglandins can reset the hypothalamic thermostat at a higher temperature, and temperature-regulating reflex mechanisms then act to bring the core body temperature up to this new setting. *Antipyretics* are agents that relieve or reduce fever. Examples include aspirin, acetaminophen (Tylenol), and ibuprofen (Advil), all of which reduce fever by inhibiting synthesis of certain prostaglandins.

Suppose that due to production of pyrogens the thermostat is reset at 39°C (103°F). Now the heat-promoting mechanisms (vasoconstriction, increased metabolism, shivering) are operating at full force. Thus, even though core temperature is climbing higher than normal—say, 38°C (101°F)—the skin remains cold, and shivering occurs. This condition, called a **chill,** is a definite sign that core temperature is rising. After several hours, core temperature reaches the setting of the thermostat, and the chills disappear. But now the body will continue to regulate temperature at 39°C (103°F). When the pyrogens disappear, the thermostat is reset at normal—37.0°C (98.6°F). Because core temperature is high in the beginning, the heat-losing mechanisms (vasodilation and sweating) go into operation to decrease core temperature. The skin becomes warm, and the person begins to sweat. This phase of the fever is called the **crisis,** and it indicates that core temperature is falling.

Although death results if core temperature rises above 44–46°C (112–114°F), up to a point, fever is beneficial. For example, a higher temperature intensifies the effects of interferons and the phagocytic activities of macrophages while hindering replication of some pathogens. Because fever increases heart rate, infection-fighting white blood cells are delivered to sites of infection more rapidly. In addition, antibody production and T cell proliferation increase. Moreover, heat speeds up the rate of chemical reactions, which may help body cells repair themselves more quickly.

Obesity

Obesity is body weight more than 20% above a desirable standard due to an excessive accumulation of adipose tissue. About one-third of the adult population in the United States is obese. (An athlete may be *overweight* due to higher-than-normal amounts of muscle tissue without being obese.) Even moderate obesity is hazardous to health; it is a risk factor in cardiovascular disease, hypertension, pulmonary disease, non-insulin-dependent diabetes mellitus, arthritis, certain cancers (breast, uterus, and colon), varicose veins, and gallbladder disease.

In a few cases, obesity may result from trauma of or tumors in the food-regulating centers in the hypothalamus. In most cases of obesity, no specific cause can be identified. Contributing factors include genetic factors, eating habits taught early in life, overeating to relieve tension, and social customs. Studies indicate that some obese people burn fewer calories during digestion and absorption of a meal, a smaller food-induced thermogenesis effect. Additionally, obese people who lose weight require about 15% fewer calories to maintain normal body weight than do people who have never been obese. Interestingly, people who gain weight easily when deliberately fed excess calories exhibit less NEAT (nonexercise activity thermogenesis, such as occurs with fidgeting) than people who resist weight gains in the face of excess calories. Although leptin suppresses appetite and produces satiety in experimental animals, it is not deficient in most obese people.

Most surplus calories in the diet are converted to triglycerides and stored in adipose cells. Initially, the adipocytes increase in size, but at a maximal size, they divide. As a result, proliferation of adipocytes occurs in extreme obesity. The enzyme endothelial lipoprotein lipase regulates triglyceride storage. The enzyme is very active in abdominal fat but less active in hip fat. Accumulation of fat in the abdomen is associated with higher blood cholesterol level and other cardiac risk factors because adipose cells in this area appear to be more metabolically active.

Treatment of obesity is difficult because most people who are successful at losing weight gain it back within two years. Yet, even modest weight loss is associated with health benefits. Treatments for obesity include behavior modification programs, very-low-calorie diets, drugs, and surgery. Behavior modification programs, offered at many hospitals, strive to alter eating behaviors and increase exercise activity. The nutrition program includes a "heart-healthy" diet that includes abundant vegetables but is low in fats, especially saturated fats. A typical exercise program suggests walking for 30 minutes a day, five to seven times a week. Regular exercise enhances both weight loss and weight-loss maintenance. Very-low-calorie (VLC) diets include 400 to 800 kcal/day in a commercially made liquid mixture. The VLC diet is usually prescribed for 12 weeks, under close medical supervision. Two drugs are available to treat obesity. Sibutramine is an appetite suppressant that works by inhibiting reuptake of serotonin and norepinephrine in brain areas that govern eating behavior. Orlistat works by inhibiting the lipases released into the lumen of the GI tract. With less lipase activity, fewer dietary triglycerides are absorbed. For those with extreme obesity who have not responded to other treatments, a surgical procedure may be considered. The two operations most commonly performed—gastric bypass and gastroplasty—both greatly reduce the stomach size so that it can hold just a tiny quantity of food.

MEDICAL TERMINOLOGY

Heat cramps Cramps that result from profuse sweating. The salt lost in sweat causes painful contractions of muscles; such cramps tend to occur in muscles used while working but do not appear until the person relaxes once the work is done. Drinking salted liquids usually leads to rapid improvement.

Heat exhaustion (heat prostration) A condition in which the core temperature is generally normal, or a little below, and the skin is cool and moist due to profuse perspiration. Heat exhaustion is usually characterized by loss of fluid and electrolytes, especially salt (NaCl). The salt loss results in muscle cramps, dizziness, vomiting, and fainting; fluid loss may cause low blood pressure. Complete rest, rehydration, and electrolyte replacement are recommended.

Heatstroke (sunstroke) A severe and often fatal disorder caused by exposure to high temperatures, especially when the relative humidity is high, which makes it difficult for the body to lose heat. Blood flow to the skin is decreased, perspiration is greatly reduced, and body temperature rises sharply because of failure of the hypothalamic thermostat. Body temperature may reach 43°C (110°F). Treatment, which must be undertaken immediately, consists of cooling the body by immersing the victim in cool water and by administering fluids and electrolytes.

Kwashiorkor (kwash-ē-OR-kor) A disorder in which protein intake is deficient despite normal or nearly normal caloric intake, character-

ized by edema of the abdomen, enlarged liver, decreased blood pressure, low pulse rate, lower-than-normal body temperature, and sometimes mental retardation. Because the main protein in corn (zein) lacks two essential amino acids, which are needed for growth and tissue repair, many African children whose diet consists largely of cornmeal develop kwashiorkor.

Malnutrition (*mal-* = bad) An imbalance of total caloric intake or intake of specific nutrients, which can be either inadequate or excessive.

Marasmus (mar-AZ-mus) A type of protein–calorie undernutrition that results from inadequate intake of both protein and calories. Its characteristics include retarded growth, low weight, muscle wasting, emaciation, dry skin, and thin, dry, dull hair.

STUDY OUTLINE

Introduction (p. 977)

1. Our only source of energy for performing biological work is the food we eat. Food also provides essential substances that we cannot synthesize.
2. Most food molecules absorbed by the gastrointestinal tract are used to supply energy for life processes, serve as building blocks during synthesis of complex molecules, or are stored for future use.

Metabolic Reactions (p. 978)

1. Metabolism refers to all chemical reactions of the body and is of two types: catabolism and anabolism.
2. Catabolism is the term for reactions that break down complex organic compounds into simple ones. Overall, catabolic reactions are exergonic; they produce more energy than they consume.
3. Chemical reactions that combine simple molecules into more complex ones that form the body's structural and functional components are collectively known as anabolism. Overall, anabolic reactions are endergonic; they consume more energy than they produce.
4. The coupling of anabolism and catabolism occurs via ATP.

Energy Transfer (p. 979)

1. Oxidation is the removal of electrons from a substance; reduction is the addition of electrons to a substance.
2. Two coenzymes that carry hydrogen atoms during coupled oxidation–reduction reactions are nicotinamide adenine dinucleotide (NAD^+) and flavin adenine dinucleotide (FAD).
3. ATP can be generated via substrate-level phosphorylation, oxidative phosphorylation, and photophosphorylation.

Carbohydrate Metabolism (p. 980)

1. During digestion, polysaccharides and disaccharides are hydrolyzed into the monosaccharides glucose (about 80%), fructose, and galactose; the latter two are then converted to glucose.
2. Some glucose is oxidized by cells to provide ATP. Glucose also can be used to synthesize amino acids, glycogen, and triglycerides.

3. Glucose moves into most body cells via facilitated diffusion through glucose transporters (GluT) and becomes phosphorylated to glucose 6-phosphate. In muscle cells, this process is stimulated by insulin. Glucose entry into neurons and hepatocytes is always "turned on."
4. Cellular respiration, the complete oxidation of glucose to CO_2 and H_2O, involves glycolysis, the Krebs cycle, and the electron transport chain.
5. Glycolysis is the breakdown of glucose into two molecules of pyruvic acid; there is a net production of two molecules of ATP.
6. When oxygen is in short supply, pyruvic acid is reduced to lactic acid; under aerobic conditions, pyruvic acid enters the Krebs cycle.
7. Pyruvic acid is prepared for entrance into the Krebs cycle by conversion to a two-carbon acetyl group followed by the addition of coenzyme A to form acetyl coenzyme A.
8. The Krebs cycle involves decarboxylations, oxidations, and reductions of various organic acids.
9. Each molecule of pyruvic acid that is converted to acetyl coenzyme A and then enters the Krebs cycle produces three molecules of CO_2, four molecules of NADH and four H^+, one molecule of $FADH_2$, and one molecule of ATP.
10. The energy originally stored in glucose and then in pyruvic acid is transferred primarily to the reduced coenzymes NADH and $FADH_2$.
11. The electron transport chain involves a series of oxidation–reduction reactions in which the energy in NADH and $FADH_2$ is liberated and transferred to ATP.
12. The electron carriers include FMN, cytochromes, iron–sulfur centers, copper atoms, and coenzyme Q.
13. The electron transport chain yields a maximum of 32 or 34 molecules of ATP and six molecules of H_2O.
14. Table 25.1 on page 988 summarizes the ATP yield during cellular respiration. The complete oxidation of glucose can be represented as follows:

$$C_6H_{12}O_6 + 6\ O_2 + 36 \text{ or } 38 \text{ ADPs} + 36 \text{ or } 38 \text{ P} \longrightarrow$$
$$6\ CO_2 + 6\ H_2O + 36 \text{ or } 38 \text{ ATPs}$$

15. The conversion of glucose to glycogen for storage in the liver and skeletal muscle is called glycogenesis. It is stimulated by insulin.

16. The conversion of glycogen to glucose is called glycogenolysis. It occurs between meals and is stimulated by glucagon and epinephrine.

17. Gluconeogenesis is the conversion of noncarbohydrate molecules into glucose. It is stimulated by cortisol and glucagon.

Lipid Metabolism (p. 990)

1. Lipoproteins transport lipids in the bloodstream. Types of lipoproteins include chylomicrons, which carry dietary lipids to adipose tissue; very-low-density lipoproteins (VLDLs), which carry triglycerides from the liver to adipose tissue; low-density lipoproteins (LDLs), which deliver cholesterol to body cells; and high-density lipoproteins (HDLs), which remove excess cholesterol from body cells and transport it to the liver for elimination.

2. Cholesterol in the blood comes from two sources: from food and from synthesis by the liver.

3. Lipids may be oxidized to produce ATP or stored as triglycerides in adipose tissue, mostly in the subcutaneous layer.

4. A few lipids are used as structural molecules or to synthesize essential molecules.

5. Adipose tissue contains lipases that catalyze the deposition of triglycerides from chylomicrons and hydrolyze triglycerides into fatty acids and glycerol.

6. In lipolysis, triglycerides are split into fatty acids and glycerol and released from adipose tissue under the influence of epinephrine, norepinephrine, cortisol, thyroid hormones, and insulinlike growth factors.

7. Glycerol can be converted into glucose by conversion into glyceraldehyde 3-phosphate.

8. In beta-oxidation of fatty acids, carbon atoms are removed in pairs from fatty acid chains; the resulting molecules of acetyl coenzyme A enter the Krebs cycle.

9. The conversion of glucose or amino acids into lipids is called lipogenesis; it is stimulated by insulin.

Protein Metabolism (p. 993)

1. During digestion, proteins are hydrolyzed into amino acids, which enter the liver via the hepatic portal vein.

2. Amino acids, under the influence of insulinlike growth factors and insulin, enter body cells via active transport.

3. Inside cells, amino acids are synthesized into proteins that function as enzymes, hormones, structural elements, and so forth; stored as fat or glycogen; or used for energy.

4. Before amino acids can be catabolized, they must be deaminated and converted to substances that can enter the Krebs cycle.

5. Amino acids may also be converted into glucose, fatty acids, and ketone bodies.

6. Protein synthesis is stimulated by insulinlike growth factors, thyroid hormones, insulin, estrogen, and testosterone.

7. Table 25.2 on page 997 summarizes carbohydrate, lipid, and protein metabolism.

Key Molecules at Metabolic Crossroads (p. 995)

1. Three molecules play a key role in metabolism: glucose 6-phosphate, pyruvic acid, and acetyl coenzyme A.

2. Glucose 6-phosphate may be converted to glucose, glycogen, ribose 5-phosphate, and pyruvic acid.

3. When ATP is low and oxygen is plentiful, pyruvic acid is converted to acetyl coenzyme A; when oxygen supply is low, pyruvic acid is converted to lactic acid. Carbohydrate and protein metabolism are linked by pyruvic acid.

4. Acetyl coenzyme A is the molecule that enters the Krebs cycle; it is also used to synthesize fatty acids, ketone bodies, and cholesterol.

Metabolic Adaptations (p. 997)

1. During the absorptive state, ingested nutrients enter the blood and lymph from the GI tract.

2. During the absorptive state, blood glucose is oxidized to form ATP, and glucose transported to the liver is converted to glycogen or triglycerides. Most triglycerides are stored in adipose tissue. Amino acids in hepatocytes are converted to carbohydrates, fats, and proteins. Table 25.3 on page 999 summarizes the hormonal regulation of metabolism during the absorptive state.

3. During the postabsorptive state, absorption is complete and the ATP needs of the body are satisfied by nutrients already present in the body. The major task is to maintain normal blood glucose level by converting glycogen in the liver and skeletal muscle into glucose, converting glycerol into glucose, and converting amino acids into glucose. Fatty acids, ketone bodies, and amino acids are oxidized to supply ATP. Table 25.4 on page 1001 summarizes the hormonal regulation of metabolism during the postabsorptive state.

4. Fasting is going without food for a few days; starvation implies weeks or months of inadequate food intake. During fasting and starvation, fatty acids and ketone bodies are increasingly utilized for ATP production.

Heat and Energy Balance (p. 1001)

1. Measurement of the metabolic rate under basal conditions is called the basal metabolic rate (BMR).

2. A calorie (cal) is the amount of energy required to raise the temperature of 1 g of water 1°C.

3. Because the calorie is a relatively small unit, the kilocalorie (kcal) or Calorie (cal) is often used to measure the body's metabolic rate and to express the energy content of foods; a kilocalorie equals 1000 calories.

4. Normal core temperature is maintained by a delicate balance between heat-producing and heat-losing mechanisms.

5. Exercise, hormones, the nervous system, body temperature, ingestion of food, age, gender, climate, sleep, and malnutrition affect metabolic rate.

6. Mechanisms of heat transfer include conduction, convection, radiation, and evaporation. Conduction is the transfer of heat between two substances or objects in contact with each other. Convection is the transfer of heat by a liquid or gas between areas of different temperatures. Radiation is the transfer of heat from a warmer object to a cooler object without physical contact. Evaporation is the conversion of a liquid to a vapor; in the process, heat is lost.

7. The hypothalamic thermostat is in the preoptic area.

8. Responses that produce, conserve, or retain heat when core temperature falls are vasoconstriction; release of epinephrine, norepinephrine, and thyroid hormones; and shivering.

9. Responses that increase heat loss when core temperature increases include vasodilation, decreased metabolic rate, and evaporation of perspiration.

10. Two nuclei in the hypothalamus that help regulate food intake are the arcuate and paraventricular nuclei. The hormone leptin, released by adipocytes, inhibits release of neuropeptide Y from the arcuate nucleus and thereby decreases food intake. Melanocortin also decreases food intake.

Nutrition (p. 1006)

1. Nutrients include water, carbohydrates, lipids, proteins, minerals, and vitamins.
2. Nutrition experts suggest dietary calories be 50–60% from carbohydrates, 30% or less from fats, and 12–15% from proteins.
3. The MyPyramid guide represents a personalized approach to making healthy food choices and maintaining regular physical activity.

4. Minerals known to perform essential functions include calcium, phosphorus, potassium, sulfur, sodium, chloride, magnesium, iron, iodide, manganese, copper, cobalt, zinc, fluoride, selenium, and chromium. Their functions are summarized in Table 25.5 on page 1008.
5. Vitamins are organic nutrients that maintain growth and normal metabolism. Many function in enzyme systems.
6. Fat-soluble vitamins are absorbed with fats and include vitamins A, D, E, and K; water-soluble vitamins include the B vitamins and vitamin C.
7. The functions and deficiency disorders of the principal vitamins are summarized in Table 25.6 on pages 1010–1011.

SELF-QUIZ QUESTIONS

Fill in the blanks in the following statements.

1. The thermostat and food intake regulating center of the body is in the _____ of the brain.
2. The three key molecules of metabolism are _____, _____, and _____.

Indicate whether the following statements are true or false.

3. Foods that we eat are used to supply energy for life processes, serve as building blocks for synthesis reactions, or are stored for future use.
4. Vitamins A, B, D, and K are fat-soluble vitamins.

Choose the one best answer to the following questions.

5. NAD^+ and FAD (1) are both derivatives of B vitamins, (2) are used to carry hydrogen atoms released during oxidation reactions, (3) become NADH and $FADH_2$ in their reduced forms, (4) act as coenzymes in the Krebs cycle, (5) are the final electron acceptors in the electron transport chain.
   (a) 1, 2, 3, 4, and 5   (b) 2, 3, and 4   (c) 2 and 4
   (d) 1, 2, and 3   (e) 1, 2, 3, and 4

6. During glycolysis, (1) a six-carbon glucose is split into two three-carbon pyruvic acids, (2) there is a net gain of two ATP molecules, (3) two NADH molecules are oxidized, (4) moderately high levels of oxygen are needed, (5) the activity of phosphofructokinase determines the rate of the chemical reactions.
   (a) 1, 2, and 3   (b) 1 and 2   (c) 1, 2, and 5
   (d) 2, 3, 4, and 5   (e) 1, 2, 3, 4, and 5

7. If glucose is not needed for immediate ATP production, it can be used for (1) vitamin synthesis, (2) amino acid synthesis, (3) gluconeogenesis, (4) glycogenesis, (5) lipogenesis.
   (a) 1, 3, and 5   (b) 2, 4, and 5   (c) 2, 3, 4, and 5
   (d) 1, 2, and 3   (e) 2 and 5

8. Which of the following is the *correct* sequence for the oxidation of glucose to produce ATP?
   (a) electron transport chain, Krebs cycle, glycolysis, formation of acetyl CoA
   (b) Krebs cycle, formation of acetyl CoA, electron transport chain, glycolysis
   (c) glycolysis, electron transport chain, Krebs cycle, formation of acetyl CoA

   (d) glycolysis, formation of acetyl CoA, Krebs cycle, electron transport chain
   (e) formation of acetyl CoA, Krebs cycle, glycolysis, electron transport chain.

9. Which of the following would you *not* expect to experience during fasting or starvation?
   (a) decrease in plasma fatty acid levels
   (b) increase in ketone body formation
   (c) lipolysis
   (d) increased use of ketones for ATP production in the brain
   (e) depletion of glycogen

10. If core body temperature rises above normal, which of the following would occur to cool the body? (1) dilation of vessels in the skin, (2) increased radiation and conduction of heat to the environment, (3) increased metabolic rate, (4) evaporation of perspiration, (5) increased secretion of thyroid hormones.
    (a) 3, 4, and 5   (b) 1, 2, and 4   (c) 1, 2, and 5
    (d) 1, 2, 3, 4 and 5   (e) 1, 2, 4, and 5

11. In which of the following situations would the metabolic rate increase? (1) sleep, (2) after ingesting food, (3) increased secretion of thyroid hormones, (4) parasympathetic nervous system stimulation, (5) fever.
    (a) 3 and 4   (b) 1, 3, and 5   (c) 2 and 3
    (d) 2, 3, and 4   (e) 2, 3, and 5

12. Which of the following are absorptive state reactions? (1) aerobic cellular respiration, (2) glycogenesis, (3) glycogenolysis, (4) gluconeogenesis using lactic acid, (5) lipolysis.
    (a) 1 and 2   (b) 2 and 3   (c) 3 and 4
    (d) 4 and 5   (e) 1 and 5

13. Match the hormones with the reactions they regulate (answers may be used more than once; some reactions have more than one answer):
    _____ (a) gluconeogenesis   (1) insulin
    _____ (b) glycogenesis   (2) cortisol
    _____ (c) glycogenolysis   (3) glucagon
    _____ (d) lipolysis   (4) thyroid
    _____ (e) lipogenesis   hormones
    _____ (f) protein catabolism   (5) epinephrine
    _____ (g) protein anabolism   (6) insulinlike
                                           growth factors

14. Match the following:

_____ (a) deliver cholesterol to body cells for use in repair of membranes and synthesis of steroid hormones and bile salts

_____ (b) remove excess cholesterol from body cells and transport it to the liver for elimination

_____ (c) organic nutrients required in small amounts for growth and normal metabolism

_____ (d) the energy-transferring molecule of the body

_____ (e) nutrient molecules that can be oxidized to produce ATP or stored in adipose tissue

_____ (f) transport endogenous lipids to adipocytes for storage

_____ (g) the body's preferred source for synthesizing ATP

_____ (h) composed of amino acids and are the primary regulatory molecules in the body

_____ (i) acetoacetic acid, beta-hydroxybutyric acid, and acetone

_____ (j) hormone secreted by adipocytes that acts to decrease total body-fat mass

_____ (k) neurotransmitter that stimulates food intake

_____ (l) inorganic substances that perform many vital functions in the body

_____ (m) carriers of electrons in the electron transport chain

(1) leptin
(2) minerals
(3) glucose
(4) lipids
(5) proteins
(6) neuropeptide Y
(7) cytochromes
(8) ketone bodies
(9) low-density lipoproteins
(10) ATP
(11) vitamins
(12) high-density lipoproteins
(13) very low-density lipoproteins

15. Match the following:

_____ (a) the mechanism of ATP generation that links chemical reactions with pumping of hydrogen ions

_____ (b) the removal of electrons from an atom or molecule resulting in a decrease in potential energy

_____ (c) the transfer of an amino group from an amino acid to a substance such as pyruvic acid

_____ (d) the formation of glucose from noncarbohydrate sources

_____ (e) refers to all the chemical reactions in the body

_____ (f) the oxidation of glucose to produce ATP

_____ (g) the splitting of a triglyceride into glycerol and fatty acids

_____ (h) the synthesis of lipids

_____ (i) the addition of electrons to a molecule resulting in an increase in potential energy content of the molecule

_____ (j) the formation of ketone bodies

_____ (k) the breakdown of glycogen back to glucose

_____ (l) exergonic chemical reactions that break down complex organic molecules into simpler ones

_____ (m) overall rate at which metabolic reactions use energy

_____ (n) the breakdown of glucose into two molecules of pyruvic acid

_____ (o) removal of CO_2 from a molecule

_____ (p) endergonic chemical reactions that combine simple molecules and monomers to make more complex ones

_____ (q) the addition of a phosphate group to a molecule

_____ (r) the removal of the amino group from an amino acid

_____ (s) the cleavage of one pair of carbon atoms at a time from a fatty acid

_____ (t) the conversion of glucose into glycogen

(1) metabolism
(2) catabolism
(3) beta oxidation
(4) lipolysis
(5) phosphorylation
(6) glycolysis
(7) cellular respiration
(8) transamination
(9) anabolism
(10) lipogenesis
(11) glycogenolysis
(12) glycogenesis
(13) metabolic rate
(14) ketogenesis
(15) oxidation
(16) reduction
(17) chemiosmosis
(18) deamination
(19) gluconeogenesis
(20) decarboxylation

CRITICAL THINKING QUESTIONS

1. Jane Doe's deceased body was found at her dining room table. Her death was considered suspicious. Lab results from the medical investigation revealed cyanide in her blood. How did the cyanide cause her death?

2. During a recent physical, 55-year-old Glenn's blood serum lab results showed the following: total cholesterol = 300mg/dL; LDL = 175 mg/dL; HDL=20 mg/dL. Interpret these results for Glenn and

indicate to him what changes, if any, he needs to make in his lifestyle. Why are these changes important?

3. Marissa has joined a weight loss program. As part of the program, she regularly submits a urine sample which is tested for ketones. She went to the clinic today, had her urine checked, and was confronted by the nurse who accused Marissa of "cheating" on her diet. How did the nurse know Marissa was not following her diet?

? ANSWERS TO FIGURE QUESTIONS

25.1 In pancreatic acinar cells, anabolism predominates because the primary activity is synthesis of complex molecules (digestive enzymes).

25.2 Glycolysis is also called anaerobic cellular respiration.

25.3 The reactions of glycolysis consume two molecules of ATP but generate four molecules of ATP, for a net gain of two.

25.4 Kinases are enzymes that phosphorylate (add phosphate to) their substrate.

25.5 Glycolysis occurs in the cytosol.

25.6 CO_2 is given off during the production of acetyl coenzyme A and during the Krebs cycle. It diffuses into the blood, is transported by the blood to the lungs, and is exhaled.

25.7 The production of reduced coenzymes is important in the Krebs cycle because they will subsequently yield ATP in the electron transport chain.

25.8 The energy source that powers the proton pumps is electrons provided by $NADH + H^+$.

25.9 The concentration of H^+ is highest in the space between the inner and outer mitochondrial membranes.

25.10 During the complete oxidation of one glucose molecule, six molecules of O_2 are used and six molecules of CO_2 are produced.

25.11 Skeletal muscle fibers can synthesize glycogen, but they cannot release glucose into the blood because they lack the enzyme

phosphatase required to remove the phosphate group from glucose.

25.12 Hepatocytes can carry out gluconeogenesis and glycogenesis.

25.13 LDLs deliver cholesterol to body cells.

25.14 Hepatocytes and adipose cells carry out lipogenesis, beta oxidation, and lipolysis; hepatocytes carry out ketogenesis.

25.15 Before an amino acid can enter the Krebs cycle, an amino group must be removed via deamination.

25.16 Acetyl coenzyme A is the gateway into the Krebs cycle for molecules being oxidized to generate ATP.

25.17 Reactions of the absorptive state are mainly anabolic.

25.18 Processes that directly elevate blood glucose during the postabsorptive state include lipolysis (in adipocytes and hepatocytes), gluconeogenesis (in hepatocytes), and glycogenolysis (in hepatocytes).

25.19 Exercise, the sympathetic nervous system, hormones (epinephrine, norepinephrine, thyroxine, testosterone, human growth hormone), elevated body temperature, and ingestion of food increase metabolic rate, which results in an increase in body temperature.

25.20 The wider base of each band represents foods with little or no solid fats or added sugars.

26 | THE URINARY SYSTEM

THE URINARY SYSTEM AND HOMEOSTASIS The urinary system contributes to homeostasis by altering blood composition, pH, volume, and pressure; maintaining blood osmolarity; excreting wastes and foreign substances; and producing hormones.

The **urinary system** consists of two kidneys, two ureters, one urinary bladder, and one urethra (Figure 26.1). After the kidneys filter blood plasma, they return most of the water and solutes to the bloodstream. The remaining water and solutes constitute **urine,** which passes through the ureters and is stored in the urinary bladder until it is excreted from the body through the urethra. **Nephrology** (nef-ROL-ō-jē; *nephr-* = kidney; *-ology* = study of) is the scientific study of the anatomy, physiology, and pathology of the kidneys. The branch of medicine that deals with the male and female urinary systems and the male reproductive system is called **urology** (ū-ROL-ō-jē; *uro-* = urine). A physician who specializes in this branch of medicine is called a **urologist** (ū-ROL-ō-jist). •

Figure 26.1 Organs of the urinary system in a female. (See Tortora, *A Photographic Atlas of the Human Body, Second Edition,* Figure 13.2.)

⑥ Urine formed by the kidneys passes first into the ureters, then to the urinary bladder for storage, and finally through the urethra for elimination from the body.

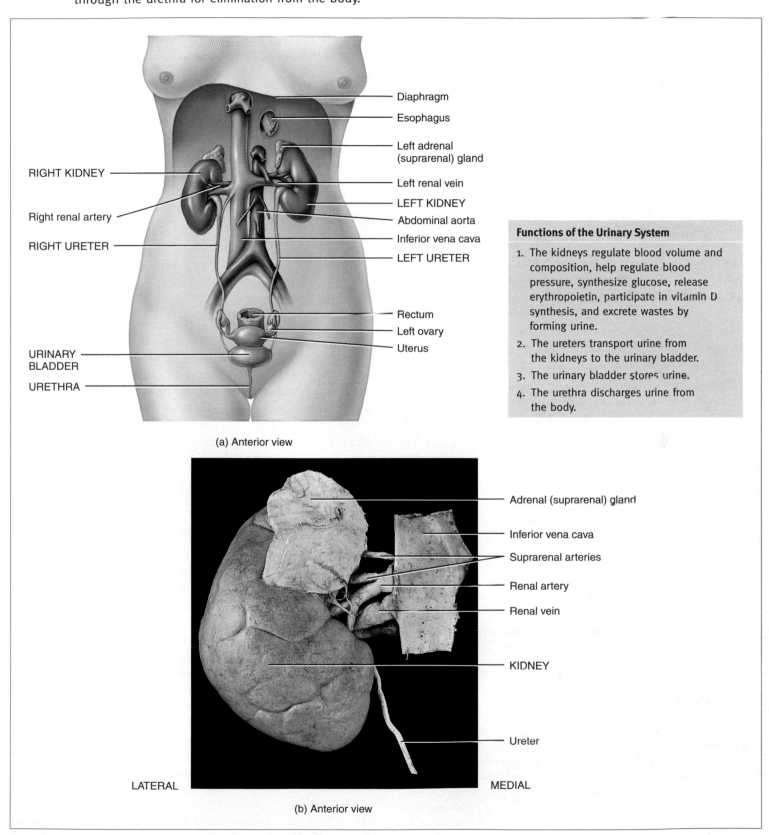

Functions of the Urinary System

1. The kidneys regulate blood volume and composition, help regulate blood pressure, synthesize glucose, release erythropoietin, participate in vitamin D synthesis, and excrete wastes by forming urine.

2. The ureters transport urine from the kidneys to the urinary bladder.

3. The urinary bladder stores urine.

4. The urethra discharges urine from the body.

(a) Anterior view

(b) Anterior view

? **Which organs constitute the urinary system?**

OVERVIEW OF KIDNEY FUNCTIONS

◉ OBJECTIVE

• List the functions of the kidneys.

The kidneys do the major work of the urinary system. The other parts of the system are mainly passageways and storage areas. Functions of the kidneys include the following:

- *Regulation of blood ionic composition.* The kidneys help regulate the blood levels of several ions, most importantly sodium ions (Na^+), potassium ions (K^+), calcium ions (Ca^{2+}), chloride ions (Cl^-), and phosphate ions (HPO_4^{2-}).

- *Regulation of blood pH.* The kidneys excrete a variable amount of hydrogen ions (H^+) into the urine and conserve bicarbonate ions (HCO_3^-), which are an important buffer of H^+ in the blood. Both of these activities help regulate blood pH.

- *Regulation of blood volume.* The kidneys adjust blood volume by conserving or eliminating water in the urine. An increase in blood volume increases blood pressure; a decrease in blood volume decreases blood pressure.

- *Regulation of blood pressure.* The kidneys also help regulate blood pressure by secreting the enzyme renin, which activates the renin–angiotensin–aldosterone pathway (see Figure 18.16 on page 667). Increased renin causes an increase in blood pressure.

- *Maintenance of blood osmolarity.* By separately regulating loss of water and loss of solutes in the urine, the kidneys maintain a relatively constant blood osmolarity close to 300 milliosmoles per liter (mOsm/liter).*

- *Production of hormones.* The kidneys produce two hormones. *Calcitriol,* the active form of vitamin D, helps regulate calcium homeostasis (see Figure 18.14 on page 664), and *erythropoietin* stimulates the production of red blood cells (see Figure 19.5 on page 697).

- *Regulation of blood glucose level.* Like the liver, the kidneys can use the amino acid glutamine in *gluconeogenesis,* the synthesis of new glucose molecules. They can then release glucose into the blood to help maintain a normal blood glucose level.

- *Excretion of wastes and foreign substances.* By forming urine, the kidneys help excrete **wastes**—substances that

---

*The **osmolarity** of a solution is a measure of the total number of dissolved particles per liter of solution. The particles may be molecules, ions, or a mixture of both. To calculate osmolarity, multiply molarity (see page 40) by the number of particles per molecule, once the molecule dissolves. A similar term, *osmolality,* is the number of particles of solute per *kilogram* of water. Because it is easier to measure volumes of solutions than to determine the mass of water they contain, osmolarity is used more commonly than osmolality. Most body fluids and solutions used clinically are dilute, in which case there is less than a 1% difference between the two measures.

have no useful function in the body. Some wastes excreted in urine result from metabolic reactions in the body. These include ammonia and urea from the deamination of amino acids; bilirubin from the catabolism of hemoglobin; creatinine from the breakdown of creatine phosphate in muscle fibers; and uric acid from the catabolism of nucleic acids. Other wastes excreted in urine are foreign substances from the diet, such as drugs and environmental toxins.

◉ CHECKPOINT

1. What are wastes, and how do the kidneys participate in their removal from the body?

ANATOMY AND HISTOLOGY OF THE KIDNEYS

◉ OBJECTIVES

- Describe the external and internal gross anatomical features of the kidneys.
- Trace the path of blood flow through the kidneys.
- Describe the structure of renal corpuscles and renal tubules.

The paired **kidneys** are reddish, kidney-bean-shaped organs located just above the waist between the peritoneum and the posterior wall of the abdomen. Because their position is posterior to the peritoneum of the abdominal cavity, they are said to be **retroperitoneal** (re′-trō-per-i-tō-NĒ-al; *retro-* = behind) organs (Figure 26.2). The kidneys are located between the levels of the last thoracic and third lumbar vertebrae, a position where they are partially protected by the eleventh and twelfth pairs of ribs. The right kidney is slightly lower than the left (see Figure 26.1) because the liver occupies considerable space on the right side superior to the kidney.

External Anatomy of the Kidneys

A typical adult kidney is 10–12 cm (4–5 in.) long, 5–7 cm (2–3 in.) wide, and 3 cm (1 in.) thick—about the size of a bar of bath soap—and has a mass of 135–150 g (4.5–5 oz). The concave medial border of each kidney faces the vertebral column (see Figure 26.1). Near the center of the concave border is an indentation called the **renal hilum** (see Figure 26.3), through which the ureter emerges from the kidney along with blood vessels, lymphatic vessels, and nerves.

Three layers of tissue surround each kidney (Figure 26.2). The deep layer, the **renal capsule** (*ren-* = kidney), is a smooth, transparent sheet of dense irregular connective tissue that is continuous with the outer coat of the ureter. It serves as a barrier against trauma and helps maintain the shape of the kidney. The middle layer, the **adipose capsule,** is a mass of fatty tissue surrounding the renal capsule. It also protects the kidney from trauma and holds it firmly in place within the abdominal cavity. The superficial layer, the **renal fascia,** is another thin layer of dense irregular connective tissue that anchors the kidney to

Figure 26.2 Position and coverings of the kidneys. (See Tortora, *A Photographic Atlas of the Human Body, Second Edition,* Figure 13.3.)

The kidneys are surrounded by a renal capsule, adipose capsule, and renal fascia.

ANTERIOR

Transverse plane

View

Stomach

Pancreas

Liver

Large intestine

Abdominal aorta

Renal artery and vein

RENAL HILUM

Body of L2

LEFT KIDNEY

Spleen

Rib

Quadratus lumborum muscle

Inferior vena cava

Peritoneum

Layers

RENAL FASCIA

ADIPOSE CAPSULE

RENAL CAPSULE

RIGHT KIDNEY

POSTERIOR

(a) Inferior view of transverse section of abdomen (L2)

SUPERIOR

Sagittal plane

Lung

Liver

Diaphragm

Adrenal (suprarenal) gland

Twelfth rib

Peritoneum

RENAL FASCIA

Right kidney

ADIPOSE CAPSULE   Layers

RENAL CAPSULE

Quadratus lumborum muscle

Large intestine

Hip bone

POSTERIOR

ANTERIOR

(b) Sagittal section through the right kidney

? **Why are the kidneys said to be retroperitoneal?**

the surrounding structures and to the abdominal wall. On the anterior surface of the kidneys, the renal fascia is deep to the peritoneum.

Internal Anatomy of the Kidneys

A frontal section through the kidney reveals two distinct regions: a superficial, light red area called the **renal cortex** (*cortex* = rind or bark) and a deep, darker reddish-brown inner region called the **renal medulla** (*medulla* = inner portion) (Figure 26.3). The renal medulla consists of several cone-shaped **renal pyramids.** The base (wider end) of each pyramid faces the renal cortex, and its apex (narrower end), called a **renal papilla,** points toward the renal hilum. The renal cortex is the smooth-textured area extending from the renal capsule to the bases of the renal pyramids and into the spaces between them. It is divided into an outer *cortical zone* and an inner *juxtamedullary zone.* Those portions of the renal cortex that extend between renal pyramids are called **renal columns.** A **renal lobe** consists of a renal pyramid, its overlying area of renal cortex, and one-half of each adjacent renal column.

Together, the renal cortex and renal pyramids of the renal medulla constitute the **parenchyma** (functional portion) of the kidney. Within the parenchyma are the functional units of the kidney—about 1 million microscopic structures called **nephrons** (NEF-rons). Urine formed by the nephrons drains into large **papillary ducts,** which extend through the renal papillae of the pyramids. The papillary ducts drain into cuplike structures called **minor** and **major calyces** (KĀ-li-sēz = cups; singular is *calyx*). Each kidney has 8 to 18 minor calyces and 2 or 3 major calyces. A minor calyx receives urine from the papillary ducts of one renal papilla and delivers it to a major calyx. From the major calyces, urine drains into a single large cavity called the **renal pelvis** (*pelv-* = basin) and then out through the ureter to the urinary bladder.

The hilum expands into a cavity within the kidney called the **renal sinus,** which contains part of the renal pelvis, the calyces, and branches of the renal blood vessels and nerves. Adipose tissue helps stabilize the position of these structures in the renal sinus.

Figure 26.3 Internal anatomy of the kidneys.

 The two main regions of the kidney parenchyma are the renal cortex and the renal pyramids in the renal medulla.

Blood and Nerve Supply of the Kidneys

Because the kidneys remove wastes from the blood and regulate its volume and ionic composition, it is not surprising that they are abundantly supplied with blood vessels. Although the kidneys constitute less than 0.5% of total body mass, they receive 20–25% of the resting cardiac output via the right and left **renal arteries** (Figure 26.4). In adults, **renal blood flow,** the blood flow through both kidneys, is about 1200 mL per minute.

Within the kidney, the renal artery divides into several **segmental arteries,** which supply different segments (areas) of the kidney. Each segmental artery gives off several branches that enter the parenchyma and pass through the renal columns between the renal pyramids as the **interlobar arteries.** At the bases of the renal pyramids, the interlobar arteries arch between the renal medulla and cortex; here they are known as the **arcuate arteries** (AR-kū-āt = shaped like a bow). Divisions of the arcuate arteries produce a series of **interlobular arteries.** These arteries are so named because they pass between renal lobules. Interlobular arteries enter the renal cortex and give off branches called **afferent arterioles** (*af-* = toward; *-ferrent* = to carry).

Each nephron receives one afferent arteriole, which divides into a tangled, ball-shaped capillary network called the **glomerulus** (glō-MER-ū-lus = little ball; plural is *glomeruli*). The glomerular capillaries then reunite to form an **efferent arteriole** (*ef-* = out) that carries blood out of the glomerulus. Glomerular capillaries are unique among capillaries in the body because they are positioned between two arterioles, rather than between an arteriole and a venule. Because they are capillary networks and they also play an important role in urine formation, the glomeruli are considered part of both the cardiovascular and the urinary systems.

The efferent arterioles divide to form the **peritubular capillaries** (*peri-* = around), which surround tubular parts of the nephron in the renal cortex. Extending from some efferent arterioles are long loop-shaped capillaries called **vasa recta** (VĀ-sa REK-ta; *vasa* = vessels; *recta* = straight) that supply tubular portions of the nephron in the renal medulla (see Figure 26.5b).

The peritubular capillaries eventually reunite to form **peritubular venules** and then **interlobular veins,** which also receive blood from the vasa recta. Then the blood drains through the **arcuate veins** to the **interlobar veins** running between the renal pyramids. Blood leaves the kidney through a single **renal vein** that exits at the renal hilum and carries venous blood to the inferior vena cava.

Many renal nerves originate in the *renal ganglion* and pass through the *renal plexus* into the kidneys along with the renal arteries. Renal nerves are part of the sympathetic division of the autonomic nervous system. Most are vasomotor nerves that

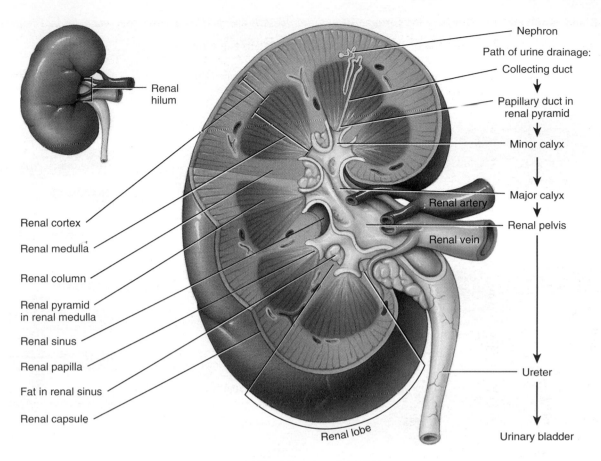

Nephron

Path of urine drainage:

Collecting duct

Papillary duct in renal pyramid

Minor calyx

Major calyx

Renal artery

Renal pelvis

Renal vein

Renal cortex

Renal medulla

Renal column

Renal pyramid in renal medulla

Renal sinus

Renal papilla

Fat in renal sinus

Renal capsule

Renal hilum

Ureter

Renal lobe

Urinary bladder

(a) Frontal section of right kidney

SUPERIOR

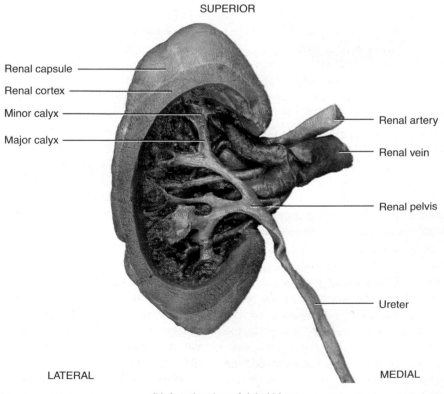

Renal capsule

Renal cortex

Minor calyx

Major calyx

Renal artery

Renal vein

Renal pelvis

Ureter

LATERAL

MEDIAL

(b) Anterior view of right kidney

❓ **What structures pass through the renal hilum?**

Figure 26.4 Blood supply of the kidneys. (See Tortora, *A Photographic Atlas of the Human Body, Second Edition,* Figure 13.6.)

The renal arteries deliver 20–25% of the resting cardiac output to the kidneys.

Blood supply of the nephron

(a) Frontal section of right kidney

(b) Path of blood flow

❓ **What volume of blood enters the renal arteries per minute?**

regulate the flow of blood through the kidney by causing vasodilation or vasoconstriction of renal arterioles.

The Nephron

Parts of a Nephron

Nephrons are the functional units of the kidneys. Each nephron (Figure 26.5) consists of two parts: a **renal corpuscle** (KOR-pus-sul = tiny body), where blood plasma is filtered, and a **renal tubule** into which the filtered fluid passes. The two components of a renal corpuscle are the **glomerulus** (capillary network) and the **glomerular (Bowman's) capsule,** a double-walled epithelial cup that surrounds the glomerular capillaries. Blood plasma is filtered in the glomerular capsule, and then the filtered fluid passes into the renal tubule, which has three main sections. In the order that fluid passes through them, the renal tubule consists of a (1) **proximal convoluted tubule,** (2) **loop of Henle (nephron loop),** and (3) **distal convoluted tubule.** *Proximal* denotes the part of the tubule attached to the glomerular capsule, and *distal* denotes the part that is further away. *Convoluted* means the tubule is tightly coiled rather than straight. The renal corpuscle and both convoluted tubules lie within the renal

Figure 26.5 **The structure of nephrons and associated blood vessels.** Note that the collecting duct and papillary duct are not part of a nephron. (a) A cortical nephron. (b) A juxtamedullary nephron.

Nephrons are the functional units of the kidneys.

FLOW OF FLUID THROUGH A
CORTICAL NEPHRON

Glomerular (Bowman's) capsule
↓
Proximal convoluted tubule
↓
Descending limb of the
loop of Henle
↓
Ascending limb of the
loop of Henle
↓
Distal convoluted tubule
(drains into collecting duct)
↓

(a) Cortical nephron and vascular supply

● FIGURE 26.5 CONTINUES ▶

cortex; the loop of Henle extends into the renal medulla, makes a hairpin turn, and then returns to the renal cortex.

The distal convoluted tubules of several nephrons empty into a single **collecting duct.** Collecting ducts then unite and converge into several hundred large **papillary ducts,** which drain into the minor calyces. The collecting ducts and papillary ducts extend from the renal cortex through the renal medulla to the renal pelvis. So one kidney has about 1 million nephrons, but a

much smaller number of collecting ducts and even fewer papillary ducts.

In a nephron, the loop of Henle connects the proximal and distal convoluted tubules. The first part of the loop of Henle dips into the renal medulla, where it is called the **descending limb of the loop of Henle** (Figure 26.5). It then makes that hairpin turn and returns to the renal cortex as the **ascending limb of the loop of Henle.** About 80–85% of the nephrons are **cortical**

● FIGURE 26.5 CONTINUED ▶

FLOW OF FLUID THROUGH A JUXTAMEDULLARY NEPHRON

Glomerular (Bowman's) capsule
↓
Proximal convoluted tubule
↓
Descending limb of the loop of Henle
↓
Thin ascending limb of the loop of Henle
↓
Thick ascending limb of the loop of Henle
↓
Distal convoluted tubule (drains into collecting duct)

(b) Juxtamedullary nephron and vascular supply

❓ **What are the basic differences between cortical and juxtamedullary nephrons?**

nephrons. Their renal corpuscles lie in the outer portion of the renal cortex, and they have *short* loops of Henle that lie mainly in the cortex and penetrate only into the outer region of the renal medulla (Figure 26.5a). The short loops of Henle receive their blood supply from peritubular capillaries that arise from efferent arterioles. The other 15–20% of the nephrons are **juxtamedullary nephrons** (*juxta-* = near to). Their renal corpus-

cles lie deep in the cortex, close to the medulla, and they have a *long* loop of Henle that extends into the deepest region of the medulla (Figure 26.5b). Long loops of Henle receive their blood supply from peritubular capillaries and from the vasa recta that arise from efferent arterioles. In addition, the ascending limb of the loop of Henle of juxtamedullary nephrons consists of two portions: a **thin ascending limb** followed by a **thick ascending**

limb (Figure 26.5b). The lumen of the thin ascending limb is the same as in other areas of the renal tubule; it is only the epithelium that is thinner. Nephrons with long loops of Henle enable the kidneys to excrete very dilute or very concentrated urine (described on pages 1042–1046).

Histology of the Nephron and Collecting Duct

A single layer of epithelial cells forms the entire wall of the glomerular capsule, renal tubule, and ducts. However, each part has distinctive histological features that reflect its particular functions. We will discuss them in the order that fluid flows through them: glomerular capsule, renal tubule, and collecting duct.

GLOMERULAR CAPSULE The glomerular (Bowman's) capsule consists of visceral and parietal layers (Figure 26.6a). The visceral layer consists of modified simple squamous epithelial cells called **podocytes** (PŌ-dō-cīts; *podo-* = foot; *-cytes* = cells).

Figure 26.6 Histology of a renal corpuscle.

A renal corpuscle consists of a glomerular (Bowman's) capsule and a glomerulus.

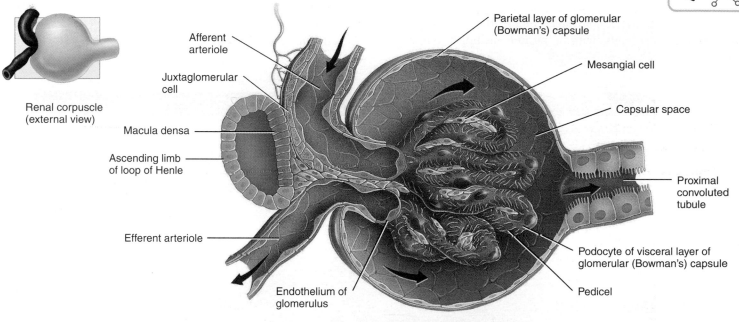

(a) Renal corpuscle (internal view)

(b) Renal corpuscle

❓ Is the photomicrograph in (b) from a section through the renal cortex or renal medulla? How can you tell?

The many footlike projections of these cells (pedicels) wrap around the single layer of endothelial cells of the glomerular capillaries and form the inner wall of the capsule. The parietal layer of the glomerular capsule consists of simple squamous epithelium and forms the outer wall of the capsule. Fluid filtered from the glomerular capillaries enters the **capsular (Bowman's) space,** the space between the two layers of the glomerular capsule. Think of the glomerulus as a fist punched into a limp balloon (the glomerular capsule) until the fist is covered by two layers of balloon (visceral and parietal layers) with a space in between (the capsular space).

RENAL TUBULE AND COLLECTING DUCT Table 26.1 illustrates the histology of the cells that form the renal tubule and collecting duct. In the proximal convoluted tubule, the cells are simple cuboidal epithelial cells with a prominent brush border of microvilli on their apical surface (surface facing the lumen). These microvilli, like those of the small intestine, increase the surface area for reabsorption and secretion. The descending limb of the loop of Henle and the first part of the ascending limb of the loop of Henle (the thin ascending limb) are composed of simple squamous epithelium. (Recall that cortical or short-loop nephrons lack the thin ascending limb.) The thick ascending limb of the loop of Henle is composed of simple cuboidal to low columnar epithelium.

In each nephron, the final part of the ascending limb of the loop of Henle makes contact with the afferent arteriole serving that renal corpuscle (Figure 26.6a). Because the columnar tubule cells in this region are crowded together, they are known as the **macula densa** (*macula* = spot; *densa* = dense). Alongside the macula densa, the wall of the afferent arteriole (and sometimes the efferent arteriole) contains modified smooth muscle fibers

---

TABLE 26.1

Histological Features of the Renal Tubule and Collecting Duct

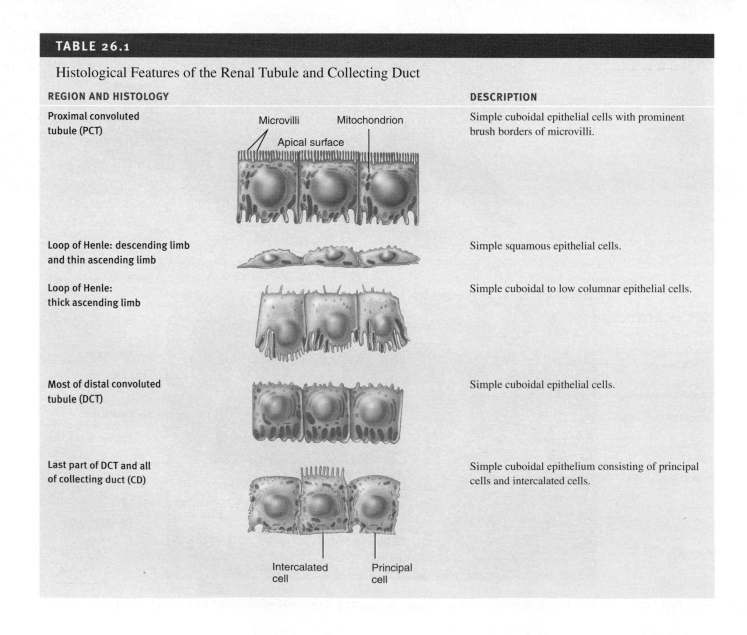

REGION AND HISTOLOGY	DESCRIPTION
Proximal convoluted tubule (PCT)	Simple cuboidal epithelial cells with prominent brush borders of microvilli.
Loop of Henle: descending limb and thin ascending limb	Simple squamous epithelial cells.
Loop of Henle: thick ascending limb	Simple cuboidal to low columnar epithelial cells.
Most of distal convoluted tubule (DCT)	Simple cuboidal epithelial cells.
Last part of DCT and all of collecting duct (CD)	Simple cuboidal epithelium consisting of principal cells and intercalated cells.

called **juxtaglomerular (JG) cells.** Together with the macula densa, they constitute the **juxtaglomerular apparatus (JGA).** As you will see later, the JGA helps regulate blood pressure within the kidneys. The distal convoluted tubule (DCT) begins a short distance past the macula densa. In the last part of the DCT and continuing into the collecting ducts, two different types of cells are present. Most are **principal cells,** which have receptors for both antidiuretic hormone (ADH) and aldosterone, two hormones that regulate their functions. A smaller number are **intercalated cells,** which play a role in the homeostasis of blood pH. The collecting ducts drain into large papillary ducts, which are lined by simple columnar epithelium.

The number of nephrons is constant from birth. Any increase in kidney size is due solely to the growth of individual nephrons. If nephrons are injured or become diseased, new ones do not form. Signs of kidney dysfunction usually do not become apparent until function declines to less than 25% of normal because the remaining functional nephrons adapt to handle a larger-than-normal load. Surgical removal of one kidney, for example, stimulates hypertrophy (enlargement) of the remaining kidney, which eventually is able to filter blood at 80% of the rate of two normal kidneys.

◉ CHECKPOINT

2. Why are the kidneys said to be retroperitoneal?
3. What are the major parts of a nephron?
4. How do cortical nephrons and juxtamedullary nephrons differ structurally?
5. Where is the juxtaglomerular apparatus (JGA) located, and what is its structure?

OVERVIEW OF RENAL PHYSIOLOGY

◉ OBJECTIVE

• Identify the three basic functions performed by nephrons and collecting ducts, and indicate where each occurs.

To produce urine, nephrons and collecting ducts perform three basic processes—glomerular filtration, tubular reabsorption, and tubular secretion (Figure 26.7):

❶ *Glomerular filtration.* In the first step of urine production, water and most solutes in blood plasma move across the wall of glomerular capillaries into the glomerular capsule and then into the renal tubule.

❷ *Tubular reabsorption.* As filtered fluid flows along the renal tubule and through the collecting duct, tubule cells reabsorb about 99% of the filtered water and many useful solutes. The water and solutes return to the blood as it flows through the peritubular capillaries and vasa recta. Note that the term *reabsorption* refers to the return of substances to the bloodstream. The term *absorption,* by contrast, means entry of new substances into the body, as occurs in the gastrointestinal tract.

❸ *Tubular secretion.* As fluid flows along the renal tubule and through the collecting duct, the tubule and duct cells secrete other materials, such as wastes, drugs, and excess ions, into the fluid. Notice that tubular secretion *removes* a substance from the blood. In other instances of secretion—for instance, secretion of hormones—cells release substances into interstitial fluid and blood.

Figure 26.7 Relation of a nephron's structure to its three basic functions: glomerular filtration, tubular reabsorption, and tubular secretion. Excreted substances remain in the urine and subsequently leave the body. For any substance S, excretion rate of S = filtration rate of S − reabsorption rate of S + secretion rate of S.

❻ ⚷ Glomerular filtration occurs in the renal corpuscle; tubular reabsorption and tubular secretion occur all along the renal tubule and collecting duct.

❓ When cells of the renal tubules secrete the drug penicillin, is the drug being added to or removed from the bloodstream?

Solutes in the fluid that drains into the renal pelvis remain in the urine and are excreted. The rate of urinary excretion of any solute is equal to its rate of glomerular filtration, plus its rate of secretion, minus its rate of reabsorption.

By filtering, reabsorbing, and secreting, nephrons help maintain homeostasis of the blood's volume and composition. The situation is somewhat analogous to a recycling center: Garbage trucks dump refuse into an input hopper, where the smaller refuse passes onto a conveyor belt (glomerular filtration of plasma). As the conveyor belt carries the garbage along, workers remove useful items, such as aluminum cans, plastics, and glass containers (reabsorption). Other workers place additional garbage left at the center and larger items onto the conveyor belt (secretion). At the end of the belt, all remaining garbage falls into a truck for transport to the landfill (excretion of wastes in urine).

GLOMERULAR FILTRATION

⊙ OBJECTIVES

- Describe the filtration membrane.
- Discuss the pressures that promote and oppose glomerular filtration.

The fluid that enters the capsular space is called the **glomerular filtrate.** The fraction of blood plasma in the afferent arterioles of the kidneys that becomes glomerular filtrate is the **filtration fraction.** Although a filtration fraction of 0.16–0.20 (16–20%) is typical, the value varies considerably in both health and disease. On average, the daily volume of glomerular filtrate in adults is 150 liters in females and 180 liters in males. More than 99% of the glomerular filtrate returns to the bloodstream via tubular reabsorption, so only 1–2 liters (about 1–2 qt) are excreted as urine.

The Filtration Membrane

Together, the endothelial cells of glomerular capillaries and the podocytes, which completely encircle the capillaries, form a leaky barrier known as the **filtration membrane.** This sandwichlike assembly permits filtration of water and small solutes but prevents filtration of most plasma proteins, blood cells, and platelets. Substances filtered from the blood cross three barriers—a glomerular endothelial cell, the basal lamina, and a filtration slit formed by a podocyte (Figure 26.8):

❶ Glomerular endothelial cells are quite leaky because they have large **fenestrations** (pores) that measure 0.07–0.1 μm in diameter. This size permits all solutes in blood plasma to exit glomerular capillaries but prevents filtration of blood cells and platelets. Located among the glomerular capillaries and in the cleft between afferent and efferent arterioles are **mesangial cells** (*mes-* = in the middle; *-angi* = blood vessel) (see Figure 26.6a). These contractile cells help regulate glomerular filtration.

❷ The **basal lamina,** a layer of acellular material between the endothelium and the podocytes, consists of minute collagen fibers and proteoglycans in a glycoprotein matrix; it prevents filtration of larger plasma proteins.

❸ Extending from each podocyte are thousands of footlike processes termed **pedicels** (PED-i-sels = little feet) that wrap around glomerular capillaries. The spaces between pedicels are the **filtration slits.** A thin membrane, the **slit membrane,** extends across each filtration slit; it permits the passage of molecules having a diameter smaller than 0.006–0.007 μm, including water, glucose, vitamins, amino acids, very small plasma proteins, ammonia, urea, and ions. Less than 1% of albumin, the most plentiful plasma protein, passes the slit membrane because, with a diameter of 0.007 μm, it is slightly too big to get through.

The principle of *filtration*—the use of pressure to force fluids and solutes through a membrane—is the same in glomerular capillaries as in capillaries elsewhere in the body (see Starling's law of the capillaries, page 770). However, the volume of fluid filtered by the renal corpuscle is much larger than in other capillaries of the body for three reasons:

1. Glomerular capillaries present a large surface area for filtration because they are long and extensive. The mesangial cells regulate how much of this surface area is available for filtration. When mesangial cells are relaxed, surface area is maximal, and glomerular filtration is very high. Contraction of mesangial cells reduces the available surface area, and glomerular filtration decreases.

2. The filtration membrane is thin and porous. Despite having several layers, the thickness of the filtration membrane is only 0.1 μm. Glomerular capillaries also are about 50 times leakier than capillaries in most other tissues, mainly because of their large fenestrations.

3. Glomerular capillary blood pressure is high. Because the efferent arteriole is smaller in diameter than the afferent arteriole, resistance to the outflow of blood from the glomerulus is high. As a result, blood pressure in glomerular capillaries is considerably higher than in capillaries elsewhere in the body.

Net Filtration Pressure

Glomerular filtration depends on three main pressures. One pressure *promotes* filtration and two pressures *oppose* filtration (Figure 26.9 on page 1032).

❶ **Glomerular blood hydrostatic pressure (GBHP)** is the blood pressure in glomerular capillaries. Generally, GBHP is about 55 mmHg. It promotes filtration by forcing water and solutes in blood plasma through the filtration membrane.

❷ **Capsular hydrostatic pressure (CHP)** is the hydrostatic pressure exerted against the filtration membrane by fluid already in the capsular space and renal tubule. CHP opposes filtration and represents a "back pressure" of about 15 mmHg.

Figure 26.8 **The filtration membrane.** The size of the endothelial fenestrations and filtration slits in (a) have been exaggerated for emphasis.

During glomerular filtration, water and solutes pass from blood plasma into the capsular space.

1 Fenestration (pore) of glomerular endothelial cell: prevents filtration of blood cells but allows all components of blood plasma to pass through

2 Basal lamina of glomerulus: prevents filtration of larger proteins

3 Slit membrane between pedicels: prevents filtration of medium-sized proteins

Filtration slit
Pedicel

Podocyte of visceral layer of glomerular (Bowman's) capsule

(a) Details of filtration membrane

Pedicel of podocyte   Filtration slit

Basal lamina

Lumen of glomerulus

Fenestration (pore) of glomerular endothelial cell

TEM 78,000x

(b) Filtration membrane

? Which part of the filtration membrane prevents red blood cells from entering the capsular space?

3 **Blood colloid osmotic pressure (BCOP),** which is due to the presence of proteins such as albumin, globulins, and fibrinogen in blood plasma, also opposes filtration. The average BCOP in glomerular capillaries is 30 mmHg.

Net filtration pressure (NFP), the total pressure that promotes filtration, is determined as follows:

Net Filtration Pressure (NFP) = GBHP − CHP − BCOP

By substituting the values just given, normal NFP may be calculated:

$$\text{NFP} = 55 \text{ mmHg} - 15 \text{ mmHg} - 30 \text{ mmHg}$$
$$= 10 \text{ mmHg}$$

Thus, a pressure of only 10 mmHg causes a normal amount of blood plasma (minus plasma proteins) to filter from the glomerulus into the capsular space.

●CLINICAL CONNECTION | **Loss of Plasma Proteins in Urine Causes Edema**

In some kidney diseases, glomerular capillaries are damaged and become so permeable that plasma proteins enter glomerular filtrate. As a result, the filtrate exerts a colloid osmotic pressure that draws water out of the blood. In this situation, the NFP increases, which means more fluid is filtered. At the same time, blood colloid osmotic pressure decreases because plasma proteins are being lost in the urine. Because more fluid filters out of blood capillaries into tissues throughout the body than returns via reabsorption, blood volume decreases and interstitial fluid volume increases. Thus, loss of plasma proteins in urine causes *edema,* an abnormally high volume of interstitial fluid. •

Glomerular Filtration Rate

The amount of filtrate formed in all the renal corpuscles of both kidneys each minute is the **glomerular filtration rate (GFR).** In adults, the GFR averages 125 mL/min in males and 105 mL/min in females. Homeostasis of body fluids requires that the kidneys maintain a relatively constant GFR. If the GFR is too high, needed substances may pass so quickly through the renal tubules that some are not reabsorbed and are lost in the urine. If the GFR is too low, nearly all the filtrate may be reabsorbed and certain waste products may not be adequately excreted.

GFR is directly related to the pressures that determine net filtration pressure; any change in net filtration pressure will affect GFR. Severe blood loss, for example, reduces mean arterial blood pressure and decreases the glomerular blood hydrostatic pressure. Filtration ceases if glomerular blood hydrostatic

Figure 26.9 The pressures that drive glomerular filtration. Taken together, these pressures determine net filtration pressure (NFP).

 Glomerular blood hydrostatic pressure promotes filtration, whereas capsular hydrostatic pressure and blood colloid osmotic pressure oppose filtration.

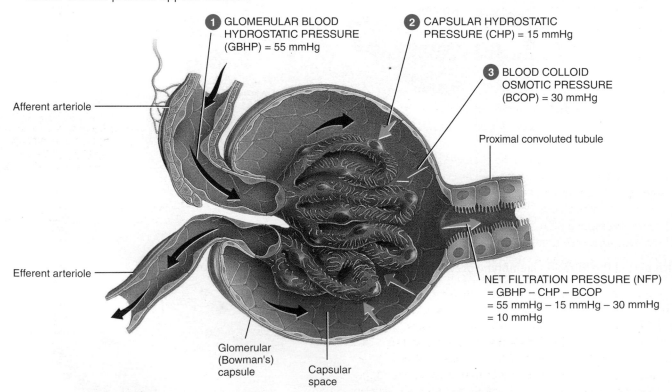

1 GLOMERULAR BLOOD HYDROSTATIC PRESSURE (GBHP) = 55 mmHg

2 CAPSULAR HYDROSTATIC PRESSURE (CHP) = 15 mmHg

3 BLOOD COLLOID OSMOTIC PRESSURE (BCOP) = 30 mmHg

Afferent arteriole

Proximal convoluted tubule

Efferent arteriole

NET FILTRATION PRESSURE (NFP)
= GBHP − CHP − BCOP
= 55 mmHg − 15 mmHg − 30 mmHg
= 10 mmHg

Glomerular (Bowman's) capsule

Capsular space

? Suppose a tumor is pressing on and obstructing the right ureter. What effect might this have on CHP and thus on NFP in the right kidney? Would the left kidney also be affected?

pressure drops to 45 mmHg because the opposing pressures add up to 45 mmHg. Amazingly, when systemic blood pressure rises above normal, net filtration pressure and GFR increase very little. GFR is nearly constant when the mean arterial blood pressure is anywhere between 80 and 180 mmHg.

The mechanisms that regulate glomerular filtration rate operate in two main ways: (1) by adjusting blood flow into and out of the glomerulus and (2) by altering the glomerular capillary surface area available for filtration. GFR increases when blood flow into the glomerular capillaries increases. Coordinated control of the diameter of both afferent and efferent arterioles regulates glomerular blood flow. Constriction of the afferent arteriole decreases blood flow into the glomerulus; dilation of the afferent arteriole increases it. Three mechanisms control GFR: renal autoregulation, neural regulation, and hormonal regulation.

Renal Autoregulation of GFR

The kidneys themselves help maintain a constant renal blood flow and GFR despite normal, everyday changes in blood pressure, like those that occur during exercise. This capability is called **renal autoregulation** and consists of two mechanisms— the myogenic mechanism and tubuloglomerular feedback. Working together, they can maintain nearly constant GFR over a wide range of systemic blood pressures.

The **myogenic mechanism** (*myo-* = muscle; *-genic* = producing) occurs when stretching triggers contraction of smooth muscle cells in the walls of afferent arterioles. As blood pressure rises, GFR also rises because renal blood flow increases. However, the elevated blood pressure stretches the walls of the afferent arterioles. In response, smooth muscle fibers in the wall of the afferent arteriole contract, which narrows the arteriole's lumen. As a result, renal blood flow decreases, thus reducing GFR to its previous level. Conversely, when arterial blood pressure drops, the smooth muscle cells are stretched less and thus relax. The afferent arterioles dilate, renal blood flow increases, and GFR increases. The myogenic mechanism normalizes renal blood flow and GFR within seconds after a change in blood pressure.

The second contributor to renal autoregulation, **tubuloglomerular feedback,** is so named because part of the renal tubules—the macula densa—provides feedback to the glomerulus (Figure 26.10). When GFR is above normal due to elevated systemic blood pressure, filtered fluid flows more rapidly along the renal tubules. As a result, the proximal convoluted tubule and loop of Henle have less time to reabsorb Na$^+$, Cl$^-$, and water. Macula densa cells are thought to detect the increased delivery of Na$^+$, Cl$^-$, and water and to inhibit release of nitric oxide (NO) from cells in the juxtaglomerular apparatus (JGA). Because NO causes vasodilation, afferent arterioles constrict when the level of NO declines. As a result, less blood flows into the glomerular capillaries, and GFR decreases. When blood pressure falls, causing GFR to be lower than normal, the opposite sequence of events occurs, although to a lesser degree. Tubuloglomerular feedback operates more slowly than the myogenic mechanism.

Neural Regulation of GFR

Like most blood vessels of the body, those of the kidneys are supplied by sympathetic ANS fibers that release norepinephrine. (NA) Norepinephrine causes vasoconstriction through the activation of α_1 receptors, which are particularly plentiful in the smooth muscle fibers of afferent arterioles. At rest, sympathetic stimulation is moderately low, the afferent and efferent arterioles are dilated, and renal autoregulation of GFR prevails. With moderate sympathetic

Figure 26.10 Tubuloglomerular feedback.

 Macula densa cells of the juxtaglomerular apparatus provide negative feedback regulation of glomerular filtration rate.

Some stimulus disrupts homeostasis by

Increasing

Glomerular filtration rate (GFR)

Receptors

Macula densa cells of JGA detect increased delivery of Na$^+$, Cl$^-$, and water

Input

Control center

Juxtaglomerular apparatus

Return to homeostasis when response brings GFR back to normal

Output — Decreased secretion of nitric oxide

Effectors

Afferent arteriole constricts, which decreases blood flow through glomerulus

Decrease in GFR

? **Why is this process termed autoregulation?**

the cells in a passive process known as **paracellular reabsorption** (*para-* = beside). In some parts of the renal tubule, the paracellular route is thought to account for up to 50% of the reabsorption of certain ions and the water that accompanies them via osmosis. In **transcellular reabsorption** (*trans-* = across), a substance passes from the fluid in the tubular lumen through the apical membrane of a tubule cell, across the cytosol, and out into interstitial fluid through the basolateral membrane.

Transport Mechanisms

When renal cells transport solutes out of or into tubular fluid, they move specific substances in one direction only. Not surprisingly, different types of transport proteins are present in the apical and basolateral membranes. The tight junctions form a barrier that prevents mixing of proteins in the apical and basolateral membrane compartments. Reabsorption of Na^+ by the renal tubules is especially important because of the large number of sodium ions that pass through the glomerular filters.

Cells lining the renal tubules, like other cells throughout the body, have a low concentration of Na^+ in their cytosol due to the activity of sodium–potassium pumps (Na^+/K^+ ATPases). These pumps are located in the basolateral membranes and eject Na^+ from the renal tubule cells (Figure 26.11). The absence of sodium–potassium pumps in the apical membrane ensures that reabsorption of Na^+ is a one-way process. Most sodium ions that cross the apical membrane will be pumped into interstitial fluid at the base and sides of the cell. The amount of ATP used by sodium–potassium pumps in the renal tubules is about 6% of the total ATP consumption of the body at rest. This may not sound like much, but it is about the same amount of energy used by the diaphragm as it contracts during quiet breathing.

As we noted in Chapter 3, transport of materials across membranes may be either active or passive. Recall that in **primary active transport** the energy derived from hydrolysis of ATP is used to "pump" a substance across a membrane; the sodium–potassium pump is one such pump. In **secondary active transport** the energy stored in an ion's electrochemical gradient, rather than hydrolysis of ATP, drives another substance across a membrane. Secondary active transport couples the movement of an ion down its electrochemical gradient to the "uphill" movement of a second substance against its electrochemical gradient. *Symporters* are membrane proteins that move two or more substances in the same direction across a membrane. *Antiporters* move two or more substances in opposite directions across a membrane. Each type of transporter has an upper limit on how fast it can work, just as an escalator has a limit on how many people it can carry from one level to another in a given period. This limit, called the **transport maximum (T_m)**, is measured in mg/min.

Solute reabsorption drives water reabsorption because all water reabsorption occurs via osmosis. About 90% of the reabsorption of water filtered by the kidneys occurs along with the reabsorption of solutes such as Na^+, Cl^-, and glucose. Water reabsorbed with solutes in tubular fluid is termed **obligatory water reabsorption** (ob-LIG-a-tor'-ē) because the water is "obliged" to follow the solutes when they are reabsorbed. This type of water reabsorption occurs in the proximal convoluted tubule and the descending limb of the loop of Henle because these segments of the nephron are always permeable to water. Reabsorption of the final 10% of the water, a total of 10–20 liters per day, is termed **facultative water reabsorption** (FAK-ul-tā'-tiv). The word *facultative* means "capable of adapting to a need." Facultative water reabsorption is regulated by antidiuretic hormone and occurs mainly in the collecting ducts.

● CLINICAL CONNECTION | Glucosuria

When the blood concentration of glucose is above 200 mg/mL, the renal symporters cannot work fast enough to reabsorb all the glucose that enters the glomerular filtrate. As a result, some glucose remains in the urine, a condition called **glucosuria** (gloo'-kō-SOO-rē-a). The most common cause of glucosuria is diabetes mellitus, in which the blood glucose level may rise far above normal because insulin activity is deficient. Rare genetic mutations in the renal Na^+–glucose symporter greatly reduce its T_m and cause glucosuria. In these cases, glucose appears in the urine even though the blood glucose level is normal. Excessive glucose in the glomerular filtrate inhibits water reabsorption by kidney tubules. This leads to increased urinary output (polyuria), decreased blood volume, and dehydration. ●

Now that we have discussed the principles of renal transport, we will follow the filtered fluid from the proximal convoluted tubule, into the loop of Henle, on to the distal convoluted tubule, and through the collecting ducts. In each segment, we will examine where and how specific substances are reabsorbed and secreted. The filtered fluid becomes *tubular fluid* once it enters the proximal convoluted tubule. The composition of tubular fluid changes as it flows along the nephron tubule and through the collecting duct due to reabsorption and secretion. The fluid that drains from papillary ducts into the renal pelvis is *urine*.

Reabsorption and Secretion in the Proximal Convoluted Tubule

The largest amount of solute and water reabsorption from filtered fluid occurs in the proximal convoluted tubules, which reabsorb 65% of the filtered water, Na^+, and K^+; 100% of most filtered organic solutes such as glucose and amino acids; 50% of the filtered Cl^-; 80–90% of the filtered HCO_3^-; 50% of the filtered urea; and a variable amount of the filtered Ca^{2+}, Mg^{2+}, and HPO_4^{2-} (phosphate). In addition, proximal convoluted tubules secrete a variable amount of H^+ ions, ammonium ions (NH_4^+), and urea.

Most solute reabsorption in the proximal convoluted tubule (PCT) involves Na^+. Na^+ transport occurs via symport and antiport mechanisms in the proximal convoluted tubule. Normally, filtered glucose, amino acids, lactic acid, water-soluble vitamins, and other nutrients are not lost in the urine. Rather, they are completely reabsorbed in the first half of the proximal convoluted tubule by several types of **Na^+ symporters** located in the apical membrane. Figure 26.12 depicts the operation of one such symporter, the **Na^+–glucose symporter** in the apical membrane of a cell in the PCT. Two Na^+ and a molecule of glucose attach

Figure 26.12 Reabsorption of glucose by Na⁺-glucose symporters in cells of the proximal convoluted tubule (PCT).

Normally, all filtered glucose is reabsorbed in the PCT.

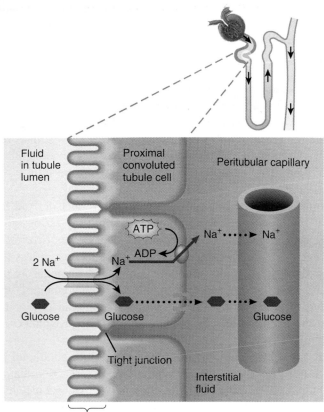

Key:

- Na⁺-glucose symporter
- Glucose facilitated diffusion transporter
- ····▶ Diffusion
- Sodium-potassium pump

? How does filtered glucose enter and leave a PCT cell?

to the symporter protein, which carries them from the tubular fluid into the tubule cell. The glucose molecules then exit the basolateral membrane via facilitated diffusion and they diffuse into peritubular capillaries. Other Na⁺ symporters in the PCT reclaim filtered HPO_4^{2-} (phosphate) and SO_4^{2-} (sulfate) ions, all amino acids, and lactic acid in a similar way.

In another secondary active transport process, the **Na⁺/H⁺ antiporters** carry filtered Na⁺ down its concentration gradient into a PCT cell as H⁺ is moved from the cytosol into the lumen (Figure 26.13a), causing Na⁺ to be reabsorbed into blood and H⁺ to be secreted into tubular fluid. PCT cells produce the H⁺ needed to keep the antiporters running in the following way. Carbon dioxide (CO_2) diffuses from peritubular blood or tubular fluid or is produced by metabolic reactions within the cells. As also occurs in red blood cells (see Figure 23.23 on page 904),

Figure 26.13 Actions of Na⁺/H⁺ antiporters in proximal convoluted tubule cells. (a) Reabsorption of sodium ions (Na⁺) and secretion of hydrogen ions (H⁺) via secondary active transport through the apical membrane; (b) reabsorption of bicarbonate ions (HCO_3^-) via facilitated diffusion through the basolateral membrane. CO_2 = carbon dioxide; H_2CO_3 = carbonic acid; CA = carbonic anhydrase.

Na⁺/H⁺ antiporters promote transcellular reabsorption of Na⁺ and secretion of H⁺.

(a) Na⁺ reabsorption and H⁺ secretion

(b) HCO_3^- reabsorption

Key:

- Na⁺/H⁺ antiporter
- HCO_3^- facilitated diffusion transporter
- ····▶ Diffusion
- Sodium-potassium pump

? Which step in Na⁺ movement in part (a) is promoted by the electrochemical gradient?

the enzyme *carbonic anhydrase (CA)* catalyzes the reaction of CO_2 with water (H_2O) to form carbonic acid (H_2CO_3), which then dissociates into H^+ and HCO_3^-:

$$CO_2 + H_2O \xrightarrow{\text{Carbonic anhydrase}} H_2CO_3 \longrightarrow H^+ + HCO_3^-$$

Most of the HCO_3^- in filtered fluid is reabsorbed in proximal convoluted tubules, thereby safeguarding the body's supply of an important buffer (Figure 26.13b). After H^+ is secreted into the fluid within the lumen of the proximal convoluted tubule, it reacts with filtered HCO_3^- to form H_2CO_3, which readily dissociates into CO_2 and H_2O. Carbon dioxide then diffuses into the tubule cells and joins with H_2O to form H_2CO_3, which dissociates into H^+ and HCO_3^-. As the level of HCO_3^- rises in the cytosol, it exits via facilitated diffusion transporters in the basolateral membrane and diffuses into the blood with Na^+. Thus, for every H^+ secreted into the tubular fluid of the proximal convoluted tubule, one HCO_3^- and one Na^+ are reabsorbed.

Solute reabsorption in proximal convoluted tubules promotes osmosis of water. Each reabsorbed solute increases the osmolarity, first inside the tubule cell, then in interstitial fluid, and finally in the blood. Water thus moves rapidly from the tubular fluid, via both the paracellular and transcellular routes, into the peritubular capillaries and restores osmotic balance (Figure 26.14). In other words, reabsorption of the solutes creates an osmotic gradient that promotes the reabsorption of water via

Figure 26.14 Passive reabsorption of Cl^-, K^+, Ca^{2+}, Mg^{2+}, urea, and water in the second half of the proximal convoluted tubule.

Electrochemical gradients promote passive reabsorption of solutes via both paracellular and transcellular routes.

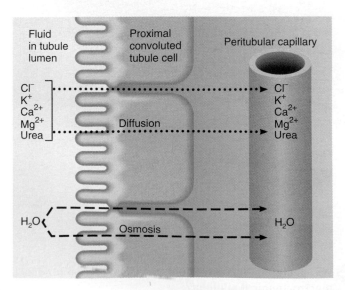

By what mechanism is water reabsorbed from tubular fluid?

osmosis. Cells lining the proximal convoluted tubule and the descending limb of the loop of Henle are especially permeable to water because they have many molecules of *aquaporin-1*. This integral protein in the plasma membrane is a water channel that greatly increases the rate of water movement across the apical and basolateral membranes.

As water leaves the tubular fluid, the concentrations of the remaining filtered solutes increase. In the second half of the PCT, electrochemical gradients for Cl^-, K^+, Ca^{2+}, Mg^{2+}, and urea promote their passive diffusion into peritubular capillaries via both paracellular and transcellular routes. Among these ions, Cl^- is present in the highest concentration. Diffusion of negatively charged Cl^- into interstitial fluid via the paracellular route makes the interstitial fluid electrically more negative than the tubular fluid. This negativity promotes passive paracellular reabsorption of cations, such as K^+, Ca^{2+}, and Mg^{2+}.

Ammonia (NH_3) is a poisonous waste product derived from the deamination (removal of an amino group) of various amino acids, a reaction that occurs mainly in hepatocytes (liver cells). Hepatocytes convert most of this ammonia to urea, a less-toxic compound. Although tiny amounts of urea and ammonia are present in sweat, most excretion of these nitrogen-containing waste products occurs via the urine. Urea and ammonia in blood are both filtered at the glomerulus and secreted by proximal convoluted tubule cells into the tubular fluid.

Proximal convoluted tubule cells can produce additional NH_3 by deaminating the amino acid glutamine in a reaction that also generates HCO_3^-. The NH_3 quickly binds H^+ to become an ammonium ion (NH_4^+), which can substitute for H^+ aboard Na^+/H^+ antiporters in the apical membrane and be secreted into the tubular fluid. The HCO_3^- generated in this reaction moves through the basolateral membrane and then diffuses into the bloodstream, providing additional buffers in blood plasma.

Reabsorption in the Loop of Henle

Because all of the proximal convoluted tubules reabsorb about 65% of the filtered water (about 80 mL/min), fluid enters the next part of the nephron, the loop of Henle, at a rate of 40–45 mL/min. The chemical composition of the tubular fluid now is quite different from that of glomerular filtrate because glucose, amino acids, and other nutrients are no longer present. The osmolarity of the tubular fluid is still close to the osmolarity of blood, however, because reabsorption of water by osmosis keeps pace with reabsorption of solutes all along the proximal convoluted tubule.

The loop of Henle reabsorbs about 15% of the filtered water; 20–30% of the filtered Na^+ and K^+; 35% of the filtered Cl^-; 10–20% of the filtered HCO_3^-; and a variable amount of the filtered Ca^{2+} and Mg^{2+}. Here, for the first time, reabsorption of water via osmosis is *not* automatically coupled to reabsorption of filtered solutes because part of the loop of Henle is relatively impermeable to water. The loop of Henle thus sets the stage for *independent* regulation of both the *volume* and *osmolarity* of body fluids.

The apical membranes of cells in the thick ascending limb of the loop of Henle have **Na⁺–K⁺–2Cl⁻ symporters** that simultaneously reclaim one Na⁺, one K⁺, and two Cl⁻ from the fluid in the tubular lumen (Figure 26.15). Na⁺ that is actively transported into interstitial fluid at the base and sides of the cell diffuses into the vasa recta. Cl⁻ moves through leakage channels in the basolateral membrane into interstitial fluid and then into the vasa recta. Because many K⁺ leakage channels are present in the apical membrane, most K⁺ brought in by the symporters moves down its concentration gradient back into the tubular fluid. Thus, the main effect of the Na⁺–K⁺–2Cl⁻ symporters is reabsorption of Na⁺ and Cl⁻.

The movement of positively charged K⁺ into the tubular fluid through the apical membrane channels leaves the interstitial fluid and blood with more negative charges relative to fluid in the ascending limb of the loop of Henle. This relative negativity promotes reabsorption of cations—Na⁺, K⁺, Ca²⁺, and Mg²⁺—via the paracellular route.

Although about 15% of the filtered water is reabsorbed in the *descending* limb of the loop of Henle, little or no water is reabsorbed in the *ascending* limb. In this segment of the tubule, the apical membranes are virtually impermeable to water. Because ions but not water molecules are reabsorbed, the osmolarity of the tubular fluid decreases progressively as fluid flows toward the end of the ascending limb.

Reabsorption in the Early Distal Convoluted Tubule

Fluid enters the distal convoluted tabules at a rate of about 25 mL/min because 80% of the filtered water has now been reabsorbed. The early or initial part of the distal convoluted tubule (DCT) reabsorbs about 10–15% of the filtered water; 5% of the filtered Na⁺; and 5% of the filtered Cl⁻. Reabsorption of Na⁺ and Cl⁻ occurs by means of **Na⁺–Cl⁻ symporters** in the apical membranes. Sodium–potassium pumps and Cl⁻ leakage channels in the basolateral membranes then permit reabsorption of Na⁺ and Cl⁻ into the peritubular capillaries. The early DCT also is a major site where parathyroid hormone (PTH) stimulates reabsorption of Ca²⁺. The amount of Ca²⁺ reabsorption in the early DCT varies depending on the body's needs.

Reabsorption and Secretion in the Late Distal Convoluted Tubule and Collecting Duct

By the time fluid reaches the end of the distal convoluted tubule, 90–95% of the filtered solutes and water have returned to the bloodstream. Recall that two different types of cells—principal cells and intercalated cells—are present at the late or terminal part of the distal convoluted tubule and throughout the collecting duct. The principal cells reabsorb Na⁺ and secrete K⁺; the intercalated cells reabsorb K⁺ and HCO₃⁻ and secrete H⁺. In the late distal convoluted tubules and collecting ducts, the amount of water and solute reabsorption and the amount of solute secretion vary depending on the body's needs.

Figure 26.15 Na⁺–K⁺–2Cl⁻ symporter in the thick ascending limb of the loop of Henle.

 Cells in the thick ascending limb have symporters that simultaneously reabsorb one Na⁺, one K⁺, and two Cl⁻.

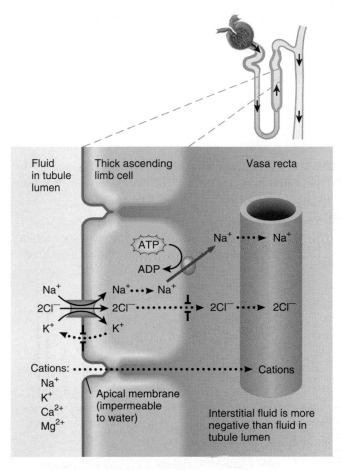

Key:

	Na⁺—K⁺—2Cl⁻ symporter
	Leakage channels
	Sodium-potassium pump
••••►	Diffusion

? Why is this process considered secondary active transport? Does water reabsorption accompany ion reabsorption in this region of the nephron?

In contrast to earlier segments of the nephron, Na^+ passes through the apical membrane of principal cells via Na^+ leakage channels rather than by means of symporters or antiporters (Figure 26.16). The concentration of Na^+ in the cytosol remains low, as usual, because the sodium–potassium pumps actively transport Na^+ across the basolateral membranes. Then Na^+ passively diffuses into the peritubular capillaries from the interstitial spaces around the tubule cells.

Normally, transcellular and paracellular reabsorption in the proximal convoluted tubule and loop of Henle return most filtered K^+ to the bloodstream. To adjust for varying dietary intake of potassium and to maintain a stable level of K^+ in body fluids, principal cells secrete a variable amount of K^+ (Figure 26.16). Because the basolateral sodium–potassium pumps continually

bring K^+ into principal cells, the intracellular concentration of K^+ remains high. K^+ leakage channels are present in both the apical and basolateral membranes. Thus, some K^+ diffuses down its concentration gradient into the tubular fluid, where the K^+ concentration is very low. This secretion mechanism is the main source of K^+ excreted in the urine.

Hormonal Regulation of Tubular Reabsorption and Tubular Secretion

Five hormones affect the extent of Na^+, Cl^-, Ca^{2+}, and water reabsorption as well as K^+ secretion by the renal tubules. These hormones include angiotensin II, aldosterone, antidiuretic hormone, atrial natriuretic peptide, and parathyroid hormone.

Renin–Angiotensin–Aldosterone System

When blood volume and blood pressure decrease, the walls of the afferent arterioles are stretched less, and the juxtaglomerular cells secrete the enzyme **renin** into the blood. Sympathetic stimulation also directly stimulates release of renin from juxtaglomerular cells. Renin clips off a 10-amino-acid peptide called angiotensin I from angiotensinogen, which is synthesized by hepatocytes. By clipping off two more amino acids, *angiotensin converting enzyme (ACE)* converts angiotensin I to **angiotensin II,** which is the active form of the hormone.

Angiotensin II affects renal physiology in three main ways:

1. It decreases the glomerular filtration rate by causing vasoconstriction of the afferent arterioles.

2. It enhances reabsorption of Na^+, Cl^-, and water in the proximal convoluted tubule by stimulating the activity of Na^+/H^+ antiporters.

3. It stimulates the adrenal cortex to release **aldosterone,** a hormone that in turn stimulates the principal cells in the collecting ducts to reabsorb more Na^+ and Cl^- and secrete more K^+. The osmotic consequence of reabsorbing more Na^+ and Cl^- is excreting less water, which increases blood volume.

Antidiuretic Hormone

Antidiuretic hormone (ADH or **vasopressin)** is released by the posterior pituitary. It regulates facultative water reabsorption by increasing the water permeability of principal cells in the last part of the distal convoluted tubule and throughout the collecting duct. In the absence of ADH, the apical membranes of principal cells have a very low permeability to water. Within principal cells are tiny vesicles containing many copies of a water channel protein known as **aquaporin-2.** * ADH stimulates insertion of the aquaporin-2–containing vesicles into the apical membranes via exocytosis. As a result, the water permeability of the principal cell's apical membrane increases, and water molecules move more rapidly from the tubular fluid into the cells. Because the

Figure 26.16 Reabsorption of Na^+ and secretion of K^+ by principal cells in the last part of the distal convoluted tubule and in the collecting duct.

 In the apical membrane of principal cells, Na^+ leakage channels allow entry of Na^+ while K^+ leakage channels allow exit of K^+ into the tubular fluid.

Key:

• • • • ▶ Diffusion

⊣⊢ Leakage channels

Sodium-potassium pump

❓ **Which hormone stimulates reabsorption and secretion by principal cells, and how does this hormone exert its effect?**

*ADH does not govern the previously mentioned water channel (aquaporin-1).

basolateral membranes are always relatively permeable to water, water molecules then move rapidly into the blood. The kidneys can produce as little as 400–500 mL of very concentrated urine each day when ADH concentration is maximal, for instance during severe dehydration. When ADH level declines, the aquaporin-2 channels are removed from the apical membrane via endocytosis. The kidneys produce a large volume of dilute urine when ADH level is low.

A negative feedback system involving ADH regulates facultative water reabsorption (Figure 26.17). When the osmolarity or osmotic pressure of plasma and interstitial fluid increases—that is, when water concentration decreases—by as little as 1%, osmoreceptors in the hypothalamus detect the change. Their nerve impulses stimulate secretion of more ADH into the blood, and the principal cells become more permeable to water. As facultative water reabsorption increases, plasma osmolarity decreases to normal. A second powerful stimulus for ADH secretion is a decrease in blood volume, as occurs in hemorrhaging or severe dehydration. In the pathological absence of ADH activity, a condition known as *diabetes insipidus,* a person may excrete up to 20 liters of very dilute urine daily.

Atrial Natriuretic Peptide

A large increase in blood volume promotes release of atrial natriuretic peptide (ANP) from the heart. Although the importance of ANP in normal regulation of tubular function is unclear, it can inhibit reabsorption of Na^+ and water in the proximal convoluted tubule and collecting duct. ANP also suppresses the secretion of aldosterone and ADH. These effects increase the excretion of Na^+ in urine (natriuresis) and increase urine output (diuresis), which decreases blood volume and blood pressure.

Parathyroid Hormone

A lower-than-normal level of Ca^{2+} in the blood stimulates the parathyroid glands to release parathyroid hormone (PTH). PTH in turn stimulates cells in the early distal convoluted tubules to reabsorb more Ca^{2+} into the blood. PTH also inhibits HPO_4^{2-} (phosphate) reabsorption in proximal convoluted tubules, thereby promoting phosphate excretion.

Table 26.4 summarizes hormonal regulation of tubular reabsorption and tubular secretion.

☉ CHECKPOINT

11. Diagram the reabsorption of substances via the transcellular and paracellular routes. Label the apical membrane and the basolateral membrane. Where are the sodium–potassium pumps located?
12. Describe two mechanisms in the PCT, one in the LOH, one in the DCT, and one in the collecting duct for reabsorption of Na^+. What other solutes are reabsorbed or secreted with Na^+ in each mechanism?
13. How do intercalated cells secrete hydrogen ions?
14. Graph the percentages of filtered water and filtered Na^+ that are reabsorbed in the PCT, LOH, DCT, and collecting duct. Indicate which hormones, if any, regulate reabsorption in each segment.

Figure 26.17 Negative feedback regulation of facultative water reabsorption by ADH.

🗝 Most water reabsorption (90%) is obligatory; 10% is facultative.

Some stimulus disrupts homeostasis by

Increasing

Osmolarity of plasma and interstitial fluid

Receptors

Osmoreceptors in hypothalamus

Input   Nerve impulses

Control center

Hypothalamus and posterior pituitary

ADH

Output   Increased release of ADH

Effectors

Principal cells become more permeable to water, which increases facultative water reabsorption

Decrease in plasma osmolarity

Return to homeostasis when response brings plasma osmolarity back to normal

❓ In addition to ADH, which other hormones contribute to the regulation of water reabsorption?

TABLE 26.4

Hormonal Regulation of Tubular Reabsorption and Tubular Secretion

HORMONE	MAJOR STIMULI THAT TRIGGER RELEASE	MECHANISM AND SITE OF ACTION	EFFECTS
Angiotensin II	Low blood volume or low blood pressure stimulates renin-induced production of angiotensin II.	Stimulates activity of Na^+/H^+ antiporters in proximal tubule cells.	Increases reabsorption of Na^+, other solutes, and water, which increases blood volume.
Aldosterone	Increased angiotensin II level and increased level of plasma K^+ promote release of aldosterone by adrenal cortex.	Enhances activity of sodium–potassium pumps in basolateral membrane and Na^+ channels in apical membrane of principal cells in collecting duct.	Increases secretion of K^+ and reabsorption of Na^+, Cl^-; increases reabsorption of water, which increases blood volume.
Antidiuretic hormone (ADH) or vasopressin	Increased osmolarity of extracellular fluid or decreased blood volume promotes release of ADH from the posterior pituitary gland.	Stimulates insertion of water-channel proteins (aquaporin-2) into the apical membranes of principal cells.	Increases facultative reabsorption of water, which decreases osmolarity of body fluids.
Atrial natriuretic peptide (ANP)	Stretching of atria of heart stimulates secretion of ANP.	Suppresses reabsorption of Na^+ and water in proximal tubule and collecting duct; also inhibits secretion of aldosterone and ADH.	Increases excretion of Na^+ in urine (natriuresis); increases urine output (diuresis) and thus decreases blood volume.
Parathyroid hormone (PTH)	Decreased level of plasma Ca^{2+} promotes release of PTH from parathyroid glands.	Stimulates opening of Ca^{2+} channels in apical membranes of early distal tubule cells.	Increases reabsorption of Ca^{2+}.

PRODUCTION OF DILUTE AND CONCENTRATED URINE

◉ OBJECTIVE

• Describe how the renal tubule and collecting ducts produce dilute and concentrated urine.

Even though your fluid intake can be highly variable, the total volume of fluid in your body normally remains stable. Homeostasis of body fluid volume depends in large part on the ability of the kidneys to regulate the rate of water loss in urine. Normally functioning kidneys produce a large volume of dilute urine when fluid intake is high, and a small volume of concentrated urine when fluid intake is low or fluid loss is large. ADH controls whether dilute urine or concentrated urine is formed. In the absence of ADH, urine is very dilute. However, a high level of ADH stimulates reabsorption of more water into blood, producing a concentrated urine.

Formation of Dilute Urine

Glomerular filtrate has the same ratio of water and solute particles as blood; its osmolarity is about 300 mOsm/liter. As previously noted, fluid leaving the proximal convoluted tubule is still isotonic to plasma. When *dilute* urine is being formed (Figure 26.18), the osmolarity of the fluid in the tubular lumen *increases*

as it flows down the descending limb of the loop of Henle, *decreases* as it flows up the ascending limb, and *decreases* still more as it flows through the rest of the nephron and collecting duct. These changes in osmolarity result from the following conditions along the path of tubular fluid:

1. Because the osmolarity of the interstitial fluid of the renal medulla becomes progressively greater, more and more water is reabsorbed by osmosis as tubular fluid flows along the descending limb toward the tip of the loop. (The source of this medullary osmotic gradient is explained shortly.) As a result, the fluid remaining in the lumen becomes progressively more concentrated.

2. Cells lining the thick ascending limb of the loop have symporters that actively reabsorb Na^+, K^+, and Cl^- from the tubular fluid (see Figure 26.15). The ions pass from the tubular fluid into thick ascending limb cells, then into interstitial fluid, and finally some diffuse into the blood inside the vasa recta.

3. Although solutes are being reabsorbed in the thick ascending limb, the water permeability of this portion of the nephron is always quite low, so water cannot follow by osmosis. As solutes—but not water molecules—are leaving the tubular fluid, its osmolarity drops to about 150 mOsm/liter. The fluid entering the distal convoluted tubule is thus more dilute than plasma.

4. While the fluid continues flowing along the distal convoluted tubule, additional solutes but only a few water molecules

are reabsorbed. The early distal convoluted tubule cells are not very permeable to water and are not regulated by ADH.

5. Finally, the principal cells of the late distal convoluted tubules and collecting ducts are impermeable to water when the ADH level is very low. Thus, tubular fluid becomes progressively more dilute as it flows onward. By the time the tubular fluid drains into the renal pelvis, its concentration can be as low as 65–70 mOsm/liter. This is four times more dilute than blood plasma or glomerular filtrate.

Figure 26.18 Formation of dilute urine. Numbers indicate osmolarity in milliosmoles per liter (mOsm/liter). Heavy brown lines in the ascending limb of the loop of Henle and in the distal convoluted tubule indicate impermeability to water; heavy blue lines indicate the last part of the distal convoluted tubule and the collecting duct, which are impermeable to water in the absence of ADH; light blue areas around the nephron represent interstitial fluid. When ADH is absent, the osmolarity of urine can be as low as 65 mOsm/liter.

When ADH level is low, urine is dilute and has an osmolarity less than the osmolarity of blood.

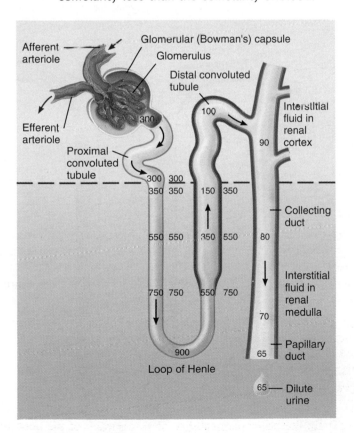

? Which portions of the renal tubule and collecting duct reabsorb more solutes than water to produce dilute urine?

Formation of Concentrated Urine

When water intake is low or water loss is high (such as during heavy sweating), the kidneys must conserve water while still eliminating wastes and excess ions. Under the influence of ADH, the kidneys produce a small volume of highly concentrated urine. Urine can be four times more concentrated (up to 1200 mOsm/liter) than blood plasma or glomerular filtrate (300 mOsm/liter).

The ability of ADH to cause excretion of concentrated urine depends on the presence of an **osmotic gradient** of solutes in the interstitial fluid of the renal medulla. Notice in Figure 26.19 on page 1044 that the solute concentration of the interstitial fluid in the kidney increases from about 300 mOsm/liter in the renal cortex to about 1200 mOsm/liter deep in the renal medulla. The three major solutes that contribute to this high osmolarity are Na^+, Cl^-, and urea. Two main factors contribute to building and maintaining this osmotic gradient: (1) differences in solute and water permeability and reabsorption in different sections of the long loops of Henle and the collecting ducts, and (2) the countercurrent flow of fluid through tube-shaped structures in the renal medulla. *Countercurrent flow* refers to the flow of fluid in opposite directions. This occurs when fluid flowing in one tube runs counter (opposite) to fluid flowing in a nearby parallel tube. Examples of countercurrent flow include the flow of tubular fluid through the descending and ascending limbs of the loop of Henle and the flow of blood through the ascending and descending parts of the vasa recta. Two types of **countercurrent mechanisms** exist in the kidneys: countercurrent multiplication and countercurrent exchange.

Countercurrent Multiplication

Countercurrent multiplication is the process by which a progressively increasing osmotic gradient is formed in the interstitial fluid of the renal medulla as a result of countercurrent flow. Countercurrent multiplication involves the long loops of Henle of juxtamedullary nephrons. Note in Figure 26.19a that the descending limb of the loop of Henle carries tubular fluid from the renal cortex deep into the medulla, and the ascending limb carries it in the opposite direction. Since countercurrent flow through the descending and ascending limbs of the long loop of Henle establishes the osmotic gradient in the renal medulla, the long loop of Henle is said to function as a **countercurrent multiplier.** The kidneys use this osmotic gradient to excrete concentrated urine.

Production of concentrated urine by the kidneys occurs in the following way (Figure 26.19):

1 *Symporters in thick ascending limb cells of the loop of Henle cause a buildup of Na^+ and Cl^- in the renal medulla.* In the thick ascending limb of the loop of Henle, the Na^+–K^+–$2Cl^-$ symporters reabsorb Na^+ and Cl^- from the tubular fluid (Figure 26.19a). Water is not reabsorbed in this segment, however, because the cells are impermeable to water. As a result, there is a buildup of Na^+ and Cl^- ions in the interstitial fluid of the medulla.

2 *Countercurrent flow through the descending and ascending limbs of the loop of Henle establishes an osmotic gradient in the renal medulla.* Since tubular fluid constantly moves from the descending limb to the thick ascending limb of the loop of Henle, the thick ascending limb is constantly reabsorbing Na⁺ and Cl⁻. Consequently, the reabsorbed Na⁺ and Cl⁻ become increasingly concentrated in the interstitial fluid of the medulla, which results in the formation of an osmotic gradient that ranges from 300 mOsm/liter in the outer medulla to 1200 mOsm/liter deep in the inner medulla. The descending limb of the loop of Henle is very permeable to water but impermeable to solutes except urea. Because the

Figure 26.19 Mechanism of urine concentration in long-loop juxtamedullary nephrons. The green line indicates the presence of Na⁺–K⁺–2Cl⁻ symporters that simultaneously reabsorb these ions into the interstitial fluid of the renal medulla; this portion of the nephron is also relatively impermeable to water and urea. All concentrations are in milliosmoles per liter (mOsm/liter).

🔑 The formation of concentrated urine depends on high concentrations of solutes in interstitial fluid in the renal medulla.

Juxtamedullary nephron and its blood supply together

(a) Reabsorption of Na⁺, Cl⁻ and water in a long-loop juxtamedullary nephron

(b) Recycling of salts and urea in the vasa recta

? Which solutes are the main contributors to the high osmolarity of interstitial fluid in the renal medulla?

osmolarity of the interstitial fluid outside the descending limb is higher than the tubular fluid within it, water moves out of the descending limb via osmosis. This causes the osmolarity of the tubular fluid to increase. As the fluid continues along the descending limb, its osmolarity increases even more: At the hairpin turn of the loop, the osmolarity can be as high as 1200 mOsm/liter in juxtamedullary nephrons. As you have already learned, the ascending limb of the loop is impermeable to water, but its symporters reabsorb Na$^+$ and Cl$^-$ from the tubular fluid into the interstitial fluid of the renal medulla, so the osmolarity of the tubular fluid progressively decreases as it flows through the ascending limb. At the junction of the medulla and cortex, the osmolarity of the tubular fluid has fallen to about 100 mOsm/liter. Overall, tubular fluid becomes progressively more concentrated as it flows along the descending limb and progressively more dilute as it moves along the ascending limb.

3 *Cells in the collecting ducts reabsorb more water and urea.* When ADH increases the water permeability of the principal cells, water quickly moves via osmosis out of the collecting duct tubular fluid, into the interstitial fluid of the inner medulla, and then into the vasa recta. With loss of water, the urea left behind in the tubular fluid of the collecting duct becomes increasingly concentrated. Because duct cells deep in the medulla are permeable to it, urea diffuses from the fluid in the duct into the interstitial fluid of the medulla.

4 *Urea recycling causes a buildup of urea in the renal medulla.* As urea accumulates in the interstitial fluid, some of it diffuses into the tubular fluid in the descending and thin ascending limbs of the long loops of Henle, which also are permeable to urea (Figure 26.19a). However, while the fluid flows through the thick ascending limb, distal convoluted tubule, and cortical portion of the collecting duct, urea remains in the lumen because cells in these segments are impermeable to it. As fluid flows along the collecting ducts, water reabsorption continues via osmosis because ADH is present. This water reabsorption *further increases* the concentration of urea in the tubular fluid, more urea diffuses into the interstitial fluid of the inner renal medulla, and the cycle repeats. The constant transfer of urea between segments of the renal tubule and the interstitial fluid of the medulla is termed *urea recycling*. In this way, reabsorption of water from the tubular fluid of the ducts promotes the buildup of urea in the interstitial fluid of the renal medulla, which in turn promotes water reabsorption. The solutes left behind in the lumen thus become very concentrated, and a small volume of concentrated urine is excreted.

Countercurrent Exchange

Countercurrent exchange is the process by which solutes and water are passively exchanged between the blood of the vasa recta and interstitial fluid of the renal medulla as a result of countercurrent flow. Note in Figure 26.19b that the vasa recta also consists of descending and ascending limbs that are parallel

to each other and to the loop of Henle. Just as tubular fluid flows in opposite directions in the loop of Henle, blood flows in opposite directions in the ascending and descending parts of the vasa recta. Since countercurrent flow between the descending and ascending limbs of the vasa recta allows for exchange of solutes and water between the blood and interstitial fluid of the renal medulla, the vasa recta is said to function as a **countercurrent exchanger.**

Blood entering the vasa recta has an osmolarity of about 300 mOsm/liter. As it flows along the descending part into the renal medulla, where the interstitial fluid becomes increasingly concentrated, Na$^+$, Cl$^-$, and urea diffuse from interstitial fluid into the blood and water diffuses from the blood into the interstitial fluid. But after its osmolarity increases, the blood flows into the ascending part of the vasa recta. Here blood flows through a region where the interstitial fluid becomes increasingly less concentrated. As a result Na$^+$, Cl$^-$, and urea diffuse from the blood back into interstitial fluid, and water diffuses from interstitial fluid back into the vasa recta. The osmolarity of blood leaving the vasa recta is only slightly higher than the osmolarity of blood entering the vasa recta. Thus, the vasa recta provides oxygen and nutrients to the renal medulla without washing out or diminishing the osmotic gradient. The long loop of Henle *establishes* the osmotic gradient in the renal medulla by countercurrent multiplication, but the vasa recta *maintains* the osmotic gradient in the renal medulla by countercurrent exchange.

Figure 26.20 summarizes the processes of filtration, reabsorption, and secretion in each segment of the nephron and collecting duct.

● CLINICAL CONNECTION | Diuretics

Diuretics are substances that slow renal reabsorption of water and thereby cause *diuresis*, an elevated urine flow rate, which in turn reduces blood volume. Diuretic drugs often are prescribed to treat *hypertension* (high blood pressure) because lowering blood volume usually reduces blood pressure. Naturally occurring diuretics include *caffeine* in coffee, tea, and sodas, which inhibits Na$^+$ reabsorption, and *alcohol* in beer, wine, and mixed drinks, which inhibits secretion of ADH. Most diuretic drugs act by interfering with a mechanism for reabsorption of filtered Na$^+$. For example, loop diuretics, such as furosemide (Lasix®), selectively inhibit the Na$^+$–K$^+$–2Cl$^-$ symporters in the thick ascending limb of the loop of Henle (see Figure 26.15). The thiazide diuretics, such as chlorthiazide (Diuril®), act in the distal convoluted tubule, where they promote loss of Na$^+$ and Cl$^-$ in the urine by inhibiting Na$^+$–Cl$^-$ symporters. ●

◉ CHECKPOINT

15. How do symporters in the ascending limb of the loop of Henle and principal cells in the collecting duct contribute to the formation of concentrated urine?
16. How does ADH regulate facultative water reabsorption?
17. What is the countercurrent mechanism? Why is it important?

Figure 26.20 Summary of filtration, reabsorption, and secretion in the nephron and collecting duct.

Filtration occurs in the renal corpuscle; reabsorption occurs all along the renal tubule and collecting ducts.

PROXIMAL CONVOLUTED TUBULE

Reabsorption (into blood) of filtered:

Water	65% (osmosis)
Na^+	65% (sodium-potassium pumps, symporters, antiporters)
K^+	65% (diffusion)
Glucose	100% (symporters and facilitated diffusion)
Amino acids	100% (symporters and facilitated diffusion)
Cl^-	50% (diffusion)
HCO_3^-	80–90% (facilitated diffusion)
Urea	50% (diffusion)
Ca^{2+}, Mg^{2+}	variable (diffusion)

Secretion (into urine) of:

H^+	variable (antiporters)
NH_4^+	variable, increases in acidosis (antiporters)
Urea	variable (diffusion)
Creatinine	small amount

At end of PCT, tubular fluid is still isotonic to blood (300 mOsm/liter).

LOOP OF HENLE

Reabsorption (into blood) of:

Water	15% (osmosis in descending limb)
Na^+	20–30% (symporters in ascending limb)
K^+	20–30% (symporters in ascending limb)
Cl^-	35% (symporters in ascending limb)
HCO_3^-	10–20% (facilitated diffusion)
Ca^{2+}, Mg^{2+}	variable (diffusion)

Secretion (into urine) of:

Urea	variable (recycling from collecting duct)

At end of loop of Henle, tubular fluid is hypotonic (100–150 mOsm/liter).

RENAL CORPUSCLE

Glomerular filtration rate:
105–125 mL/min of fluid that is isotonic to blood

Filtered substances: water and all solutes present in blood (except proteins) including ions, glucose, amino acids, creatinine, uric acid

EARLY DISTAL CONVOLUTED TUBULE

Reabsorption (into blood) of:

Water	10–15% (osmosis)
Na^+	5% (symporters)
Cl^-	5% (symporters)
Ca^{2+}	variable (stimulated by parathyroid hormone)

LATE DISTAL CONVOLUTED TUBULE AND COLLECTING DUCT

Reabsorption (into blood) of:

Water	5–9% (insertion of water channels stimulated by ADH)
Na^+	1–4% (sodium-potassium pumps and sodium channels stimulated by aldosterone)
HCO_3^-	variable amount, depends on H^+ secretion (antiporters)
Urea	variable (recycling to loop of Henle)

Secretion (into urine) of:

K^+	variable amount to adjust for dietary intake (leakage channels)
H^+	variable amounts to maintain acid-base homeostasis (H^+ pumps)

Tubular fluid leaving the collecting duct is dilute when ADH level is low and concentrated when ADH level is high.

Urine

? In which segments of the nephron and collecting duct does secretion occur?

TABLE 26.5

Characteristics of Normal Urine

CHARACTERISTIC	DESCRIPTION
Volume	One to two liters in 24 hours but varies considerably.
Color	Yellow or amber but varies with urine concentration and diet. Color is due to urochrome (pigment produced from breakdown of bile) and urobilin (from breakdown of hemoglobin). Concentrated urine is darker in color. Diet (reddish-colored urine from beets), medications, and certain diseases affect color. Kidney stones may produce blood in urine.
Turbidity	Transparent when freshly voided but becomes turbid (cloudy) upon standing.
Odor	Mildly aromatic but becomes ammonia-like upon standing. Some people inherit the ability to form methylmercaptan from digested asparagus that gives urine a characteristic odor. Urine of diabetics has a fruity odor due to presence of ketone bodies.
pH	Ranges between 4.6 and 8.0; average 6.0; varies considerably with diet. High-protein diets increase acidity; vegetarian diets increase alkalinity.
Specific gravity	Specific gravity (density) is the ratio of the weight of a volume of a substance to the weight of an equal volume of distilled water. In urine, it ranges from 1.001 to 1.035. The higher the concentration of solutes, the higher the specific gravity.

EVALUATION OF KIDNEY FUNCTION

◉ OBJECTIVES

- Define urinalysis and describe its importance.
- Define renal plasma clearance and describe its importance.

Routine assessment of kidney function involves evaluating both the quantity and quality of urine and the levels of wastes in the blood.

Urinalysis

An analysis of the volume and physical, chemical, and microscopic properties of urine, called a **urinalysis,** reveals much about the state of the body. Table 26.5 summarizes the major characteristics of normal urine. The volume of urine eliminated per day in a normal adult is 1–2 liters (about 1–2 qt). Fluid intake, blood pressure, blood osmolarity, diet, body temperature, diuretics, mental state, and general health influence urine volume. For example, low blood pressure triggers the renin–angiotensin–aldosterone pathway. Aldosterone increases reabsorption of water and salts in the renal tubules and decreases urine volume. By contrast, when blood osmolarity decreases—for example, after drinking a large volume of water—secretion of ADH is inhibited and a larger volume of urine is excreted.

Water accounts for about 95% of the total volume of urine. The remaining 5% consists of electrolytes, solutes derived from cellular metabolism, and exogenous substances such as drugs. Normal urine is virtually protein-free. Typical solutes normally present in urine include filtered and secreted electrolytes that are not reabsorbed, urea (from breakdown of proteins), creatinine (from breakdown of creatine phosphate in muscle fibers), uric acid (from breakdown of nucleic acids), urobilinogen (from breakdown of hemoglobin), and small quantities of other substances, such as fatty acids, pigments, enzymes, and hormones.

If disease alters body metabolism or kidney function, traces of substances not normally present may appear in the urine, or normal constituents may appear in abnormal amounts. Table 26.6 lists several abnormal constituents in urine that may be detected as part of a urinalysis. Normal values of urine components and the clinical implications of deviations from normal are listed in Appendix D.

Blood Tests

Two blood-screening tests can provide information about kidney function. One is the **blood urea nitrogen (BUN)** test, which measures the blood nitrogen that is part of the urea resulting from catabolism and deamination of amino acids. When glomerular filtration rate decreases severely, as may occur with renal disease or obstruction of the urinary tract, BUN rises steeply. One strategy in treating such patients is to minimize their protein intake, thereby reducing the rate of urea production.

Another test often used to evaluate kidney function is measurement of **plasma creatinine,** which results from catabolism of creatine phosphate in skeletal muscle. Normally, the blood creatinine level remains steady because the rate of creatinine excretion in the urine equals its discharge from muscle. A creatinine level above 1.5 mg/dL (135 mmol/liter) usually is an indication of poor renal function. Normal values for selected blood tests are listed in Appendix C along with situations that may cause the values to increase or decrease.

Renal Plasma Clearance

Even more useful than BUN and blood creatinine values in the diagnosis of kidney problems is an evaluation of how effectively the kidneys are removing a given substance from blood plasma. **Renal plasma clearance** is the volume of blood that is "cleaned" or cleared of a substance per unit of time, usually expressed in units of *milliliters per minute.* High renal plasma clearance indicates efficient excretion of a substance in the urine; low clearance indicates inefficient excretion. For example, the clearance of glucose normally is zero because it is completely reabsorbed (see Table 26.3); therefore, glucose is not excreted at all. Knowing a drug's clearance is essential for determining the correct dosage. If clearance is high (one example is penicillin), then the dosage must also be high, and the drug must be given several times a day to maintain an adequate therapeutic level in the blood.

The following equation is used to calculate clearance:

$$\text{Renal plasma clearance of substance S} = \left(\frac{U \times V}{P}\right)$$

TABLE 26.6

Summary of Abnormal Constituents in Urine

ABNORMAL CONSTITUENT	COMMENTS
Albumin	A normal constituent of plasma, it usually appears in only very small amounts in urine because it is too large to pass through capillary fenestrations. The presence of excessive albumin in the urine—**albuminuria** (al′-bū-mi-NOO-rē-a)—indicates an increase in the permeability of filtration membranes due to injury or disease, increased blood pressure, or irritation of kidney cells by substances such as bacterial toxins, ether, or heavy metals.
Glucose	The presence of glucose in the urine is called **glucosuria** (gloo-kō-SOO-rē-a) and usually indicates diabetes mellitus. Occasionally it may be caused by stress, which can cause excessive amounts of epinephrine to be secreted. Epinephrine stimulates the breakdown of glycogen and liberation of glucose from the liver.
Red blood cells (erythrocytes)	The presence of red blood cells in the urine is called **hematuria** (hēm-a-TOO-rē-a) and generally indicates a pathological condition. One cause is acute inflammation of the urinary organs as a result of disease or irritation from kidney stones. Other causes include tumors, trauma, and kidney disease, or possible contamination of the sample by menstrual blood.
Ketone bodies	High levels of ketone bodies in the urine, called **ketonuria** (kē-tō-NOO-rē-a), may indicate diabetes mellitus, anorexia, starvation, or simply too little carbohydrate in the diet.
Bilirubin	When red blood cells are destroyed by macrophages, the globin portion of hemoglobin is split off and the heme is converted to biliverdin. Most of the biliverdin is converted to bilirubin, which gives bile its major pigmentation. An above-normal level of bilirubin in urine is called **bilirubinuria** (bil′-ē-roo-bi-NOO-rē-a).
Urobilinogen	The presence of urobilinogen (breakdown product of hemoglobin) in urine is called **urobilinogenuria** (ū′-rō-bi-lin′-ō-je-NOO-rē-a). Trace amounts are normal, but elevated urobilinogen may be due to hemolytic or pernicious anemia, infectious hepatitis, biliary obstruction, jaundice, cirrhosis, congestive heart failure, or infectious mononucleosis.
Casts	**Casts** are tiny masses of material that have hardened and assumed the shape of the lumen of the tubule in which they formed. They are then flushed out of the tubule when filtrate builds up behind them. Casts are named after the cells or substances that compose them or based on their appearance. For example, there are white blood cell casts, red blood cell casts, and epithelial cell casts that contain cells from the walls of the tubules.
Microbes	The number and type of bacteria vary with specific infections in the urinary tract. One of the most common is *E. coli*. The most common fungus to appear in urine is the yeast *Candida albicans*, a cause of vaginitis. The most frequent protozoan seen is *Trichomonas vaginalis*, a cause of vaginitis in females and urethritis in males.

where U and P are the concentrations of the substance in urine and plasma, respectively (both expressed in the same units, such as mg/mL), and V is the urine flow rate in mL/min.

The clearance of a solute depends on the three basic processes of a nephron: glomerular filtration, tubular reabsorption, and tubular secretion. Consider a substance that is filtered but neither reabsorbed nor secreted. Its clearance equals the glomerular filtration rate because all the molecules that pass the filtration membrane appear in the urine. This is very nearly the situation for creatinine; it easily passes the filter, it is not reabsorbed, and it is secreted only to a very small extent. Measuring the creatinine clearance, which normally is 120–140 mL/min, is the easiest way to assess glomerular filtration rate. The waste product urea is filtered, reabsorbed, and secreted to varying extents. Its clearance typically is less than the GFR, about 70 mL/min.

The clearance of the organic anion **para-aminohippuric acid (PAH)** is also of clinical importance. After PAH is administered intravenously, it is filtered and secreted in a single pass through the kidneys. Thus, the clearance of PAH is used to measure **renal plasma flow,** the amount of plasma that passes through the kidneys in one minute. Typically, the renal plasma flow is 650 mL per minute, which is about 55% of the renal blood flow (1200 mL per minute).

● CLINICAL CONNECTION | Dialysis

If a person's kidneys are so impaired by disease or injury that he or she is unable to function adequately, then blood must be cleansed artificially by **dialysis** (dī-AL-i-sis; *dialyo* = to separate), the separation of large solutes from smaller ones by diffusion through a selectively permeable membrane. One method of dialysis is **hemodialysis** (hē-mō-dī-AL-i-sis; *hemo-* = blood), which directly filters the patient's blood by removing wastes and excess electrolytes and fluid and then returning the cleansed blood to the patient. Blood removed from the body is delivered to a *hemodialyzer* (artificial kidney). Inside the hemodialyzer, blood flows through a *dialysis membrane*, which contains pores large enough to permit the diffusion of small solutes. A special solution, called the *dialysate* (dī-AL-i-sāt), is pumped into the hemodialyzer so that it surrounds the dialysis membrane. The dialysate is specially formulated to maintain diffusion gradients that remove wastes from the blood (for example, urea, creatinine, uric acid, excess phosphate, potassium, and sulfate ions) and add needed substances (for example, glucose and bicarbonate ions) to it. The cleansed blood is passed through an air embolus detector to remove air and then returned to the body. An anticoagulant (heparin) is added to prevent blood from clotting in the hemodialyzer. As a rule, most people on hemodialysis require about 6–12 hours a week, typically divided into three sessions.

Another method of dialysis, called **peritoneal dialysis,** uses the peritoneum of the abdominal cavity as the dialysis membrane to filter the blood. The peritoneum has a large surface area and numerous blood vessels, and is a very effective filter. A catheter is inserted into the peritoneal cavity and connected to a bag of dialysate. The fluid flows into the peritoneal cavity by gravity and is left there for sufficient time to permit wastes and excess electrolytes and fluids to diffuse into the dialysate. Then the dialysate is drained out into a bag, discarded, and replaced with fresh dialysate.

Each cycle is called an *exchange*. One variation of peritoneal dialysis, called **continuous ambulatory peritoneal dialysis (CAPD),** can be performed at home. Usually, the dialysate is drained and replenished four times a day and once at night during sleep. Between exchanges the person can move about freely with the dialysate in the peritoneal cavity. •

◉ CHECKPOINT

18. What are the characteristics of normal urine?
19. What chemical substances normally are present in urine?
20. How may kidney function be evaluated?
21. Why are the renal plasma clearances of glucose, urea, and creatinine different? How does each clearance compare to glomerular filtration rate?

URINE TRANSPORTATION, STORAGE, AND ELIMINATION

◉ OBJECTIVE

• Describe the anatomy, histology, and physiology of the ureters, urinary bladder, and urethra.

From collecting ducts, urine drains through papillary ducts into the minor calyces, which join to become major calyces that unite to form the renal pelvis (see Figure 26.3). From the renal pelvis, urine first drains into the ureters and then into the urinary bladder. Urine is then discharged from the body through the single urethra (see Figure 26.1).

Ureters

Each of the two **ureters** (Ū-rē-ters) transports urine from the renal pelvis of one kidney to the urinary bladder. Peristaltic contractions of the muscular walls of the ureters push urine toward the urinary bladder, but hydrostatic pressure and gravity also contribute. Peristaltic waves that pass from the renal pelvis to the urinary bladder vary in frequency from one to five per minute, depending on how fast urine is being formed.

The ureters are 25–30 cm (10–12 in.) long and are thick-walled, narrow tubes that vary in diameter from 1 mm to 10 mm along their course between the renal pelvis and the urinary bladder. Like the kidneys, the ureters are retroperitoneal. At the base of the urinary bladder, the ureters curve medially and pass obliquely through the wall of the posterior aspect of the urinary bladder (Figure 26.21).

Even though there is no anatomical valve at the opening of each ureter into the urinary bladder, a physiological one is quite effective. As the urinary bladder fills with urine, pressure within it compresses the oblique openings into the ureters and prevents the backflow of urine. When this physiological valve is not operating properly, it is possible for microbes to travel up the ureters from the urinary bladder to infect one or both kidneys.

Figure 26.21 Ureters, urinary bladder, and urethra in a female. (See Tortora, *A Photographic Atlas of the Human Body, Second Edition,* Figures 13.8, 13.9.)

⑥ Urine is stored in the urinary bladder before being expelled by micturition.

Anterior view of frontal section

❓ What is a lack of voluntary control over micturition called?

Three layers of tissue form the wall of the ureters. The deepest coat, the **mucosa,** is a mucous membrane with **transitional epithelium** (see Table 4.1I on page 119) and an underlying **lamina propria** of areolar connective tissue with considerable collagen, elastic fibers, and lymphatic tissue. Transitional epithelium is able to stretch—a marked advantage for any organ that must accommodate a variable volume of fluid. Mucus secreted by the goblet cells of the mucosa prevents the cells from coming in contact with urine, the solute concentration and pH of which may differ drastically from the cytosol of cells that form the wall of the ureters.

Throughout most of the length of the ureters, the intermediate coat, the **muscularis,** is composed of inner longitudinal and outer circular layers of smooth muscle fibers. This arrangement is opposite to that of the gastrointestinal tract, which contains inner circular and outer longitudinal layers. The muscularis of the distal third of the ureters also contains an outer layer of longitudinal muscle fibers. Thus, the muscularis in the distal third of the ureter is inner longitudinal, middle circular, and outer longitudinal. Peristalsis is the major function of the muscularis.

The superficial coat of the ureters is the **adventitia,** a layer of areolar connective tissue containing blood vessels, lymphatic vessels, and nerves that serve the muscularis and mucosa. The adventitia blends in with surrounding connective tissue and anchors the ureters in place.

Urinary Bladder

The **urinary bladder** is a hollow, distensible muscular organ situated in the pelvic cavity posterior to the pubic symphysis. In males, it is directly anterior to the rectum; in females, it is anterior to the vagina and inferior to the uterus (see Figure 26.22). Folds of the peritoneum hold the urinary bladder in position. When slightly distended due to the accumulation of urine, the urinary bladder is spherical. When it is empty, it collapses. As urine volume increases, it becomes pear-shaped and rises into the abdominal cavity. Urinary bladder capacity averages 700–800 mL. It is smaller in females because the uterus occupies the space just superior to the urinary bladder.

Anatomy and Histology of the Urinary Bladder

In the floor of the urinary bladder is a small triangular area called the **trigone** (TRĪ-gōn = triangle). The two posterior corners of the trigone contain the two ureteral openings; the opening into the urethra, the **internal urethral orifice,** lies in the anterior corner (Figure 26.21). Because its mucosa is firmly bound to the muscularis, the trigone has a smooth appearance.

Three coats make up the wall of the urinary bladder. The deepest is the **mucosa,** a mucous membrane composed of **transitional epithelium** and an underlying **lamina propria** similar to that of the ureters. Rugae (the folds in the mucosa) are also present to permit expansion of the urinary bladder. Surrounding the mucosa is the intermediate **muscularis,** also called the **detrusor muscle** (de-TROO-ser = to push down), which consists of three layers of smooth muscle fibers: the inner longitudinal, middle circular, and outer longitudinal layers. Around

the opening to the urethra the circular fibers form an **internal urethral sphincter;** inferior to it is the **external urethral sphincter,** which is composed of skeletal muscle and is a modification of the deep muscles of the perineum (see Figure 11.12 on page 367). The most superficial coat of the urinary bladder on the posterior and inferior surfaces is the **adventitia,** a layer of areolar connective tissue that is continuous with that of the ureters. Over the superior surface of the urinary bladder is the **serosa,** a layer of visceral peritoneum.

The Micturition Reflex

Discharge of urine from the urinary bladder, called **micturition** (mik′-too-RISH-un; *mictur-* = urinate), is also known as *urination* or *voiding*. Micturition occurs via a combination of involuntary and voluntary muscle contractions. When the volume of urine in the urinary bladder exceeds 200–400 mL, pressure within the bladder increases considerably, and stretch receptors in its wall transmit nerve impulses into the spinal cord. These impulses propagate to the **micturition center** in sacral spinal cord segments S2 and S3 and trigger a spinal reflex called the **micturition reflex.** In this reflex arc, parasympathetic impulses from the micturition center propagate to the urinary bladder wall and internal urethral sphincter. The nerve impulses cause *contraction* of the detrusor muscle and *relaxation* of the internal urethral sphincter muscle. Simultaneously, the micturition center inhibits somatic motor neurons that innervate skeletal muscle in the external urethral sphincter. Upon contraction of the urinary bladder wall and relaxation of the sphincters, urination takes place. Urinary bladder filling causes a sensation of fullness that initiates a conscious desire to urinate before the micturition reflex actually occurs. Although emptying of the urinary bladder is a reflex, in early childhood we learn to initiate it and stop it voluntarily. Through learned control of the external urethral sphincter muscle and certain muscles of the pelvic floor, the cerebral cortex can initiate micturition or delay its occurrence for a limited period.

• CLINICAL CONNECTION | **Cystoscopy**

Cystoscopy (sis-TOS-kō-pē; *cysto-* = bladder; *-skopy* = to examine) is a very important procedure for direct examination of the mucosa of the urethra and urinary bladder and prostate in males. In the procedure, a *cystoscope* (a flexible narrow tube with a light) is inserted into the urethra to examine the structures through which it passes. With special attachments, tissue samples can be removed for examination (biopsy) and small stones can be removed. Cystoscopy is useful for evaluating urinary bladder problems such as cancer and infections. It can also evaluate the degree of obstruction resulting from an enlarged prostate. •

Urethra

The **urethra** (ū-RĒ-thra) is a small tube leading from the internal urethral orifice in the floor of the urinary bladder to the exterior of the body. In both males and females, the urethra is the terminal portion of the urinary system and the passageway

Figure 26.22 **Comparison between female and male urethras.**

The female urethra is about 4 cm (1.5 in.) in length, while the male urethra is about 20 cm (8 in.) in length.

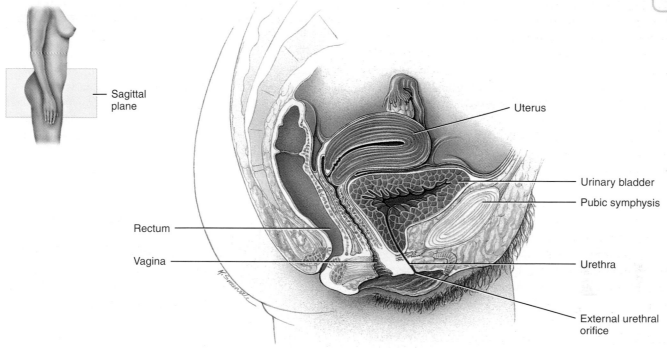

Sagittal plane

Uterus

Urinary bladder

Pubic symphysis

Rectum

Vagina

Urethra

External urethral orifice

(a) Sagittal section

Sagittal plane

Urinary bladder

Rectum

Pubic symphysis

Prostatic urethra

Prostate

Membranous urethra

Deep muscles of perineum

Penis

Spongy urethra

Testis

Scrotum

External urethral orifice

(b) Sagittal section

? **What are the three subdivisions of the male urethra?**

for discharging urine from the body. In males, it discharges semen (fluid that contains sperm) as well.

In females, the urethra lies directly posterior to the pubic symphysis, is directed obliquely, inferiorly, and anteriorly, and has a length of 4 cm (1.5 in.) (Figure 26.22a). The opening of the urethra to the exterior, the **external urethral orifice,** is located between the clitoris and the vaginal opening (see Figure 28.11a on page 1096). The wall of the female urethra consists of a deep **mucosa** and a superficial **muscularis.** The mucosa is a mucous membrane composed of **epithelium** and **lamina propria** (areolar connective tissue with elastic fibers and a plexus of veins). The muscularis consists of circularly arranged smooth muscle fibers and is continuous with that of the urinary bladder. Near the urinary bladder, the mucosa contains transitional epithelium that is continuous with that of the urinary bladder; near the external urethral orifice, the epithelium is nonkeratinized stratified squamous epithelium. Between these areas, the mucosa contains stratified columnar or pseudostratified columnar epithelium.

In males, the urethra also extends from the internal urethral orifice to the exterior, but its length and passage through the body are considerably different than in females (Figure 26.22b). The male urethra first passes through the prostate, then through the deep muscles of the perineum, and finally through the penis, a distance of about 20 cm (8 in.).

The male urethra, which also consists of a deep **mucosa** and a superficial **muscularis,** is subdivided into three anatomical regions: (1) The **prostatic urethra** passes through the prostate. (2) The **membranous (intermediate) urethra,** the shortest portion, passes through the deep muscles of the perineum. (3) The **spongy urethra,** the longest portion, passes through the penis. The epithelium of the prostatic urethra is continuous with that of the urinary bladder and consists of transitional epithelium that becomes stratified columnar or pseudostratified columnar epithelium more distally. The mucosa of the membranous urethra contains stratified columnar or pseudostratified columnar epithelium. The epithelium of the spongy urethra is stratified columnar or pseudostratified columnar epithelium, except near the external urethral orifice. There it is nonkeratinized stratified squamous epithelium. The **lamina propria** of the male urethra is areolar connective tissue with elastic fibers and a plexus of veins.

The muscularis of the prostatic urethra is composed of mostly circular smooth muscle fibers superficial to the lamina propria; these circular fibers help form the internal urethral sphincter of the urinary bladder. The muscularis of the membranous urethra consists of circularly arranged skeletal muscle fibers of the deep muscles of the perineum that help form the external urethral sphincter of the urinary bladder.

Several glands and other structures associated with reproduction (described in detail in Chapter 28) deliver their contents into the male urethra. The prostatic urethra receives secretions that contain sperm, neutralize the acidity of the female reproductive tract, and contribute to sperm motility and viability. The spongy urethra receives an alkaline substance before ejaculation that neutralizes the acidity of the urethra, and mucus, which lubricates the end of the penis during sexual arousal. The entire urethra, but especially the spongy urethra, receives mucus during sexual arousal or ejaculation.

• CLINICAL CONNECTION | Urinary Incontinence

A lack of voluntary control over micturition is called **urinary incontinence.** In infants and children under 2–3 years old, incontinence is normal because neurons to the external urethral sphincter muscle are not completely developed; voiding occurs whenever the urinary bladder is sufficiently distended to stimulate the micturition reflex. Urinary incontinence also occurs in adults. There are four types of urinary incontinence—stress, urge, overflow, and functional. **Stress incontinence** is the most common type of incontinence in young and middle-aged females, and results from weakness of the deep muscles of the pelvic floor. As a result, any physical stress that increases abdominal pressure, such as coughing, sneezing, laughing, exercising, straining, lifting heavy objects, and pregnancy, causes leakage of urine from the urinary bladder. **Urge incontinence** is most common in older people and is characterized by an abrupt and intense urge to urinate followed by an involuntary loss of urine. It may be caused by irritation of the urinary bladder wall by infection or kidney stones, stroke, multiple sclerosis, spinal cord injury, or anxiety. **Overflow incontinence** refers to the involuntary leakage of small amounts of urine caused by some type of blockage or weak contractions of the musculature of the urinary bladder. When urine flow is blocked (for example, from an enlarged prostate or kidney stones) or the urinary bladder muscles can no longer contract, the urinary bladder becomes overfilled and the pressure inside increases until small amounts of urine dribble out. **Functional incontinence** is urine loss resulting from the inability to get to a toilet facility in time as a result of conditions such as stroke, severe arthritis, and Alzheimer disease. Choosing the right treatment option depends on correct diagnosis of the type of incontinence. Treatments include Kegel exercises (see page 367), urinary bladder training, medication, and possibly even surgery. •

■ CHECKPOINT

22. What forces help propel urine from the renal pelvis to the urinary bladder?
23. What is micturition? How does the micturition reflex occur?
24. How do the location, length, and histology of the urethra compare in males and females?

WASTE MANAGEMENT IN OTHER BODY SYSTEMS

■ OBJECTIVE
• Describe the ways that body wastes are handled.

As we have seen, just one of the many functions of the urinary system is to help rid the body of some kinds of waste materials. Besides the kidneys, several other tissues, organs, and processes contribute to the temporary confinement of wastes, the transport of waste materials for disposal, the recycling of materials, and the excretion of excess or toxic substances in the body. These waste management systems include the following:

- **Body buffers.** Buffers in body fluids bind excess hydrogen ions (H^+), thereby preventing an increase in the acidity of body fluids. Buffers, like wastebaskets, have a limited capacity; eventually the H^+, like the paper in a wastebasket, must be eliminated from the body by excretion.

- **Blood.** The bloodstream provides pickup and delivery services for the transport of wastes, in much the same way that garbage trucks and sewer lines serve a community.

- **Liver.** The liver is the primary site for metabolic recycling, as occurs, for example, in the conversion of amino acids into glucose or of glucose into fatty acids. The liver also converts toxic substances into less toxic ones, such as ammonia into urea. These functions of the liver are described in Chapters 24 and 25.

- **Lungs.** With each exhalation, the lungs excrete CO_2, and expel heat and a little water vapor.

- **Sweat (sudoriferous) glands.** Especially during exercise, sweat glands in the skin help eliminate excess heat, water, and CO_2, plus small quantities of salts and urea as well.

- **Gastrointestinal tract.** Through defecation, the gastrointestinal tract excretes solid, undigested foods; wastes; some CO_2; water; salts; and heat.

◉ **CHECKPOINT**

25. What roles do the liver and lungs play in the elimination of wastes?

DEVELOPMENT OF THE URINARY SYSTEM

◉ **OBJECTIVE**

- Describe the development of the urinary system.

Starting in the third week of fetal development, a portion of the mesoderm along the posterior aspect of the embryo, the **intermediate mesoderm,** differentiates into the kidneys. The intermediate mesoderm is located in paired elevations called **urogenital ridges.** Three pairs of kidneys form within the intermediate mesoderm in succession: the pronephros, the mesonephros, and the metanephros (Figure 26.23 on page 1054). Only the last pair remains as the functional kidneys of the newborn.

The first kidney to form, the **pronephros** (prō-NEF-rōs; *pro- = before; -nephros = kidney*), is the most superior of the three and has an associated **pronephric duct.** This duct empties into the **cloaca,** the expanded terminal part of the hindgut, which functions as a common outlet for the urinary, digestive, and reproductive ducts. The pronephros begins to degenerate during the fourth week and is completely gone by the sixth week.

The second kidney, the **mesonephros** (mez′-ō-NEF-rōs; *meso- = middle*), replaces the pronephros. The retained portion of the pronephric duct, which connects to the mesonephros, develops into the **mesonephric duct.** The mesonephros begins to degenerate by the sixth week and is almost gone by the eighth week.

At about the fifth week, a mesodermal outgrowth, called a **ureteric bud** (ū-rē-TER-ik), develops from the distal portion of the mesonephric duct near the cloaca. The **metanephros** (met-a-NEF-rōs; *meta = after*), or ultimate kidney, develops from the ureteric bud and metanephric mesoderm. The ureteric bud forms the *collecting ducts, calyces, renal pelvis,* and *ureter.* The **metanephric mesoderm** forms the *nephrons* of the kidneys. By the third month, the fetal kidneys begin excreting urine into the surrounding amniotic fluid; indeed, fetal urine makes up most of the amniotic fluid.

During development, the cloaca divides into a **urogenital sinus,** into which urinary and genital ducts empty, and a *rectum* that discharges into the anal canal. The *urinary bladder* develops from the urogenital sinus. In females, the *urethra* develops as a result of lengthening of the short duct that extends from the urinary bladder to the urogenital sinus. In males, the urethra is considerably longer and more complicated, but it is also derived from the urogenital sinus.

Although the metanephric kidneys form in the pelvis, they ascend to their ultimate destination in the abdomen. As they do so, they receive renal blood vessels. Although the inferior blood vessels usually degenerate as superior ones appear, sometimes the inferior vessels do not degenerate. Consequently, some individuals (about 30%) develop multiple renal vessels.

◉ **CHECKPOINT**

26. Which type of embryonic tissue develops into nephrons?
27. Which tissue gives rise to collecting ducts, calyces, renal pelves, and ureters?

AGING AND THE URINARY SYSTEM

◉ **OBJECTIVE**

- Describe the effects of aging on the urinary system.

With aging, the kidneys shrink in size, have a decreased blood flow, and filter less blood. These age-related changes in kidney size and function seem to be linked to a progressive reduction in blood supply to the kidneys as an individual gets older; for example, blood vessels such as the glomeruli become damaged or decrease in number. The mass of the two kidneys decreases from an average of nearly 300 g in 20-year-olds to less than 200 g by age 80, a decrease of about one-third. Likewise, renal blood flow and filtration rate decline by 50% between ages 40 and 70. By age 80, about 40% of glomeruli are not functioning and thus filtration, reabsorption, and secretion decrease. Kidney diseases that become more common with age include acute and chronic kidney inflammations and renal calculi (kidney stones). Because the sensation of thirst diminishes with age, older individuals also are susceptible to dehydration. Urinary bladder changes that occur with aging include a reduction

Figure 26.23 Development of the urinary system.

Three pairs of kidneys form within intermediate mesoderm in succession: pronephros, mesonephros, and metanephros.

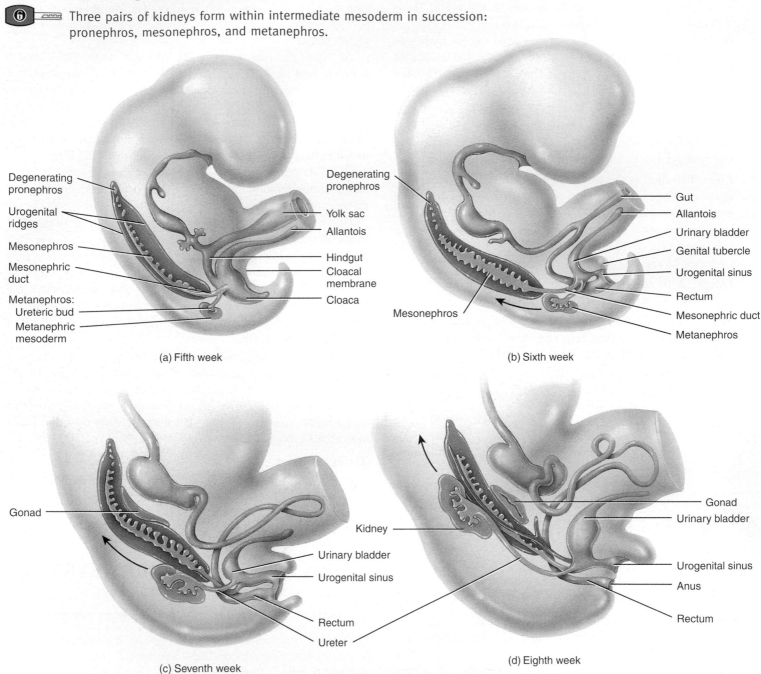

(a) Fifth week

(b) Sixth week

(c) Seventh week

(d) Eighth week

? **When do the kidneys begin to develop?**

in size and capacity and weakening of the muscles. Urinary tract infections are more common among the elderly, as are polyuria (excessive urine production), nocturia (excessive urination at night), increased frequency of urination, dysuria (painful urination), urinary retention or incontinence, and hematuria (blood in the urine).

● CHECKPOINT

28. To what extent do kidney mass and filtration rate decrease with age?

• • •

To appreciate the many ways that the urinary system contributes to homeostasis of other body systems, examine *Focus on Homeostasis: The Urinary System*. Next, in Chapter 27, we will see how the kidneys and lungs contribute to maintenance of homeostasis of body fluid volume, electrolyte levels in body fluids, and acid–base balance.

Focus on Homeostasis

BODY SYSTEM	CONTRIBUTION OF THE URINARY SYSTEM

For all body systems

Kidneys regulate the volume, composition, and pH of body fluids by removing wastes and excess substances from blood and excreting them in urine; the ureters transport urine from the kidneys to the urinary bladder, which stores urine until it is eliminated through the urethra.

Integumentary system

Kidneys and skin both contribute to synthesis of calcitriol, the active form of vitamin D.

Skeletal system

Kidneys help adjust levels of blood calcium and phosphates, needed for building extracellular bone matrix.

Muscular system

Kidneys help adjust level of blood calcium, needed for contraction of muscle.

Nervous system

Kidneys perform gluconeogenesis, which provides glucose for ATP production in neurons, especially during fasting or starvation.

Endocrine system

Kidneys participate in synthesis of calcitriol, the active form of vitamin D, and release erythropoietin, the hormone that stimulates production of red blood cells.

THE URINARY SYSTEM

Cardiovascular system
By increasing or decreasing their reabsorption of water filtered from blood, the kidneys help adjust blood volume and blood pressure; renin released by juxtaglomerular cells in the kidneys raises blood pressure; some bilirubin from hemoglobin breakdown is converted to a yellow pigment (urobilin), which is excreted in urine.

Lymphatic system and immunity
By increasing or decreasing their reabsorption of water filtered from blood, the kidneys help adjust the volume of interstitial fluid and lymph; urine flushes microbes out of the urethra.

Respiratory system

Kidneys and lungs cooperate in adjusting pH of body fluids.

Digestive system

Kidneys help synthesize calcitriol, the active form of vitamin D, which is needed for absorption of dietary calcium.

Reproductive systems

In males, the portion of the urethra that extends through the prostate and penis is a passageway for semen as well as urine.

DISORDERS: HOMEOSTATIC IMBALANCES

Renal Calculi

The crystals of salts present in urine occasionally precipitate and solidify into insoluble stones called **renal calculi** (*calculi* = pebbles) or **kidney stones.** They commonly contain crystals of calcium oxalate, uric acid, or calcium phosphate. Conditions leading to calculus formation include the ingestion of excessive calcium, low water intake, abnormally alkaline or acidic urine, and overactivity of the parathyroid glands. When a stone lodges in a narrow passage, such as a ureter, the pain can be intense. **Shock-wave lithotripsy** (LITH-ō-trip'-sē; *litho* = stone) is a procedure that uses high-energy shock waves to disintegrate kidney stones and offers an alternative to surgical removal. Once the kidney stone is located using x-rays, a device called a *lithotripter* delivers brief, high-intensity sound waves through a water- or gel-filled cushion placed under the back. Over a period of 30 to 60 minutes, 1000 or more shock waves pulverize the stone, creating fragments that are small enough to wash out in the urine.

Urinary Tract Infections

The term **urinary tract infection (UTI)** is used to describe either an infection of a part of the urinary system or the presence of large numbers of microbes in urine. UTIs are more common in females due to the shorter length of the urethra. Symptoms include painful or burning urination, urgent and frequent urination, low back pain, and bedwetting. UTIs include *urethritis* (inflammation of the urethra), *cystitis* (inflammation of the urinary bladder), and *pyelonephritis* (inflammation of the kidneys). If pyelonephritis becomes chronic, scar tissue can form in the kidneys and severely impair their function. Drinking cranberry juice can prevent the attachment of *E. coli* bacteria to the lining of the urinary bladder so that they are more readily flushed away during urination.

Glomerular Diseases

A variety of conditions may damage the kidney glomeruli, either directly or indirectly because of disease elsewhere in the body. Typically, the filtration membrane sustains damage, and its permeability increases.

Glomerulonephritis is an inflammation of the kidney that involves the glomeruli. One of the most common causes is an allergic reaction to the toxins produced by streptococcal bacteria that have recently infected another part of the body, especially the throat. The glomeruli become so inflamed, swollen, and engorged with blood that the filtration membranes allow blood cells and plasma proteins to enter the filtrate. As a result, the urine contains many erythrocytes (hematuria) and a lot of protein. The glomeruli may be permanently damaged, leading to chronic renal failure.

Nephrotic syndrome is a condition characterized by *proteinuria* (protein in the urine) and *hyperlipidemia* (high blood levels of cholesterol, phospholipids, and triglycerides). The proteinuria is due to an increased permeability of the filtration membrane, which permits proteins, especially albumin, to escape from blood into urine. Loss of albumin results in *hypoalbuminemia* (low blood albumin level) once liver production of albumin fails to meet increased urinary losses. Edema, usually seen around the eyes, ankles, feet, and abdomen, occurs in nephrotic syndrome because loss of albumin from the blood decreases blood colloid osmotic pressure. Nephrotic syndrome is associated with several glomerular diseases of unknown cause, as well as with systemic disorders such as diabetes mellitus, systemic lupus erythematosus (SLE), a variety of cancers, and AIDS.

Renal Failure

Renal failure is a decrease or cessation of glomerular filtration. In **acute renal failure (ARF),** the kidneys abruptly stop working entirely (or almost entirely). The main feature of ARF is the suppression of urine flow, usually characterized either by *oliguria* (daily urine output between 50 mL and 250 mL), or by *anuria* (daily urine output less than 50 mL). Causes include low blood volume (for example, due to hemorrhage), decreased cardiac output, damaged renal tubules, kidney stones, the dyes used to visualize blood vessels in angiograms, nonsteroidal anti-inflammatory drugs, and some antibiotic drugs. It is also common in people who suffer a devastating illness or overwhelming traumatic injury; in such cases it may be related to a more general organ failure known as *multiple organ dysfunction syndrome (MODS).*

Renal failure causes a multitude of problems. There is edema due to salt and water retention and metabolic acidosis due to an inability of the kidneys to excrete acidic substances. In the blood, urea builds up due to impaired renal excretion of metabolic waste products and potassium level rises, which can lead to cardiac arrest. Often, there is anemia because the kidneys no longer produce enough erythropoietin for adequate red blood cell production. Because the kidneys are no longer able to convert vitamin D to calcitriol, which is needed for adequate calcium absorption from the small intestine, osteomalacia also may occur.

Chronic renal failure (CRF) refers to a progressive and usually irreversible decline in glomerular filtration rate (GFR). CRF may result from chronic glomerulonephritis, pyelonephritis, polycystic kidney disease, or traumatic loss of kidney tissue. CRF develops in three stages. In the first stage, *diminished renal reserve,* nephrons are destroyed until about 75% of the functioning nephrons are lost. At this stage, a person may have no signs or symptoms because the remaining nephrons enlarge and take over the function of those that have been lost. Once 75% of the nephrons are lost, the person enters the second stage, called *renal insufficiency,* characterized by a decrease in GFR and increased blood levels of nitrogen-containing wastes and creatinine. Also, the kidneys cannot effectively concentrate or dilute the urine. The final stage, called *end-stage renal failure,* occurs when about 90% of the nephrons have been lost. At this stage, GFR diminishes to 10–15% of normal, oliguria is present, and blood levels of nitrogen-containing wastes and creatinine increase further. People with end-stage renal failure need dialysis therapy and are possible candidates for a kidney transplant operation.

Polycystic Kidney Disease

Polycystic kidney disease (PKD) is one of the most common inherited disorders. In PKD, the kidney tubules become riddled with hundreds or thousands of cysts (fluid-filled cavities). In addition, inappropriate apoptosis (programmed cell death) of cells in noncystic tubules leads to progressive impairment of renal function and eventually to end-stage renal failure.

People with PKD also may have cysts and apoptosis in the liver, pancreas, spleen, and gonads; increased risk of cerebral aneurysms; heart valve defects; and diverticuli in the colon. Typically, symptoms are not noticed until adulthood, when patients may have back pain, urinary tract infections, blood in the urine, hypertension, and large abdominal masses. Using drugs to restore normal blood pressure, restricting protein and salt in the diet, and controlling urinary tract infections may slow progression to renal failure.

Urinary Bladder Cancer

Each year, nearly 12,000 Americans die from **urinary bladder cancer.** It generally strikes people over 50 years of age and is three times more likely to develop in males than females. The disease is typically painless as it develops, but in most cases blood in the urine is a primary sign of the disease. Less often, people experience painful and/or frequent urination.

As long as the disease is identified early and treated promptly, the prognosis is favorable. Fortunately, about 75% of the urinary bladder cancers are confined to the epithelium of the urinary bladder and are easily removed by surgery. The lesions tend to be low-grade, meaning that they have only a small potential for metastasis.

Urinary bladder cancer is frequently the result of a carcinogen. About half of all cases occur in people who smoke or have at some time smoked cigarettes. The cancer also tends to develop in people who are exposed to chemicals called aromatic amines. Workers in the leather, dye, rubber, and aluminum industries, as well as painters, are often exposed to these chemicals.

MEDICAL TERMINOLOGY

Azotemia (az-ō-TĒ-mē-a; *azot-* = nitrogen; *-emia* = condition of blood) Presence of urea or other nitrogen-containing substances in the blood.

Cystocele (SIS-tō-sēl; *cysto-* = bladder; *-cele* = hernia or rupture) Hernia of the urinary bladder.

Diabetic kidney disease A disorder caused by diabetes mellitus in which glomeruli are damaged. The result is the leakage of proteins into the urine and a reduction in the ability of the kidney to remove water and waste.

Dysuria (dis-Ū-rē-a; *dys-* = painful; *uria* = urine) Painful urination.

Enuresis (en'-ū-RĒ-sis = to void urine) Involuntary voiding of urine after the age at which voluntary control has typically been attained.

Hydronephrosis (hī'-drō-ne-FRŌ-sis; *hydro-* = water; *nephros* = kidney; *-osis* = condition) Swelling of the kidney due to dilation of the renal pelvis and calyces as a result of an obstruction to the flow of urine. It may be due to a congenital abnormality, a narrowing of the ureter, a kidney stone, or an enlarged prostate.

Intravenous pyelogram (in'-tra-VĒ-nus PĪ e-lō-gram'; *intra-* = within; *veno-* = vein; *pyelo-* = pelvis of kidney; *-gram* = record) (or **IVP**) Radiograph (x-ray) of the kidneys, ureters, and urinary bladder after venous injection of a radiopaque contrast medium.

Nephropathy (ne-FROP-a-thē; *neph-* = kidney; *-pathos* = suffering) Any disease of the kidneys. Types include analgesic (from long-term and excessive use of drugs such as ibuprofen), lead (from ingestion of lead-based paint), and solvent (from carbon tetrachloride and other solvents).

Nocturnal enuresis (nok-TUR-nal en'-ū-RĒ-sis) Discharge of urine during sleep, resulting in bed-wetting; occurs in about 15% of 5-year-old children and generally resolves spontaneously, afflicting only about 1% of adults. It may have a genetic basis, as bed-wetting occurs more often in identical twins than in fraternal twins and more often in children whose parents or siblings were bed-wetters. Possible causes include smaller-than-normal bladder capacity, failure to awaken in response to a full bladder, and above-normal production of urine at night. Also referred to as **nocturia.**

Polyuria (pol'-ē-Ū-rē-a; *poly-* = too much) Excessive urine formation. It may occur in conditions such as diabetes mellitus and glomerulonephritis.

Stricture (STRIK-chur) Narrowing of the lumen of a canal or hollow organ, as may occur in the ureter, urethra, or any other tubular structure in the body.

Uremia (ū-RĒ-mē-a; *emia* = condition of blood) Toxic levels of urea in the blood resulting from severe malfunction of the kidneys.

Urinary retention A failure to completely or normally void urine; may be due to an obstruction in the urethra or neck of the urinary bladder, to nervous contraction of the urethra, or to lack of urge to urinate. In men, an enlarged prostate may constrict the urethra and cause urinary retention. If urinary retention is prolonged, a catheter (slender rubber drainage tube) must be placed into the urethra to drain the urine.

STUDY OUTLINE

Introduction (p. 1018)

1. The organs of the urinary system are the kidneys, ureters, urinary bladder, and urethra.
2. After the kidneys filter blood and return most water and many solutes to the bloodstream, the remaining water and solutes constitute urine.

Overview of Kidney Functions (p. 1020)

1. The kidneys regulate blood ionic composition, blood osmolarity, blood volume, blood pressure, and blood pH.
2. The kidneys also perform gluconeogenesis, release calcitriol and erythropoietin, and excrete wastes and foreign substances.

Anatomy and Histology of the Kidneys (p. 1020)

1. The kidneys are retroperitoneal organs attached to the posterior abdominal wall.
2. Three layers of tissue surround the kidneys: renal capsule, adipose capsule, and renal fascia.
3. Internally, the kidneys consist of a renal cortex, a renal medulla, renal pyramids, renal papillae, renal columns, major and minor calyces, and a renal pelvis.
4. Blood flows into the kidney through the renal artery and successively into segmental, interlobar, arcuate, and interlobular arteries; afferent arterioles; glomerular capillaries; efferent arterioles; peritubular capillaries and vasa recta; and interlobular, arcuate, and interlobar veins before flowing out of the kidney through the renal vein.

5. Vasomotor nerves from the sympathetic division of the autonomic nervous system supply kidney blood vessels; they help regulate the flow of blood through the kidney.

6. The nephron is the functional unit of the kidneys. A nephron consists of a renal corpuscle (glomerulus and glomerular or Bowman's capsule) and a renal tubule.

7. A renal tubule consists of a proximal convoluted tubule, a loop of Henle, and a distal convoluted tubule, which drains into a collecting duct (shared by several nephrons). The loop of Henle consists of a descending limb and an ascending limb.

8. A cortical nephron has a short loop that dips only into the superficial region of the renal medulla; a juxtamedullary nephron has a long loop of Henle that stretches through the renal medulla almost to the renal papilla.

9. The wall of the entire glomerular capsule, renal tubule, and ducts consists of a single layer of epithelial cells. The epithelium has distinctive histological features in different parts of the tubule. Table 26.1 on page 1028 summarizes the histological features of the renal tubule and collecting duct.

10. The juxtaglomerular apparatus (JGA) consists of the juxtaglomerular cells of an afferent arteriole and the macula densa of the final portion of the ascending limb of the loop of Henle.

Overview of Renal Physiology (p. 1029)

1. Nephrons perform three basic tasks: glomerular filtration, tubular secretion, and tubular reabsorption.

Glomerular Filtration (p. 1030)

1. Fluid that enters the capsular space is glomerular filtrate.

2. The filtration membrane consists of the glomerular endothelium, basal lamina, and filtration slits between pedicels of podocytes.

3. Most substances in blood plasma easily pass through the glomerular filter. However, blood cells and most proteins normally are not filtered.

4. Glomerular filtrate amounts to up to 180 liters of fluid per day. This large amount of fluid is filtered because the filter is porous and thin, the glomerular capillaries are long, and the capillary blood pressure is high.

5. Glomerular blood hydrostatic pressure (GBHP) promotes filtration; capsular hydrostatic pressure (CHP) and blood colloid osmotic pressure (BCOP) oppose filtration. Net filtration pressure (NFP) = GBHP − CHP − BCOP. NFP is about 10 mmHg.

6. Glomerular filtration rate (GFR) is the amount of filtrate formed in both kidneys per minute; it is normally 105–125 mL/min.

7. Glomerular filtration rate depends on renal autoregulation, neural regulation, and hormonal regulation. Table 26.2 on page 1034 summarizes regulation of GFR.

Tubular Reabsorption and Tubular Secretion (p. 1034)

1. Tubular reabsorption is a selective process that reclaims materials from tubular fluid and returns them to the bloodstream. Reabsorbed substances include water, glucose, amino acids, urea, and ions, such as sodium, chloride, potassium, bicarbonate, and phosphate (Table 26.3 on page 1035).

2. Some substances not needed by the body are removed from the blood and discharged into the urine via tubular secretion. Included are ions (K^+, H^+, and NH_4^+), urea, creatinine, and certain drugs.

3. Reabsorption routes include both paracellular (between tubule cells) and transcellular (across tubule cells) routes.

4. The maximum amount of a substance that can be reabsorbed per unit time is called the transport maximum (T_m).

5. About 90% of water reabsorption is obligatory; it occurs via osmosis, together with reabsorption of solutes, and is not hormonally regulated. The remaining 10% is facultative water reabsorption, which varies according to body needs and is regulated by ADH.

6. Na^+ ions are reabsorbed throughout the basolateral membrane via primary active transport.

7. In the proximal convoluted tubule, sodium ions are reabsorbed through the apical membranes via Na^+–glucose symporters and Na^+/H^+ antiporters; water is reabsorbed via osmosis; Cl^-, K^+, Ca^{2+}, Mg^{2+}, and urea are reabsorbed via passive diffusion; and NH_3 and NH_4^+ are secreted.

8. The loop of Henle reabsorbs 20–30% of the filtered Na^+, K^+, Ca^{2+}, and HCO_3^-; 35% of the filtered Cl^-; and 15% of the filtered water.

9. The distal convoluted tubule reabsorbs sodium and chloride ions via Na^+–Cl^- symporters.

10. In the collecting duct, principal cells reabsorb Na^+ and secrete K^+; intercalated cells reabsorb K^+ and HCO_3^- and secrete H^+.

11. Angiotensin II, aldosterone, antidiuretic hormone, atrial natriuretic peptide, and parathyroid hormone regulate solute and water reabsorption, as summarized in Table 26.4 on page 1042.

Production of Dilute and Concentrated Urine (p. 1042)

1. In the absence of ADH, the kidneys produce dilute urine; renal tubules absorb more solutes than water.

2. In the presence of ADH, the kidneys produce concentrated urine; large amounts of water are reabsorbed from the tubular fluid into interstitial fluid, increasing solute concentration of the urine.

3. The countercurrent multiplier establishes an osmotic gradient in the interstitial fluid of the renal medulla that enables production of concentrated urine when ADH is present.

Evaluation of Kidney Function (p. 1047)

1. A urinalysis is an analysis of the volume and physical, chemical, and microscopic properties of a urine sample. Table 26.5 on page 1047 summarizes the principal physical characteristics of normal urine.

2. Chemically, normal urine contains about 95% water and 5% solutes. The solutes normally include urea, creatinine, uric acid, urobilinogen, and various ions.

3. Table 26.6 on page 1048 lists several abnormal components that can be detected in a urinalysis, including albumin, glucose, red and white blood cells, ketone bodies, bilirubin, excessive urobilinogen, casts, and microbes.

4. Renal clearance refers to the ability of the kidneys to clear (remove) a specific substance from blood.

Urine Transportation, Storage, and Elimination (p. 1049)

1. The ureters are retroperitoneal and consist of a mucosa, muscularis, and adventitia. They transport urine from the renal pelvis to the urinary bladder, primarily via peristalsis.

2. The urinary bladder is located in the pelvic cavity posterior to the pubic symphysis; its function is to store urine before micturition.

3. The urinary bladder consists of a mucosa with rugae, a muscularis (detrusor muscle), and an adventitia (serosa over the superior surface).

4. The micturition reflex discharges urine from the urinary bladder via parasympathetic impulses that cause contraction of the detrusor muscle and relaxation of the internal urethral sphincter muscle and via inhibition of impulses in somatic motor neurons to the external urethral sphincter.

5. The urethra is a tube leading from the floor of the urinary bladder to the exterior. Its anatomy and histology differ in females and males. In both sexes, the urethra functions to discharge urine from the body; in males, it discharges semen as well.

Waste Management in Other Body Systems (p. 1052)

1. Besides the kidneys, several other tissues, organs, and processes temporarily confine wastes, transport waste materials for disposal, recycle materials, and excrete excess or toxic substances.

2. Buffers bind excess H^+, the blood transports wastes, the liver converts toxic substances into less toxic ones, the lungs exhale CO_2, sweat glands help eliminate excess heat, and the gastrointestinal tract eliminates solid wastes.

Development of the Urinary System (p. 1053)

1. The kidneys develop from intermediate mesoderm.

2. The kidneys develop in the following sequence: pronephros, mesonephros, and metanephros. Only the metanephros remains and develops into a functional kidney.

Aging and the Urinary System (p. 1053)

1. With aging, the kidneys shrink in size, have a decreased blood flow, and filter less blood.

2. Common problems related to aging include urinary tract infections, increased frequency of urination, urinary retention or incontinence, and renal calculi.

Q SELF-QUIZ QUESTIONS

Fill in the blanks in the following statements.

1. The renal corpuscle consists of the _____ and _____.
2. Discharge of urine from the urinary bladder is called _____.

Indicate whether the following statements are true or false.

3. The most superficial region of the internal kidney is the renal medulla.
4. When dilute urine is being formed, the osmolarity of the fluid in the tubular lumen increases as it flows down the descending limb of the loop of Henle, decreases as it flows up the ascending limb, and continues to decrease as it flows through the rest of the nephron and collecting duct.

Choose the one best answer to the following questions.

5. Which of the following statements are *correct?* (1) Glomerular filtration rate (GFR) is directly related to the pressures that determine net filtration pressure. (2) Angiotensin II and atrial natriuretic peptide help regulate GFR. (3) Mechanisms that regulate GFR work by adjusting blood flow into and out of the glomerulus and by altering the glomerular capillary surface area available for filtration. (4) GFR increases when blood flow into glomerular capillaries decreases. (5) Normally, GFR increases very little when systemic blood pressure rises.
   (a) 1, 2, and 3   (b) 2, 3, and 4   (c) 3, 4, and 5
   (d) 1, 2, 3, and 5   (e) 2, 3, 4, and 5

6. Which of the following hormones affect Na^+, Cl^-, Ca^{2+}, and water reabsorption and K^+ secretion by the renal tubules? (1) angiotensin II, (2) aldosterone, (3) ADH, (4) atrial natriuretic peptide, (5) thyroid hormone, (6) parathyroid hormone.
   (a) 1, 3, and 5   (b) 2, 3, and 6   (c) 2, 4, and 5
   (d) 1, 2, 4, and 5   (e) 1, 2, 3, 4, and 6

7. Which of the following are features of the renal corpuscle that enhance its filtering capacity? (1) large glomerular capillary surface area, (2) thick, selectively permeable filtration membrane, (3) high capsular hydrostatic pressure, (4) high glomerular capillary pressure, (5) mesangial cells regulating the filtering surface area.
   (a) 1, 2, and 3   (b) 2, 4, and 5   (c) 1, 4, and 5
   (d) 2, 3, and 4   (e) 2, 3, and 5

8. Given the following values, calculate the net filtration pressure: (1) glomerular blood hydrostatic pressure = 40 mmHg, (2) capsular hydrostatic pressure = 10 mmHg, (3) blood colloid osmotic pressure = 30 mmHg.
   (a) 20 mmHg   (b) 0 mmHg   (c) 20 mmHg
   (d) 60 mmHg   (e) 80 mmHg

9. The micturition reflex (1) is initiated by stretch receptors in the ureters, (2) relies on parasympathetic impulses from the micturition center in S2 and S3, (3) results in contraction of the detrusor muscle, (4) results in contraction of the internal urethral sphincter muscle, (5) inhibits motor neurons in the external urethral sphincter.
   (a) 1, 2, 3, 4, and 5   (b) 1, 3, and 4   (c) 2, 3, 4, and 5
   (d) 2 and 5   (e) 2, 3, and 5

10. Which of the following are mechanisms that control GFR? (1) renal autoregulation, (2) neural regulation, (3) hormonal regulation, (4) chemical regulation of ions, (5) presence or absence of a transporter.
   (a) 1, 2, and 3   (b) 2, 3, and 4   (c) 3, 4, and 5
   (d) 1, 3, and 5   (e) 1, 3, and 4

11. Place the route of blood flow through the kidney in the correct order:
   (a) segmental arteries   (b) vasa recta
   (c) arcuate arteries   (d) peritubular venules
   (e) interlobular veins   (f) renal vein
   (g) renal artery   (h) interlobar arteries
   (i) peritubular capillaries   (j) efferent arterioles
   (k) interlobar veins   (l) glomeruli
   (m) arcuate veins   (n) afferent arterioles
   (o) interlobular arteries

12. Place the route of filtrate flow in the correct order from its origin to the ureter:
    (a) minor calyx
    (b) ascending limb of loop of Henle
    (c) papillary duct
    (d) distal convoluted tubule
    (e) major calyx
    (f) descending limb of loop of Henle
    (g) proximal convoluted tubule
    (h) collecting duct
    (i) renal pelvis

13. Match the following:
    ____(a) cells in the last portion of the distal convoluted tubule and in the collecting ducts; regulated by ADH and aldosterone
    ____(b) a capillary network lying in the glomerular capsule and functioning in filtration
    ____(c) the functional unit of the kidney
    ____(d) drains into a collecting duct
    ____(e) combined glomerulus and glomerular capsule; where plasma is filtered
    ____(f) the visceral layer of the glomerular capsule consisting of modified simple squamous epithelial cells
    ____(g) cells of the final portion of the ascending limb of the loop of Henle that make contact with the afferent arteriole
    ____(h) site of obligatory water reabsorption
    ____(i) pores in the glomerular endothelial cells that allow filtration of blood solutes but not blood cells and platelets
    ____(j) can secrete H^+ against a concentration gradient
    ____(k) modified smooth muscle cells in the wall of the afferent arteriole

    (1) podocytes
    (2) glomerulus
    (3) renal corpuscle
    (4) proximal convoluted tubule
    (5) distal convoluted tubule
    (6) juxtaglomerular cells
    (7) macula densa
    (8) principal cells
    (9) intercalated cells
    (10) nephron
    (11) fenestrations

14. Match the following:
    ____(a) measure of blood nitrogen resulting from the catabolism and deamination of amino acids
    ____(b) produced from the catabolism of creatine phosphate in skeletal muscle
    ____(c) volume of blood that is cleared of a substance per unit of time
    ____(d) can result from diabetes mellitus
    ____(e) insoluble stones of crystallized salts
    ____(f) usually indicates a pathological condition
    ____(g) lack of voluntary control of micturition
    ____(h) can be caused by damage to the filtration membranes

    (1) incontinence
    (2) renal calculi
    (3) plasma creatinine
    (4) BUN test
    (5) albuminuria
    (6) glucosuria
    (7) renal plasma clearance
    (8) hematuria

15. Match the following:
    ____(a) membrane proteins that function as water channels
    ____(b) a secondary active transport process that achieves Na^+ reabsorption, returns filtered HCO_3^- and water to the peritubular capillaries, and secretes H^+
    ____(c) stimulates principal cells to secrete more K^+ into tubular fluid and absorb more Na^+ and Cl^- into tubular fluid
    ____(d) enzyme secreted by juxtaglomerular cells
    ____(e) reduces glomerular filtration rate; increases blood volume and pressure
    ____(f) inhibits Na^+ and H_2O reabsorption in the proximal convoluted tubules and collecting ducts
    ____(g) regulates facultative water reabsorption by increasing the water permeability of principal cells in the distal convoluted tubules and collecting ducts
    ____(h) reabsorb Na^+ together with a variety of other solutes
    ____(i) stimulates cells in the distal convoluted tubule to reabsorb more calcium into the blood

    (1) angiotensin II
    (2) atrial natriuretic peptide
    (3) Na^+ symporters
    (4) Na^+/H^+ antiporters
    (5) aquaporins
    (6) aldosterone
    (7) ADH
    (8) renin
    (9) parathyroid hormone

CRITICAL THINKING QUESTIONS

1. Imagine the discovery of a new toxin that blocks renal tubule reabsorption but does not affect filtration. Predict the short-term effects of this toxin.

2. For each of the following urinalysis results, indicate whether you should be concerned or not and why: (a) dark yellow urine that is turbid; (b) ammonia-like odor of the urine; (c) presence of excessive albumin; (d) presence of epithelial cell casts, (e) pH of 5.5; (f) hematuria.

3. Bruce is experiencing sudden, rhythmic waves of pain in his groin area. He has noticed that, although he is consuming fluids, his urine output has decreased. From what condition is Bruce suffering? How is it treated? How can he prevent future episodes?

? ANSWERS TO FIGURE QUESTIONS

26.1 The kidneys, ureters, urinary bladder, and urethra are the components of the urinary system.

26.2 The kidneys are retroperitoneal because they are posterior to the peritoneum.

26.3 Blood vessels, lymphatic vessels, nerves, and a ureter pass through the renal hilum.

26.4 About 1200 mL of blood enters the renal arteries each minute.

26.5 Cortical nephrons have glomeruli in the superficial renal cortex, and their short loops of Henle penetrate only into the superficial renal medulla. Juxtamedullary nephrons have glomeruli deep in the renal cortex, and their long loops of Henle extend through the renal medulla nearly to the renal papilla.

26.6 This section must pass through the renal cortex because there are no renal corpuscles in the renal medulla.

26.7 Secreted penicillin is being removed from the bloodstream.

26.8 Endothelial fenestrations (pores) in glomerular capillaries are too small for red blood cells to pass through.

26.9 Obstruction of the right ureter would increase CHP and thus decrease NFP in the right kidney; the obstruction would have no effect on the left kidney.

26.10 *Auto* means self; tubuloglomerular feedback is an example of autoregulation because it takes place entirely within the kidneys.

26.11 The tight junctions between tubule cells form a barrier that prevents diffusion of transporter, channel, and pump proteins between the apical and basolateral membranes.

26.12 Glucose enters a PCT cell via a Na^+–glucose symporter in the apical membrane and leaves via facilitated diffusion through the basolateral membrane.

26.13 The electrochemical gradient promotes movement of Na^+ into the tubule cell through the apical membrane antiporters.

26.14 Reabsorption of the solutes creates an osmotic gradient that promotes the reabsorption of water via osmosis.

26.15 This is considered secondary active transport because the symporter uses the energy stored in the concentration gradient of Na^+ between extracellular fluid and the cytosol. No water is reabsorbed here because the thick ascending limb of the loop of Henle is virtually impermeable to water.

26.16 In principal cells, aldosterone stimulates secretion of K^+ and reabsorption of Na^+ by increasing the activity of sodium-potassium pumps and number of leakage channels for Na^+ and K^+.

26.17 Aldosterone and atrial natriuretic peptide influence renal water reabsorption along with ADH.

26.18 Dilute urine is produced when the thick ascending limb of the loop of Henle, the distal convoluted tubule, and the collecting duct reabsorb more solutes than water.

26.19 The high osmolarity of interstitial fluid in the renal medulla is due mainly to Na^+, Cl^-, and urea.

26.20 Secretion occurs in the proximal convoluted tubule, the loop of Henle, and the collecting duct.

26.21 Lack of voluntary control over micturition is termed urinary incontinence.

26.22 The three subdivisions of the male urethra are the prostatic urethra, membranous urethra, and spongy urethra.

26.23 The kidneys start to form during the third week of development.

The processes of filtration, reabsorption, diffusion, and osmosis allow continual exchange of water and solutes among body fluid compartments (Figure 27.1b). Yet the volume of fluid in each compartment remains remarkably stable. The pressures that promote filtration of fluid from blood capillaries and reabsorption of fluid back into capillaries can be reviewed in Figure 21.7 on page 771. Because osmosis is the primary means of water movement between intracellular fluid and interstitial fluid, the concentration of solutes in these fluids determines the *direction* of water movement. Because most solutes in body fluids are **electrolytes,** inorganic compounds that dissociate into ions, fluid balance is closely related to electrolyte balance. Because intake of water and electrolytes rarely occurs in exactly the same proportions as their presence in body fluids, the ability of the kidneys to excrete excess water by producing dilute urine, or to excrete excess electrolytes by producing concentrated urine, is of utmost importance in the maintenance of homeostasis.

Sources of Body Water Gain and Loss

The body can gain water by ingestion and by metabolic synthesis (Figure 27.2). The main sources of body water are ingested liquids (about 1600 mL) and moist foods (about 700 mL)

absorbed from the gastrointestinal (GI) tract, which total about 2300 mL/day. The other source of water is **metabolic water** that is produced in the body mainly when electrons are accepted by oxygen during aerobic cellular respiration (see Figure 25.2 on page 981) and to a smaller extent during dehydration synthesis reactions (see Figure 2.15 on page 45). Metabolic water gain accounts for only 200 mL/day. Daily water gain from these two sources totals about 2500 mL.

Normally, body fluid volume remains constant because water loss equals water gain. Water loss occurs in four ways (Figure 27.2). Each day the kidneys excrete about 1500 mL in urine, the skin evaporates about 600 mL (400 mL through insensible perspiration, sweat that evaporates before it is perceived as moisture, and 200 mL as sweat), the lungs exhale about 300 mL as water vapor, and the gastrointestinal tract eliminates about 100 mL in feces. In women of reproductive age, additional water is lost in menstrual flow. On average, daily water loss totals about 2500 mL. The amount of water lost by a given route can vary considerably over time. For example, water may literally pour from the skin in the form of sweat during strenuous exertion. In other cases, water may be lost in diarrhea during a GI tract infection.

Regulation of Body Water Gain

The volume of metabolic water formed in the body depends entirely on the level of aerobic cellular respiration, which reflects the demand for ATP in body cells. When more ATP is produced, more water is formed. Body water gain is regulated mainly by the volume of water intake, or how much fluid you drink. An area in the hypothalamus known as the **thirst center** governs the urge to drink.

When water loss is greater than water gain, **dehydration**—a decrease in volume and an increase in osmolarity of body fluids—stimulates thirst (Figure 27.3). When body mass decreases by 2% due to fluid loss, mild dehydration exists. A decrease in blood volume causes blood pressure to fall. This change stimulates the kidneys to release renin, which promotes the formation of angiotensin II. Increased nerve impulses from osmoreceptors in the hypothalamus, triggered by increased blood osmolarity, and increased angiotensin II in the blood both stimulate the thirst center in the hypothalamus. Other signals that stimulate thirst come from (1) neurons in the mouth that detect dryness due to a decreased flow of saliva and (2) baroreceptors that detect lowered blood pressure in the heart and blood vessels. As a result, the sensation of thirst increases, which usually leads to increased fluid intake (if fluids are available) and restoration of normal fluid volume. Overall, fluid gain balances fluid loss. Sometimes, however, the sensation of thirst does not occur quickly enough or access to fluids is restricted, and significant dehydration ensues. This happens most often in elderly people, in infants, and in those who are in a confused mental state. When heavy sweating or fluid loss from diarrhea or vomiting occurs, it is wise to start replacing body fluids by drinking fluids even before the sensation of thirst occurs.

Figure 27.2 Sources of daily water gain and loss under normal conditions. Numbers are average volumes for adults.

Normally, daily water loss equals daily water gain.

WATER GAIN

- Metabolic water (200mL)
- Ingested foods (700 mL)
- Ingested liquids (1600 mL)

WATER LOSS

- GI tract (100mL)
- Lungs (300 mL)
- Skin (600 mL)
- Kidneys (1500 mL)

Volume of water (mL)

❓ **How does each of the following affect fluid balance: Hyperventilation? Vomiting? Fever? Diuretics?**

Figure 27.3 Pathways through which dehydration stimulates thirst.

Dehydration occurs when water loss is greater than water gain.

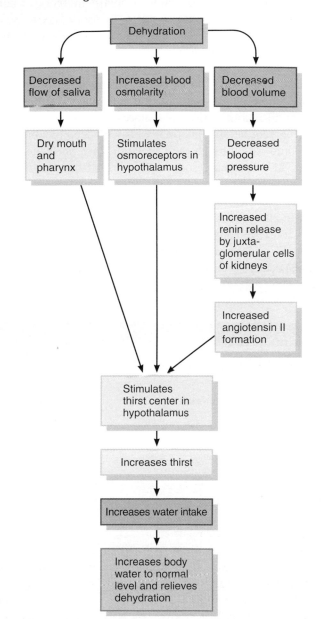

Does regulation of these pathways occur via negative or positive feedback? Why?

Regulation of Water and Solute Loss

Even though the loss of water and solutes through sweating and exhalation increases during exercise, elimination of *excess* body water or solutes occurs mainly by control of their loss in urine. The extent of *urinary salt (NaCl) loss* is the main factor that determines body fluid *volume.* The reason for this is that "water follows solutes" in osmosis, and the two main solutes in extra-cellular fluid (and in urine) are sodium ions (Na^+) and chloride ions (Cl^-). In a similar way, the main factor that determines body fluid *osmolarity* is the extent of *urinary water loss.*

Because our daily diet contains a highly variable amount of NaCl, urinary excretion of Na^+ and Cl^- must also vary to maintain homeostasis. Hormonal changes regulate the urinary loss of these ions, which in turn affects blood volume. Figure 27.4 depicts the sequence of changes that occur after a salty meal. The increased intake of NaCl produces an increase in plasma levels of Na^+ and Cl^- (the major contributors to osmolarity of extracellular fluid). As a result, the osmolarity of interstitial fluid increases, which causes movement of water from intracellular fluid into interstitial fluid and then into plasma. Such water movement increases blood volume.

The three most important hormones that regulate the extent of renal Na^+ and Cl^- reabsorption (and thus how much is lost in the urine) are **angiotensin II, aldosterone,** and **atrial natriuretic peptide (ANP).** When your body is dehydrated, angiotensin II and aldosterone promote urinary reabsorption of Na^+ and Cl^- (and water by osmosis with the electrolytes), conserving the volume of body fluids by reducing urinary loss. An increase in blood volume, as might occur after you finish one or more supersized drinks, stretches the atria of the heart and promotes release of atrial natriuretic peptide. ANP promotes **natriuresis,** elevated urinary excretion of Na^+ (and Cl^-) followed by water excretion, which decreases blood volume. An increase in blood volume also slows release of renin from juxtaglomerular cells of the kidneys. When renin level declines, less angiotensin II is formed. Decline in angiotensin II from a moderate level to a low level increases glomerular filtration rate and reduces Na^+, Cl^-, and water reabsorption in the kidney tubules. In addition, less angiotensin II leads to lower levels of aldosterone, which causes reabsorption of filtered Na^+ and Cl^- to slow in the renal collecting ducts. More filtered Na^+ and Cl^- thus remain in the tubular fluid to be excreted in the urine. The osmotic consequence of excreting more Na^+ and Cl^- is loss of more water in urine, which decreases blood volume and blood pressure.

The major hormone that regulates water loss is **antidiuretic hormone (ADH).** This hormone, also known as **vasopressin,** is produced by neurosecretory cells that extend from the hypothalamus to the posterior pituitary. In addition to stimulating the thirst mechanism, an increase in the osmolarity of body fluids stimulates release of ADH (see Figure 26.17 on page 1041). ADH promotes the insertion of water-channel proteins (aquaporin-2) into the apical membranes of principal cells in the collecting ducts of the kidneys. As a result, the permeability of these cells to water increases. Water molecules move by osmosis from the renal tubular fluid into the cells and then from the cells into the bloodstream. The result is production of a small volume of very concentrated urine (see page 1043). Intake of water in response to the thirst mechanism decreases the osmolarity of blood and interstitial fluid. Within minutes, ADH secretion shuts down, and soon its blood level is close to zero. When the principal cells are not stimulated by ADH, aquaporin-2 molecules are removed

Figure 27.4 Hormonal regulation of renal Na⁺ and Cl⁻ reabsorption.

The three main hormones that regulate renal Na⁺ and Cl⁻ reabsorption (and thus the amount lost in the urine) are angiotensin II, aldosterone, and atrial natriuretic peptide.

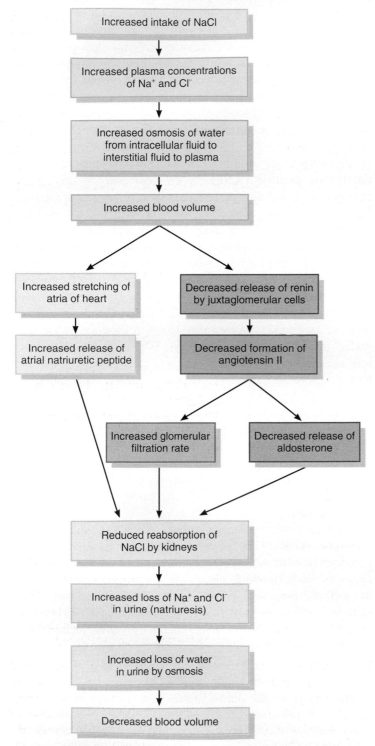

? How does hyperaldosteronism (excessive aldosterone secretion) cause edema?

from the apical membrane by endocytosis. As the number of water channels decreases, the water permeability of the principal cells' apical membrane falls, and more water is lost in the urine.

Under some conditions, factors other than blood osmolarity influence ADH secretion. A large decrease in blood volume, which is detected by baroreceptors (sensory neurons that respond to stretching) in the left atrium and in blood vessel walls, also stimulates ADH release. In severe dehydration, glomerular filtration rate decreases because blood pressure falls, so that less water is lost in the urine. Conversely, the intake of too much water increases blood pressure, causing the rate of glomerular filtration to rise, and more water to be lost in the urine. Hyperventilation (abnormally fast and deep breathing) can increase fluid loss through the exhalation of more water vapor. Vomiting and diarrhea result in fluid loss from the GI tract. Finally, fever, heavy sweating, and destruction of extensive areas of the skin from burns can cause excessive water loss through the skin. In all of these conditions, an increase in ADH secretion will help conserve body fluids.

Table 27.1 summarizes the factors that maintain body water balance.

TABLE 27.1

Summary of Factors That Maintain Body Water Balance

FACTOR	MECHANISM	EFFECT
Thirst center in hypothalamus	Stimulates desire to drink fluids.	Water gain if thirst is quenched.
Angiotensin II	Stimulates secretion of aldosterone.	Reduces loss of water in urine.
Aldosterone	By promoting urinary reabsorption of Na⁺ and Cl⁻, increases water reabsorption via osmosis.	Reduces loss of water in urine.
Atrial natriuretic peptide (ANP)	Promotes natriuresis, elevated urinary excretion of Na⁺ (and Cl⁻), accompanied by water.	Increases loss of water in urine.
Antidiuretic hormone (ADH), also known as vasopressin	Promotes insertion of water-channel proteins (aquaporin-2) into the apical membranes of principal cells in the collecting ducts of the kidneys. As a result, the water permeability of these cells increases and more water is reabsorbed.	Reduces loss of water in urine.

Movement of Water Between Body Fluid Compartments

Normally, cells neither shrink nor swell because intracellular and interstitial fluids have the same osmolarity. Changes in the osmolarity of interstitial fluid, however, cause fluid imbalances. An increase in the osmolarity of interstitial fluid draws water out of cells, and they shrink slightly. A decrease in the osmolarity of interstitial fluid, by contrast, causes cells to swell. Changes in osmolarity most often result from changes in the concentration of Na^+.

A decrease in the osmolarity of interstitial fluid, as may occur after drinking a large volume of water, inhibits secretion of ADH. Normally, the kidneys then excrete a large volume of dilute urine, which restores the osmotic pressure of body fluids to normal. As a result, body cells swell only slightly, and only for a brief period. But when a person steadily consumes water faster than the kidneys can excrete it (the maximum urine flow rate is about 15 mL/min) or when renal function is poor, the result may be **water intoxication,** a state in which excessive body water causes cells to swell dangerously (Figure 27.5). If the body wa-

ter and Na^+ lost during blood loss or excessive sweating, vomiting, or diarrhea is replaced by drinking plain water, then body fluids become more dilute. This dilution can cause the Na^+ concentration of plasma and then of interstitial fluid to fall below the normal range. When the Na^+ concentration of interstitial fluid decreases, its osmolarity also falls. The net result is osmosis of water from interstitial fluid into the cytosol. Water entering the cells causes them to swell, producing convulsions, coma, and possibly death. To prevent this dire sequence of events in cases of severe electrolyte and water loss, solutions given for intravenous or oral rehydration therapy (ORT) include a small amount of table salt (NaCl).

• CLINICAL CONNECTION | Enemas and Fluid Balance

An **enema** (EN-e-ma) is the introduction of a solution into the rectum to draw water (and electrolytes) into the colon osmotically. The increased volume increases peristalsis, which evacuates feces. Enemas are used to treat constipation. Repeated enemas, especially in young children, increase the risk of fluid and electrolyte imbalances. •

◉ CHECKPOINT

1. What is the approximate volume of each of your body fluid compartments?
2. How are the routes of water gain and loss from the body regulated?
3. By what mechanism does thirst help regulate water intake?
4. How do angiotensin II, aldosterone, atrial natriuretic peptide, and antidiuretic hormone regulate the volume and osmolarity of body fluids?
5. What factors control the movement of water between interstitial fluid and intracellular fluid?

Figure 27.5 Series of events in water intoxication.

Water intoxication is a state in which excessive body water causes cells to swell.

? Why do solutions used for oral rehydration therapy contain a small amount of table salt (NaCl)?

ELECTROLYTES IN BODY FLUIDS

OBJECTIVES
• Compare the electrolyte composition of the three major fluid compartments: plasma, interstitial fluid, and intracellular fluid.
• Discuss the functions of sodium, chloride, potassium, bicarbonate, calcium, phosphate, and magnesium ions, and explain how their concentrations are regulated.

The ions formed when electrolytes dissolve and dissociate serve four general functions in the body. (1) Because they are largely confined to particular fluid compartments and are more numerous than nonelectrolytes, certain ions *control the osmosis of water between fluid compartments.* (2) Ions *help maintain the acid–base balance* required for normal cellular activities. (3) Ions *carry electrical current,* which allows production of action potentials and graded potentials. (4) Several ions *serve as cofactors* needed for optimal activity of enzymes.

Concentrations of Electrolytes in Body Fluids

To compare the charge carried by ions in different solutions, the concentration of ions is typically expressed in units of **milliequivalents per liter (mEq/liter).** These units give the concentration of cations or anions in a given volume of solution. One equivalent is the positive or negative charge equal to the amount of charge in one mole of H^+; a milliequivalent is one-thousandth of an equivalent. Recall that a mole of a substance is its molecular weight expressed in grams. For ions such as sodium (Na^+), potassium (K^+), and bicarbonate (HCO_3^-), which have a single positive or negative charge, the number of mEq/liter is equal to the number of mmol/liter. For ions such as calcium (Ca^{2+}) or phosphate (HPO_4^{2-}), which have two positive or negative charges, the number of mEq/liter is twice the number of mmol/liter.

Figure 27.6 compares the concentrations of the main electrolytes and protein anions in blood plasma, interstitial fluid, and intracellular fluid. The chief difference between the two extracellular fluids—blood plasma and interstitial fluid—is that blood plasma contains many protein anions, in contrast to interstitial fluid, which has very few. Because normal capillary membranes are virtually impermeable to proteins, only a few plasma proteins leak out of blood vessels into the interstitial fluid. This difference in protein concentration is largely responsible for the blood colloid osmotic pressure exerted by blood plasma. In other respects, the two fluids are similar.

The electrolyte content of intracellular fluid differs considerably from that of extracellular fluid. In extracellular fluid, the most abundant cation is Na^+, and the most abundant anion is Cl^-. In intracellular fluid, the most abundant cation is K^+, and the most abundant anions are proteins and phosphates (HPO_4^{2-}). By actively transporting Na^+ out of cells and K^+ into cells, sodium–potassium pumps (Na^+/K^+ ATPase) play a major role in maintaining the high intracellular concentration of K^+ and high extracellular concentration of Na^+.

Sodium

Sodium ions (Na^+) are the most abundant ions in extracellular fluid, accounting for 90% of the extracellular cations. The normal blood plasma Na^+ concentration is 136–148 mEq/liter. As we have already seen, Na^+ plays a pivotal role in fluid and electrolyte balance because it accounts for almost half of the osmolarity of extracellular fluid (142 of about 300 mOsm/liter). The flow of Na^+ through voltage-gated channels in the plasma membrane also is necessary for the generation and conduction of action potentials in neurons and muscle fibers. The typical daily intake of Na^+ in North America often far exceeds the body's normal daily requirements, due largely to excess dietary salt.

Figure 27.6 Electrolyte and protein anion concentrations in plasma, interstitial fluid, and intracellular fluid. The height of each column represents the milliequivalents per liter (mEq/liter).

The electrolytes present in extracellular fluids are different from those present in intracellular fluid.

? What cation and two anions are present in the highest concentrations in ECF and ICF?

The kidneys excrete excess Na^+, but they also can conserve it during periods of shortage.

The Na^+ level in the blood is controlled by aldosterone, antidiuretic hormone (ADH), and atrial natriuretic peptide (ANP). Aldosterone increases renal reabsorption of Na^+. When the blood plasma concentration of Na^+ drops below 135 mEq/liter, a condition called *hyponatremia,* ADH release ceases. The lack of ADH in turn permits greater excretion of water in urine and restoration of the normal Na^+ level in ECF. Atrial natriuretic peptide (ANP) increases Na^+ excretion by the kidneys when Na^+ level is above normal, a condition called *hypernatremia.*

• CLINICAL **CONNECTION** | Indicators of Na^+ Imbalance

If excess sodium ions remain in the body because the kidneys fail to excrete enough of them, water is also osmotically retained. The result is increased blood volume, increased blood pressure, and **edema,** an abnormal accumulation of interstitial fluid. Renal failure and hyperaldosteronism (excessive aldosterone secretion) are two causes of Na^+ retention. Excessive urinary loss of Na^+, by contrast, causes excessive water loss, which results in **hypovolemia,** an abnormally low blood volume. Hypovolemia related to Na^+ loss is most frequently due to the inadequate secretion of aldosterone associated with adrenal insufficiency or overly vigorous therapy with diuretic drugs. •

Chloride

Chloride ions (Cl^-) are the most prevalent anions in extracellular fluid. The normal blood plasma Cl^- concentration is 95–105 mEq/liter. Cl^- moves relatively easily between the extracellular and intracellular compartments because most plasma membranes contain many Cl^- leakage channels and antiporters. For this reason, Cl^- can help balance the level of anions in different fluid compartments. One example is the chloride shift that occurs between red blood cells and blood plasma as the blood level of carbon dioxide either increases or decreases (see Figure 23.23 on page 904). In this case, the antiporter exchange of Cl^- for HCO_3^- maintains the correct balance of anions between ECF and ICF. Chloride ions also are part of the hydrochloric acid secreted into gastric juice. ADH helps regulate Cl^- balance in body fluids because it governs the extent of water loss in urine. Processes that increase or decrease renal reabsorption of sodium ions also affect reabsorption of chloride ions. (Recall that reabsorption of Na^+ and Cl^- occurs by means of Na^+–Cl^- symporters.)

Potassium

Potassium ions (K^+) are the most abundant cations in intracellular fluid (140 mEq/liter). K^+ plays a key role in establishing the resting membrane potential and in the repolarization phase of action potentials in neurons and muscle fibers; K^+ also helps maintain normal intracellular fluid volume. When K^+ moves into or out of cells, it often is exchanged for H^+ and thereby helps regulate the pH of body fluids.

The normal blood plasma K^+ concentration is 3.5–5.0 mEq/liter and is controlled mainly by aldosterone. When blood plasma K^+ concentration is high, more aldosterone is secreted into the blood. Aldosterone then stimulates principal cells of the renal collecting ducts to secrete more K^+ so excess K^+ is lost in the urine. Conversely, when blood plasma K^+ concentration is low, aldosterone secretion decreases and less K^+ is excreted in urine. Because K^+ is needed during the repolarization phase of action potentials, abnormal K^+ levels can be lethal. For instance, *hyperkalemia* (above-normal concentration of K^+ in blood) can cause death due to ventricular fibrillation.

Bicarbonate

Bicarbonate ions (HCO_3^-) are the second most prevalent extracellular anions. Normal blood plasma HCO_3^- concentration is 22–26 mEq/liter in systemic arterial blood and 23–27 mEq/liter in systemic venous blood. HCO_3^- concentration increases as blood flows through systemic capillaries because the carbon dioxide released by metabolically active cells combines with water to form carbonic acid; the carbonic acid then dissociates into H^+ and HCO_3^-. As blood flows through pulmonary capillaries, however, the concentration of HCO_3^- decreases again as carbon dioxide is exhaled. (Figure 23.23 on page 904 shows these reactions.) Intracellular fluid also contains a small amount of HCO_3^-. As previously noted, the exchange of Cl^- for HCO_3^- helps maintain the correct balance of anions in extracellular fluid and intracellular fluid.

The kidneys are the main regulators of blood HCO_3^- concentration. The intercalated cells of the renal tubule can either form HCO_3^- and release it into the blood when the blood level is low (see Figure 27.8) or excrete excess HCO_3^- in the urine when the level in blood is too high. Changes in the blood level of HCO_3^- are considered later in this chapter in the section on acid–base balance.

Calcium

Because such a large amount of calcium is stored in bone, it is the most abundant mineral in the body. About 98% of the calcium in adults is located in the skeleton and teeth, where it is combined with phosphates to form a crystal lattice of mineral salts. In body fluids, calcium is mainly an extracellular cation (Ca^{2+}). The normal concentration of free or unattached Ca^{2+} in blood plasma is 4.5–5.5 mEq/liter. About the same amount of Ca^{2+} is attached to various plasma proteins. Besides contributing to the hardness of bones and teeth, Ca^{2+} plays important roles in blood clotting, neurotransmitter release, maintenance of muscle tone, and excitability of nervous and muscle tissue.

The most important regulator of Ca^{2+} concentration in blood plasma is parathyroid hormone (PTH) (see Figure 18.14 on page 664). A low level of Ca^{2+} in blood plasma promotes release of more PTH, which stimulates osteoclasts in bone tissue to release calcium (and phosphate) from bone extracellular matrix. Thus, PTH increases bone *resorption.* Parathyroid hormone also enhances *reabsorption* of Ca^{2+} from glomerular filtrate through re-

1. *Buffer systems.* Buffers act quickly to temporarily bind H^+, removing the highly reactive, excess H^+ from solution. Buffers thus raise pH of body fluids but do not remove H^+ from the body.

2. *Exhalation of carbon dioxide.* By increasing the rate and depth of breathing, more carbon dioxide can be exhaled. Within minutes this reduces the level of carbonic acid in blood, which raises the blood pH (reduces blood H^+ level).

3. *Kidney excretion of H^+.* The slowest mechanism, but the only way to eliminate acids other than carbonic acid, is through their excretion in urine.

We will examine each of these mechanisms in more detail in the following sections.

The Actions of Buffer Systems

Most buffer systems in the body consist of a weak acid and the salt of that acid, which functions as a weak base. Buffers prevent rapid, drastic changes in the pH of body fluids by converting strong acids and bases into weak acids and weak bases within fractions of a second. Strong acids lower pH more than weak acids because strong acids release H^+ more readily and thus contribute more free hydrogen ions. Similarly, strong bases raise pH more than weak ones. The principal buffer systems of the body fluids are the protein buffer system, the carbonic acid–bicarbonate buffer system, and the phosphate buffer system.

Protein Buffer System

The **protein buffer system** is the most abundant buffer in intracellular fluid and blood plasma. For example, the protein hemoglobin is an especially good buffer within red blood cells, and albumin is the main protein buffer in blood plasma. Proteins are composed of amino acids, organic molecules that contain at least one carboxyl group (—COOH) and at least one amino group (—NH_2); these groups are the functional components of the protein buffer system. The free carboxyl group at one end of a protein acts like an acid by releasing H^+ when pH rises; it dissociates as follows:

$$NH_2-\overset{\overset{\displaystyle R}{|}}{\underset{\underset{\displaystyle H}{|}}{C}}-COOH \longrightarrow NH_2-\overset{\overset{\displaystyle R}{|}}{\underset{\underset{\displaystyle H}{|}}{C}}-COO^- + H^+$$

The H^+ is then able to react with any excess OH^- in the solution to form water. The free amino group at the other end of a protein can act as a base by combining with H^+ when pH falls, as follows:

$$NH_2-\overset{\overset{\displaystyle R}{|}}{\underset{\underset{\displaystyle H}{|}}{C}}-COOH + H^+ \longrightarrow {}^+NH_3-\overset{\overset{\displaystyle R}{|}}{\underset{\underset{\displaystyle H}{|}}{C}}-COOH$$

So proteins can buffer both acids and bases. In addition to the terminal carboxyl and amino groups, side chains that can buffer H^+ are present on seven of the 20 amino acids.

As we have already noted, the protein hemoglobin is an important buffer of H^+ in red blood cells (see Figure 23.23 on page 904). As blood flows through the systemic capillaries, carbon dioxide (CO_2) passes from tissue cells into red blood cells, where it combines with water (H_2O) to form carbonic acid (H_2CO_3). Once formed, H_2CO_3 dissociates into H^+ and HCO_3^-. At the same time that CO_2 is entering red blood cells, oxyhemoglobin (Hb—O_2) is giving up its oxygen to tissue cells. Reduced hemoglobin (deoxyhemoglobin) picks up most of the H^+. For this reason, reduced hemoglobin usually is written as Hb—H. The following reactions summarize these relations:

$$\underset{\substack{\text{Water}}}{H_2O} + \underset{\substack{\text{Carbon dioxide} \\ \text{(entering RBCs)}}}{CO_2} \longrightarrow \underset{\substack{\text{Carbonic acid}}}{H_2CO_3}$$

$$\underset{\substack{\text{Carbonic acid}}}{H_2CO_3} \longrightarrow \underset{\substack{\text{Hydrogen ion}}}{H^+} + \underset{\substack{\text{Bicarbonate ion}}}{HCO_3^-}$$

$$\underset{\substack{\text{Oxyhemoglobin} \\ \text{(in RBCs)}}}{Hb\text{–}O_2} + \underset{\substack{\text{Hydrogen ion} \\ \text{(from carbonic} \\ \text{acid)}}}{H^+} \longrightarrow \underset{\substack{\text{Reduced} \\ \text{hemoglobin}}}{Hb\text{–}H} + \underset{\substack{\text{Oxygen} \\ \text{(released to} \\ \text{tissue cells)}}}{O_2}$$

Carbonic Acid–Bicarbonate Buffer System

The **carbonic acid–bicarbonate buffer system** is based on the *bicarbonate ion* (HCO_3^-), which can act as a weak base, and *carbonic acid* (H_2CO_3), which can act as a weak acid. As you have already learned, HCO_3^- is a significant anion in both intracellular and extracellular fluids (see Figure 27.6). Because the kidneys also synthesize new HCO_3^- and reabsorb filtered HCO_3^-, this important buffer is not lost in the urine. If there is an excess of H^+, the HCO_3^- can function as a weak base and remove the excess H^+ as follows:

$$\underset{\substack{\text{Hydrogen ion}}}{H^+} + \underset{\substack{\text{Bicarbonate ion} \\ \text{(weak base)}}}{HCO_3^-} \longrightarrow \underset{\substack{\text{Carbonic acid}}}{H_2CO_3}$$

Then, H_2CO_3 dissociates into water and carbon dioxide, and the CO_2 is exhaled from the lungs.

Conversely, if there is a shortage of H^+, the H_2CO_3 can function as a weak acid and provide H^+ as follows:

$$\underset{\substack{\text{Carbonic acid} \\ \text{(weak acid)}}}{H_2CO_3} \longrightarrow \underset{\substack{\text{Hydrogen ion}}}{H^+} + \underset{\substack{\text{Bicarbonate ion}}}{HCO_3^-}$$

At a pH of 7.4, HCO_3^- concentration is about 24 mEq/liter and H_2CO_3 concentration is about 1.2 mmol/liter, so bicarbonate ions outnumber carbonic acid molecules by 20 to 1. Because CO_2 and H_2O combine to form H_2CO_3, this buffer system cannot protect against pH changes due to respiratory problems in which there is an excess or shortage of CO_2.

Phosphate Buffer System

The **phosphate buffer system** acts via a mechanism similar to the one for the carbonic acid–bicarbonate buffer system. The components of the phosphate buffer system are the ions *dihydrogen phosphate* ($H_2PO_4^-$) and *monohydrogen phosphate* (HPO_4^{2-}). Recall that phosphates are major anions in intracellular fluid and minor ones in extracellular fluids (Figure 27.6). The dihydrogen phosphate ion acts as a weak acid and is capable of buffering strong bases such as OH^-, as follows:

$$OH^- + H_2PO_4^- \longrightarrow H_2O + HPO_4^{2-}$$

Hydroxide ion	Dihydrogen	Water	Monohydrogen
(strong base)	phosphate		phosphate
	(weak acid)		(weak base)

The monohydrogen phosphate ion is capable of buffering the H^+ released by a strong acid such as hydrochloric acid (HCl) by acting as a weak base:

$$H^+ + HPO_4^{2-} \longrightarrow H_2PO_4^-$$

Hydrogen ion	Monohydrogen	Dihydrogen
(strong acid)	phosphate	phosphate
	(weak base)	(weak acid)

Because the concentration of phosphates is highest in intracellular fluid, the phosphate buffer system is an important regulator of pH in the cytosol. It also acts to a smaller degree in extracellular fluids and buffers acids in urine. $H_2PO_4^{2-}$ is formed when excess H^+ in the kidney tubule fluid combines with HPO_4^{2-} (see Figure 27.8). The H^+ that becomes part of the $H_2PO_4^-$ passes into the urine. This reaction is one way the kidneys help maintain blood pH by excreting H^+ in the urine.

Exhalation of Carbon Dioxide

The simple act of breathing also plays an important role in maintaining the pH of body fluids. An increase in the carbon dioxide (CO_2) concentration in body fluids increases H^+ concentration and thus lowers the pH (makes body fluids more acidic). Because H_2CO_3 can be eliminated by exhaling CO_2, it is called a **volatile acid.** Conversely, a decrease in the CO_2 concentration of body fluids raises the pH (makes body fluids more alkaline). This chemical interaction is illustrated by the following reversible reactions:

$$CO_2 + H_2O \rightleftharpoons H_2CO_3 \rightleftharpoons H^+ + HCO_3^-$$

| Carbon | Water | Carbonic | Hydrogen | Bicarbonate |
| dioxide | | acid | ion | ion |

Changes in the rate and depth of breathing can alter the pH of body fluids within a couple of minutes. With increased ventilation, more CO_2 is exhaled. When CO_2 levels decrease, the reaction is driven to the left (blue arrows), H^+ concentration falls, and blood pH increases. Doubling the ventilation increases pH by about 0.23 units, from 7.4 to 7.63. If ventilation is slower than normal, less carbon dioxide is exhaled. When CO_2 levels increase, the reaction is driven to the right (red arrows), the H^+ concentration increases, and blood pH decreases. Reducing ventilation to one-quarter of normal lowers the pH by 0.4 units, from 7.4 to 7.0. These examples show the powerful effect of alterations in breathing on the pH of body fluids.

The pH of body fluids and the rate and depth of breathing interact via a negative feedback loop (Figure 27.7). When the blood acidity increases, the decrease in pH (increase in concentration of H^+) is detected by central chemoreceptors in the medulla oblongata and peripheral chemoreceptors in the aortic

Figure 27.7 Negative feedback regulation of blood pH by the respiratory system.

Exhalation of carbon dioxide lowers the H^+ concentration of blood.

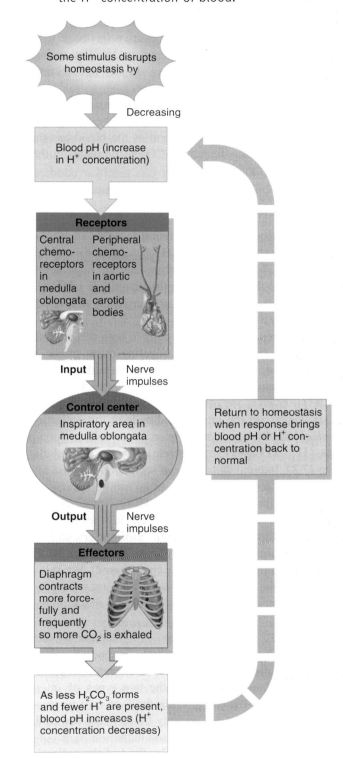

If you hold your breath for 30 seconds, what is likely to happen to your blood pH?

may occur with severe diarrhea or renal dysfunction; (2) accumulation of an acid other than carbonic acid, as may occur in ketosis (described on page 993); or (3) failure of the kidneys to excrete H^+ from metabolism of dietary proteins. If the problem is not too severe, hyperventilation can help bring blood pH into the normal range (respiratory compensation). Treatment of metabolic acidosis consists of administering intravenous solutions of sodium bicarbonate and correcting the cause of the acidosis.

Metabolic Alkalosis

In **metabolic alkalosis,** the systemic arterial blood HCO_3^- concentration is above 26 mEq/liter. A nonrespiratory loss of acid or excessive intake of alkaline drugs causes the blood pH to increase above 7.45. Excessive vomiting of gastric contents, which results in a substantial loss of hydrochloric acid, is probably the most frequent cause of metabolic alkalosis. Other causes include gastric suctioning, use of certain diuretics, endocrine disorders, excessive intake of alkaline drugs (antacids), and severe dehydration. Respiratory compensation through hypoventilation may bring blood pH into the normal range. Treatment of metabolic alkalosis consists of giving fluid solutions to correct Cl^-, K^+, and other electrolyte deficiencies plus correcting the cause of alkalosis.

Table 27.4 summarizes respiratory and metabolic acidosis and alkalosis.

• CLINICAL CONNECTION | Diagnosis of Acid–Base Imbalances

One can often pinpoint the cause of an acid–base imbalance by careful evaluation of three factors in a sample of systemic arterial blood: pH, concentration of HCO_3^-, and P_{CO_2}. These three blood chemistry values are examined in the following four-step sequence:

1. Note whether the pH is high (alkalosis) or low (acidosis).
2. Then decide which value—P_{CO_2} or HCO_3^-—is out of the normal range and could be the *cause* of the pH change. For example, *elevated pH* could be caused by *low* P_{CO_2} or *high* HCO_3^-.
3. If the cause is a *change in* P_{CO_2}, the problem is *respiratory*; if the cause is a *change in* HCO_3^-, the problem is *metabolic*.
4. Now look at the value that doesn't correspond with the observed pH change. If it is within its normal range, there is no compensation. If it is outside the normal range, compensation is occurring and partially correcting the pH imbalance. •

⊙ CHECKPOINT

8. Explain how each of the following buffer systems helps to maintain the pH of body fluids: proteins, carbonic acid–bicarbonate buffers, and phosphates.
9. Define acidosis and alkalosis. Distinguish among respiratory and metabolic acidosis and alkalosis.
10. What are the principal physiological effects of acidosis and alkalosis?

TABLE 27.4

Summary of Acidosis and Alkalosis

CONDITION	DEFINITION	COMMON CAUSES	COMPENSATORY MECHANISM
Respiratory acidosis	Increased P_{CO_2} (above 45 mmHg) and decreased pH (below 7.35) if there is no compensation.	Hypoventilation due to emphysema, pulmonary edema, trauma to respiratory center, airway obstructions, or dysfunction of muscles of respiration.	Renal: increased excretion of H^+; increased reabsorption of HCO_3^-. If compensation is complete, pH will be within the normal range but P_{CO_2} will be high.
Respiratory alkalosis	Decreased P_{CO_2} (below 35 mmHg) and increased pH (above 7.45) if there is no compensation.	Hyperventilation due to oxygen deficiency, pulmonary disease, cerebrovascular accident (CVA), or severe anxiety.	Renal: decreased excretion of H^+; decreased reabsorption of HCO_3^-. If compensation is complete, pH will be within the normal range but P_{CO_2} will be low.
Metabolic acidosis	Decreased HCO_3^- (below 22 mEq/liter) and decreased pH (below 7.35) if there is no compensation.	Loss of bicarbonate ions due to diarrhea, accumulation of acid (ketosis), renal dysfunction.	Respiratory: hyperventilation, which increases loss of CO_2. If compensation is complete, pH will be within the normal range but HCO_3^- will be low.
Metabolic alkalosis	Increased HCO_3^- (above 26 mEq/liter) and increased pH (above 7.45) if there is no compensation.	Loss of acid due to vomiting, gastric suctioning, or use of certain diuretics; excessive intake of alkaline drugs.	Respiratory: hypoventilation, which slows loss of CO_2. If compensation is complete, pH will be within the normal range but HCO_3^- will be high.

AGING AND FLUID, ELECTROLYTE, AND ACID–BASE BALANCE

◉ OBJECTIVE

- Describe the changes in fluid, electrolyte, and acid–base balance that may occur with aging.

There are significant differences between adults and infants, especially premature infants, with respect to fluid distribution, regulation of fluid and electrolyte balance, and acid–base homeostasis. Accordingly, infants experience more problems than adults in these areas. The differences are related to the following conditions:

- *Proportion and distribution of water.* A newborn's total body mass is about 75% water (and can be as high as 90% in a premature infant); an adult's total body mass is about 55–60% water. (The "adult" percentage is achieved at about 2 years of age.) Adults have twice as much water in ICF as ECF, but the opposite is true in premature infants. Because ECF is subject to more changes than ICF, rapid losses or gains of body water are much more critical in infants. Given that the rate of fluid intake and output is about seven times higher in infants than in adults, the slightest changes in fluid balance can result in severe abnormalities.

- *Metabolic rate.* The metabolic rate of infants is about double that of adults. This results in the production of more metabolic wastes and acids, which can lead to the development of acidosis in infants.

- *Functional development of the kidneys.* Infant kidneys are only about half as efficient in concentrating urine as those of adults. (Functional development is not complete until close to the end of the first month after birth.) As a result, the kidneys of newborns can neither concentrate urine nor rid the body of excess acids as effectively as those of adults.

- *Body surface area.* The ratio of body surface area to body volume of infants is about three times greater than that of adults. Water loss through the skin is significantly higher in infants than in adults.

- *Breathing rate.* The higher breathing rate of infants (about 30 to 80 times a minute) causes greater water loss from the lungs. Respiratory alkalosis may occur because greater ventilation eliminates more CO_2 and lowers the P_{CO_2}.

- *Ion concentrations.* Newborns have higher K^+ and Cl^- concentrations than adults. This creates a tendency toward metabolic acidosis.

By comparison with children and younger adults, older adults often have an impaired ability to maintain fluid, electrolyte, and acid–base balance. With increasing age, many people have a decreased volume of intracellular fluid and decreased total body K^+ due to declining skeletal muscle mass and increasing mass of adipose tissue (which contains very little water). Age-related decreases in respiratory and renal functioning may compromise acid–base balance by slowing the exhalation of CO_2 and the excretion of excess acids in urine. Other kidney changes, such as decreased blood flow, decreased glomerular filtration rate, and reduced sensitivity to antidiuretic hormone, have an adverse effect on the ability to maintain fluid and electrolyte balance. Due to a decrease in the number and efficiency of sweat glands, water loss from the skin declines with age. Because of these age-related changes, older adults are susceptible to several fluid and electrolyte disorders:

- *Dehydration* and *hypernatremia* often occur due to inadequate fluid intake or loss of more water than Na^+ in vomit, feces, or urine.

- *Hyponatremia* may occur due to inadequate intake of Na^+; elevated loss of Na^+ in urine, vomit, or diarrhea; or impaired ability of the kidneys to produce dilute urine.

- *Hypokalemia* often occurs in older adults who chronically use laxatives to relieve constipation or who take K^+-depleting diuretic drugs for treatment of hypertension or heart disease.

- *Acidosis* may occur due to impaired ability of the lungs and kidneys to compensate for acid–base imbalances. One cause of acidosis is decreased production of ammonia (NH_3) by renal tubule cells, which then is not available to combine with H^+ and be excreted in urine as NH_4^+; another cause is reduced exhalation of CO_2.

◉ CHECKPOINT

11. Why do infants experience greater problems with fluid, electrolyte, and acid–base balance than adults?

STUDY OUTLINE

Fluid Compartments and Fluid Balance (p. 1063)

1. Body fluid includes water and dissolved solutes.
2. About two-thirds of the body's fluid is located within cells and is called intracellular fluid (ICF). The other one-third, called extracellular fluid (ECF), includes interstitial fluid; blood plasma and lymph; cerebrospinal fluid; gastrointestinal tract fluids; synovial fluid; fluids of the eyes and ears; pleural, pericardial, and peritoneal fluids; and glomerular filtrate.

3. Fluid balance means that the required amounts of water and solutes are present and are correctly proportioned among the various compartments.
4. An inorganic substance that dissociates into ions in solution is called an electrolyte.
5. Water is the largest single constituent in the body. It makes up 45–75% of total body mass, depending on age, gender, and the amount of adipose tissue present.

6. Daily water gain and loss are each about 2500 mL. Sources of water gain are ingested liquids and foods, and water produced by cellular respiration and dehydration synthesis reactions (metabolic water). Water is lost from the body via urination, evaporation from the skin surface, exhalation of water vapor, and defecation. In women, menstrual flow is an additional route for loss of body water.

7. Body water gain is regulated by adjusting the volume of water intake, mainly by drinking more or less fluid. The thirst center in the hypothalamus governs the urge to drink.

8. Although increased amounts of water and solutes are lost through sweating and exhalation during exercise, loss of excess body water or excess solutes depends mainly on regulating excretion in the urine. The extent of urinary NaCl loss is the main determinant of body fluid volume; the extent of urinary water loss is the main determinant of body fluid osmolarity.

9. Table 27.1 on page 1066 summarizes the factors that regulate water gain and water loss in the body.

10. Angiotensin II and aldosterone reduce urinary loss of Na^+ and Cl^- and thereby increase the volume of body fluids. ANP promotes natriuresis, elevated excretion of Na^+ (and Cl^-), which decreases blood volume.

11. The major hormone that regulates water loss and thus body fluid osmolarity is antidiuretic hormone (ADH).

12. An increase in the osmolarity of interstitial fluid draws water out of cells, and they shrink slightly. A decrease in the osmolarity of interstitial fluid causes cells to swell. Most often a change in osmolarity is due to a change in the concentration of Na^+, the dominant solute in interstitial fluid.

13. When a person consumes water faster than the kidneys can excrete it or when renal function is poor, the result may be water intoxication, in which cells swell dangerously.

Electrolytes in Body Fluids (p. 1067)

1. Ions formed when electrolytes dissolve in body fluids control the osmosis of water between fluid compartments, help maintain acid–base balance, and carry electrical current.

2. The concentrations of cations and anions is expressed in units of milliequivalents/liter (mEq/liter).

3. Blood plasma, interstitial fluid, and intracellular fluid contain varying types and amounts of ions.

4. Sodium ions (Na^+) are the most abundant extracellular ions. They are involved in impulse transmission, muscle contraction, and fluid and electrolyte balance. Na^+ level is controlled by aldosterone, antidiuretic hormone, and atrial natriuretic peptide.

5. Chloride ions (Cl^-) are the major extracellular anions. They play a role in regulating osmotic pressure and forming HCl in gastric juice. Cl^- level is controlled indirectly by antidiuretic hormone and by processes that increase or decrease renal reabsorption of Na^+.

6. Potassium ions (K^+) are the most abundant cations in intracellular fluid. They play a key role in the resting membrane potential and action potential of neurons and muscle fibers; help maintain intracellular fluid volume; and contribute to regulation of pH. K^+ level is controlled by aldosterone.

7. Bicarbonate ions (HCO_3^-) are the second most abundant anions in extracellular fluid. They are the most important buffer in blood plasma.

8. Calcium is the most abundant mineral in the body. Calcium salts are structural components of bones and teeth. Ca^{2+}, which are principally extracellular cations, function in blood clotting, neurotransmitter release, and contraction of muscle. Ca^{2+} level is controlled mainly by parathyroid hormone and calcitriol.

9. Phosphate ions ($H_2PO_4^-$, HPO_4^{2-}, and PO_4^{3-}) are principally intracellular anions, and their salts are structural components of bones and teeth. They are also required for the synthesis of nucleic acids and ATP and participate in buffer reactions. Their level is controlled by parathyroid hormone and calcitriol.

10. Magnesium ions (Mg^{2+}) are primarily intracellular cations. They act as cofactors in several enzyme systems.

11. Table 27.2 on page 1071 describes the imbalances that result from deficiency or excess of important body electrolytes.

Acid–Base Balance (p. 1070)

1. The overall acid–base balance of the body is maintained by controlling the H^+ concentration of body fluids, especially extracellular fluid.

2. The normal pH of systemic arterial blood is 7.35–7.45.

3. Homeostasis of pH is maintained by buffer systems, via exhalation of carbon dioxide, and via kidney excretion of H^+ and reabsorption of HCO_3^-.

4. The important buffer systems include proteins, carbonic acid–bicarbonate buffers, and phosphates.

5. An increase in exhalation of carbon dioxide increases blood pH; a decrease in exhalation of CO_2 decreases blood pH.

6. In the proximal convoluted tubules of the kidneys, Na^+/H^+ antiporters secrete H^+ as they reabsorb Na^+. In the collecting ducts of the kidneys, some intercalated cells reabsorb K^+ and HCO_3^- and secrete H^+; other intercalated cells secrete HCO_3^-. In these ways, the kidneys can increase or decrease the pH of body fluids.

7. Table 27.3 on page 1075 summarizes the mechanisms that maintain pH of body fluids.

8. Acidosis is a systemic arterial blood pH below 7.35; its principal effect is depression of the central nervous system (CNS). Alkalosis is a systemic arterial blood pH above 7.45; its principal effect is overexcitability of the CNS.

9. Respiratory acidosis and alkalosis are disorders due to changes in blood P_{CO_2}; metabolic acidosis and alkalosis are disorders associated with changes in blood HCO_3^- concentration.

10. Metabolic acidosis or alkalosis can be compensated by respiratory mechanisms (respiratory compensation); respiratory acidosis or alkalosis can be compensated by renal mechanisms (renal compensation).

11. Table 27.4 on page 1076 summarizes the effects of respiratory and metabolic acidosis and alkalosis.

12. By examining systemic arterial blood pH, HCO_3^-, and P_{CO_2} values, it is possible to pinpoint the cause of an acid–base imbalance.

Aging and Fluid, Electrolyte, and Acid–Base Balance (p. 1077)

1. With increasing age, there is decreased intracellular fluid volume and decreased K^+ due to declining skeletal muscle mass.

2. Decreased kidney function with aging adversely affects fluid and electrolyte balance.

Q SELF-QUIZ QUESTIONS

Fill in the blanks in the following statements.

1. The source of water that is derived from aerobic cellular respiration and dehydration synthesis reactions is _____ water.

2. In the carbonic acid–bicarbonate buffer system, the _____ acts as a weak base, and _____ acts as a weak acid.

Indicate whether the following statements are true or false.

3. The phosphate buffer system is an important regulator of pH in the cytosol.

4. The two compartments in which water can be found are plasma and cytosol.

Choose the one best answer to the following questions.

5. The primary means of regulating body water gain is adjusting
   (a) the volume of water intake.
   (b) the rate of cellular respiration.
   (c) the formation of metabolic water.
   (d) the volume of metabolic water.
   (e) the metabolic use of water.

6. Which of the following stimulate thirst? (1) a decreased production of saliva, (2) a decrease in nerve impulses from hypothalamic osmoreceptors, (3) an increase in osmolarity of body fluids, (4) angiotensin II release, (5) release of atrial natriuretic peptide, (6) an increase in blood volume.
   (a) 1, 2, 4, and 6   (b) 1, 3, 5, and 6   (c) 1, 3, and 4
   (d) 2, 4, and 6   (e) 1, 4, 5, and 6

7. Which of the following is *not* true concerning the protein buffer system?
   (a) Albumin is considered the main protein buffer in blood plasma.
   (b) Albumin is the most abundant buffer in blood plasma and intracellular fluid.
   (c) The functional components of a protein buffer system are the carboxyl group and the amino group.
   (d) Protein buffers are the primary buffers of acids in urine.
   (e) Proteins can buffer both acids and bases.

8. Which of the following statements are *true*? (1) Buffers prevent rapid, drastic changes in pH of a body fluid. (2) Buffers work slowly. (3) Strong acids lower pH more than weak acids because strong acids contribute fewer H^+. (4) Most buffers consist of a weak acid and the salt of that acid, which acts as weak base. (5) Hemoglobin is an important buffer.
   (a) 1, 2, 3, and 5   (b) 1, 3, 4, and 5   (c) 1, 3, and 5
   (d) 1, 4, and 5   (e) 2, 3, and 5

9. Which of the following hormones regulate fluid loss? (1) antidiuretic hormone, (2) aldosterone, (3) atrial natriuretic peptide, (4) thyroxine, (5) cortisol.
   (a) 1, 3, and 5   (b) 1, 2, and 3   (c) 2, 4, and 5
   (d) 2, 3, and 4   (e) 1, 3, and 4

10. Which of the following are true concerning ions in the body? (1) They control osmosis of water between fluid compartments. (2) They help maintain acid–base balance. (3) They carry electrical current. (4) They serve as cofactors for enzyme activity. (5) They serve as neurotransmitters under special circumstances.
    (a) 1, 3, and 5   (b) 2, 4, and 5   (c) 1, 4, and 5
    (d) 1, 2, and 4   (e) 1, 2, 3, and 4

11. Which of the following statements are *true*? (1) An increase in the carbon dioxide concentration in body fluids increases H^+ concentration and thus lowers pH. (2) Breath holding results in a decline in blood pH. (3) The respiratory buffer mechanism can eliminate a single volatile acid: carbonic acid. (4) The only way to eliminate nonvolatile acids is to excrete H^+ in the urine. (5) When the diet contains a large amount of protein, normal metabolism produces more acids than bases.
    (a) 1, 2, 3, 4, and 5   (b) 1, 3, 4, and 5   (c) 1, 2, 3, and 4
    (d) 1, 2, 4, and 5   (e) 1, 3, and 4

12. Concerning acid–base imbalances: (1) Acidosis can cause depression of the central nervous system through depression of synaptic transmission. (2) Renal compensation can resolve respiratory alkalosis or acidosis. (3) A major physiological effect of alkalosis is lack of excitability in the central nervous system and peripheral nerves. (4) Resolution of metabolic acidosis and alkalosis occurs through renal compensation. (5) In adjusting blood pH, renal compensation occurs quickly, but respiratory compensation takes days.
    (a) 1, 2, and 5   (b) 1 and 2   (c) 2, 3, and 4
    (d) 2, 3, and 5   (e) 1, 2, 3, and 5

13. Which of the following are *mismatched*?
    (a) hypoventilation: respiratory alkalosis
    (b) severe diarrhea: metabolic acidosis
    (c) excessive vomiting: metabolic alkalosis
    (d) airway obstruction: respiratory acidosis
    (e) inability of kidneys to excrete H^+ from dietary protein metabolism: metabolic acidosis

14. Match the following:
    ____(a) the most abundant cation in intracellular fluid; plays a key role in establishing the resting membrane potential
    ____(b) the most abundant mineral in the body; plays important roles in blood clotting, neurotransmitter release, maintenance of muscle tone, and excitability of nervous and muscle tissue
    ____(c) second most common intracellular cation; is a cofactor for enzymes involved in carbohydrate, protein, and Na^+/K^+ ATPase metabolism
    ____(d) the most abundant extracellular cation; essential in fluid and electrolyte balance
    ____(e) ions that are mostly combined with lipids, proteins, carbohydrates, nucleic acids, and ATP inside cells
    ____(f) most prevalent extracellular anion; can help balance the level of anions in different fluid compartments
    ____(g) second most prevalent extracellular anion; mainly regulated by the kidneys; important for acid–base balance
    ____(h) substances that act to prevent rapid, drastic changes in the pH of a body fluid
    ____(i) inorganic substances that dissociate into ions when in solution

    (1) sodium
    (2) chloride
    (3) electrolytes
    (4) bicarbonate
    (5) buffers
    (6) phosphate
    (7) magnesium
    (8) potassium
    (9) calcium

MALE REPRODUCTIVE SYSTEM

◉ OBJECTIVES

- Describe the location, structure, and functions of the organs of the male reproductive system.
- Discuss the process of spermatogenesis in the testes.

The organs of the **male reproductive system** include the testes, a system of ducts (including the epididymis, ductus deferens, ejaculatory ducts, and urethra), accessory sex glands (seminal vesicles, prostate, and bulbourethral glands), and several supporting structures, including the scrotum and the penis (Figure 28.1). The testes (male gonads) produce sperm and secrete hormones. The duct system transports and stores sperm, assists in their maturation, and conveys them to the exterior. Semen contains sperm plus the secretions provided by the accessory sex glands. The supporting structures have various functions. The penis delivers sperm into the female reproductive tract and the scrotum supports the testes.

Scrotum

The **scrotum** (SKRŌ-tum = bag), the supporting structure for the testes, consists of loose skin and underlying subcutaneous layer that hangs from the root (attached portion) of the penis (Figure 28.1a). Externally, the scrotum looks like a single pouch of skin separated into lateral portions by a median ridge called the **raphe** (RĀ-fē = seam). Internally, the **scrotal septum** divides the scrotum into two sacs, each containing a single testis (Figure 28.2 on page 1084). The septum is made up of a subcutaneous layer and muscle tissue called the **dartos muscle** (DAR-tōs = skinned), which is composed of bundles of smooth muscle fibers.

Figure 28.1 Male organs of reproduction and surrounding structures.

 Reproductive organs are adapted for producing new individuals and passing on genetic material from one generation to the next.

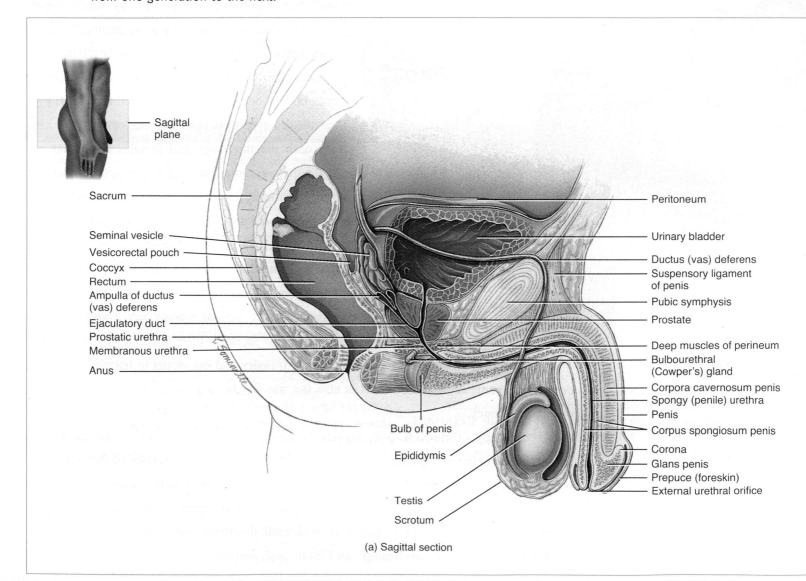

(a) Sagittal section

The dartos muscle is also found in the subcutaneous layer of the scrotum. Associated with each testis in the scrotum is the **cremaster muscle** (krē-MAS-ter = suspender), a series of small bands of skeletal muscle that descend as an extension of the internal oblique muscle through the spermatic cord to surround the testes.

The location of the scrotum and the contraction of its muscle fibers regulate the temperature of the testes. Normal sperm production requires a temperature about 2–3°C below core body temperature. This lowered temperature is maintained within the scrotum because it is outside the pelvic cavity. In response to cold temperatures, the cremaster and dartos muscles contract. Contraction of the cremaster muscles moves the testes closer to the body, where they can absorb body heat. Contraction of the dartos muscle causes the scrotum to become tight (wrinkled in appearance), which reduces heat loss. Exposure to warmth reverses these actions.

Testes

The **testes** (TES-tēz), or **testicles,** are paired oval glands in the scrotum measuring about 5 cm (2 in.) long and 2.5 cm (1 in.) in diameter (Figure 28.3 on page 1085). Each testis (singular) has a mass of 10–15 grams. The testes develop near the kidneys, in the posterior portion of the abdomen, and they usually begin their descent into the scrotum through the inguinal canals (passageways in the anterior abdominal wall; see Figure 28.2) during the latter half of the seventh month of fetal development.

A serous membrane called the **tunica vaginalis** (*tunica* = sheath), which is derived from the peritoneum and forms during the descent of the testes, partially covers the testes. A collection of serous fluid in the tunica vaginalis is called a **hydrocele** (HĪ-drō-sēl; *hydro-* = water; *-kele* = hernia). It may be caused by injury to the testes or inflammation of the epididymis. Usually, no

---

Functions of the Male Reproductive System

1. The testes produce sperm and the male sex hormone testosterone.
2. The ducts transport, store, and assist in maturation of sperm.
3. The accessory sex glands secrete most of the liquid portion of semen.
4. The penis contains the urethra, a passageway for ejaculation of semen and excretion of urine.

SUPERIOR

Seminal vesicle

Ductus (vas) deferens

Prostatic urethra

Ejaculatory duct

Membranous urethra

Rectum

Bulb of penis

Bulbospongiosus muscle

Urinary bladder (opened)

Prostate

Pubic symphysis

Corpora cavernosum penis

Corpus spongiosum penis

Spongy (penile) urethra

Testis

Corona

Glans penis

POSTERIOR

ANTERIOR

(b) Sagittal section

❓ What are the groups of reproductive organs in males, and what are the functions of each group?

Figure 28.2 **The scrotum, the supporting structure for the testes.**

🔓 ⚿ The scrotum consists of loose skin and an underlying subcutaneous layer and supports the testes.

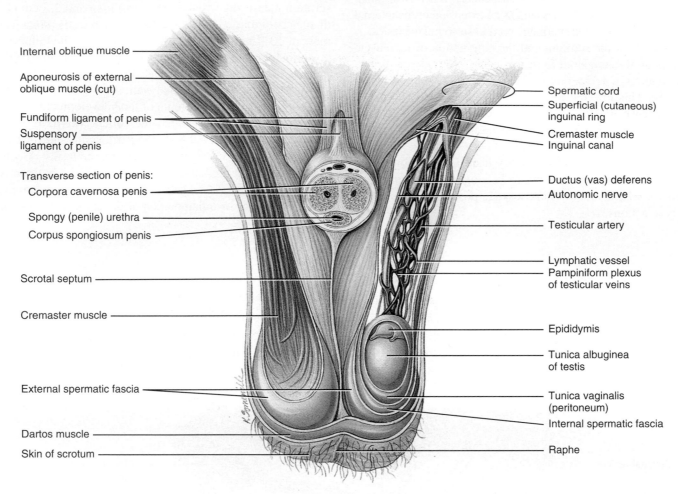

Internal oblique muscle

Aponeurosis of external oblique muscle (cut)

Fundiform ligament of penis

Suspensory ligament of penis

Transverse section of penis:
   Corpora cavernosa penis

   Spongy (penile) urethra
   Corpus spongiosum penis

Scrotal septum

Cremaster muscle

External spermatic fascia

Dartos muscle

Skin of scrotum

Spermatic cord
Superficial (cutaneous) inguinal ring
Cremaster muscle
Inguinal canal

Ductus (vas) deferens
Autonomic nerve

Testicular artery

Lymphatic vessel
Pampiniform plexus of testicular veins

Epididymis

Tunica albuginea of testis

Tunica vaginalis (peritoneum)
Internal spermatic fascia

Raphe

Anterior view of scrotum and testes and transverse section of penis

❓ **Which muscles help regulate the temperature of the testes?**

treatment is required. Internal to the tunica vaginalis is a white fibrous capsule composed of dense irregular connective tissue, the **tunica albuginea** (al′-bū-JIN-ē-a; *albu-* = white); it extends inward, forming septa that divide the testis into a series of internal compartments called **lobules.** Each of the 200–300 lobules contains one to three tightly coiled tubules, the **seminiferous tubules** (*semin-* = seed; *fer-* = to carry), where sperm are produced. The process by which the seminiferous tubules of the testes produce sperm is called spermatogenesis (sper′-ma-tō-JEN-e-sis; *genesis* = beginning process or production).

The seminiferous tubules contain two types of cells: **spermatogenic cells,** the sperm-forming cells, and **Sertoli cells,** which have several functions in supporting spermatogenesis (Figure 28.4 on page 1086). Stem cells called **spermatogonia** (sper′-ma-tō-GŌ-nē-a; *-gonia* = offspring; singular is *spermatogonium*) develop from **primordial germ cells** (*primordi-* = primitive or early form) that arise from the yolk sac and enter the testes during the fifth week of development. In the embryonic testes, the primordial germ cells differentiate into spermatogonia, which remain dormant during childhood and actively begin producing

sperm at puberty. Toward the lumen of the seminiferous tubule are layers of progressively more mature cells. In order of advancing maturity, these are primary spermatocytes, secondary spermatocytes, spermatids, and sperm cells. After a **sperm cell,** or **spermatozoon** (sper′-ma-tō-ZŌ-on; *-zoon* = life) has formed, it is released into the lumen of the seminiferous tubule. (The plural terms are *sperm* and *spermatozoa*.)

Embedded among the spermatogenic cells in the seminiferous tubules are large **Sertoli cells** or *sustentacular cells* (sus′-ten-TAK-ū-lar), which extend from the basement membrane to the lumen of the tubule. Internal to the basement membrane and spermatogonia, tight junctions join neighboring Sertoli cells to one another. These junctions form an obstruction known as the **blood–testis barrier** because substances must first pass through the Sertoli cells before they can reach the developing sperm. By isolating the developing gametes from the blood, the blood–testis barrier prevents an immune response against the spermatogenic cell's surface antigens, which are recognized as "foreign" by the immune system. The blood–testis barrier does not include spermatogonia.

Figure 28.3 Internal and external anatomy of a testis.

The testes are the male gonads, which produce haploid sperm.

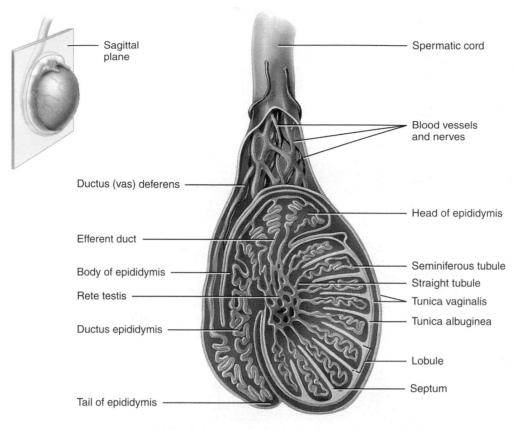

- Sagittal plane
- Spermatic cord
- Blood vessels and nerves
- Ductus (vas) deferens
- Head of epididymis
- Efferent duct
- Body of epididymis
- Seminiferous tubule
- Straight tubule
- Rete testis
- Tunica vaginalis
- Tunica albuginea
- Ductus epididymis
- Lobule
- Septum
- Tail of epididymis

(a) Sagittal section of a testis showing seminiferous tubules

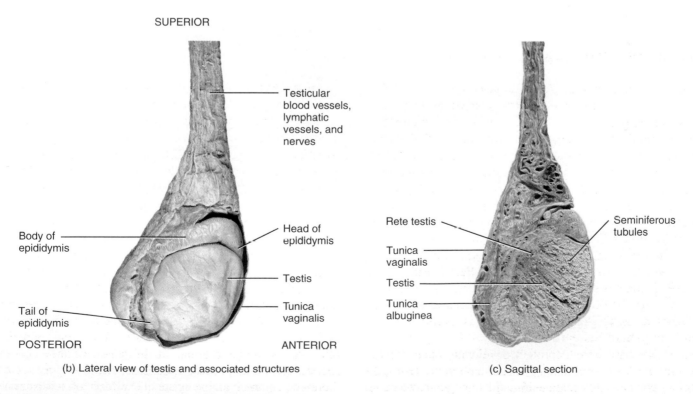

SUPERIOR

- Testicular blood vessels, lymphatic vessels, and nerves
- Body of epididymis
- Head of epididymis
- Tail of epididymis
- Testis
- Tunica vaginalis

POSTERIOR   ANTERIOR

(b) Lateral view of testis and associated structures

- Rete testis
- Seminiferous tubules
- Tunica vaginalis
- Testis
- Tunica albuginea

(c) Sagittal section

? **What tissue layers cover and protect the testes?**

Figure 28.4 Microscopic anatomy of the seminiferous tubules and stages of sperm production (spermatogenesis). Arrows in (b) indicate the progression of spermatogenic cells from least mature to most mature. The (*n*) and (2*n*) refer to haploid and diploid numbers of chromosomes, respectively.

🔑 Spermatogenesis occurs in the seminiferous tubules of the testes.

Transverse plane

Basement membrane

Sertoli cell

Spermatid (*n*)

Secondary spermatocyte (*n*)

Primary spermatocyte (2*n*)

Spermatogonium (2*n*) (stem cell)

LM 160x

(a) Transverse section of several seminiferous tubules

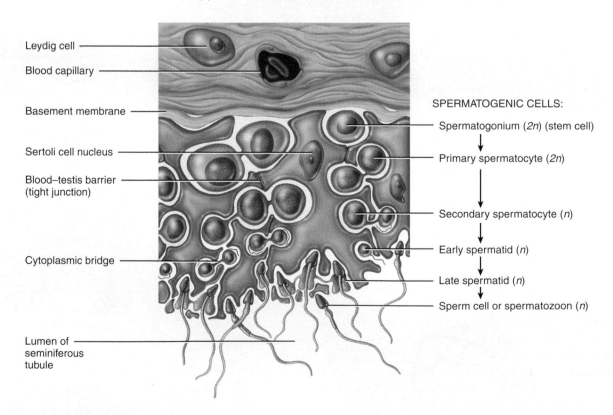

Leydig cell

Blood capillary

Basement membrane

Sertoli cell nucleus

Blood–testis barrier (tight junction)

Cytoplasmic bridge

Lumen of seminiferous tubule

SPERMATOGENIC CELLS:

Spermatogonium (2*n*) (stem cell)
↓
Primary spermatocyte (2*n*)
↓
Secondary spermatocyte (*n*)
↓
Early spermatid (*n*)
↓
Late spermatid (*n*)
↓
Sperm cell or spermatozoon (*n*)

(b) Transverse section of a portion of a seminiferous tubule

❓ **Which cells secrete testosterone?**

Sertoli cells support and protect developing spermatogenic cells in several ways. They nourish spermatocytes, spermatids, and sperm; phagocytize excess spermatid cytoplasm as development proceeds; and control movements of spermatogenic cells and the release of sperm into the lumen of the seminiferous tubule. They also produce fluid for sperm transport, secrete the hormone inhibin, and regulate the effects of testosterone and FSH (follicle-stimulating hormone).

In the spaces between adjacent seminiferous tubules are clusters of cells called **Leydig (interstitial) cells** (Figure 28.4). These cells secrete testosterone, the most prevalent androgen. An **androgen** is a hormone that promotes the development of masculine characteristics. Testosterone also promotes a man's *libido* (sexual drive).

Spermatogenesis

Before you read this section, please review the topic of reproductive cell division in Chapter 3 on pages 97–99. Pay particular attention to Figures 3.33 and 3.34, which appear on pages 98 and 99, respectively.

In humans, spermatogenesis takes 65–75 days. It begins with the spermatogonia, which contain the diploid (2*n*) number of chromosomes (Figure 28.5). Spermatogonia are types of *stem cells;* when they undergo mitosis, some spermatogonia remain near the basement membrane of the seminiferous tubule in an undifferentiated state to serve as a reservoir of cells for future cell division and subsequent sperm production. The rest of the spermatogonia lose contact with the basement membrane, squeeze through the tight junctions of the blood–testis barrier, undergo developmental changes, and differentiate into **primary spermatocytes** (SPER-ma-tō-sītz′). Primary spermatocytes, like spermatogonia, are diploid (2*n*); that is, they have 46 chromosomes.

Shortly after it forms, each primary spermatocyte replicates its DNA and then meiosis begins (Figure 28.5). In meiosis I, homologous pairs of chromosomes line up at the metaphase plate, and crossing-over occurs. Then, the meiotic spindle pulls one (duplicated) chromosome of each pair to an opposite pole of the dividing cell. The two cells formed by meiosis I are called **secondary spermatocytes.** Each secondary spermatocyte has 23 chromosomes, the haploid number (*n*). Each chromosome within a secondary spermatocyte, however, is made up of two chromatids (two copies of the DNA) still attached by a centromere. No replication of DNA occurs in the secondary spermatocytes.

In meiosis II, the chromosomes line up in single file along the metaphase plate, and the two chromatids of each chromosome separate. The four haploid cells resulting from meiosis II are called **spermatids.** A single primary spermatocyte therefore produces four spermatids via two rounds of cell division (meiosis I and meiosis II).

A unique process occurs during spermatogenesis. As spermatogenic cells proliferate, they fail to complete cytoplasmic separation (cytokinesis). The cells remain in contact via cytoplasmic bridges through their entire development (see Figures 28.4b and 28.5). This pattern of development most likely accounts for the synchronized production of sperm in any given area of seminiferous tubule. It may also have survival value in that half of the sperm contain an X chromosome and half contain a Y chromosome. The larger X chromosome may carry genes needed for spermatogenesis that are lacking on the smaller Y chromosome.

The final stage of spermatogenesis, **spermiogenesis** (sper′-mē-ō-JEN-e-sis), is the development of haploid spermatids into sperm. No cell division occurs in spermiogenesis; each spermatid becomes a single **sperm cell.** During this process, spherical spermatids transform into elongated, slender sperm. An acrosome (described shortly) forms atop the nucleus, which condenses and elongates, a flagellum develops, and mitochondria multiply. Sertoli cells dispose of the excess cytoplasm that sloughs off. Finally, sperm are released from their connections to

Figure 28.5 Events in spermatogenesis. Diploid cells (2*n*) have 46 chromosomes; haploid cells (*n*) have 23 chromosomes.

Spermiogenesis involves the maturation of spermatids into sperm.

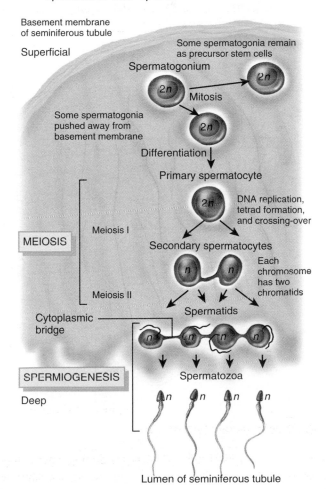

What is "reduced" during meiosis I?

Sertoli cells, an event known as **spermiation.** Sperm then enter the lumen of the seminiferous tubule. Fluid secreted by Sertoli cells pushes sperm along their way, toward the ducts of the testes. At this point, sperm are not yet able to swim.

Sperm

Each day about 300 million sperm complete the process of spermatogenesis. A sperm is about 60 μm long and contains several structures that are highly adapted for reaching and penetrating a secondary oocyte (Figure 28.6). The major parts of a sperm are the head and the tail. The flattened, pointed **head** of the sperm is about 4–5 μm long. It contains a **nucleus** with 23 highly condensed chromosomes. Covering the anterior two-thirds of the nucleus is the **acrosome** (*acro-* = atop; *-some* = body), a caplike vesicle filled with enzymes that help a sperm to penetrate a secondary oocyte to bring about fertilization. Among the enzymes are hyaluronidase and proteases. The **tail** of a sperm is subdivided into four parts: neck, middle piece, principal piece, and end piece. The **neck** is the constricted region just behind the head that contains centrioles. The centrioles form the microtubules that comprise the remainder of the tail. The **middle piece** contains mitochondria arranged in a spiral, which provide the energy (ATP) for locomotion of sperm to the site of fertilization and for sperm metabolism. The **principal piece** is the longest portion of the tail, and the **end piece** is the terminal, tapering portion of the tail. Once ejaculated, most sperm do not survive more than 48 hours within the female reproductive tract.

Figure 28.6 Parts of a sperm cell.

About 300 million sperm mature each day.

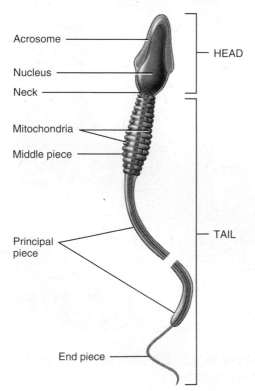

Acrosome

Nucleus

Neck

Mitochondria

Middle piece

Principal piece

End piece

HEAD

TAIL

? What are the functions of each part of a sperm cell?

Hormonal Control of the Testes

Although the initiating factors are unknown, at puberty certain hypothalamic neurosecretory cells increase their secretion of **gonadotropin-releasing hormone (GnRH).** This hormone in turn stimulates gonadotrophs in the anterior pituitary to increase their secretion of the two gonadotropins, **luteinizing hormone (LH)** and **follicle-stimulating hormone (FSH).** Figure 28.7 shows the hormones and negative feedback loops that control secretion of testosterone and spermatogenesis.

LH stimulates Leydig cells, which are located between seminiferous tubules, to secrete the hormone **testosterone** (tes-TOS-te-rōn). This steroid hormone is synthesized from cholesterol in the testes and is the principal androgen. It is lipid-soluble and readily diffuses out of Leydig cells into the interstitial fluid and then into blood. Via negative feedback, testosterone suppresses secretion of LH by anterior pituitary gonadotrophs and suppresses secretion of GnRH by hypothalamic neurosecretory cells. In some target cells, such as those in the external genitals and prostate, the enzyme 5 alpha-reductase converts testosterone to another androgen called **dihydrotestosterone (DHT).**

FSH acts indirectly to stimulate spermatogenesis (Figure 28.7). FSH and testosterone act synergistically on the Sertoli cells to stimulate secretion of **androgen-binding protein (ABP)** into the lumen of the seminiferous tubules and into the interstitial fluid around the spermatogenic cells. ABP binds to testosterone, keeping its concentration high. Testosterone stimulates the final steps of spermatogenesis in the seminiferous tubules. Once the degree of spermatogenesis required for male reproductive functions has been achieved, Sertoli cells release **inhibin,** a protein hormone named for its role in inhibiting FSH secretion by the anterior pituitary (Figure 28.7). If spermatogenesis is proceeding too slowly, less inhibin is released, which permits more FSH secretion and an increased rate of spermatogenesis.

Testosterone and dihydrotestosterone both bind to the same androgen receptors, which are found within the nuclei of target cells. The hormone–receptor complex regulates gene expression, turning some genes on and others off. Because of these changes, the androgens produce several effects:

- *Prenatal development.* Before birth, testosterone stimulates the male pattern of development of reproductive system ducts and the descent of the testes. Dihydrotestosterone stimulates development of the external genitals (described on page 1121). Testosterone also is converted in the brain to estrogens (feminizing hormones), which may play a role in the development of certain regions of the brain in males.

- *Development of male sexual characteristics.* At puberty, testosterone and dihydrotestosterone bring about development and enlargement of the male sex organs and the development of masculine secondary sexual characteristics. *Secondary sex characteristics* are traits that distinguish males and females but do not have a direct role in reproduction. These include muscular and skeletal growth that results in wide shoulders and narrow hips; facial and chest hair (within hereditary limits) and more hair on other parts of the body; thickening of the skin; increased sebaceous (oil) gland secretion; and enlargement of the larynx and consequent deepening of the voice.

- *Development of sexual function.* Androgens contribute to male sexual behavior and spermatogenesis and to sex drive (libido) in both males and females. Recall that the adrenal cortex is the main source of androgens in females.

- *Stimulation of anabolism.* Androgens are anabolic hor-

mones; that is, they stimulate protein synthesis. This effect is obvious in the heavier muscle and bone mass of most men as compared to women.

A negative feedback system regulates testosterone production (Figure 28.8). When testosterone concentration in the blood increases to a certain level, it inhibits the release of GnRH by cells in the hypothalamus. As a result, there is less GnRH in the portal

Figure 28.7 **Hormonal control of spermatogenesis and actions of testosterone and dihydrotestosterone (DHT).** In response to stimulation by FSH and testosterone, Sertoli cells secrete androgen-binding protein (ABP). Dashed red lines indicate negative feedback inhibition.

Release of FSH is stimulated by GnRH and inhibited by inhibin; release of LH is stimulated by GnRH and inhibited by testosterone.

Figure 28.8 **Negative feedback control of blood level of testosterone.**

Gonadotrophs of the anterior pituitary produce luteinizing hormone (LH).

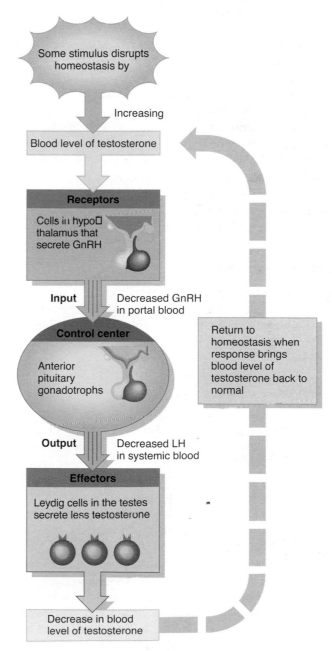

? **Which hormones inhibit secretion of FSH and LH by the anterior pituitary?**

? **Which cells secrete inhibin?**

blood that flows from the hypothalamus to the anterior pituitary. Gonadotrophs in the anterior pituitary then release less LH, so the concentration of LH in systemic blood falls. With less stimulation by LH, the Leydig cells in the testes secrete less testosterone, and there is a return to homeostasis. If the testosterone concentration in the blood falls too low, however, GnRH is again released by the hypothalamus and stimulates secretion of LH by the anterior pituitary. LH in turn stimulates testosterone production by the testes.

◉ **CHECKPOINT**

1. Describe the function of the scrotum in protecting the testes from temperature fluctuations.
2. Describe the internal structure of a testis. Where are sperm cells produced? What are the functions of Sertoli cells and Leydig cells?
3. Describe the principal events of spermatogenesis.
4. Identify the parts of a sperm cell, and list the functions of each.
5. What are the roles of FSH, LH, testosterone, and inhibin in the male reproductive system? How is secretion of these hormones controlled?

Reproductive System Ducts in Males

Ducts of the Testis

Pressure generated by the fluid secreted by Sertoli cells pushes sperm and fluid along the lumen of seminiferous tubules and then into a series of very short ducts called **straight tubules** (see Figure 28.3a). The straight tubules lead to a network of ducts in the testis called the **rete testis** (RĒ-tē = network). From the rete testis, sperm move into a series of coiled **efferent ducts** in the epididymis that empty into a single tube called the **ductus epididymis.**

Epididymis

The **epididymis** (ep′-i-DID-i-mis; *epi-* = above or over; *-didymis* = testis) is a comma-shaped organ about 4 cm (1.5 in.) long that lies along the posterior border of each testis (see Figure 28.3a). The plural is **epididymides** (ep′-i-did-ĪM-i-dēs). Each epididymis consists mostly of the tightly coiled **ductus epididymis.** The efferent ducts from the testis join the ductus epididymis at the larger, superior portion of the epididymis called the **head.** The **body** is the narrow midportion of the epididymis, and the **tail** is the smaller, inferior portion. At its distal end, the tail of the epididymis continues as the ductus (vas) deferens (discussed shortly).

The ductus epididymis would measure about 6 m (20 ft) in length if it were uncoiled. It is lined with pseudostratified columnar epithelium and encircled by layers of smooth muscle. The free surfaces of the columnar cells contain **stereocilia,** which despite their name are long, branching microvilli (not cilia) that increase the surface area for the reabsorption of degenerated sperm. Connective tissue around the muscle layer

attaches the loops of the ductus epididymis and carries blood vessels and nerves.

Functionally, the epididymis is the site of **sperm maturation,** the process by which sperm acquire motility and the ability to fertilize an ovum. This occurs over a period of about 14 days. The epididymis also helps propel sperm into the ductus (vas) deferens during sexual arousal by peristaltic contraction of its smooth muscle. In addition, the epididymis stores sperm, which remain viable here for up to several months. Any stored sperm that are not ejaculated by that time are eventually reabsorbed.

Ductus Deferens

Within the tail of the epididymis, the ductus epididymis becomes less convoluted, and its diameter increases. Beyond this point, the duct is known as the **ductus deferens** or **vas deferens** (see Figure 28.3a). The ductus deferens, which is about 45 cm (18 in.) long, ascends along the posterior border of the epididymis through the spermatic cord and then enters the pelvic cavity. There it loops over the ureter and passes over the side and down the posterior surface of the urinary bladder (see Figure 28.1a). The dilated terminal portion of the ductus deferens is the **ampulla** (am-PUL-la = little jar; Figure 28.9). The mucosa of the ductus deferens consists of pseudostratified columnar epithelium and lamina propria (areolar connective tissue). The muscularis is composed of three layers of smooth muscle; the inner and outer layers are longitudinal, and the middle layer is circular.

Functionally, the ductus deferens conveys sperm during sexual arousal from the epididymis toward the urethra by peristaltic contractions of its muscular coat. Like the epididymis, the ductus deferens also can store sperm for several months. Any stored sperm that are not ejaculated by that time are eventually reabsorbed.

Spermatic Cord

The **spermatic cord** is a supporting structure of the male reproductive system that ascends out of the scrotum (see Figure 28.2). It consists of the ductus (vas) deferens as it ascends through the scrotum, the testicular artery, veins that drain the testes and carry testosterone into circulation (the pampiniform plexus), autonomic nerves, lymphatic vessels, and the cremaster muscle. The term **varicocele** (VAR-i-kō-sēl; *varico-* = varicose; *-kele* = hernia) refers to a swelling in the scrotum due to a dilation of the veins that drain the testes. It is usually more apparent when the person is standing and typically does not require treatment. The spermatic cord and ilioinguinal nerve pass through the **inguinal canal** (IN-gwin-al = groin), an oblique passageway in the anterior abdominal wall just superior and parallel to the medial half of the inguinal ligament. The canal, which is about 4–5 cm (about 2 in.) long, originates at the **deep (abdominal) inguinal ring,** a slitlike opening in the aponeurosis of the transversus abdominis muscle; the canal ends at the **superficial (subcutaneous) inguinal ring** (see Figure 28.2), a somewhat triangular opening in the aponeurosis of the external oblique muscle. In females, the round ligament of the uterus and ilioinguinal nerve pass through the inguinal canal.

The principal method for sterilization of males is a **vasectomy** (vas-EK-tō-mē; *-ectomy* = cut out), in which a portion of each ductus deferens is removed. An incision is made on either side of the scrotum, the ducts are located and cut, each is tied (ligated) in two places with stitches, and the portion between the ties is removed. Although sperm produc-tion continues in the testes, sperm can no longer reach the exterior. The sperm degenerate and are destroyed by phagocytosis. Because the blood vessels are not cut, testosterone levels in the blood remain normal, so vasectomy has no effect on sexual desire, performance, and ejaculation. If done correctly, it is close to 100% effective. The procedure can be reversed, but the chance of regaining fertility is only 30–40%. ●

Figure 28.9 Locations of several accessory reproductive organs in males. The prostate, urethra, and penis have been sectioned to show internal details.

The male urethra has three subdivisions: the prostatic, membranous, and spongy (penile) urethra.

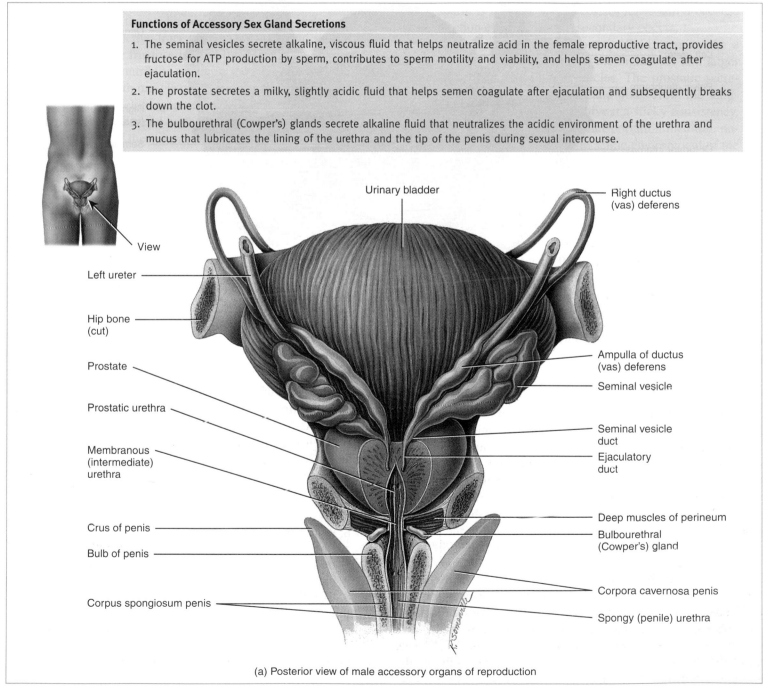

Functions of Accessory Sex Gland Secretions

1. The seminal vesicles secrete alkaline, viscous fluid that helps neutralize acid in the female reproductive tract, provides fructose for ATP production by sperm, contributes to sperm motility and viability, and helps semen coagulate after ejaculation.
2. The prostate secretes a milky, slightly acidic fluid that helps semen coagulate after ejaculation and subsequently breaks down the clot.
3. The bulbourethral (Cowper's) glands secrete alkaline fluid that neutralizes the acidic environment of the urethra and mucus that lubricates the lining of the urethra and the tip of the penis during sexual intercourse.

View

Urinary bladder

Right ductus (vas) deferens

Left ureter

Hip bone (cut)

Ampulla of ductus (vas) deferens

Seminal vesicle

Prostate

Prostatic urethra

Seminal vesicle duct

Ejaculatory duct

Membranous (intermediate) urethra

Deep muscles of perineum

Crus of penis

Bulbourethral (Cowper's) gland

Bulb of penis

Corpora cavernosa penis

Corpus spongiosum penis

Spongy (penile) urethra

(a) Posterior view of male accessory organs of reproduction

Figure 28.11 Female organs of reproduction and surrounding structures.

The organs of reproduction in females include the ovaries, uterine (fallopian) tubes, uterus, vagina, vulva, and mammary glands.

Functions of the Female Reproductive System

1. The ovaries produce secondary oocytes and hormones, including progesterone and estrogens (female sex hormones), inhibin, and relaxin.
2. The uterine tubes transport a secondary oocyte to the uterus and normally are the sites where fertilization occurs.
3. The uterus is the site of implantation of a fertilized ovum, development of the fetus during pregnancy, and labor.
4. The vagina receives the penis during sexual intercourse and is a passageway for childbirth.
5. The mammary glands synthesize, secrete, and eject milk for nourishment of the newborn.

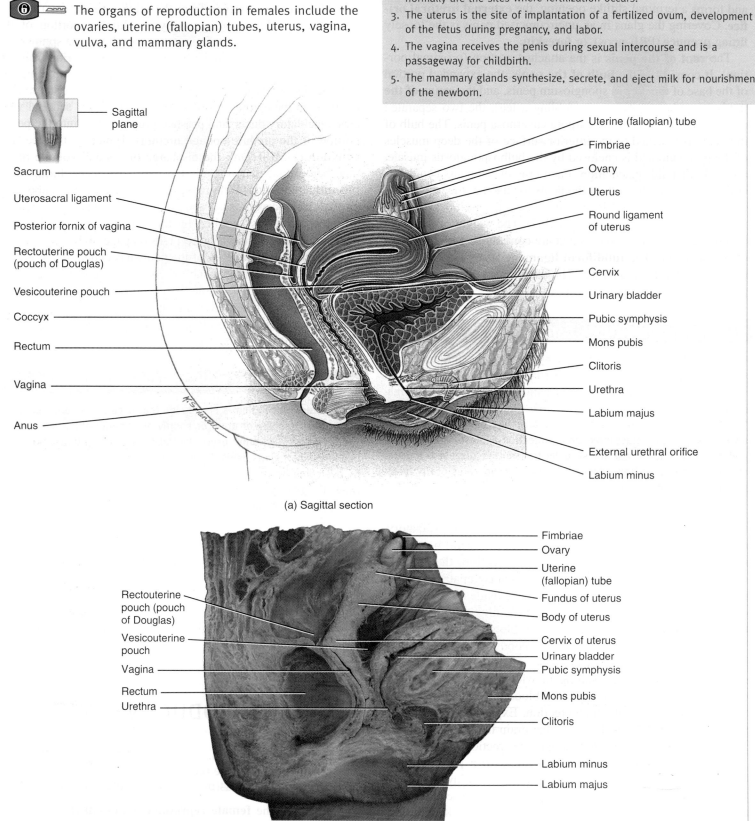

Sagittal plane

Sacrum

Uterosacral ligament

Posterior fornix of vagina

Rectouterine pouch (pouch of Douglas)

Vesicouterine pouch

Coccyx

Rectum

Vagina

Anus

Uterine (fallopian) tube

Fimbriae

Ovary

Uterus

Round ligament of uterus

Cervix

Urinary bladder

Pubic symphysis

Mons pubis

Clitoris

Urethra

Labium majus

External urethral orifice

Labium minus

(a) Sagittal section

Rectouterine pouch (pouch of Douglas)

Vesicouterine pouch

Vagina

Rectum

Urethra

Fimbriae

Ovary

Uterine (fallopian) tube

Fundus of uterus

Body of uterus

Cervix of uterus

Urinary bladder

Pubic symphysis

Mons pubis

Clitoris

Labium minus

Labium majus

(b) Sagittal section

? Which structures in males are homologous to the ovaries, the clitoris, the paraurethral glands, and the greater vestibular glands?

• **Ovarian** and con plus the form a s developr ferred to the deve follicle g

• A **matu** that is r process

• A **corpu** of a mat duces pi degenera **cans** (=

Oogenesis

The formatic (ō-ō-JEN-e-s begins in ma they are eve manner as sp and the resul

During ea cells migrate differentiate gular is **oogt**

Figure 28.:

 A

Follicular c

Basement membrane

Stromal cell

Collagen fiber

which are collectively called the vulva, or pudendum. The mammary glands are considered part of both the integumentary system and the female reproductive system.

Ovaries

The **ovaries** (= egg receptacles), which are the female gonads, are paired glands that resemble unshelled almonds in size and shape; they are homologous to the testes. (Here *homologous* means that two organs have the same embryonic origin.) The ovaries produce (1) gametes, secondary oocytes that develop into mature ova (eggs) after fertilization, and (2) hormones, including progesterone and estrogens (the female sex hormones), inhibin, and relaxin.

The ovaries, one on either side of the uterus, descend to the brim of the superior portion of the pelvic cavity during the third month of development. A series of ligaments holds them in position (Figure 28.12). The **broad ligament** of the uterus (see also Figure 28.11b), which is itself part of the parietal peritoneum, attaches to the ovaries by a double-layered fold of peritoneum called the **mesovarium.** The **ovarian ligament** anchors the ovaries to the uterus, and the **suspensory ligament** attaches them to the pelvic wall. Each ovary contains a **hilum,** the point of entrance and exit for blood vessels and nerves along which the mesovarium is attached.

Histology of the Ovary

Each ovary consists of the following parts (Figure 28.13):

• The **germinal epithelium** (*germen* = sprout or bud) is a layer of simple epithelium (low cuboidal or squamous) that covers the surface of the ovary. We now know that the term germinal epithelium in humans is not accurate because it does not give rise to ova; the name came about because, at one time, people believed that it did. We have since learned that the cells that produce ova arise from the yolk sac and migrate to the ovaries during embryonic development.

• The **tunica albuginea** is a whitish capsule of dense irregular connective tissue located immediately deep to the germinal epithelium.

• The **ovarian cortex** is a region just deep to the tunica albuginea. It consists of ovarian follicles (described shortly) surrounded by dense irregular connective tissue that contains collagen fibers and fibroblast-like cells called *stromal cells.*

• The **ovarian medulla** is deep to the ovarian cortex. The border between the cortex and medulla is indistinct, but the medulla consists of more loosely arranged connective tissue and contains blood vessels, lymphatic vessels, and nerves.

Figure 28.12 Relative positions of the ovaries, the uterus, and the ligaments that support them.

Ligaments holding the ovaries in position are the mesovarium, the ovarian ligament, and the suspensory ligament.

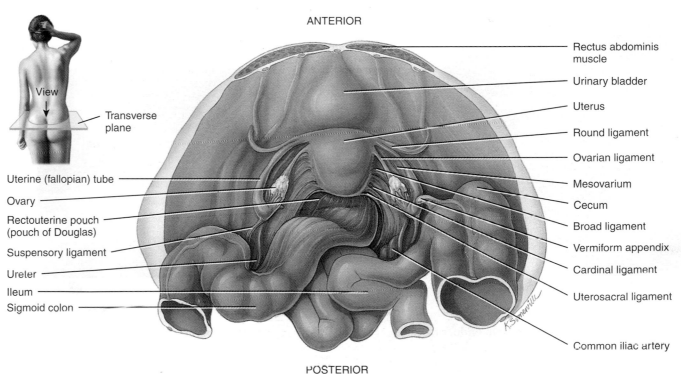

Superior view of transverse section

? To which structures do the mesovarium, ovarian ligament, and suspensory ligament anchor the ovary?

Figure 28.13

The

● FIGURE 28.14 CONTINUED ▶

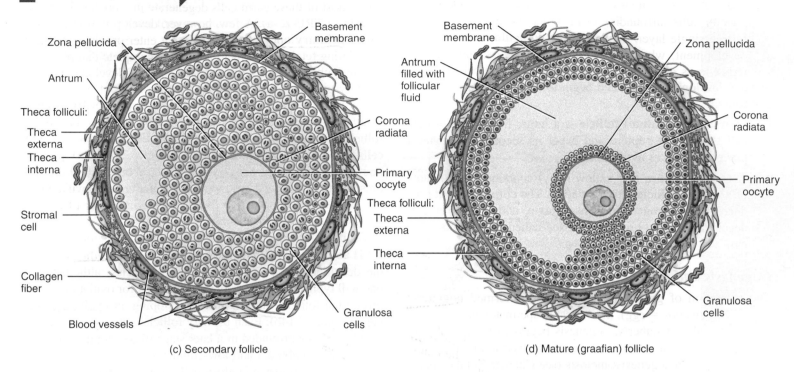

(c) Secondary follicle

(d) Mature (graafian) follicle

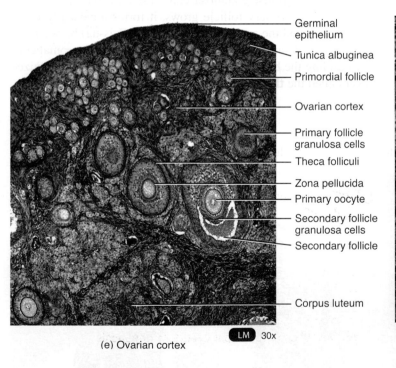

(e) Ovarian cortex

LM 30x

LM 70x

(f) Secondary follicle

? What happens to most ovarian follicles?

? What struc

With continuing maturation, a primary follicle develops into a secondary follicle (Figure 28.14c). In a **secondary follicle,** the theca differentiates into two layers: (1) the **theca interna,** a highly vascularized internal layer of cuboidal secretory cells that secrete estrogens and (2) the **theca externa,** an outer layer of stromal cells and collagen fibers. In addition, the granulosa cells begin to secrete follicular fluid, which builds up in a cavity called the **antrum** in the center of the secondary follicle. The innermost layer of granulosa cells becomes firmly attached to the zona pellucida and is now called the **corona radiata** (*corona* = crown; *radiata* = radiation) (Figure 28.14c).

The secondary follicle eventually becomes larger, turning into a **mature (graafian) follicle** (Figure 28.14d). While in this follicle, and just before ovulation, the diploid primary oocyte completes meiosis I, producing two haploid (*n*) cells of unequal size—each with 23 chromosomes (Figure 28.15). The smaller cell produced by meiosis I, called the **first polar body,** is essentially a packet of discarded nuclear material. The larger cell, known as the **secondary oocyte,** receives most of the cytoplasm. Once a secondary oocyte is formed, it begins meiosis II but then stops in metaphase. The mature (graafian) follicle soon ruptures and releases its secondary oocyte, a process known as **ovulation.**

At ovulation, the secondary oocyte is expelled into the pelvic cavity together with the first polar body and corona radiata. Normally these cells are swept into the uterine tube. If fertilization does not occur, the cells degenerate. If sperm are present in the uterine tube and one penetrates the secondary oocyte, however, meiosis II resumes. The secondary oocyte splits into two haploid cells, again of unequal size. The larger cell is the **ovum,** or mature egg; the smaller one is the **second polar body.** The nuclei of the sperm cell and the ovum then unite, forming a diploid **zygote.** If the first polar body undergoes another division to produce two polar bodies, then the primary oocyte ultimately gives rise to three haploid polar bodies, which all degenerate, and a single haploid ovum. Thus, one primary oocyte gives rise to a single gamete (an ovum). By contrast, recall that in males one primary spermatocyte produces four gametes (sperm).

Table 28.1 on page 1102 summarizes the events of oogenesis and follicular development.

Figure 28.15 Oogenesis. Diploid cells (2*n*) have 46 chromosomes; haploid cells (*n*) have 23 chromosomes.

In a secondary oocyte, meiosis II is completed only if fertilization occurs.

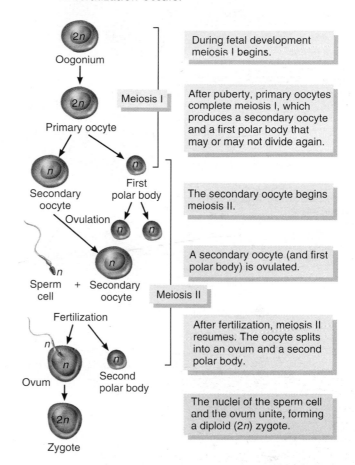

During fetal development meiosis I begins.

After puberty, primary oocytes complete meiosis I, which produces a secondary oocyte and a first polar body that may or may not divide again.

The secondary oocyte begins meiosis II.

A secondary oocyte (and first polar body) is ovulated.

After fertilization, meiosis II resumes. The oocyte splits into an ovum and a second polar body.

The nuclei of the sperm cell and the ovum unite, forming a diploid (2*n*) zygote.

? How does the age of a primary oocyte in a female compare with the age of a primary spermatocyte in a male?

◉ **CHECKPOINT**
14. How are the ovaries held in position in the pelvic cavity?
15. Describe the microscopic structure and functions of an ovary.
16. Describe the principal events of oogenesis.

● CLINICAL **CONNECTION** | Ovarian Cysts

An **ovarian cyst** is a fluid-filled sac in or on an ovary. Such cysts are relatively common, are usually noncancerous, and frequently disappear on their own. Cancerous cysts are more likely to occur in women over 40. Ovarian cysts may cause pain, pressure, a dull ache, or fullness in the abdomen; pain during sexual intercourse; delayed, painful, or irregular menstrual periods; abrupt onset of sharp pain in the lower abdomen; and/or vaginal bleeding. Most ovarian cysts require no treatment, but larger ones (more than 5 cm or 2 in.) may be removed surgically. ●

TABLE 28.1

Summary of Oogenesis and Follicular Development

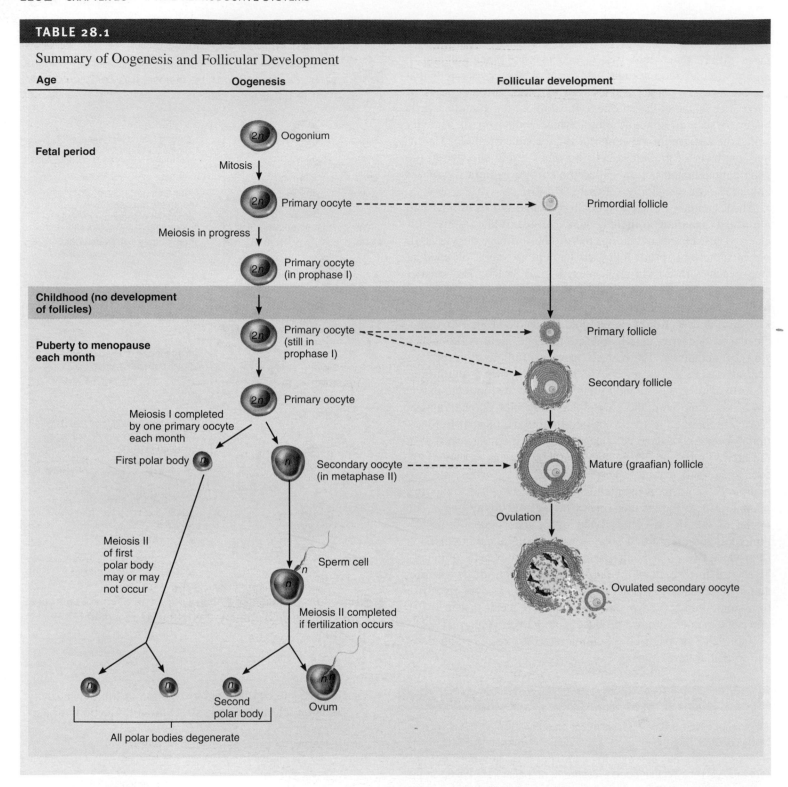

Uterine Tubes

Females have two **uterine (fallopian) tubes,** or **oviducts,** that extend laterally from the uterus (Figure 28.16). The tubes, which measure about 10 cm (4 in.) long, lie between the folds of the broad ligaments of the uterus. They provide a route for sperm to reach an ovum and transport secondary oocytes and fertilized ova from the ovaries to the uterus. The funnel-shaped portion of each tube, called the **infundibulum,** is close to the ovary but is open to the pelvic cavity. It ends in a fringe of fingerlike projections called **fimbriae** (FIM-brē-ē = fringe), one of which is attached to the lateral end of the ovary. From the infundibulum, the uterine tube extends medially and eventually inferiorly and attaches to the superior lateral angle of the uterus. The **ampulla** of the uterine tube is the widest, longest portion, making up about the lateral two-thirds of its length. The **isthmus** of the uterine tube is the more medial, short, narrow, thick-walled portion that joins the uterus.

Histologically, the uterine tubes are composed of three layers: mucosa, muscularis, and serosa. The mucosa consists of epithelium and lamina propria (areolar connective tissue). The epithelium contains ciliated simple columnar cells, which function as a "ciliary conveyor belt" to help move a fertilized ovum (or secondary oocyte) within the uterine tube toward the uterus, and nonciliated cells called *peg cells,* which have microvilli and secrete a fluid that provides nutrition for the ovum (Figure 28.17). The

middle layer, the muscularis, is composed of an inner, thick, circular ring of smooth muscle and an outer, thin region of longitudinal smooth muscle. Peristaltic contractions of the muscularis and the ciliary action of the mucosa help move the oocyte or fertilized ovum toward the uterus. The outer layer of the uterine tubes is a serous membrane, the serosa.

Local currents produced by movements of the fimbriae, which surround the ovary during ovulation, sweep the ovulated secondary oocyte from the pelvic cavity into the uterine tube. A sperm cell usually encounters and fertilizes a secondary oocyte in the ampulla of the uterine tube, although fertilization in the pelvic cavity is not uncommon. Fertilization can occur at any time up to about 24 hours after ovulation. Some hours after fertilization, the nuclear materials of the haploid ovum and sperm

Figure 28.16 **Relationship of the uterine (fallopian) tubes to the ovaries, uterus, and associated structures.** In the left side of the drawing the uterine tube and uterus have been sectioned to show internal structures. (See Tortora, *A Photographic Atlas of the Human Body, Second Edition,* Figure 14.9a.)

After ovulation, a secondary oocyte and its corona radiata move from the pelvic cavity into the infundibulum of the uterine tube. The uterus is the site of menstruation, implantation of a fertilized ovum, development of the fetus, and labor.

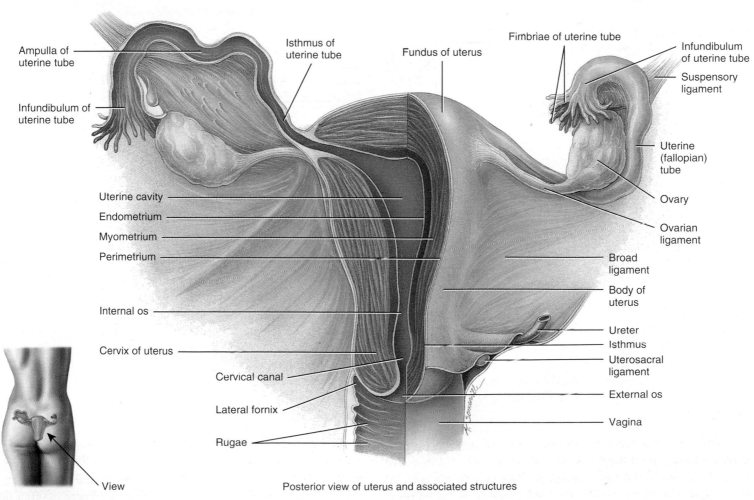

Posterior view of uterus and associated structures

? **Where does fertilization usually occur?**

Figure 28.17 **Histology of the uterine (fallopian) tube.** (See Tortora, *A Photographic Atlas of the Human Body, Second Edition*, Figure 14.11a.)

Peristaltic contractions of the muscularis and ciliary action of the mucosa of the uterine tube help move the oocyte or fertilized ovum toward the uterus.

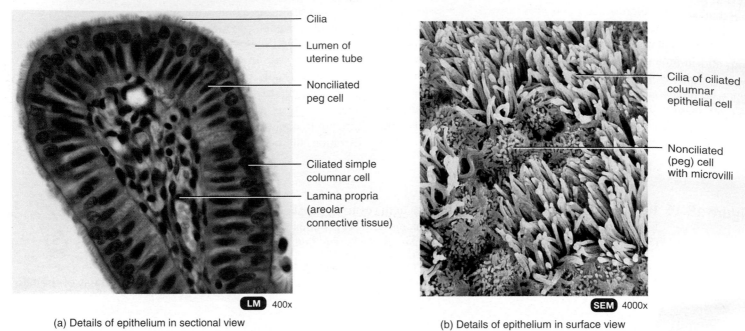

(a) Details of epithelium in sectional view

(b) Details of epithelium in surface view

? **What types of cells line the uterine tubes?**

unite. The diploid fertilized ovum is now called a **zygote** and begins to undergo cell divisions while moving toward the uterus. It arrives at the uterus 6 to 7 days after ovulation.

Uterus

The **uterus** (womb) serves as part of the pathway for sperm deposited in the vagina to reach the uterine tubes. It is also the site of implantation of a fertilized ovum, development of the fetus during pregnancy, and labor. During reproductive cycles when implantation does not occur, the uterus is the source of menstrual flow.

Anatomy of the Uterus

Situated between the urinary bladder and the rectum, the uterus is the size and shape of an inverted pear (see Figure 28.16). In females who have never been pregnant, it is about 7.5 cm (3 in.) long, 5 cm (2 in.) wide, and 2.5 cm (1 in.) thick. The uterus is larger in females who have recently been pregnant, and smaller (atrophied) when sex hormone levels are low, as occurs after menopause.

Anatomical subdivisions of the uterus include: (1) a dome-shaped portion superior to the uterine tubes called the **fundus,** (2) a tapering central portion called the **body,** and (3) an inferior narrow portion called the **cervix** that opens into the vagina. Between the body of the uterus and the cervix is the **isthmus** (IS-mus), a constricted region about 1 cm (0.5 in.) long. The in-

terior of the body of the uterus is called the **uterine cavity,** and the interior of the cervix is called the **cervical canal.** The cervical canal opens into the uterine cavity at the **internal os** (*os* = mouthlike opening) and into the vagina at the **external os.**

Normally, the body of the uterus projects anteriorly and superiorly over the urinary bladder in a position called **anteflexion.** The cervix projects inferiorly and posteriorly and enters the anterior wall of the vagina at nearly a right angle (see Figure 28.11). Several ligaments that are either extensions of the parietal peritoneum or fibromuscular cords maintain the position of the uterus (see Figure 28.12). The paired **broad ligaments** are double folds of peritoneum attaching the uterus to either side of the pelvic cavity. The paired **uterosacral ligaments,** also peritoneal extensions, lie on either side of the rectum and connect the uterus to the sacrum. The **cardinal (lateral cervical) ligaments** are located inferior to the bases of the broad ligaments and extend from the pelvic wall to the cervix and vagina. The **round ligaments** are bands of fibrous connective tissue between the layers of the broad ligament; they extend from a point on the uterus just inferior to the uterine tubes to a portion of the labia majora of the external genitalia. Although the ligaments normally maintain the anteflexed position of the uterus, they also allow the uterine body enough movement such that the uterus may become malpositioned. A posterior tilting of the uterus, called **retroflexion** (*retro-* = backward or behind), is a harmless variation of the normal position of the uterus. There is often no cause for the condition, but it may occur after childbirth.

A condition called **uterine prolapse** (*prolapse* = falling down or downward displacement) may result from weakening of supporting ligaments and pelvic musculature associated with age or disease, traumatic vaginal delivery, chronic straining from coughing or difficult bowel movements, or pelvic tumors. The prolapse may be characterized as *first degree (mild)*, in which the cervix remains within the vagina; *second degree (marked)*, in which the cervix protrudes through the vagina to the exterior; and *third degree (complete)*, in which the entire uterus is outside the vagina. Depending on the degree of prolapse, treatment may involve pelvic exercises, dieting if a patient is overweight, a stool softener to minimize straining during defecation, pessary therapy (placement of a rubber device around the uterine cervix that helps prop up the uterus), or surgery. ●

Histology of the Uterus

Histologically, the uterus consists of three layers of tissue: perimetrium, myometrium, and endometrium (Figure 28.18). The outer layer—the **perimetrium** (*peri-* = around; *-metrium* = uterus) or serosa—is part of the visceral peritoneum; it is composed of simple squamous epithelium and areolar connective tissue. Laterally, it becomes the broad ligament. Anteriorly, it covers the urinary bladder and forms a shallow pouch, the **vesicouterine pouch** (ves′-i-kō-Û-ter-in; *vesico-* = bladder; see Figure 28.11). Posteriorly, it covers the rectum and forms a deep pouch between the uterus and urinary bladder, the **rectouterine pouch** (rek-tō-Û-ter-in; *recto-* = rectum) or *pouch of Douglas*—the most inferior point in the pelvic cavity.

Figure 28.18 Histology of the uterus. (See Tortora, *A Photographic Atlas of the Human Body, Second Edition,* Figure 14.12.)

🔑 The three layers of the uterus from superficial to deep are the perimetrium (serosa), the myometrium, and the endometrium.

Transverse plane

Uterus

Lumen of uterus

Endometrium:
Stratum functionalis
Stratum basalis

Myometrium:
Inner longitudinal

Middle circular

Outer longitudinal

Perimetrium

LM 4x

(a) Transverse section through the uterus

Lumen of uterus

Simple columnar epithelium

Stratum functionalis

Endometrial gland

Stratum basalis

LM 115x

(b) Details of endometrium

 What structural features of the endometrium and myometrium contribute to their functions?

The middle layer of the uterus, the **myometrium** (*myo-* = muscle), consists of three layers of smooth muscle fibers that are thickest in the fundus and thinnest in the cervix. The thicker middle layer is circular; the inner and outer layers are longitudinal or oblique. During labor and childbirth, coordinated contractions of the myometrium in response to oxytocin from the posterior pituitary help expel the fetus from the uterus.

The inner layer of the uterus, the **endometrium** (*endo-* = within), is highly vascularized and has three components: (1) An innermost layer composed of simple columnar epithelium (ciliated and secretory cells) lines the lumen. (2) An underlying endometrial stroma is a very thick region of lamina propria (areolar connective tissue). (3) Endometrial (uterine) glands develop as invaginations of the luminal epithelium and extend almost to the myometrium. The endometrium is divided into two layers. The **stratum functionalis** (*functional layer*) lines the uterine cavity and sloughs off during menstruation. The deeper layer, the **stratum basalis** (*basal layer),* is permanent and gives rise to a new stratum functionalis after each menstruation.

Branches of the internal iliac artery called **uterine arteries** (Figure 28.19) supply blood to the uterus. Uterine arteries give off branches called **arcuate arteries** (= shaped like a bow) that are arranged in a circular fashion in the myometrium. These arteries branch into **radial arteries** that penetrate deeply into the myometrium. Just before the branches enter the endometrium, they divide into two kinds of arterioles: **Straight arterioles** supply the stratum basalis with the materials needed to regenerate the stratum functionalis; **spiral arterioles** supply the stratum functionalis and change markedly during the menstrual cycle. Blood leaving the uterus is drained by the **uterine veins** into the internal iliac veins. The extensive blood supply of the uterus is essential to support regrowth of a new stratum functionalis after menstruation, implantation of a fertilized ovum, and development of the placenta.

Cervical Mucus

The secretory cells of the mucosa of the cervix produce a secretion called **cervical mucus,** a mixture of water, glycoproteins, lipids, enzymes, and inorganic salts. During their reproductive years, females secrete 20–60 mL of cervical mucus per day. Cervical mucus is more hospitable to sperm at or near the time of ovulation because it is then less viscous and more alkaline (pH 8.5). At other times, a more viscous mucus forms a cervical plug that physically impedes sperm penetration. Cervical mucus supplements the energy needs of sperm, and both the cervix and cervical mucus protect sperm from phagocytes and the hostile environment of the vagina and uterus. Cervical mucus may also play a role in *capacitation*—a series of functional changes that

Figure 28.19 Blood supply of the uterus. The inset shows histological details of the blood vessels of the endometrium.

Straight arterioles supply the materials needed for regeneration of the stratum functionalis.

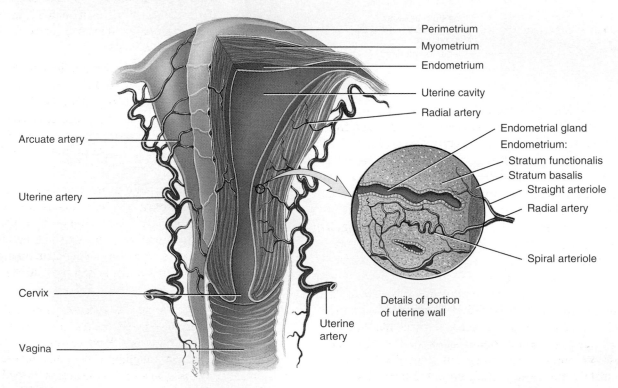

Anterior view with left side of uterus partially sectioned

? What is the functional significance of the stratum basalis of the endometrium?

sperm undergo in the female reproductive tract before they are able to fertilize a secondary oocyte. Capacitation causes a sperm cell's tail to beat even more vigorously, and it prepares the sperm cell's plasma membrane to fuse with the oocyte's plasma membrane.

◉ CHECKPOINT

17. Where are the uterine tubes located, and what is their function?
18. What are the principal parts of the uterus? Where are they located in relation to one another?
19. Describe the arrangement of ligaments that hold the uterus in its normal position.
20. Describe the histology of the uterus.
21. Why is an abundant blood supply important to the uterus?

Vagina

The **vagina** (= sheath) is a tubular, 10-cm (4-in.) long fibromuscular canal lined with mucous membrane that extends from the exterior of the body to the uterine cervix (see Figures 28.11 and 28.16). It is the receptacle for the penis during sexual intercourse, the outlet for menstrual flow, and the passageway for childbirth. Situated between the urinary bladder and the rectum, the vagina is directed superiorly and posteriorly, where it attaches to the uterus. A recess called the **fornix** (= arch or vault) surrounds the vaginal attachment to the cervix. When properly inserted, a contraceptive diaphragm rests in the fornix, where it is held in place as it covers the cervix.

The **mucosa** of the vagina is continuous with that of the uterus. Histologically, it consists of nonkeratinized stratified squamous epithelium and areolar connective tissue that lies in a series of transverse folds called **rugae** (ROO-gē). Dendritic cells in the mucosa are antigen-presenting cells (described on page 850). Unfortunately, they also participate in the transmission of viruses—for example, HIV (the virus that causes AIDS)—to a female during intercourse with an infected male. The mucosa of the vagina contains large stores of glycogen, the decomposition

of which produces organic acids. The resulting acidic environment retards microbial growth, but it also is harmful to sperm. Alkaline components of semen, mainly from the seminal vesicles, raise the pH of fluid in the vagina and increase viability of the sperm.

The **muscularis** is composed of an outer circular layer and an inner longitudinal layer of smooth muscle that can stretch considerably to accommodate the penis during sexual intercourse and a child during birth.

The **adventitia,** the superficial layer of the vagina, consists of areolar connective tissue. It anchors the vagina to adjacent organs such as the urethra and urinary bladder anteriorly and the rectum and anal canal posteriorly.

A thin fold of vascularized mucous membrane, called the **hymen** (= membrane), forms a border around and partially closes the inferior end of the vaginal opening to the exterior, the **vaginal orifice** (see Figure 28.20). After its rupture, usually following the first sexual intercourse, only remnants of the hymen remain. Sometimes the hymen completely covers the orifice, a condition called **imperforate hymen** (im-PER-fō-rāt). Surgery may be needed to open the orifice and permit the discharge of menstrual flow.

Vulva

The term **vulva** (VUL-va = to wrap around), or **pudendum** (pū-DEN-dum), refers to the external genitals of the female (Figure 28.20). The following components comprise the vulva:

• Anterior to the vaginal and urethral openings is the **mons pubis** (MONZ PŪ-bis; *mons* = mountain), an elevation of adipose tissue covered by skin and coarse pubic hair that cushions the pubic symphysis.

• From the mons pubis, two longitudinal folds of skin, the **labia majora** (LĀ-bē-a ma-JŌ-ra; *labia* = lips; *majora* = larger), extend inferiorly and posteriorly. The singular term is *labium majus.* The labia majora are covered by pubic hair and contain an abundance of adipose tissue, sebaceous (oil) glands, and apocrine sudoriferous (sweat) glands. They are homologous to the scrotum.

• Medial to the labia majora are two smaller folds of skin called the **labia minora** (mī-NŌ-ra; *minora* = smaller). The singular term is *labium minus.* Unlike the labia majora, the labia minora are devoid of pubic hair and fat and have few sudoriferous glands, but they do contain many sebaceous glands. The labia minora are homologous to the spongy (penile) urethra.

• The **clitoris** (KLI-to-ris) is a small cylindrical mass composed of two small erectile bodies, the *corpora cavernosa,* and numerous nerves and blood vessels. The clitoris is located at the anterior junction of the labia minora. A layer of skin called the **prepuce of the clitoris** is formed at the point where the labia minora unite and covers the body of the clitoris. The exposed portion of the clitoris is the **glans clitoris.** The clitoris is homologous to the glans penis in

● CHECKPOINT

22. How does the histology of the vagina contribute to its function?
23. What are the structures and functions of each part of the vulva?
24. Describe the components of the mammary glands and the structures that support them.
25. Outline the route milk takes from the alveoli of the mammary gland to the nipple.

THE FEMALE REPRODUCTIVE CYCLE

● OBJECTIVE

• Compare the major events of the ovarian and uterine cycles.

During their reproductive years, nonpregnant females normally exhibit cyclical changes in the ovaries and uterus. Each cycle takes about a month and involves both oogenesis and preparation of the uterus to receive a fertilized ovum. Hormones secreted by the hypothalamus, anterior pituitary, and ovaries control the main events. The **ovarian cycle** is a series of events in the ovaries that occur during and after the maturation of an oocyte. The **uterine (menstrual) cycle** is a concurrent series of changes in the endometrium of the uterus to prepare it for the arrival of a fertilized ovum that will develop there until birth. If fertilization does not occur, ovarian hormones wane, which causes the stratum functionalis of the endometrium to slough off. The general term **female reproductive cycle** encompasses the ovarian and uterine cycles, the hormonal changes that regulate them, and the related cyclical changes in the breasts and cervix.

Hormonal Regulation of the Female Reproductive Cycle

Gonadotropin-releasing hormone (GnRH) secreted by the hypothalamus controls the ovarian and uterine cycles (Figure 28.23). GnRH stimulates the release of **follicle-stimulating hormone (FSH)** and **luteinizing hormone (LH)** from the anterior pituitary. FSH initiates follicular growth, while LH stimulates further development of the ovarian follicles. In addition, both FSH and LH stimulate the ovarian follicles to secrete estrogens.

LH stimulates the theca cells of a developing follicle to produce androgens. Under the influence of FSH, the androgens are taken up by the granulosa cells of the follicle and then converted into estrogens. At midcycle, LH triggers ovulation and then promotes formation of the corpus luteum, the reason for the name luteinizing hormone. Stimulated by LH, the corpus luteum produces and secretes estrogens, progesterone, relaxin, and inhibin.

At least six different estrogens have been isolated from the plasma of human females, but only three are present in significant quantities: *beta (β)-estradiol, estrone,* and *estriol.* In a nonpregnant woman, the most abundant estrogen is β-estradiol, which is synthesized from cholesterol in the ovaries.

Estrogens secreted by ovarian follicles have several important functions (Figure 28.23).

• Estrogens promote the development and maintenance of female reproductive structures, secondary sex characteristics, and the breasts. The secondary sex characteristics include distribution of adipose tissue in the breasts, abdomen, mons pubis, and hips; voice pitch; a broad pelvis; and pattern of hair growth on the head and body.

• Estrogens increase protein anabolism, including the building of strong bones. In this regard, estrogens are synergistic with human growth hormone (hGH).

• Estrogens lower blood cholesterol level, which is probably the reason that women under age 50 have a much lower risk of coronary artery disease than do men of comparable age.

• Moderate levels of estrogens in the blood inhibit both the release of GnRH by the hypothalamus and secretion of LH and FSH by the anterior pituitary.

Progesterone, secreted mainly by cells of the corpus luteum, cooperates with estrogens to prepare and maintain the endometrium for implantation of a fertilized ovum and to prepare the mammary glands for milk secretion. High levels of progesterone also inhibit secretion of GnRH and LH.

The small quantity of **relaxin** produced by the corpus luteum during each monthly cycle relaxes the uterus by inhibiting contractions of the myometrium. Presumably, implantation of a fertilized ovum occurs more readily in a "quiet" uterus. During pregnancy, the placenta produces much more relaxin, and it continues to relax uterine smooth muscle. At the end of pregnancy, relaxin also increases the flexibility of the pubic symphysis and may help dilate the uterine cervix, both of which ease delivery of the baby.

Inhibin is secreted by granulosa cells of growing follicles and by the corpus luteum after ovulation. It inhibits secretion of FSH and, to a lesser extent, LH.

Phases of the Female Reproductive Cycle

The duration of the female reproductive cycle typically ranges from 24 to 35 days. For this discussion, we assume a duration of 28 days and divide it into four phases: the menstrual phase, the preovulatory phase, ovulation, and the postovulatory phase (Figure 28.24 on page 1114).

Figure 28.23 **Secretion and physiological effects of estrogens, progesterone, relaxin, and inhibin in the female reproductive cycle.**
Dashed red lines indicate negative feedback inhibition.

 The uterine and ovarian cycles are controlled by gonadotropin-releasing hormone (GnRH) and ovarian hormones (estrogens and progesterone).

? Of the several estrogens, which one exerts the major effect?

Menstrual Phase

The **menstrual phase** (MEN-stroo-al), also called **menstruation** (men′-stroo-Ā-shun) or **menses** (MEN-sēz = month), lasts for roughly the first 5 days of the cycle. (By convention, the first day of menstruation is day one of a new cycle.)

EVENTS IN THE OVARIES Under the influence of FSH, several primordial follicles develop into primary follicles and then into secondary follicles. This developmental process may take several months to occur. Therefore, a follicle that begins to develop at the beginning of a particular menstrual cycle may not reach maturity and ovulate until several menstrual cycles later.

EVENTS IN THE UTERUS Menstrual flow from the uterus consists of 50–150 mL of blood, tissue fluid, mucus, and epithelial cells shed from the endometrium. This discharge occurs because the declining levels of progesterone and estrogens stimulate release of prostaglandins that cause the uterine spiral arterioles to constrict. As a result, the cells they supply become oxygen-deprived and start to die. Eventually, the entire stratum functionalis sloughs off. At this time the endometrium is very thin, about

2–5 mm, because only the stratum basalis remains. The menstrual flow passes from the uterine cavity through the cervix and vagina to the exterior.

Preovulatory Phase

The **preovulatory phase** is the time between the end of menstruation and ovulation. The preovulatory phase of the cycle is more variable in length than the other phases and accounts for most of the differences in length of the cycle. It lasts from days 6 to 13 in a 28-day cycle.

EVENTS IN THE OVARIES Some of the secondary follicles in the ovaries begin to secrete estrogens and inhibin. By about day 6, a single secondary follicle in one of the two ovaries has outgrown all the others to become the **dominant follicle.** Estrogens and inhibin secreted by the dominant follicle decrease the secretion of FSH, which causes other, less well-developed follicles to stop growing and undergo atresia. Fraternal (nonidentical) twins or triplets result when two or three secondary follicles become codominant and later are ovulated and fertilized at about the same time.

Figure 28.24 The female reproductive cycle. The length of the female reproductive cycle typically is 24 to 36 days; the preovulatory phase is more variable in length than the other phases. (a) Events in the ovarian and uterine cycles and the release of anterior pituitary hormones are correlated with the sequence of the cycle's four phases. In the cycle shown, fertilization and implantation have not occurred. (b) Relative concentrations of anterior pituitary hormones (FSH and LH) and ovarian hormones (estrogens and progesterone) during the phases of a normal female reproductive cycle.

🔒 Estrogens are the primary ovarian hormones before ovulation; after ovulation, both progesterone and estrogens are secreted by the corpus luteum.

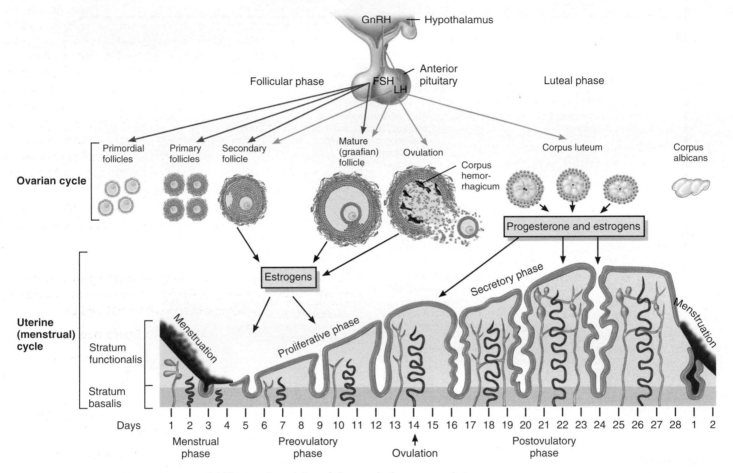

(a) Hormonal regulation of changes in the ovary and uterus

(b) Changes in concentration of anterior pituitary and ovarian hormones

❓ **Which hormones are responsible for the proliferative phase of endometrial growth, for ovulation, for growth of the corpus luteum, and for the surge of LH at midcycle?** ·

The female reproductive cycle can be disrupted by many factors, including weight loss, low body weight, disordered eating, and vigorous physical activity. The observation that three conditions—disordered eating, amenorrhea, and osteoporosis—often occur together in female athletes led researchers to coin the term **female athlete triad.**

Many athletes experience intense pressure from coaches, parents, peers, and themselves to lose weight to improve performance. Hence, they may develop disordered eating behaviors and engage in other harmful weight-loss practices in a struggle to maintain a very low body weight. **Amenorrhea** (ā-men′-ō-RĒ-a; *a-* = without; *men-* = month; *-rrhea* = a flow) is the absence of menstruation. The most common causes of amenorrhea are pregnancy and menopause. In female athletes, amenorrhea results from reduced secretion of gonadotropin-releasing hormone, which decreases the release of LH and FSH. As a result, ovarian follicles fail to develop, ovulation does not occur, synthesis of estrogens and progesterone wanes, and monthly menstrual bleeding ceases. Most cases of the female athlete triad occur in young women with very low amounts of body fat. Low levels of the hormone leptin, secreted by adipose cells, may be a contributing factor.

Because estrogens help bones retain calcium and other minerals, chronically low levels of estrogens are associated with loss of bone mineral density. The female athlete triad causes "old bones in young women." In one study, amenorrheic runners in their twenties had low bone mineral densities, similar to those of postmenopausal women 50 to 70 years old! Short periods of amenorrhea in young athletes may cause no lasting harm. However, long-term cessation of the reproductive cycle may be accompanied by a loss of bone mass, and adolescent athletes may fail to achieve an adequate bone mass; both of these situations can lead to premature osteoporosis and irreversible bone damage. •

CHECKPOINT

26. Describe the function of each of the following hormones in the uterine and ovarian cycles: GnRH, FSH, LH, estrogens, progesterone, and inhibin.
27. Briefly outline the major events of each phase of the uterine cycle, and correlate them with the events of the ovarian cycle.
28. Prepare a labeled diagram of the major hormonal changes that occur during the uterine and ovarian cycles.

BIRTH CONTROL METHODS AND ABORTION

OBJECTIVE

• Explain the differences among the various types of birth control methods and compare their effectiveness.

Birth control refers to restricting the number of children by various methods designed to control fertility and prevent conception. No single, ideal method of birth control exists. The only

method of preventing pregnancy that is 100% reliable is **complete abstinence,** the avoidance of sexual intercourse. Several other methods are available; each has its advantages and disadvantages. These include surgical sterilization, hormonal methods, intrauterine devices, spermicides, barrier methods, and periodic abstinence. Table 28.3 provides the failure rates for various methods of birth control. Although it is not a form of birth control, in this section we will also discuss abortion, the premature expulsion of the products of conception from the uterus.

Surgical Sterilization

Sterilization is a procedure that renders an individual incapable of further reproduction. The principal method for sterilization of males is a **vasectomy** (described in the Clinical Connection on page 1091). Sterilization in females most often is achieved by performing a **tubal ligation** (lī-GĀ-shun), in which both uterine tubes are tied closed and then cut. This can be achieved in a few

TABLE 28.3

Failure Rates for Several Birth Control Methods

METHOD	FAILURE RATES* (%) PERFECT USE†	TYPICAL USE
Complete abstinence	0	0
Surgical sterilization		
Vasectomy	0.10	0.15
Tubal ligation	0.5	0.5
Hormonal methods		
Oral contraceptives		
Combined pill	0.3	1–2
Seasonale®	0.3	1–2
Minipill	0.5	2
Non-oral contraceptives		
Contraceptive skin patch	0.1	1–2
Vaginal contraceptive ring	0.1	1–2
Emergency contraception	25	25
Hormone injections	0.3	1–2
Intrauterine devices (copper T 380 A®)	0.6	0.8
Spermicides (alone)	15	29
Barrier methods		
Male condom	2	15
Vaginal pouch	5	21
Diaphragm (with spermicide)	6	16
Cervical cap (with spermicide)	9	16
Periodic abstinence		
Rhythm	9	25
Sympto-thermal	2	20
No method	85	85

* Defined as percentage of women having an unintended pregnancy during the first year of use.

† Failure rate when the method is used correctly and consistently.

different ways. "Clips" or "clamps" can be placed on the uterine tubes, the tubes can be tied and/or cut, and sometimes they are cauterized. In any case the result is that the secondary oocyte cannot pass through the uterine tubes, and sperm cannot reach the oocyte. Tubal ligation reduces the risk of pelvic inflammatory disease in women who are exposed to sexually transmitted infections; it may also reduce the risk of ovarian cancer.

Hormonal Methods

Aside from complete abstinence or surgical sterilization, hormonal methods are the most effective means of birth control. **Oral contraceptives** (the pill) contain hormones designed to prevent pregnancy. Some, called *combined oral contraceptives (COCs)*, contain both progestin (hormone with actions similar to progesterone) and estrogens. The primary action of COCs is to inhibit ovulation by suppressing the gonadotropins FSH and LH. The low levels of FSH and LH usually prevent the development of a dominant follicle in the ovary. As a result, levels of estrogens do not rise, the midcycle LH surge does not occur, and ovulation does not take place. Even if ovulation does occur, as it does in some cases, COCs may also block implantation in the uterus and inhibit the transport of ova and sperm in the uterine tubes.

Progestins thicken cervical mucus and make it more difficult for sperm to enter the uterus. *Progestin-only pills* thicken cervical mucus and may block implantation in the uterus, but they do not consistently inhibit ovulation.

Among the noncontraceptive benefits of oral contraceptives are regulation of the length of menstrual cycle and decreased menstrual flow (and therefore decreased risk of anemia). The pill also provides protection against endometrial and ovarian cancers and reduces the risk of endometriosis. However, oral contraceptives may not be advised for women with a history of blood clotting disorders, cerebral blood vessel damage, migraine headaches, hypertension, liver malfunction, or heart disease. Women who take the pill and smoke face far higher odds of having a heart attack or stroke than do nonsmoking pill users. Smokers should quit smoking or use an alternative method of birth control.

Following are several variations of *oral* hormonal methods of contraception:

- **Combined pill.** Contains both progestin and estrogens and is typically taken once a day for three weeks to prevent pregnancy and regulate the menstrual cycle. The pills taken during the fourth week are inactive (do not contain hormones) and permit menstruation to occur.
- **Seasonale®.** Contains both progestin and estrogens and is taken once a day in 3-month cycles of 12 weeks of hormone-containing pills followed by one week of inactive pills. Menstruation occurs during the thirteenth week.
- **Minipill.** Contains progestin only and is taken every day of the month.

Non-oral hormonal methods of contraception are also available. Among these are the following:

- **Contraceptive skin patch (Ortho Evra®).** Contains both progestin and estrogens delivered in a skin patch placed on the skin (upper outer arm, back, lower abdomen, or buttocks) once a week for three weeks. After one week, the patch is removed from one location and then a new one is placed elsewhere. During the fourth week no patch is used.
- **Vaginal contraceptive ring (NuvaRing®).** A flexible doughnut-shaped ring about 5 cm (2 in.) in diameter that contains estrogens and progesterone and is inserted by the female herself into the vagina. It is left in the vagina for three weeks to prevent conception and then removed for one week to permit menstruation.
- **Emergency contraception (EC) (morning after pill).** Consists of progestin and estrogens or progestin alone to prevent pregnancy following unprotected sexual intercourse. The relatively high levels of progestin and estrogens in EC pills provide inhibition of FSH and LH secretion. Loss of the stimulating effects of these gonadtropic hormones causes the ovaries to cease secretion of their own estrogens and progesterone. In turn, declining levels of estrogens and progesterone induce shedding of the uterine lining, thereby blocking implantation. One pill is taken as soon as possible but within 72 hours of unprotected sexual intercourse. The second pill must be taken 12 hours after the first. The pills work in the same way as regular birth control pills.
- **Hormone injections (Depo-provera®).** An injectable progestin given intramuscularly by a health-care practitioner once every three months.

Intrauterine Devices

An **intrauterine device (IUD)** is a small object made of plastic, copper, or stainless steel that is inserted by a health-care professional into the cavity of the uterus. IUDs prevent fertilization from taking place by blocking sperm from entering the uterine tubes. The IUD most commonly used in the United States today is the Copper T 380A®, which is approved for up to 10 years of use and has long-term effectiveness comparable to that of tubal ligation. Some women cannot use IUDs because of expulsion, bleeding, or discomfort.

Spermicides

Various foams, creams, jellies, suppositories, and douches that contain sperm-killing agents, or **spermicides,** make the vagina and cervix unfavorable for sperm survival and are available without prescription. They are placed in the vagina before sexual intercourse. The most widely used spermicide is *nonoxynol-9,* which kills sperm by disrupting their plasma membranes. A spermicide is more effective when used with a barrier method such as a male condom, vaginal pouch, diaphragm, or cervical cap.

Barrier Methods

Barrier methods use a physical barrier and are designed to prevent sperm from gaining access to the uterine cavity and uterine tubes. In addition to preventing pregnancy, certain barrier methods (male condom and vaginal pouch) may also provide some protection against sexually transmitted diseases (STDs) such as AIDS. In contrast, oral contraceptives and IUDs confer no such protection, Among the barrier methods are the male condom, vaginal pouch, diaphragm, and cervical cap.

A **male condom** is a nonporous, latex covering placed over the penis that prevents deposition of sperm in the female reproductive tract. A **vaginal pouch,** sometimes called a **female condom,** is designed to prevent sperm from entering the uterus. It is made of two flexible rings connected by a polyurethane sheath. One ring lies inside the sheath and is inserted to fit over the cervix; the other ring remains outside the vagina and covers the female external genitals. A **diaphragm** is a rubber, dome-shaped structure that fits over the cervix and is used in conjunction with a spermicide. It can be inserted by the female up to 6 hours before intercourse. The diaphragm stops most sperm from passing into the cervix and the spermicide kills most sperm that do get by. Although diaphragm use does decrease the risk of some STDs, it does not fully protect against HIV infection because the vagina is still exposed. **A cervical cap** resembles a diaphragm but is smaller and more rigid. It fits snugly over the cervix and must be fitted by a health-care professional. Spermicides should be used with the cervical cap.

Periodic Abstinence

A couple can use their knowledge of the physiological changes that occur during the female reproductive cycle to decide either to abstain from intercourse on those days when pregnancy is a likely result, or to plan intercourse on those days if they wish to conceive a child. In females with normal and regular menstrual cycles, these physiological events help to predict the day on which ovulation is likely to occur.

The first physiologically based method, developed in the 1930s, is known as the **rhythm method.** It involves abstaining from sexual activity on the days that ovulation is likely to occur in each reproductive cycle. During this time (3 days before ovulation, the day of ovulation, and 3 days after ovulation) the couple abstains from intercourse. The effectiveness of the rhythm method for birth control is poor in many women due to the irregularity of the female reproductive cycle.

Another system is the **sympto-thermal method,** in which couples are instructed to know and understand certain signs of fertility. The signs of ovulation include increased basal body temperature; the production of abundant clear, stretchy cervical mucus; and pain associated with ovulation (*mittelschmerz*). If a couple abstains from sexual intercourse when the signs of ovulation are present and for 3 days afterward, the chance of pregnancy is decreased. A big problem with this method is that fertilization is very likely if intercourse occurs one or two days *before* ovulation.

Abortion

Abortion refers to the premature expulsion of the products of conception from the uterus, usually before the twentieth week of pregnancy. An abortion may be **spontaneous** (naturally occurring; also called a *miscarriage*) or **induced** (intentionally performed).

There are several types of induced abortions. One involves **mifepristone,** called **miniprex** in the United States and **RU 486** in Europe. It is a hormone approved only for pregnancies 9 weeks or less when taken with misoprostol (a prostaglandin).

Mifepristone is an antiprogestin; it blocks the action of progesterone by binding to and blocking progesterone receptors. Progesterone prepares the uterine endometrium for implantation and then maintains the uterine lining after implantation. If the level of progesterone falls during pregnancy or if the action of the hormone is blocked, menstruation occurs, and the embryo sloughs off along with the uterine lining. Within 12 hours after taking mifepristone, the endometrium starts to degenerate, and within 72 hours it begins to slough off. Misoprostol stimulates uterine contractions, and is given after mifepristone to aid in expulsion of the endometrium.

Another type of induced abortion is called **vacuum aspiration** (suction) and can be performed up to the sixteenth week of pregnancy. A small, flexible tube attached to a vacuum source is inserted into the uterus through the vagina. The embryo or fetus, placenta, and lining of the uterus are then removed by suction. For pregnancies between 13 and 16 weeks, a technique called **dilation and evacuation** is commonly used. After the cervix is dilated, suction and forceps are used to remove the fetus, placenta, and uterine lining. From the 16th to 24th week, a **late-stage abortion** may be employed using surgical methods similar to dilation and evacuation or through nonsurgical methods using a saline solution or medications to induce abortion. Labor may be induced by using vaginal suppositories, intravenous infusion, or injections into the amniotic fluid through the uterus.

◉ CHECKPOINT

29. How do oral contraceptives reduce the likelihood of pregnancy?
30. How do some methods of birth control protect against sexually transmitted diseases?
31. What is the problem with developing an oral contraceptive pill for males?

DEVELOPMENT OF THE REPRODUCTIVE SYSTEMS

◉ OBJECTIVE

• Describe the development of the male and female reproductive systems.

The *gonads* develop from the **intermediate mesoderm** during the fifth week of development, and they appear as bulges (Figure 28.27). Adjacent to the gonads are the **mesonephric (Wolffian) ducts,** which eventually develop into structures of the reproductive system in males. A second pair of ducts, the **paramesonephric (Müllerian) ducts,** develop lateral to the

Figure 28.27 Development of the internal reproductive systems.

The gonads develop from intermediate mesoderm.

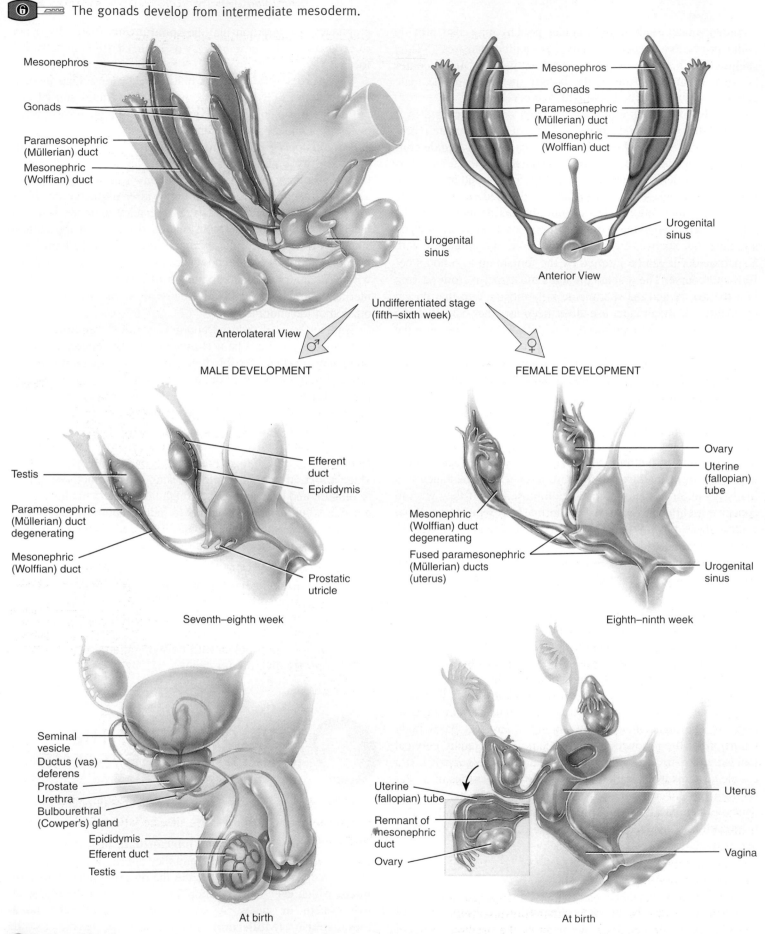

Mesonephros

Gonads

Paramesonephric (Müllerian) duct

Mesonephric (Wolffian) duct

Anterolateral View

Mesonephros

Gonads

Paramesonephric (Müllerian) duct

Mesonephric (Wolffian) duct

Urogenital sinus

Urogenital sinus

Anterior View

Undifferentiated stage (fifth–sixth week)

MALE DEVELOPMENT

FEMALE DEVELOPMENT

Testis

Paramesonephric (Müllerian) duct degenerating

Mesonephric (Wolffian) duct

Efferent duct

Epididymis

Prostatic utricle

Seventh–eighth week

Ovary

Uterine (fallopian) tube

Mesonephric (Wolffian) duct degenerating

Fused paramesonephric (Müllerian) ducts (uterus)

Urogenital sinus

Eighth–ninth week

Seminal vesicle

Ductus (vas) deferens

Prostate

Urethra

Bulbourethral (Cowper's) gland

Epididymis

Efferent duct

Testis

At birth

Uterine (fallopian) tube

Remnant of mesonephric duct

Ovary

Uterus

Vagina

At birth

Which gene is responsible for the development of the gonads into testes?

1120

mesonephric ducts and eventually form structures of the reproductive system in females. Both sets of ducts empty into the urogenital sinus. An early embryo has the potential to follow either the male or the female pattern of development because it contains both sets of ducts and primitive gonads that can differentiate into either testes or ovaries.

Cells of a male embryo have one X chromosome and one Y chromosome. The male pattern of development is initiated by a "master switch" gene on the Y chromosome named *SRY,* which stands for *Sex-determining Region of the Y chromosome.* When the *SRY* gene is expressed during development, its protein product causes the primitive Sertoli cells to begin to differentiate in the gonadal tissues during the seventh week. The developing Sertoli cells secrete a hormone called **Müllerian-inhibiting hormone (MIH),** which causes apoptosis of cells within the paramesonephric (Müllerian) ducts. As a result, those cells do not contribute any functional structures to the male reproductive system. Stimulated by human chorionic gonadotropin (hCG), primitive Leydig cells in the gonadal tissue begin to secrete the androgen **testosterone** during the eighth week. Testosterone then stimulates development of the mesonephric duct on each side into the *epididymis, ductus (vas) deferens, ejaculatory duct,* and *seminal vesicle.* The *testes* connect to the mesonephric duct through a series of tubules that eventually become the *seminiferous tubules.* The *prostate* and *bulbourethral glands* are **endodermal** outgrowths of the urethra.

Cells of a female embryo have two X chromosomes and no Y chromosome. Because *SRY* is absent, the gonads develop into *ovaries,* and because **MIH** is not produced, the paramesonephric ducts flourish. The distal ends of the paramesonephric ducts fuse to form the *uterus* and *vagina;* the unfused proximal portions of the ducts become the *uterine (fallopian) tubes.* The mesonephric ducts degenerate without contributing any functional structures to the female reproductive system because of the absence of testosterone. The *greater* and *lesser vestibular glands* develop from **endodermal** outgrowths of the vestibule.

The *external genitals* of both male and female embryos (penis and scrotum in males and clitoris, labia, and vaginal orifice in females) also remain undifferentiated until about the eighth week. Before differentiation, all embryos have an elevated midline swelling called the **genital tubercle** (Figure 28.28). The tubercle consists of the **urethral groove** (opening into the urogenital sinus), paired **urethral folds,** and paired **labioscrotal swellings.**

In male embryos, some testosterone is converted to a second androgen called **dihydrotestosterone (DHT).** DHT stimulates development of the urethra, prostate, and external genitals (scrotum and penis).

Figure 28.28 Development of the external genitals.

 The external genitals of male and female embryos remain undifferentiated until about the eighth week.

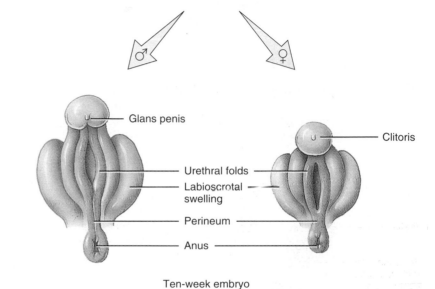

MALE DEVELOPMENT   FEMALE DEVELOPMENT

? **Which hormone is responsible for the differentiation of the external genitals?**

Part of the genital tubercle elongates and develops into a penis. Fusion of the urethral folds forms the *spongy (penile) urethra* and leaves an opening to the exterior only at the distal end of the penis, the *external urethral orifice.* The labioscrotal swellings develop into the *scrotum.* In the absence of DHT, the genital tubercle gives rise to the *clitoris* in female embryos. The urethral folds remain open as the *labia minora,* and the labioscrotal swellings become the *labia majora.* The urethral groove becomes the *vestibule.* After birth, androgen levels decline because hCG is no longer present to stimulate secretion of testosterone.

◉ CHECKPOINT

32. Describe the role of hormones in differentiation of the gonads, the mesonephric ducts, the paramesonephric ducts, and the external genitals.

AGING AND THE REPRODUCTIVE SYSTEMS

◉ OBJECTIVE

• Describe the effects of aging on the reproductive systems.

During the first decade of life, the reproductive system is in a juvenile state. At about age 10, hormone-directed changes start to occur in both sexes. **Puberty** (PŪ-ber-tē = a ripe age) is the period when secondary sexual characteristics begin to develop and the potential for sexual reproduction is reached. The onset of puberty is marked by pulses or bursts of LH and FSH secretion, each triggered by a pulse of GnRH. Most pulses occur during sleep. As puberty advances, the hormone pulses occur during the day as well as at night. The pulses increase in frequency during a three- to four-year period until the adult pattern is established. The stimuli that cause the GnRH pulses are still unclear, but a role for the hormone leptin is starting to unfold. Just before puberty, leptin levels rise in proportion to adipose tissue mass. Interestingly, leptin receptors are present in both the hypothalamus and anterior pituitary. Mice that lack a functional leptin gene from birth are sterile and remain in a prepubertal state. Giving leptin to such mice elicits secretion of gonadotropins, and they become fertile. Leptin may signal the hypothalamus that long-term energy stores (triglycerides in adipose tissue) are adequate for reproductive functions to begin.

In females, the reproductive cycle normally occurs once each month from **menarche** (me-NAR-kē), the first menses, to **menopause,** the permanent cessation of menses. Thus, the female reproductive system has a time-limited span of fertility between menarche and menopause. For the first 1 to 2 years after menarche, ovulation only occurs in about 10% of the cycles and the luteal phase is short. Gradually, the percentage of ovulatory cycles increases, and the luteal phase reaches its normal duration of 14 days. With age, fertility declines. Between the ages of 40 and 50 the pool of remaining ovarian follicles becomes exhausted. As a result, the ovaries become less responsive to hormonal stimulation. The production of estrogens declines, despite copious secretion of FSH and LH by the anterior pituitary. Many women experience hot flashes and heavy sweating, which coincide with bursts of GnRH release. Other symptoms of menopause are headache, hair loss, muscular pains, vaginal dryness, insomnia, depression, weight gain, and mood swings. Some atrophy of the ovaries, uterine tubes, uterus, vagina, external genitalia, and breasts occurs in postmenopausal women. Due to loss of estrogens, most women experience a decline in bone mineral density after menopause. Sexual desire (libido) does not show a parallel decline; it may be maintained by adrenal sex steroids. The risk of having uterine cancer peaks at about 65 years of age, but cervical cancer is more common in younger women.

In males, declining reproductive function is much more subtle than in females. Healthy men often retain reproductive capacity into their eighties or nineties. At about age 55 a decline in testosterone synthesis leads to reduced muscle strength, fewer viable sperm, and decreased sexual desire. Although sperm production decreases 50–70% between ages 60 and 80, abundant sperm may still be present even in old age.

Enlargement of the prostate to two to four times its normal size occurs in most males over age 60. This condition, called **benign prostatic hyperplasia (BPH),** decreases the size of the prostatic urethra and is characterized by frequent urination, nocturia (bed-wetting), hesitancy in urination, decreased force of urinary stream, postvoiding dribbling, and a sensation of incomplete emptying.

◉ CHECKPOINT

33. What changes occur in males and females at puberty?
34. What do the terms menarche and menopause mean?

• • •

To appreciate the many ways that the reproductive systems contribute to homeostasis of other body systems, examine *Focus on Homeostasis: The Reproductive Systems.* Next, in Chapter 29, you will explore the major events that occur during pregnancy and you will discover how genetics (inheritance) plays a role in the development of a child.

BODY SYSTEM	CONTRIBUTION OF THE REPRODUCTIVE SYSTEM
For all body systems	The male and female reproductive systems produce gametes (oocytes and sperm) that unite to form embryos and fetuses, which contain cells that divide and differentiate to form all of the organ systems of the body.
Integumentary systems	Androgens promote the growth of body hair. Estrogens stimulate the deposition of fat in the breasts, abdomen, and hips. Mammary glands produce milk. Skin stretches during pregnancy as the fetus enlarges.
Skeletal system	Androgens and estrogens stimulate the growth and maintenance of bones of the skeletal system.
Muscular system	Androgens stimulate the growth of skeletal muscles.
Nervous system	Androgens influence libido (sex drive). Estrogens may play a role in the development of certain regions of the brain in males.
Endocrine system	Testosterone and estrogens exert feedback effects on the hypothalamus and anterior pituitary gland.
Cardiovascular system	Estrogens lower blood cholesterol level and may reduce the risk of coronary artery disease in women under age 50.
Lymphatic systems and Immunity	The presence of an antibiotic-like chemical in semen and the acidic pH of vaginal fluid provide innate immunity against microbes in the reproductive tract.
Respiratory system	Sexual arousal increases the rate and depth of breathing.
Digestive system	The presence of the fetus during pregnancy crowds the digestive organs, which leads to heartburn and constipation.
Urinary system	In males, the portion of the urethra that extends through the prostate and penis is a passageway for urine as well as semen.

THE REPRODUCTIVE SYSTEMS

Reproductive System Disorders in Males

Testicular Cancer

Testicular cancer is the most common cancer in males between the ages of 20 and 35. More than 95% of testicular cancers arise from spermatogenic cells within the seminiferous tubules. An early sign of testicular cancer is a mass in the testis, often associated with a sensation of testicular heaviness or a dull ache in the lower abdomen; pain usually does not occur. To increase the chance for early detection of a testicular cancer, all males should perform regular self-examinations of the testes. The examination should be done starting in the teen years and once each month thereafter. After a warm bath or shower (when the scrotal skin is loose and relaxed) each testicle should be examined as follows. The testicle is grasped and gently rolled between the index finger and thumb, feeling for lumps, swellings, hardness, or other changes. If a lump or other change is detected, a physician should be consulted as soon as possible.

Prostate Disorders

Because the prostate surrounds part of the urethra, any infection, enlargement, or tumor can obstruct the flow of urine. Acute and chronic infections of the prostate are common in postpubescent males, often in association with inflammation of the urethra. Symptoms may include fever, chills, urinary frequency, frequent urination at night, difficulty in urinating, burning or painful urination, low back pain, joint and muscle pain, blood in the urine, or painful ejaculation. However, often there are no symptoms. Antibiotics are used to treat most cases that result from a bacterial infection. In **acute prostatitis,** the prostate becomes swollen and tender. **Chronic prostatitis** is one of the most common chronic infections in men of the middle and later years. On examination, the prostate feels enlarged, soft, and very tender, and its surface outline is irregular.

Prostate cancer is the leading cause of death from cancer in men in the United States, having surpassed lung cancer in 1991. Each year it is diagnosed in almost 200,000 U.S. men and causes nearly 40,000 deaths. The amount of PSA (prostate-specific antigen), which is produced only by prostate epithelial cells, increases with enlargement of the prostate and may indicate infection, benign enlargement, or prostate cancer. A blood test can measure the level of PSA in the blood. Males over the age of 40 should have an annual examination of the prostate gland. In a **digital rectal exam,** a physician palpates the gland through the rectum with the fingers (digits). Many physicians also recommend an annual PSA test for males over age 50. Treatment for prostate cancer may involve surgery, cryotherapy, radiation, hormonal therapy, and chemotherapy. Because many prostate cancers grow very slowly, some urologists recommend "watchful waiting" before treating small tumors in men over age 70.

Erectile Dysfunction

Erectile dysfunction (ED), previously termed *impotence,* is the consistent inability of an adult male to ejaculate or to attain or hold an erection long enough for sexual intercourse. Many cases of impotence are caused by insufficient release of nitric oxide (NO), which relaxes the smooth muscle of the penile arterioles and erectile tissue. The drug *Viagra*® (sildenafil) enhances smooth muscle relaxation by nitric oxide in the penis. Other causes of erectile dysfunction include diabetes mellitus, physical abnormalities of the penis, systemic disorders such as syphilis, vascular disturbances (arterial or venous obstructions), neurological disorders, surgery, testosterone deficiency, and

drugs (alcohol, antidepressants, antihistamines, antihypertensives, narcotics, nicotine, and tranquilizers). Psychological factors such as anxiety or depression, fear of causing pregnancy, fear of sexually transmitted diseases, religious inhibitions, and emotional immaturity may also cause ED.

Reproductive System Disorders in Females

Premenstrual Syndrome and Premenstrual Dysphoric Disorder

Premenstrual syndrome (PMS) is a cyclical disorder of severe physical and emotional distress. It appears during the postovulatory (luteal) phase of the female reproductive cycle and dramatically disappears when menstruation begins. The signs and symptoms are highly variable from one woman to another. They may include edema, weight gain, breast swelling and tenderness, abdominal distension, backache, joint pain, constipation, skin eruptions, fatigue and lethargy, greater need for sleep, depression or anxiety, irritability, mood swings, headache, poor coordination and clumsiness, and cravings for sweet or salty foods. The cause of PMS is unknown. For some women, getting regular exercise; avoiding caffeine, salt, and alcohol; and eating a diet that is high in complex carbohydrates and lean proteins can bring considerable relief.

Premenstrual dysphoric disorder (PMDD) is a more severe syndrome in which PMS-like signs and symptoms do not resolve after the onset of menstruation. Clinical research studies have found that suppression of the reproductive cycle by a drug that interferes with GnRH (leuprolide) decreases symptoms significantly. Because symptoms reappear when estradiol or progesterone is given together with leuprolide, researchers propose that PMDD is caused by abnormal responses to normal levels of these ovarian hormones. *SSRIs* (selective serotonin receptor inhibitors) have shown promise in treating both PMS and PMDD.

Endometriosis

Endometriosis (en-dō-mē-trē-Ō-sis; *endo-* = within; *metri-* = uterus; *osis* = condition) is characterized by the growth of endometrial tissue outside the uterus. The tissue enters the pelvic cavity via the open uterine tubes and may be found in any of several sites—on the ovaries, the rectouterine pouch, the outer surface of the uterus, the sigmoid colon, pelvic and abdominal lymph nodes, the cervix, the abdominal wall, the kidneys, and the urinary bladder. Endometrial tissue responds to hormonal fluctuations, whether it is inside or outside the uterus. With each reproductive cycle, the tissue proliferates and then breaks down and bleeds. When this occurs outside the uterus, it can cause inflammation, pain, scarring, and infertility. Symptoms include premenstrual pain or unusually severe menstrual pain.

Breast Cancer

One in eight women in the United States faces the prospect of **breast cancer.** After lung cancer, it is the second-leading cause of death from cancer in U.S. women. Breast cancer can occur in males but is rare. In females, breast cancer is seldom seen before age 30; its incidence rises rapidly after menopause. An estimated 5% of the 180,000 cases diagnosed each year in the United States, particularly those that arise in younger women, stem from inherited genetic mutations (changes in the DNA). Researchers have now identified two genes that increase susceptibility to breast cancer: *BRCA*1 (*br*east *ca*ncer 1) and *BRCA*2. Mutation of *BRCA*1 also confers a high risk for ovarian cancer. In addition, mutations of the *p*53 gene increase the risk of breast cancer in both males and females, and mutations of the androgen receptor gene are associated with the occurrence of breast cancer in some males. Because breast cancer generally is not painful until it becomes quite advanced, any lump, no matter how small, should be reported to a physician at once. Early detection—by breast self-examination and mammograms—is the best way to increase the chance of survival.

The most effective technique for detecting tumors less than 1 cm (0.4 in.) in diameter is **mammography** (mam-OG-ra-fē; -*graphy* = to record), a type of radiography using very sensitive x-ray film. The image of the breast, called a **mammogram** (see Table 1.3 on page 21), is best obtained by compressing the breasts, one at a time, using flat plates. A supplementary procedure for evaluating breast abnormalities is **ultrasound.** Although ultrasound cannot detect tumors smaller than 1 cm in diameter (which mammography can detect), it can be used to determine whether a lump is a benign, fluid-filled cyst or a solid (and therefore possibly malignant) tumor.

Among the factors that increase the risk of developing breast cancer are (1) a family history of breast cancer, especially in a mother or sister; (2) nulliparity (never having borne a child) or having a first child after age 35; (3) previous cancer in one breast; (4) exposure to ionizing radiation, such as x-rays; (5) excessive alcohol intake; and (6) cigarette smoking.

The American Cancer Society recommends the following steps to help diagnose breast cancer as early as possible:

- All women over 20 should develop the habit of monthly breast self-examination.

- A physician should examine the breasts every 3 years when a woman is between the ages of 20 and 40, and every year after age 40.

- A mammogram should be taken in women between the ages of 35 and 39, to be used later for comparison (baseline mammogram).

- Women with no symptoms should have a mammogram every year after age 40.

- Women of any age with a history of breast cancer, a strong family history of the disease, or other risk factors should consult a physician to determine a schedule for mammography.

Treatment for breast cancer may involve hormone therapy, chemotherapy, radiation therapy, **lumpectomy** (removal of the tumor and the immediate surrounding tissue), a modified or radical mastectomy, or a combination of these approaches. A **radical mastectomy** (*mast-* = breast) involves removal of the affected breast along with the underlying pectoral muscles and the axillary lymph nodes. (Lymph nodes are removed because metastasis of cancerous cells usually occurs through lymphatic or blood vessels.) Radiation treatment and chemotherapy may follow the surgery to ensure the destruction of any stray cancer cells. Several types of chemotherapeutic drugs are used to decrease the risk of relapse or disease progression. *Nolvadex®* (*tamoxifen*) is an estrogen antagonist that binds to and blocks estrogen receptors, thus decreasing the stimulating effect of estrogen on breast cancer cells. Tamoxifen has been used for 20 years and greatly reduces the risk of cancer recurrence. *Herceptin®*, a monoclonal antibody drug, targets an antigen on the surface of breast cancer cells. It is effective in causing regression of tumors and retarding progression of the disease. The early data from clinical trials of two new drugs, *Femara®* and *Amimidex®*, show relapse rates that are lower than those for tamoxifen. These drugs are inhibitors of aromatase, the enzyme needed for the final step in synthesis of estrogens. Finally, two drugs—tamoxifen and *Evista®* (*raloxifene*)—are being marketed for breast cancer *prevention.*

Interestingly, raloxifene blocks estrogen receptors in the breasts and uterus but activates estrogen receptors in bone, providing effective treatment for osteoporosis while possibly decreasing the risk of breast or endometrial (uterine) cancer.

Ovarian Cancer

Even though **ovarian cancer** is the sixth most common form of cancer in females, it is the leading cause of death from all gynecological malignancies (excluding breast cancer) because it is difficult to detect before it metastasizes (spreads) beyond the ovaries. Risk factors associated with ovarian cancer include age (usually over age 50); race (whites are at highest risk); family history of ovarian cancer; more than 40 years of active ovulation; nulliparity or first pregnancy after age 30; a high-fat, low-fiber, vitamin A–deficient diet; and prolonged exposure to asbestos or talc. Early ovarian cancer has no symptoms or only mild ones associated with other common problems, such as abdominal discomfort, heartburn, nausea, loss of appetite, bloating, and flatulence. Later-stage signs and symptoms include an enlarged abdomen, abdominal and/or pelvic pain, persistent gastrointestinal disturbances, urinary complications, menstrual irregularities, and heavy menstrual bleeding.

Cervical Cancer

Cervical cancer, carcinoma of the cervix of the uterus, starts with **cervical dysplasia** (dis-PLĀ-sē-a), a change in the shape, growth, and number of cervical cells. The cells may either return to normal or progress to cancer. In most cases, cervical cancer may be detected in its earliest stages by a Pap test (see page 120). Some evidence links cervical cancer to the virus that causes genital warts, human papillomavirus (HPV). Increased risk is associated with having a large number of sexual partners, having first intercourse at a young age, and smoking cigarettes.

Vulvovaginal Candidiasis

Candida albicans is a yeastlike fungus that commonly grows on mucous membranes of the gastrointestinal and genitourinary tracts. The organism is responsible for **vulvovaginal candidiasis** (vul-vō-VAJ-i-nal can-di-DĪ-a-sis), the most common form of **vaginitis** (vaj-i-NĪ-tis), inflammation of the vagina. Candidiasis is characterized by severe itching; a thick, yellow, cheesy discharge; a yeasty odor; and pain. The disorder, experienced at least once by about 75% of females, is usually a result of proliferation of the fungus following antibiotic therapy for another condition. Predisposing conditions include the use of oral contraceptives or cortisone-like medications, pregnancy, and diabetes.

Sexually Transmitted Diseases

A **sexually transmitted disease (STD)** is one that is spread by sexual contact. In most developed countries of the world, such as those of Western Europe, Japan, Australia, and New Zealand, the incidence of STDs has declined markedly during the past 25 years. In the United States, by contrast, STDs have been rising to near-epidemic proportions; they currently affect more than 65 million people. AIDS and hepatitis B, which are sexually transmitted diseases that also may be contracted in other ways, are discussed in Chapters 22 and 24, respectively.

Chlamydia

Chlamydia (kla-MID-ē-a) is a sexually transmitted disease caused by the bacterium *Chlamydia trachomatis* (*chlamy-* = cloak). This unusual bacterium cannot reproduce outside body cells; it "cloaks" itself inside cells, where it divides. At present, chlamydia is the most prevalent sexually transmitted disease in the United States. In most cases,

Female Reproductive System (p. 1095)

1. The female organs of reproduction include the ovaries (gonads), uterine (fallopian) tubes or oviducts, uterus, vagina, and vulva.
2. The mammary glands are part of the integumentary system and also are considered part of the reproductive system in females.
3. The ovaries, the female gonads, are located in the superior portion of the pelvic cavity, lateral to the uterus.
4. Ovaries produce secondary oocytes, discharge secondary oocytes (the process of ovulation), and secrete estrogens, progesterone, relaxin, and inhibin.
5. Oogenesis (the production of haploid secondary oocytes) begins in the ovaries. The oogenesis sequence includes meiosis I and meiosis II, which goes to completion only after an ovulated secondary oocyte is fertilized by a sperm cell.
6. The uterine (fallopian) tubes transport secondary oocytes from the ovaries to the uterus and are the normal sites of fertilization. Ciliated cells and peristaltic contractions help move a secondary oocyte or fertilized ovum toward the uterus.
7. The uterus is an organ the size and shape of an inverted pear that functions in menstruation, implantation of a fertilized ovum, development of a fetus during pregnancy, and labor. It also is part of the pathway for sperm to reach the uterine tubes to fertilize a secondary oocyte. Normally, the uterus is held in position by a series of ligaments.
8. Histologically, the layers of the uterus are an outer perimetrium (serosa), a middle myometrium, and an inner endometrium.
9. The vagina is a passageway for sperm and the menstrual flow, the receptacle of the penis during sexual intercourse, and the inferior portion of the birth canal. It is capable of considerable stretching.
10. The vulva, a collective term for the external genitals of the female, consists of the mons pubis, labia majora, labia minora, clitoris, vestibule, vaginal and urethral orifices, hymen, bulb of the vestibule, and three sets of glands: the paraurethral (Skene's), greater vestibular (Bartholin's), and lesser vestibular glands.
11. The perineum is a diamond-shaped area at the inferior end of the trunk medial to the thighs and buttocks.
12. The mammary glands are modified sweat glands lying superficial to the pectoralis major muscles. Their function is to synthesize, secrete, and eject milk (lactation).
13. Mammary gland development depends on estrogens and progesterone.
14. Milk production is stimulated by prolactin, estrogens, and progesterone; milk ejection is stimulated by oxytocin.

The Female Reproductive Cycle (p. 1112)

1. The function of the ovarian cycle is to develop a secondary oocyte; the function of the uterine (menstrual) cycle is to prepare the endometrium each month to receive a fertilized egg. The female reproductive cycle includes both the ovarian and uterine cycles.
2. The uterine and ovarian cycles are controlled by GnRH from the hypothalamus, which stimulates the release of FSH and LH by the anterior pituitary.
3. FSH and LH stimulate development of follicles and secretion of estrogens by the follicles. LH also stimulates ovulation, formation of the corpus luteum, and the secretion of progesterone and estrogens by the corpus luteum.
4. Estrogens stimulate the growth, development, and maintenance of female reproductive structures; stimulate the development of secondary sex characteristics; and stimulate protein synthesis.

5. Progesterone works with estrogens to prepare the endometrium for implantation and the mammary glands for milk synthesis.
6. Relaxin relaxes the myometrium at the time of possible implantation. At the end of a pregnancy, relaxin increases the flexibility of the pubic symphysis and helps dilate the uterine cervix to facilitate delivery.
7. During the menstrual phase, the stratum functionalis of the endometrium is shed, discharging blood, tissue fluid, mucus, and epithelial cells.
8. During the preovulatory phase, a group of follicles in the ovaries begins to undergo final maturation. One follicle outgrows the others and becomes dominant while the others degenerate. At the same time, endometrial repair occurs in the uterus. Estrogens are the dominant ovarian hormones during the preovulatory phase.
9. Ovulation is the rupture of the mature (graafian) follicle and the release of a secondary oocyte into the pelvic cavity. It is brought about by a surge of LH. Signs and symptoms of ovulation include increased basal body temperature; clear, stretchy cervical mucus; changes in the uterine cervix; and abdominal pain.
10. During the postovulatory phase, both progesterone and estrogens are secreted in large quantity by the corpus luteum of the ovary, and the uterine endometrium thickens in readiness for implantation.
11. If fertilization and implantation do not occur, the corpus luteum degenerates, and the resulting low levels of progesterone and estrogens allow discharge of the endometrium followed by the initiation of another reproductive cycle.
12. If fertilization and implantation do occur, the corpus luteum is maintained by hCG. The corpus luteum and later the placenta secrete progesterone and estrogens to support pregnancy and breast development for lactation.

Birth Control Methods and Abortion (p. 1117)

1. Methods include complete abstinence, surgical sterilization (vasectomy, tubal ligation), hormonal methods (combined pill, minipill, contraceptive skin patch, vaginal contraceptive ring, emergency contraception, hormonal injections), intrauterine devices, spermicides, barrier methods (male condom, vaginal pouch, diaphragm, cervical cap), and periodic abstinence (rhythm and sympto-thermal methods).
2. Contraceptive pills of the combination type contain progestin and estrogens in concentrations that decrease the secretion of FSH and LH and thereby inhibit development of ovarian follicles and ovulation, inhibit transport of ova and sperm in the uterine tubes, and block implantation in the uterus.
3. An abortion is the premature expulsion from the uterus of the products of conception; it may be spontaneous or induced.

Development of the Reproductive Systems (p. 1119)

1. The gonads develop from intermediate mesoderm. In the presence of the *SRY* gene, the gonads begin to differentiate into testes during the seventh week. The gonads differentiate into ovaries when the *SRY* gene is absent.
2. In males, testosterone stimulates development of each mesonephric duct into an epididymis, ductus (vas) deferens, ejaculatory duct, and seminal vesicle, and Müllerian-inhibiting hormone (MIH) causes the paramesonephric duct cells to die. In females, testosterone and MIS are absent; the paramesonephric ducts develop into the uterine tubes, uterus, and vagina and the mesonephric ducts degenerate.

3. The external genitals develop from the genital tubercle and are stimulated to develop into typical male structures by the hormone dihydrotestosterone (DHT). The external genitals develop into female structures when DHT is not produced, the normal situation in female embryos.

Aging and the Reproductive Systems (p. 1122)

1. Puberty is the period when secondary sex characteristics begin to develop and the potential for sexual reproduction is reached.
2. The onset of puberty is marked by pulses or bursts of LH and FSH secretion, each triggered by a pulse of GnRH. The hormone leptin, released by adipose tissue, may signal the hypothalamus that long-term energy stores (triglycerides in adipose tissue) are adequate for reproductive functions to begin.

3. In females, the reproductive cycle normally occurs once each month from menarche, the first menses, to menopause, the permanent cessation of menses.
4. Between the ages of 40 and 50, the pool of remaining ovarian follicles becomes exhausted and levels of progesterone and estrogens decline. Most women experience a decline in bone mineral density after menopause, together with some atrophy of the ovaries, uterine tubes, uterus, vagina, external genitalia, and breasts. Uterine and breast cancer increase in incidence with age.
5. In older males, decreased levels of testosterone are associated with decreased muscle strength, waning sexual desire, and fewer viable sperm; prostate disorders are common.

SELF-QUIZ QUESTIONS

Fill in the blanks in the following statement.

1. The period of time when secondary sexual characteristics begin to develop and the potential for sexual reproduction is reached is called _____. The first menses is called _____, and the permanent cessation of menses is called _____.

Indicate whether the following statements are true or false.

2. Spermatogenesis does not occur at normal core body temperature.
3. The route of sperm from the production in the testes to the exterior of the body is: seminiferous tubules, straight tubules, rete testes, epididymis, ductus (vas) deferens, ejaculatory duct, prostatic urethra, membranous urethra, spongy urethra, external urethral orifice.

Choose the one best answer to the following questions.

4. Which of the following are functions of Sertoli cells? (1) protection of developing spermatogenic cells, (2) nourishment of spermatocytes, spermatids, and sperm, (3) phagocytosis of excess sperm cytoplasm as development proceeds, (4) mediation of the effects of testosterone and FSH, (5) control of movements of spermatogenic cells and release of sperm into the lumen of seminiferous tubules.
   (a) 1, 2, 4, and 5   (b) 1, 2, 3, and 5   (c) 2, 3, 4, and 5
   (d) 1, 2, 3, and 4   (e) 1, 2, 3, 4, and 5

5. Which of the following are *true?* (1) An erection is a sympathetic response initiated by sexual stimulation. (2) Dilation of blood vessels supplying erectile tissue results in erection. (3) Nitric oxide causes smooth muscle within erectile tissue to relax, which results in widening of blood sinuses. (4) Ejaculation is a sympathetic reflex coordinated by the sacral region of the spinal cord. (5) The purpose of the corpus cavernosa penis is to keep the spongy urethra open during ejaculation.
   (a) 1, 2, and 3   (b) 1, 2, 3, 4, and 5   (c) 2 and 3
   (d) 2, 4, and 5   (e) 1, 2, 3, and 4

6. Which of the following are *true* concerning estrogens? (1) They promote development and maintenance of female reproductive structures and secondary sex characteristics. (2) They help control fluid and electrolyte balance. (3) They increase protein catabolism. (4) They lower blood cholesterol. (5) In moderate levels, they inhibit the release of GnRH and the secretion of LH and FSH.
   (a) 1, 4, and 5   (b) 1, 3, 4, and 5   (c) 1, 2, 3, and 5
   (d) 1, 2, 3, and 4   (e) 1, 2, 3, 4, and 5

7. Which of the following statements are *correct?* (1) A sperm head contains DNA and an acrosome. (2) An acrosome is a specialized lysosome that contains enzymes that enable sperm to produce the ATP needed to propel themselves out of the male reproductive tract. (3) Mitochondria in the midpiece of a sperm produce ATP for sperm motility. (4) A sperm's tail, a flagellum, propels it along its way. (5) Once ejaculated, sperm are viable and normally are able to fertilize a secondary oocyte for 5 days.
   (a) 1, 2, 3, and 4   (b) 2, 3, 4, and 5   (c) 1, 3, and 4
   (d) 2, 4, and 5   (e) 2, 3, and 4

8. Which of the following statements are *correct?* (1) Spermatogonia are stem cells because when they undergo mitosis, some of the daughter cells remain to serve as a reservoir of cells for future mitosis. (2) Meiosis I is a division of pairs of chromosomes resulting in daughter cells with only one member of each chromosome pair. (3) Meiosis II separates the chromatids of each chromosome. (4) Spermiogenesis involves the maturation of spermatids into sperm. (5) The process by which the seminiferous tubules produce haploid sperm is called spermatogenesis.
   (a) 1, 2, 3, and 5   (b) 1, 2, 3, 4, and 5   (c) 1, 3, 4, and 5
   (d) 1, 2, 3, and 4   (e) 1, 3, and 5

9. Which of the following statements are *correct?* (1) Cells from the yolk sac give rise to oogonia. (2) Ova arise from the germinal epithelium of the ovary. (3) Primary oocytes enter prophase of meiosis I during fetal development but do not complete it until after puberty. (4) Once a secondary oocyte is formed, it proceeds to metaphase of meiosis II and stops at this stage. (5) The secondary oocyte resumes meiosis II and forms the ovum and a polar body only if fertilization occurs. (6) A primary oocyte gives rise to an ovum and four polar bodies.
   (a) 1, 3, 4, and 5   (b) 1, 3, 4, and 6   (c) 1, 2, 4, and 6
   (d) 1, 2, 4, and 5   (e) 1, 2, 5, and 6

10. Which of the following statements are *correct?* (1) The female reproductive cycle consists of a menstrual phase, a preovulatory phase, ovulation, and a postovulatory phase. (2) During the menstrual phase, small secondary follicles in the ovary begin to enlarge while the uterus is shedding its lining. (3) During the preovulatory phase, a dominant follicle continues to grow and begins to secrete estrogens and inhibin while the uterine lining begins to rebuild.

gametes; forms part of the gut (gastrointestinal tract); functions as a shock absorber; and helps prevent drying out of the embryo.

Development of Sinusoids

On the ninth day after fertilization, the blastocyst becomes completely embedded in the endometrium. As the syncytiotrophoblast expands, small spaces called **lacunae** (la-KOO-nē = little lakes) develop within it (Figure 29.6b).

By the twelfth day of development, the lacunae fuse to form larger, interconnecting spaces called **lacunar networks** (Figure 29.6c). Endometrial capillaries around the developing embryo become dilated and are referred to as **maternal sinusoids.** As the syncytiotrophoblast erodes some of the maternal sinusoids and endometrial glands, maternal blood and secretions from the glands enter the lacunar networks and flow through them. Maternal blood is both a rich source of materials for embryonic nutrition and a disposal site for the embryo's wastes.

Development of the Extraembryonic Coelom

About the twelfth day after fertilization, the **extraembryonic mesoderm** develops. These mesodermal cells are derived from the yolk sac and form a connective tissue layer (mesenchyme) around the amnion and yolk sac (Figure 29.6c). Soon a number of large cavities develop in the extraembryonic mesoderm, which then fuse to form a single, larger cavity called the **extraembryonic coelom** (SĒ-lōm).

Development of the Chorion

The extraembryonic mesoderm, together with the two layers of the trophoblast (the cytotrophoblast and syncytiotrophoblast), together form the **chorion** (KOR-ē-on = membrane) (Figure 29.6c). The chorion surrounds the embryo and, later, the fetus (see Figure 29.11a inset). Eventually it becomes the principal embryonic part of the placenta, the structure for exchange of materials between mother and fetus. The chorion also protects the embryo and fetus from the immune responses of the mother in two ways: (1) It secretes proteins that block antibody production by the mother. (2) It promotes the production of T lymphocytes that suppress the normal immune response in the uterus. Finally, the chorion produces human chorionic gonadotropin (hCG), an important hormone of pregnancy (see Figure 29.16).

The inner layer of the chorion eventually fuses with the amnion. With the development of the chorion, the extraembryonic coelom is now referred to as the **chorionic cavity.** By the end of the second week of development, the bilaminar embryonic disc becomes connected to the trophoblast by a band of extraembryonic mesoderm called the **connecting (body) stalk** (see Figure 29.7, inset). The connecting stalk is the future umbilical cord.

◉ CHECKPOINT

6. What are the functions of the trophoblast?
7. How is the bilaminar embryonic disc formed?
8. Describe the formation of the amnion, yolk sac, and chorion and explain their functions.
9. Why are sinusoids important during embryonic development?

Third Week of Development

The third embryonic week begins a six-week period of very rapid development and differentiation. During the third week, the three primary germ layers are established and lay the groundwork for organ development in weeks four through eight.

Gastrulation

The first major event of the third week of development, **gastrulation** (gas'-troo-LĀ-shun), occurs about 15 days after fertilization. In this process, the bilaminar (two-layered) embryonic disc, consisting of epiblast and hypoblast, transforms into a **trilaminar** (three-layered) **embryonic disc** consisting of three primary germ layers: the ectoderm, mesoderm, and endoderm. The primary germ layers are the major embryonic tissues from which the various tissues and organs of the body develop.

Gastrulation involves the rearrangement and migration of cells from the epiblast. The first evidence of gastrulation is the formation of the **primitive streak,** a faint groove on the dorsal surface of the epiblast that elongates from the posterior to the anterior part of the embryo (Figure 29.7a). The primitive streak clearly establishes the head and tail ends of the embryo, as well as its right and left sides. At the head end of the primitive streak a small group of epiblastic cells forms a rounded structure called the **primitive node.**

Following formation of the primitive streak, cells of the epiblast move inward below the primitive streak and detach from the epiblast (Figure 29.7b) in a process called **invagination** (in-vaj-i-NĀ-shun). Once the cells have invaginated, some of them displace the hypoblast, forming the **endoderm** (endo- = inside; -derm = skin). Other cells remain between the epiblast and newly formed endoderm to form the **mesoderm** (meso- = middle). Cells remaining in the epiblast then form the **ectoderm** (ecto- = outside). The ectoderm and endoderm are epithelia composed of tightly packed cells; the mesoderm is a loosely organized connective tissue (mesenchyme). As the embryo develops, the endoderm ultimately becomes the epithelial lining of the gastrointestinal tract, respiratory tract, and several other organs. The mesoderm gives rise to muscles, bones, and other connective tissues, and the peritoneum. The ectoderm develops into the epidermis of the skin and the nervous system. Table 29.1 on page 1142 provides more details about the fates of these primary germ layers.

About 16 days after fertilization, mesodermal cells from the primitive node migrate toward the head end of the embryo and form a hollow tube of cells in the midline called the **notochordal process** (nō-tō-KOR-dal) (Figure 29.8 on page 1142). By days 22–24, the notochordal process becomes a solid cylinder of cells called the **notochord** (nō-tō-KORD; noto- = back; -chord = cord). This structure plays an extremely important role in **induction** (in-DUK-shun), the process by which one tissue (inducing tissue) stimulates the development of an adjacent unspecialized tissue (responding tissue) into a specialized one. An inducing tissue usually produces a chemical substance that influences the responding tissue. The notochord induces certain mesodermal cells to develop

Figure 29.7 Gastrulation.

Gastrulation involves the rearrangement and migration of cells from the epiblast.

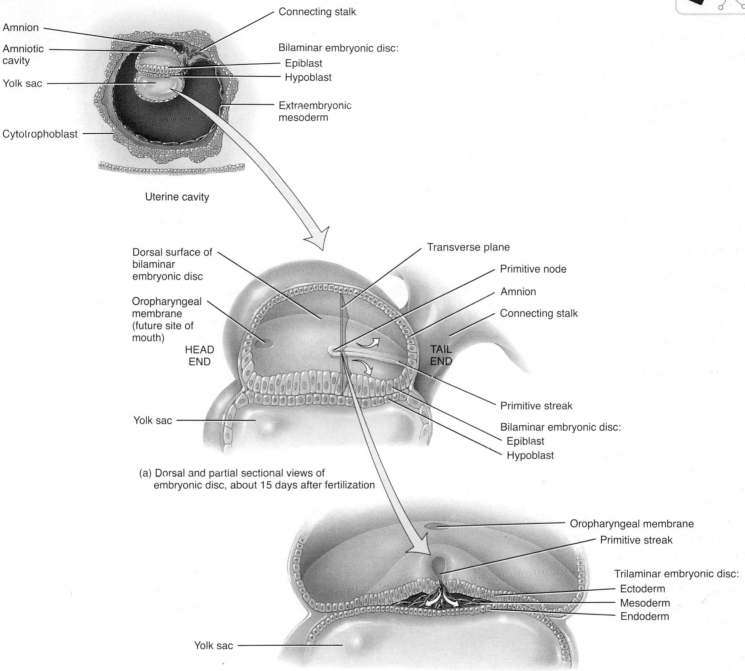

Amnion

Amniotic cavity

Yolk sac

Cytotrophoblast

Connecting stalk

Bilaminar embryonic disc:
Epiblast
Hypoblast

Extraembryonic mesoderm

Uterine cavity

Dorsal surface of bilaminar embryonic disc

Oropharyngeal membrane (future site of mouth)

HEAD END

Yolk sac

Transverse plane

Primitive node

Amnion

Connecting stalk

TAIL END

Primitive streak

Bilaminar embryonic disc:
Epiblast
Hypoblast

(a) Dorsal and partial sectional views of embryonic disc, about 15 days after fertilization

Oropharyngeal membrane

Primitive streak

Trilaminar embryonic disc:
Ectoderm
Mesoderm
Endoderm

Yolk sac

(b) Transverse section of trilaminar embryonic disc, about 16 days after fertilization

? What is the significance of gastrulation?

TABLE 29.1

Structures Produced by the Three Primary Germ Layers

ENDODERM	MESODERM	ECTODERM
Epithelial lining of gastrointestinal tract (except the oral cavity and anal canal) and the epithelium of its glands.	All skeletal and cardiac muscle tissue and most smooth muscle tissue.	All nervous tissue.
Epithelial lining of urinary bladder, gallbladder, and liver.	Cartilage, bone, and other connective tissues.	Epidermis of skin.
Epithelial lining of pharynx, auditory (eustachian) tubes, tonsils, larynx, trachea, bronchi, and lungs.	Blood, red bone marrow, and lymphatic tissue.	Hair follicles, arrector pili muscles, nails, epithelium of skin glands (sebaceous and sudoriferous), and mammary glands.
Epithelium of thyroid gland, parathyroid glands, pancreas, and thymus.	Endothelium of blood vessels and lymphatic vessels.	Lens, cornea, and internal eye muscles.
Epithelial lining of prostate and bulbourethral (Cowper's) glands, vagina, vestibule, urethra, and associated glands such as the greater (Bartholin's) vestibular and lesser vestibular glands.	Dermis of skin.	Internal and external ear.
	Fibrous tunic and vascular tunic of eye.	Neuroepithelium of sense organs.
	Middle ear.	Epithelium of oral cavity, nasal cavity, paranasal sinuses, salivary glands, and anal canal.
	Mesothelium of thoracic, abdominal, and pelvic cavities.	Epithelium of pineal gland, pituitary gland, and adrenal medullae.
	Epithelium of kidneys and ureters.	
	Epithelium of adrenal cortex.	
	Epithelium of gonads and genital ducts.	

into the vertebral bodies. It also forms the nucleus pulposus of the intervertebral discs (see Figure 7.24 on page 229).

Also during the third week of development, two faint depressions appear on the dorsal surface of the embryo. The structure closer to the head end is called the **oropharyngeal membrane** (or-ō-fa-RIN-jē-al; *oro-* = mouth; *-pharyngeal* = pertaining to the pharynx) (Figure 29.8a, b). It breaks down during the fourth week to connect the mouth cavity to the pharynx and the remainder of the gastrointestinal tract. The structure

closer to the tail end is called the **cloacal membrane** (klō-Ā-kul = sewer), which degenerates in the seventh week to form the openings of the anus and urinary and reproductive tracts.

When the cloacal membrane appears, the wall of the yolk sac forms a small vascularized outpouching called the **allantois** (a-LAN-tō-is; *allant-* = sausage) that extends into the connecting stalk (Figure 29.8b). In most other mammals, the allantois is used for gas exchange and waste removal. Because of the role of the human placenta in these activities, the allantois is not a

Figure 29.8 Development of the notochordal process.

 The notochordal process develops from the primitive node and later becomes the notochord.

(a) Dorsal and partial sectional views of trilaminar embryonic disc, about 16 days after fertilization

(b) Sagittal section of trilaminar embryonic disc, about 16 days after fertilization

? **What is the significance of the notochord?**

prominent structure in humans (see Figure 29.11a, inset). Nevertheless, it does function in the early formation of blood and blood vessels, and it is associated with the development of the urinary bladder.

Neurulation

In addition to inducing mesodermal cells to develop into vertebral bodies, the notochord also induces ectodermal cells over it to form the **neural plate** (Figure 29.9a). (Also see Figure 14.27 on page 537.) By the end of the third week, the lateral edges of the neural plate become more elevated and form the **neural fold** (Figure 29.9b). The depressed midregion is called the **neural groove** (Figure 29.9c). Generally, the neural folds approach each other and fuse, thus converting the neural plate into a **neural tube** (Figure 29.9d). This occurs first near the middle of the embryo and then progresses toward the head and tail ends. Neural

Figure 29.9 Neurulation and the development of somites.

Neurulation is the process by which the neural plate, neural folds, and neural tube form.

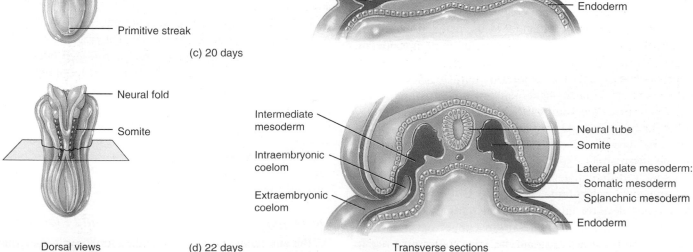

(a) 17 days

(b) 19 days

(c) 20 days

(d) 22 days

Dorsal views   Transverse sections

? **Which structures develop from the neural tube and somites?**

tube cells then develop into the brain and spinal cord. The process by which the neural plate, neural folds, and neural tube form is called **neurulation** (noor-oo-LĀ-shun).

As the neural tube forms, some of the ectodermal cells from the tube migrate to form several layers of cells called the **neural crest** (see Figure 14.27b on page 537). Neural crest cells give rise to spinal and cranial nerves and their ganglia, autonomic nervous system ganglia, the meninges of the brain and spinal cord, the adrenal medullae, and several skeletal and muscular components of the head.

At about four weeks after fertilization, the head end of the neural tube develops into three enlarged areas called **primary brain vesicles** (see Figure 14.28 on page 538): the **prosencephalon (forebrain), mesencephalon (midbrain),** and **rhombencephalon (hindbrain).** At about five weeks, the prosencephalon develops into **secondary brain vesicles** called the **telencephalon** and **diencephalon,** and the rhombencephalon develops into secondary brain vesicles called the **metencephalon** and **myelencephalon.** The areas of the neural tube adjacent to the myelencephalon develop into the spinal cord. The parts of the brain that develop from the various brain vesicles are described on page 496 in Chapter 14.

> ● CLINICAL **CONNECTION** | Anencephaly
>
> **Neural tube defects (NTDs)** are caused by arrest of the normal development and closure of the neural tube. These include spina bifida (discussed on page 230) and **anencephaly** (an′-en-SEPH-a-lē; *an-* = without; *encephal* = brain). In anencephaly, the cranial bones fail to develop and certain parts of the brain remain in contact with amniotic fluid and degenerate. Usually, a part of the brain that controls vital functions such as breathing and regulation of the heart is also affected. Infants with anencephaly are stillborn or die within a few days after birth. The condition occurs about once in every 1000 births and is 2 to 4 times more common in female infants than males. ●

Development of Somites

By about the 17th day after fertilization, the mesoderm adjacent to the notochord and neural tube forms paired longitudinal columns of **paraxial mesoderm** (par-AK-sē-al; *para-* = near) (Figure 29.9b). The mesoderm lateral to the paraxial mesoderm forms paired cylindrical masses called **intermediate mesoderm.** The mesoderm lateral to the intermediate mesoderm consists of a pair of flattened sheets called **lateral plate mesoderm.** The paraxial mesoderm soon segments into a series of paired, cube-shaped structures called **somites** (SŌ-mīts = little bodies). By the end of the fifth week, 42–44 pairs of somites are present. The number of somites that develop over a given period can be correlated to the approximate age of the embryo.

Each somite differentiates into three regions: a **myotome,** a **dermatome,** and a **sclerotome** (see Figure 10.19b on page 329). The myotomes develop into the skeletal muscles of the neck, trunk, and limbs; the dermatomes form connective tissue, including the dermis of the skin; and the sclerotomes give rise to the vertebrae and ribs.

Development of the Intraembryonic Coelom

In the third week of development, small spaces appear in the lateral plate mesoderm. These spaces soon merge to form a larger cavity called the **intraembryonic coelom** (SĒ-lom = cavity). This cavity splits the lateral plate mesoderm into two parts called the splanchnic mesoderm and somatic mesoderm (Figure 29.9d). **Splanchnic mesoderm** (SPLANGK-nik = visceral) forms the heart and the visceral layer of the serous pericardium, blood vessels, the smooth muscle and connective tissues of the respiratory and digestive organs, and the visceral layer of the serous membrane of the pleurae and peritoneum. **Somatic mesoderm** (sō-MAT-ik; *soma-* = body) gives rise to the bones, ligaments, and dermis of the limbs and the parietal layer of the serous membrane of the pericardium, pleurae, and peritoneum.

Development of the Cardiovascular System

At the beginning of the third week, **angiogenesis** (an-jē-ō-JEN-e-sis; *angio-* = vessel; *-genesis* = production), the formation of blood vessels, begins in the extraembryonic mesoderm in the yolk sac, connecting stalk, and chorion. This early development is necessary because there is insufficient yolk in the yolk sac and ovum to provide adequate nutrition for the rapidly developing embryo. Angiogenesis is initiated when mesodermal cells differentiate into **hemangioblasts.** These then develop into cells called **angioblasts,** which aggregate to form isolated masses of cells referred to as **blood islands** (see Figure 21.31 on page 823). Spaces soon develop in the blood islands and form the lumens of blood vessels. Some angioblasts arrange themselves around each space to form the endothelium and the tunics (layers) of the developing blood vessels. As the blood islands grow and fuse, they soon form an extensive system of blood vessels throughout the embryo.

About 3 weeks after fertilization, blood cells and blood plasma begin to develop *outside* the embryo from hemoangioblasts in the blood vessels in the walls of the yolk sac, allantois, and chorion. These then develop into pluripotent stem cells that form blood cells. Blood formation begins *within* the embryo at about the fifth week in the liver and the twelfth week in the spleen, red bone marrow, and thymus.

The heart forms from splanchnic mesoderm in the head end of the embryo on days 18 and 19. This region of mesodermal cells is called the **cardiogenic area** (kar-dē-ō-JEN-ik; *cardio-* = heart; *-genic* = producing). In response to induction signals from the underlying endoderm, these mesodermal cells form a pair of **endocardial tubes** (see Figure 20.19 on page 748). The tubes then fuse to form a single **primitive heart tube.** By the end of the third week, the primitive heart tube bends on itself, becomes S-shaped, and begins to beat. It then joins blood vessels in other parts of the embryo, connecting stalk, chorion, and yolk sac to form a primitive cardiovascular system.

Development of the Chorionic Villi and Placenta

As the embryonic tissue invades the uterine wall, maternal uterine vessels are eroded and maternal blood fills spaces, called

lacunae, within the invading tissue. By the end of the second week of development, **chorionic villi** (ko-rē-ON-ik VIL-ī) begin to develop. These fingerlike projections consist of chorion (syncytiotrophoblast surrounded by cytotrophoblast) that projects into the endometrial wall of the uterus (Figure 29.10a). By the end of the third week, blood capillaries develop in the chorionic villi (Figure 29.10b). Blood vessels in the chorionic villi connect to the embryonic heart by way of the umbilical arteries and umbilical vein through the connecting (body) stalk, which will eventually become the umbilical cord (Figure 29.10c). The fetal blood capillaries within the chorionic villi project into the lacunae, which unite to form the **intervillous spaces** that bathe the chorionic villi with maternal blood. As a result, maternal and fetal blood vessels are in close proximity. Note, however, that maternal and fetal blood vessels do not join, and the blood they carry does not normally mix. Instead, oxygen and nutrients in the blood of the mother's intervillous spaces, which are the spaces between chorionic villi, diffuse across the cell membranes into the capillaries of the villi. Waste products such as carbon dioxide diffuse in the opposite direction.

Figure 29.10 Development of chorionic villi.

Blood vessels in chorionic villi connect to the embryonic heart via the umbilical arteries and umbilical vein.

(a) Frontal section through uterus showing blastocyst, about 13 days after fertilization

(b) Details of two chorionic villi, about 21 days after fertilization

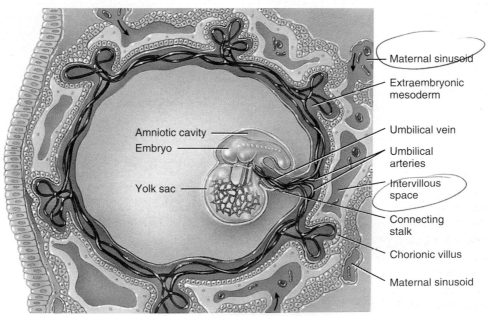

(c) Frontal section through uterus showing an embryo and its vascular supply, about 21 days after fertilization

? Why is development of chorionic villi important?

Placentation (plas'-en-TĀ-shun) is the process of forming the **placenta** (pla-SEN-ta = flat cake), the site of exchange of nutrients and wastes between the mother and fetus. The placenta also produces hormones needed to sustain the pregnancy (see Figure 29.16). The placenta is unique because it develops from two separate individuals, the mother and the fetus.

By the beginning of the twelfth week, the placenta has two distinct parts: (1) the fetal portion formed by the chorionic villi of the chorion and (2) the maternal portion formed by the decidua basalis of the endometrium (Figure 29.11a). When fully developed, the placenta is shaped like a pancake (Figure 29.11b). Functionally, the placenta allows oxygen and nutrients to diffuse from maternal blood into fetal blood while carbon dioxide and wastes diffuse from fetal blood into maternal blood. The placenta also is a protective barrier because most microorganisms cannot pass through it. However, certain viruses, such as those that cause AIDS, German measles, chickenpox, measles, encephalitis, and poliomyelitis, can cross the placenta. Many drugs, alcohol, and some substances that can cause birth defects also pass freely. The placenta stores nutrients such as carbohydrates, proteins, calcium, and iron, which are released into fetal circulation as required.

The actual connection between the placenta and embryo, and later the fetus, is through the **umbilical cord** (um-BIL-i-kul = navel), which develops from the connecting stalk and is usually about 2 cm (1 in.) wide and about 50–60 cm (20–24 in.) in

Figure 29.11 Placenta and umbilical cord.

The placenta is formed by the chorionic villi of the embryo and the decidua basalis of the endometrium of the mother.

(a) Details of placenta and umbilical cord

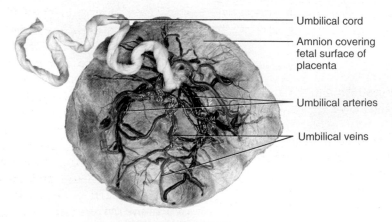

(b) Fetal aspect of placenta

? What is the function of the placenta?

length. The umbilical cord consists of two umbilical arteries that carry deoxygenated fetal blood to the placenta, one umbilical vein that carries oxygen and nutrients acquired from the mother's intervillous spaces into the fetus, and supporting mucous connective tissue called **Wharton's jelly** derived from the allantois. A layer of amnion surrounds the entire umbilical cord and gives it a shiny appearance (Figure 29.11a). In some cases, the umbilical vein is used to transfuse blood into a fetus or to introduce drugs for various medical treatments.

In about 1 in 200 newborns, only one of the two umbilical arteries is present in the umbilical cord. It may be due to failure of the artery to develop or degeneration of the vessel early in development. Nearly 20% of infants with this condition develop cardiovascular defects.

After the birth of the baby, the placenta detaches from the uterus and is therefore termed the **afterbirth.** At this time, the umbilical cord is tied off and then severed. The small portion (about an inch) of the cord that remains attached to the infant begins to wither and falls off, usually within 12 to 15 days after birth. The area where the cord was attached becomes covered by a thin layer of skin, and scar tissue forms. The scar is the **umbilicus** (navel).

Pharmaceutical companies use human placentas as a source of hormones, drugs, and blood; portions of placentas are also used for burn coverage. The placental and umbilical cord veins can also be used in blood vessel grafts, and cord blood can be frozen to provide a future source of pluripotent stem cells, for example, to repopulate red bone marrow following radiotherapy for cancer.

● CLINICAL **CONNECTION** | Placenta Previa

In some cases, the entire placenta or part of it may become implanted in the inferior portion of the uterus, near or covering the internal os of the cervix. This condition is called **placenta previa** (PRE-vē-a = before or in front of). Although placenta previa may lead to spontaneous abortion, it also occurs in approximately 1 in 250 live births. It is dangerous to the fetus because it may cause premature birth and intrauterine hypoxia due to maternal bleeding. Maternal mortality is increased due to hemorrhage and infection. The most important symptom is sudden, painless, bright-red vaginal bleeding in the third trimester. Cesarean section is the preferred method of delivery in placenta previa. ●

◉ CHECKPOINT

10. What is the significance of gastrulation?
11. How do the three primary germ layers form? Why are they important?
12. What is the function of the notochord?
13. Describe how neurulation occurs. Why is it important?
14. What are the functions of somites?
15. How does the cardiovascular system develop?
16. How does the placenta form, and what is its importance?

Fourth Week of Development

The fourth through eighth weeks of development are very significant in embryonic development because all major organs appear during this time. The term **organogenesis** (or'-ga-nō-JEN-e-sis) refers to the formation of body organs and systems. By the end of the eighth week, all the major body systems have begun to develop, although their functions for the most part are minimal. Organogenesis requires the presence of blood vessels to supply developing organs with oxygen and other nutrients. However, recent studies suggest that blood vessels play a significant role in organogenesis even before blood begins to flow within them. The endothelial cells of blood vessels apparently provide some type of developmental signal, either a secreted substance or a direct cell-to-cell interaction, that is necessary for organogenesis.

During the fourth week after fertilization, the embryo undergoes very dramatic changes in shape and size, nearly tripling its size. It is essentially converted from a flat, two-dimensional trilaminar embryonic disc to a three-dimensional cylinder, a process called **embryonic folding** (Figure 29.12a–d). The cylinder consists of endoderm in the center (gut), ectoderm on the outside (epidermis), and mesoderm in between. The main force responsible for embryonic folding is the different rates of growth of various parts of the embryo, especially the rapid longitudinal growth of the nervous system (neural tube). Folding in the median plane produces a **head fold** and a **tail fold;** folding in the horizontal plane results in the two **lateral folds.** Overall, due to the foldings, the embryo curves into a C-shape.

The head fold brings the developing heart and mouth into their eventual adult positions. The tail fold brings the developing anus into its eventual adult position. The lateral folds form as the lateral margins of the trilaminar embryonic disc bend ventrally. As they move toward the midline, the lateral folds incorporate the dorsal part of the yolk sac into the embryo as the **primitive gut,** the forerunner of the gastrointestinal tract (Figure 29.12b). The primitive gut differentiates into an anterior **foregut,** an intermediate **midgut,** and a posterior **hindgut** (Figure 29.12c). The fates of the foregut, midgut, and hindgut are described on page 967. Recall that the oropharyngeal membrane is located in the head end of the embryo (see Figure 29.8). It separates the future pharyngeal (throat) region of the foregut from the **stomodeum** (stō-mō-DĒ-um; *stomo-* = mouth), the future oral cavity. Because of head folding, the oropharyngeal membrane moves downward and the foregut and stomodeum move closer to their final positions. When the oropharyngeal membrane ruptures during the fourth week, the pharyngeal region of the pharynx is brought into contact with the stomodeum.

In a developing embryo, the last part of the hindgut expands into a cavity called the **cloaca** (see Figure 26.23 on page 1054). On the outside of the embryo is a small cavity in the tail region called the **proctodeum** (prok-tō-DĒ-um; *procto-* = anus) (Figure 29.12c). Separating the cloaca from the proctodeum is the **cloacal membrane** (see Figure 29.8). During embryonic development, the cloaca divides into a ventral urogenital sinus and a dorsal anorectal canal. As a result of tail folding, the cloacal

Figure 29.12 Embryonic folding.

Embryonic folding converts the two-dimensional trilaminar embryonic disc into a three-dimensional cylinder.

(a) 22 days

(b) 24 days

(c) 26 days

(d) 28 days

Sagittal sections

Transverse sections

? What are the results of embryonic folding?

membrane moves downward and the urogenital sinus, anorectal canal, and proctodeum move closer to their final positions. When the cloacal membrane ruptures during the seventh week of development, the urogenital and anal openings are created.

Along with embryonic folding, development of somites and the neural tube (previously described) occurs during the fourth week of development. In addition, several **pharyngeal (branchial) arches** develop on each side of the future head and neck regions (Figure 29.13). These five paired structures form swellings on the surface of the embryo beginning on the 22nd day after fertilization. Each pharyngeal arch consists of an outer covering of ectoderm, an inner covering of endoderm, and mesoderm in between. The pharyngeal arches also contain an artery, a cranial nerve, cartilage, and muscular tissue. Also on the outside of the embryo are a series of grooves between the pharyngeal arches called **pharyngeal clefts,** which separate the pharyngeal arches (Figure 29.13a). Simultaneous with the development of the pharyngeal arches and clefts, four distinct pairs of **pharyngeal pouches** develop inside the embryo (Figure 29.13b).

Figure 29.13 Development of pharyngeal arches, pharyngeal clefts, and pharyngeal pouches.

 The five pairs of pharyngeal pouches consist of ectoderm, mesoderm, and endoderm and contain blood vessels, cranial nerves, cartilage, and muscle tissue.

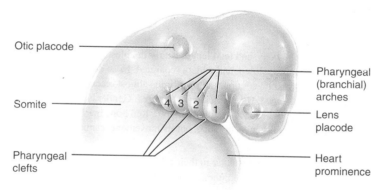

(a) External view, about 28-day embryo

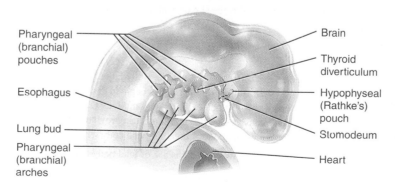

(b) Sagittal section, about 28-day embryo

? Why are pharyngeal arches, clefts, and pouches important?

The pharyngeal pouches are endoderm-lined, balloon-like outgrowths of the primitive pharynx, the most cranial part of the foregut.

Just as the somite gives rise to specified structures in the body wall, each pharyngeal arch, cleft, and pouch gives rise to specified structures in the head and neck. Each pharyngeal arch is a developmental unit and includes a skeletal component, muscle, nerve, and blood vessels. In the human embryo, there are four obvious pharyngeal arches and two less distinct arches. Each of these arches develops into a specific and unique component of the head and neck region. For example, the first pharyngeal arch is often called the *mandibular arch* because it forms the jaws (the *mandible* is the lower jawbone).

The first sign of a developing ear is a thickened area of ectoderm, the **otic placode** (future internal ear), which can be distinguished about 22 days after fertilization (see Figure 29.13a). A thickened area of ectoderm called the **lens placode** (see Figure 29.13a), which will become the eye, also appears at this time.

By the middle of the fourth week, the upper limbs begin their development as outgrowths of mesoderm covered by ectoderm called **upper limb buds** (see Figure 8.18b on page 259). By the end of the fourth week, the **lower limb buds** develop. The heart also forms a distinct projection on the ventral surface of the embryo called the **heart prominence** (see Figure 8.18a). At the end of the fourth week the embryo has a distinctive **tail** (see Figure 8.18b).

Fifth Through Eighth Weeks of Development

During the fifth week of development, there is a very rapid development of the brain, so growth of the head is considerable. By the end of the sixth week, the head grows even larger relative to the trunk, and the limbs show substantial development (see Figure 8.18c on page 259). In addition, the neck and trunk begin to straighten, and the heart is now four-chambered. By the seventh week, the various regions of the limbs become distinct and the beginnings of digits appear (see Figure 8.18d). At the start of the eighth week (the final week of the embryonic period), the digits of the hands are short and webbed, the tail is shorter but still visible, the eyes are open, and the auricles of the ears are visible (see Figure 8.18c). By the end of the eighth week, all regions of limbs are apparent; the digits are distinct and no longer webbed due to removal of cells via apoptosis. Also, the eyelids come together and may fuse, the tail disappears, and the external genitals begin to differentiate. The embryo now has clearly human characteristics.

⊙ CHECKPOINT

17. How does embryonic folding occur?
18. How does the primitive gut form, and what is its significance?
19. What is the origin of the structures of the head and neck?
20. What are limb buds?
21. What changes occur in the limbs during the second half of the embryonic period?

◉ CHECKPOINT

22. What are the general developmental trends during the fetal period?

23. Using Table 29.2 as a guide, select any one body structure in weeks 9 through 12 and trace its development through the remainder of the fetal period.

Figure 29.14 Summary of representative developmental events of the embryonic and fetal periods. The embryos and fetuses are not shown at their actual sizes.

🔑 Development during the fetal period is mostly concerned with the growth and differentiation of tissues and organs formed during the embryonic period.

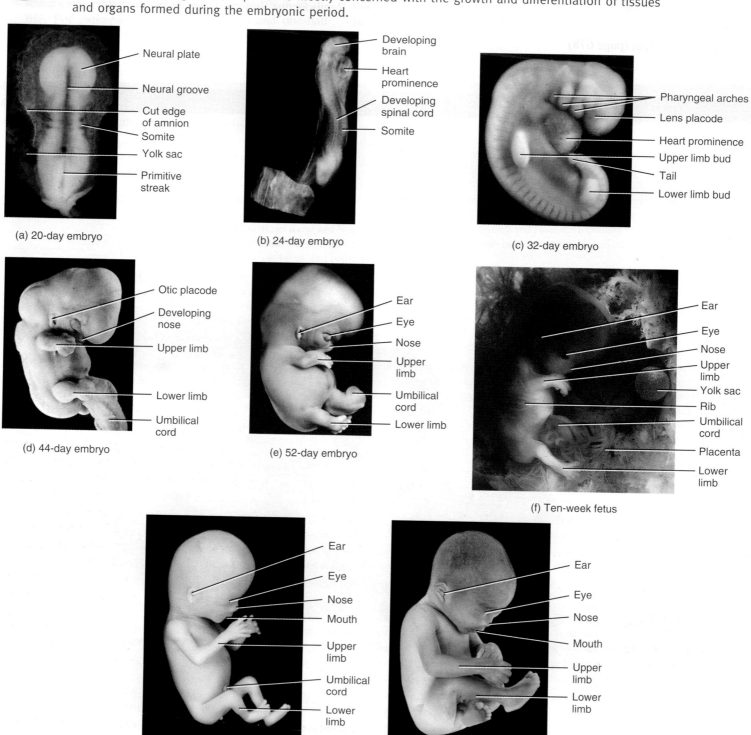

(a) 20-day embryo

- Neural plate
- Neural groove
- Cut edge of amnion
- Somite
- Yolk sac
- Primitive streak

(b) 24-day embryo

- Developing brain
- Heart prominence
- Developing spinal cord
- Somite

(c) 32-day embryo

- Pharyngeal arches
- Lens placode
- Heart prominence
- Upper limb bud
- Tail
- Lower limb bud

(d) 44-day embryo

- Otic placode
- Developing nose
- Upper limb
- Lower limb
- Umbilical cord

(e) 52-day embryo

- Ear
- Eye
- Nose
- Upper limb
- Umbilical cord
- Lower limb

(f) Ten-week fetus

- Ear
- Eye
- Nose
- Upper limb
- Yolk sac
- Rib
- Umbilical cord
- Placenta
- Lower limb

(g) Thirteen-week fetus

- Ear
- Eye
- Nose
- Mouth
- Upper limb
- Umbilical cord
- Lower limb

(h) Twenty-six-week fetus

- Ear
- Eye
- Nose
- Mouth
- Upper limb
- Lower limb

❓ How does mid-fetal weight compare to end-fetal weight?

TERATOGENS

Exposure of a developing embryo or fetus to certain environmental factors can damage the developing organism or even cause death. A **teratogen** (TER-a-tō-jen; *terato-* = monster; *-gen* = creating) is any agent or influence that causes developmental defects in the embryo. In the following sections we briefly discuss several examples.

Chemicals and Drugs

Because the placenta is not an absolute barrier between the maternal and fetal circulations, any drug or chemical that is dangerous to an infant should be considered potentially dangerous to the fetus when given to the mother. Alcohol is by far the number-one fetal teratogen. Intrauterine exposure to even a small amount of alcohol may result in **fetal alcohol syndrome (FAS),** one of the most common causes of mental retardation and the most common preventable cause of birth defects in the United States. The symptoms of FAS may include slow growth before and after birth, characteristic facial features (short palpebral fissures, a thin upper lip, and sunken nasal bridge), defective heart and other organs, malformed limbs, genital abnormalities, and central nervous system damage. Behavioral problems, such as hyperactivity, extreme nervousness, reduced ability to concentrate, and an inability to appreciate cause-and-effect relationships, are common.

Other teratogens include certain viruses (hepatitis B and C and certain papilloma viruses that cause sexually transmitted diseases); pesticides; defoliants (chemicals that cause plants to shed their leaves prematurely); industrial chemicals; some hormones; antibiotics; oral anticoagulants, anticonvulsants, antitumor agents, thyroid drugs, thalidomide, diethylstilbestrol (DES), and numerous other prescription drugs; LSD; and cocaine. A pregnant woman who uses cocaine, for example, subjects the fetus to higher risk of retarded growth, attention and orientation problems, hyperirritability, a tendency to stop breathing, malformed or missing organs, strokes, and seizures. The risks of spontaneous abortion, premature birth, and stillbirth also increase with fetal exposure to cocaine.

Cigarette Smoking

Strong evidence implicates cigarette smoking during pregnancy as a cause of low infant birth weight; there is also a strong association between smoking and a higher fetal and infant mortality rate. Women who smoke have a much higher risk of an ectopic pregnancy. Cigarette smoke may be teratogenic and may cause cardiac abnormalities as well as anencephaly (see page 1144). Maternal smoking also is a significant factor in the development of cleft lip and palate and has been linked with sudden infant death syndrome (SIDS). Infants nursing from smoking mothers have also been found to have an increased incidence of gastrointestinal disturbances. Even a mother's exposure to secondhand cigarette smoke (breathing air containing tobacco smoke) during pregnancy or while nursing predisposes her baby to increased incidence of respiratory problems, including bronchitis and pneumonia, during the first year of life.

Irradiation

Ionizing radiation of various kinds is a potent teratogen. Exposure of pregnant mothers to x-rays or radioactive isotopes during the embryo's susceptible period of development may cause microcephaly (small head size relative to the rest of the body), mental retardation, and skeletal malformations. Caution is advised, especially during the first trimester of pregnancy.

◉ **CHECKPOINT**

24. What are some of the symptoms of fetal alcohol syndrome?
25. How does cigarette smoking affect embryonic and fetal development?

PRENATAL DIAGNOSTIC TESTS

Several tests are available to detect genetic disorders and assess fetal well-being. Here we describe fetal ultrasonography, amniocentesis, and chorionic villi sampling (CVS).

Fetal Ultrasonography

If there is a question about the normal progress of a pregnancy, **fetal ultrasonography** (ul'-tra-son-OG-ra-fē) may be performed. By far the most common use of diagnostic ultrasound is to determine a more accurate fetal age when the date of conception is unclear. It is also used to confirm pregnancy, evaluate fetal viability and growth, determine fetal position, identify multiple pregnancies, identify fetal–maternal abnormalities, and serve as an adjunct to special procedures such as amniocentesis. During fetal ultrasonography, a transducer, an instrument that emits high-frequency sound waves, is passed back and forth over the abdomen. The reflected sound waves from the developing fetus are picked up by the transducer and converted to an on-screen image called a **sonogram** (see Table 1.3 on page 21). Because the urinary bladder serves as a landmark during the procedure, the patient needs to drink liquids before the procedure and not void urine to maintain a full bladder.

Amniocentesis

Amniocentesis (am'-nē-ō-sen-TĒ-sis; *amnio-* = amnion; *-centesis* = puncture to remove fluid) involves withdrawing some of the amniotic fluid that bathes the developing fetus and analyzing the fetal cells and dissolved substances. It is used to test for the presence of certain genetic disorders, such as Down syndrome

(DS), hemophilia, Tay-Sachs disease, sickle-cell disease, and certain muscular dystrophies. It is also used to help determine survivability of the fetus. The test is usually done at 14–18 weeks of gestation. All gross chromosomal abnormalities and over 50 biochemical defects can be detected through amniocentesis. It can also reveal the baby's gender, which is important information for the diagnosis of sex-linked disorders, in which an abnormal gene carried by the mother affects her male offspring only (described on page 1167).

During amniocentesis, the position of the fetus and placenta is first identified using ultrasound and palpation. After the skin is prepared with an antiseptic and a local anesthetic is given, a hypodermic needle is inserted through the mother's abdominal wall and into the amniotic cavity within the uterus. Then, 10 to 30 mL of fluid and suspended cells are aspirated (Figure 29.15a) for microscopic examination and biochemical testing. Elevated levels of alpha-fetoprotein (AFP) and acetylcholinesterase may indicate failure of the nervous system to develop properly, as occurs in spina bifida or anencephaly (absence of the cerebrum) or may be due to other developmental or chromosomal problems. Chromosome studies, which require growing the cells for 2–4 weeks in a culture medium, may reveal rearranged, missing, or extra chromosomes. Amniocentesis is performed only when a risk for genetic defects is suspected, because there is about a 0.5% chance of spontaneous abortion after the procedure.

Chorionic Villi Sampling

In **chorionic villi sampling** (ko-rē-ON-ik VIL-ī) or **CVS,** a catheter is guided through the vagina and cervix of the uterus and then advanced to the chorionic villi under ultrasound guidance (Figure 29.15b). About 30 milligrams of tissue are suctioned out and prepared for chromosomal analysis. Alternatively, the chorionic villi can be sampled by inserting a needle through the abdominal cavity, as performed in amniocentesis.

CVS can identify the same defects as amniocentesis because chorion cells and fetal cells contain the same genome. CVS offers several advantages over amniocentesis: It can be performed as early as eight weeks of gestation, and test results are available in only a few days, permitting an earlier decision on whether to continue the pregnancy. However, CVS is slightly riskier than amniocentesis; after the procedure there is a 1–2% chance of spontaneous abortion.

Noninvasive Prenatal Tests

Currently, chorionic villi testing and amniocentesis are the only useful ways to obtain fetal tissue for prenatal testing of gene defects. While these invasive procedures pose relatively little risk when performed by experts, much work has been done to develop **noninvasive prenatal tests,** which do not require the penetration of any embryonic structure. The goal is to develop

Figure 29.15 Amniocentesis and chorionic villi sampling.

To detect genetic abnormalities, amniocentesis is performed at 14–16 weeks of gestation; chorionic villi sampling may be performed as early as 8 weeks of gestation.

(a) Amniocentesis

(b) Chorionic villi sampling (CVS)

? What information can be provided by amniocentesis?

accurate, safe, more efficient, and less expensive tests for screening a large population.

The first such test developed was the **maternal alpha-fetoprotein (AFP) test.** In this test, the mother's blood is analyzed for the presence of AFP, a protein synthesized in the fetus that passes into the maternal circulation. The highest levels of AFP normally occur during weeks 12 through 15 of pregnancy. Later, AFP is not produced, and its concentration decreases to a very low level both in the fetus and in maternal blood. A high level of AFP after week 16 usually indicates that the fetus has a neural tube defect, such as spina bifida or anencephaly. Because the test is 95% accurate, it is now recommended that all pregnant women be tested for AFP. A newer test (Quad AFP Plus) probes maternal blood for AFP and three other molecules. The test permits prenatal screening for Down syndrome, trisomy 18, and neural tube defects; it also helps predict the delivery date and may reveal the presence of twins.

◉ CHECKPOINT

26. What conditions can be detected using fetal ultrasonography, amniocentesis, and chorionic villi sampling? What are the advantages of noninvasive prenatal tests?

MATERNAL CHANGES DURING PREGNANCY

◉ OBJECTIVES
- Describe the sources and functions of the hormones secreted during pregnancy.
- Describe the hormonal, anatomical, and physiological changes in the mother during pregnancy.

Hormones of Pregnancy

During the first 3 to 4 months of pregnancy, the corpus luteum in the ovary continues to secrete **progesterone** and **estrogens,** which maintain the lining of the uterus during pregnancy and prepare the mammary glands to secrete milk. The amounts secreted by the corpus luteum, however, are only slightly more than those produced after ovulation in a normal menstrual cycle. From the third month through the remainder of the pregnancy, the placenta itself provides the high levels of progesterone and estrogens required. As noted previously, the chorion secretes **human chorionic gonadotropin (hCG)** into the blood. In turn, hCG stimulates the corpus luteum to continue production of progesterone and estrogens—an activity required to prevent menstruation and for the continued attachment of the embryo and fetus to the lining of the uterus (Figure 29.16a). By the eighth day after fertilization, hCG can be detected in the blood and urine of a pregnant woman. Peak secretion of hCG occurs at about the ninth week of pregnancy (Figure 29.16b). During the fourth and fifth months the hCG level decreases sharply and then levels off until childbirth.

The chorion begins to secrete estrogens after the first 3 or 4 weeks of pregnancy and progesterone by the sixth week. These hormones are secreted in increasing quantities until the time of birth (Figure 29.16b). By the fourth month, when the placenta is fully established, the secretion of hCG is greatly reduced, and the secretions of the corpus luteum are no longer essential. A high level of progesterone ensures that the uterine myometrium is relaxed and that the cervix is tightly closed. After delivery, estrogens and progesterone in the blood decrease to normal levels.

Relaxin, a hormone produced first by the corpus luteum of the ovary and later by the placenta, increases the flexibility of the pubic symphysis and ligaments of the sacroiliac and sacrococcygeal joints and helps dilate the uterine cervix during labor. Both of these actions ease delivery of the baby.

A third hormone produced by the chorion of the placenta is **human chorionic somatomammotropin (hCS),** also known as **human placental lactogen (hPL).** The rate of secretion of hCS increases in proportion to placental mass, reaching maximum levels after 32 weeks and remaining relatively constant after that. It is thought to help prepare the mammary glands for lactation, enhance maternal growth by increasing protein synthesis, and regulate certain aspects of metabolism in both mother and fetus. For example, hCS decreases the use of glucose by the mother and promotes the release of fatty acids from her adipose tissue, making more glucose available to the fetus.

The hormone most recently found to be produced by the placenta is **corticotropin-releasing hormone (CRH),** which in nonpregnant people is secreted only by neurosecretory cells in the hypothalamus. CRH is now thought to be part of the "clock" that establishes the timing of birth. Secretion of CRH by the placenta begins at about 12 weeks and increases enormously toward the end of pregnancy. Women who have higher levels of CRH earlier in pregnancy are more likely to deliver prematurely; those who have low levels are more likely to deliver after their due date. CRH from the placenta has a second important effect: It increases secretion of cortisol, which is needed for maturation of the fetal lungs and the production of surfactant (see page 888).

● CLINICAL **CONNECTION** | Early Pregnancy Tests

Early pregnancy tests detect the tiny amounts of human chorionic gonadotropin (hCG) in the urine that begin to be excreted about 8 days after fertilization. The test kits can detect pregnancy as early as the first day of a missed menstrual period—that is, at about 14 days after fertilization. Chemicals in the kits produce a color change if a reaction occurs between hCG in the urine and hCG antibodies included in the kit.

Several of the test kits available at pharmacies are as sensitive and accurate as test methods used in many hospitals. Still, false-negative and false-positive results can occur. A false-negative result (the test is negative, but the woman is pregnant) may be due to testing too soon or to an ectopic pregnancy. A false-positive result (the test is positive, but the woman is not pregnant) may be due to excess protein or blood in the urine or to hCG production due to a rare type of uterine cancer. Thiazide diuretics, hormones, steroids, and thyroid drugs may also affect the outcome of an early pregnancy test. •

Figure 29.16 Hormones during pregnancy.

The corpus luteum produces progesterone and estrogens during the first 3–4 months of pregnancy, after which time the placenta assumes this function.

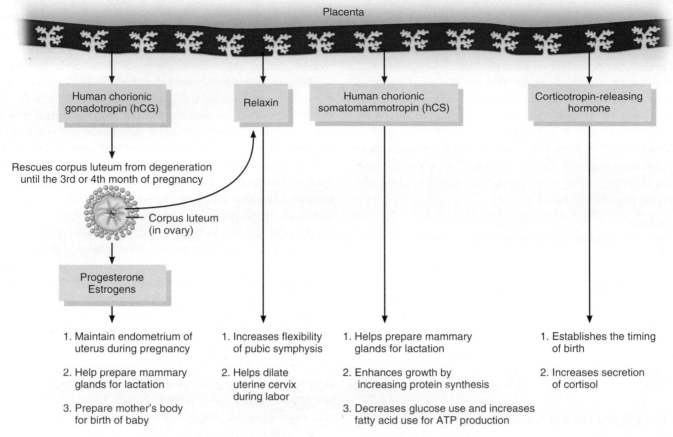

(a) Sources and functions of hormones

(b) Blood levels of hormones during pregnancy

? **Which hormone is detected by early pregnancy tests?**

Changes During Pregnancy

Near the end of the third month of pregnancy, the uterus occupies most of the pelvic cavity. As the fetus continues to grow, the uterus extends higher and higher into the abdominal cavity. Toward the end of a full-term pregnancy, the uterus fills nearly the entire abdominal cavity, reaching above the costal margin nearly to the xiphoid process of the sternum (Figure 29.17). It pushes the maternal intestines, liver, and stomach superiorly, elevates the diaphragm, and widens the thoracic cavity. Pressure on the stomach may force the stomach contents superiorly into the esophagus, resulting in heartburn. In the pelvic cavity, compression of the ureters and urinary bladder occurs.

Pregnancy-induced physiological changes also occur, including weight gain due to the fetus, amniotic fluid, the placenta,

uterine enlargement, and increased total body water; increased storage of proteins, triglycerides, and minerals; marked breast enlargement in preparation for lactation; and lower back pain due to lordosis (hollow back).

Several changes occur in the maternal cardiovascular system. Stroke volume increases by about 30% and cardiac output rises by 20–30% due to increased maternal blood flow to the placenta and increased metabolism. Heart rate increases 10–15% and blood volume increases 30–50%, mostly during the second half of pregnancy. These increases are necessary to meet the additional demands of the fetus for nutrients and oxygen. When a pregnant woman is lying on her back, the enlarged uterus may compress the aorta, resulting in diminished blood flow to the uterus. Compression of the inferior vena cava also decreases venous return, which leads to edema in the lower limbs and may

Figure 29.17 Normal fetal location and position at the end of a full-term pregnancy.

The gestation period is the time interval (about 38 weeks) from fertilization to birth.

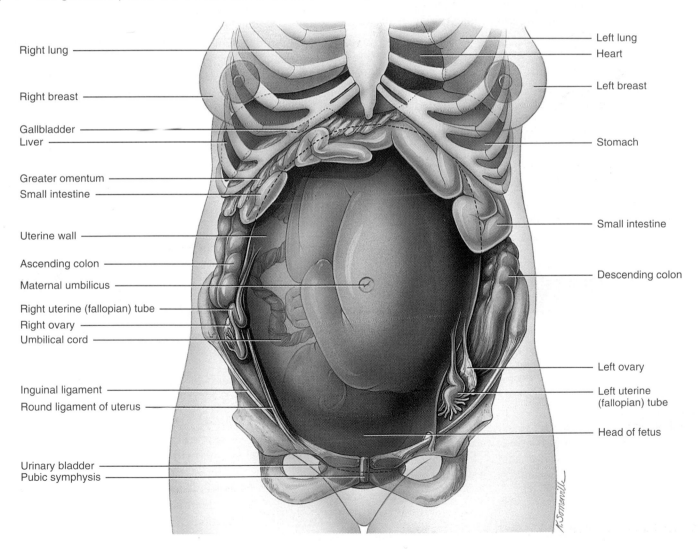

Anterior view of position of organs at end of full-term pregnancy

? **What hormone increases the flexibility of the pubic symphysis and helps dilate the cervix of the uterus to ease delivery of the baby?**

produce varicose veins. Compression of the renal artery can lead to renal hypertension.

Respiratory function is also altered during pregnancy to meet the added oxygen demands of the fetus. Tidal volume can increase by 30–40%, expiratory reserve volume can be reduced by up to 40%, functional residual capacity can decline by up to 25%, minute ventilation (the total volume of air inhaled and exhaled each minute) can increase by up to 40%, airway resistance in the bronchial tree can decline by 30–40%, and total body oxygen consumption can increase by about 10–20%. Dyspnea (difficult breathing) also occurs.

The digestive system also undergoes changes. Pregnant women experience an increase in appetite due to the added nutritional demands of the fetus. A general decrease in GI tract motility can cause constipation, delay gastric emptying time, and produce nausea, vomiting, and heartburn.

Pressure on the urinary bladder by the enlarging uterus can produce urinary symptoms, such as increased frequency and urgency of urination, and stress incontinence. An increase in renal plasma flow up to 35% and an increase in glomerular filtration rate up to 40% increase the renal filtering capacity, which allows faster elimination of the extra wastes produced by the fetus.

Changes in the skin during pregnancy are more apparent in some women than in others. Some women experience increased pigmentation around the eyes and cheekbones in a masklike pattern (chloasma), in the areolae of the breasts, and in the linea alba of the lower abdomen (linea nigra). Striae (stretch marks) over the abdomen can occur as the uterus enlarges, and hair loss increases.

Changes in the reproductive system include edema and increased vascularity of the vulva and increased pliability and vascularity of the vagina. The uterus increases from its nonpregnant mass of 60–80 g to 900–1200 g at term because of hyperplasia of muscle fibers in the myometrium in early pregnancy and hypertrophy of muscle fibers during the second and third trimesters.

• CLINICAL CONNECTION | Pregnancy-Induced Hypertension

About 10–15% of all pregnant women in the United States experience **pregnancy-induced hypertension (PIH),** an elevated blood pressure that is associated with pregnancy. The major cause is **preeclampsia** (prē-ē-KLAMP-sē-a), an abnormal condition of pregnancy characterized by sudden hypertension, large amounts of protein in the urine, and generalized edema that typically appears after the 20th week of pregnancy. Other signs and symptoms are generalized edema, blurred vision, and headaches. Preeclampsia might be related to an autoimmune or allergic reaction resulting from the presence of a fetus. Treatment involves bed rest and various drugs. When the condition is also associated with convulsions and coma, it is termed **eclampsia.** •

◉ CHECKPOINT

27. List the hormones involved in pregnancy, and describe the functions of each.
28. What structural and functional changes occur in the mother during pregnancy?

EXERCISE AND PREGNANCY

◉ OBJECTIVE

• Explain the effects of pregnancy on exercise and of exercise on pregnancy.

Only a few changes in early pregnancy affect the ability to exercise. A pregnant woman may tire more easily than usual, or morning sickness may interfere with regular exercise. As the pregnancy progresses, weight is gained and posture changes, so more energy is needed to perform activities, and certain maneuvers (sudden stopping, changes in direction, rapid movements) are more difficult to execute. In addition, certain joints, especially the pubic symphysis, become less stable in response to the increased level of the hormone relaxin. As compensation, many mothers-to-be walk with widely spread legs and a shuffling motion.

Although blood shifts from viscera (including the uterus) to the muscles and skin during exercise, there is no evidence of inadequate blood flow to the placenta. The heat generated during exercise may cause dehydration and further increase body temperature. Especially during early pregnancy, excessive exercise and heat buildup should be avoided because elevated body temperature has been implicated in neural tube defects. Exercise has no known effect on lactation, provided a woman remains hydrated and wears a bra that provides good support. Overall, moderate physical activity does not endanger the fetus of a healthy woman who has a normal pregnancy. However, any physical activity that might endanger the fetus should be avoided.

Among the benefits of exercise to the mother during pregnancy are a greater sense of well-being and fewer physical complaints.

◉ CHECKPOINT

29. Which changes in pregnancy have an effect on the ability to exercise?

LABOR

◉ OBJECTIVE

• Explain the events associated with the three stages of labor.

Labor is the process by which the fetus is expelled from the uterus through the vagina, also referred to as giving birth. A synonym for labor is **parturition** (par'-toor-ISH-un; *parturit-* = childbirth).

The onset of labor is determined by complex interactions of several placental and fetal hormones. Because progesterone inhibits uterine contractions, labor cannot take place until its effects are diminished. Toward the end of gestation, the levels of estrogens in the mother's blood rise sharply, producing changes that overcome the inhibiting effects of progesterone. The rise in estrogens results from increasing secretion by the placenta of corticotropin-releasing hormone, which stimulates the anterior pituitary gland of the fetus to secrete ACTH (adrenocorticotropic hormone). In turn, ACTH stimulates the fetal adrenal gland to secrete cortisol and dehydroepiandrosterone (DHEA), the major adrenal androgen. The placenta then converts DHEA into an estrogen. High levels of estrogens cause the number of receptors for oxytocin on uterine muscle fibers to increase, and cause uterine muscle fibers to form gap junctions with one another. Oxytocin released by the posterior pituitary stimulates uterine contractions, and relaxin from the placenta assists by increasing the flexibility of the pubic symphysis and helping dilate the uterine cervix. Estrogen also stimulates the placenta to release prostaglandins, which induce production of enzymes that digest collagen fibers in the cervix, causing it to soften.

Control of labor contractions during parturition occurs via a positive feedback cycle (see Figure 1.4 on page 11). Contractions of the uterine myometrium force the baby's head or body into the cervix, distending (stretching) the cervix. Stretch receptors in the cervix send nerve impulses to neurosecretory cells in the hypothalamus, causing them to release oxytocin into blood capillaries of the posterior pituitary gland. Oxytocin then is carried by the blood to the uterus, where it stimulates the myometrium to contract more forcefully. As the contractions intensify, the baby's body stretches the cervix still more, and the resulting nerve impulses stimulate the secretion of yet more oxytocin. With birth of the infant, the positive feedback cycle is broken because cervical distension suddenly lessens.

Uterine contractions occur in waves (quite similar to the peristaltic waves of the gastrointestinal tract) that start at the top of the uterus and move downward, eventually expelling the fetus. **True labor** begins when uterine contractions occur at regular intervals, usually producing pain. As the interval between contractions shortens, the contractions intensify. Another symptom of true labor in some women is localization of pain in the back that is intensified by walking. The most reliable indicator of true labor is dilation of the cervix and the "show," a discharge of a blood-containing mucus into the cervical canal. In **false labor,** pain is felt in the abdomen at irregular intervals, but it does not intensify and walking does not alter it significantly. There is no "show" and no cervical dilation.

True labor can be divided into three stages (Figure 29.18):

❶ *Stage of dilation.* The time from the onset of labor to the complete dilation of the cervix is the **stage of dilation.** This stage, which typically lasts 6–12 hours, features regular contractions of the uterus, usually a rupturing of the amniotic sac, and complete dilation (to 10 cm) of the cervix. If the amniotic sac does not rupture spontaneously, it is ruptured intentionally.

❷ *Stage of expulsion.* The time (10 minutes to several hours) from complete cervical dilation to delivery of the baby is the **stage of expulsion.**

❸ *Placental stage.* The time (5–30 minutes or more) after delivery until the placenta or "afterbirth" is expelled by powerful uterine contractions is the **placental stage.** These contrac-

Figure 29.18 **Stages of true labor.**

The term *parturition* refers to birth.

- Urinary bladder
- Vagina
- Ruptured amniotic sac
- Rectum

❶ Stage of dilation

- Placenta

❷ Stage of expulsion

- Uterus
- Placenta
- Umbilical cord

❸ Placental stage

? **What event marks the beginning of the stage of expulsion?**

tions also constrict blood vessels that were torn during delivery, reducing the likelihood of hemorrhage.

As a rule, labor lasts longer with first babies, typically about 14 hours. For women who have previously given birth, the average duration of labor is about 8 hours—although the time varies enormously among births. Because the fetus may be squeezed through the birth canal (cervix and vagina) for up to several hours, the fetus is stressed during childbirth: The fetal head is compressed, and the fetus undergoes some degree of intermittent hypoxia due to compression of the umbilical cord and the placenta during uterine contractions. In response to this stress, the fetal adrenal medullae secrete very high levels of epinephrine and norepinephrine, the "fight-or-flight" hormones. Much of the protection against the stresses of parturition, as well as preparation of the infant for surviving extrauterine life, is provided by these hormones. Among other functions, epinephrine and norepinephrine clear the lungs and alter their physiology in readiness for breathing air, mobilize readily usable nutrients for cellular metabolism, and promote an increased flow of blood to the brain and heart.

About 7% of pregnant women do not deliver by 2 weeks after their due date. Such cases carry an increased risk of brain damage to the fetus, and even fetal death, due to inadequate supplies of oxygen and nutrients from an aging placenta. Post-term deliveries may be facilitated by inducing labor, initiated by administration of oxytocin (Pitocin®), or by surgical delivery (cesarean section).

Following the delivery of the baby and placenta is a 6-week period during which the maternal reproductive organs and physiology return to the prepregnancy state. This period is called the **puerperium** (pū′-er-PER-ē-um). Through a process of tissue catabolism, the uterus undergoes a remarkable reduction in size, called **involution** (in′-vō-LOO-shun), especially in lactating women. The cervix loses its elasticity and regains its prepregnancy firmness. For 2–4 weeks after delivery, women have a uterine discharge called **lochia** (LŌ-kē-a), which consists initially of blood and later of serous fluid derived from the former site of the placenta.

• CLINICAL CONNECTION | **Dystocia and Cesarean Section**

Dystocia (dis-TŌ-sē-a; *dys-* = painful or difficult; *toc-* = birth), or difficult labor, may result either from an abnormal position (presentation) of the fetus or a birth canal of inadequate size to permit vaginal delivery. In a **breech presentation,** for example, the fetal buttocks or lower limbs, rather than the head, enter the birth canal first; this occurs most often in premature births. If fetal or maternal distress prevents a vaginal birth, the baby may be delivered surgically through an abdominal incision. A low, horizontal cut is made through the abdominal wall and lower portion of the uterus, through which the baby and placenta are removed. Even though it is popularly associated with the birth of Julius Caesar, the true reason this procedure is termed a **cesarean section (C-section)** is because it was described in Roman Law, *lex cesarea,* about 600 years before Julius Caesar was born. Even a history of multiple C-sections need not exclude a pregnant woman from attempting a vaginal delivery. •

◉ CHECKPOINT

30. What hormonal changes induce labor?
31. What is the difference between false labor and true labor?
32. What happens during the stage of dilation, the stage of expulsion, and the placental stage of true labor?

ADJUSTMENTS OF THE INFANT AT BIRTH

◉ OBJECTIVE

• Explain the respiratory and cardiovascular adjustments that occur in an infant at birth.

During pregnancy, the embryo (and later the fetus) is totally dependent on the mother for its existence. The mother supplies the fetus with oxygen and nutrients, eliminates its carbon dioxide and other wastes, protects it against shocks and temperature changes, and provides antibodies that confer protection against certain harmful microbes. At birth, a physiologically mature baby becomes much more self-supporting, and the newborn's body systems must make various adjustments. The most dramatic changes occur in the respiratory and cardiovascular systems.

Respiratory Adjustments

The reason that the fetus depends entirely on the mother for obtaining oxygen and eliminating carbon dioxide is that the fetal lungs are either collapsed or partially filled with amniotic fluid. The production of surfactant begins by the end of the sixth month of development. Because the respiratory system is fairly well developed at least 2 months before birth, premature babies delivered at 7 months are able to breathe and cry. After delivery, the baby's supply of oxygen from the mother ceases, and any amniotic fluid in the fetal lungs is absorbed. Because carbon dioxide is no longer being removed, it builds up in the blood. A rising CO_2 level stimulates the respiratory center in the medulla oblongata, causing the respiratory muscles to contract, and the baby to draw his or her first breath. Because the first inspiration is unusually deep, as the lungs contain no air, the baby also exhales vigorously and naturally cries. A full-term baby may breathe 45 times a minute for the first 2 weeks after birth. Breathing rate gradually declines until it approaches a normal rate of 12 breaths per minute.

Cardiovascular Adjustments

After the baby's first inspiration, the cardiovascular system must make several adjustments (see Figure 21.30 on page 821). Closure of the foramen ovale between the atria of the fetal heart, which occurs at the moment of birth, diverts deoxygenated blood to the lungs for the first time. The foramen ovale is closed by two flaps of septal heart tissue that fold together and permanently fuse. The remnant of the foramen ovale is the fossa ovalis.

Once the lungs begin to function, the ductus arteriosus shuts off due to contractions of smooth muscle in its wall, and it becomes the ligamentum arteriosum. The muscle contraction is probably mediated by the polypeptide bradykinin, released from the lungs during their initial inflation. The ductus arteriosus generally does not close completely until about 3 months after birth. Prolonged incomplete closure results in a condition called **patent ductus arteriosus** (see Figure 20.23b on page 753).

After the umbilical cord is tied off and severed and blood no longer flows through the umbilical arteries, they fill with connective tissue, and their distal portions become the medial umbilical ligaments. The umbilical vein then becomes the ligamentum teres (round ligament) of the liver.

In the fetus, the ductus venosus connects the umbilical vein directly with the inferior vena cava, allowing blood from the placenta to bypass the fetal liver. When the umbilical cord is severed, the ductus venosus collapses, and venous blood from the viscera of the fetus flows into the hepatic portal vein to the liver and then via the hepatic vein to the inferior vena cava. The remnant of the ductus venosus becomes the ligamentum venosum.

At birth, an infant's pulse may range from 120 to 160 beats per minute and may go as high as 180 upon excitation. After birth, oxygen use increases, which stimulates an increase in the rate of red blood cell and hemoglobin production. The white blood cell count at birth is very high—sometimes as much as 45,000 cells per microliter—but the count decreases rapidly by the seventh day. Recall that the white blood cell count of an adult is 5000–10,000 cells per microliter.

● CLINICAL **CONNECTION** | Premature Infants

Delivery of a physiologically immature baby carries certain risks. A **premature infant** or "preemie" is generally considered a baby who weighs less than 2500 g (5.5 lb) at birth. Poor prenatal care, drug abuse, history of a previous premature delivery, and mother's age below 16 or above 35 increase the chance of premature delivery. The body of a premature infant is not yet ready to sustain some critical functions, and thus its survival is uncertain without medical intervention. The major problem after delivery of an infant under 36 weeks of gestation is respiratory distress syndrome (RDS) of the newborn due to insufficient surfactant. RDS can be eased by use of artificial surfactant and a ventilator that delivers oxygen until the lungs can operate on their own. •

◉ CHECKPOINT

33. Why are respiratory and cardiovascular adjustments so important at birth?

THE PHYSIOLOGY OF LACTATION

◉ OBJECTIVE

• Discuss the physiology and hormonal control of lactation.

Lactation (lak'-TĀ-shun) is the secretion and ejection of milk from the mammary glands. A principal hormone in promoting milk synthesis and secretion is **prolactin (PRL),** which is secreted from the anterior pituitary gland. Even though prolactin levels increase as the pregnancy progresses, no milk secretion occurs because progesterone inhibits the effects of prolactin. After delivery, the levels of estrogens and progesterone in the mother's blood decrease, and the inhibition is removed. The principal stimulus in maintaining prolactin secretion during lactation is the sucking action of the infant. Suckling initiates nerve impulses from stretch receptors in the nipples to the hypothalamus; the impulses decrease hypothalamic release of prolactin-

inhibiting hormone (PIH) and increase release of prolactin-releasing hormone (PRH), so more prolactin is released by the anterior pituitary.

Oxytocin causes release of milk into the mammary ducts via the **milk ejection reflex** (Figure 29.19). Milk formed by the

Figure 29.19 The milk ejection reflex, a positive feedback cycle.

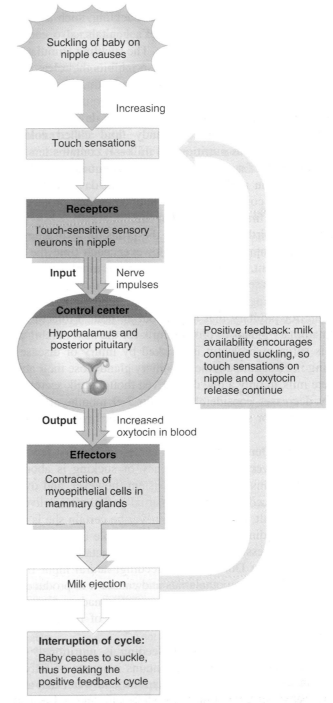

Oxytocin stimulates contraction of myoepithelial cells in the breasts, which squeezes the glandular and duct cells and causes milk ejection.

 What is another function of oxytocin?

to their offspring. Such individuals are called **carriers** of the recessive gene.

Most genes give rise to the same phenotype whether they are inherited from the mother or the father. In a few cases, however, the phenotype is dramatically different, depending on the parental origin. This surprising phenomenon, first appreciated in the 1980s, is called **genomic imprinting.** In humans, the abnormalities most clearly associated with mutation of an imprinted gene are *Angelman syndrome* (mental retardation, ataxia, seizures, and minimal speech), which results when the gene for a particular abnormal trait is inherited from the mother, and *Prader-Willi syndrome* (short stature, mental retardation, obesity, poor responsiveness to external stimuli, and sexual immaturity), which results when it is inherited from the father.

Alleles that code for normal traits do not always dominate over those that code for abnormal ones, but dominant alleles for severe disorders usually are lethal and cause death of the embryo or fetus. One exception is Huntington disease (HD) (see page 588), which is caused by a dominant allele whose effects do not become manifest until adulthood. Both homozygous dominant and heterozygous people exhibit the disease; homozygous recessive people are normal. HD causes progressive degeneration of the nervous system and eventual death, but because symptoms typically do not appear until after age 30 or 40, many afflicted individuals will already have passed on the allele for the condition to their children by the time they discover they have the disease.

Occasionally an error in cell division, called **nondisjunction,** results in an abnormal number of chromosomes. In this situation, homologous chromosomes (during meiosis I) or sister chromatids (during anaphase of mitosis or meiosis II) fail to separate properly. See Figure 3.34 on page 99. A cell from which one or more chromosomes has been added or deleted is called an **aneuploid** (AN-ū-ployd). A monosomic cell ($2n - 1$) is missing a chromosome; a trisomic cell ($2n + 1$) has an extra chromosome. Most cases of Down syndrome (see page 1169) are aneuploid disorders in which there is trisomy of chromosome 21. Nondisjunction usually occurs during gametogenesis (meiosis), but about 2% of Down syndrome cases result from nondisjunction during mitotic divisions in early embryonic development.

Another error in meiosis is a **translocation.** In this case, two chromosomes that are *not* homologous break and interchange portions of their chromosomes. The individual who has a translocation may be perfectly normal if no loss of genetic material took place when the rearrangement occurred. However, some of the person's gametes may not contain the correct amount and type of genetic material. About 3% of Down syndrome cases result from a translocation of part of chromosome 21 to another chromosome, usually chromosome 14 or 15. The individual who has this translocation is normal and does not even know that he or she is a "carrier." When such a carrier produces gametes, however, some gametes end up with a whole chromosome 21 plus another chromosome with the translocated fragment of chromosome 21. Upon fertilization, the zygote then has three, rather than two, copies of that part of chromosome 21.

Table 29.3 lists some dominant and recessive inherited structural and functional traits in humans.

TABLE 29.3

Selected Hereditary Traits in Humans

DOMINANT	RECESSIVE
Normal skin pigmentation	Albinism
Near- or farsightedness	Normal vision
PTC taster*	PTC nontaster
Polydactyly (extra digits)	Normal digits
Brachydactyly (short digits)	Normal digits
Syndactylism (webbed digits)	Normal digits
Diabetes insipidus	Normal urine excretion
Huntington disease	Normal nervous system
Widow's peak	Straight hairline
Curved (hyperextended) thumb	Straight thumb
Normal Cl⁻ transport	Cystic fibrosis
Hypercholesterolemia (familial)	Normal cholesterol level

*Ability to taste a chemical compound called phenylthiocarbamide (PTC).

Variations on Dominant–Recessive Inheritance

Most patterns of inheritance do not conform to the simple **dominant–recessive inheritance** we have just described, in which only dominant and recessive alleles interact. The phenotypic expression of a particular gene may be influenced not only by which alleles are present, but also by other genes and by the environment. Most inherited traits are influenced by more than one gene, and, to complicate matters, most genes can influence more than one trait. Variations on dominant–recessive inheritance include incomplete dominance, multiple-allele inheritance, and complex inheritance.

Incomplete Dominance

In **incomplete dominance,** neither member of a pair of alleles is dominant over the other, and the heterozygote has a phenotype intermediate between the homozygous dominant and the homozygous recessive phenotypes. An example of incomplete dominance in humans is the inheritance of **sickle-cell disease (SCD)** (Figure 29.21). People with the homozygous dominant genotype $Hb^A Hb^A$ form normal hemoglobin; those with the homozygous recessive genotype $Hb^S Hb^S$ have sickle-cell disease and severe anemia. Although they are usually healthy, those with the heterozygous genotype $Hb^A Hb^S$ have minor problems with anemia because half their hemoglobin is normal and half is not. Heterozygotes are carriers, and they are said to have *sickle-cell trait.*

Multiple-Allele Inheritance

Although a single individual inherits only two alleles for each gene, some genes may have more than two alternative forms;

Figure 29.21 Inheritance of sickle-cell disease.

 Sickle-cell disease is an example of incomplete dominance.

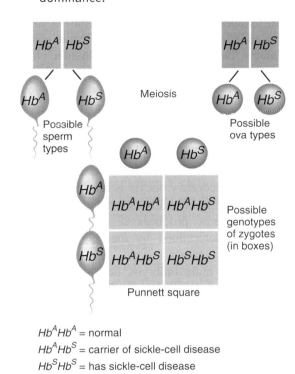

$Hb^A Hb^A$ = normal
$Hb^A Hb^S$ = carrier of sickle-cell disease
$Hb^S Hb^S$ = has sickle-cell disease

? What are the distinguishing features of incomplete dominance?

this is the basis for **multiple-allele inheritance.** One example of multiple-allele inheritance is the inheritance of the ABO blood group. The four blood types (phenotypes) of the ABO group—A, B, AB, and O—result from the inheritance of six combinations of three different alleles of a single gene called the *I* gene: (1) allele I^A produces the A antigen, (2) allele I^B produces the B antigen, and (3) allele *i* produces neither A nor B antigen. Each person inherits two *I*-gene alleles, one from each parent, that give rise to the various phenotypes. The six possible genotypes produce four blood types, as follows:

Genotype	Blood type (phenotype)
$I^A I^A$ or $I^A i$	A
$I^B I^B$ or $I^B i$	B
$I^A I^B$	AB
ii	O

Notice that both I^A and I^B are inherited as dominant traits, and *i* is inherited as a recessive trait. Because an individual with type AB blood has characteristics of both type A and type B red blood cells expressed in the phenotype, alleles I^A and I^B are said to be **codominant.** In other words, both genes are expressed equally in the heterozygote. Depending on the parental blood types, different offspring may have blood types different from each other. Figure 29.22 shows the blood types offspring could inherit, given the blood types of their parents.

Figure 29.22 The 10 possible combinations of parental ABO blood types and the blood types their offspring could inherit.
For each possible set of parents, the blue letters represent the blood types their offspring could inherit.

 Inheritance of ABO blood types is an example of multiple-allele inheritance.

? How is it possible for a baby to have type O blood if neither parent is type O?

Complex Inheritance

Most inherited traits are not controlled by one gene, but instead by the combined effects of two or more genes, a situation referred to as **polygenic inheritance** (*poly-* = many), or the combined effects of many genes and environmental factors, a situation referred to as **complex inheritance.** Examples of complex traits include skin color, hair color, eye color, height, metabolism rate, and body build. In complex inheritance, one genotype can have many possible phenotypes, depending on the environment, or one phenotype can include many possible genotypes. For example, even though a person inherits several genes for tallness, full height potential may not be reached due to environmental factors, such as disease or malnutrition during the growth years. You have already learned that the risk of having a child with a neural tube defect is greater in pregnant women who lack adequate folic acid in their diet; this is also considered an environmental factor. Because neural tube defects are more prevalent in some families than in others, however, one or more genes may also contribute.

Often, a complex trait shows a continuous gradation of small differences between extremes among individuals. It is relatively easy to predict the risk of passing on an undesirable trait that is due to a single dominant or recessive gene, but it is very difficult to make this prediction when the trait is complex. Such traits are difficult to follow in a family because the range of variation is large, the number of different genes involved usually is not known, and the impact of environmental factors may be incompletely understood.

Choose the one best answer to the following questions.

5. Which of the following are *true*? (1) During implantation the outer cell mass of the blastocyst orients toward the endometrium. (2) The decidua basalis provides glycogen and lipids for the developing fetus. (3) The decidua parietalis becomes the maternal part of the placenta. (4) During implantation, the syncytiotrophoblast secretes enzymes that allow the blastocyst to penetrate the uterine lining. (5) After fetal delivery, the decidua separates from the endometrium and is released from the uterus.
   (a) 2, 4, and 5   (b) 1, 2, and 3   (c) 2, 3, 4, and 5
   (d) 1, 2, 3, 4, and 5   (e) 1, 3, and 5

6. Which of the following are maternal changes that occur during pregnancy? (1) altered pulmonary function; (2) increased stroke volume, cardiac output, and heart rate, and decreased blood volume; (3) weight gain; (4) increased gastric motility, causing a delay in gastric emptying time; (5) edema and possible varicose veins.
   (a) 1, 2, 3, and 4   (b) 2, 3, 4, and 5   (c) 1, 3, 4, and 5
   (d) 1, 3, and 5   (e) 2, 4, and 5

7. Which of the following statements is *correct*? (a) Normal traits always dominate over abnormal traits. (b) Occasionally an error in meiosis called nondisjunction results in an abnormal number of chromosomes. (c) The mother always determines the sex of the child because she has either an X or Y gene in her oocytes. (d) Most patterns of inheritance are simple dominant-recessive inheritances. (e) Genes are expressed normally regardless of any outside influence such as chemicals or radiation.

8. Which of the following are *true* concerning fertilization? (1) The sperm first penetrate the zona pellucida and then the corona radiata. (2) The binding of specific membrane proteins in the sperm head to ZP3 causes the release of acrosomal contents. (3) Sperm are able to fertilize the oocyte within minutes after ejaculation. (4) Depolarization of the cell membrane of the secondary oocyte inhibits fertilization by more than one sperm. (5) The oocyte completes meiosis II after fertilization.
   (a) 1, 2, 4, and 5   (b) 1, 3, and 5   (c) 1, 2, 3, and 4
   (d) 1, 4, and 5   (e) 2, 4, and 5

9. Amniotic fluid (1) is derived entirely from a filtrate of maternal blood, (2) acts as a fetal shock absorber, (3) provides nutrients to the fetus, (4) helps regulate fetal body temperature, (5) prevents adhesions between the skin of the fetus and surrounding tissues.
   (a) 1, 2, 3, 4, and 5   (b) 2, 4, and 5   (c) 2, 3, 4, and 5
   (d) 1, 4, and 5   (e) 1, 2, 4, and 5

10. Which of the following structures develop during the fourth week after fertilization? (1) embryonic folding, (2) the neural tube, (3) otic placode (beginning of the ear), (4) beginning of the eyes, (5) upper and lower limb buds.
    (a) 1 and 2   (b) 1, 2, and 5   (c) 1, 2, 3, 4, and 5
    (d) 2, 3, and 5   (e) 1, 3, 4, and 5

11. Match the following:
    ___ (a) a fluid-filled sphere of cells that enters the uterine cavity
    ___ (b) cells produced by cleavage
    ___ (c) the developing individual from week nine of pregnancy until birth
    ___ (d) the outer covering of cells of the blastocyst
    ___ (e) membrane derived from trophoblast
    ___ (f) early divisions of the zygote
    ___ (g) a solid sphere of cells still surrounded by the zona pellucida
    ___ (h) event in which differentiation into the three primary germ layers occurs
    ___ (i) embryonic development of structures that will become the nervous system
    ___ (j) the formation of blood vessels to support the developing embryo
    ___ (k) result of the fusion of female and male pronuclei

    (1) cleavage
    (2) blastomeres
    (3) morula
    (4) angiogenesis
    (5) trophoblast
    (6) blastocyst
    (7) zygote
    (8) gastrulation
    (9) neurulation
    (10) chorion
    (11) fetus

12. Match the following:
    (a) stimulates the corpus luteum to continue production of progesterone and estrogens
    (b) increases the flexibility of the pubic symphysis and helps dilate the uterine cervix during labor
    (c) secreted by the placenta; helps establish the timing of birth and increases the secretion of cortisol for fetal lung maturation
    (d) helps prepare mammary glands for lactation; regulates certain aspects of maternal and fetal metabolism
    (e) stimulates uterine contractions; responsible for the milk ejection reflex
    (f) promotes milk synthesis and secretion; inhibited by progesterone during pregnancy

    (1) oxytocin
    (2) human chorionic somatomammotropin
    (3) human chorionic gonadotropin
    (4) prolactin
    (5) corticotropin-releasing hormone
    (6) relaxin

13. Match the following:

____ (a) the penetration of a secondary oocyte by a single sperm cell

____ (b) fertilization of a secondary oocyte by more than one sperm

____ (c) the attachment of a blastocyst to the endometrium

____ (d) the fusion of the genetic material from a haploid sperm and a haploid secondary oocyte into a single diploid nucleus

____ (e) the induction by the female reproductive tract of functional changes in sperm that allow them to fertilize a secondary oocyte

____ (f) the examination of embryonic or fetal cells sloughed off into the amniotic fluid

____ (g) an abnormal condition of pregnancy characterized by sudden hypertension, large amounts of protein in urine, and generalized edema

____ (h) noninvasive test that can detect fetal neural tube defects

____ (i) the process of giving birth

____ (j) the period of time (about 6 weeks) during which the maternal reproductive organs and physiology return to the prepregnancy state

(1) fertilization
(2) capacitation
(3) syngamy
(4) polyspermy
(5) implantation
(6) amniocentesis
(7) preeclampsia
(8) parturition
(9) puerperium
(10) maternal AFP test

14. Match the following:

____ (a) the control of inherited traits by the combined effects of many genes

____ (b) the two alternative forms of a gene that code for the same trait and are at the same location on homologous chromosomes

____ (c) abnormal number of chromosomes due to failure of homologous chromosomes or chromatids to separate

____ (d) inheritance based on genes that have more than two alternative forms; an example is the inheritance of blood type

____ (e) a cell in which one or more chromosomes of a set is added or deleted

____ (f) refers to an individual with different alleles on homologous chromosomes

____ (g) traits linked to the X chromosome

____ (h) permanent inheritable change in an allele that produces a different variant of the same trait

____ (i) neither member of the allelic pair is dominant over the other, and the heterozygote has a phenotype intermediate between the homozygous dominant and the homozygous recessive

____ (j) refers to how the genetic makeup is expressed in the body; the physical or outward expression of a gene

____ (k) a homozygous dominant, homozygous recessive, or heterozygous genetic makeup; the actual gene arrangement

____ (l) refers to a person with the same alleles on homologous chromosomes

____ (m) inactivated X chromosome in females

____ (n) heterozygous individuals who possess a recessive gene (but do not express it) and can pass the gene on to their offspring

____ (o) interchange of portions of nonhomologous chromosomes

____ (p) an allele that masks the presence of another allele and is fully expressed

(1) genotype
(2) phenotype
(3) alleles
(4) aneuploid
(5) incomplete dominance
(6) multiple-allele inheritance
(7) polygenic inheritance
(8) sex-linked inheritance
(9) homozygous
(10) heterozygous
(11) carriers
(12) dominant trait
(13) mutation
(14) nondisjunction
(15) translocation
(16) Barr body

15. Match the following:

_____ (a) the embryonic membrane that entirely surrounds the embryo

_____ (b) functions as an early site of blood formation; contains cells that migrate into the gonads and differentiate into the primitive germ cells

_____ (c) becomes the principal part of the embryonic placenta; produces human chorionic gonadotropin

_____ (d) modified endometrium after implantation has occurred; separates from the endometrium after the fetus is delivered

_____ (e) contains the vascular connections between mother and fetus

_____ (f) the fetal portion is formed by the chorionic villi and the maternal portion is formed by the decidua basalis of the endometrium; allows oxygen and nutrients to diffuse from maternal blood into fetal blood

_____ (g) serves as an early site of blood vessel formation

_____ (h) finger-like projections of the chorion that bring maternal and fetal blood vessels into close proximity

_____ (i) plays an important role in induction whereby an inducing tissue stimulates the development of an unspecialized responding tissue into a specialized tissue

(1) decidua
(2) placenta
(3) amnion
(4) chorion
(5) allantois
(6) yolk sac
(7) notochord
(8) chorionic villi
(9) umbilical cord

CRITICAL THINKING QUESTIONS

1. Kathy is breast-feeding her infant and is experiencing what feels like early labor pains. What is causing these painful feelings? Is there a benefit to them?

2. Jack has hemophilia, which is a sex-linked blood clotting disorder. He blames his father for passing on the gene for hemophilia. Explain to Jack why his reasoning is wrong. How can Jack have hemophilia if his parents do not?

3. Alisa has asked her obstetrician to save and freeze her baby's cord blood after delivery in case the child needs a future bone marrow transplant. What is in the baby's cord blood that could be used to treat future disorders in the child?

? ANSWERS TO FIGURE QUESTIONS

29.1 Capacitation is the group of functional changes in sperm that enable them to fertilize a secondary oocyte, which occur after they have been deposited in the female reproductive tract.

29.2 A morula is a solid ball of cells; a blastocyst consists of a rim of cells (trophoblast) surrounding a cavity (blastocyst cavity) and an inner cell mass.

29.3 The blastocyst secretes digestive enzymes that eat away the endometrial lining at the site of implantation.

29.4 The decidua basalis helps form the maternal part of the placenta.

29.5 Implantation occurs during the secretory phase of the uterine cycle.

29.6 The bilaminar embryonic disc is attached to the trophoblast by the connecting stalk.

29.7 Gastrulation converts a bilaminar embryonic disc into a trilaminar embryonic disc.

29.8 The notochord induces mesodermal cells to develop into vertebral bodies and forms the nucleus pulposus of intervertebral discs.

29.9 The neural tube forms the brain and spinal cord; somites develop into skeletal muscles, connective tissue, and the vertebrae.

29.10 Chorionic villi help to bring the fetal and maternal blood vessels into close proximity.

29.11 The placenta participates in the exchange of materials between fetus and mother, serves as a protective barrier against many microbes, and stores nutrients.

29.12 As a result of embryonic folding, the embryo curves into a C-shape, various organs are brought into their eventual adult positions, and the primitive gut is formed.

29.13 Pharyngeal arches, clefts, and pouches give rise to structures of the head and neck.

29.14 Fetal weight doubles between the mid-fetal period and birth.

29.15 Amniocentesis is used primarily to detect genetic disorders, but it also provides information concerning the maturity (and survivability) of the fetus.

29.16 Early pregnancy tests detect elevated levels of human chorionic gonadotropin (hCG).

29.17 Relaxin increases the flexibility of the pubic symphysis and helps dilate the cervix of the uterus to ease delivery.

29.18 Complete dilation of the cervix marks the onset of the stage of expulsion.

29.19 Oxytocin also stimulates contraction of the uterus during delivery of a baby.

29.20 The odds that a child will have PKU are the same for each child—25%.

29.21 In incomplete dominance, neither member of an allelic pair is dominant; the heterozygote has a phenotype intermediate between the homozygous dominant and the homozygous recessive phenotypes.

29.22 A baby can have blood type O if each parent is heterozygous and has one i allele.

29.23 Hair color, height, and body build, among others, are traits passed on by complex inheritance.

29.24 The female sex chromosomes are XX, and the male sex chromosomes are XY.

29.25 The chromosomes that are not sex chromosomes are called autosomes.

29.26 A red–green color-blind female has an $X^c X^c$ genotype.

Measurements

U.S. Customary System

PARAMETER	UNIT	RELATION TO OTHER U.S. UNITS	SI (METRIC) EQUIVALENT
Length	inch	1/12 foot	2.54 centimeters
foot	12 inches	0.305 meter	
yard	36 inches	0.914 meters	
mile	5,280 feet	1.609 kilometers	
Mass	grain	1/1000 pound	64.799 milligrams
dram	1/16 ounce	1.772 grams	
ounce	16 drams	28.350 grams	
pound	16 ounces	453.6 grams	
ton	2,000 pounds	907.18 kilograms	
Volume (Liquid)	ounce	1/16 pint	29.574 milliliters
pint	16 ounces	0.473 liter	
quart	2 pints	0.946 liter	
gallon	4 quarts	3.785 liters	
Volume (Dry)	pint	1/2 quart	0.551 liter
quart	2 pints	1.101 liters	
peck	8 quarts	8.810 liters	
bushel	4 pecks	35.239 liters	

International System (SI)

BASE UNITS			PREFIXES		
UNIT	QUANTITY	SYMBOL	PREFIX	MULTIPLIER	SYMBOL
meter	length	m	tera-	$10^{12} = 1,000,000,000,000$	T
kilogram	mass	kg	giga-	$10^{9} = 1,000,000,000$	G
second	time	s	mega-	$10^{6} = 1,000,000$	M
liter	volume	L	kilo-	$10^{3} = 1,000$	k
mole	amount of matter	mol	hecto-	$10^{2} = 100$	h
			deca-	$10^{1} = 10$	da
			deci-	$10^{-1} = 0.1$	d
			centi-	$10^{-2} = 0.01$	c
			milli-	$10^{-3} = 0.001$	m
			micro-	$10^{-6} = 0.000,001$	μ
			nano-	$10^{-9} = 0.000,000,001$	n
			pico-	$10^{-12} = 0.000,000,000,001$	p

Temperature Conversion

FAHRENHEIT (F) TO CELSIUS (C)

$$°C = (°F - 32) \div 1.8$$

CELSIUS (C) TO FAHRENHEIT (F)

$$°F = (°C \times 1.8) + 32$$

U.S To SI (Metric) Conversion

WHEN YOU KNOW	MULTIPLY BY	TO FIND
inches	2.54	centimeters
feet	30.48	centimeters
yards	0.91	meters
miles	1.61	kilometers
ounces	28.35	grams
pounds	0.45	kilograms
tons	0.91	metric tons
fluid ounces	29.57	milliliters
pints	0.47	liters
quarts	0.95	liters
gallons	3.79	liters

SI (Metric) To U.S. Conversion

WHEN YOU KNOW	MULTIPLY BY	TO FIND
millimeters	0.04	inches
centimeters	0.39	inches
meters	3.28	feet
kilometers	0.62	miles
liters	1.06	quarts
cubic meters	35.32	cubic feet
grams	0.035	ounces
kilograms	2.21	pounds

Periodic Table

The periodic table lists the known **chemical elements,** the basic units of matter. The elements in the table are arranged left-to-right in rows in order of their **atomic number,** the number of protons in the nucleus. Each horizontal row, numbered from 1 to 7, is a **period.** All elements in a given period have the same number of electron shells as their period number. For example, an atom of hydrogen or helium each has one electron shell, while an atom of potassium or calcium each has four electron shells. The elements in each column, or **group,** share chemical properties. For example, the elements in column IA are very chemically reactive, whereas the elements in column VIIIA have full electron shells and thus are chemically inert.

Scientists now recognize 117 different elements; 92 occur naturally on Earth, and the rest are produced from the natural elements using particle accelerators or nuclear reactors. Elements are designated by **chemical symbols,** which are the first one or two letters of the element's name in English, Latin, or another language.

Twenty-six of the 92 naturally occurring elements normally are present in your body. Of these, just four elements—oxygen (O), carbon (C), hydrogen (H), and nitrogen (N) (coded blue)—constitute about 96% of the body's mass. Eight others—calcium (Ca), phosphorus (P), potassium (K), sulfur (S), sodium (Na), chlorine (Cl), magnesium (Mg), and iron (Fe) (coded pink)—contribute 3.8% of the body's mass. An additional 14 elements, called **trace elements** because they are present in tiny amounts, account for the remaining 0.2% of the body's mass. The trace elements are aluminum, boron, chromium, cobalt, copper, fluorine, iodine, manganese, molybdenum, selenium, silicon, tin, vanadium, and zinc (coded yellow). Table 2.1 on page 29 provides information about the main chemical elements in the body.

Normal Values for Selected Blood Tests

The system of international (SI) units (Système Internationale d'Unités) is used in most countries and in many medical and scientific journals. Clinical laboratories in the United States, by contrast, usually report values for blood and urine tests in conventional units. The laboratory values in this Appendix give conventional units first, followed by SI equivalents in parentheses. Values listed for various blood tests should be viewed as reference values rather than absolute "normal" values for all well people. Values may vary due to age, gender, diet, and environment of the subject or the equipment, methods, and standards of the lab performing the measurement.

KEY TO SYMBOLS

g = gram	mL = milliliter
mg = milligram = 10^{-3} gram	μL = microliter
μg = microgram = 10^{-6} gram	mEq/L = milliequivalents per liter
U = units	mmol/L = millimoles per liter
L = liter	μmol/L = micromoles per liter
dL = deciliter	> = greater than; < = less than

Blood Tests

TEST (SPECIMEN)	U.S. REFERENCE VALUES (SI UNITS)	VALUES INCREASE IN	VALUES DECREASE IN
Aminotransferases (serum)			
Alanine aminotransferase (ALT)	0–35 U/L (same)	Liver disease or liver damage due to toxic drugs.	
Aspartate aminotransferase (AST)	0–35 U/L (same)	Myocardial infarction, liver disease, trauma to skeletal muscles, severe burns.	Beriberi, uncontrolled diabetes mellitus with acidosis, pregnancy.
Ammonia (plasma)	20–120 μg/dL (12–55 μmol/L)	Liver disease, heart failure, emphysema, pneumonia, hemolytic disease of newborn.	Hypertension.
Bilirubin (serum)	Conjugated: <0.5 mg/dL (<5.0 μmol/L)	Conjugated bilirubin: liver dysfunction or gallstones.	
	Unconjugated: 0.2–1.0 mg/dL (18–20 μmol/L) Newborn: 1.0–12.0 mg/dL (<200 μmol/L)	Unconjugated bilirubin: excessive hemolysis of red blood cells.	
Blood urea nitrogen (BUN) (serum)	8–26 mg/dL (2.9–9.3 mmol/L)	Kidney disease, urinary tract obstruction, shock, diabetes, burns, dehydration, myocardial infarction.	Liver failure, malnutrition, overhydration, pregnancy.
Carbon dioxide content (bicarbonate + dissolved CO_2) (whole blood)	Arterial: 19–24 mEq/L (19–24 mmol/L) Venous: 22–26 mEq/L (22–26 mmol/L)	Severe diarrhea, severe vomiting, starvation, emphysema, aldosteronism.	Renal failure, diabetic ketoacidosis, shock.

Blood Tests CONTINUED

TEST (SPECIMEN)	U.S. REFERENCE VALUES (SI UNITS)	VALUES INCREASE IN	VALUES DECREASE IN
Cholesterol, total (plasma) **HDL cholesterol** (plasma) **LDL cholesterol** (plasma)	<200 mg/dL (<5.2 mmol/L) is desirable >40 mg/dL (>1.0 mmol/L) is desirable <130 mg/dL (<3.2 mmol/L) is desirable	Hypercholesterolemia, uncontrolled diabetes mellitus, hypothyroidism, hypertension, atherosclerosis, nephrosis.	Liver disease, hyperthyroidism, fat malabsorption, pernicious or hemolytic anemia, severe infections.
Creatine (serum)	Males: 0.15–0.5 mg/dL (10–40 μmol/L) Females: 0.35–0.9 mg/dL (30–70 μmol/L)	Muscular dystrophy, damage to muscle tissue, electric shock, chronic alcoholism.	
Creatine kinase (CK), also known as **creatine phosphokinase (CPK)** (serum)	0–130 U/L (same)	Myocardial infarction, progressive muscular dystrophy, hypothyroidism, pulmonary edema.	
Creatinine (serum)	0.5–1.2 mg/dL (45–105 μmol/L)	Impaired renal function, urinary tract obstruction, giantism, acromegaly.	Decreased muscle mass, as occurs in muscular dystrophy or myasthenia gravis.
Electrolytes (plasma)	See Table 27.2 on page 1071		
Gamma-glutamyl transferase (GGT) (serum)	0–30 U/L (same)	Bile duct obstruction, cirrhosis, alcoholism, metastatic liver cancer, congestive heart failure.	
Glucose (plasma)	70–110 mg/dL (3.9–6.1 mmol/L)	Diabetes mellitus, acute stress, hyperthyroidism, chronic liver disease, Cushing's disease.	Addison's disease, hypothyroidism, hyperinsulinism.
Hemoglobin (whole blood)	Males: 14–18 g/100 mL (140–180 g/L) Females: 12–16 g/100 mL (120–160 g/L) Newborns: 14–20 g/100 mL (140–200 g/L)	Polycythemia, congestive heart failure, chronic obstructive pulmonary disease, living at high altitude.	Anemia, severe hemorrhage, cancer, hemolysis, Hodgkin's disease, nutritional deficiency of vitamin B_{12}, systemic lupus erythematosus, kidney disease.
Iron, total (serum)	Males: 80–180 μg/dL (14–32 μmol/L) Females: 60–160 μg/dL (11–29 μmol/L)	Liver disease, hemolytic anemia, iron poisoning.	Iron-deficiency anemia, chronic blood loss, pregnancy (late), chronic heavy menstruation.
Lactic dehydrogenase (LDH) (serum)	71–207 U/L (same)	Myocardial infarction, liver disease, skeletal muscle necrosis, extensive cancer.	
Lipids (serum) **Total** **Triglycerides**	 400–850 mg/dL (4.0–8.5 g/L) 10–190 mg/dL (0.1–1.9 g/L)	Hyperlipidemia, diabetes mellitus.	Fat malabsorption, hypothyroidism.
Platelet (thrombocyte) count (whole blood)	150,000–400,000/μL	Cancer, trauma, leukemia, cirrhosis.	Anemias, allergic conditions, hemorrhage.
Protein (serum) **Total** **Albumin** **Globulin**	 6–8 g/dL (60–80 g/L) 4–6 g/dL (40–60 g/L) 2.3–3.5 g/dL (23–35 g/L)	Dehydration, shock, chronic infections.	Liver disease, poor protein intake, hemorrhage, diarrhea, malabsorption, chronic renal failure, severe burns.
Red blood cell (erythrocyte) count (whole blood)	Males: 4.5–6.5 million/μL Females: 3.9–5.6 million/μL	Polycythemia, dehydration, living at high altitude.	Hemorrhage, hemolysis, anemias, cancer, overhydration.
Uric acid (urate) (serum)	2.0–7.0 mg/dL (120–420) μmol/L	Impaired renal function, gout, metastatic cancer, shock, starvation.	
White blood cell (leukocyte) count, total (whole blood)	5,000–10,000/μL (See Table 19.3 on page 702 for relative percentages of different types of WBCs.)	Acute infections, trauma, malignant diseases, cardiovascular diseases. (See also Table 19.2 on page 701.)	Diabetes mellitus, anemia. (See also Table 19.2 on page 701.)

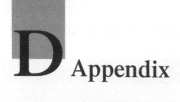

D Appendix

Normal Values for Selected Urine Tests

Urine Tests

TEST (SPECIMEN)	U.S. REFERENCE VALUES (SI UNITS)	CLINICAL IMPLICATIONS
Amylase (2 hour)	35–260 Somogyi units/hr (6.5–48.1 units/hr)	Values increase in inflammation of the pancreas (pancreatitis) or salivary glands, obstruction of the pancreatic duct, and perforated peptic ulcer.
Bilirubin* (random)	Negative	Values increase in liver disease and obstructive biliary disease.
Blood* (random)	Negative	Values increase in renal disease, extensive burns, transfusion reactions, and hemolytic anemia.
Calcium (Ca^{2+}) (random)	10 mg/dL (2.5 mmol/liter); up to 300 mg/24 hr (7.5 mmol/24 hr)	Amount depends on dietary intake; values increase in hyperparathyroidism, metastatic malignancies, and primary cancer of breasts and lungs; values decrease in hypoparathyroidism and vitamin D deficiency.
Casts (24 hour)		
Epithelial	Occasional	Values increase in nephrosis and heavy metal poisoning.
Granular	Occasional	Values increase in nephritis and pyelonephritis.
Hyaline	Occasional	Values increase in kidney infections.
Red blood cell	Occasional	Values increase in glomerular membrane damage and fever.
White blood cell	Occasional	Values increase in pyelonephritis, kidney stones, and cystitis.
Chloride (Cl$^-$) (24 hour)	140–250 mEq/24 hr (140–250 mmol/24 hr)	Amount depends on dietary salt intake; values increase in Addison's disease, dehydration, and starvation; values decrease in pyloric obstruction, diarrhea, and emphysema.
Color (random)	Yellow, straw, amber	Varies with many disease states, hydration, and diet.
Creatinine (24 hour)	Males: 1.0–2.0 g/24 hr (9–18 mmol/24 hr) Females: 0.8–1.8 g/24 hr (7–16 mmol/24 hr)	Values increase in infections; values decrease in muscular atrophy, anemia, and kidney diseases.
Glucose*	Negative	Values increase in diabetes mellitus, brain injury, and myocardial infarction.
Hydroxycorticosteroids (17-hydroxysteroids) (24 hour)	Males: 5–15 mg/24 hr (13–41 μmol/24 hr) Females: 2–13 mg/24 hr (5–36 μmol/24 hr)	Values increase in Cushing's syndrome, burns, and infections; values decrease in Addison's disease.
Ketone bodies* (random)	Negative	Values increase in diabetic acidosis, fever, anorexia, fasting, and starvation.
17-ketosteroids (24 hour)	Males: 8–25 mg/24 hr (28–87 μmol/24 hr) Females: 5–15 mg/24 hr (17–53 μmol/24 hr)	Values decrease in surgery, burns, infections, adrenogenital syndrome, and Cushing's syndrome.

TEST (SPECIMEN)	U.S. REFERENCE VALUES (SI UNITS)	CLINICAL IMPLICATIONS
Odor (random)	Aromatic	Becomes acetonelike in diabetic ketosis.
Osmolality (24 hour)	500–1400 mOsm/kg water (500–1400 mmol/kg water)	Values increase in cirrhosis, congestive heart failure (CHF), and high-protein diets; values decrease in aldosteronism, diabetes insipidus, and hypokalemia.
pH* (random)	4.6–8.0	Values increase in urinary tract infections and severe alkalosis; values decrease in acidosis, emphysema, starvation, and dehydration.
Phenylpyruvic acid (random)	Negative	Values increase in phenylketonuria (PKU).
Potassium (K⁺) (24 hour)	40–80 mEq/24 hr (40–80 mmol/24 hr)	Values increase in chronic renal failure, dehydration, starvation, and Cushing's syndrome; values decrease in diarrhea, malabsorption syndrome, and adrenal cortical insufficiency.
Protein* (albumin) (random)	Negative	Values increase in nephritis, fever, severe anemias, trauma, and hyperthyroidism.
Sodium (Na⁺) (24 hour)	75–200 mg/24 hr (75–200 mmol/24 hr)	Amount depends on dietary salt intake; values increase in dehydration, starvation, and diabetic acidosis; values decrease in diarrhea, acute renal failure, emphysema, and Cushing's syndrome.
Specific gravity* (random)	1.001–1.035 (same)	Values increase in diabetes mellitus and excessive water loss; values decrease in absence of antidiuretic hormone (ADH) and severe renal damage.
Urea (random)	25–35 g/24 hr (420–580 mmol/24 hr)	Values increase in response to increased protein intake; values decrease in impaired renal function.
Uric acid (24 hour)	0.4–1.0 g/24 hr (1.5–4.0 mmol/24 hr)	Values increase in gout, leukemia, and liver disease; values decrease in kidney disease.
Urobilinogen* (2 hour)	0.3–1.0 Ehrlich units (1.7–6.0 μmol/24 hr)	Values increase in anemias, hepatitis A (infectious), biliary disease, and cirrhosis; values decrease in cholelithiasis and renal insufficiency.
Volume, total (24 hour)	1000–2000 mL/24 hr (1.0–2.0 liters/24 hr)	Varies with many factors.

* Test often performed using a **dipstick,** a plastic strip impregnated with chemicals that is dipped into a urine specimen to detect particular substances. Certain colors indicate the presence or absence of a substance and sometimes give a rough estimate of the amount(s) present.

Answers

Answers to Self-Quiz Questions

Chapter 1 1. tissue 2. metabolism, anabolism, catabolism 3. intracellular fluid (ICF), extracellular fluid (ECF) 4. true 5. false 6. false 7. e 8. d 9. a 10. c 11. c 12. (a) 4, (b) 6, (c) 8 , (d) 1, (e) 9, (f) 5, (g) 2, (h) 7, (i) 3, (j) 10 13. (a) 1, (b) 12, (c) 1, 6, (d) 6, (e) 4, (f) 8, (g) 7 (h) 3, (i) 2, (j) 10 14. (a) 4, (b) 1, (c) 3, (d) 6, (e) 5, (f) 7, (g) 2 15. (a) 6, (b) 1, (c) 11, (d) 5, (e) 10, (f) 8, (g) 7, (h) 9, (i) 4, (j) 3, (k) 2

Chapter 2 1. 8 2. solid, liquid, gas 3. monosaccharides, amino acids 4. true 5. false 6. true 7. c 8. a 9. d 10. b 11. e 12. a 13. e 14. (a) 1, (b), 2, (c) 1, (d) 4, (e) 3 15. (a) 11, (b) 1, (c) 8, (d) 3, (e) 7, (f) 4, (g) 5, (h) 9, (i) 10, (j) 12, (k) 6, (l) 2

Chapter 3 1. plasma membrane, cytoplasm, nucleus 2. apoptosis, necrosis 3. Telomeres 4. UAG 5. false 6. true 7. true 8. e 9. c 10. c, g, i, b, d, k, f, j, a, e, h 11. a 12. c 13. (a) 2, (b) 3, (c) 5, (d) 7, (e) 6, (f) 8, (g) 1, (h) 4 14. (a) 2, (b) 9, (c) 3, (d) 5, (e) 11, (f) 8, (g) 1, (h) 6, (i) 10, (j) 7, (k) 13, (l) 4, (m) 12 15. (a) 3, (b) 9, (c) 1, (d) 5, (e) 11, (f) 4, (g) 8, (h) 7, (i) 2, (j) 10, (k) 6

Chapter 4 1. epithelial, connective, muscle, nervous 2. arrangement of cells in layers, cell shape 3. true 4. true 5. e 6. b 7. a 8. c 9. e 10. b 11. d 12. c 13. (a) C, (b) M, (c) N, (d) E, (e) C, (f) E, (g) M, (h) E, (i) C, (j) M, (k) N, (l) E, (m) C, (n) E, (o) M and N 14. (a) 4, (b) 8, (c) 5, (d) 2, (e) 6, (f) 3, (g) 1, (h) 7 15. (a) 3, (b) 5, (c) 8, (d) 13, (e) 9, (f) 7, (g) 11, (h) 6, (i) 2, (j) 4, (k) 10, (l) 12, (m) 1

Chapter 5 1. stratum lucidum 2. eccrine, ceruminous, apocrine 3. false 4. true 5. c 6. e 7. a 8. c 9. b 10. e 11. a 12. c 13. (a) 3, (b) 5, (c) 4, (d) 1, (e) 6, (f) 11, (g) 2, (h) 8, (i) 9, (j) 10, (k) 7 14. (a) 3, (b) 4, (c) 1, (d) 2 15. (a) 4, (b) 3, (c) 2, (d) 1, inflammatory, migratory, proliferative, maturation

Chapter 6 1. interstitial, appositional 2. hardness, tensile strength 3. true 4. true 5. true 6. d 7. a 8. e 9. c 10. a 11. (a) 1, (b) 4, (c) 3, (d) 2 12. (a) 3, (b) 9, (c) 8, (d) 1, (e) 5, (f) 4, (g) 6, (h) 7, (i) 12, (j) 2, (k) 11 (l) 10 13. (a) 2, (b) 6, (c) 4, (d) 5, (e) 7, (f) 3, (g) 1 14. (a) 12, (b) 4, (c) 8, (d) 6, (e) 3, (f) 9, (g) 13, (h) 10, (i) 7, (j) 5, (k) 2, (l) 11, (m) 1 15. (a) 3, (b) 7, (c) 6, (d) 1, (e) 4, (f) 2, (g) 5, (h) 9, (i) 8, (j) 10

Chapter 7 1. fontanels 2. pituitary gland 3. sacrum, coccyx 4. false 5. false 6. c 7. c 8. a 9. e 10. d 11. e 12. (a) 4, (b) 9, (c) 7, (d) 5, (e) 3, (f) 1, (g) 2, (h) 8, (i) 6 13. (a) 7, (b) 5, (c) 1, (d) 6, (e) 2, (f) 4, (g) 8, (h) 9, (i) 3, (j) 10, (k) 11, (l) 13, (m) 12 14. (a) 2, (b) 3, (c) 5, (d) 6, (e) 4, (f) 1, (g) 5, (h) 4, (i) 2, (j) 4, (k) 3 15. (a) 3, (b) 1, (c) 6, (d) 9, (e) 13, (f) 12, (g) 2, (h) 4, (i) 5, (j) 7, (k) 10, (l) 15, (m) 8, (n) 11, (o) 14

Chapter 8 1. metacarpals 2. ilium, ischium, pubis 3. true (lesser), false (greater) 4. false 5. true 6. b 7. c 8. e 9. c 10. a 11. b and e 12. a 13. (a) 2, (b) 6, (c) 9, (d) 7, (e) 4, (f) 5, (g) 8, (h) 10, (i) 1, (j) 3 14. (a) 3, (b) 8, (c) 4, (d) 11, (e) 9, (f) 13, (g) 5, (h) 6, (i) 10, (j) 14, (k) 2, (l) 1, (m) 7, (n) 12 15. (t) 9, (b) 3, (c) 3, (d) 6, (e) 7, (f) 1, (g) 3, (h) 2, (i) 5, (j) 9, (k) 8, (l) 2, (m) 4, (n) 6, (o) 7, (p) 9, (q) 6, (r) 3, (s) 4, (t) 4 and 5, (u) 4 and 5

Chapter 9 1. joint, articulation or arthrosis 2. arthroplasty 3. false 4. false 5. false 6. e 7. d 8. b 9. c 10. a 11. c 12. e 13. (a) 5, (b) 3, (c) 7, (d) 2, (e) 6, (f) 4, (g) 1 14. (a) 6, (b) 4, (c) 5, (d) 1, (e) 3, (f) 2 15. (a) 8, (b) 11, (b) 10, (d) 13, (e) 15, (f) 9, (g) 6, (h) 12, (i) 3, (j) 4, (k) 16, (l) 2, (m) 18, (n) 1, (o) 7, (p) 14, (q) 17, (r) 5

Chapter 10 1. motor unit 2. muscular atrophy, fibrosis 3. acetylcholine 4. true 5. true 6. e 7. a 8. c 9. e 10. d 11. b 12. (a) 5, (b) 6, (c) 9, (d) 7, (e) 2, (f) 4, (g) 10, (h) 3, (i) 1, (j) 8 13. (a) 7, (b) 10, (c) 9, (d) 12, (e) 8, (f) 11, (g) 6, (h) 1, (i) 2, (j) 3, (k) 4, (l) 13, (m) 5 14. (a) 2, (b) 3, (c) 1, (d) 1 and 2, (e) 3, (f) 2, (g) 1, (h) 3, (i) 1 and 2, (j) 3, (k) 2 and 3, (l) 3 15. (a) 10, (b) 2, (c) 4, (d) 3, (e) 6, (f) 5, (g) 1, (h) 12, (i) 7, (j) 9, (k) 11, (l) 8

Chapter 11 1. buccinator 2. gastrocnemius, soleus, plantaris, calcaneus 3. true 4. true 5. b 6. c 7. d 8. a 9. e 10. e 11. (a) 6, (b) 2, (c) 8, (d) 5, (e) 3, (f) 1, (g) 7, (h) 4 12. (a) 13, (b) 9, (c) 8, (d) 6, (e) 3, (f) 11, (g) 10, (h) 1, (i) 2, (j) 7, (k) 12, (l) 4, (m) 5 13. (a) 6, (b) 3, (c) 7, (d) 4, (e) 2, (f) 9, (g) 5, (h) 1, (i) 8 14. (a) 10, (b) 1, (c) 9, (d) 8, (e) 12, (f) 17, (g) 2, (h) 6, (i) 8, (j) 14, (k) 5, (l) 4, (m) 2, (n) 15, (o) 1, (p) 11, (q) 13, (r) 12, (s) 7, (t) 16, (u) 11, (v) 17, (w) 16, (x) 15, (y) 3, (z) 10 15. (a) 3, (b) 1, (c) 2, (d) 1, (e) 2, (f) 3, (g) 3

Chapter 12 1. somatic, autonomic, enteric 2. sympathetic, parasympathetic 3. false 4. false 5. c 6. d 7. c 8. e 9. e 10. d 11. e 12. b 13. (a) 6, (b) 12, (c) 1, (d) 2, (e) 9, (f) 14, (g) 4, (h) 8, (i) 7, (j) 13, (k) 5, (l) 3, (m) 10, (n) 15, (o) 11 14. (a) 2, (b) 1, (c) 10, (d) 9, (e) 6, (f) 3, (g) 4, (h) 5, (i) 12, (j) 8, (k) 7, (l) 13, (m) 11 15. (a) 4, (b) 5, (c) 16, (d) 8, (e) 7, (f) 1, (g) 2, (h) 10, (i) 15, (j) 6, (k) 3, (l) 13, (m) 9, (n) 11, (o) 14, (p) 12

Chapter 13 1. mixed 2. sensory receptor, sensory neuron, integrating center, motor neuron, effector 3. true 4. false 5. c 6. c 7. a 8. c 9. d 10. e 11. a 12. d 13. (a) 1, (b) 8, (c) 4, (d) 2, (e) 11, (f) 1, (g) 6, (h) 5, (i) 3, (j) 9, (k) 1, (l) 12, (m) 7, (n) 2, (o) 10 14. (a) 14, (b) 12, (c) 13, (d) 1, (e) 2, (f) 5, (g) 11, (h) 8, (i) 10, (j) 9, (k) 15, (l) 4, (m) 7, (n) 3, (o) 6 15. (a) 2, (b) 1, (c) 3, (d) 4, (e) 1, (f) 5, (g) 3, (h) 2, (i) 4, (j) 1, (k) 2, (l) 4, (m) 3, (n) 5, (o) 1

Chapter 14 1. corpus callosum 2. frontal, temporal, parietal, occipital, insula 3. longitudinal fissure 4. false 5. true 6. d 7. c 8. d 9. e 10. d 11. e 12. (a) 3, (b) 5, (c) 6, (d) 8, (e) 11, (f) 10, (g) 7, (h) 9, (i) 1, (j) 4, (k) 2, (l) 12, (m) 1, (n) 8, (o) 5, (p) 7, (q) 12, (r) 10, (s) 9, (t) 1 and 2, (u) 3, 4, and 6, (v) 11 13. (a) 9, (b) 2, (c) 6, (d) 10, (e) 4, (f) 11, (g) 1, (h) 2, (i) 5, (j) 8, (k) 12, (l) 7, (m) 3, (n) 6 and 8, (o) 13, (p) 7, (q) 1 14. (a) 5, (b) 9, (c) 11, (d) 6, (e) 3, (f) 1, (g) 10, (h) 8, (i) 2, (j) 4, (k) 7 15. (a) 10, (b) 2, (c) 6, (d) 8, (e) 7, (f) 5, (g) 3, (h) 11, (i) 14, (j) 13, (k) 4, (l) 1, (m) 12, (n) 9 (o) 15

Chapter 15 1. acetylcholine, epinephrine or norepinephrine 2. thoracolumbar, craniosacral 3. true 4. true 5. d 6. e 7. b 8. c 9. e 10. a 11. a 12. c 13. e, b, g, f, d, a, c14. (a) 3, (b) 2, (c) 1, (d) 1, (e) 2, (f) 3, (g) 3, (h) 1, (i) 4, (j) 2, (k) 5 15. (a) 2, (b) 1, (c) 1, (d) 2, (e) 1, (f) 1, (g) 2, (h) 2

Chapter 16 1. sensation, perception 2. decussation 3. false 4. true 5. c 6. a 7. d 8. b 9. d 10. e 11. e 12. d 13. (a) 9, (b) 8, (c) 4, (d) 7, (e) 10, (f) 2, (g) 3, (h) 1, (i) 5, (j) 6, (k) 11 (l) 12 14. (a) 3, (b) 2, (c) 5, (d) 7, (e) 1, (f) 9, (g) 3, (h) 11, (i) 8, (j) 4, (k) 6, (l) 10 15. (a) 10, (b) 8, (c) 7, (d) 1, (e) 4, (f) 3, (g) 5, (h) 6, (i) 9, (j) 2

Chapter 17 1. sweet, sour, salty, bitter, umami 2. static, dynamic 3. true 4. false 5. d 6. a 7. d 8. b 9. b 10. c, j, k, d, h, l, e, b, f, i, a, m, g 11. c 12. a 13. (a) 1, (b) 5, (c) 7, (d) 6, (e) 8, (f) 2, (g) 4, (h) 3 14. (a) 3, (b) 6, (c) 9, (d) 14, (e) 1, (f) 5, (g) 10, (h) 13, (i) 7, (j) 15, (k) 2, (l) 11, (m) 12, (n) 4, (o) 8 15. (a) 2, (b) 11, (c) 14, (d) 13, (e) 3, (f) 10, (g) 6, (h) 12, (i) 4, (j) 5, (k) 9, (l) 1, (m) 7, (n) 8

Chapter 18 1. fight-or-flight response, resistance reaction, exhaustion 2. hypothalamus 3. less, more 4. false 5. true 6. b 7. e 8. d 9. a 10. e 11. c 12. a 13. (a) 8, (b) 2, (c) 7, (d) 1, (e) 12, (f) 20, (g) 5, (h) 18, (i) 22, (j) 15, (k) 3, (l) 17, (m) 21, (n) 6, (o) 13, (p) 11, (q) 4, (r) 10, (s) 14, (t) 9, (u) 16, (v) 19 14. (a) 10, (b) 8, (c) 2, (d) 12, (e) 15, (f) 4, (g) 1, (h) 16, (i) 6, (j) 9, (k) 13, (l) 7, (m) 5, (n) 14, (o) 3, (p) 11 15. (a) 12, (b) 1, (c) 11, (d) 7, (e) 3, (f) 10, (g) 2, (h) 9, (i) 4, (j) 8, (k) 5, (l) 6

Chapter 19 1. serum 2. clot retraction 3. true 4. true 5. e 6. a 7. b 8. c 9. d 10. a 11. d 12. e 13. (a) 8, (b) 14, (c) 7, (d) 3, (e) 16, (f) 2, (g) 11, (h) 4, (i) 1, (j) 10, (k) 9, (l) 12, (m) 13, (n) 18, (o) 15, (p) 5, (q) 19, (r) 6, (s) 17 14. (a) 4, (b) 6, (c) 8, (d) 1, (e) 7, (f) 3, (g) 5, (h) 2 15. (a) 4, (b) 7, (c) 6, (d) 1, (e) 3, (f) 5, (g) 2

Chapter 20 1. left ventricle 2. systole, diastole 3. false 4. true 5. a 6. c 7. d 8. b 9. b 10. e 11. c 12. (a) 3, (b) 6, (c) 1, (d) 5, (e) 2, (f) 4 13. (a) 8, (b) 4, (c) 11, (d) 5, (e) 1, (f) 9, (g) 7, (h) 2, (i) 10, (j) 6, (k) 3 14. (a) 3, (b) 2, (c) 9, (d) 14, (e) 8, (f) 7, (g) 11, (h) 12, (i) 15, (j) 4, (k) 5, (l) 1, (m) 6, (n) 21, (o) 22, (p) 19, (q) 17, (r) 18, (s) 20, (t) 16, (u) 13, (v) 10 15. (a) 3, (b) 7, (c) 2, (d) 5, (e) 1, (f) 6, (g) 4 and 7

Chapter 21 1. carotid sinus, aortic 2. skeletal muscle pump, respiratory pump 3. true 4. true 5. b 6. a 7. c 8. e 9. a 10. d 11. (a) D, (b) C, (c) C, (d) D, (e) D, (f) C, (g) C, (h) C, (i) D, (j) D, (k) C 12. (a) 2, (b) 5, (c) 1, (d) 4, (e) 3 13. (a) 11, (b) 1, (c) 4, (d) 9, (e) 3, (f) 8, (g) 6, (h) 2, (i) 7, (j) 5, (k) 10, (l) 12, (m) 13 14. (a) 2, (b) 6, (c) 4, (d) 1, (e) 3, (f) 5 15. (a) 5, (b) 3, (c) 1, (d) 4, (e) 2, (f) 4, (g) 1, (h) 5, (i) 3

Chapter 22 1. skin, mucous membranes, antimicrobial proteins, natural killer cells, phagocytes 2. antigens 3. true 4. true 5. c 6. d 7. e 8. d 9. e 10. b 11. c 12. e, h, b, f, a, d, g, i, c 13. (a) 3, (b) 1, (c) 7, (d) 4, (e) 2, (f) 5, (g) 6 14. (a) 2, (b) 3, (c) 4, (d) 7, (e) 1, (f) 6, (g) 5 15. (a) 12, (b) 12, (c) 8, (d) 1, (e) 2, (f) 5, (g) 4, (h) 7, (i) 9, (j) 14, (k) 13, (l) 3, (m) 6, (n) 11, (o) 10

Chapter 23 1. oxyhemoglobin; dissolved CO_2, carbamino compounds (primarily carbaminohemoglobin), and bicarbonate ion 2. $CO_2 + H_2O \rightarrow H_2CO_3 \rightarrow H^+ + HCO_3^-$ 3. false 4. true 5. c 6. e 7. b 8. d 9. a 10. e 11. e, g, b, h, a, d, f, c 12. (a) 2, (b) 11, (c) 3, (d) 9, (e) 1, (f) 12, (g) 10, (h) 5, (i) 13, (j) 8, (k) 4, (l) 6, (m) 7 13. (a) 7, (b) 8, (c) 1, (d) 5, (e) 6, (f) 9, (g) 2, (h) 3, (i) 4 14. (a) 3, (b) 8, (c) 5, (d) 9, (e) 2, (f) 7, (g) 10, (h) 1, (i) 4, (j) 6 15. (a) 9, (b) 11, (c) 3, (d) 4, (e) 6, (f) 1, (g) 5, (h) 10, (i) 7, (j) 8, (k) 2

Chapter 24 1. monosaccharides; amino acids; monoglycerides, fatty acids; pentoses, phosphates, nitrogenous bases 2. diffusion, facilitated diffusion, osmosis, active transport 3. true 4. true 5. b 6. d 7. e 8. c 9. a 10. b 11. d 12. a 13. (a) 5, (b) 13, (c) 8, (d) 11, (e) 9, (f) 1, (g) 7, (h) 12, (i) 14, (j) 2, (k) 4, (l) 10, (m) 6, (n) 3 14. (a) 4, (b) 6, (c) 7, (d) 1, (e) 5, (f) 3, (g) 9, (h) 2, (i) 11, (j) 10, (k) 8 15. (a) 4, (b) 8, (c) 2, (d) 10, (e) 11, (f) 7, (g) 1, (h) 13, (i) 12, (j) 6, (k) 3, (l) 9, (m) 5

Chapter 25 1. hypothalamus 2. glucose 6-phosphate, pyruvic acid, acetyl coenzyme A 3. true 4. false 5. e 6. c 7. b 8. d 9. a 10. b 11. e 12. a 13. (a) 2 and 3, (b) 1, (c) 3 and 5, (d) 2, 4, 5, and 6, (e) 1, (f) 2, (g) 1, 4, and 6 14. (a) 9, (b) 12, (c) 11, (d) 10, (e) 4, (f) 13, (g) 3, (h) 5, (i) 8, (j) 1, (k) 6, (l) 2, (m) 7 15. (a) 17, (b) 15, (c) 8, (d) 19, (e) 1, (f) 7, (g) 4, (h) 10, (i) 16, (j) 14, (k) 11, (l) 2, (m) 13, (n) 6, (o) 20, (p) 9, (q) 5, (r) 18, (s) 3, (t) 12

Chapter 26 1. glomerulus, glomerular (Bowman's) capsule 2. micturition 3. false 4. true 5. d 6. e 7. c 8. b 9. e 10. a 11. g, a, h, c, o, n, l, j, i, b, d, e, m, k, f

Chapter 26 12. g, f, b, d, h, c, a, e, i 13. (a) 8, (b) 2, (c) 10, (d) 5, (e) 3, (f) 1, (g) 7, (h) 4, (i) 11, (j) 9, (k) 6 14. (a) 4, (b) 3, (c) 7, (d) 6, (e) 2, (f) 8, (g) 1, (h) 5 15. (a) 5, (b) 4, (c) 6, (d) 8, (e) 1, (f) 2, (g) 7, (h) 3, (i) 9

Chapter 27 1. metabolic 2. bicarbonate ion, carbonic acid 3. true 4. false 5. a 6. c 7. d 8. d 9. b 10. e 11. a 12. b 13. a 14. (a) 8, (b) 9, (c) 7, (d) 1, (e) 6, (f) 2, (g) 4, (h) 5, (i) 3 15. (a) 8, (b) 12, (c) 7, (d) 5, (e) 10, (f) 6, (g) 9, (h) 13, (i) 1, (j) 11, (k) 3, (l) 4, (m) 2

Chapter 28 1. puberty, menarche, menopause 2. true 3. true 4. e 5. c 6. a 7. c 8. b 9. a 10. e 11. d 12. (a) 13, (b) 10, (c) 12, (d) 1, (e) 5, (f) 2, (g) 4, (h) 6, (i) 14, (j) 11, (k) 8, (l) 3, (m) 9, (n) 7 13. (a) 6, (b) 4, (c) 1, (d) 12, (e) 8, (f) 5, (g) 7, (h) 13, (i) 11, (j) 3, (k) 14, (l) 2, (m) 10, (n) 15, (o) 9 14. (a) 7, (b) 2, (c) 1, (d) 10, (e) 4, (f) 12, (g) 8, (h) 11, (i) 5, (j) 3, (k) 6, (l) 9 15. (a) 4, (b) 2, (c) 1, (d) 6, (e) 5, (f) 3

Chapter 29 1. dilation, expulsion, placental 2. corpus luteum, human chorionic gonadotropin 3. mesoderm, ectoderm, endoderm 4. false 5. a 6. d 7. b 8. e 9. b 10. c 11. (a) 6, (b) 2, (c) 11, (d) 5, (e) 10, (f) 1, (g) 3, (h) 8, (i) 9, (j) 4, (k) 7 12. (a) 3, (b) 6, (c) 5, (d) 2, (e) 1, (f) 4 13. (a) 3, (b) 4, (c) 5, (d) 1, (e) 2, (f) 6, (g) 7, (h) 10, (i) 8, (j) 9 14. (a) 7, (b) 3, (c) 14, (d) 6, (e) 4, (f) 10, (g) 8, (h) 13, (i) 5, (j) 2, (k) 1, (l) 9, (m) 16, (n) 11, (o) 15, (p) 12 15. (a) 3, (b) 6, (c) 4, (d) 1, (e) 9, (f) 2, (g) 5, (h) 8, (i) 7

Answers to Critical Thinking Questions

Chapter 1

1. No. Computed tomography is used to look at differences in tissue density. To assess activity in an organ such as the brain, a positron emission tomography (PET) scan or a single-photo-emission computerized tomography (SPECT) scan would both provide a colorized visual assessment of brain activity.

2. Stem cells are undifferentiated cells. Research using stem cells has shown that these undifferentiated cells may be prompted to differentiate into the specific cells needed to replace those which are damaged or malfunctioning.

3. Homeostasis is the relative constancy of the body's internal environment. Homeostasis is maintained as the body changes in response to shifting external and internal conditions. Body temperature should vary within a narrow range around normal body temperature (37°C or 98.6°F), which is above normal room temperature (usually around 72°F (25°C).

Chapter 2

1. Neither butter nor margarine are particularly good choices for frying eggs. Butter contains saturated fats that are associated with heart disease. However, many margarines contain hydrogenated or partially hydrogenated trans-fatty acids that also increase the risk of heart disease. An alternative would be frying the eggs in any of the mono- or polyunsaturated fats such as olive oil, peanut oil, or corn oil.

2. High body temperatures can be life-threatening, especially in infants. The increased temperature can cause denaturing of structural proteins and vital enzymes. When this happens, the proteins become nonfunctional. If the denatured enzymes are required for reactions that are necessary for life, then the infant could die.

3. Simply adding water to the table sugar does not cause it to break apart into monosaccharides. The water acts as a solvent, dissolving the sucrose, and forming a sugar-water solution. To complete the breakdown of table sugar to glucose and fructose would require the presence of the enzyme sucrase.

Chapter 3

1. Synthesis of mucin by ribosomes on rough endoplasmic reticulum, to transport vesicle, to entry face of Golgi complex, to transfer vesicle, to medial cisternae where protein is modified, to transfer vesicle, to exit face, to secretory vesicle, to plasma membrane where it undergoes exocytosis.

2. Since smooth ER inactivates or detoxifies drugs, and peroxisomes also destroy harmful substances such as alcohol, we would expect to see increased numbers of these organelles in Sebastian's liver cells.

3. Removing part of the small intestine will drastically reduce the cellular membrane surface area for absorption of digested nutrients. Although it can cause weight loss, it also can result in a lack of absorption of essential vitamins and minerals; dietary supplements are generally required.

4. In order to restore water balance to the cells, the runners need to consume hypotonic solutions. The water in the hypotonic solution will move from the blood, into the interstitial fluid, and then into the cells. Plain water works well; sports drinks contain water and some electrolytes (which may have been lost due to sweating) but will still be hypotonic in relation to the body cells.

Chapter 4

1. There are many possible adaptations, including: more adipose tissue for insulation; thicker bones for support; more red blood cells for oxygen transport; increased thickness of skin to prevent water loss; etc.

2. Infants tend to have a high proportion of brown fat, which contains many mitochondria and is highly vascularized. When broken down, brown fat produces heat that helps to maintain infants' body temperatures. This heat can also warm the blood, which then distributes the heat throughout the body.

3. Your bread-and-water diet is not providing you with the necessary nutrients to encourage tissue repair. You need proper amounts of many essential vitamins, especially vitamin C, which is required for repair of the matrix and blood vessels. Vitamin A is needed to help properly maintain epithelial tissue. Adequate protein is also needed in order to synthesize the structural proteins of the damaged tissue.

Chapter 5

1. The dust particles are primarily keratinocytes that are shed from the stratum corneum of the skin.

2. Tattoos are created by depositing ink into the dermis, which does not undergo shedding as the epidermis does. Although the tattoo will fade due to exposure to sunlight and the flushing away of ink particles by the lymphatic system, the tattoo is indeed permanent.

3. Chef Eduardo has damaged the nail matrix—the part of the nail that produces growth. Because the damaged area has not regrown properly, the nail matrix may be permanently damaged.

Chapter 6

1. Due to the strenuous, repetitive activity, Taryn has probably developed a stress fracture of her right tibia (lower leg bone). Stress fractures are due to repeated stress on a bone that causes microscopic breaks in the bone without any evidence of injury to other tissue. An x-ray would not reveal the stress fracture, but a bone scan would. Thus the bone scan would either confirm or negate the physician's diagnosis.

2. When Marcus broke his arm as a child, he injured his epiphyseal (growth) plate. Damage to the cartilage in the epiphyseal plate resulted in premature closure of the plate, which interfered with the lengthwise growth of the arm bone.

3. Exercise causes mechanical stress on bones, but because there is effectively zero gravity in space, the pull of gravity on bones is missing. The lack of stress from gravity results in bone demineralization and weakness.

Chapter 7

1. Inability to open mouth—damage to the mandible, probably at temporomandibular joint; black eye—trauma to the ridge over the supraorbital margin; broken nose—probably damage to the nasal septum (includes the vomer, septal cartilage, and perpendicular plate of the ethmoid) and possibly the nasal bones; broken cheek—fracture of zygomatic bone; broken upper jaw—fracture of maxilla; damaged eye socket—fracture of parts of the sphenoid, frontal, ethmoid, palatine, zygomatic, lacrimal and maxilla (all compose the eye socket); punctured lung—damage to the thoracic vertebrae, which have punctured the lung.

2. Due to the repeated and extensive tension on his bone surfaces, Bubba would experience deposition of new bone tissue. His arm bones would be thicker and with increased raised areas (projections) where the tendons attach his muscles to bone.

3. The "soft area" being referred to is the anterior fontanel, located between the parietal and frontal bones. This is one of several areas of fibrous connective tissue in the skull that has not ossified; it should complete its ossification at 18–24 months after birth. Fontanels allow flexibility of the skull for childbirth

and for brain growth after birth. The connective tissue will not allow passage of water; thus no brain damage will occur through simply washing the baby's hair.

Chapter 8

1. There are several characteristics of the bony pelves that can be used to differentiate male from female: (1) The pelvis in the female is wider and more shallow that the male's; (2) the pelvic brim of the female is larger and more oval; (3) the pubic arch has an angle greater than 90°; (4) the pelvic outlet is wider than in a male's; (5) the female's iliac crest is less curved and the ilium less vertical. Table 8.1 provides additional differences between female and male pelves. Age of the skeleton can be determined by the size of the bones, the presence or absence of epiphyseal plates, the degree of demineralization of the bones, and the general appearance of the "bumps" and ridges of bones.

2. Infants do have "flat feet" because their arches have not yet developed. As they begin to stand and walk, the arches should begin to develop in order to accommodate and support their body weight. The arches are usually fully developed by age 12 or 13, so Dad doesn't need to worry yet!

3. There are 14 phalanges in each hand: two bones in the thumb and three in each of the other fingers. Farmer White has lost five phalanges on his left hand (two in his thumb and three in his index finger), so he has nine remaining on his left and 14 remaining on his right for a total of 23.

Chapter 9

1. Katie's vertebral column, head, thighs, lower legs, lower arms, and fingers are flexed. Her lower arms and shoulders are medially rotated. Her thighs and arms are adducted.

2. The knee joint is commonly injured, especially among athletes. The twisting of Jeremiah's leg could have resulted in a multitude of internal injuries to the knee joint but often football players suffer tearing of the anterior cruciate ligament and medial meniscus. The immediate swelling is due to blood from damaged blood vessels, damaged synovial membranes, and the torn meniscus. Continued swelling is a result of a buildup of synovial fluid, which can result in pain and decreased mobility. Jeremiah's doctor may aspirate some of the fluid ("draining the water off his knee") and might want to perform arthroscopy to check for the extent of the knee damage.

3. The condylar processes of the mandible passed anteriorly to the articular tubercles of the temporal bones and this dislocated Antonio's mandible. It could be corrected by pressing the thumbs downward on the lower molar teeth and pushing the mandible backward.

Chapter 10

1. Muscle cells lose their ability to undergo cell division after birth. Therefore, the increase in size is not due to an increase in the number of muscle cells but rather is due to enlargement of the existing muscle fibers (hypertrophy). This enlargement can occur from forceful, repetitive muscular activity. It will cause the muscle fibers to increase their production of internal structures such as mitochondria and myofibrils and produce an increase in the muscle fiber diameter.

2. The "dark meat" of both chickens and ducks is composed primarily of slow oxidative (SO) muscle fibers. These fibers contain large amounts of myoglobin and capillaries, which accounts for their dark color. In addition, these fibers contain large numbers of mitochondria and generate ATP by aerobic respiration. SO fibers are resistant to fatigue and can produce sustained contractions for many hours. The legs of chickens and ducks are used for support, walking, and swimming (in ducks), all activities in which endurance is needed. In addition, migrating ducks require SO fibers in their breasts to enable them to have enough energy to fly for extremely long distances while migrating. There may be some fast oxidative-glycolytic (FOG) fibers in the dark meat. FOG fibers also contain large amounts of myoglobin and capillaries, contributing to the dark color. They can use aerobic or anaerobic cellular respiration to generate ATP and have high-to-moderate resistance to fatigue. These fibers would be good for the occasional "sprint" that ducks and chickens undergo to escape dangerous situations. In contrast, the white meat of a chicken breast is composed primarily of fast glycolytic (FG) fibers. FG fibers have lower amounts of myoglobin and capillaries that give the meat its white color. There are also few mitochondria in FG fibers so these fibers generate ATP mainly by glycolysis.

2. In emphysema, there is destruction of the alveolar walls, producing abnormally large air spaces that remain filled with air during exhalation. The destruction of alveoli decreases the surface area for gas exchange across the respiratory membrane, resulting in a decreased blood O_2 level. Damage to the alveolar walls also causes a loss of elasticity, making exhalation more difficult. This can result in a buildup of CO_2. Emphysema causes constriction of bronchioles. This narrowing of the breathing airways increases resistance and results in more pressure required to maintain airflow. Cigarette smoke contains nicotine, carbon monoxide, and a variety of irritants, all of which affect the lungs. Nicotine constricts terminal bronchioles, decreasing the air flow into and out of the lungs; carbon monoxide binds to hemoglobin, reducing its ability to carry oxygen; irritants such as tar and fine particulate matter destroy cilia and increase mucus secretion, interfering with the ability of the respiratory passages to cleanse themselves.

3. The squirrel's nest blocked the passage of exhaust gas from the furnace, causing an accumulation of carbon monoxide (CO), a colorless, odorless gas, in the home. As they were sleeping, their blood was saturated with CO, which has a stronger affinity for hemoglobin than oxygen. As a result, the Robinsons became oxygen deficient. Without adequate oxygenation of the brain, the Robinsons died during their sleep.

Chapter 24

1. HCl has several important roles in digestion. HCl stimulates the secretion of hormones that promote the flow of bile and pancreatic juice. The presence of HCl destroys certain microbes that may have been ingested with food. HCl begins denaturing proteins in food, and provides the proper chemical environment for activating pepsinogen into pepsin, which breaks apart certain peptide bonds in proteins. It also helps in the action of gastric lipase, which splits triglycerides in fat molecules found in milk into fatty acids and monoglycerides.

2. Blockage of the pancreatic and bile ducts prevent pancreatic digestive enzymes and bile from reaching the duodenum. As a consequence, there will be problems digesting carbohydrates, proteins, nucleic acids, and lipids. Of particular concern is lipid digestion since the pancreatic juices contain the primary lipid-digesting enzyme. Fats will not be adequately digested, and Trey's feces will contain larger than normal amounts of lipids. In addition, the lack of bile salts will affect the body's ability to emulsify lipids and to form micelles required for absorption of fatty acids and monoglycerides (from lipid breakdown). When lipids are not absorbed properly, then there will be malabsorption of the lipid-soluble vitamins (A, D, E, and K).

3. Antonio experienced gastroesophageal reflux. The stomach contents backed up (refluxed) into Antonio's esophagus due to a failure of the lower esophageal sphincter to fully close. The HCl from the stomach irritated the esophageal wall, which resulted in the burning sensation he felt; this is commonly known as "heartburn," even though it is not related to the heart. Antonio's recent meal worsened the problem. Alcohol and smoking both can cause the sphincter to relax, while certain foods such tomatoes, chocolate, and coffee can stimulate stomach acid secretion. In addition, lying down immediately after a meal can exacerbate the problem.

Chapter 25

1. Ingestion of cyanide affects cellular respiration. The cyanide binds to the cytochrome oxidase complex in the inner membrane of mitochondria. Blocking this complex interferes with the last step in electron transport in aerobic ATP production. Jane Doe's body would quickly run out of energy to perform vital functions, resulting in her death.

2. Glenn's total cholesterol and LDL levels are very high, while his HDL levels are low. Total cholesterol above 239mg/dL and LDL above 159mg/dL are considered high. The ratio of total cholesterol (TC) to HDL-cholesterol is a predictor of the risk of developing coronary artery disease. Glenn's TC to HDL is 15; a ratio above 4 is undesirable. His ratio places him at high risk of developing coronary artery disease. In addition, for every 50mg/dL TC over 200mg/dL, the risk of a heart attack doubles. Glenn needs to reduce his TC and LDL-cholesterol while raising his HDL-cholesterol levels. LDLs contribute to fatty plaque formation on coronary artery walls. On the other hand, HDLs help remove excess cholesterol from the blood, which helps decrease the risk of coronary artery disease. Glenn will need to reduce his dietary intake of total fat, saturated fats, and cholesterol, all of which contribute to raising LDL levels. Exercise will raise HDL levels. If those changes are not successful, drug therapy may be required.

3. The goal of weight loss programs is to reduce caloric intake so that the body utilizes stored lipids as an energy source. As part of that desired lipid metabolism, ketone bodies are produced. Some of these ketone bodies will be excreted in the urine. If no ketones are present, then Marissa's body is not breaking down lipids. Only through using fewer calories than needed will her body break down the stored fat and release ketones. Thus, she must be eating more calories than needed to support her daily activities—she is "cheating."

Chapter 26

1. Without reabsorption, initially 105–125mL of filtrate would be lost per minute, assuming normal glomerular filtration rate. Fluid loss from the blood would cause a decrease in blood pressure, and therefore a decrease in GBHP. When GBHP dropped below 45mmHg, filtration would stop (assuming normal CHP and BCOP) because NFP would be zero.

2. a. Although normally pale yellow, urine color can vary based upon concentration, diet, drugs, and disease. A dark yellow color would not necessarily indicate a problem, but further investigation may be needed. Turbidity or cloudiness can be caused by urine that has been standing for a period of time, from certain foods, or from bacterial infections. Further investigation is needed. b. Ammonia-like odor occurs when the urine sample is allowed to stand. c. Albumin should not be present in urine (or present in very small amounts) because it is too large to pass through the filtration membranes. The presence of high levels of albumin is cause for concern as it indicates damage to the filtration membranes. d. Casts are hardened masses of material that are flushed out in the urine. The presence of casts is not normal and indicates a pathology. e. The pH of normal urine ranges from 4.8 to 8.0. A pH of 5.5 is in normal range. f. Hematuria is the presence of red blood cells in the urine. It can occur with certain pathological conditions or from kidney trauma. Hematuria may occur if the urine sample was contaminated with menstrual blood.

3. Bruce has developed renal calculi (kidney stones), which are blocking his ureters and interfering with the flow of urine from the kidneys to the bladder. The rhythmic pains are a result of the peristaltic contractions of the ureters as they attempt to move the stones toward the bladder. Bruce can wait for the stones to pass, can have them surgically removed, or can use shock-wave lithotripsy to break apart the stones into smaller fragments that can be eliminated with urine. To prevent future episodes, Bruce needs to watch his diet (limit calcium), and drink fluids, and may need drug intervention.

Chapter 27

1. The loss of stomach acids can result in metabolic alkalosis. Robin's HCO_3^- levels would be higher than normal. She would be hypoventilating in order to decrease her pH by slowing the loss of CO_2. Excessive vomiting can result in hyponatremia, hypokalemia, and hypochloremia. Both hyponatremia and hypokalemia can cause mental confusion.

2. (Step 1) pH=7.30 indicates slight acidosis, which could be caused by elevated PCO_2 or lowered HCO_3^-. (Step 2) The HCO_3^- is lower than normal (20mEq/liter), so (Step 3) the cause is metabolic. (Step 4) The PCO_2 is lower than normal (32mmHg), so hyperventilation is providing some compensation. Diagnosis: Henry has partially compensated metabolic acidosis. A possible cause is kidney damage that resulted from interruption of blood flow during the heart attack.

3. Sam will experience increased fluid loss through increased evaporation from the skin and water vapor from the respiratory system through his increased respiratory rate. His insensible water loss will also increase (loss of water from mucous membranes of the mouth and respiratory system). Sam will have a decrease in urine formation.

Chapter 28

1. Monica's excessive training has resulted in an abnormally low amount of body fat. A certain amount of body fat is needed in order to produce the hormones required for the ovarian cycle. Several hormones are involved. Her amennorhea is due to a lack of gonadotropin-releasing hormone, which in turn reduces the release of LH and FSH. Her follicles with their enclosed ova fail to develop and ovulation will not occur. In addition, synthesis of estrogens and progesterone declines from the lack of hormonal feedback. Usually a gain of weight will allow normal hormonal feedback mechanisms to return.

2. Along with estrogens, progesterone helps to prepare the endometrium for possible implantation of a zygote by promoting growth of the endometrium. The endometrial glands secrete glycogen, which will help sustain an embryo if one should implant. If implantation occurs, progesterone helps maintain the endometrium for the developing fetus. In addition, it helps prepare mammary glands to secrete milk. It inhibits the release of GnRH and LH, which stops a new ovarian cycle from occurring.

3. The ductus deferens is cut and tied in a vasectomy. This stops the release of sperm into the ejaculatory duct and urethra. Mark will still produce the secretions from his accessory glands (prostate, seminal vesicles, bulbourethral glands) in his ejaculate. In addition, a vasectomy does not affect sexual performance; he will be able to achieve erection and ejaculation as those events are nervous system responses. Mark and Isabella will need to use additional protection for a period of time after the vasectomy as sperm can be stored in the ductus deferens and remain viable for several months. The physician will want to monitor his ejaculate for sperm levels for a short period of time.

Chapter 29

1. As part of the feedback mechanism for lactation, oxytocin is released from the posterior pituitary. It is carried to the mammary glands where it causes milk to be released into the mammary ducts (milk ejection). The oxytocin is also transported in the blood to the uterus, which contains oxytocin receptors on the myometrium. The oxytocin causes contraction of the myometrium, resulting in the painful sensations that Kathy is experiencing. The uterine contractions can help return the uterus back to its prepregnancy size.

2. Sex-linked genetic traits, such as hemophilia, are present on the X chromosomes but not on the Y chromosomes. In males, the X chromosome is always inherited from the mother, and the Y chromosome from the father. Thus, Jack's hemophilia gene was inherited from his mother on his X chromosome. The gene for hemophilia is a recessive gene. His mother would need two recessive genes, one on each of her X chromosomes, to be hemophiliac. His father must carry the dominant (nonhemophiliac) gene on his X chromosome, so he also would not have hemophilia.

3. The cord blood is a source of pluripotent stem cells, which are unspecialized cells that have the potential to specialize into cells with specific functions. The hope is that stem cells can be used to generate cells and tissues to treat a variety of disorders. It is assumed that the tissues would not be rejected since they would contain the same genetic material as the patient—in this case Alisa's baby.

GLOSSARY

Pronunciation Key

1. The most strongly accented syllable appears in capital letters, for example, bilateral (bī-LAT-er-al) and diagnosis (dī-ag-NŌ-sis).

2. If there is a secondary accent, it is noted by a prime ('), for example, constitution (kon′-sti-TOO-shun) and physiology (fiz′-ē-OL-ō-jē). Any additional secondary accents are also noted by a prime, for example, decarboxylation (dē′-kar-bok′-si-LĀ-shun).

3. Vowels marked by a line above the letter are pronounced with the long sound, as in the following common words:

    ā as in *māke*   ō as in *pōle*
    ē as in *bē*   ū as in *cūte*
    ī as in *īvy*

4. Vowels not marked by a line above the letter are pronounced with the short sound, as in the following words:

    a as in *above* or *at*   o as in *not*
    e as in *bet*   u as in *bud*
    i as in *sip*

5. Other vowel sounds are indicated as follows:

    oy as in *oil*
    oo as in *root*

6. Consonant sounds are pronounced as in the following words:

    b as in *bat*   m as in *mother*
    ch as in *chair*   n as in *no*
    d as in *dog*   p as in *pick*
    f as in *father*   r as in *rib*
    g as in *get*   s as in *so*
    h as in *hat*   t as in *tea*
    j as in *jump*   v as in *very*
    k as in *can*   w as in *welcome*
    ks as in *tax*   z as in *zero*
    kw as in *quit*   zh as in *lesion*
    l as in *let*

A

Abdomen (ab-DŌ-men *or* AB-dō-men) The area between the diaphragm and pelvis.

Abdominal (ab-DŌM-i-nal) **cavity** Superior portion of the abdominopelvic cavity that contains the stomach, spleen, liver, gallbladder, most of the small intestine, and part of the large intestine.

Abdominal thrust maneuver A first-aid procedure for choking. Employs a quick, upward thrust against the diaphragm that forces air out of the lungs with sufficient force to eject any lodged material.

Abdominopelvic (ab-dom′-i-nō-PEL-vik) **cavity** Inferior to the diaphragm that is subdivided into a superior abdominal cavity and an inferior pelvic cavity.

Abduction (ab-DUK-shun) Movement away from the midline of the body.

Abortion (a-BOR-shun) The premature loss (spontaneous) or removal (induced) of the embryo or nonviable fetus; miscarriage due to a failure in the normal process of developing or maturing.

Abscess (AB-ses) A localized collection of pus and liquefied tissue in a cavity.

Absorption (ab-SORP-shun) Intake of fluids or other substances by cells of the skin or mucous membranes; the passage of digested foods from the gastrointestinal tract into blood or lymph.

Accessory duct A duct of the pancreas that empties into the duodenum about 2.5 cm (1 in.) superior to the ampulla of Vater (hepatopancreatic ampulla). Also called the **duct of Santorini** (san′-tō-RĒ-nē).

Acetabulum (as′-e-TAB-ū-lum) The rounded cavity on the external surface of the hip bone that receives the head of the femur.

Acetylcholine (as′-ē-til-KŌ-lēn) **(ACh)** A neurotransmitter liberated by many peripheral nervous system neurons and some central nervous system neurons. It is excitatory at neuromuscular junctions but inhibitory at some other synapses (for example, it slows heart rate).

Achalasia (ak′-a-LĀ-zē-a) A condition, caused by malfunction of the myenteric plexus, in which the lower esophageal sphincter fails to relax normally as food approaches. A whole meal may become lodged in the esophagus and enter the stomach very slowly. Distension of the esophagus results in chest pain that is often confused with pain originating from the heart.

Achilles tendon *See* **Calcaneal tendon.**

Acini (AS-i-nē) Groups of cells in the pancreas that secrete digestive enzymes.

Acoustic (a-KOOS-tik) Pertaining to sound or the sense of hearing.

Acquired immunodeficiency syndrome (AIDS) A fatal disease caused by the human immunodeficiency virus (HIV). Characterized by a positive HIV-antibody test, low helper T cell count, and certain indicator diseases (for example Kaposi's sarcoma, pneumocystis carinii pneumonia, tuberculosis, fungal diseases). Other symptoms include fever or night sweats, coughing, sore throat, fatigue, body aches, weight loss, and enlarged lymph nodes.

Acrosome (AK-rō-sōm) A lysosomelike organelle in the head of a sperm cell containing enzymes that facilitate the penetration of a sperm cell into a secondary oocyte.

Actin (AK-tin) A contractile protein that is part of thin filaments in muscle fibers.

Action potential (AP) An electrical signal that propagates along the membrane of a neuron or muscle fiber (cell); a rapid change in membrane potential that involves a depolarization followed by a repolarization. Also called a **nerve action potential** or **nerve impulse** as it relates to a neuron, and a **muscle action potential** as it relates to a muscle fiber.

Activation (ak′-ti-VĀ-shun) **energy** The minimum amount of energy required for a chemical reaction to occur.

Active transport The movement of substances across cell membranes against a concentration gradient, requiring the expenditure of cellular energy (ATP).

Acute (a-KŪT) Having rapid onset, severe symptoms, and a short course; not chronic.

Adaptation (ad′-ap-TĀ shun) The adjustment of the pupil of the eye to changes in light intensity. The property by which a sensory neuron relays a decreased frequency of action potentials from a receptor, even though the strength of the stimulus remains constant; the decrease in perception of a sensation over time while the stimulus is still present.

Adduction (ad-DUK-shun) Movement toward the midline of the body.

Adenoids (AD-e-noyds) The pharyngeal tonsils.

Adenosine triphosphate (a-DEN-ō-sēn trī-FOS-fāt) **(ATP)** The main energy currency in living cells; used to transfer the chemical energy needed for metabolic reactions. ATP consists of the purine base *adenine* and the five-carbon sugar *ribose,* to which are added, in linear array, three *phosphate* groups.

Adhesion (ad-HĒ-zhun) Abnormal joining of parts to each other.

Adipocyte (AD-i-pō-sīt) Fat cell, derived from a fibroblast.

Adipose (AD-i-pōz) **tissue** Tissue composed of adipocytes specialized for triglyceride storage and present in the form of soft pads between various organs for support, protection, and insulation.

Adrenal cortex (a-DRĒ-nal KOR-teks) The outer portion of an adrenal gland, divided into three zones; the zona glomerulosa secretes mineralocorticoids, the zona fasciculata secretes glucocorticoids, and the zona reticularis secretes androgens.

Adrenal glands Two glands located superior to each kidney. Also called the **suprarenal** (soo′-pra-RĒ-nal) **glands.**

Adrenal medulla (me-DUL-a) The inner part of an adrenal gland, consisting of cells that secrete epinephrine, norepinephrine, and a small amount of dopamine in response to stimulation by sympathetic preganglionic neurons.

Adrenergic (ad′-ren-ER-jik) **neuron** A neuron that releases epinephrine (adrenaline) or norepinephrine (noradrenaline) as its neurotransmitter.

Adrenocorticotropic (ad-rē′-nō-kor-ti-kō-TRŌP-ik) **hormone (ACTH)** A hormone produced by the anterior pituitary that influences the production and secretion of certain hormones of the adrenal cortex.

Adventitia (ad-ven-TISH-a) The outermost covering of a structure or organ.

Aerobic (air-Ō-bik) Requiring molecular oxygen.

Afferent arteriole (AF-er-ent ar-TĒ-rē-ōl) A blood vessel of a kidney that divides into the capillary network called a glomerulus; there is one afferent arteriole for each glomerulus.

Agglutination (a-gloo-ti-NĀ-shun) Clumping of microorganisms or blood cells, typically due to an antigen–antibody reaction.

Aggregated lymphatic follicles Clusters of lymph nodules that are most numerous in the ileum. Also called **Peyer's** (PĪ-erz) **patches.**

Albinism (AL-bin-izm) Abnormal, nonpathological, partial, or total absence of pigment in skin, hair, and eyes.

Aldosterone (al-DOS-ter-ōn) A mineralocorticoid produced by the adrenal cortex that promotes sodium and water reabsorption by the kidneys and potassium excretion in urine.

Allantois (a-LAN-tō-is) A small, vascularized outpouching of the yolk sac that serves as an early site for blood formation and development of the urinary bladder.

Alleles (a-LĒLZ) Alternate forms of a single gene that control the same inherited trait (such as type A blood) and are located at the same position on homologous chromosomes.

Allergen (AL-er-jen) An antigen that evokes a hypersensitivity reaction.

Alopecia (al′-ō-PĒ-shē-a) The partial or complete lack of hair as a result of factors such as genetics, aging, endocrine disorders, chemotherapy, and skin diseases.

Alpha (AL-fa) **cell** A type of cell in the pancreatic islets (islets of Langerhans) in the pancreas that secretes the hormone glucagon. Also termed an **A cell.**

Alpha receptor A type of receptor for norepinephrine and epinephrine; present on visceral effectors innervated by sympathetic postganglionic neurons.

Alveolar duct Branch of a respiratory bronchiole around which alveoli and alveolar sacs are arranged.

Alveolar macrophage (MAK-rō-fāj) Highly phagocytic cell found in the alveolar walls of the lungs. Also called a **dust cell.**

Alveolar sac A cluster of alveoli that share a common opening.

Alveolus (al-VĒ-ō-lus) A small hollow or cavity; an air sac in the lungs; milk-secreting portion of a mammary gland. *Plural* is **alveoli** (al-VĒ-ol-ī).

Alzheimer (ALTZ-hī-mer) **disease (AD)** Disabling neurological disorder characterized by dysfunction and death of specific cerebral neurons, resulting in widespread intellectual impairment, personality changes, and fluctuations in alertness.

Amenorrhea (ā-men-ō-RĒ-a) Absence of menstruation.

Amnesia (am-NĒ-zē-a) A lack or loss of memory.

Amnion (AM-nē-on) A thin, protective fetal membrane that develops from the epiblast; holds the fetus suspended in amniotic fluid. Also called the "bag of waters."

Amniotic (am′-nē-OT-ik) **fluid** Fluid in the amniotic cavity, the space between the developing embryo (or fetus) and amnion; the fluid is initially produced as a filtrate from maternal blood and later includes fetal urine. It functions as a shock absorber, helps regulate fetal body temperature, and helps prevent desiccation.

Amphiarthrosis (am′-fē-ar-THRŌ-sis) A slightly movable joint, in which the articulating bony surfaces are separated by fibrous connective tissue or fibrocartilage to which both are attached; types are syndesmosis and symphysis.

Ampulla (am-PUL-la) A saclike dilation of a canal or duct.

Ampulla of Vater *See* **Hepatopancreatic ampulla.**

Anabolism (a-NAB-ō-lizm) Synthetic, energy-requiring reactions whereby small molecules are built up into larger ones.

Anaerobic (an-ar-Ō-bik) Not requiring oxygen.

Anal (Ā-nal) **canal** The last 2 or 3 cm (1 in.) of the rectum; opens to the exterior through the anus.

Anal column A longitudinal fold in the mucous membrane of the anal canal that contains a network of arteries and veins.

Anal triangle The subdivision of the female or male perineum that contains the anus.

Analgesia (an-al-JĒ-zē-a) Pain relief; absence of the sensation of pain.

Anaphase (AN-a-fāz) The third stage of mitosis in which the chromatids that have separated at the centromeres move to opposite poles of the cell.

Anaphylaxis (an′-a-fi-LAK-sis) A hypersensitivity (allergic) reaction in which IgE antibodies attach to mast cells and basophils, causing them to produce mediators of anaphylaxis (histamine, leukotrienes, kinins, and prostaglandins) that bring about increased blood permeability, increased smooth muscle contraction, and increased mucus production. Examples are hay fever, hives, and anaphylactic shock.

Anastomosis (a-nas-tō-MŌ-sis) An end-to-end union or joining of blood vessels, lymphatic vessels, or nerves.

Anatomic dead space Spaces of the nose, pharynx, larynx, trachea, bronchi, and bronchioles totaling about 150 mL of the 500 mL in a quiet breath (tidal volume); air in the anatomic dead space does not reach the alveoli to participate in gas exchange.

Anatomical (an′-a-TOM-i-kal) **position** A position of the body universally used in anatomical descriptions in which the body is erect, the head is level, the eyes face forward, the upper limbs are at the sides, the palms face forward, and the feet are flat on the floor.

Anatomy (a-NAT-ō-mē) The structure or study of the structure of the body and the relation of its parts to each other.

Androgens (AN-drō-jenz) Masculinizing sex hormones produced by the testes in males and the adrenal cortex in both sexes; responsible for libido (sexual desire); the two main androgens are testosterone and dihydrotestosterone.

Anemia (a-NĒ-mē-a) Condition of the blood in which the number of functional red blood cells or their hemoglobin content is below normal.

Anesthesia (an′-es-THĒ-zē-a) A total or partial loss of feeling or sensation; may be general or local.

Aneurysm (AN-ū-rizm) A saclike enlargement of a blood vessel caused by a weakening of its wall.

Angina pectoris (an-JĪ-na *or* AN-ji-na PEK-tō-ris) A pain in the chest related to reduced coronary circulation due to coronary artery disease (CAD) or spasms of vascular smooth muscle in coronary arteries.

Angiogenesis (an′-jē-ō-JEN-e-sis) The formation of blood vessels in the extraembryonic mesoderm of the yolk sac, connecting stalk, and chorion at the beginning of the third week of development.

Ankylosis (ang′-ki-LŌ-sis) Severe or complete loss of movement at a joint as the result of a disease process.

Antagonist (an-TAG-ō-nist) A muscle that has an action opposite that of the prime mover (agonist) and yields to the movement of the prime mover.

Antagonistic (an-tag-ō-NIST-ik) **effect** A hormonal interaction in which the effect of one hormone on a target cell is opposed by another hormone. For example, calcitonin (CT) lowers blood calcium level, whereas parathyroid hormone (PTH) raises it.

Anterior (an-TĒR-ē-or) Nearer to or at the front of the body. Equivalent to **ventral** in bipeds.

Anterior pituitary Anterior lobe of the pituitary gland. Also called the **adenohypophysis** (ad′-e-nō-hī-POF-i-sis).

Anterior root The structure composed of axons of motor (efferent) neurons that emerges from the anterior aspect of the spinal cord and extends laterally to join a posterior root, forming a spinal nerve. Also called a **ventral root.**

Anterolateral (an′-ter-ō-LAT-er-al) **pathway** Sensory pathway that conveys information related to pain, temperature, tickle, and itch.

Antibody (AN-ti-bod'-ē) A protein produced by plasma cells in response to a specific antigen; the antibody combines with that antigen to neutralize, inhibit, or destroy it. Also called an **immunoglobulin** (im-ū-nō-GLOB-ū-lin) or **Ig.**

Anticoagulant (an-tī-cō-AG-ū-lant) A substance that can delay, suppress, or prevent the clotting of blood.

Antidiuretic (an'-ti-dī-ū-RET-ik) Substance that inhibits urine formation.

Antidiuretic hormone (ADH) Hormone produced by neurosecretory cells in the paraventricular and supraoptic nuclei of the hypothalamus that stimulates water reabsorption from kidney tubule cells into the blood and vasoconstriction of arterioles. Also called **vasopressin** (vāz-ō-PRES-in).

Antigen (AN-ti-jen) A substance that has immunogenicity (the ability to provoke an immune response) and reactivity (the ability to react with the antibodies or cells that result from the immune response); contraction of *anti*body *gen*erator. Also termed a **complete antigen.**

Antigen-presenting cell (APC) Special class of migratory cell that processes and presents antigens to T cells during an immune response; APCs include macrophages, B cells, and dendritic cells, which are present in the skin, mucous membranes, and lymph nodes.

Antrum (AN-trum) Any nearly closed cavity or chamber, especially one within a bone, such as a sinus.

Anuria (an-Ū-rē-a) Absence of urine formation or daily urine output of less than 50 mL.

Anus (Ā-nus) The distal end and outlet of the rectum.

Aorta (ā-OR-ta) The main systemic trunk of the arterial system of the body that emerges from the left ventricle.

Aortic (ā-OR-tik) **body** Cluster of chemoreceptors on or near the arch of the aorta that respond to changes in blood levels of oxygen, carbon dioxide, and hydrogen ions (H^+).

Aortic reflex A reflex that helps maintain normal systemic blood pressure; initiated by baroreceptors in the wall of the ascending aorta and arch of the aorta. Nerve impulses from aortic baroreceptors reach the cardiovascular center via sensory axons of the vagus (X) nerves.

Apex (Ā-peks) The pointed end of a conical structure, such as the apex of the heart.

Aphasia (a-FĀ-zē-a) Loss of ability to express oneself properly through speech or loss of verbal comprehension.

Apnea (AP-nē-a) Temporary cessation of breathing.

Apneustic (ap-NOO-stik) **area** A part of the respiratory center in the pons that sends stimulatory nerve impulses to the inspiratory area that activate and prolong inhalation and inhibit exhalation.

Apocrine (AP-ō-krin) **gland** A type of gland in which the secretory products gather at the free end of the secreting cell and are pinched off, along with some of the cytoplasm, to become the secretion, as in mammary glands.

Aponeurosis (ap'-ō-noo-RŌ-sis) A sheetlike tendon joining one muscle with another or with bone.

Apoptosis (ap-ō-TŌ-sis *or* ap'-ōp-TŌ-sis) Programmed cell death; a normal type of cell death that removes unneeded cells during embryological development, regulates the number of cells in tissues, and eliminates many potentially dangerous cells such as cancer cells. During apoptosis, the DNA fragments, the nucleus condenses, mitochondria cease to function, and the cytoplasm shrinks, but the plasma membrane remains intact. Phagocytes engulf and digest the apoptotic cells, and an inflammatory response does not occur.

Appositional (a-pō-ZISH-o-nal) **growth** Growth due to surface deposition of material, as in the growth in diameter of cartilage and bone. Also called **exogenous** (eks-OJ-e-nus) **growth.**

Aqueous humor (AK-wē-us HŪ-mer) The watery fluid, similar in composition to cerebrospinal fluid, that fills the anterior cavity of the eye.

Arachnoid (a-RAK-noyd) **mater** The middle of the three meninges (coverings) of the brain and spinal cord. Also termed the **arachnoid.**

Arachnoid villus (VIL-us) Berrylike tuft of the arachnoid mater that protrudes into the superior sagittal sinus and through which cerebrospinal fluid is reabsorbed into the bloodstream.

Arbor vitae (AR-bor VĪ-tē) The white matter tracts of the cerebellum, which have a treelike appearance when seen in midsagittal section.

Arch of the aorta The most superior portion of the aorta, lying between the ascending and descending segments of the aorta.

Areola (a-RĒ-ō-la) Any tiny space in a tissue. The pigmented ring around the nipple of the breast.

Arm The part of the upper limb from the shoulder to the elbow.

Arousal (a-ROW-zal) Awakening from sleep, a response due to stimulation of the reticular activating system (RAS).

Arrector pili (a-REK-tor PI-lē) Smooth muscles attached to hairs; contraction pulls the hairs into a vertical position, resulting in "goose bumps."

Arrhythmia (a-RITH-mē-a) An irregular heart rhythm. Also called a **dysrhythmia.**

Arteriole (ar-TĒ-rē-ōl) A small, almost microscopic, artery that delivers blood to a capillary.

Arteriosclerosis (ar-tē-rē-ō-skle-RŌ-sis) Group of diseases characterized by thickening of the walls of arteries and loss of elasticity.

Artery (AR-ter-ē) A blood vessel that carries blood away from the heart.

Arthritis (ar-THRĪ-tis) Inflammation of a joint.

Arthrology (ar-THROL-ō-jē) The study or description of joints.

Arthroplasty (AR-thrō-plas'-tē) Surgical replacement of joints, for example, the hip and knee joints.

Arthroscopy (ar-THROS-kō-pē) A procedure for examining the interior of a joint, usually the knee, by inserting an arthroscope into a small incision; used to determine extent of damage, remove torn cartilage, repair cruciate ligaments, and obtain samples for analysis.

Arthrosis (ar-THRŌ-sis) A joint or articulation.

Articular (ar-TIK-ū-lar) **capsule** Sleevelike structure around a synovial joint composed of a fibrous capsule and a synovial membrane.

Articular cartilage (KAR-ti-lij) Hyaline cartilage attached to articular bone surfaces.

Articular disc Fibrocartilage pad between articular surfaces of bones of some synovial joints. Also called a **meniscus** (men-IS-kus).

Articulation (ar-tik-ū-LĀ-shun) A joint; a point of contact between bones, cartilage and bones, or teeth and bones.

Arytenoid (ar'-i-TĒ-noyd) **cartilages** A pair of small, pyramidal cartilages of the larynx that attach to the vocal folds and intrinsic pharyngeal muscles and can move the vocal folds.

Ascending colon (KŌ-lon) The part of the large intestine that passes superiorly from the cecum to the inferior border of the liver, where it bends at the right colic (hepatic) flexure to become the transverse colon.

Ascites (as-SĪ-tēz) Abnormal accumulation of serous fluid in the peritoneal cavity.

Association areas Large cortical regions on the lateral surfaces of the occipital, parietal, and temporal lobes and on the frontal lobes anterior to the motor areas connected by many motor and sensory axons to other parts of the cortex. The association areas are concerned with motor patterns, memory, concepts of word-hearing and word seeing, reasoning, will, judgment, and personality traits.

Asthma (AZ-ma) Usually allergic reaction characterized by smooth muscle spasms in bronchi resulting in wheezing and difficult breathing. Also called **bronchial asthma.**

Astigmatism (a-STIG-ma-tizm) An irregularity of the lens or cornea of the eye causing the image to be out of focus and producing faulty vision.

Astrocyte (AS-trō-sīt) A neuroglial cell having a star shape that participates in brain development and the metabolism of neurotransmitters, helps form the blood–brain barrier, helps maintain the proper balance of K^+ for generation of nerve impulses, and provides a link between neurons and blood vessels.

Ataxia (a-TAK-sē-a) A lack of muscular coordination, lack of precision.

Atherosclerotic (ath'-er-ō-skle-RO-tic) **plaque** (PLAK) A lesion that results from accumulated cholesterol and smooth muscle fibers (cells) of the tunica media of an artery; may become obstructive.

Atom Unit of matter that makes up a chemical element; consists of a nucleus (containing positively charged protons and uncharged neutrons) and negatively charged electrons that orbit the nucleus.

Atresia (a-TRĒ-zē-a) Degeneration and reabsorption of an ovarian follicle before it fully matures and ruptures; abnormal closure of a passage, or absence of a normal body opening.

Atrial fibrillation (Ā-trē-al fib-ri-LĀ-shun) Asynchronous contraction of cardiac muscle fibers in the atria that results in the cessation of atrial pumping.

Atrial natriuretic (na′-trē-ū-RET-ik) **peptide (ANP)** Peptide hormone, produced by the atria of the heart in response to stretching, that inhibits aldosterone production and thus lowers blood pressure; causes natriuresis, increased urinary excretion of sodium.

Atrioventricular (AV) (ā′-trē-ō-ven-TRIK-ū-lar) **bundle** The part of the conduction system of the heart that begins at the atrioventricular (AV) node, passes through the cardiac skeleton separating the atria and the ventricles, then extends a short distance down the interventricular septum before splitting into right and left bundle branches. Also called the **bundle of His** (HISS).

Atrioventricular (AV) node The part of the conduction system of the heart made up of a compact mass of conducting cells located in the septum between the two atria.

Atrioventricular (AV) valve A heart valve made up of membranous flaps or cusps that allows blood to flow in one direction only, from an atrium into a ventricle.

Atrium (Ā-trē-um) A superior chamber of the heart.

Atrophy (AT-rō-fē) Wasting away or decrease in size of a part, due to a failure, abnormality of nutrition, or lack of use.

Auditory ossicle (AW-di-tō-rē OS-si-kul) One of the three small bones of the middle ear called the malleus, incus, and stapes.

Auditory tube The tube that connects the middle ear with the nose and nasopharynx region of the throat. Also called the **eustachian** (ū-STĀ-shun *or* ū-STĀ-kē-an) **tube** or **pharyngotympanic tube.**

Auscultation (aws-kul-TĀ-shun) Examination by listening to sounds in the body.

Autoimmunity An immunological response against a person's own tissues.

Autolysis (aw-TOL-i-sis) Self-destruction of cells by their own lysosomal digestive enzymes after death or in a pathological process.

Autonomic ganglion (aw′-tō-NOM-ik GANG-lē-on) A cluster of cell bodies of sympathetic or parasympathetic neurons located outside the central nervous system.

Autonomic nervous system (ANS) Visceral sensory (afferent) and visceral motor (efferent) neurons. Autonomic motor neurons, both sympathetic and parasympathetic, conduct nerve impulses from the central nervous system to smooth muscle, cardiac muscle, and glands. So named because this part of the nervous system was thought to be self-governing or spontaneous.

Autonomic plexus (PLEK-sus) A network of sympathetic and parasympathetic axons; examples are the cardiac, celiac, and pelvic plexuses, which are located in the thorax, abdomen, and pelvis, respectively.

Autophagy (aw-TOF-a-jē) Process by which worn-out organelles are digested within lysosomes.

Autopsy (AW-top-sē) The examination of the body after death.

Autorhythmic cells Cardiac or smooth muscle fibers that are self-excitable (generate impulses without an external stimulus); act as the heart's pacemaker and conduct the pacing impulse through the conduction system of the heart; self-excitable neurons in the central nervous system, as in the inspiratory area of the brain stem.

Autosome (AW-tō-sōm) Any chromosome other than the X and Y chromosomes (sex chromosomes).

Axilla (ak-SIL-a) The small hollow beneath the arm where it joins the body at the shoulders. Also called the **armpit.**

Axon (AK-son) The usually single, long process of a nerve cell that propagates a nerve impulse toward the axon terminals.

Axon terminal Terminal branch of an axon where synaptic vesicles undergo exocytosis to release neurotransmitter molecules.

B

B cell A lymphocyte that can develop into a clone of antibody-producing plasma cells or memory cells when properly stimulated by a specific antigen.

Babinski (ba-BIN-skē) **sign** Extension of the great toe, with or without fanning of the other toes, in response to stimulation of the outer margin of the sole; normal up to 18 months of age and indicative of damage to descending motor pathways such as the corticospinal tracts after that.

Back The posterior part of the body; the dorsum.

Ball-and-socket joint A synovial joint in which the rounded surface of one bone moves within a cup-shaped depression or socket of another bone, as in the shoulder or hip joint. Also called a **spheroid** (SFĒ-royd) **joint.**

Baroreceptor (bar′-ō-re-SEP-tor) Neuron capable of responding to changes in blood, air, or fluid pressure. Also called a **pressoreceptor.**

Basal ganglia (GANG-glē-a) Paired clusters of gray matter deep in each cerebral hemisphere including the globus pallidus, putamen, and caudate nucleus. Together, the caudate nucleus and putamen are known as the **corpus striatum.** Nearby structures that are functionally linked to the basal ganglia are the substantia nigra of the midbrain and the subthalamic nuclei of the diencephalon.

Basement membrane Thin, extracellular layer between epithelium and connective tissue consisting of a basal lamina and a reticular lamina.

Basilar (BĀS-i-lar) **membrane** A membrane in the cochlea of the internal ear that separates the cochlear duct from the scala tympani and on which the spiral organ (organ of Corti) rests.

Basophil (BĀ-sō-fil) A type of white blood cell characterized by a pale nucleus and large granules that stain blue-purple with basic dyes.

Belly The abdomen. The gaster or prominent, fleshy part of a skeletal muscle. Also called a **B Cell.**

Beta (BĀ-ta) **cell** A type of cell in the pancreatic islets (islets of Langerhans) in the pancreas that secretes the hormone insulin. Also called a **B cell.**

Beta receptor A type of adrenergic receptor for epinephrine and norepinephrine; found on visceral effectors innervated by sympathetic postganglionic neurons.

Bicuspid (bī-KUS-pid) **valve** Atrioventricular (AV) valve on the left side of the heart. Also called the **mitral valve.**

Bilateral (bī-LAT-er-al) Pertaining to two sides of the body.

Bile (BĪL) A secretion of the liver consisting of water, bile salts, bile pigments, cholesterol, lecithin, and several ions; it emulsifies lipids prior to their digestion.

Bilirubin (bil-ē-ROO-bin) An orange pigment that is one of the end products of hemoglobin breakdown in the hepatocytes and is excreted as a waste material in bile.

Blastocyst (BLAS-tō-sist) In the development of an embryo, a hollow ball of cells that consists of a blastocele (the internal cavity), trophoblast (outer cells), and embryoblast (inner cell mass).

Blastocyst (BLAS-tō-sist) **cavity** The fluid-filled cavity within the blastocyst. Also called the **blastocele.**

Blastomere (BLAS-tō-mēr) One of the cells resulting from the cleavage of a fertilized ovum.

Blastula (BLAS-tyū-la) An early stage in the development of a zygote.

Blind spot Area in the retina at the end of the optic (II) nerve in which there are no photoreceptors.

Blood The fluid that circulates through the heart, arteries, capillaries, and veins and that constitutes the chief means of transport within the body.

Blood–brain barrier (BBB) A barrier consisting of specialized brain capillaries and astrocytes that prevents the passage of materials from the blood to the cerebrospinal fluid and brain.

Blood island Isolated mass of mesoderm derived from angioblasts and from which blood vessels develop.

Blood pressure (BP) Force exerted by blood against the walls of blood vessels due to contraction of the heart and influenced by the elasticity of the vessel walls; clinically, a measure of the pressure in arteries during ventricular systole and ventricular diastole.

Blood reservoir (REZ-er-vwar) Systemic veins and venules that contain large amounts of blood that can be moved quickly to parts of the body requiring the blood.

Blood–testis barrier (BTB) A barrier formed by Sertoli cells that prevents an immune response against antigens produced by spermatogenic cells by isolating the cells from the blood.

Body cavity A space within the body that contains various internal organs.

Bolus (BŌ-lus) A soft, rounded mass, usually food, that is swallowed.

Bony labyrinth (LAB-i-rinth) A series of cavities within the petrous portion of the temporal bone forming the vestibule, cochlea, and semicircular canals of the inner ear.

Bowman's capsule *See* **Glomerular capsule.**

Brachial plexus (BRĀ-kē-al PLEK-sus) A network of nerve axons of the ventral rami of spinal nerves C5, C6, C7, C8, and T1. The nerves that emerge from the brachial plexus supply the upper limb.

Bradycardia (brād′-i-KAR-dē-a) A slow resting heart or pulse rate (under 50 beats per minute).

Brain The part of the central nervous system contained within the cranial cavity.

Brain stem The portion of the brain immediately superior to the spinal cord, made up of the medulla oblongata, pons, and midbrain.

Brain waves Electrical signals that can be recorded from the skin of the head due to electrical activity of brain neurons.

Broad ligament A double fold of parietal peritoneum attaching the uterus to the side of the pelvic cavity.

Broca's (BRŌ-kaz) **area** Motor area of the brain in the frontal lobe that translates thoughts into speech. Also called the **motor speech area.**

Bronchi (BRON-kī) Branches of the respiratory passageway including primary bronchi (the two divisions of the trachea), secondary or lobar bronchi (divisions of the primary bronchi that are distributed to the lobes of the lung), and tertiary or segmental bronchi (divisions of the secondary bronchi that are distributed to bronchopulmonary segments of the lung). *Singular* is **bronchus.**

Bronchial tree The trachea, bronchi, and their branching structures up to and including the terminal bronchioles.

Bronchiole (BRONG-kē-ōl) Branch of a tertiary bronchus further dividing into terminal bronchioles (distributed to lobules of the lung), which divide into respiratory bronchioles (distributed to alveolar sacs).

Bronchitis (brong-KĪ-tis) Inflammation of the mucous membrane of the bronchial tree; characterized by hypertrophy and hyperplasia of seromucous glands and goblet cells that line the bronchi which results in a productive cough.

Bronchopulmonary (brong′-kō-PUL-mō-ner-ē) **segment** One of the smaller divisions of a lobe of a lung supplied by its own branches of a bronchus.

Brunner's gland *See* **Duodenal gland.**

Buccal (BUK-al) Pertaining to the cheek or mouth.

Bulb of penis Expanded portion of the base of the corpus spongiosum penis.

Bulbourethral (bul′-bō-ū-RĒ-thral) **gland** One of a pair of glands located inferior to the prostate on either side of the urethra that secretes an alkaline fluid into the cavernous urethra. Also called a **Cowper's** (KOW-perz) **gland.**

Bulimia (boo-LIM-ē-a *or* boo-LĒ-mē-a) A disorder characterized by overeating at least twice a week followed by purging by self-induced vomiting, strict dieting or fasting, vigorous exercise, or use of laxatives or diuretics. Also called **binge–purge syndrome.**

Bulk-phase endocytosis A process by which most body cells can ingest membrane-surrounded droplets of interstitial fluid.

Bundle branch One of the two branches of the atrioventricular (AV) bundle made up of specialized muscle fibers (cells) that transmit electrical impulses to the ventricles.

Bundle of His *See* **Atrioventricular (AV) bundle.**

Bursa (BUR-sa) A sac or pouch of synovial fluid located at friction points, especially about joints.

Bursitis (bur-SĪ-tis) Inflammation of a bursa.

Buttocks (BUT-oks) The two fleshy masses on the posterior aspect of the inferior trunk, formed by the gluteal muscles.

C

Calcaneal (kal-KĀ-nē-al) **tendon** The tendon of the soleus, gastrocnemius, and plantaris muscles at the back of the heel. Also called the **Achilles** (a-KIL-ēz) **tendon.**

Calcification (kal′-si-fi-KĀ-shun) Deposition of mineral salts, primarily hydroxyapatite, in a framework formed by collagen fibers in which the tissue hardens. Also called **mineralization** (min′-e-ral-i-ZĀ-shun).

Calcitonin (kal-si-TŌ-nin) **(CT)** A hormone produced by the parafollicular cells of the thyroid gland that can lower the amount of blood calcium and phosphates by inhibiting bone resorption (breakdown of bone extracellular matrix) and by accelerating uptake of calcium and phosphates into bone matrix.

Calculus (KAL-kū-lus) A stone, or insoluble mass of crystallized salts or other material, formed within the body, as in the gallbladder, kidney, or urinary bladder.

Callus (KAL-lus) A growth of new bone tissue in and around a fractured area, ultimately replaced by mature bone. An acquired, localized thickening.

Calyx (KĀL-iks) Any cuplike division of the kidney pelvis. *Plural* is **calyces** (KĀ-li-sēz).

Canal (ka-NAL) A narrow tube, channel, or passageway.

Canaliculus (kan′-a-LIK-ū-lus) A small channel or canal, as in bones, where they connect lacunae. *Plural* is **canaliculi** (kan′-a-LIK-ū-lī).

Canal of Schlemm *See* **Scleral venous sinus.**

Capacitation (ka′-pas-i-TĀ-shun) The functional changes that sperm undergo in the female reproductive tract that allow them to fertilize a secondary oocyte.

Capillary (KAP-i-lar′-ē) A microscopic blood vessel located between an arteriole and venule through which materials are exchanged between blood and interstitial fluid.

Carcinogen (car-SIN-ō-jen) A chemical substance or radiation that causes cancer.

Cardiac (KAR-dē-ak) **arrest** Cessation of an effective heartbeat in which the heart is completely stopped or in ventricular fibrillation.

Cardiac cycle A complete heartbeat consisting of systole (contraction) and diastole (relaxation) of both atria plus systole and diastole of both ventricles.

Cardiac muscle Striated muscle fibers (cells) that form the wall of the heart; stimulated by an intrinsic conduction system and autonomic motor neurons.

Cardiac notch An angular notch in the anterior border of the left lung into which part of the heart fits.

Cardinal ligament A ligament of the uterus, extending laterally from the cervix and vagina as a continuation of the broad ligament.

Cardiogenic area (kar-dē-ō-JEN-ik) A group of mesodermal cells in the head end of an embryo that gives rise to the heart.

Cardiology (kar-dē-OL-ō-jē) The study of the heart and diseases associated with it.

Cardiovascular (kar-dē-ō-VAS-kū-lar) **center** Groups of neurons scattered within the medulla oblongata that regulate heart rate, force of contraction, and blood vessel diameter.

Carotene (KAR-ō-tēn) Antioxidant precursor of vitamin A, which is needed for synthesis of photopigments; yellow-orange pigment present in the stratum corneum of the epidermis. Accounts for the yellowish coloration of skin. Also termed **beta-carotene.**

Carotid (ka-ROT-id) **body** Cluster of chemoreceptors on or near the carotid sinus that respond to changes in blood levels of oxygen, carbon dioxide, and hydrogen ions.

Carotid sinus A dilated region of the internal carotid artery just superior to where it branches from the common carotid artery; it contains baroreceptors that monitor blood pressure.

Carpal bones The eight bones of the wrist. Also called **carpals.**

Carpus (KAR-pus) A collective term for the eight bones of the wrist.

Cartilage (KAR-ti-lij) A type of connective tissue consisting of chondrocytes in lacunae embedded in a dense network of collagen and elastic fibers and an extracellular matrix of chondroitin sulfate.

Cartilaginous (kar-ti-LAJ-i-nus) **joint** A joint without a synovial (joint) cavity where the articulating bones are held tightly together by cartilage, allowing little or no movement.

Catabolism (ka-TAB-ō-lizm) Chemical reactions that break down complex organic compounds into simple ones, with the net release of energy.

Cataract (KAT-a-rakt) Loss of transparency of the lens of the eye or its capsule or both.

Cauda equina (KAW-da ē-KWĪ-na) A tail like array of roots of spinal nerves at the inferior end of the spinal cord.

Caudal (KAW-dal) Pertaining to any tail-like structure; inferior in position.

Cecum (SĒ-kum) A blind pouch at the proximal end of the large intestine that attaches to the ileum.

Celiac plexus (PLEK-sus) A large mass of autonomic ganglia and axons located at the level of the superior part of the first lumbar vertebra. Also called the **solar plexus.**

Cell The basic structural and functional unit of all organisms; the smallest structure capable of performing all the activities vital to life.

Cell cycle Growth and division of a single cell into two identical cells; consists of interphase and cell division.

Cell division Process by which a cell reproduces itself that consists of a nuclear division (mitosis) and a cytoplasmic division (cytokinesis); types include somatic and reproductive cell division.

Cell junction Point of contact between plasma membranes of tissue cells.

Cementum (se-MEN-tum) Calcified tissue covering the root of a tooth.

Central canal A microscopic tube running the length of the spinal cord in the gray commissure. A circular channel running longitudinally in the center of an osteon (haversian system) of mature compact bone, containing blood and lymphatic vessels and nerves. Also called an **haversian** (ha-VER-shun) **canal.**

Central fovea (FŌ-vē-a) A depression in the center of the macula lutea of the retina, containing cones only and lacking blood vessels; the area of highest visual acuity (sharpness of vision).

Central nervous system (CNS) That portion of the nervous system that consists of the brain and spinal cord.

Centrioles (SEN-trē-ōlz) Paired, cylindrical structures of a centrosome, each consisting of a ring of microtubules and arranged at right angles to each other.

Centromere (SEN-trō-mēr) The constricted portion of a chromo-some where the two chromatids are joined; serves as the point of attachment for the microtubules that pull chromatids during anaphase of cell division.

Centrosome (SEN-trō-sōm) A dense network of small protein fibers near the nucleus of a cell, containing a pair of centrioles and pericentriolar material.

Cephalic (se-FAL-ik) Pertaining to the head; superior in position.

Cerebellar peduncle (ser-e-BEL-ar pe-DUNG-kul) A bundle of nerve axons connecting the cerebellum with the brain stem.

Cerebellum (ser'-e-BEL-um) The part of the brain lying posterior to the medulla oblongata and pons; governs balance and coordinates skilled movements.

Cerebral aqueduct (SER-ē-bral AK-we-dukt) A channel through the midbrain connecting the third and fourth ventricles and containing cerebrospinal fluid. Also termed the **aqueduct of Sylvius.**

Cerebral arterial circle A ring of arteries forming an anastomosis at the base of the brain between the internal carotid and basilar arteries and arteries supplying the cerebral cortex. Also called the **circle of Willis.**

Cerebral cortex The surface of the cerebral hemispheres, 2–4 mm thick, consisting of gray matter; arranged in six layers of neuronal cell bodies in most areas.

Cerebral peduncle One of a pair of nerve axon bundles located on the anterior surface of the midbrain, conducting nerve impulses between the pons and the cerebral hemispheres.

Cerebrospinal (se-rē'-brō-SPĪ-nal) **fluid (CSF)** A fluid produced by ependymal cells that cover choroid plexuses in the ventricles of the brain; the fluid circulates in the ventricles, the central canal, and the subarachnoid space around the brain and spinal cord.

Cerebrovascular (se rē'-brō-VAS-kū-lar) **accident (CVA)** Destruction of brain tissue (infarction) resulting from obstruction or rupture of blood vessels that supply the brain. Also called a **stroke** or **brain attack.**

Cerebrum (SER-e-brum *or* se-RĒ-brum) The two hemispheres of the forebrain (derived from the telencephalon), making up the largest part of the brain.

Cerumen (se-ROO-men) Waxlike secretion produced by ceruminous glands in the external auditory meatus (ear canal). Also termed **ear wax.**

Ceruminous (se-RŪ-mi-nus) **gland** A modified sudoriferous (sweat) gland in the external auditory meatus that secretes cerumen (ear wax).

Cervical ganglion (SER-vi-kul GANG-glē-on) A cluster of cell bodies of postganglionic sympathetic neurons located in the neck, near the vertebral column.

Cervical plexus (PLEK-sus) A network formed by nerve axons from the ventral rami of the first four cervical nerves and receiving gray rami communicantes from the superior cervical ganglion.

Cervix (SER-viks) Neck; any constricted portion of an organ, such as the inferior cylindrical part of the uterus.

Chemoreceptor (kē'-mō-rē-SEP-tor) Sensory receptor that detects the presence of a specific chemical.

Chiasm (KĪ-azm) A crossing; especially the crossing of axons in the optic (II) nerve.

Chief cell The secreting cell of a gastric gland that produces pepsinogen, the precursor of the enzyme pepsin, and the enzyme gastric lipase. Also called a **zymogenic** (zī'-mō-JEN-ik) **cell.** Cell in the parathyroid glands that secretes parathyroid hormone (PTH). Also called a **principal cell.**

Cholecystectomy (kō'-lē-sis-TEK-tō-mē) Surgical removal of the gallbladder.

Cholecystitis (kō'-lē-sis-TĪ-tis) Inflammation of the gallbladder.

Cholesterol (kō-LES-te-rol) Classified as a lipid, the most abundant steroid in animal tissues; located in cell membranes and used for the synthesis of steroid hormones and bile salts.

Cholinergic (kō'-lin-ER-jik) **neuron** A neuron that liberates acetylcholine as its neurotransmitter.

Chondrocyte (KON-drō-sīt) Cell of mature cartilage.

Chondroitin (kon-DROY-tin) **sulfate** An amorphous extracellular matrix material found outside connective tissue cells.

Chordae tendineae (KOR-dē TEN-di-nē-ē) Tendonlike, fibrous cords that connect atrioventricular valves of the heart with papillary muscles.

Chorion (KŌ-rē-on) The most superficial fetal membrane that becomes the principal embryonic portion of the placenta; serves a protective and nutritive function.

Chorionic villi (kō-rē-ON-ik VIL-lī) Fingerlike projections of the chorion that grow into the decidua basalis of the endometrium and contain fetal blood vessels.

Chorionic villi sampling (CVS) The removal of a sample of chorionic villus tissue by means of a catheter to analyze the tissue for prenatal genetic defects.

Choroid (KŌ-royd) One of the vascular coats of the eyeball.

Choroid plexus (PLEK-sus) A network of capillaries located in the roof of each of the four ventricles of the brain; ependymal cells around choroid plexuses produce cerebrospinal fluid.

Chromaffin (KRŌ-maf-in) **cell** Cell that has an affinity for chrome salts, due in part to the presence of the precursors of the neurotransmitter epinephrine; found, among other places, in the adrenal medulla.

Chromatid (KRŌ-ma-tid) One of a pair of identical connected nucleoprotein strands that are joined at the centromere and separate during cell division, each becoming a chromosome of one of the two daughter cells.

Chromatin (KRŌ-ma-tin) The threadlike mass of genetic material, consisting of DNA and histone proteins, that is present in the nucleus of a nondividing or interphase cell.

Chromatolysis (krō'-ma-TOL-i-sis) The breakdown of Nissl bodies into finely granular masses in the cell body of a neuron whose axon has been damaged.

Chromosome (KRŌ-mō-sōm) One of the small, threadlike structures in the nucleus of a cell, normally 46 in a human diploid cell, that bears the genetic material; composed of DNA and proteins (histones) that form a delicate chromatin thread during interphase; becomes packaged into compact rodlike structures that are visible under the light microscope during cell division.

Chronic (KRON-ik) Long term or frequently recurring; applied to a disease that is not acute.

Chronic obstructive pulmonary disease (COPD) A disease, such as bronchitis or emphysema, in which there is some degree of obstruction of airways and consequent increase in airway resistance.

Chyle (KĪL) The milky-appearing fluid found in the lacteals of the small intestine after absorption of lipids in food.

Chyme (KĪM) The semifluid mixture of partly digested food and digestive secretions found in the stomach and small intestine during digestion of a meal.

Ciliary (SIL-ē-ar'-ē) **body** One of the three parts of the vascular tunic of the eyeball, the others being the choroid and the iris; includes the ciliary muscle and the ciliary processes.

Ciliary ganglion (GANG-glē-on) A very small parasympathetic ganglion whose preganglionic axons come from the oculomotor (III) nerve and whose postganglionic axons carry nerve impulses to the ciliary muscle and the sphincter muscle of the iris.

Cilium (SIL-ē-um) A hair or hairlike process projecting from a cell that may be used to move the entire cell or to move substances along the surface of the cell. *Plural* is **cilia.**

Circle of Willis *See* **Cerebral arterial circle.**

Circular folds Permanent, deep, transverse folds in the mucosa and submucosa of the small intestine that increase the surface area for absorption. Also called **plicae circulares** (PLĪ-kē SER-kū-lar-ēs).

Circumduction (ser-kum-DUK-shun) A movement at a synovial joint in which the distal end of a bone moves in a circle while the proximal end remains relatively stable.

Cirrhosis (si-RŌ-sis) A liver disorder in which the parenchymal cells are destroyed and replaced by connective tissue.

Cisterna chyli (sis-TER-na KĪ-lē) The origin of the thoracic duct.

Cleavage The rapid mitotic divisions following the fertilization of a secondary oocyte, resulting in an increased number of progressively smaller cells, called blastomeres.

Clitoris (KLI-to-ris) An erectile organ of the female, located at the anterior junction of the labia minora, that is homologous to the male penis.

Clone (KLŌN) A population of identical cells.

Coarctation (kō′-ark-TĀ-shun) **of the aorta** A congenital heart defect in which a segment of the aorta is too narrow. As a result, the flow of oxygenated blood to the body is reduced, the left ventricle is forced to pump harder, and high blood pressure develops.

Coccyx (KOK-siks) The fused bones at the inferior end of the vertebral column.

Cochlea (KOK-lē-a) A winding, cone-shaped tube forming a portion of the inner ear and containing the spiral organ (organ of Corti).

Cochlear duct The membranous cochlea consisting of a spirally arranged tube enclosed in the bony cochlea and lying along its outer wall. Also called the **scala media** (SCA-la MĒ-dē-a).

Collagen (KOL-a-jen) A protein that is the main organic constituent of connective tissue.

Collateral circulation The alternate route taken by blood through an anastomosis.

Colliculus (ko-LIK-ū-lus) A small elevation.

Colon The portion of the large intestine consisting of ascending, transverse, descending, and sigmoid portions.

Colony-stimulating factor (CSF) One of a group of molecules that stimulates development of white blood cells. Examples are macrophage CSF and granulocyte CSF.

Colostrum (kō-LOS-trum) A thin, cloudy fluid secreted by the mammary glands a few days prior to or after delivery before true milk is produced.

Column (KOL-um) Group of white matter tracts in the spinal cord.

Common bile duct A tube formed by the union of the common hepatic duct and the cystic duct that empties bile into the duodenum at the hepatopancreatic ampulla (ampulla of Vater).

Compact (dense) bone tissue Bone tissue that contains few spaces between osteons (haversian systems); forms the external portion of all bones and the bulk of the diaphysis (shaft) of long bones; is found immediately deep to the periosteum and external to spongy bone.

Concha (KONG-ka) A scroll-like bone found in the skull. *Plural is* **conchae** (KONG-kē).

Concussion (kon-KUSH-un) Traumatic injury to the brain that produces no visible bruising but may result in abrupt, temporary loss of consciousness.

Conduction system A group of autorhythmic cardiac muscle fibers that generates and distributes electrical impulses to stimulate coordinated contraction of the heart chambers; includes the sinoatrial (SA) node, the atrioventricular (AV) node, the atrioventricular (AV) bundle, the right and left bundle branches, and the Purkinje fibers.

Condyloid (KON-di-loyd) **joint** A synovial joint structured so that an oval-shaped condyle of one bone fits into an elliptical cavity of another bone, permitting side-to-side and back-and-forth movements, such as the joint at the wrist between the radius and carpals. Also called an **ellipsoidal** (ē-lip-SOYD-al) **joint.**

Cone (KŌN) The type of photoreceptor in the retina that is specialized for highly acute color vision in bright light.

Congenital (kon-JEN-i-tal) Present at the time of birth.

Conjunctiva (kon′-junk-TĪ-va) The delicate membrane covering the eyeball and lining the eyes.

Connective tissue One of the most abundant of the four basic tissue types in the body, performing the functions of binding and supporting; consists of relatively few cells in a generous extracellular matrix (the ground substance and fibers between the cells).

Consciousness (KON-shus-nes) A state of wakefulness in which an individual is fully alert, aware, and oriented, partly as a result of feedback between the cerebral cortex and reticular activating system.

Continuous conduction (kon-DUK-shun) Propagation of an action potential (nerve impulse) in a step-by-step depolarization of each adjacent area of an axon membrane.

Contraception (kon′-tra-SEP-shun) The prevention of fertilization or impregnation without destroying fertility.

Contractility (kon′-trak-TIL-i-tē) The ability of cells or parts of cells to actively generate force to undergo shortening for movements. Muscle fibers (cells) exhibit a high degree of contractility.

Contralateral (CON-tra-lat-er-al) On the opposite side; affecting the opposite side of the body.

Conus medullaris (KŌ-nus med-ū-LAR-is) The tapered portion of the spinal cord inferior to the lumbar enlargement.

Convergence (con-VER-jens) A synaptic arrangement in which the synaptic end bulbs of several presynaptic neurons terminate on one postsynaptic neuron. The medial movement of the two eyeballs so that both are directed toward a near object being viewed in order to produce a single image.

Cornea (KOR-nē-a) The nonvascular, transparent fibrous coat through which the iris of the eye can be seen.

Corona (kō-RŌ-na) Margin of the glans penis.

Corona radiata The innermost layer of granulosa cells that is firmly attached to the zona pellucida around a secondary oocyte.

Coronary artery disease (CAD) A condition such as atherosclerosis that causes narrowing of coronary arteries so that blood flow to the heart is reduced. The result is **coronary heart disease (CHD),** in which the heart muscle receives inadequate blood flow due to an interruption of its blood supply.

Coronary circulation The pathway followed by the blood from the ascending aorta through the blood vessels supplying the heart and returning to the right atrium. Also called **cardiac circulation.**

Coronary sinus (SĪ-nus) A wide venous channel on the posterior surface of the heart that collects the blood from the coronary circulation and returns it to the right atrium.

Corpus albicans (KOR-pus AL-bi-kanz) A white fibrous patch in the ovary that forms after the corpus luteum regresses.

Corpus callosum (kal-LŌ-sum) The great commissure of the brain between the cerebral hemispheres.

Corpuscle of touch See **Meissner corpuscle.**

Corpus luteum (LOO-tē-um) A yellowish body in the ovary formed when a follicle has discharged its secondary oocyte; secretes estrogens, progesterone, relaxin, and inhibin.

Corpus striatum (strī-Ā-tum) An area in the interior of each cerebral hemisphere composed of the caudate and putamen of the basal ganglia and white matter of the internal capsule, arranged in a striated manner.

Cortex (KOR-teks) An outer layer of an organ. The convoluted layer of gray matter covering each cerebral hemisphere.

Costal (KOS-tal) Pertaining to a rib.

Cramp A spasmodic, usually painful contraction of a muscle.

Cranial (KRĀ-ne-al) **cavity** A body cavity formed by the cranial bones and containing the brain.

Cranial nerve One of 12 pairs of nerves that leave the brain; pass through foramina in the skull; and supply sensory and motor neurons to the head, neck, part of the trunk, and viscera of the thorax and abdomen. Each is designated by a Roman numeral and a name.

Craniosacral (krā-nē-ō-SĀK-ral) **outflow** The axons of parasympathetic preganglionic neurons, which have their cell bodies located in nuclei in the brain stem and in the lateral gray matter of the sacral portion of the spinal cord.

Cranium (KRĀ-nē-um) The skeleton of the skull that protects the brain and the organs of sight, hearing, and balance; includes the frontal, parietal, temporal, occipital, sphenoid, and ethmoid bones.

Crista (KRIS-ta) A crest or ridged structure. A small elevation in the ampulla of each semicircular duct that contains receptors for dynamic equilibrium. *Plural is* **cristae.**

Crossing-over The exchange of a portion of one chromatid with another during meiosis. It permits an exchange of genes among chromatids and is one factor that results in genetic variation of progeny.

Crus (KRUS) **of penis** Separated, tapered portion of the corpora cavernosa penis. *Plural is* **crura** (KROO-ra).

Crypt of Lieberkühn *See* **Intestinal gland.**

Cryptorchidism (krip-TOR-ki-dizm) The condition of undescended testes.

Cuneate (KŪ-nē-āt) **nucleus** A group of neurons in the inferior part of the medulla oblongata in which axons of the cuneate fasciculus terminate.

Cupula (KU-pū-la) A mass of gelatinous material covering the hair cells of a crista; a sensory receptor in the ampulla of a semicircular canal stimulated when the head moves.

Cushing's syndrome Condition caused by a hypersecretion of glucocorticoids characterized by spindly legs, "moon face," "buffalo hump," pendulous abdomen, flushed facial skin, poor wound healing, hyperglycemia, osteoporosis, hypertension, and increased susceptibility to disease.

Cutaneous (kū-TĀ-nē-us) Pertaining to the skin.

Cyanosis (sī-a-NŌ-sis) A blue or dark purple discoloration, most easily seen in nail beds and mucous membranes, that results from an increased concentration of deoxygenated (reduced) hemoglobin (more than 5 gm/dL).

Cyst (SIST) A sac with a distinct connective tissue wall, containing a fluid or other material.

Cystic (SIS-tik) **duct** The duct that carries bile from the gallbladder to the common bile duct.

Cystitis (sis-TĪ-tis) Inflammation of the urinary bladder.

Cytokinesis (sī′-tō-ki-NĒ-sis) Distribution of the cytoplasm into two separate cells during cell division; coordinated with nuclear division (mitosis).

Cytolysis (sī-TOL-i-sis) The rupture of living cells in which the contents leak out.

Cytoplasm (SĪ-tō-plasm) Cytosol plus all organelles except the nucleus.

Cytoskeleton Complex internal structure of cytoplasm consisting of microfilaments, microtubules, and intermediate filaments.

Cytosol (SĪ-tō-sol) Semifluid portion of cytoplasm in which organelles and inclusions are suspended and solutes are dissolved. Also called **intracellular fluid.**

D

Dartos (DAR-tōs) The contractile tissue deep to the skin of the scrotum.

Decidua (dē-SID-ū-a) That portion of the endometrium of the uterus (all but the deepest layer) that is modified during pregnancy and shed after childbirth.

Deciduous (dē-SID-ū-us) Falling off or being shed seasonally or at a particular stage of development. In the body, referring to the first set of teeth.

Decussation (dē′-ku-SĀ-shun) A crossing-over to the opposite (contralateral) side; an example is the crossing of 90% of the axons in the large motor tracts to opposite sides in the medullary pyramids.

Deep Away from the surface of the body or an organ.

Deep inguinal (IN-gwi-nal) **ring** A slitlike opening in the aponeurosis of the transversus abdominis muscle that represents the origin of the inguinal canal.

Deep-venous thrombosis (DVT) The presence of a thrombus in a vein, usually a deep vein of the lower limbs.

Defecation (def-e-KĀ-shun) The discharge of feces from the rectum.

Deglutition (dē-gloo-TISH-un) The act of swallowing.

Dehydration (dē-hī-DRĀ-shun) Excessive loss of water from the body or its parts.

Delta cell A cell in the pancreatic islets (islets of Langerhans) in the pancreas that secretes somatostatin. Also termed a **D cell.**

Demineralization (de-min′-er-al-i-ZĀ-shun) Loss of calcium and phosphorus from bones.

Dendrite (DEN-drīt) A neuronal process that carries electrical signals, usually graded potentials, toward the cell body.

Dendritic (den-DRIT-ik) **cell** One type of antigen-presenting cell with long branchlike projections that commonly is present in mucosal linings such as the vagina, in the skin (Langerhans cells in the epidermis), and in lymph nodes (follicular dendritic cells).

Dental caries (KA-rēz) Gradual demineralization of the enamel and dentin of a tooth that may invade the pulp and alveolar bone. Also called **tooth decay.**

Denticulate (den-TIK-ū-lāt) Finely toothed or serrated; characterized by a series of small, pointed projections.

Dentin (DEN-tin) The bony tissues of a tooth enclosing the pulp cavity.

Dentition (den-TI-shun) The eruption of teeth. The number, shape, and arrangement of teeth.

Deoxyribonucleic (dē-ok′-sē-rī-bō-nū-KLĒ-ik) **acid (DNA)** A nucleic acid constructed of nucleotides consisting of one of four bases (adenine, cytosine, guanine, or thymine), deoxyribose, and a phosphate group; encoded in the nucleotides is genetic information.

Depression (de-PRESH-un) Movement in which a part of the body moves inferiorly.

Dermal papilla (pa-PILL-a) Fingerlike projection of the papillary region of the dermis that may contain blood capillaries or corpuscles of touch (Meissner corpuscles).

Dermatology (der′-ma-TOL-ō-jē) The medical specialty dealing with diseases of the skin.

Dermatome (DER-ma-tōm) The cutaneous area developed from one embryonic spinal cord segment and receiving most of its sensory innervation from one spinal nerve. An instrument for incising the skin or cutting thin transplants of skin.

Dermis (DER-mis) A layer of dense irregular connective tissue lying deep to the epidermis.

Descending colon (KŌ-lon) The part of the large intestine descending from the left colic (splenic) flexure to the level of the left iliac crest.

Detrusor (de-TROO-ser) **muscle** Smooth muscle that forms the wall of the urinary bladder.

Developmental biology The study of development from the fertilized egg to the adult form.

Diagnosis (dī′-ag-NŌ-sis) Distinguishing one disease from another or determining the nature of a disease from signs and symptoms by inspection, palpation, laboratory tests, and other means.

Dialysis (dī-AL-i-sis) The removal of waste products from blood by diffusion through a selectively permeable membrane.

Diaphragm (DĪ-a-fram) Any partition that separates one area from another, especially the dome-shaped skeletal muscle between the thoracic and abdominal cavities. Also a dome-shaped device that is placed over the cervix, usually with a spermicide, to prevent conception.

Diaphysis (dī-AF-i-sis) The shaft of a long bone.

Diarrhea (dī-a-RE-a) Frequent defecation of liquid feces caused by increased motility of the intestines.

Diarthrosis (dī-ar-THRŌ-sis) A freely movable joint; types are gliding, hinge, pivot, condyloid, saddle, and ball-and-socket.

Diastole (dī-AS-tō-lē) In the cardiac cycle, the phase of relaxation or dilation of the heart muscle, especially of the ventricles.

Diastolic (dī-as-TOL-ik) **blood pressure** The force exerted by blood on arterial walls during ventricular relaxation; the lowest blood pressure measured in the large arteries, normally about 80 mmHg in a young adult.

Diencephalon (dī′-en-SEF-a-lon) A part of the brain consisting of the thalamus, hypothalamus, and epithalamus.

Diffusion (di-FŪ-zhun) A passive process in which there is a net or greater movement of molecules or ions from a region of high concentration to a region of low concentration until equilibrium is reached.

Digestion (dī-JES-chun) The mechanical and chemical breakdown of food to simple molecules that can be absorbed and used by body cells.

Dilate (DĪ-lāt) To expand or swell.

Diploid (DIP-loid) Having the number of chromosomes characteristically found in the somatic cells of an organism; having two haploid sets of chromosomes, one each from the mother and father. Symbolized 2*n.*

Direct motor pathways Collections of upper motor neurons with cell bodies in the motor cortex that project axons into the spinal cord, where they synapse with lower motor neurons or interneurons in the anterior horns. Also called the **pyramidal pathways.**

Disease Any change from a state of health.

Dislocation (dis′-lō-KĀ-shun) Displacement of a bone from a joint with tearing of ligaments, tendons, and articular capsules. Also called **luxation** (luks-Ā-shun).

Dissect (di-SEKT) To separate tissues and parts of a cadaver or an organ for anatomical study.

Distal (DIS-tal) Farther from the attachment of a limb to the trunk; farther from the point of origin or attachment.

Diuretic (dī-ū-RET-ik) A chemical that increases urine volume by decreasing reabsorption of water, usually by inhibiting sodium reabsorption.

Divergence (dī-VER-jens) A synaptic arrangement in which the synaptic end bulbs of one presynaptic neuron terminate on several postsynaptic neurons.

Diverticulum (dī-ver-TIK-ū-lum) A sac or pouch in the wall of a canal or organ, especially in the colon.

Dorsal ramus (RĀ-mus) A branch of a spinal nerve containing motor and sensory axons supplying the muscles, skin, and bones of the posterior part of the head, neck, and trunk.

Dorsiflexion (dor-si-FLEK-shun) Bending the foot in the direction of the dorsum (upper surface).

Down-regulation Phenomenon in which there is a decrease in the number of receptors in response to an excess of a hormone or neurotransmitter.

Duct of Santorini *See* **Accessory duct.**

Duct of Wirsung *See* **Pancreatic duct.**

Ductus arteriosus (DUK-tus ar-tē-rē-O-sus) A small vessel connecting the pulmonary trunk with the aorta; found only in the fetus.

Ductus (vas) deferens (DEF-er-ens) The duct that carries sperm from the epididymis to the ejaculatory duct. Also called the **seminal duct.**

Ductus epididymis (ep′-i-DID-i-mis) A tightly coiled tube inside the epididymis, distinguished into a head, body, and tail, in which sperm undergo maturation.

Ductus venosus (ve-NŌ-sus) A small vessel in the fetus that helps the circulation bypass the liver.

Duodenal (doo-ō-DĒ-nal) **gland** Gland in the submucosa of the duodenum that secretes an alkaline mucus to protect the lining of the small intestine from the action of enzymes and to help neutralize the acid in chyme. Also called a **Brunner's** (BRUN-erz) **gland.**

Duodenal papilla (pa-PILL-a) An elevation on the duodenal mucosa that receives the hepatopancreatic ampulla (ampulla of Vater).

Duodenum (doo′-ō-DĒ-num *or* doo-OD-e-num) The first 25cm (10 in.) of the small intestine, which connects the stomach and the ileum.

Dura mater (DOO-ra MĀ-ter) The outermost of the three meninges (coverings) of the brain and spinal cord.

Dynamic equilibrium (ē-kwi-LIB-rē-um) The maintenance of body position, mainly the head, in response to sudden movements such as rotation.

Dysmenorrhea (dis′-men-ō-RĒ-a) Painful menstruation.

Dysplasia (dis-PLĀ-zē-a) Change in the size, shape, and organization of cells due to chronic irritation or inflammation; may either revert to normal if stress is removed or progress to neoplasia.

Dyspnea (DISP-nē-a) Shortness of breath; painful or labored breathing.

E

Ectoderm The primary germ layer that gives rise to the nervous system and the epidermis of skin and its derivatives.

Ectopic (ek-TOP-ik) Out of the normal location, as in ectopic pregnancy.

Edema (e-DĒ-ma) An abnormal accumulation of interstitial fluid.

Effector (e-FEK-tor) An organ of the body, either a muscle or a gland, that is innervated by somatic or autonomic motor neurons.

Efferent arteriole (EF-er-ent ar-TĒ-rē-ōl) A vessel of the renal vascular system that carries blood from a glomerulus to a peritubular capillary.

Efferent (EF-er-ent) **ducts** A series of coiled tubes that transport sperm from the rete testis to the epididymis.

Ejaculation (e-jak-ū-LĀ-shun) The reflex ejection or expulsion of semen from the penis.

Ejaculatory (e-JAK-ū-la-tō-rē) **duct** A tube that transports sperm from the ductus (vas) deferens to the prostatic urethra.

Elasticity (e-las-TIS-i-tē) The ability of tissue to return to its original shape after contraction or extension.

Electrocardiogram (e-lek′-trō-KAR-dē-ō-gram) **(ECG** or **EKG)** A recording of the electrical changes that accompany the cardiac cycle that can be detected at the surface of the body; may

be resting, stress, or ambulatory.

Elevation (el-e-VĀ-shun) Movement in which a part of the body moves superiorly.

Embolus (EM-bō-lus) A blood clot, bubble of air or fat from broken bones, mass of bacteria, or other debris or foreign material transported by the blood.

Embryo (EM-brē-ō) The young of any organism in an early stage of development; in humans, the developing organism from fertilization to the end of the eighth week of development.

Embryoblast A region of cells of a blastocyst that differentiates into the three primary germ layers— ectoderm, mesoderm, and endoderm—from which all tissues and organs develop; also called an **inner cell mass.**

Embryology (em′-brē-OL-ō-jē) The study of development from the fertilized egg to the end of the eighth week of development.

Emesis (EM-e-sis) Vomiting.

Emigration (em′-i-GRĀ-shun) Process whereby white blood cells (WBCs) leave the bloodstream by rolling along the endothelium, sticking to it, and squeezing between the endothelial cells. Adhesion molecules help WBCs stick to the endothelium. Also known as **migration** or **extravasation.**

Emission (ē-MISH-un) Propulsion of sperm into the urethra due to peristaltic contractions of the ducts of the testes, epididymides, and ductus (vas) deferens as a result of sympathetic stimulation.

Emphysema (em-fi-SĒ-ma) A lung disorder in which alveolar walls disintegrate, producing abnormally large air spaces and loss of elasticity in the lungs; typically caused by exposure to cigarette smoke.

Emulsification (ē-mul′-si-fi-KĀ-shun) The dispersion of large lipid globules into smaller, uniformly distributed particles in the presence of bile.

Enamel (e-NAM-el) The hard, white substance covering the crown of a tooth.

Endocardium (en-dō-KAR-dē-um) The layer of the heart wall, composed of endothelium and smooth muscle, that lines the inside of the heart and covers the valves and tendons that hold the valves open.

Endochondral ossification (en′-dō-KON-dral os′-i-fi-KĀ-shun) The replacement of cartilage by bone. Also called **intracartilaginous** (in′-tra-kar′-ti-LAJ-i-nus) **ossification.**

Endocrine (EN-dō-krin) **gland** A gland that secretes hormones into interstitial fluid and then the blood; a ductless gland.

Endocrinology (en′-dō-kri-NOL-ō-jē) The science concerned with the structure and functions of endocrine glands and the diagnosis and treatment of disorders of the endocrine system.

Endocytosis (en′-dō-sī-TŌ-sis) The uptake into a cell of large molecules and particles in which a segment of plasma membrane surrounds the substance, encloses it, and brings it in; includes phagocytosis, pinocytosis, and receptor-mediated endocytosis.

Endoderm (EN-dō-derm) A primary germ layer of the developing embryo; gives rise to the gastrointestinal tract, urinary bladder, urethra, and respiratory tract.

Endodontics (en′-dō-DON-tiks) The branch of dentistry concerned with the prevention, diagnosis, and treatment of diseases that affect the pulp, root, periodontal ligament, and alveolar bone.

Endolymph (EN-dō-limf′) The fluid within the membranous labyrinth of the internal ear.

Endometriosis (en′-dō-MĒ-trē-ō′-sis) The growth of endometrial tissue outside the uterus.

Endometrium (en′-dō-MĒ-trē-um) The mucous membrane lining the uterus.

Endomysium (en′-dō-MĪZ-ē-um) Invagination of the perimysium separating each individual muscle fiber (cell).

Endoneurium (en′-dō-NOO-rē-um) Connective tissue wrapping around individual nerve axons.

Endoplasmic reticulum (en′-dō-PLAS-mik re-TIK-ū-lum) **(ER)** A network of channels running through the cytoplasm of a cell that serves in intracellular transportation, support, storage, synthesis, and packaging of molecules. Portions of ER where ribosomes are attached to the outer surface are called **rough ER;** portions that have no ribosomes are called **smooth ER.**

End organ of Ruffini *See* **Type II cutaneous mechanoreceptor.**

Endosteum (end-OS-tē-um) The membrane that lines the medullary (marrow) cavity of bones, consisting of osteogenic cells and scattered osteoclasts.

Endothelium (en′-dō-THĒ-lē-um) The layer of simple squamous epithelium that lines the cavities of the heart, blood vessels, and lymphatic vessels.

Enteric (EN-ter-ik) **nervous system** The part of the nervous system that is embedded in the submucosa and muscularis of the gastrointestinal (GI) tract; governs motility and secretions of the GI tract.

Enteroendocrine (en-ter-ō-EN-dō-krin) **cell** A cell of the mucosa of the gastrointestinal tract that secretes a hormone that governs function of the GI tract; hormones secreted include gastrin, cholecystokinin, glucose-dependent insulinotropic peptide (GIP), and secretin.

Enzyme (EN-zīm) A substance that accelerates chemical reactions; an organic catalyst, usually a protein.

Eosinophil (ē-ō-SIN-ō-fil) A type of white blood cell characterized by granules that stain red or pink with acid dyes.

Ependymal (ep-EN-de-mal) **cells** Neuroglial cells that cover choroid plexuses and produce cerebrospinal fluid (CSF); they also line the ventricles of the brain and probably assist in the circulation of CSF.

Epicardium (ep′-i-KAR-dē-um) The thin outer layer of the heart wall, composed of serous tissue and mesothelium. Also called the **visceral pericardium.**

Epidemiology (ep′-i-dē-mē-OL-ō-jē) Study of the occurrence and transmission of diseases and disorders in human populations.

Epidermis (ep′-i-DERM-is) The superficial, thinner layer of skin, composed of keratinized stratified squamous epithelium.

Epididymis (ep′-i-DID-i-mis) A comma-shaped organ that lies along the posterior border of the testis and contains the ductus epididymis, in which sperm undergo maturation. *Plural* is **epididymides** (ep′-i-di-DIM-i-dēz).

Epidural (ep′-i-DOO-ral) **space** A space between the spinal dura mater and the vertebral canal, containing areolar connective tissue and a plexus of veins.

Epiglottis (ep′-i-GLOT-is) A large, leaf-shaped piece of cartilage lying on top of the larynx, attached to the thyroid cartilage; its unattached portion is free to move up and down to cover the glottis (vocal folds and rima glottidis) during swallowing.

Epimysium (ep-i-MĪZ-ē-um) Fibrous connective tissue around muscles.

Epinephrine (ep-ē-NEF-rin) Hormone secreted by the adrenal medulla that produces actions similar to those that result from sympathetic stimulation. Also called **adrenaline** (a-DREN-a-lin).

Epineurium (ep′-i-NOO-rē-um) The superficial connective tissue covering around an entire nerve.

Epiphyseal (ep′-i-FIZ-ē-al) **line** The remnant of the epiphyseal plate in the metaphysis of a long bone.

Epiphyseal plate The hyaline cartilage plate in the metaphysis of a long bone; site of lengthwise growth of long bones.

Epiphysis (e-PIF-i-sis) The end of a long bone, usually larger in diameter than the shaft (diaphysis).

Epiphysis cerebri (se-RĒ-brē) Pineal gland.

Episiotomy (e-piz′-ē-OT-ō-mē) A cut made with surgical scissors to avoid tearing of the perineum at the end of the second stage of labor.

Epistaxis (ep′-i-STAK-sis) Loss of blood from the nose due to trauma, infection, allergy, neoplasm, and bleeding disorders. Also called **nosebleed.**

Epithalamus (ep′-i-THAL-a-mus) Part of the diencephalon superior and posterior to the thalamus, comprising the pineal gland and associated structures.

Epithelial (ep-i-THĒ-lē-al) **tissue** The tissue that forms the innermost and outermost surfaces of body structures and forms glands.

Eponychium (ep′-o-NIK-ē-um) Narrow band of stratum corneum at the proximal border of a nail that extends from the margin of the nail wall. Also called the **cuticle.**

Erectile dysfunction Failure to maintain an erection long enough for sexual intercourse. Also known as **impotence** (IM-pō-tens).

Erection (ē-REK-shun) The enlarged and stiff state of the penis or clitoris resulting from the engorgement of the spongy erectile tissue with blood.

Eructation (e-ruk′-TĀ-shun) The forceful expulsion of gas from the stomach. Also called **belching.**

Erythema (er-e-THĒ-ma) Skin redness usually caused by dilation of the capillaries.

Erythrocyte (e-RITH-rō-sīt) A mature red blood cell.

Erythropoietin (e-rith′-rō-POY-e-tin) A hormone released by the juxtaglomerular cells of the kidneys that stimulates red blood cell production.

Esophagus (e-SOF-a-gus) The hollow muscular tube that connects the pharynx and the stomach.

Estrogens (ES-tro-jenz) Feminizing sex hormones produced by the ovaries; govern development of oocytes, maintenance of female reproductive structures, and appearance of secondary sex characteristics; also affect fluid and electrolyte balance, and protein anabolism. Examples are β-estradiol, estrone, and estriol.

Eupnea (ŪP-nē-a) Normal quiet breathing.

Eustachian tube *See* **Auditory tube.**

Eversion (ē-VER-zhun) The movement of the sole laterally at the ankle joint or of an atrioventricular valve into an atrium during ventricular contraction.

Excitability (ek-sīt′-a-BIL-i-tē) The ability of muscle fibers to receive and respond to stimuli; the ability of neurons to respond to stimuli and generate nerve impulses.

Excretion (eks-KRĒ-shun) The process of eliminating waste products from the body; also the products excreted.

Exocrine (EK-sō-krin) **gland** A gland that secretes its products into ducts that carry the secretions into body cavities, into the lumen of an organ, or to the outer surface of the body.

Exocytosis (ex′-ō-sī-TŌ-sis) A process in which membrane-enclosed secretory vesicles form inside the cell, fuse with the plasma membrane, and release their contents into the interstitial fluid; achieves secretion of materials from a cell.

Exhalation (eks-ha-LĀ-shun) Breathing out; expelling air from the lungs into the atmosphere. Also called **expiration.**

Extensibility (ek-sten′-si-BIL-i-tē) The ability of muscle tissue to stretch when it is pulled.

Extension (eks-TEN-shun) An increase in the angle between two bones; restoring a body part to its anatomical position after flexion.

External Located on or near the surface.

External auditory (AW-di-tōr-ē) **canal** or **meatus** (mē-Ā-tus) A curved tube in the temporal bone that leads to the middle ear.

External ear The outer ear, consisting of the pinna, external auditory canal, and tympanic membrane (eardrum).

External nares (NĀ-rez) The openings into the nasal cavity on the exterior of the body. Also called the **nostrils.**

External respiration The exchange of respiratory gases between the lungs and blood. Also called **pulmonary respiration.**

Exteroceptor (EKS-ter-ō-sep′-tor) A sensory receptor adapted for the reception of stimuli from outside the body.

Extracellular fluid (ECF) Fluid outside body cells, such as interstitial fluid and plasma.

Extracellular matrix (MĀ-triks) The ground substance and fibers between cells in a connective tissue.

Eyebrow The hairy ridge superior to the eye.

F

F cell A cell in the pancreatic islets (islets of Langerhans) that secretes pancreatic polypeptide.

Face The anterior aspect of the head.

Falciform ligament (FAL-si-form LIG-a-ment) A sheet of parietal peritoneum between the two principal lobes of the liver. The ligamentum teres, or remnant of the umbilical vein, lies within its fold.

Falx cerebelli (FALKS cer-e-BEL-li) A small triangular process of the dura mater attached to the occipital bone in the posterior cranial fossa and projecting inward between the two cerebellar hemispheres.

Falx cerebri (FALKS CER-e-brē) A fold of the dura mater extending deep into the longitudinal fissure between the two cerebral hemispheres.

Fascia (FASH-ē-a) Large connective tissue sheet that wraps around groups of muscles.

Fascicle (FAS-i-kul) A small bundle or cluster, especially of nerve or muscle fibers (cells). Also called a **fasciculus** (fa-SIK-ū-lus). *Plural* is **fasciculi** (fa-SIK-yoo-lī).

Fasciculation (fa-sik-ū-LĀ-shun) Abnormal, spontaneous twitch of all skeletal muscle fibers in one motor unit that is visible at the skin surface; not associated with movement of the affected muscle; present in progressive diseases of motor neurons, for example, poliomyelitis.

Fauces (FAW-sēs) The opening from the mouth into the pharynx.

Feces (FĒ-sēz) Material discharged from the rectum and made up of bacteria, excretions, and food residue. Also called **stool.**

Female reproductive cycle General term for the ovarian and uterine cycles, the hormonal changes that accompany them, and cyclic changes in the breasts and cervix; includes changes in the endometrium of a nonpregnant female that prepares the lining of the uterus to receive a fertilized ovum. Less correctly termed the **menstrual cycle.**

Fertilization (fer′-ti-li-ZĀ-shun) Penetration of a secondary oocyte by a sperm cell, meiotic division of secondary oocyte to form an ovum, and subsequent union of the nuclei of the gametes.

Fetal circulation The cardiovascular system of the fetus, including the placenta and special blood vessels involved in the exchange of materials between fetus and mother.

Fetus (FE-tus) In humans, the developing organism *in utero* from the beginning of the third month to birth.

Fever An elevation in body temperature above the normal temperature of 37 °C (98.6 °F) due to a resetting of the hypothalamic thermostat.

Fibroblast (FĪ-brō-blast) A large, flat cell that secretes most of the extracellular matrix of areolar and dense connective tissues.

Fibrous (FĪ-brus) **joint** A joint that allows little or no movement, such as a suture or a syndesmosis.

Fibrous tunic (TOO-nik) The superficial coat of the eyeball, made up of the posterior sclera and the anterior cornea.

Fight-or-flight response The effects produced upon stimulation of the sympathetic division of the autonomic nervous system.

Filiform papilla (FIL-i-form pa-PIL-a) One of the conical projections that are distributed in parallel rows over the anterior two-thirds of the tongue and lack taste buds.

Filtration (fil-TRĀ-shun) The flow of a liquid through a filter (or membrane that acts like a filter) due to a hydrostatic pressure; occurs in capillaries due to blood pressure.

Filum terminale (FĪ-lum ter-mi-NAL-ē) Nonnervous fibrous tissue of the spinal cord that extends inferiorly from the conus medullaris to the coccyx.

Fimbriae (FIM-brē-ē) Fingerlike structures, especially the lateral ends of the uterine (Fallopian) tubes.

Fissure (FISH-ur) A groove, fold, or slit that may be normal or abnormal.

Fixator A muscle that stabilizes the origin of the prime mover so that the prime mover can act more efficiently.

Fixed macrophage (MAK-rō-fāj) Stationary phagocytic cell found in the liver, lungs, brain, spleen, lymph nodes, subcutaneous tissue, and red bone marrow. Also called a **histiocyte** (HIS-tē-ō-sīt).

Flaccid (FLAS-sid) Relaxed, flabby, or soft; lacking muscle tone.

Flagellum (fla-JEL-um) A hairlike, motile process on the extremity of a bacterium, protozoan, or sperm cell. *Plural* is **flagella** (fla-JEL-a).

Flatus (FLĀ-tus) Gas in the stomach or intestines; commonly used to denote expulsion of gas through the anus.

Flexion (FLEK-shun) Movement in which there is a decrease in the angle between two bones.

Follicle (FOL-i-kul) A small secretory sac or cavity; the group of cells that contains a developing oocyte in the ovaries.

Follicle-stimulating hormone (FSH) Hormone secreted by the anterior pituitary; it initiates development of ova and stimulates the ovaries to secrete estrogens in females, and initiates sperm production in males.

Fontanel (fon-ta-NEL) A mesenchyme-filled space where bone formation is not yet complete, especially between the cranial bones of an infant's skull.

Foot The terminal part of the lower limb, from the ankle to the toes.

Foramen (fō-RĀ-men) A passage or opening; a communication between two cavities of an organ, or a hole in a bone for passage of vessels or nerves. *Plural* is **foramina** (fō-RAM-i-na).

Foramen ovale (fō-RĀ-men ō-VAL-ē) An opening in the fetal heart in the septum between the right and left atria. A hole in the greater wing of the sphenoid bone that transmits the mandibular branch of the trigeminal (V) nerve.

Forearm (FOR-arm) The part of the upper limb between the elbow and the wrist.

Fornix (FOR-niks) An arch or fold; a tract in the brain made up of association fibers, connecting the hippocampus with the mammillary bodies; a recess around the cervix of the uterus where it protrudes into the vagina.

Fossa (FOS-a) A furrow or shallow depression.

Fourth ventricle (VEN-tri-kul) A cavity filled with cerebrospinal fluid within the brain lying between the cerebellum and the medulla oblongata and pons.

Fracture (FRAK-choor) Any break in a bone.

Frontal plane A plane at a right angle to a midsagittal plane that divides the body or organs into anterior and posterior portions. Also called a **coronal** (kō-RŌ-nal) **plane.**

Fundus (FUN-dus) The part of a hollow organ farthest from the opening.

Fungiform papilla (FUN-ji-form pa-PIL-a) A mushroomlike elevation on the upper surface of the tongue appearing as a red dot; most contain taste buds.

Furuncle (FŪ-rung-kul) A boil; painful nodule caused by bacterial infection and inflammation of a hair follicle or sebaceous (oil) gland.

G

Gallbladder A small pouch, located inferior to the liver, that stores bile and empties by means of the cystic duct.

Gallstone A solid mass, usually containing cholesterol, in the gallbladder or a bile-containing duct; formed anywhere between bile canaliculi in the liver and the hepatopancreatic ampulla (ampulla of Vater), where bile enters the duodenum. Also called a **biliary calculus.**

Gamete (GAM-ēt) A male or female reproductive cell; a sperm cell or secondary oocyte.

Ganglion (GANG-glē-on) Usually, a group of neuronal cell bodies lying outside the central nervous system (CNS). *Plural* is **ganglia** (GANG-glē-a).

Gastric (GAS-trik) **glands** Glands in the mucosa of the stomach composed of cells that empty their secretions into narrow channels called gastric pits. Types of cells are chief cells (secrete pepsinogen), parietal cells (secrete hydrochloric acid and intrinsic factor), surface mucous and mucous neck cells (secrete mucus), and G cells (secrete gastrin).

Gastroenterology (gas′-trō-en′-ter-OL-ō-jē) The medical specialty that deals with the structure, function, diagnosis, and treatment of diseases of the stomach and intestines.

Gastrointestinal (gas-trō-in-TES-ti-nal) **(GI) tract** A continuous tube running through the ventral body cavity extending from the mouth to the anus. Also called the **alimentary** (al′-i-MEN-tar-ē) **canal.**

Gastrulation (gas′-troo-LĀ-shun) The migration of groups of cells from the epiblast that transform a bilaminar embryonic disc into a trilaminar embryonic disc with three primary germ layers; transformation of the blastula into the gastrula.

Gene (JĒN) Biological unit of heredity; a segment of DNA located in a definite position on a particular chromosome; a sequence of DNA that codes for a particular mRNA, rRNA, or tRNA.

Genetic engineering The manufacture and manipulation of genetic material.

Genetics The study of genes and heredity.

Genome (JĒ-nōm) The complete set of genes of an organism.

Genotype (JĒ-nō-tīp) The genetic makeup of an individual; the combination of alleles present at one or more chromosomal locations, as distinguished from the appearance, or phenotype, that results from those alleles.

Geriatrics (jer′-ē-AT-riks) The branch of medicine devoted to the medical problems and care of elderly persons.

Gestation (jes-TĀ-shun) The period of development from fertilization to birth.

Gingivae (jin-JI-vē) Gums. They cover the alveolar processes of the mandible and maxilla and extend slightly into each socket.

Gland Specialized epithelial cell or cells that secrete substances; may be exocrine or endocrine.

Glans penis (glanz PĒ-nis) The slightly enlarged region at the distal end of the penis.

Glaucoma (glaw-KŌ-ma) An eye disorder in which there is increased intraocular pressure due to an excess of aqueous humor.

Gliding joint A synovial joint having articulating surfaces that are usually flat, permitting only side-to-side and back-and-forth movements, as between carpal bones, tarsal bones, and the scapula and clavicle. Also called an **arthrodial** (ar-THRŌ-dē-al) **joint.**

Glomerular (glō-MER-ū-lar) **capsule** A double-walled globe at the proximal end of a nephron that encloses the glomerular capillaries. Also called **Bowman's** (BŌ-manz) **capsule.**

Glomerular filtrate (glō-MER-ū-lar FIL-trāt) The fluid produced when blood is filtered by the filtration membrane in the glomeruli of the kidneys.

Glomerular filtration The first step in urine formation in which substances in blood pass through the filtration membrane and the filtrate enters the proximal convoluted tubule of a nephron.

Glomerulus (glō-MER-ū-lus) A rounded mass of nerves or blood vessels, especially the microscopic tuft of capillaries that is surrounded by the glomerular (Bowman's) capsule of each kidney tubule. *Plural* is **glomeruli.**

Glottis (GLOT-is) The vocal folds (true vocal cords) in the larynx plus the space between them (rima glottidis).

Glucagon (GLOO-ka-gon) A hormone produced by the alpha cells of the pancreatic islets (islets of Langerhans) that increases blood glucose level.

Glucocorticoids (gloo′-kō-KOR-ti-koyds) Hormones secreted by the cortex of the adrenal gland, especially cortisol, that influence glucose metabolism.

Glucose (GLOO-kōs) A hexose (six-carbon sugar), $C_6H_{12}O_6$, that is a major energy source for the production of ATP by body cells.

Glucosuria (gloo′-kō-SOO-rē-a) The presence of glucose in the urine; may be temporary or pathological. Also called **glycosuria.**

Glycogen (GLĪ-kō-jen) A highly branched polymer of glucose containing thousands of subunits; functions as a compact store of glucose molecules in liver and muscle fibers (cells).

Goblet cell A goblet-shaped unicellular gland that secretes mucus; present in epithelium of the airways and intestines.

Goiter (GOY-ter) An enlarged thyroid gland.

Golgi (GOL-jē) **complex** An organelle in the cytoplasm of cells consisting of four to six flattened sacs (cisternae), stacked on one another, with expanded areas at their ends; functions in processing, sorting, packaging, and delivering proteins and lipids to the plasma membrane, lysosomes, and secretory vesicles.

Golgi tendon organ *See* **Tendon organ.**

Gomphosis (gom-FŌ-sis) A fibrous joint in which a cone-shaped peg fits into a socket.

Gonad (GŌ-nad) A gland that produces gametes and hormones; the ovary in the female and the testis in the male.

Gonadotropic hormone Anterior pituitary hormone that affects the gonads.

Gout (GOWT) Hereditary condition associated with excessive uric acid in the blood; the acid crystallizes and deposits in joints, kidneys, and soft tissue.

Graafian follicle *See* **Vesicular ovarian follicle.**

Gracile (GRAS-il) **nucleus** A group of nerve cells in the inferior part of the medulla oblongata in which axons of the gracile fasciculus terminate.

Gray commissure (KOM-mi-shur) A narrow strip of gray matter connecting the two lateral gray masses within the spinal cord.

Gray matter Areas in the central nervous system and ganglia containing neuronal cell bodies, dendrites, unmyelinated axons, axon terminals, and neuroglia; Nissl bodies impart a gray color and there is little or no myelin in gray matter.

Gray ramus communicans (RĀ-mus kō-MŪ-ni-kans) A short nerve containing axons of sympathetic postganglionic neurons; the cell bodies of the neurons are in a sympathetic chain ganglion, and the unmyelinated axons extend via the gray ramus to a spinal nerve and then to the periphery to supply smooth muscle in blood vessels, arrector pili muscles, and sweat glands. *Plural* is **rami communicantes** (RĀ-mē kō-mū-ni-KAN-tēz).

Greater omentum (ō-MEN-tum) A large fold in the serosa of the stomach that hangs down like an apron anterior to the intestines.

Greater vestibular (ves-TIB-ū-lar) **glands** A pair of glands on either side of the vaginal orifice that open by a duct into the space between the hymen and the labia minora. Also called **Bartholin's** (BAR-to-linz) **glands.**

Groin (GROYN) The depression between the thigh and the trunk; the inguinal region.

Gross anatomy The branch of anatomy that deals with structures that can be studied without using a microscope. Also called **macroscopic anatomy.**

Growth An increase in size due to an increase in (1) the number of cells, (2) the size of existing cells as internal components increase in size, or (3) the size of intercellular substances.

Gustatory (GUS-ta-tō′-rē) Pertaining to taste.

Gynecology (gī′-ne-KOL-ō-jē) The branch of medicine dealing with the study and treatment of disorders of the female reproductive system.

Gynecomastia (gīn′e-kō-MAS-tē-a) Excessive growth (benign) of the male mammary glands due to secretion of estrogens by an adrenal gland tumor (feminizing adenoma).

Gyrus (JĪ-rus) One of the folds of the cerebral cortex of the brain. *Plural* is **gyri** (JĪ-rī). Also called a **convolution.**

H

Hair A threadlike structure produced by hair follicles that develops in the dermis. Also called a **pilus** (PĪ-lus).

Hair follicle (FOL-li-kul) Structure composed of epithelium and surrounding the root of a hair from which hair develops.

Hair root plexus (PLEK-sus) A network of dendrites arranged around the root of a hair as free or naked nerve endings that are stimulated when a hair shaft is moved.

Hand The terminal portion of an upper limb, including the carpus, metacarpus, and phalanges.

Haploid (HAP-loyd) **cell** Having half the number of chromosomes characteristically found in the somatic cells of an organism; characteristic of mature gametes. Symbolized *n.*

Hard palate (PAL-at) The anterior portion of the roof of the mouth, formed by the maxillae and palatine bones and lined by mucous membrane.

Haustra (HAWS-tra) A series of pouches that characterize the colon; caused by tonic contractions of the teniae coli. *Singular is* **haustrum.**

Haversian canal *See* **Central canal.**

Haversian system *See* **Osteon.**

Head The superior part of a human, cephalic to the neck. The superior or proximal part of a structure.

Heart A hollow muscular organ lying slightly to the left of the midline of the chest that pumps the blood through the cardiovascular system.

Heart block An arrhythmia (dysrhythmia) of the heart in which the atria and ventricles contract independently because of a blocking of electrical impulses through the heart at some point in the conduction system.

Heart murmur (MER-mer) An abnormal sound that consists of a flow noise that is heard before, between, or after the normal heart sounds, or that may mask normal heart sounds.

Hemangioblast (hē-MAN-jē-ō-blast) A precursor mesodermal cell that develops into blood and blood vessels.

Hematocrit (hē-MAT-ō-krit) **(Hct)** The percentage of blood made up of red blood cells. Usually measured by centrifuging a blood sample in a graduated tube and then reading the volume of red blood cells and dividing it by the total volume of blood in the sample.

Hematology (hēm-a-TOL-ō-jē) The study of blood.

Hematoma (hē′-ma-TŌ-ma) A tumor or swelling filled with blood.

Hemiplegia (hem-i-PLĒ-jē-a) Paralysis of the upper limb, trunk, and lower limb on one side of the body.

Hemoglobin (hē′-mō-GLŌ-bin) **(Hb)** A substance in red blood cells consisting of the protein globin and the iron-containing red pigment heme that transports most of the oxygen and some carbon dioxide in blood.

Hemolysis (hē-MOL-i-sis) The escape of hemoglobin from the interior of a red blood cell into the surrounding medium; results from disruption of the cell membrane by toxins or drugs, freezing or thawing, or hypotonic solutions.

Hemolytic disease of the newborn A hemolytic anemia of a newborn child that results from the destruction of the infant's erythrocytes (red blood cells) by antibodies produced by the mother; usually the antibodies are due to an Rh blood type incompatibility. Also called **erythroblastosis fetalis** (e-rith′-rō-blas-TŌ-sis fe-TAL-is).

Hemophilia (hē-mō-FIL-ē-a) A hereditary blood disorder where there is a deficient production of certain factors involved in blood clotting, resulting in excessive bleeding into joints, deep tissues, and elsewhere.

Hemopoiesis (hēm-ō-poy-Ē-sis) Blood cell production, which occurs in red bone marrow after birth. Also called **hematopoiesis** (hem′-a-tō-poy-Ē-sis).

Hemorrhage (HEM-o-rij) Bleeding; the escape of blood from blood vessels, especially when the loss is profuse.

Hemorrhoids (HEM-ō-royds) Dilated or varicosed blood vessels (usually veins) in the anal region. Also called **piles.**

Hepatic (he-PAT-ik) Refers to the liver.

Hepatic duct A duct that receives bile from the bile capillaries. Small hepatic ducts merge to form the larger right and left hepatic ducts that unite to leave the liver as the common hepatic duct.

Hepatic portal circulation The flow of blood from the gastrointestinal organs to the liver before returning to the heart.

Hepatocyte (he-PAT-ō-cyte) A liver cell.

Hepatopancreatic (hep′-a-tō-pan′-krē-A-tik) **ampulla** A small, raised area in the duodenum where the combined common bile duct and main pancreatic duct empty into the duodenum. Also called the **ampulla of Vater** (VA-ter).

Hernia (HER-nē-a) The protrusion or projection of an organ or part of an organ through a membrane or cavity wall, usually the abdominal cavity.

Herniated (HER-nē-ā′-ted) **disc** A rupture of an intervertebral disc so that the nucleus pulposus protrudes into the vertebral cavity. Also called a **slipped disc.**

Hiatus (hī-Ā-tus) An opening; a foramen.

Hilum (HĪ-lum) An area, depression, or pit where blood vessels and nerves enter or leave an organ. Also called a **hilus.**

Hinge joint A synovial joint in which a convex surface of one bone fits into a concave surface of another bone, such as the elbow, knee, ankle, and interphalangeal joints. Also called a **ginglymus** (JIN-gli-mus) **joint.**

Hirsutism (HER-soo-tizm) An excessive growth of hair in females and children, with a distribution similar to that in adult males, due to the conversion of vellus hairs into large terminal hairs in response to higher-than-normal levels of androgens.

Histamine (HISS-ta-mēn) Substance found in many cells, especially mast cells, basophils, and platelets, that is released when the cells are injured; results in vasodilation, increased permeability of blood vessels, and constriction of bronchioles.

Histology (hiss′-TOL-ō-jē) Microscopic study of the structure of tissues.

Holocrine (HŌ-lō-krin) **gland** A type of gland in which entire secretory cells, along with their accumulated secretions, make up the secretory product of the gland, as in the sebaceous (oil) glands.

Homeostasis (hō′-mē-ō-STĀ-sis) The condition in which the body's internal environment remains relatively constant within physiological limits.

Homologous chromosomes Two chromosomes that belong to a pair. Also called **homologs.**

Hormone (HOR-mōn) A secretion of endocrine cells that alters the physiological activity of target cells of the body.

Horn An area of gray matter (anterior, lateral, or posterior) in the spinal cord.

Human chorionic gonadotropin (kō-rē-ON-ik gō-nad-ō-TRŌ-pin) **(hCG)** A hormone produced by the developing placenta that maintains the corpus luteum.

Human chorionic somatomammotropin (sō-mat-ō-mam-ō-TRŌ-pin) **(hCS)** Hormone produced by the chorion of the placenta that stimulates breast tissue for lactation, enhances body growth, and regulates metabolism. Also called **human placental lactogen (hPL).**

Human growth hormone (hGH) Hormone secreted by the anterior pituitary that stimulates growth of body tissues, especially skeletal and muscular tissues. Also known as **somatotropin** and **somatotropic hormone (STH).**

Hyaluronic (hī′-a-loo-RON-ik) **acid** A viscous, amorphous extracellular material that binds cells together, lubricates joints, and maintains the shape of the eyeballs.

Hymen (HĪ-men) A thin fold of vascularized mucous membrane at the vaginal orifice.

Hyperextension (hī′-per-ek-STEN-shun) Continuation of extension beyond the anatomical position, as in bending the head backward.

Hyperplasia (hī-per-PLĀ-zē-a) An abnormal increase in the number of normal cells in a tissue or organ, increasing its size.

Hypersecretion (hī′-per-se-KRĒ-shun) Overactivity of glands resulting in excessive secretion.

Hypersensitivity (hī′-per-sen-si-TI-vi-tē) Overreaction to an allergen that results in pathological changes in tissues. Also called **allergy.**

Hypertension (hī′-per-TEN-shun) High blood pressure.

Hyperthermia (hī′-per-THERM-ē-a) An elevated body temperature.

Hypertonia (hī′-per-TŌ-nē-a) Increased muscle tone that is expressed as spasticity or rigidity.

Hypertonic (hī′-per-TON-ik) Solution that causes cells to shrink due to loss of water by osmosis.

Hypertrophy (hī-PER-trō-fē) An excessive enlargement or overgrowth of tissue without cell division.

Hyperventilation (hī′-per-ven-ti-LĀ-shun) A rate of inhalation and exhalation higher than that required to maintain a normal partial pressure of carbon dioxide in the blood.

Hyponychium (hī′-pō-NIK-ē-um) Free edge of the fingernail.

Hypophyseal fossa (hī′-pō-FIZ-ē-al FOS-a) A depression on the superior surface of the sphenoid bone that houses the pituitary gland.

Hypophyseal (hī′-pō-FIZ-ē-al) **pouch** An outgrowth of ectoderm from the roof of the mouth from which the anterior pituitary develops.

Hypophysis (hī-POF-i-sis) Pituitary gland.

Hyposecretion (hī′-pō-se-KRĒ-shun) Underactivity of glands resulting in diminished secretion.

Hypothalamohypophyseal (hī′-pō-thal′-a-mō-hī′-pō-FIZ-ē-al) **tract** A bundle of axons containing secretory vesicles filled with oxytocin or antidiuretic hormone that extend from the hypothalamus to the posterior pituitary.

Hypothalamus (hī′-pō-THAL-a-mus) A portion of the diencephalon, lying beneath the thalamus and forming the floor and part of the wall of the third ventricle.

Hypothermia (hī′-pō-THER-mē-a) Lowering of body temperature below 35 °C (95 °F); in surgical procedures, it refers to deliberate cooling of the body to slow down metabolism and reduce oxygen needs of tissues.

Hypotonia (hī′-pō-TŌ-nē-a) Decreased or lost muscle tone in which muscles appear flaccid.

Hypotonic (hī′-pō-TON-ik) Solution that causes cells to swell and perhaps rupture due to gain of water by osmosis.

Hypoventilation (hī-pō-ven-ti-LĀ-shun) A rate of inhalation and exhalation lower than that required to maintain a normal partial pressure of carbon dioxide in plasma.

Hypoxia (hī-POKS-ē-a) Lack of adequate oxygen at the tissue level.

Hysterectomy (hiss-te-REK-tō-mē) The surgical removal of the uterus.

I

Ileocecal (il-ē-ō-SĒ-kal) **sphincter** A fold of mucous membrane that guards the opening from the ileum into the large intestine. Also called the **ileocecal valve.**

Ileum (IL-ē-um) The terminal part of the small intestine.

Immunity (im-Ū-ni-tē) The state of being resistant to injury, particularly by poisons, foreign proteins, and invading pathogens.

Immunoglobulin (im-ū-nō-GLOB-ū-lin) **(Ig)** An antibody synthesized by plasma cells derived from B lymphocytes in response to the introduction of an antigen. Immunoglobulins are divided into five kinds (IgG, IgM, IgA, IgD, IgE).

Immunology (im′-ū-NOL-ō-jē) The study of the responses of the body when challenged by antigens.

Imperforate (im-PER-fō-rāt) Abnormally closed.

Implantation (im-plan-TĀ-shun) The insertion of a tissue or a part into the body. The attachment of the blastocyst to the stratum basalis of the endometrium about 6 days after fertilization.

Incontinence (in-KON-ti-nens) Inability to retain urine, semen, or feces through loss of sphincter control.

Indirect motor pathways Motor tracts that convey information from the brain down the spinal cord for automatic movements, coordination of body movements with visual stimuli, skeletal muscle tone and posture, and balance. Also known as **extrapyramidal pathways.**

Induction (in-DUK-shun) The process by which one tissue (inducting tissue) stimulates the development of an adjacent unspecialized tissue (responding tissue) into a specialized one.

Infarction (in-FARK-shun) A localized area of necrotic tissue, produced by inadequate oxygenation of the tissue.

Infection (in-FEK-shun) Invasion and multiplication of microorganisms in body tissues, which may be inapparent or characterized by cellular injury.

Inferior (in-FĒR-ē-or) Away from the head or toward the lower part of a structure. Also called **caudad** (KAW-dad).

Inferior vena cava (VĒ-na CĀ-va) **(IVC)** Large vein that collects blood from parts of the body inferior to the heart and returns it to the right atrium.

Infertility Inability to conceive or to cause conception. Also called **sterility.**

Inflammation (in′-fla-MĀ-shun) Localized, protective response to tissue injury designed to destroy, dilute, or wall off the infecting agent or injured tissue; characterized by redness, pain, heat, swelling, and sometimes loss of function.

Infundibulum (in-fun-DIB-ū-lum) The stalklike structure that attaches the pituitary gland to the hypothalamus of the brain. The funnel-shaped, open, distal end of the uterine (Fallopian) tube.

Ingestion (in-JES-chun) The taking in of food, liquids, or drugs, by mouth.

Inguinal (IN-gwi-nal) Pertaining to the groin.

Inguinal canal An oblique passageway in the anterior abdominal wall just superior and parallel to the medial half of the inguinal ligament that transmits the spermatic cord and ilioinguinal nerve in the male and round ligament of the uterus and ilioinguinal nerve in the female.

Inhalation (in-ha-LĀ-shun) The act of drawing air into the lungs. Also termed **inspiration.**

Inheritance The acquisition of body traits by transmission of genetic information from parents to offspring.

Inhibin A hormone secreted by the gonads that inhibits release of follicle-stimulating hormone (FSH) by the anterior pituitary.

Inhibiting hormone Hormone secreted by the hypothalamus that can suppress secretion of hormones by the anterior pituitary.

Insertion (in-SER-shun) The attachment of a muscle tendon to a movable bone or the end opposite the origin.

Insula (IN-soo-la) A triangular area of the cerebral cortex that lies deep within the lateral cerebral fissue, under the parietal, frontal, and temporal lobes.

Insulin (IN-soo-lin) A hormone produced by the beta cells of a pancreatic islet (islet of Langerhans) that decreases the blood glucose level.

Integrins (IN-te-grinz) A family of transmembrane glycoproteins in plasma membranes that function in cell adhesion; they are present in hemidesmosomes, which anchor cells to a basement membrane, and they mediate adhesion of neutrophils to endothelial cells during emigration.

Integumentary (in-teg-ū-MEN-tar-ē) Relating to the skin.

Intercalated (in-TER-ka-lā t-ed) **disc** An irregular transverse thickening of sarcolemma that contains desmosomes, which hold cardiac muscle fibers (cells) together, and gap junctions, which aid in conduction of muscle action potentials from one fiber to the next.

Intercostal (in′-ter-KOS-tal) **nerve** A nerve supplying a muscle located between the ribs.

Intermediate Between two structures, one of which is medial and one of which is lateral.

Intermediate filament Protein filament, ranging from 8 to 12 nm in diameter, that may provide structural reinforcement, hold organelles in place, and give shape to a cell.

Internal Away from the surface of the body.

Internal capsule A large tract of projection fibers lateral to the thalamus that is the major connection between the cerebral cortex and the brain stem and spinal cord; contains axons of sensory neurons carrying auditory, visual, and somatic sensory signals to the cerebral cortex plus axons of motor neurons descending from the cerebral cortex to the thalamus, subthalamus, brain stem, and spinal cord.

Internal ear The inner ear or labyrinth, lying inside the temporal bone, containing the organs of hearing and balance.

Internal nares (NĀ-rez) The two openings posterior to the nasal cavities opening into the nasopharynx. Also called the **choanae** (kō-Ā-nē).

Internal respiration The exchange of respiratory gases between blood and body cells. Also called **tissue respiration.**

Interneurons (in′-ter-NOO-ronz) Neurons whose axons extend only for a short distance and contact nearby neurons in the brain, spinal cord, or a ganglion; they comprise the vast majority of neurons in the body.

Interoceptor (IN-ter-ō-sep′-tor) Sensory receptor located in blood vessels and viscera that provides information about the body's internal environment.

Interphase (IN-ter-fāz) The period of the cell cycle between cell divisions, consisting of the G_1-(gap or growth) phase, when the cell is engaged in growth, metabolism, and production of substances required for division; S-(synthesis) phase, during which chromosomes are replicated; and G_2-phase.

Interstitial cell of Leydig *See* **Interstitial endocrinocyte.**

Interstitial (in′-ter-STISH-al) **endocrinocyte** A cell that is located in the connective tissue between seminiferous tubules in a mature testis that secretes testosterone. Also called an **interstitial cell of Leydig** (LĪ-dig).

Interstitial (in′-ter-STISH-al) **fluid** The portion of extracellular fluid that fills the microscopic spaces between the cells of tissues; the internal environment of the body. Also called **intercellular** or **tissue fluid.**

Interstitial growth Growth from within, as in the growth of cartilage. Also called **endogenous** (en-DOJ-e-nus) **growth.**

Interventricular (in′-ter-ven-TRIK-ū-lar) **foramen** A narrow, oval opening through which the lateral ventricles of the brain communicate with the third ventricle. Also called the **foramen of Monro.**

Intervertebral (in′-ter-VER-te-bral) **disc** A pad of fibrocartilage located between the bodies of two vertebrae.

Intestinal gland A gland that opens onto the surface of the intestinal mucosa and secretes digestive enzymes. Also called a **crypt of Lieberkühn** (LĒ-ber-kūn).

Intracellular (in′-tra-SEL-yū-lar) **fluid (ICF)** Fluid located within cells.

Intrafusal (in′-tra-FŪ-sal) **fibers** Three to ten specialized muscle fibers (cells), partially enclosed in a spindle-shaped connective tissue capsule, that make up a muscle spindle.

Intramembranous ossification (in′-tra-MEM-bra-nus os′-i-fi-KĀ-shun) The method of bone formation in which the bone is formed directly in mesenchyme arranged sheet like layers that resemble membranes.

Intraocular (in′-tra-OK-ū-lar) **pressure (IOP)** Pressure in the eyeball, produced mainly by aqueous humor.

Intrinsic factor (IF) A glycoprotein, synthesized and secreted by the parietal cells of the gastric mucosa, that facilitates vitamin B_{12} absorption in the small intestine.

Invagination (in-vaj′-i-NĀ-shun) The pushing of the wall of a cavity into the cavity itself.

Inversion (in-VER-zhun) The movement of the sole medially at the ankle joint.

In vitro (VĒ-trō) Literally, in glass; outside the living body and in an artificial environment such as a laboratory test tube.

Ipsilateral (ip-si-LAT-er-al) On the same side, affecting the same side of the body.

Iris The colored portion of the vascular tunic of the eyeball seen through the cornea that contains circular and radial smooth muscle; the hole in the center of the iris is the pupil.

Irritable bowel syndrome (IBS) Disease of the entire gastrointestinal tract in which a person reacts to stress by developing symptoms (such as cramping and abdominal pain) associated with alternating patterns of diarrhea and constipation. Excessive amounts of mucus may appear in feces, and other symptoms include flatulence, nausea, and loss of appetite. Also known as **irritable colon** or **spastic colitis.**

Ischemia (is-KĒ-mē-a) A lack of sufficient blood to a body part due to obstruction or constriction of a blood vessel.

Islet of Langerhans *See* **Pancreatic islet.**

Isotonic (ī′-sō-TON-ik) Having equal tension or tone. A solution having the same concentration of impermeable solutes as cytosol.

Isthmus (IS-mus) A narrow strip of tissue or narrow passage connecting two larger parts.

J

Jaundice (JON-dis) A condition characterized by yellowness of the skin, the white of the eyes, mucous membranes, and body fluids because of a buildup of bilirubin.

Jejunum (je-JOO-num) The middle part of the small intestine.

Joint kinesthetic (kin′-es-THET-ik) **receptor** A proprioceptive receptor located in a joint, stimulated by joint movement.

Juxtaglomerular (juks-ta-glō-MER-ū-lar) **apparatus (JGA)** Con-sists of the macula densa (cells of the distal convoluted tubule adjacent to the afferent and efferent arteriole) and juxtaglomerular cells (modified cells of the afferent and sometimes efferent arteriole); secretes renin when blood pressure starts to fall.

K

Keratin (KER-a-tin) An insoluble protein found in the hair, nails, and other keratinized tissues of the epidermis.

Keratinocyte (ker-a-TIN-ō-sīt) The most numerous of the epidermal cells; produces keratin.

Kidney (KID-nē) One of the paired reddish organs located in the lumbar region that regulates the composition, volume, and pressure of blood and produces urine.

Kidney stone A solid mass, usually consisting of calcium oxalate, uric acid, or calcium phosphate crystals, that may form in any portion of the urinary tract. Also called a **renal calculus.**

Kinesiology (ki-nē-sē′-OL-ō-jē) The study of the movement of body parts.

Kinesthesia (kin′-es-THĒ-zē-a) The perception of the extent and direction of movement of body parts; this sense is possible due to nerve impulses generated by proprioceptors.

Kinetochore (ki-NET-ō-kor) Protein complex attached to the outside of a centromere to which kinetochore microtubules attach.

Kupffer's cell *See* **Stellate reticuloendothelial cell.**

Kyphosis (kī-FŌ-sis) An exaggeration of the thoracic curve of the vertebral column, resulting in a "round-shouldered" appearance. Also called **hunchback.**

L

Labial frenulum (LĀ-bē-al FREN-ū-lum) A medial fold of mucous membrane between the inner surface of the lip and the gums.

Labia majora (LĀ-bē-a ma-JŌ-ra) Two longitudinal folds of skin extending downward and backward from the mons pubis of the female.

Labia minora (min-OR-a) Two small folds of mucous membrane lying medial to the labia majora of the female.

Labium (LĀ-bē-um) A lip. A liplike structure. *Plural* is **labia** (LA-bē-a).

Labor The process of giving birth in which a fetus is expelled from the uterus through the vagina.

Labyrinth (LAB-i-rinth) Intricate communicating passageway, especially in the internal ear.

Lacrimal canal A duct, one on each eyelid, beginning at the punctum at the medial margin of an eyelid and conveying tears medially into the nasolacrimal sac.

Lacrimal gland Secretory cells, located at the superior anterolateral portion of each orbit, that secrete tears into excretory ducts that open onto the surface of the conjunctiva.

Lacrimal sac The superior expanded portion of the nasolacrimal duct that receives the tears from a lacrimal canal.

Lactation (lak-TĀ-shun) The secretion and ejection of milk by the mammary glands.

Lacteal (LAK-tē-al) One of many lymphatic vessels in villi of the intestines that absorb triglycerides and other lipids from digested food.

Lacuna (la-KOO-na) A small, hollow space, such as that found in bones in which the osteocytes lie. *Plural* is **lacunae** (la-KOO-nē).

Lambdoid (LAM-doyd) **suture** The joint in the skull between the parietal bones and the occipital bone; sometimes contains sutural (Wormian) bones.

Lamellae (la-MEL-ē) Concentric rings of hard, calcified extracellular matrix found in compact bone.

Lamellated corpuscle See **pacinian corpuscle.**

Lamina (LAM-i-na) A thin, flat layer or membrane, as the flattened part of either side of the arch of a vertebra. *Plural* is **laminae** (LAM-i-nē).

Lamina propria (PRŌ-prē-a) The connective tissue layer of a mucosa.

Langerhans (LANG-er-hans) **cell** Epidermal dendritic cell that functions as an antigen-presenting cell (APC) during an immune response.

Lanugo (la-NOO-gō) Fine downy hairs that cover the fetus.

Large intestine The portion of the gastrointestinal tract extending from the ileum of the small intestine to the anus, divided structurally into the cecum, colon, rectum, and anal canal.

Laryngopharynx (la-rin′-gō-FAR-inks) The inferior portion of the pharynx, extending downward from the level of the hyoid bone that divides posteriorly into the esophagus and anteriorly into the larynx. Also called the **hypopharynx.**

Larynx (LAR-inks) The voice box, a short passageway that connects the pharynx with the trachea.

Lateral (LAT-er-al) Farther from the midline of the body or a structure.

Lateral ventricle (VEN-tri-kul) A cavity within a cerebral hemisphere that communicates with the lateral ventricle in the other cerebral hemisphere and with the third ventricle by way of the interventricular foramen.

Leg The part of the lower limb between the knee and the ankle.

Lens A transparent organ constructed of proteins (crystallins) lying posterior to the pupil and iris of the eyeball and anterior to the vitreous body.

Lesion (LĒ-zhun) Any localized, abnormal change in a body tissue.

Lesser omentum (ō-MEN-tum) A fold of the peritoneum that extends from the liver to the lesser curvature of the stomach and the first part of the duodenum.

Lesser vestibular (ves-TIB-ū-lar) **gland** One of the paired mucussecreting glands with ducts that open on either side of the urethral orifice in the vestibule of the female.

Leukemia (loo-KĒ-mē-a) A malignant disease of the blood-forming tissues characterized by either uncontrolled production and accumulation of immature leukocytes in which many cells fail to reach maturity (acute) or an accumulation of mature leukocytes in the blood because they do not die at the end of their normal life span (chronic).

Leukocyte (LOO-kō-sīt) A white blood cell.

Leydig (LĪ-dig) **cell** A type of cell that secretes testosterone; located in the connective tissue between seminiferous tubules in a mature testis. Also known as **interstitial cell of Leydig** or **interstitial endocrinocyte.**

Ligament (LIG-a-ment) Dense regular connective tissue that attaches bone to bone.

Ligand (LĪ-gand) A chemical substance that binds to a specific receptor.

Limbic system A part of the forebrain, sometimes termed the visceral brain, concerned with various aspects of emotion and behavior; includes the limbic lobe, dentate gyrus, amygdala, septal nuclei, mammillary bodies, anterior thalamic nucleus, olfactory bulbs, and bundles of myelinated axons.

Lingual frenulum (LIN-gwal FREN-ū-lum) A fold of mucous membrane that connects the tongue to the floor of the mouth.

Lipase An enzyme that splits fatty acids from triglycerides and phospholipids.

Lipid (LIP-id) An organic compound composed of carbon, hydrogen, and oxygen that is usually insoluble in water, but soluble in alcohol, ether, and chloroform; examples include triglycerides (fats and oils), phospholipids, steroids, and eicosanoids.

Lipid bilayer Arrangement of phospholipid, glycolipid, and cholesterol molecules in two parallel sheets in which the hydrophilic "heads" face outward and the hydrophobic "tails" face inward; found in cellular membranes.

Lipoprotein (lip′-ō-PRŌ-tēn) One of several types of particles containing lipids (cholesterol and triglycerides) and proteins that make it water soluble for transport in the blood; high levels of **low-density lipoproteins (LDLs)** are associated with increased risk of atherosclerosis, whereas high levels of **high-density lipoproteins (HDLs)** are associated with decreased risk of atherosclerosis.

Liver Large organ under the diaphragm that occupies most of the right hypochondriac region and part of the epigastric region. Functionally, it produces bile and synthesizes most plasma proteins; interconverts nutrients; detoxifies substances; stores glycogen, iron, and vitamins; carries on phagocytosis of worn-out blood cells and bacteria; and helps synthesize the active form of vitamin D.

Long-term potentiation (LTP) Prolonged, enhanced synaptic transmission that occurs at certain synapses within the hippocampus of the brain; believed to underlie some aspects of memory.

Lordosis (lor-DŌ-sis) An exaggeration of the lumbar curve of the vertebral column. Also called **hollow back.**

Lower limb The appendage attached at the pelvic (hip) girdle, consisting of the thigh, knee, leg, ankle, foot, and toes. Also called the **lower extremity.**

Lumbar (LUM-bar) Region of the back and side between the ribs and pelvis; loin.

Lumbar plexus (PLEK-sus) A network formed by the anterior (ventral) branches of spinal nerves L1 through L4.

Lumen (LOO-men) The space within an artery, vein, intestine, renal tubule, or other tubular structure.

Lungs Main organs of respiration that lie on either side of the heart in the thoracic cavity.

Lunula (LOO-noo-la) The moon-shaped white area at the base of a nail.

Luteinizing (LOO-tē-in′-īz-ing) **hormone (LH)** A hormone secreted by the anterior pituitary that stimulates ovulation, stimulates pro-gesterone secretion by the corpus luteum, and readies the mammary glands for milk secretion in females; stimulates testosterone secretion by the testes in males.

Lymph (LIMF) Fluid confined in lymphatic vessels and flowing through the lymphatic system until it is returned to the blood.

Lymph node An oval or bean-shaped structure located along lymphatic vessels.

Lymphatic (lim-FAT-ik) **capillary** Closed-ended microscopic lymphatic vessel that begins in spaces between cells and converges with other lymphatic capillaries to form lymphatic vessels.

Lymphatic tissue A specialized form of reticular tissue that contains large numbers of lymphocytes.

Lymphatic vessel A large vessel that collects lymph from lymphatic capillaries and converges with other lymphatic vessels to form the thoracic and right lymphatic ducts.

Lymphocyte (LIM-fō-sīt) A type of white blood cell that helps carry out cell-mediated and antibody-mediated immune responses; found in blood and in lymphatic tissues.

Lysosome (LĪ-sō-sōm) An organelle in the cytoplasm of a cell, enclosed by a single membrane and containing powerful digestive enzymes.

Lysozyme (LĪ-sō-zīm) A bactericidal enzyme found in tears, saliva, and perspiration.

M

Macrophage (MAK-rō-fāj) Phagocytic cell derived from a monocyte; may be fixed or wandering.

Macula (MAK-ū-la) A discolored spot or a colored area. A small, thickened region on the wall of the utricle and saccule that contains receptors for static equilibrium.

Macula lutea (LOO-tē-a) The yellow spot in the center of the retina.

Major histocompatibility (MHC) antigens Surface proteins on white blood cells and other nucleated cells that are unique for each person (except for identical siblings); used to type tissues and help prevent rejection of transplanted tissues. Also known as **human leukocyte antigens (HLA).**

Malignant (ma-LIG-nant) Referring to diseases that tend to become worse and cause death, especially the invasion and spreading of cancer.

Mammary (MAM-ar-ē) **gland** Modified sudoriferous (sweat) gland of the female that produces milk for the nourishment of the young.

Mammillary (MAM-i-ler-ē) **bodies** Two small rounded bodies on the inferior aspect of the hypothalamus that are involved in reflexes related to the sense of smell.

Marrow (MAR-ō) Soft, spongelike material in the cavities of bone. Red bone marrow produces blood cells; yellow bone marrow contains adipose tissue that stores triglycerides.

Mast cell A cell found in areolar connective tissue that releases histamine, a dilator of small blood vessels, during inflammation.

Mastication (mas′-ti-KĀ-shun) Chewing.

Mature follicle A large, fluid-filled follicle containing a secondary oocyte and surrounding granulosa cells that secrete estrogens. Also called a **graafian** (GRAF-ē-an) **follicle.**

Meatus (mē-Ā-tus) A passage or opening, especially the external portion of a canal.

Mechanoreceptor (me-KAN-ō-rē-sep-tor) Sensory receptor that detects mechanical deformation of the receptor itself or adjacent cells; stimuli so detected include those related to touch, pressure, vibration, proprioception, hearing, equilibrium, and blood pressure.

Medial (MĒ-dē-al) Nearer the midline of the body or a structure.

Medial lemniscus (lem-NIS-kus) A white matter tract that originates in the gracile and cuneate nuclei of the medulla oblongata and extends to the thalamus on the same side; sensory axons in this tract conduct nerve impulses for the sensations of proprioception, fine touch, vibration, hearing, and equilibrium.

Median aperture (AP-er-choor) One of the three openings in the roof of the fourth ventricle through which cerebrospinal fluid enters the subarachnoid space of the brain and cord. Also called the **foramen of Magendie.**

Median plane A vertical plane dividing the body into right and left halves. Situated in the middle.

Mediastinum (mē′-dē-as-TĪ-num) The anatomical region on the thoracic cavity between the pleurae of the lungs that extends from the sternum to the vertebral column and from the first rib to the diaphragm.

Medulla (me-DOOL-la) An inner layer of an organ, such as the medulla of the kidneys.

Medulla oblongata (me-DOOL-la ob′-long-GA-ta) The most inferior part of the brain stem. Also termed the **medulla.**

Medullary (MED-ū-lar′-ē) **cavity** The space within the diaphysis of a bone that contains yellow bone marrow. Also called the **marrow cavity.**

Medullary rhythmicity (rith-MIS-i-tē) **area** The neurons of the respiratory center in the medulla oblongata that control the basic rhythm of respiration.

Meibomian gland *See* **Tarsal gland.**

Meiosis (mī-Ō-sis) A type of cell division that occurs during production of gametes, involving two successive nuclear divisions that result in cells with the haploid *(n)* number of chromosomes.

Meissner (mīs-ner) **corpuscle** A sensory receptor for touch; found in dermal papillae, especially in the palms and soles. Also called a **corpuscle of touch.**

Melanin (MEL-a-nin) A dark black, brown, or yellow pigment found in some parts of the body such as the skin, hair, and pigmented layer of the retina.

Melanocyte (MEL-a-nō-sīt′) A pigmented cell, located between or beneath cells of the deepest layer of the epidermis, that synthesizes melanin.

Melanocyte-stimulating hormone (MSH) A hormone secreted by the anterior pituitary that stimulates the dispersion of melanin granules in melanocytes in amphibians; continued administration produces darkening of skin in humans.

Melatonin (mel-a-TŌN-in) A hormone secreted by the pineal gland that helps set the timing of the body's biological clock.

Membrane A thin, flexible sheet of tissue composed of an epithelial layer and an underlying connective tissue layer, as in an epithelial membrane, or of areolar connective tissue only, as in a synovial membrane.

Membranous labyrinth (mem-BRA-nus LAB-i-rinth) The part of the labyrinth of the internal ear that is located inside the bony labyrinth and separated from it by the perilymph; made up of the semicircular ducts, the saccule and utricle, and the cochlear duct.

Memory The ability to recall thoughts; commonly classifed as short-term (activated) and long-term.

Menarche (me-NAR-kē) The first menses (menstrual flow) and beginning of ovarian and uterine cycles.

Meninges (me-NIN-jēz) Three membranes covering the brain and spinal cord, called the dura mater, arachnoid mater, and pia mater. *Singular* is **meninx** (MEN-inks).

Menopause (MEN-ō-pawz) The termination of the menstrual cycles.

Menstrual (MEN-strū-al) **cycle** A series of changes in the endometrium of a nonpregnant female that prepares the lining of the uterus to receive a fertilized ovum.

Menstruation (men′-stroo-Ā-shun) Periodic discharge of blood, tissue fluid, mucus, and epithelial cells that usually lasts for 5 days; caused by a sudden reduction in estrogens and progesterone. Also called the **menstrual phase** or **menses.**

Merkel (MER-kel) **cell** Type of cell in the epidermis of hairless skin that makes contact with a Merkel (tactile) disc, which functions in touch.

Merkel Disc Saucer-shaped free nerve endings that make contact with Merkel cells in the epidermis and function as touch receptors. Also called a **tactile disc.**

Merocrine (MER-ō-krin) **gland** Gland made up of secretory cells that remain intact throughout the process of formation and discharge of the secretory product, as in the salivary and pancreatic glands.

Mesenchyme (MEZ-en-kīm) An embryonic connective tissue from which all other connective tissues arise.

Mesentery (MEZ-en-ter′-ē) A fold of peritoneum attaching the small intestine to the posterior abdominal wall.

Mesocolon (mez′-ō-KŌ-lon) A fold of peritoneum attaching the colon to the posterior abdominal wall.

Mesoderm The middle primary germ layer that gives rise to connective tissues, blood and blood vessels, and muscles.

Mesothelium (mez′-ō-THĒ-lē-um) The layer of simple squamous epithelium that lines serous membranes.

Mesovarium (mez′-ō-VAR-ē-um) A short fold of peritoneum that attaches an ovary to the broad ligament of the uterus.

Metabolism (me-TAB-ō-lizm) All the biochemical reactions that occur within an organism, including the synthetic (anabolic) reactions and decomposition (catabolic) reactions.

Metacarpus (met′-a-KAR-pus) A collective term for the five bones that make up the palm.

Metaphase (MET-a-phāz) The second stage of mitosis, in which chromatid pairs line up on the metaphase plate of the cell.

Metaphysis (me-TAF-i-sis) Region of a long bone between the diaphysis and epiphysis that contains the epiphyseal plate in a growing bone.

Metarteriole (met′-ar-TĒ-rē-ōl) A blood vessel that emerges from an arteriole, traverses a capillary network, and empties into a venule.

Metastasis (me-TAS-ta-sis) The spread of cancer to surrounding tissues (local) or to other body sites (distant).

Metatarsus (met′-a-TAR-sus) A collective term for the five bones located in the foot between the tarsals and the phalanges.

Microfilament (mī-krō-FIL-a-ment) Rodlike protein filament about 6 nm in diameter; constitutes contractile units in muscle fibers (cells) and provides support, shape, and movement in nonmuscle cells.

Microglia (mī-KROG-lē-a) Neuroglial cells that carry on phagocytosis.

Microtubule (mī-krō-TOO-būl′) Cylindrical protein filament, from 18 to 30 nm in diameter, consisting of the protein tubulin; provides support, structure, and transportation.

Microvilli (mī′-krō-VIL-ē) Microscopic, fingerlike projections of the plasma membranes of cells that increase surface area for absorption, especially in the small intestine and proximal convoluted tubules of the kidneys.

Micturition (mik′-choo-RISH-un) The act of expelling urine from the urinary bladder. Also called **urination** (ū-ri-NĀ-shun).

Midbrain The part of the brain between the pons and the diencephalon. Also called the **mesencephalon** (mes′-en-SEF-a-lon).

Middle ear A small, epithelial-lined cavity hollowed out of the temporal bone, separated from the external ear by the eardrum and from the internal ear by a thin bony partition containing the oval and round windows; extending across the middle ear are the three auditory ossicles. Also called the **tympanic** (tim-PAN-ik) **cavity.**

Midline An imaginary vertical line that divides the body into equal left and right sides.

Midsagittal plane A vertical plane through the midline of the body that divides the body or organs into *equal* right and left sides. Also called a **median plane.**

Mineralocorticoids (min′-er-al-ō-KOR-ti-koyds) A group of hormones of the adrenal cortex that help regulate sodium and potassium balance.

Mitochondrion (mī-tō-KON-drē-on) A double-membraned organelle that plays a central role in the production of ATP; known as the "power-house" of the cell. *Plural* is **mitochondria.**

Mitosis (mī-TŌ-sis) The orderly division of the nucleus of a cell that ensures that each new nucleus has the same number and kind of chromosomes as the original nucleus. The process includes the replication of chromosomes and the distribution of the two sets of chromosomes into two separate and equal nuclei.

Mitotic spindle Collective term for a football-shaped assembly of micro-tubules (nonkinetochore, kinetochore, and aster) that is responsible for the movement of chromosomes during cell division.

Modality (mō-DAL-i-tē) Any of the specific sensory entities, such as vision, smell, taste, or touch.

Modiolus (mō-DĪ-ō′-lus) The central pillar or column of the cochlea.

Monocyte (MON-ō-sit′) The largest type of white blood cell, characterized by agranular cytoplasm.

Monounsaturated fat A fatty acid that contains one double covalent bond between its carbon atoms; it is not completely saturated with hydrogen atoms. Plentiful in triglycerides of olive and peanut oils.

Mons pubis (MONZ PŪ-bis) The rounded, fatty prominence over the pubic symphysis, covered by coarse pubic hair.

Morula (MOR-ū-la) A solid sphere of cells produced by successive cleavages of a fertilized ovum about four days after fertilization.

Motor area The region of the cerebral cortex that governs muscular movement, particularly the precentral gyrus of the frontal lobe.

Motor end plate Region of the sarcolemma of a muscle fiber (cell) that includes acetylcholine (ACh) receptors, which bind ACh released by synaptic end bulbs of somatic motor neurons.

Motor neurons (NOO-ronz) Neurons that conduct impulses from the brain toward the spinal cord or out of the brain and spinal cord into cranial or spinal nerves to effectors that may be either muscles or glands. Also called **efferent neurons.**

Motor unit A motor neuron together with the muscle fibers (cells) it stimulates.

Mucosa-associated lymphatic tissue (MALT) Lymphatic nodules scattered throughout the lamina propria (connective tissue) of mucous membranes lining the gastrointestinal tract, res-

piratory airways, urinary tract, and reproductive tract.

Mucous (MŪ-kus) **cell** A unicellular gland that secretes mucus. Two types are mucous neck cells and surface mucous cells in the stomach.

Mucous membrane A membrane that lines a body cavity that opens to the exterior. Also called the **mucosa** (mū-KŌ-sa).

Mucus The thick fluid secretion of goblet cells, mucous cells, mucous glands, and mucous membranes.

Muscarinic (mus′ ka RIN-ik) **receptor** Receptor for the neurotransmitter acetylcholine found on all effectors innervated by parasympathetic postganglionic axons and on sweat glands innervated by cholinergic sympathetic postganglionic axons; so named because muscarine activates these receptors but does not activate nicotinic receptors for acetylcholine.

Muscle An organ composed of one of three types of muscle tissue (skeletal, cardiac, or smooth), specialized for contraction to produce voluntary or involuntary movement of parts of the body.

Muscle action potential A stimulating impulse that propagates along the sarcolemma and transverse tubules; in skeletal muscle, it is generated by acetylcholine, which increases the permeability of the sarcolemma to cations, especially sodium ions (Na⁺).

Muscle fatigue (fa-TĒG) Inability of a muscle to maintain its strength of contraction or tension; may be related to insufficient oxygen, depletion of glycogen, and/or lactic acid buildup.

Muscle spindle An encapsulated proprioceptor in a skeletal muscle, consisting of specialized intrafusal muscle fibers and nerve endings; stimulated by changes in length or tension of muscle fibers.

Muscle tone A sustained, partial contraction of portions of a skeletal or smooth muscle in response to activation of stretch receptors or a baseline level of action potentials in the innervating motor neurons.

Muscular dystrophies (DIS-trō-fēz′) Inherited muscle-destroying diseases, characterized by degeneration of muscle fibers (cells), which causes progressive atrophy of the skeletal muscle.

Muscularis (MUS-kū-la′-ris) A muscular layer (coat or tunic) of an organ.

Muscularis mucosae (mū-KŌ-sē) A thin layer of smooth muscle fibers that underlie the lamina propria of the mucosa of the gastrointestinal tract.

Muscular tissue A tissue specialized to produce motion in response to muscle action potentials by its qualities of contractility, extensibility, elasticity, and excitability; types include skeletal, cardiac, and smooth.

Mutation (mū-TĀ-shun) Any change in the sequence of bases in a DNA molecule resulting in a permanent alteration in some inheritable trait.

Myasthenia (mī-as-THĒ-nē-a) **gravis** Weakness and fatigue of skeletal muscles caused by antibodies directed against acetylcholine receptors.

Myelin (MĪ-e-lin) **sheath** Multilayered lipid and protein covering, formed by Schwann cells and

oligodendrocytes, around axons of many peripheral and central nervous system neurons.

Myenteric plexus A network of autonomic axons and postganglionic cell bodies located in the muscularis of the gastrointestinal tract. Also called the **plexus of Auerbach** (OW-er-bak).

Myocardial infarction (mī′-ō-KAR-dē-al in-FARK-shun) **(MI)** Gross necrosis of myocardial tissue due to interrupted blood supply. Also called a **heart attack.**

Myocardium (mī′-ō-KAR-dē-um) The middle layer of the heart wall, made up of cardiac muscle tissue, lying between the epicardium and the endocardium and constituting the bulk of the heart.

Myofibril (mī-ō-FĪ-bril) A threadlike structure, extending longitudinally through a muscle fiber (cell) consisting mainly of thick filaments (myosin) and thin filaments (actin, troponin, and tropomyosin).

Myoglobin (mī-ō-GLŌB-in) The oxygen-binding, iron-containing protein present in the sarcoplasm of muscle fibers (cells); contributes the red color to muscle.

Myogram (MĪ-ō-gram) The record or tracing produced by a myograph, an apparatus that measures and records the force of muscular contractions.

Myology (mī-OL-ō-jē) The study of muscles.

Myometrium (mī′-ō-MĒ-trē-um) The smooth muscle layer of the uterus.

Myopathy (mī-OP-a-thē) Any abnormal condition or disease of muscle tissue.

Myopia (mī-Ō-pē-a) Defect in vision in which objects can be seen distinctly only when very close to the eyes; nearsightedness.

Myosin (MĪ-ō-sin) The contractile protein that makes up the thick filaments of muscle fibers.

Myotome (MĪ-ō-tōm) A group of muscles innervated by the motor neurons of a single spinal segment. In an embryo, the portion of a somite that develops into some skeletal muscles.

N

Nail A hard plate, composed largely of keratin, that develops from the epidermis of the skin to form a protective covering on the dorsal surface of the distal phalanges of the fingers and toes.

Nail matrix (MĀ-triks) The part of the nail beneath the body and root from which the nail is produced.

Nasal (NĀ-zal) **cavity** A mucosa-lined cavity on either side of the nasal septum that opens onto the face at the external nares and into the nasopharynx at the internal nares.

Nasal septum (SEP-tum) A vertical partition composed of bone (perpendicular plate of ethmoid and vomer) and cartilage, covered with a mucous membrane, separating the nasal cavity into left and right sides.

Nasolacrimal (nā′-zō-LAK-ri-mal) **duct** A canal that transports the lacrimal secretion (tears) from the nasolacrimal sac into the nose.

Nasopharynx (nā′-zō-FAR-inks) The superior portion of the pharynx, lying posterior to the nose and extending inferiorly to the soft palate.

Neck The part of the body connecting the head and the trunk. A constricted portion of an organ, such as the neck of the femur or uterus.

Necrosis (ne-KRŌ-sis) A pathological type of cell death that results from disease, injury, or lack of blood supply in which many adjacent cells swell, burst, and spill their contents into the interstitial fluid, triggering an inflammatory response.

Neonatal (nē-ō-NĀ-tal) Pertaining to the first four weeks after birth.

Neoplasm (NĒ-ō-plazm) A new growth that may be benign or malignant.

Nephron (NEF-ron) The functional unit of the kidney.

Nerve A cordlike bundle of neuronal axons and/or dendrites and associated connective tissue coursing together outside the central nervous system.

Nerve fiber General term for any process (axon or dendrite) projecting from the cell body of a neuron.

Nerve impulse A wave of depolarization and repolarization that self-propagates along the plasma membrane of a neuron; also called a **nerve action potential.**

Nervous tissue Tissue containing neurons that initiate and conduct nerve impulses to coordinate homeostasis, and neuroglia that provide support and nourishment to neurons.

Neuralgia (noo-RŌG-lē-a) Attacks of pain along the entire course or branch of a peripheral sensory nerve.

Neural plate A thickening of ectoderm, induced by the notochord, that forms early in the third week of development and represents the beginning of the development of the nervous system.

Neural tube defect (NTD) A developmental abnormality in which the neural tube does not close properly. Examples are spina bifida and anencephaly.

Neuritis (noo-RĪ-tis) Inflammation of one or more nerves.

Neurofibral node *See* **Node of Ranvier.**

Neuroglia (noo-RŌG-lē-a) Cells of the nervous system that perform various supportive functions. The neuroglia of the central nervous system are the astrocytes, oligodendrocytes, microglia, and ependymal cells; neuroglia of the peripheral nervous system include Schwann cells and satellite cells. Also called **glial** (GLĒ-al) **cells.**

Neurohypophyseal (noo′-rō-hī′-pō-FIZ-ē-al) **bud** An outgrowth of ectoderm located on the floor of the hypothalamus that gives rise to the posterior pituitary.

Neurolemma (noo-rō-LEM-ma) The peripheral, nucleated cytoplasmic layer of the Schwann cell. Also called **sheath of Schwann** (SCHWON).

Neurology (noo-ROL-ō-jē) The study of the normal functioning and disorders of the nervous system.

Neuromuscular (noo-rō-MUS-kū-lar) **junction (NMJ)** A synapse between the axon terminals of a motor neuron and the sarcolemma of a muscle fiber (cell).

Neuron (NOO-ron) A nerve cell, consisting of a cell body, dendrites, and an axon.

Neurosecretory (noo-rō-SĒC-re-tō-rē) **cell** A neuron that secretes a hypothalamic releasing hormone or inhibiting hormone into blood capillaries of the hypothalmus; a neuron that secretes oxytocin or antidiuretic hormone into blood capillaries of the posterior pituitary.

Neurotransmitter One of a variety of molecules within axon terminals that are released into the synaptic cleft in response to a nerve impulse and that change the membrane potential of the postsynaptic neuron.

Neurulation (noor-oo-LĀ-shun) The process by which the neural plate, neural folds, and neural tube develop.

Neutrophil (NOO-trō-fil) A type of white blood cell characterized by granules that stain pale lilac with a combination of acidic and basic dyes.

Nicotinic (nik′-ō-TIN-ik) **receptor** Receptor for the neurotransmitter acetylcholine found on both sympathetic and parasympathetic postganglionic neurons and on skeletal muscle in the motor end plate; so named because nicotine activates these receptors but does not activate muscarinic receptors for acetylcholine.

Nipple A pigmented, wrinkled projection on the surface of the breast that is the location of the openings of the lactiferous ducts for milk release.

Nociceptor (nō′-sē-SEP-tor) A free (naked) nerve ending that detects painful stimuli.

Node of Ranvier (RON-vē-ā) A space along a myelinated axon between the individual Schwann cells that form the myelin sheath and the neurolemma. Also called a **neurofibral node.**

Norepinephrine (nor′-ep-ē-NEF-rin) **(NE)** A hormone secreted by the adrenal medulla that produces actions similar to those that result from sympathetic stimulation. Also called **noradrenaline** (nor-a-DREN-a-lin).

Notochord (NŌ-tō-cord) A flexible rod of mesodermal tissue that lies where the future vertebral column will develop and plays a role in induction.

Nucleic (noo-KLĒ-ic) **acid** An organic compound that is a long polymer of nucleotides, with each nucleotide containing a pentose sugar, a phosphate group, and one of four possible nitrogenous bases (adenine, cytosine, guanine, and thymine or uracil).

Nucleolus (noo′-KLĒ-ō-lus) Spherical body within a cell nucleus composed of protein, DNA, and RNA that is the site of the assembly of small and large ribosomal subunits. *Plural* is **nucleoli.**

Nucleosome (NOO-klē-ō-sōm) Structural subunit of a chromosome consisting of histones and DNA.

Nucleus (NOO-klē-us) A spherical or oval organelle of a cell that contains the hereditary factors of the cell, called genes. A cluster of unmyelinated nerve cell bodies in the central nervous system. The central part of an atom made up of protons and neutrons.

Nucleus pulposus (pul-PŌ-sus) A soft, pulpy, highly elastic substance in the center of an intervertebral disc; a remnant of the notochord.

Nutrient A chemical substance in food that provides energy, forms new body components, or assists in various body functions.

O

Obesity (ō-BĒS-i-tē) Body weight more than 20% above a desirable standard due to excessive accumulation of fat.

Oblique (ō-BLĒK) **plane** A plane that passes through the body or an organ at an angle between the transverse plane and either the midsagittal, parasagittal, or frontal plane.

Obstetrics (ob-STET-riks) The specialized branch of medicine that deals with pregnancy, labor, and the period of time immediately after delivery (about 6 weeks).

Olfactory (ōl-FAK-tō-rē) Pertaining to smell.

Olfactory bulb A mass of gray matter containing cell bodies of neurons that form synapses with neurons of the olfactory (I) nerve, lying inferior to the frontal lobe of the cerebrum on either side of the crista galli of the ethmoid bone.

Olfactory receptor A bipolar neuron with its cell body lying between supporting cells located in the mucous membrane lining the superior portion of each nasal cavity; transduces odors into neural signals.

Olfactory tract A bundle of axons that extends from the olfactory bulb posteriorly to olfactory regions of the cerebral cortex.

Oligodendrocyte (OL-i-gō-den′-drō-sīt) A neuroglial cell that supports neurons and produces a myelin sheath around axons of neurons of the central nervous system.

Oliguria (ol′-i-GŪ-rē-a) Daily urinary output usually less than 250 ml.

Olive A prominent oval mass on each lateral surface of the superior part of the medulla oblongata.

Oncogenes (ON-kō-jēnz) Cancer-causing genes; they derive from normal genes, termed protooncogenes, that encode proteins involved in cell growth or cell regulation but have the ability to transform a normal cell into a cancerous cell when they are mutated or inappropriately activated. One example is *p53*.

Oncology (on-KOL-ō-jē) The study of tumors.

Oogenesis (ō′-ō-JEN-e-sis) Formation and development of female gametes (oocytes).

Oophorectomy (ō′-of-ō-REK-tō-me) Surgical removal of the ovaries.

Ophthalmic (of-THAL-mik) Pertaining to the eye.

Ophthalmologist (of′-thal-MOL-ō-jist) A physician who specializes in the diagnosis and treatment of eye disorders using drugs, surgery, and corrective lenses.

Ophthalmology (of-thal-MOL-ō-jē) The study of the structure, function, and diseases of the eye.

Optic (OP-tik) Refers to the eye, vision, or properties of light.

Optic chiasm (kī-AZ-m) A crossing point of the two branches of the optic (II) nerve, anterior to the pituitary gland. Also called **optic chiasma.**

Optic disc A small area of the retina containing openings through which the axons of the ganglion cells emerge as the optic (II) nerve. Also called the **blind spot.**

Optic tract A bundle of axons that carry nerve impulses from the retina of the eye between the optic chiasm and the thalamus.

Ora serrata (Ō-ra ser-RĀ-ta) The irregular margin of the retina lying internal and slightly posterior to the junction of the choroid and ciliary body.

Orbit (OR-bit) The bony, pyramidal-shaped cavity of the skull that holds the eyeball.

Organ A structure composed of two or more different kinds of tissues with a specific function and usually a recognizable shape.

Organelle (or-gan-EL) A permanent structure within a cell with characteristic morphology that is specialized to serve a specific function in cellular activities.

Organism (OR-ga-nizm) A total living form; one individual.

Organogenesis (or′-ga-nō-JEN-e-sis) The formation of body organs and systems. By the end of the eighth week of development, all major body systems have begun to develop.

Orifice (OR-i-fis) Any aperture or opening.

Origin (OR-i-jin) The attachment of a muscle tendon to a stationary bone or the end opposite the insertion.

Oropharynx (or′-ō-FAR-inks) The intermediate portion of the pharynx, lying posterior to the mouth and extending from the soft palate to the hyoid bone.

Orthopedics (or′-thō-PĒ-diks) The branch of medicine that deals with the preservation and restoration of the skeletal system, articulations, and associated structures.

Osmoreceptor (oz′-mō-re-CEP-tor) Receptor in the hypothalamus that is sensitive to changes in blood osmolarity and, in response to high osmolarity (low water concentration), stimulates synthesis and release of antidiuretic hormone (ADH).

Osmosis (oz-MŌ-sis) The net movement of water molecules through a selectively permeable membrane from an area of higher water concentration to an area of lower water concentration until equilibrium is reached.

Osseous (OS-ē-us) Bony.

Ossicle (OS-si-kul) One of the small bones of the middle ear (malleus, incus, stapes).

Ossification (os′-i-fi-KĀ-shun) Formation of bone. Also called **osteogenesis.**

Ossification (os′-i-fi-KĀ-shun) **center** An area in the cartilage model of a future bone where the cartilage cells hypertrophy, secrete enzymes that calcify their extracellular matrix, and die, and the area they occupied is invaded by osteoblasts that then lay down bone.

Osteoblast (OS-tē-ō-blast′) Cell formed from an osteogenic cell that participates in bone formation by secreting some organic components and inorganic salts.

Osteoclast (OS-tē-ō-clast′) A large, multinuclear cell that resorbs (destroys) bone matrix.

Osteocyte (OS-tē-ō-sīt′) A mature bone cell that maintains the daily activities of bone tissue.

Osteogenic (os′-tē-ō-JEN-ik) **cell** Stem cell derived from mesenchyme that has mitotic potential and the ability to differentiate into an osteoblast.

Osteogenic layer The inner layer of the periosteum that contains cells responsible for forming new bone during growth and repair.

Osteology (os-tē-OL-ō-jē) The study of bones.

Osteon (OS-tē-on) The basic unit of structure in adult compact bone, consisting of a central (haversian) canal with its concentrically arranged lamellae, lacunae, osteocytes, and canaliculi. Also called a **haversian** (ha-VER-shan) **system.**

Osteoporosis (os′-tē-ō-pō-RŌ-sis) Age-related disorder characterized by decreased bone mass and increased susceptibility to fractures, often as a result of decreased levels of estrogens.

Otic (Ō-tik) Pertaining to the ear.

Otolith (Ō-tō-lith) A particle of calcium carbonate embedded in the otolithic membrane that functions in maintaining static equilibrium.

Otolithic (ō-tō-LITH-ik) **membrane** Thick, gelatinous, glycoprotein layer located directly over hair cells of the macula in the saccule and utricle of the internal ear.

Otorhinolaryngology (ō-tō-rī′-nō-lar-in-GOL-ō-jē) The branch of medicine that deals with the diagnosis and treatment of diseases of the ears, nose, and throat.

Oval window A small, membrane-covered opening between the middle ear and inner ear into which the footplate of the stapes fits.

Ovarian (ō-VAR-ē-an) **cycle** A monthly series of events in the ovary associated with the maturation of a secondary oocyte.

Ovarian follicle (FOL-i-kul) A general name for oocytes (immature ova) in any stage of development, along with their surrounding epithelial cells.

Ovarian ligament (LIG-a-ment) A rounded cord of connective tissue that attaches the ovary to the uterus.

Ovary (Ō-var-ē) Female gonad that produces oocytes and the estrogens, progesterone, inhibin, and relaxin hormones.

Ovulation (ov-ū-LĀ-shun) The rupture of a mature ovarian (Graafian) follicle with discharge of a secondary oocyte into the pelvic cavity.

Ovum (Ō-vum) The female reproductive or germ cell; an egg cell; arises through completion of meiosis in a secondary oocyte after penetration by a sperm.

Oxyhemoglobin (ok′-sē-HĒ-mō-glō-bin) $(Hb-O_2)$ Hemoglobin combined with oxygen.

Oxytocin (ok′-sē-TŌ-sin) **(OT)** A hormone secreted by neurosecretory cells in the paraventricular and supraoptic nuclei of the hypothalamus that stimulates contraction of smooth muscle in the pregnant uterus and myoepithelial cells around the ducts of mammary glands.

P

P wave The deflection wave of an electrocardiogram that signifies atrial depolarization.

Pacinian corpuscle Oval-shaped pressure receptor located in the dermis or subcutaneous tissue and consisting of concentric layers of a connective tissue wrapped around the dendrites of a sensory neuron. Also called a **lamellated corpuscle.**

Palate (PAL-at) The horizontal structure separating the oral and the nasal cavities; the roof of the mouth.

Palpate (PAL-pāt) To examine by touch; to feel.

Pancreas (PAN-krē-as) A soft, oblong organ lying along the greater curvature of the stomach and connected by a duct to the duodenum. It is both an exocrine gland (secreting pancreatic juice) and an endocrine gland (secreting insulin, glucagon, somatostatin, and pancreatic polypeptide).

Pancreatic (pan′-krē-AT-ik) **duct** A single large tube that unites with the common bile duct from the liver and gallbladder and drains pancreatic juice into the duodenum at the hepatopancreatic ampulla (ampulla of Vater). Also called the **duct of Wirsung.**

Pancreatic islet A cluster of endocrine gland cells in the pancreas that secretes insulin, glucagon, somatostatin, and pancreatic polypeptide. Also called an **islet of Langerhans** (LANG-er-hanz).

Papanicolaou (pa-pa-NI-kō-lō) **test** A cytological staining test for the detection and diagnosis of premalignant and malignant conditions of the female genital tract. Cells scraped from the epithelium of the cervix of the uterus are examined microscopically. Also called a **Pap test** or **Pap smear.**

Papilla (pa-PIL-a) A small nipple-shaped projection or elevation.

Paralysis (pa-RAL-a-sis) Loss or impairment of motor function due to a lesion of nervous or muscular origin.

Paranasal sinus (par′-a-NĀ-zal SĪ-nus) A mucus-lined air cavity in a skull bone that communicates with the nasal cavity. Paranasal sinuses are located in the frontal, maxillary, ethmoid, and sphenoid bones.

Paraplegia (par-a-PLĒ-jē-a) Paralysis of both lower limbs.

Parasagittal plane (par-a-SAJ-i-tal) A vertical plane that does not pass through the midline and that divides the body or organs into *unequal* left and right portions.

Parasympathetic (par′-a-sim-pa-THET-ik) **division** One of the two subdivisions of the autonomic nervous system, having cell bodies of preganglionic neurons in nuclei in the brain stem and in the lateral gray horn of the sacral portion of the spinal cord; primarily concerned with activities that conserve and restore body energy.

Parathyroid (par′-a-THĪ-royd) **gland** One of usually four small endocrine glands embedded in the posterior surfaces of the lateral lobes of the thyroid gland.

Parathyroid hormone (PTH) A hormone secreted by the chief (principal) cells of the parathyroid glands that increases blood calcium level and decreases blood phosphate level.

Paraurethral (par′-a-ū-RĒ-thral) **gland** Gland embedded in the wall of the urethra whose duct opens on either side of the urethral orifice and secretes mucus. Also called **Skene's** (SKĒNZ) **gland.**

Parenchyma (par-EN-ki-ma) The functional parts of any organ, as opposed to tissue that forms its stroma or framework.

Parietal (pa-RĪ-e-tal) Pertaining to or forming the outer wall of a body cavity.

Parietal cell A type of secretory cell in gastric glands that produces hydrochloric acid and intrinsic factor. Also called an **oxyntic cell.**

Parietal pleura (PLOO-ra) The outer layer of the serous pleural membrane that encloses and protects the lungs; the layer that is attached to the wall of the pleural cavity.

Parkinson disease (PD) Progressive degeneration of the basal ganglia and substantia nigra of the cerebrum resulting in decreased production of dopamine (DA) that leads to tremor, slowing of voluntary movements, and muscle weakness.

Parotid (pa-ROT-id) **gland** One of the paired salivary glands located inferior and anterior to the ears and connected to the oral cavity via a duct (Stensen's) that opens into the inside of the cheek opposite the maxillary (upper) second molar tooth.

Pars intermedia A small avascular zone between the anterior and posterior pituitary glands.

Parturition (par′-too-RISH-un) Act of giving birth to young; childbirth, delivery.

Patent ductus arteriosus A congenital heart defect in which the ductus arteriosus remains open. As a result, aortic blood flows into the lower-pressure pulmonary trunk, increasing pulmonary trunk pressure and overworking both ventricles.

Pathogen (PATH-ō-jen) A disease-producing microbe.

Pathological (path′-ō-LOJ-i-kal) **anatomy** The study of structural changes caused by disease.

Pectinate (PEK-ti-nāt) **muscles** Projecting muscle bundles of the anterior atrial walls and the lining of the auricles.

Pectoral (PEK-tō-ral) Pertaining to the chest or breast.

Pedicel (PED-i-sel) Footlike structure, as on podocytes of a glomerulus.

Pelvic (PEL-vik) **cavity** Inferior portion of the abdominopelvic cavity that contains the urinary bladder, sigmoid colon, rectum, and internal female and male reproductive structures.

Pelvic splanchnic (PEL-vik SPLANGK-nik) **nerves** Consist of preganglionic parasympathetic axons from the levels of S2, S3, and S4 that supply the urinary bladder, reproductive organs, and the descending and sigmoid colon and rectum.

Pelvis The basinlike structure formed by the two hip bones, the sacrum, and the coccyx. The expanded, proximal portion of the ureter, lying within the kidney and into which the major calyces open.

Penis (PĒ-nis) The organ of urination and copulation in males; used to deposit semen into the female vagina.

Pepsin Protein-digesting enzyme secreted by chief cells of the stomach in the inactive form pepsinogen, which is converted to active pepsin by hydrochloric acid.

Peptic ulcer An ulcer that develops in areas of the gastrointestinal tract exposed to hydrochloric acid; classified as a gastric ulcer if in the lesser curvature of the stomach and as a duodenal ulcer if in the first part of the duodenum.

Percussion (pur-KUSH-un) The act of striking (percussing) an underlying part of the body with short, sharp taps as an aid in diagnosing the part by the quality of the sound produced.

Perforating canal A minute passageway by means of which blood vessels and nerves from the periosteum penetrate into compact bone. Also called **Volkmann's** (FŌLK-mans) **canal.**

Pericardial (per′-i-KAR-dē-al) **cavity** Small potential space between the visceral and parietal layers of the serous pericardium that contains pericardial fluid.

Pericardium (per-i-KAR-dē-um) A loose-fitting membrane that encloses the heart, consisting of a superficial fibrous layer and a deep serous layer.

Perichondrium (per′-i-KON-drē-um) A covering of dense irregular connective tissue that surrounds the surface of most cartilage.

Perilymph (PER-i-limf) The fluid contained between the bony and membranous labyrinths of the inner ear.

Perimetrium (per′-i-MĒ-trē-um) The serosa of the uterus.

Perimysium (per-i-MĪZ-ē-um) Invagination of the epimysium that divides muscles into bundles.

Perineum (per′-i-NĒ-um) The pelvic floor; the space between the anus and the scrotum in the male and between the anus and the vulva in the female.

Perineurium (per′-i-NOO-rē-um) Connective tissue wrapping around fascicles in a nerve.

Periodontal (per-ē-ō-DON-tal) **disease** A collective term for conditions characterized by degeneration of gingivae, alveolar bone, periodontal ligament, and cementum.

Periodontal ligament The periosteum lining the alveoli (sockets) for the teeth in the alveolar processes of the mandible and maxillae.

Periosteum (per′-ē-OS-tē-um) The covering of a bone that consists of connective tissue, osteogenic cells, and osteoblasts; is essential for bone growth, repair, and nutrition.

Peripheral (pe-RIF-er-al) Located on the outer part or a surface of the body.

Peripheral nervous system (PNS) The part of the nervous system that lies outside the central nervous system, consisting of nerves and ganglia.

Peristalsis (per′-i-STAL-sis) Successive muscular contractions along the wall of a hollow muscular structure.

Peritoneum (per-i-tō-NĒ-um) The largest serous membrane of the body that lines the abdominal cavity and covers the viscera within it.

Peritonitis (per′-i-tō-NĪ-tis) Inflammation of the peritoneum.

Peroxisome (pe-ROKS-i-sōm) Organelle similar in structure to a lysosome that contains enzymes that use molecular oxygen to oxidize various organic compounds; such reactions produce hydrogen peroxide; abundant in liver cells.

Perspiration Sweat; produced by sudoriferous (sweat) glands and containing water, salts, urea, uric acid, amino acids, ammonia, sugar, lactic acid, and ascorbic acid. Helps maintain body temperature and eliminate wastes.

Peyer's patches *See* **Aggregated lymphatic follicles.**

pH A measure of the concentration of hydrogen ions (H^+) in a solution. The pH scale extends from 0 to 14, with a value of 7 expressing neutrality, values lower than 7 expressing increasing acidity, and values higher than 7 expressing increasing alkalinity.

Phagocytosis (fag′-ō-sī-TŌ-sis) The process by which phagocytes ingest and destroy microbes, cell debris, and other foreign matter.

Phalanx (FĀ-lanks) The bone of a finger or toe. *Plural* is **phalanges** (fa-LAN-jēz).

Pharmacology (far′-ma-KOL-ō-jē) The science of the effects and uses of drugs in the treatment of disease.

Pharynx (FAR-inks) The throat; a tube that starts at the internal nares and runs partway down the neck, where it opens into the esophagus posteriorly and the larynx anteriorly.

Phenotype (FĒ-nō-tīp) The observable expression of genotype; physical characteristics of an organism determined by genetic makeup and influenced by interaction between genes and internal and external environmental factors.

Phlebitis (fle-BĪ-tis) Inflammation of a vein, usually in a lower limb.

Photopigment A substance that can absorb light and undergo structural changes that can lead to the development of a receptor potential. An example is rhodopsin. In the eye, also called **visual pigment.**

Photoreceptor Receptor that detects light shining on the retina of the eye.

Physiology (fiz′-ē-OL-o-jē) Science that deals with the functions of an organism or its parts.

Pia mater (PĪ-a MĀ-ter *or* PĒ-a MA-ter) The innermost of the three meninges (coverings) of the brain and spinal cord.

Pineal (PĪN-ē-al) **gland** A cone-shaped gland located in the roof of the third ventricle that secretes melatonin. Also called the **epiphysis cerebri** (ē-PIF-i-sis se-RĒ-brē).

Pinealocyte (pin-ē-AL-ō-sīt) Secretory cell of the pineal gland that releases melatonin.

Pinna (PIN-na) The projecting part of the external ear composed of elastic cartilage and covered by skin and shaped like the flared end of a trumpet. Also called the **auricle** (OR-i-kul).

Pituicyte (pi-TOO-i-sīt) Supporting cell of the posterior pituitary.

Pituitary (pi-TOO-i-tā r-ē) **gland** A small endocrine gland occupying the hypophyseal fossa of the sphenoid bone and attached to the hypothalamus by the infundibulum. Also called the **hypophysis** (hī-POF-i-sis).

Pivot joint A synovial joint in which a rounded, pointed, or conical surface of one bone articulates with a ring formed partly by another bone and partly by a ligament, as in the joint between the atlas and axis and between the proximal ends of the radius and ulna. Also called a **trochoid** (TRŌ-koyd) **joint.**

Placenta (pla-SEN-ta) The special structure through which the exchange of materials between fetal and maternal circulations occurs. Also called the **afterbirth.**

Plantar flexion (PLAN-tar FLEK-shun) Bending the foot in the direction of the plantar surface (sole).

Plaque (PLAK) A layer of dense proteins on the inside of a plasma membrane in adherens junctions and desmosomes. A mass of bacterial cells, dextran (polysaccharide), and other debris that adheres to teeth (dental plaque). *See* also **Atherosclerotic plaque.**

Plasma (PLAZ-ma) The extracellular fluid found in blood vessels; blood minus the formed elements.

Plasma cell Cell that develops from a B cell (lymphocyte) and produces antibodies.

Plasma (cell) membrane Outer, limiting membrane that separates the cell's internal parts from extracellular fluid or the external environment.

Platelet (PLĀT-let) A fragment of cytoplasm enclosed in a cell membrane and lacking a nucleus; found in the circulating blood; plays a role in hemostasis. Also called a **thrombocyte** (THROM-bō-sīt).

Platelet plug Aggregation of platelets (thrombocytes) at a site where a blood vessel is damaged that helps stop or slow blood loss.

Pleura (PLOO-ra) The serous membrane that covers the lungs and lines the walls of the chest and the diaphragm.

Pleural cavity Small potential space between the visceral and parietal pleurae.

Plexus (PLEK-sus) A network of nerves, veins, or lymphatic vessels.

Plexus of Auerbach *See* **Myenteric plexus.**

Plexus of Meissner *See* **Submucosal plexus.**

Pluripotent stem cell Immature stem cell in red bone marrow that gives rise to precursors of all the different mature blood cells.

Pneumotaxic (noo-mō-TAK-sik) **area** A part of the respiratory center in the pons that continually sends inhibitory nerve impulses to the inspiratory area, limiting inhalation and facilitating exhalation.

Polycythemia (pol′-ē-sī-THĒ-mē-a) Disorder characterized by an above-normal hematocrit (above 55%) in which hypertension, thrombosis, and hemorrhage can occur.

Polyunsaturated fat A fatty acid that contains more than one double covalent bond between its carbon atoms; abundant in triglycerides of corn oil, safflower oil, and cottonseed oil.

Polyuria (pol′-ē-Ū-rē-a) An excessive production of urine.

Pons (PONZ) The part of the brain stem that forms a "bridge" between the medulla oblongata and the midbrain, anterior to the cerebellum.

Portal system The circulation of blood from one capillary network into another through a vein.

Postcentral gyrus Gyrus of cerebral cortex located immediately posterior to the central sulcus; contains the primary somatosensory area.

Posterior (pos-TĒR-ē-or) Nearer to or at the back of the body. Equivalent to **dorsal** in bipeds.

Posterior column–medial lemniscus pathways Sensory pathways that carry information related to proprioception, fine touch, two-point discrimination, pressure, and vibration. First-order neurons project from the spinal cord to the ipsilateral medulla in the posterior columns (gracile fasciculus and cuneate fasciculus). Second-order neurons project from the medulla to the contralateral thalamus in the medial lemniscus. Third-order neurons project from the thalamus to the somatosensory cortex (postcentral gyrus) on the same side.

Posterior pituitary Posterior lobe of the pituitary gland. Also called the **neurohypophysis** (noo-rō-hī-POF-i-sis).

Posterior root The structure composed of sensory axons lying between a spinal nerve and the dorsolateral aspect of the spinal cord. Also called the **dorsal (sensory) root.**

Posterior root ganglion (GANG-glē-on) A group of cell bodies of sensory neurons and their supporting cells located along the posterior root of a spinal nerve. Also called a **dorsal (sensory) root ganglion.**

Postganglionic neuron (pōst′-gang-lē-ON-ik NOO-ron) The second autonomic motor neuron in an autonomic pathway, having its cell body and dendrites located in an autonomic ganglion and its unmyelinated axon ending at cardiac muscle, smooth muscle, or a gland.

Postsynaptic (pōst-sin-AP-tik) **neuron** The nerve cell that is activated by the release of a neurotransmitter from another neuron and carries nerve impulses away from the synapse.

Pouch of Douglas *See* **Rectouterine pouch.**

Precapillary sphincter (SFINGK-ter) The destal most muscle fiber (cell) at the metanteriole-capillary junction that regulates blood flow into capillaries.

Precentral gyrus Gyrus of cerebral cortex located immediately anterior to the central sulcus; contains the primary motor area.

Preganglionic (prē′-gang-lē-ON-ik) **neuron** The first autonomic motor neuron in an autonomic pathway, with its cell body and dendrites in the brain or spinal cord and its myelinated axon ending at an autonomic ganglion, where it synapses with a postganglionic neuron.

Pregnancy Sequence of events that normally includes fertilization, implantation, embryonic growth, and fetal growth and terminates in birth.

Premenstrual syndrome (PMS) Severe physical and emotional stress ocurring late in the postovulatory phase of the menstrual cycle and sometimes overlapping with menstruation.

Prepuce (PRĒ-poos) The loose-fitting skin covering the glans of the penis and clitoris. Also called the **foreskin.**

Presbyopia (prez-bē-Ō-pē-a) A loss of elasticity of the lens of the eye due to advancing age with resulting inability to focus clearly on near objects.

Presynaptic (prē-sin-AP-tik) **neuron** A neuron that propagates nerve impulses toward a synapse.

Prevertebral ganglion (prē-VER te bral GANG-glē-on) A cluster of cell bodies of postganglionic sympathetic neurons anterior to the spinal column and close to large abdominal arteries. Also called a **collateral ganglion.**

Primary germ layer One of three layers of embryonic tissue, called ectoderm, mesoderm, and endoderm, that give rise to all tissues and organs of the body.

Primary motor area A region of the cerebral cortex in the precentral gyrus of the frontal lobe of the cerebrum that controls specific muscles or groups of muscles.

Primary somatosensory area A region of the cerebral cortex posterior to the central sulcus in the postcentral gyrus of the parietal lobe of the cerebrum that localizes exactly the points of the body where somatic sensations originate.

Prime mover The muscle directly responsible for producing a desired motion. Also called an **agonist** (AG-ō-nist).

Primitive gut Embryonic structure formed from the dorsal part of the yolk sac that gives rise to most of the gastrointestinal tract.

Primordial (prī-MŌR-dē-al) Existing first; especially primordial egg cells in the ovary.

Principal cell Cell type in the distal convoluted tubules and collecting ducts of the kidneys that is stimulated by aldosterone and antidiuretic hormone.

Proctology (prok-TOL-ō-jē) The branch of medicine concerned with the rectum and its disorders.

Progeny (PROJ-e-nē) Offspring or descendants.

Progesterone (prō-JES-te-rōn) A female sex hormone produced by the ovaries that helps prepare the endometrium of the uterus for implantation of a fertilized ovum and the mammary glands for milk secretion.

Prognosis (prog-NŌ-sis) A forecast of the probable results of a disorder; the outlook for recovery.

Prolactin (prō-LAK-tin) **(PRL)** A hormone secreted by the anterior pituitary that initiates and maintains milk secretion by the mammary glands.

Prolapse (PRŌ-laps) A dropping or falling down of an organ, especially the uterus or rectum.

Proliferation (prō-lif′-er-Ā-shun) Rapid and repeated reproduction of new parts, especially cells.

Pronation (prō-NĀ-shun) A movement of the forearm in which the palm is turned posteriorly.

Prophase (PRŌ-fāz) The first stage of mitosis during which chromatid pairs are formed and aggregate around the metaphase plate of the cell.

Proprioception (prō-prē-ō-SEP-shun) The perception of the position of body parts, especially the limbs, independent of vision; this sense is possible due to nerve impulses generated by proprioceptors.

Proprioceptor (PRŌ-prē-ō-sep′-tor) A receptor located in muscles, tendons, joints, or the internal ear (muscle spindles, tendon organs, joint kines thetic receptors, and hair cells of the vestibular apparatus) that provides information about body position and movements.

Prostaglandin (pros′-ta-GLAN-din) **(PG)** A membrane-associated lipid; released in small quantities and acts as a local hormone.

Prostate (PROS-tā t) A doughnut-shaped gland inferior to the urinary bladder that surrounds the superior portion of the male urethra and secretes a slightly acidic solution that contributes to sperm motility and viability.

Proteasome (PRŌ-tē-a-sōm) Tiny cellular organelle in cytosol and nucleus containing proteases that destroy unneeded, damaged, or faulty proteins.

Protein An organic compound consisting of carbon, hydrogen, oxygen, nitrogen, and sometimes sulfur and phosphorus; synthesized on ribosomes and made up of amino acids linked by peptide bonds.

Gustatory pathway, 604, 604f
Gustatory receptor cell, 571, 602
Gynecomastia, 683
Gyri, 514

H
H zone, 308, 309t
Habenular nuclei, 513
Hair, 155–157, 156f
  anatomy of, 155
  buds, 164
  bulbs, 164
  bundle, 630
  cells, 571, 623, 626, 630, 631
  chemotherapy and, 157
  color of, 157
  coloring, 157
  follicles, 155, 164
  growth, 155–156
  hormones and, 157
  matrix, 155
  removal of, 155
  root plexuses, 155, 574
  types of, 157
Haldane effect, 905
Half-life, 30–31
Halitosis, 970
Hallux valgus, 261
Hamate carpal, 242
Hamstrings, 399
Hand
  intrinsic muscles of, 384–387, 386f–387f
  muscles that move, 380–383, 382f–383f
Hand plate, 259
Haploid cell, 97
Hapten, 849
Hard palate, 928
Hassall's corpuscles, 837
Haustra, 961
Haustral churning, 963
Haversian canal, 179
Haversian system, 134, 179
Head, muscles that move, 358–359, 359f
Head and neck veins, 804–805, 805f
Head fold, 1147
Healthy eating guidelines, 1006–1007. See also Nutrition
Hearing. See also Ears
  auditory pathway and, 627–628, 628f
  deafness and, 626, 636
  ear anatomy, 620–625, 621f
  physiology of, 626–627
  sound waves and, 626
Heart. See also Cardiac entries
  action potential and, 734–735, 734f
  artificial pacemakers and, 734
  ATP production and, 735
  autorhythmic fibers and, 732–733, 733f
  cardiac assist devices, 746t–747t
  cardiac cycle and, 738–740, 739f
  cardiac muscle tissue histology, 731–732, 731f
  cardiac output and, 741–744

cell regeneration, 732
chambers of, 720–724
conduction system of, 732–733, 733f
congestive heart failure, 742
contractile fiber contraction and, 734–735, 734f
coronary circulation, 728–730, 729f
development of, 748–749, 748f
ECG wave correlation and, 736–738, 737f
electrocardiograms and, 735–736, 735f
exercise and, 745
fibrous skeleton of, 725, 725f
heart cell regeneration, 732
heart wall layers, 720, 720f
help for failing hearts, 745–746
internal anatomy of, 723f–724f
location of, 718–719, 718f
murmurs, 740
myocardial infarction and, 730
myocardial ischemia and, 730
myocardial thickness and function, 724–725
nervous system control of, 743f
partitioning of, 749f
pericardium and, 719, 719f–720f
prominence, 1149
sounds, 740, 740f
surface features of, 721f–722f
sympathetic nerves to, 555
systemic and pulmonary circulations, 728, 728f
wall layers, 720, 720f
Heart block, 754
Heart chamber, 720–724
  left atrium, 724
  left ventricle, 724
  right atrium, 722
  right ventricle, 722–724
Heart rate regulation, 742–744
  autonomic regulation, 742–743
  chemical regulation, 743
  factors in, 743–744
Heart valve
  atrioventricular valve operation, 725
  disorders of, 727
  responses to heart pumping, 726f–727f
  semilunar valve operation, 727
Heartburn, 936, 970
Heat and energy balance, 1001–1006
  body temperature homeostasis and, 1002–1004
  energy homeostasis and, 1005
  food intake regulation and, 1005
  heat production and, 1002, 1004f
  heat transfer and, 1002–1003
  hypothalamic thermostat and, 1003
  metabolic rate and, 1002
  thermoregulation and, 1003
Heat capacity of water, 40
Heat cramp, 1012
Heat exhaustion, 1012–1013
Heat generation, muscular tissue and, 302
Heat of vaporization, 40
Heat production, 1002, 1004f
Heat transfer, 1002–1003

Heat-losing center, hypothalamus, 1003
Heat-promoting center, hypothalamus, 1003
Heatstroke, 1013
Heel stick, 690
Height, hormonal abnormalities and, 187
Heimlich maneuver, 915
Helicotrema, 623
Helix, 620
Helper T cell, 848, 853–854, 853f, 863t
Hemangioblast, 823, 1144
Hemangiomas, 170
Hematocrit, 692
Hematoma, fracture, 189
Hematuria, 1048t
Heme, 696
Hemiazygos vein, 810
Hemidesmosome, 110–111
Hemiplegia, 489
Hemisection, 489
Hemispheric lateralization, 521–522
Hemochromatosis, 712
Hemocytoblasts, 693
Hemodialysis, 1048
Hemodynamics, 772–775
  blood flow velocity and, 774
  blood pressure and, 772–773, 772f
  vascular resistance and, 773
  venous return and, 773–774, 774f
Hemoglobin
  functions of, 695–696
  recycling of, 697f
  saturation, 901, 901f
  shape of, 696f
  skin color and, 154
Hemoglobin affinity for oxygen, 901–903, 902f
  acidity and, 901–902
  bisphosphoglycerate and, 902
  carbon dioxide pressure and, 902
  temperature and, 902
Hemolysis, 70, 708
Hemolytic disease of the newborn (HDN), 710, 710f
Hemophilia, 712, 1167
Hemopoiesis, 176, 693
Hemopoietic growth factors, 695
Hemorrhage, 703, 712
Hemorrhagic anemia, 711
Hemorrhoids, 970
Hemospermia, 1093
Hemostasis, 703–707. See also Blood clotting
  anticoagulants and, 707
  aspirin and, 707
  blood clotting and, 704–706, 705f
  control mechanisms, 706–707
  intravascular clotting and, 707
  platelet plug formation and, 703–704
  thrombolytic agents and, 707
  vascular spasm and, 703
Hemothorax, 885
Henry's law, 897
Heparin, 707
Hepatic acinus, 946
Hepatic jaundice, 948
Hepatic laminae, 945
Hepatic lobule, 945

Hepatic portal circulation, 818–819, 818f–819f
Hepatic portal vein, 818
Hepatic sinusoid, 945
Hepatic veins, 813, 945
Hepatitis, 969–970
Hepatitis A, 969
Hepatitis B, 969
Hepatitis C, 969
Hepatitis D, 969–970
Hepatitis E, 970
Hepatocytes, 945
Hepatopancreatic ampulla, 942
Hering-Breuer reflex, 909
Hermaphroditism, 1126
Hernia, 970
Herniated (slipped) disc, 229, 229f
Heterozygous trait, 1163
Hiatus hernia, 934
Hiccupping, 503, 895t
High altitude sickness, 899
High-density lipoproteins, 750, 991
Highly active antiretroviral therapy, 867
Hilum, 839, 885, 1097
Hindbrain, 1144
Hindgut, 967, 1147
Hinge joint, 276–277, 276f
Hinge region, antigens, 858
Hip joint, 288–289, 288f–289f
  abduction and adduction at, 272f
  angular movements at, 271f
  bones, 245–249, 245f
  circumduction at, 273f
  fracture, 261
  rotation at, 273f
Hip pointer, 246
Hip replacements, 294, 295f
Hippocampus, 518
Hirsutism, 157, 683
Histamine, 845
Histocompatibility testing, 856
Histology, 2t, 109
Histone, 87–88
Histotoxic hypoxia, 908
Hives, 171
Hodgkin disease, 868
Holocrine gland, 122
Homeostasis, 8–11. See also Homeostatic imbalances
  blood pressure and, 10t
  body fluids and, 8
  control of, 8–11
  feedback systems and, 9–11
  imbalances and, 11
  labor contractions and, 11f
Homeostatic imbalances
  abnormal contractions of skeletal muscle, 332
  abnormal curves of vertebral column, 229–230
  acquired immunodeficiency syndrome (AIDS), 866–867, 886f
  adrenal gland disorders, 682–683
  allergic reactions, 867
  Alzheimer's disease, 539–540
  anemia, 711
  ankylosing spondylitis, 296
  anorexia nervosa, 970